Some Astronomical Facts

$R_{Earth} = 6.38 \times 10^6$ m, $M_{Earth} = 5.97 \times 10^{24}$ kg, *earth-sun* distance = astronomical unit (AU) = 1.496×10^{11} m, orbit period = 365.3 days
$R_{Sun} = 6.96 \times 10^8$ m, $M_{Sun} = 1.99 \times 10^{30}$ kg
$R_{moon} = 1.74 \times 10^6$ m, $M_{moon} = 7.35 \times 10^{22}$ kg, *moon-earth* distance 3.84×10^8 m, orbit period = 27.3 days
Distance to *nearest star* is 4.0×10^{16} m; to *galactic center* is 2.2×10^{20} m; to *Andromeda galaxy* is 2.1×10^{22} m; to *edge of observable universe* is $\sim 1.3 \times 10^{26}$ m.

SI Prefixes

10^{18} = exa (E) 10^{15} = peta (P) 10^{12} = tera (T)
10^9 = giga (G) 10^6 = mega (M) 10^3 = kilo (k)
10^{-3} = milli (m) 10^{-6} = micro (μ) 10^{-9} = nano (n)
10^{-12} = pico (p) 10^{-15} = femto (f) 10^{-18} = atto (a)

Greek Alphabet

A, α – alpha   I, ι – iota   P, ρ – rho
B, β – beta   K, κ – kappa Σ, σ – sigma
Γ, γ – gamma Λ, λ – lambda   T, τ – tau
Δ, δ – delta   M, μ – mu Υ, υ – upsilon
E, ϵ – epsilon   N, ν – nu Φ, ϕ – phi
Z, ζ – zeta Ξ, ξ – xi   X, χ – chi
H, η – eta   O, o – omicron Ψ, ψ – psi
Θ, θ – theta Π, π – pi Ω, ω – omega

Some Mathematical Symbols

$\approx$ is approximately equal to
$\sim$ is proportional to
$\equiv$ is defined as (equivalent to)
$>$ ($<$) is greater (less) than
$\geq$ ($\leq$) is greater (less) than or equal to
$\gg$ ($\ll$) is much greater (less) than
Δx is the change in x
$\sum_{i=1}^{N} x_i$ is $x_1 + x_2 + \cdots + x_N$
$|x|$ is the magnitude, or absolute value, of x
$x \to O$ is x approaches zero
$n!$ is $n(n-1)(n-2)\cdots(2)(1)$
$\int$ is the integral symbol
dx/dt and $\frac{dx}{dt}$ are the derivative of x with respect to t
∇ – nabla, used for the gradient operator $\vec{\nabla}$ (pp. 132–153)

Electricity, Magnetism, and Light

Wayne M. Saslow
Texas A&M University

BROOKS/COLE

THOMSON LEARNING

Australia • Canada • Mexico • Singapore • Spain
United Kingdom • United States

Physics Editor: Angus McDonald
Production Manager: Alicia Jackson
Print/Media Buyer: Nancy Panziera
Production Service: TechBooks
Text Designer: Rebecca Evans & Associates
Illustrator: Dartmouth Publishing, Inc.
Cover Designer: Ross Carron Design
Cover Image: Lightning at Night, Texas
   © Jeremy Woodhouse/Photodisc
Cover Printer: Transcontinental Gagne
Compositor: TechBooks
Printer: Transcontinental Gagne

Printed in Canada
1 2 3 4 5 6 7   05 04 03 02 01

**For more information about our products,
contact us at:**
Thomson Learning Academic Resource Center
1-800-423-0563

**For permission to use material from this text,
contact us by:**
Phone: **1-800-730-2214**
Fax: **1-800-731-2215**
Web: **www.thomsonrights.com**

Electricity, Magnetism, and Light
ISBN 0-12619455-6
Library of Congress Control Number:
2001094351

Asia
Thomson Learning
60 Albert Complex, #15-01
Albert Complex
Singapore 189969

Australia
Nelson Thomson Learning
102 Dodds Street
South Street
South Melbourne, Victoria 3205
Australia

Canada
Nelson Thomson Learning
1120 Birchmount Road
Toronto, Ontario M1K 5G4
Canada

Europe/Middle East/South Africa
Thomson Learning
Berkshire House
168-173 High Holborn
London WC1 V7AA
United Kingdom

Latin America
Thomson Learning
Seneca, 53
Colonia Polanco
11560 Mexico D.F.
Mexico

Spain
Paraninfo Thomson Learning
Calle/Magallanes, 25
28015 Madrid, Spain

Contents

Preface for the Lecturer

This text is a calculus-based introduction to electricity, magnetism, and light. My original motivation for writing it was to provide: (1) a simple and practical approach to magnets (every student to study magnetism should be taught how to calculate the lifting strength of a bar magnet); and (2) a unified approach to the magnetism of electric currents (the Heaviside-Lorentz force law, the Biot-Savart law, and Ampère's law all can be deduced from Ampère's equivalence, at a distance, between a magnet and a current loop). Together, these give a unified viewpoint toward the subject of magnetism, contained in Chapters 9 to 11.

The text organization closely parallels the historical evolution of the subject matter and, where relevant, the text contains extensive amounts of history. By relating experimental measurement and theoretical concepts, it emphasizes the operational nature of physics. Modern concepts about matter are introduced from the start and are used to reflect on the applicability and limitations of a strictly classical approach.

As remarked by more than one reviewer, most topics have been rethought. One worth noting is the voltaic cell. Whereas typical introductory texts give the false impression that a voltaic cell is a volume pump with an external resistance, a voltaic cell should be thought of as two surface pumps and an internal resistance, as discussed in Chapter 7. For more detail, see *The American Journal of Physics* **67**, 574 (1999).

A major deviation from the standard structure is the addition of two chapters at the outset. The Review/Preview chapter, written in part to eliminate common misconceptions like "voltage passes through the body," provides students a simple but reasonably accurate mental picture of electricity and electrical phenomena. It discusses critically the electric fluid model and invokes the idea of electron orbitals, both localized and extended, and concludes with a detailed discussion of vectors. Chapter 1 takes a historical approach to electric charge and its effects, in order to ease students into more quantitative matters, and concludes with calculations of charge from charge densities.

If these chapters are omitted or neglected, there is a real danger that student understanding will degenerate rapidly, because they provide the conceptual foundation for the more quantitative considerations that follow. Students should be made to realize that they should treat these chapters like old family friends, who can be relied upon for help in a pinch.

Using the Text

Assuming a total of 39 50-minute lectures, there is a core of material from Chapters 2 to 12 inclusive, which I give in 30 lectures. With a typical class of largely engineering students I omit the optional sections and spend one lecture on about half of Chapter 9. I can lecture on about 90% of the chapters included on my syllabus (with the exception of Review/Preview). Here is the core:

> Chap. 2—2, Chap. 3—3, Chap. 4—2.5, Chap. 5—3, Chap. 6—2.5,
> Chap. 7—3.5, Chap. 8—2.5, Chap. 9—1, Chap. 10—3,
> Chap. 11—3, Chap. 12—4.

The remaining lectures depend on the class syllabus.
For ac circuits but no physical optics:

> R/P—1, Chap. 1—2, Chap. 13—0, Chap. 14—3, Chap. 15—3, Chap. 16—0.

For some physical optics and motors but no ac circuits:

> R/P—1, Chap. 1—2, Chap. 13—1, Chap. 14—1, Chap. 15—3, Chap. 16—1.

For physical optics but no ac circuits:

> R/P—0.5, Chap. 1—1.5, Chap. 13—0, Chap. 14—1, Chap. 15—3, Chap. 16—3.

For a class that is extremely well-prepared, so that the math review and the material on waves in Chapter 15 would be unnecessary:

> R/P—0, Chap. 1—1, Chap. 13—0, Chap. 14—3, Chap. 15—2, Chap. 16—3.

To include material on image formation and optical instruments (e.g., a pdf file available from the author at wsaslow@tamu.edu) requires replacing 3 lectures from the above possible sets of lectures. The choice should of course be made by the lecturer, according to the needs of the class.

Note the Nutshell Summary following the Preface for Students, to which the student can be referred for a more global picture during the course of the term. Also note that the optional sections really are optional, but the lecturer may want to spice up the course with a few of them.

The Role of the Lecturer

Although this manuscript was written with the possibility of self-study in mind (thus the home experiments, especially at the beginning), no student can help but benefit from a good lecturer, who can provide memorable demonstrations and motivation in many forms. The lecturer can take advantage of the fact that students perk up when they are given a juicy historical tidbit; discuss real-life examples like charging a battery; or are shown surprising demonstrations like the jumping rings or the (Genecon) generator run as a motor. They also pay close attention to reminders that they've seen something before, but with a different name—what I tell them is the "method of old friends."

Perhaps most important, many of them can be convinced to accept the idea that it is important to check that a complex result reduces to a well-known case in the appropriate limit. I close with a recent email from a student: "At the beginning the stupidity checks seemed so ... well ... so stupid, but that was when I really began to buy into the system, and it saved my butt on more than one occasion, and in more than one class."

Acknowledgments

I would like to acknowledge the following people:

Peter Heller for encouraging me to undertake this project. A visit to his lair at Brandeis, filled with wonderful demonstrations, showed me why he has been so effective with his students. His suggestions have added greatly to the text and the problems.

Ian Lindevald for going over the text and problems with a fine-toothed comb, and sending me the comments of his students during the fall of 2000. I came to rely on his judgment of what was appropriate and what wasn't, what was clear and what wasn't, and what worked and what didn't.

Andrew Zangwill for reading and thinking about every chapter. He consistently provided a bracing tonic of constructive criticism.

For many and varied contributions, Glenn Agnolet, Art Belmonte, Karl Berkelman, Richard Brower, Jim Eckert, Serkan Erdin, Ed Fry, Marc Gabay, Nicholas Gauthier, Harvey Gould, David Halliday, J. D. Jackson, Eric Mazur, Mark Heald, Chia-Ren Hu, Eckhard Krotscheck, Albert Kuhfelt, Antonio Mondragon, Robert Morse, Don Naugle, Robert Park, Carl Patton, Phil Platzman, Bob Romer, Joe Ross, Alan E. Shapiro, John Shoemaker, Roger Smith, Lincoln Taiz, Darin Zimmerman, Michael Weimer, and Edmund Weiss.

Anonymous readers of early versions of the manuscript.

Mat Merchan, Igor Rudychev, and especially David Miller for working out the odd-numbered problems.

My students over the years, both the unnamed and the named, for their suggestions, encouragement, and corrections. These include Laudelino Mulero, Joseph Tufts, Greg Merchan, Kurt Decker, Jonathan Claridge, Tammy Mills, Sara Rumbaugh, and Ross Dillon.

Mike Cohen, Eli Burstein, and Ralph Amado for, in that chronological order, teaching and encouraging me in my undergraduate years.

Richard Benning and the staff at Duddley's Draw for effectively providing a study carrel where I could write, rewrite, edit, and re-edit, free from the interruptions of visitors, e-mail, and phone calls.

The many anonymous authors of the Friday afternoon Liquid Physics Seminar announcements, including Charlie Albert, C. J. Tymczak, Jim Adams, Brad Joelson, Marcus Drew, Steve Johnson, and Hatcher Tynes.

I also would like to acknowledge the following works:

Arnold Arons' *A Guide to Introductory Physics Teaching* strongly influenced me in its discussion of Ohm's law and its admonition not to introduce a term until an operational means is given to measure it (a refrain of Galileo's).

Maxwell's *Treatise on Electricity and Magnetism*, which I read in my trusty Dover paperback.

R. A. R. Tricker's valuable books, *The Contributions of Faraday and Maxwell to Electrical Science* and *Early Electrodynamics*, in which are found excerpts from Faraday, Maxwell, and Ampère.

J. J. Thomson's *Mathematical Theory of Electricity and Magnetism*, for its use of the electric fluid model and magnetic poles, and its discussion of the energy, tension, and pressure of electric field lines. These powerful conceptual tools serve as great inspiration and should not be slighted for lack of rigor. Isaac Newton favored geometrical reasoning, once remarking to a contemporary that "algebra is for bunglers."

Wilhelm Ostwald's remarkable two-volume reference, *Electrochemistry: History and Theory*, was indispensable to my understanding of the work of Galvani and Volta, and provided both valuable quotations and illuminating insights.

E. Pearce Williams' biography, *Michael Faraday*, was a useful history and helped me to appreciate the significance of the nineteenth century terminology of "ponderomotive force" and "electromotive force."

J. Heilbron's *Electricity in the 17th and 18th Centuries* was a valuable resource for Chapter 1.

I would also like to thank John Mathews for bringing the manuscript to the attention of Harcourt/Academic; Jeremy Hayhurst for his editorial decision to support this manuscript; and Ed Wade for locating the cover photo, and Zachary Dorsey for his work in book production.

Finally, I would like to thank my wife, Mary, the artist, for unwavering support and reminders to be visual; my daughters, Joan and Laura, for their tolerance and patience; and my parents, Abraham and Henrietta, for their unconditional love.

Despite the critical readings that this work has received, any remaining errors are the responsibility of the author. I encourage readers not only to call to the author's attention misprints and corrections, but also to suggest improvements.

Wayne M. Saslow
Department of Physics
Texas A&M University
College Station, TX 77843-4242
wsaslow@tamu.edu
January 3, 2002

Preface for the Student

Prerequisites

Just as mechanics is the course in which many students learn how to apply differential calculus, so electricity and magnetism (E&M) is the course in which many students learn how to apply integral calculus. For completeness, therefore, the Review/Preview chapter concludes with a review of vectors and Chapter 1 contains a review, by way of an electrical application, of integral calculus.

Physics Prerequisites: A calculus-based mechanics course.
Mathematics Prerequisites: Algebra, geometry, trigonometry, differential calculus, and the vectors that you will have learned in a calculus-based mechanics course.
Concurrent Mathematics: Integral calculus. Of course, the more math you have had the better.

Concepts before Equations

In this book, whenever a fundamental law is introduced, it first will be discussed qualitatively, and then will be expanded upon quantitatively. A knowledge of the equations with no understanding of how, when, and where they apply is of little value. In short, your mind should be in Conceptland rather than in Equationland. Do *not* go diving for the equations at the first conceptual difficulty you confront. Even the brightest students don't get everything the first time.

How Should You Read This Book?

1. Optional sections are really optional. They typically are of two types. One type of optional section contains additional factual material; if you are interested in a given topic (e.g. lightning in Section 6.9.3), then you might want to read it. The other type contains historical information; if you like to figure out things for yourself, then this will help you learn how other people figured out things for themselves.

2. The insets are *not* optional. They provide parenthetical material or emphasis.

3. Examples are not enough. You cannot master this material simply by learning the examples. The text contains material that is not suited to simple numerical computation, either because it is more conceptual or because it is more complex. Only a subset of the important concepts and principles is in the examples. If this subset were enough, it would have been the book.

4. The Review/Preview was written for those who did *not* have a really good preparation in high school. Nevertheless, even the well-prepared are recommended to at least skim it, since they may find a few topics that are new to them.

5. Equations have been given identification numbers to make it easier to refer to them later, as when in giving derivations. The most important equations, such as (1.1), are boxed and shaded.

6. Don't make up your own rules. Albert Einstein once wrote: "Things should be made as simple as possible, but not any simpler." This text provides the laws of electricity and magnetism, simplified as much as possible by thousands of scientists working for hundreds of years.

Some Details

1. Appendix D gives the answers to the odd-numbered problems associated with specific sections, but not to the general problems.

 2. Especially difficult problems are marked with a lightning bolt icon.

The Big Picture

Michael Faraday, through his remarkable experiments and clear theoretical reasoning, provided a kernel of ideas that James Clerk Maxwell employed to give mathematical precision to the subject of electricity and magnetism. Both of these nineteenth century scientists thought visually:

> *"You seem to see the lines of force curving round obstacles and driving plumb at conductors and swerving towards certain directions in crystals, and carrying with them everywhere the same amount of attractive power spread wider or denser as the lines widen or contract."*
>
> James Clerk Maxwell (Nov 9, 1857), letter to Michael Faraday

Richard Feynman was a twentieth century scientist whose visual way of thinking helped him merge the laws of electricity and magnetism with those of quantum mechanics. Both Maxwell and Feynman had fertile imaginations, deep insights into physics, well-developed senses of humor, and capabilities in the arts. And both wrote textbooks of enduring value:

> *"Why repeat all this? Because there are new generations born every day. Because there are great ideas developed in the history of man, and these ideas do not last unless they are passed purposely and clearly from generation to generation."*
>
> Richard Feynman, *The Meaning of It All* (publ. 1998, Addison-Wesley-Longman)

It is now your turn to study some of these great ideas. In doing so keep in mind that physics is not a disconnected set of equations to be memorized. Rather, physics is a set of facts about the real world, and a coherent set of relationships between these facts, whose natural expression is through the language of mathematics. Therefore we must study, with equal care, both the measurements that yield these facts and the mathematics that relates them.

E&M and Light in a Nutshell

These pages are an attempt to give you a quick overview. For the time being, only look over the questions in bold face. As you read, look at the references to the chapters you have studied.

A. What Is Quantitatively Measureable?

Review/Preview   Not quantitative.

Chap. 1   Semiquantitative (we state charge conservation, and we detect charge qualitatively with electroscopes).

Chap. 2   Force $\vec{F}$ is used to measure electric charge Q. Charge electrometer.

Chap. 3 $\vec{F}$ on Q and Q are used to define electric field $\vec{E}$.

Chap. 4 $\vec{E}$ is used to define electric flux Φ_E; Q and Φ_E are then related by Gauss's law.

Chap. 5 $\vec{F}$ between plates of a capacitor is used to measure voltage difference ΔV. Voltage electrometer.

Chap. 6 Q and ΔV are used to define capacitance $C = Q = \Delta V$.

Chap. 7   Electric current I is measured by deflection of a magnetic needle due to I (magnetic force, Chap. 10). ΔV is measured with a voltage electrometer. One can then measure electrical resistance R of resistors, and internal resistance r and emf $\mathcal{E}$ of sources of electromotive force.

Chap. 8   No new measureable quantity; applications.

Chap. 9   Force $\vec{F}$ on a magnet yields magnetic pole strength Q_m; $\vec{F}$ and Q_m yield magnetic field $\vec{B}$.

Chap. 10   Measured $\vec{F}$ on a wire, and measured $\vec{B}$ yield I for galvanometer.

Chap. 11   Standard of the ampere from force between parallel wires.

Chap. 12   Measure I and R, deduce Faraday emf $\mathcal{E}$; $\mathcal{E}$ and dI/dt then define inductance L.

Chap. 13   No new measureable quantity; applications.

Chap. 14   No new measureable quantity; applications.

Chap. 15   Displacement current; new relationships; applications.

Chap. 16   Physical optics.

B. How Do We Know about Microscopic Objects?

Chap. 7   Quantum of charge $q \equiv e$ is measured by Millikan oil drop experiment.

Chap. 7 q/m for ions is measured by Faraday's law of electrolysis.

Chap. 10 q/m for all charges is measured by electric and magnetic deflection (cathode ray tube, mass spectrometer).

C. What Are the Fundamental Laws?

Chap. 1   Conservation of charge (always valid).

Chap. 2   Coulomb's law (valid only for fixed charges).

Chap. 4   Gauss's law: electric charge is the flux source for $\vec{E}$ (always valid).

Chap. 5   No circulation of $\vec{E}$ (valid only for fixed charges).

Chap. 9   Magnets, no true flux source for $\vec{B}$ (always valid).

Chap. 10 Magnetic force on currents.

Chap. 11 How currents make $\vec{B}$; circulation of $\vec{B}$ proportional to current enclosed (valid only for fixed currents and fixed $\vec{E}$).

Chap. 12 Faraday's law: time-varying $\vec{B}$ produces circulating $\vec{E}$ (always valid).

Chap. 15 Maxwell's new term in Ampère's law: time-varying $\vec{E}$ produces circulating $\vec{B}$ (always valid). Related to conservation of charge.

D. What Are the Nonuniversal Material Properties?

Chap. 6   Polarization $\vec{P}$. Dielectric constant ϵ and dielectric breakdown field E_d.

Chap. 7   Electrical resistance R. Resistivity ρ and conductivity σ.

Chap. 9   Magnetization $\vec{M}$. Magnetic susceptibility χ, remanent magnetization M_r, saturation magnetization M_s, coercive field.

Chap. 10 Hall resistance and Hall coefficient.

E. What Right-Hand Rules Are Employed?

Review/Preview   Vector-product right-hand rule (mathematics convention).

Chap. 10 Oersted's right-hand rule and Ampère's right-hand rule (physics conventions).

Chap. 11 Circuit-normal right-hand rule (mathematics convention).

Reason

"Anyone who conducts an argument by appealing to authority is not using his intelligence, he is using his memory."

Experiment

"There is no higher or lower knowledge, but one only, flowing from experimentation."

Understand

"He who loves practice without theory is like the sailor who boards ship without a rudder or compass, and never knows where he may cast."

Leonardo da Vinci, April 15, 1452–May 2, 1519

Review/Preview

Electricity: Its Uses and Its Visualization

Chapter Overview

Section R.1 provides a brief introduction. Sections R.2–R.4 discuss electricity at home and elsewhere, including automobiles and computers. Section R.5 poses two electrical questions, and Section R.6 presents the electric fluid model and R.7 applies it to answer them. Section R.8 discusses why the electric fluid model must be extended, and how to reconcile the collective nature of the electric fluid with the behavior of individual electrons. Section R.9 reviews vectors, primarily addition and subtraction, and Section R.10 discusses two rules for multiplying vectors, the scalar product and the vector product. ∎

R.1 Introduction

Without electricity, modern life would be impossible. Almost every item on your person—from your shoes to your sunglasses—owes its manufacture to electrical power. Indeed, since this is also true of your clothing, without electricity you might well be completely naked.

This chapter discusses electricity in the home. Most importantly, it tries to make physical and perceptible that difficult-to-visualize stuff called electricity. The next chapter reports the struggles of early scientists—even as they were learning to ask the right questions—to grasp the elusive electricity. Together, both chapters provide a foundation of ideas and concepts, expressed mainly without equations.

R.1.1 *The Electric Fluid Model Serves as a Conceptual Guide*

Once the rules to produce and detect static electricity were established, the major advances were (1) Stephen Gray's 1729 discovery of two classes of materials (conductors, which transport electricity, and insulators, which do not transport electricity); (2) Charles Dufay's 1733 discovery of two classes of electric charges and the rule that "opposites attract and likes repel"; and (3) Benjamin Franklin's 1750 development of the *electric fluid model*. This model implies the first quantitative law of electricity—the Law of Conservation of

Electric Charge. Once this law was understood, it became easier to manipulate electricity, and to study other electrical phenomena in a quantitative fashion. The amount of electric fluid is known as electric charge Q; its unit is the *coulomb*, or C.

The electric fluid model serves as a conceptual guide through Chapter 8, which deal with static electricity and electric currents. The mathematical theory of electricity in electrical conductors, although not strictly analogous to the mathematical theory of ordinary fluids, nevertheless describes a type of fluid. As for air and water, the amount of the electric fluid is conserved. However, compared to air and water, the electric fluid has some special properties. Thus, two blobs with an excess (or a deficit) of electric fluid repel each other (a consequence of Dufay's discovery), whereas two drops of water are indifferent to each other.

Our modern view of ordinary matter is that it has relatively light and mobile negatively charged electrons, and relatively heavy and immobile positively charged nuclei. This view can be made consistent with the electric fluid model and can explain Gray's two classes of materials, conductors and insulators.

R.2   Electricity at Home: A Presumed Common Experience

R.2.1   *Hot, Neutral, and Ground*

Modern buildings are equipped with three-hole electrical wall outlets (or receptacles, or sockets), where the plugs of electrical devices must be inserted to obtain electrical power, as in Figure R.1(a). The holes of the outlet are called *hot*, *neutral*, and *ground*. In normal operation, electric current is carried only by the hot and neutral wires. Electric current is measured in *amperes* (A), or amps.

If you stood on the ground in bare feet and accidentally touched the neutral or the ground wire of a properly wired electrical outlet, you would not be shocked. However, you would receive a shock if you touched the hot wire: your

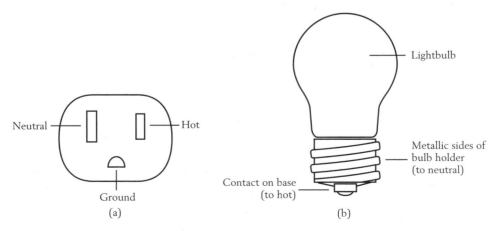

Figure R.1 (a) Grounded three-prong wall outlet. (b) Lightbulb with electrical contacts on its base and on sides.

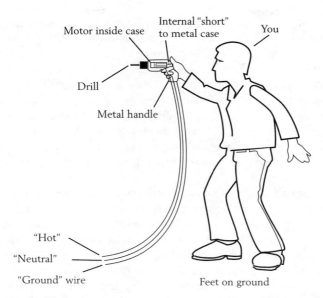

Figure R.2 How grounded wiring protects you when there is a "short."

feet, touching the ground, would provide a path for the current to flow, thus completing an electric circuit.

The size difference in Figure R.1(a) between the neutral and hot holes—the neutral hole is visibly longer than the hot one—is to ensure that only one of the two possible types of connection takes place in devices like a lamp. Figure R.1(b) illustrates the connections for a lightbulb inserted in a lamp with a modern, asymmetrical ("polarized") two-prong plug, which in turn is plugged into a correctly wired wall-socket. The wall-socket's hot wire is connected to the (relatively inaccessible) base of the bulb-holder. For an old-fashioned symmetrical ("unpolarized") two-prong plug, the hot wire could just as likely be connected to the more accidentally touched threaded end of the bulb-holder; the first cartoon characters, of the prepolarized plug 1930s, regularly received shocks in this manner. To help ensure proper wiring, inside the lamp the screws for the two electrical connections typically have different colors, one like copper (the hot wire) and one that is silver-gray (the neutral wire).

The round prong, or ground wire, is employed for safety purposes. Figure R.2 depicts a "short" between the hot wire and the electrically conducting case of an electric drill. (A "short" is a connection that shouldn't be there; shorts are undesirable.) Without the ground wire, the drill operator would provide the only path from the hot wire to ground: hot wire to short to case to person to ground. With the ground wire, there is an alternate "path of least resistance" through which most of the electric current can pass: hot wire to short to case to ground wire to ground.

R.2.2   *Voltage and Frequency of Electrical Power in the House*

Voltage bears much the same relationship to electricity as pressure does to water. We write V as an abbreviation for the unit of voltage, the *volt*. The electrical power in a house in the United States is provided at 120 V. The voltage oscillates from minimum to maximum and back again in 1/60 of a second. This corresponds to a frequency of 60 cps (cycles per second) or, more technically, 60 Hz (*hertz*). The power is provided by an electric company, which uses huge electrical generators to convert mechanical energy from turbines to electrical energy. The turbines are driven by water or by steam. The mechanical energy of the churning waters of the Columbia River (recall the quote at the beginning of the chapter) provides a large fraction of the power needs of the Pacific Northwest. On the other hand, the chemical energy released by burning coal or oil, or the nuclear energy released in a nuclear reactor, vaporizes water into steam and drives the steam that turns a turbine.

The electric light had an extraordinary influence on human society. American children learn that, in the 1830s, the young president-to-be Abraham Lincoln stayed up late reading by candlelight. However, by the 1890s, house lighting by electricity was becoming available in large cities. Nevertheless, not until the Rural Electrification Project of the 1930s did many parts of the United States finally become freed of the fire hazards of oil lamps and candles. In the year 2000, many people in the United States were still alive who could remember not having electrical lighting.

R.2.3 *Watts and Impedance Matching*

When you turn on a light switch, light is produced by bulbs that are rated in units of the *watt*, or W. (The watt is the SI, or *Système Internationale*, unit of power, or energy per second; it is a joule per second, or J/s.) If there is an electrical power failure, for illumination you use a flashlight, with power provided by one or more *voltaic cells*; for a car the power is provided by six voltaic cells in series, which truly constitutes a *battery*.

If you have ever tried to power a house lightbulb with a car battery, you noticed that it did not light. This is due to poor *impedance matching*. A mechanical example of poor impedance matching is the use of a regular tennis ball with (relatively small) table-tennis rackets. *Impedance mismatch* of another type occurs when a bulb intended for a low voltage application (flashlights, automobiles, some external house lighting) is used in a house application; the bulb then gets so much power that it burns up. A mechanical example is the use of a table-tennis ball with (relatively large) regular tennis rackets. Proper impedance matching is a fundamental design principle.

We now preview a few simple but important equations in order to apply them to some real-life situations.

R.2.4 *Current Is the Rate of Charge Flow*

If a constant current I flows for a time t, then it transfers a charge Q given by

$$Q = It. \qquad \text{(relating charge and current)} \qquad \text{(R.1)}$$

Hence the unit of charge, the coulomb (C), has the same units as the ampere-second, so C = A-s. If 0.2 C of charge is transferred in 5 s, by (R.1) this corresponds to a current $I = 0.2/5 = 0.04$ A.

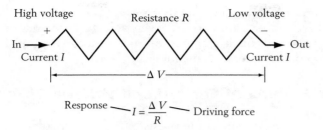

Figure R.3 A resistor, a voltage difference, and an electric current: Ohm's law.

R.2.5 *Ohm's Law: When Current Is Proportional to the Driving Voltage*

Ohm's law is an experimentally determined relation that holds for most materials (e.g., sea water or copper wire), but not all materials (e.g., the important semiconductors silicon and germanium). When the two ends of a wire are connected, respectively, to the high- and low-voltage terminals of a voltaic cell or of an electrical outlet, there is a voltage difference ΔV across the wire. Associated with ΔV is the electric current I passing through the wire. See Figure R.3, where the wire is represented by a jagged line.

Ohm's law says that (1) current I flows in the direction from high voltage to low voltage; (2) I is proportional to the *voltage difference* ΔV across the object:

$$I = \frac{\Delta V}{R}. \qquad \text{(Ohm's law)} \qquad \text{(R.2)}$$

In (2), *proportional* means that R, called the *electrical resistance*, is independent of the value of ΔV. Ohm's law holds for copper, but not for silicon. Equation (R.2) can be made to apply to silicon if we let R depend on ΔV.

Here's how to "read" (R.2). Knowing how to "read" an equation is an important skill. Equation (R.2) implies that if we measure both the "input" ΔV and the "output" I, and then we employ $R = \Delta V/I$, then we can obtain the electrical resistance R. The unit of electrical resistance, the *ohm*, or Ω (the Greek letter Omega), is the same as a volt/amp $= V/A$. Thus $\Omega = V/A$. Equation (R.2) does not apply to objects that store appreciable amounts of electrical energy (capacitors) or magnetic energy (inductors).

Equation (R.2) also implies that if you increase the "input" ΔV, then you also increase the "output" I; and at fixed ΔV if you increase the resistance R, then you decrease "output" I. See Figure R.3. An equation like (R.2) holds for the *water current* through a pipe with a *fluid resistance*, driven by the *pressure difference* between one end of the pipe and the other. Of course, the units of voltage and pressure are different, as are the units of electric current and water current, and as are the units of electrical resistance and fluid resistance. Note that it is pressure *difference* that drives a water current; water will not flow through a pipe whose ends are connected to two reservoirs at the same pressure. Similarly, it is voltage *difference* that drives electric current through a wire; electric charge Q will not flow through a wire whose ends are connected to two charge reservoirs at the same voltage. For water, we also can drive water current with water pumps. For electricity, we also can drive an electric current with voltaic cells, thermoelectric devices, electromagnetic induction, and by a variety of other means. Any source of energy that drives an electric current (even voltage difference) is called an *electromotive force*, or *emf*. Such a source of energy does work on the electric charge, so it also provides a force that causes an electric current to flow.

R.2.6 *Power Is the Product of Current and Voltage Difference*

The power $\mathcal{P}$ (in watts, or W) going into a resistor (in the form of heat) is given by

$$\mathcal{P} = I \Delta V. \tag{R.3}$$

When (R.2) and (R.3) are combined, they yield another equation,

$$\mathcal{P} = I^2 R, \qquad \text{(rate of joule heating)} \tag{R.4}$$

first obtained by Joule, and for that reason sometimes called *Joule's law*.

From (R.3), the greater the current I at fixed ΔV, the greater the power $\mathcal{P}$; and the greater the voltage difference ΔV at fixed current I, the greater the power $\mathcal{P}$: 4 A at 120 V provides 480 W; 8 A at 120 V provides 960 W; and 8 A at 240 V provides 1920 W. We will now employ equations (R.1–R.4) to answer some basic questions about power, voltage, current, electrical resistance, and electrical safety.

R.2.7 *Applications: Toasters and Power Cords*

Consider a toaster, one of the simplest of electrical devices. Its working element is a heat-resistant wire. Assuming that it produces $\mathcal{P} = 720$ W, and using $\Delta V = 120$ V, (R.3) yields a current of $I = 6$ A. Putting this into (R.2) yields a value of $R = 20\,\Omega$ for the electrical resistance of the toaster. Hence, from the current rating or power rating of a household appliance, we can deduce its electrical resistance. Similarly, a 50-foot-long, 16-gauge extension cord that is rated at 13 A for 125 V must also be rated at $13 \cdot 125 = 1625$ W for 125 V. Excess power will start to melt the wire's insulation.

R.2.8 *Overload: Fuses and Circuit Breakers*

Most modern house wiring is rated to carry safely a current of either 15 A or 20 A. Circuit breakers (found in a fuse or breaker box, often located in some obscure part of the house) protect the house wiring from carrying too large an electric current; they "trip" if the current exceeds the rated value. Overload can occur by using too many appliances on the same outlet; if a 1000 W hairdryer, a 600 W toaster, and a 1200 W microwave oven were all to use the same outlet, the total power consumption would be 2800 W, corresponding to 23.3 A, an overload even on a 20 A circuit.

Extension cords provide a way to exceed a rated value, even for a house that is properly wired. In the summer of 1992, a fraternity house in Bryan, Texas, burned down; someone had operated an air conditioner using an extension cord with too low a current rating. The extension cord, under the overload of current, began heating up like toaster wire, ultimately setting on fire the insulation or a nearby object. For an air conditioner rated at $\mathcal{P} = 2400$ W, and $\Delta V = 120$ V, (R.3) yields $I = 20$ A. Circuit breakers can safely carry such a current, but a 10 A or 12 A extension cord cannot. Note that, by (R.2), the air conditioner, when running, has an effective electrical resistance of $R = 120/20 = 6\,\Omega$.

Electrical motors, such as those employed in air conditioners, have different electrical properties when they are turning than when they have not yet started to turn. When turning, electrical motors produce a *back emf* that opposes the driving emf, and this causes the current to be less when it is running than when it is starting up. When an electrical motor is prevented from turning, no back emf is produced, so a larger current goes to the motor, which can cause it to burn out.

Fuses are intended to burn out if excessively large currents flow through them, thus protecting electrical devices and electrical wiring from too large a current flow. Circuit breakers, on the other hand, do not burn out, and can be reset, and for that reason they have supplanted fuses in modern buildings. In the 1940s and earlier, when fuses were used instead of circuit breakers, many a house burned down because, on overloaded circuits, people "cleverly" replaced fuses by pennies, which permitted a much higher current flow than the fuses they replaced. (Those who knew that pennies would serve to pass current, like a fuse, but didn't know that they wouldn't protect the house wiring, unlike a fuse, illustrate the maxim that "a little knowledge is a dangerous thing.") Fuses (which typically are used in automobiles) must be replaced, once the cause of the electrical problem has been fixed.

R.3 Some Uses of Electrical Power

Fans, blenders, and many other appliances employ electric motors to convert electrical energy to mechanical energy. Electric motors use the magnetism of electric currents to provide the torque needed to turn the fan or blender blade.

Automobiles use the chemical energy of a car battery to provide electrical energy to start the starting motor, which in turn provides the mechanical energy to start the gasoline engine. (The earliest automobiles employed no starting motors and no batteries; cars were started by the driver turning a crank to provide the mechanical energy to start the engine. That is the origin of the term *crank over* used to describe how the starting motor gets the gasoline engine turning.) The chemical energy in gasoline (released as explosions within the cylinders of the engine) is converted to mechanical energy (the pistons move, and this causes the crankshaft to turn). Some of this mechanical energy gets converted into electrical energy by an electrical generator—also called an alternator. This goes into recharging the chemical energy of the battery, which then has enough chemical energy to start the car later. Motors and generators are discussed in Chapter 13.

Radios and TVs receive, tune, demodulate (i.e., extract the useful signal), filter, and amplify weak and scrambled electromagnetic signals (Chapter 15), making them intelligible and clear. Stereo systems do much the same for unintelligible signals embedded in plastic on a record or compact disk. We discuss many aspects of the operation of these devices in Chapter 14.

A "walkman" employed to play compact disks (CDs) or cassette tapes uses up the chemical energy in its voltaic cells much more rapidly than one used only to listen to radio stations. This is because it takes much more energy to turn a CD and to amplify the signal from the CD than to simply amplify the signal from the radio station. In terms of (R.3), the voltage difference ΔV is the same in both cases, but the current I is much greater when the CD, rather than the radio, is used.

When you use a computer, the keyboard may actuate by detecting the effect of electric charge rushing back and forth when you exert pressure on a key. Inside the computer is a "hard disk" made of a magnetic recording material. This records information according to whether the magnetizations of some tiny magnetic particles are pointing along or opposite to a given direction in the plane of the disk. Most important of all to the computer are its integrated circuits, which contain miniaturized versions of circuit devices called resistors (Chapter 7), capacitors (Chapter 6), and transistors (Chapter 7). A monitor using a vacuum tube yields images from light produced by electrons that have been guided, either by electric forces (Chapters 2 and 3) or magnetic forces (Chapters 10 and 11), to the screen, on which special materials called phosphors (Chapter 5) have been deposited. (Phosphors absorb energy from the electrons and quickly release that energy as light.) Many portable computers use monitors with a liquid crystal screen; the images on the screen are determined by electric forces acting on the molecules of the liquid crystal material. Ink-jet printers are not only powered and controlled by electricity, but even the ink is guided by electrical forces (Chapters 2 and 3) as it moves down toward the paper. Electrical forces hold the ink to the paper, just as they hold together the atoms and molecules of our own bodies.

R.4 Electricity from Voltaic Cells: DC Power in an Automobile

Voltaic cells cause electric current in an electric circuit to flow in only one direction, the direction being determined by how it is wired. This is called *direct current*, or dc, in distinction to the oscillating current provided by the electric company, *alternating current*, or ac. Here are some questions and answers about electrical power associated with an automobile's use of voltaic cells.

1. How much power do car headlights use? The ratings on the packages reveal that each low beam uses 35 W and each high beam uses 65 W. Ordinary house lightbulbs usually exceed this, but are not as bright; unlike headlights, house lightbulbs do not send off their light in a relatively narrow beam.

2. What is the voltage of a car battery? Typical car batteries are rated at 12 V.

3. What does "charge" mean for voltaic cells? Here charge has two meanings. In the chemical sense, charge is the amount of chemicals available for electricity-producing chemical reactions. In the electrical sense, charge is the actual electricity released by electricity-producing chemical reactions. These need not be the same, because the chemical charge can be used up by non-electricity-producing reactions. A car battery might have a charge specified as 50 A-hr, which by (R.1) is equivalent to $50 \text{ A} \cdot 3600 \text{ s} = 1.8 \times 10^5$ C.

4. How long does it take to discharge a car battery? A car battery with a charge of 50 A-hr will discharge after it has produced either 50 A for one hour, or 10 A for 5 hours, and so forth. We now find the time to discharge it if we leave the headlights on. Using the 35 W value for one low beam, we must use a power of 70 W for both. From (R.3), with 12 V, that means a current of 5.83 A. Hence, by (R.1) it takes 50 A-hr/5.83 A = 8.6 hr to fully discharge a 50 A-hr battery.

5. What maximum current can a car battery provide? Ads for batteries tell us that: "600 cold-cranking amps." Some batteries can produce as much as 1000 A. They are intended for use either at very high temperatures (where they very readily discharge due to non-electricity-producing chemical reactions) or at very low temperatures (where all chemical reactions—even those producing electricity—are suppressed).

6. What is the electrical resistance r of a car battery? We can estimate it, using the maximum current and the concept of impedance matching (of the battery and the starting motor). In (R.2) we use 12 V for the battery, a current of 600 A, and a total resistance of the impedance-matched battery and starting motor of $R = r + r = 2r$. This leads to $r = R/2 = \Delta V/2I = 0.01\ \Omega$. A battery may be characterized by its emf, its "charge," and its internal resistance.

7. Which provides electrical energy more cheaply, the battery company or the electric company? The electric company—by about a factor of 100! This is why we don't light our houses with giant flashlights. We use batteries primarily because they are portable sources of electrical energy.

8. Why do voltaic cells run down? They obtain their energy from chemical reactions at the terminals (called the *electrodes*). Material at the electrodes and in the interior of the voltaic cell (called the *electrolyte*) are consumed by the chemical reactions.

9. What is the difference between the way electric charge is carried in a wire and in a voltaic cell? In a wire, negatively charged electrons carry the electric charge. In a voltaic cell, ions (which are much more massive than electrons and can be either positive or negative) carry the electric charge.

10. If a battery is thought of as a "pump" for electricity, where in the battery is the pump located? Each of the two electrode–electrolyte interfaces, at the positive and negative terminals, serves as a pump. Small AA cells and larger D cells have the same chemistry, and thus the same pump strength per unit surface area of their electrodes. But the much larger D cell has a much larger charge in its electrolytes and electrodes. D and AA cells have about the same internal resistance.

11. Does a 9 V battery work on a different principle than a 1.5 V AA cell? No. Open up a 9 V battery and you will find six 1.5 V cells connected in series (the positive terminal of one is connected to the negative terminal of the next). Note that not all 1.5 V cells employ the same chemical reactions.

12. How do jumper cables work? They connect two batteries in parallel (both of the positive terminals are connected, and both of the negative terminals are connected). Then both batteries can provide electric power to the starting motor.

R.5 Two Practice Exam Questions

The two questions that follow have consistently produced an enormous variety of individually unique incorrect answers, revealing an equally enormous variety of individually unique incorrect conceptions about the subject of electricity. After posing them, we will answer these questions explicitly, and in the process (we hope) establish a common language.

1. A rock 'n' roll band is playing a concert on an electrically conducting platform (e.g., steel) that is electrically insulated (e.g., dry wood) from the rest of the concert hall. During the concert, the platform is accidentally connected to a high-voltage ac source; perhaps there is a short connecting the platform to the "hot" lead of a frayed wire from a guitar amplifier. The band members' hair stands on end. Why? (*Note:* at 120 V, unless the power is direct current, the surface of the body is unlikely to charge up enough for hair to stand on end; let's assume it happens anyway, for argument's sake.)

2. At the end of the concert, the band must descend from the platform. Although it would be safer simply to cut off the power, let us assume this cannot be done. Should the band members jump down from or step off the platform?

To answer these questions, which require no quantitative knowledge of electricity, we set up the mental picture of the electric fluid model, based on an analogy between electricity and water, both of them fluids. Although this analogy is not perfect, it was taught with enthusiasm by no less a practitioner of the electrical science than J. J. Thomson. Thomson's careful measurement, in 1897, of the ratio of electrical charge to inertial mass for the emissions from diverse cathode materials, convinced scientists that there was a unique common component to all "cathode rays"—the electron. In 1936, as an elder statesman of physics, Thomson wrote:

> [*The service of the electric fluid concept*] *to the science of electricity, by suggesting and co-ordinating experiments, can hardly be overestimated.* [*For, in the laboratory,*] *if we move a piece of brass or decrease the effect we are observing, we do not fly to the higher mathematics, but use the simple conception of the electric fluid which would tell us as much as we wanted to know in a few seconds.*

In order to avoid having to "fly to the higher mathematics" (which even mathematically sophisticated scientists sometimes would like to avoid), we too will employ the "conception of the electric fluid." The reader will be warned when the analogy breaks down, at which time the model will be modified to produce a more precise physical picture of the phenomenon of electricity. *Science constantly develops and refines its most important ideas.*

R.6 The Electric Fluid Model

The version of the electric fluid model presented here is an extension of the original version by Franklin. The present one is more accurate since it is based on our

Table R.1 Equivalences for the electric fluid model

Water	Electricity, or electric fluid
1. Amount (mass)	Amount (electric charge)
2. Pressure	Electrical potential or voltage
3. Reservoir	Capacitor
4. Mass flow (mass current)	Charge flow (electric current)
5. Pipes and resistance to mass flow	Wires and resistance to charge flow
6. Pumps	Batteries or the electric company
7. Reservoir breakdown	Electrical breakdown (e.g., sparking)

current knowledge of the microscopic constitution of matter. Under the circumstances, Franklin's conception was remarkably good; moreover, in the scientific tradition, he modified his views, as new facts became available.

Broadly speaking, we can set up the equivalences shown in Table R.1.

1. **Amount: Mass and Electric Charge**   Ordinary matter has a number of properties, the most important of which for our purposes are mass and electric charge. Mass M (measured in kilograms, kg) is a positive quantity, and mass density (mass per unit volume, kg/m^3) is also a positive quantity. Water has a background mass density that is always positive, and increases only slightly when the system is put under pressure. Moreover, the addition of matter can only increase the amount of mass.

On the other hand, electric charge can be either positive or negative. The *net* electric charge Q is the algebraic sum of the positive and negative electric charges. Increases in the net electric charge can occur either by increasing the amount of positive charge or by decreasing the amount of negative charge. The addition of matter can increase, decrease, or leave unchanged the amount of electric charge. Ordinary atoms have zero net charge: they have positive charge in their relatively massive nuclei and an equal amount of negative charge in the relatively light electrons surrounding the nucleus. The less massive electrons can be stripped off or added to an atom or collection of atoms (e.g., a solid).

Franklin's Electric Fluid

Franklin, on the other hand, considered ordinary matter to be immobile and uncharged, sort of a sponge for the positively charged electric fluid. The most important aspect of Franklin's conception is that there *is* an electric fluid that can be neither created nor destroyed, and therefore it is *conserved*.

Franklin's Contributions

To Franklin we owe the concept of an excess or deficit of electrical fluid, the idea of connecting electrical storage devices in series and in parallel, a deep appreciation of the distinction between conductors and insulators, and the lightning rod. He is also responsible for bifocals, the rocking chair, the heat-retaining Franklin stove, daylight-saving time, and a host of other inventive ideas. It was Franklin's fame as a scientist—many of his electrical experiments were performed in the court of the French King Louis XV—that later gave him the credibility in France to plead the case of the American Colonies against the British.

Recall from (R.2) that the unit of charge, the coulomb (C), is based upon the unit of electric current, the ampere (A). If one amp of dc current flows for one second, it transfers one coulomb of charge, so $C = A\text{-s}$. Recall that a 50 A-hr battery has a "charge" Q of It, or $(50\ A)(3600\ s) = 180,000\ C$. This vastly exceeds the amount that can be held by the nonchemical charge storage devices called capacitors, to be discussed shortly. (Batteries can provide currents as large as their emf divided by their internal resistance, and for a considerable period of time. Capacitors can provide very large currents, but only for very short periods of time.) The charge on the electron is $-e$, where $e \approx 1.6 \times 10^{-19}\ C$. This is very small; billions of electrons are transferred to your comb when you run it through your hair. (This sounds like a lot of electrons. However, your hair contains billions of billions of billions of electrons. Don't worry about rubbing too many electrons out of your hair.)

2. **Pressure and Electrical Potential, or Voltage**   Pressure P is force per unit area. This can be rewritten as

$$\text{pressure} = \frac{\text{force}}{\text{area}} = \frac{\text{force-distance}}{\text{area-distance}} = \frac{\text{energy}}{\text{volume}}.$$

Dividing both the far left-hand and the far right-hand sides of this equation by mass per unit volume yields

$$\frac{\text{pressure}}{\text{mass per unit volume}} = \frac{\text{energy}}{\text{mass}}.$$

This quantity, an energy per unit mass, has the same units as gravitational potential and is analogous to electrical potential, which is an energy per unit charge. However, because a fluid like water is nearly incompressible, pressure and pressure divided by mass per unit volume are nearly proportional to each other. Hence pressure and voltage are nearly analogous.

With an excess of positive or negative charge, we can produce either positive or negative voltage. On the other hand, it is not possible to produce stable negative pressures, although small negative pressures can be produced temporarily (e.g., when a motor blade turns quickly through the water). At negative pressures, bubbles spontaneously form—the technical term is *cavitation*.

Pressure in water depends on the collisions of neutral atoms with a wall or with one another. Pressure changes are communicated similarly by collisions between water molecules. This leads to the generation of sound, which in water travels at a speed of about 1435 m/s. On the other hand, voltage changes are communicated by the long-range electrical force, which propagates at the velocity of light (about 3×10^8 m/s). This implies that, along an ordinary electric circuit, voltage changes are nearly instantaneously transmitted, and that throughout a "charge reservoir" the voltage is nearly uniform. However, for a long communications cable, the signal is precisely a nonuniform voltage, which travels along the cable in a short but finite time.

3. **Reservoirs and Capacitors**   Water is often stored in a reservoir or a water tower. For our purposes, we will think of a reservoir as being a tank of water

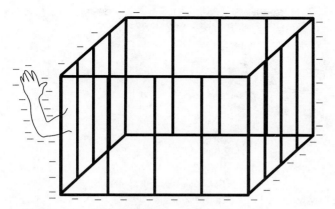

Figure R.4 Schematic of how electric charge goes to the outside of a conductor. The invisible body associated with the ghostly hand is in electrical contact with the cage, which itself is both charged up and on an insulating platform.

that is under pressure. Electric charge, too, can be stored in electrical reservoirs, called *capacitors*.

Although a water reservoir has all the water uniformly distributed throughout its *volume*, a charge reservoir, such as a metallic plate or bar, has any excess charge distributed (not necessarily uniformly!) over its *surface*. A person standing within a charged-up metal cage will not be charged up, but on passing his hand through the bars, charge will flow from the outside of the cage to the person's hand, which now serves as part of the outside surface of the cage. See Figure R.4. In Chapter 4, we will study why this occurs. It is related to the fact that there are two types of electric charge, that "opposites attract and likes repel", and to the specific way in which the electrical force between two charges falls off with distance.

Before electrical reservoirs, experimenters worked directly with the electricity produced by an electrical source (such as a silk-rubbed glass rod). The first type of electrical reservoir was a bar of metal, called a prime conductor, placed upon an insulating surface to keep it from losing its charge. (Because, in equilibrium, the net charge on a conductor resides on its surface, a cannonball and a metalized balloon of the same radius are equally effective at storing charge.) The second type of storage device was the Leyden jar (see Figure R.5). It has two surfaces—the outer tinfoil surrounding the glass and the inner water surface touching the glass—that hold equal and opposite charges $\pm Q$. Later the Leyden jar was given the name *condenser* (this usage survives today in the language of automobile ignition systems) because it "condensed" the electricity. There is a voltage difference ΔV across the two plates, with $+Q$ at the higher voltage. The modern name for the condenser, the *capacitor*, indicates that these devices have the *capacity* to store electric charge. This is expressed in terms of what is called the capacitance $C = Q/\Delta V$, whose unit is the *farad* (F), which is the same as a coulomb/volt, so F = C/V.

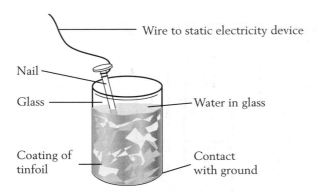

Figure R.5 The Leyden jar. The glass prevents charges on the inner and outer surfaces from making contact.

Modern capacitors consist of two pieces ("plates") of electrical conductor separated by an electrical insulator, in geometries that maximize the area of each plate and minimize the plate separation (subject to no plate contact, which would cause discharge). In this way, large amounts of electric charge, equal in quantity but opposite in sign, can be collected. Capacitors are crucial for the proper operation of nearly all types of electrical equipment. Just as the pressure in a pressurized water reservoir is uniform, so the voltage on a single capacitor plate is uniform. The two plates of a capacitor are like two distinct reservoirs of water.

4. **Mass and Charge Flow** A mass current to a region corresponds to an increase in the mass of that region. Mass current, or mass flow, is measured in units of kg/sec. Water current is sometimes expressed in terms of volume flow, in units of m^3/sec.

To repeat, electric current, or charge flow, is measured in units of the ampere A, or coulomb/sec = C/s, sometimes called the amp. An electric current to a region corresponds to an increase in the *net* charge of that region. A net increase can be due to a flow of positive charge into the region, a flow of negative charge out of the region, or a combination of the two. In a voltaic cell, or in salt water, the charge carriers are both positively and negatively charged ions (e.g., in salt water, the Na$^+$ is a positive ion, and the Cl$^-$ is a negative ion). Ions lead to *ionic conduction*. In a metal, the charge carriers are electrons. However, not all electrons in a metal are able to move freely; the ones that move freely are called *conduction electrons*. Using the modern conception of the *orbital*, we would say that the conduction electrons are in *delocalized* orbitals, which can extend throughout the metal. Other electrons in the metal are in *localized* states and cannot move freely. Electrons lead to *electronic conduction*. In insulators, all the orbitals are localized. This explains Gray's observation that some materials are conductors and some are insulators.

Electron Orbitals

Newton's laws do not correctly describe the orbital motion of electrons within atoms and molecules. However, modern physics (properly, what is called quantum mechanics) permits electron orbits to be described in terms of orbitals. These specify the probability of finding an electron at any given position in space.

Electrical *insulators*, such as air or glass or wood, conduct electricity, but much less effectively than good electrical conductors, such as copper. For example, glass is about 20 orders of magnitude less effective than copper at carrying electricity. This extraordinary variation in material properties is why, as a first approximation, we can make a simple distinction between conductors and insulators. The *electrical conductivity* and its inverse, the *electrical resistivity*, provide a continuous scale for the conducting abilities of materials. Semiconductors, like silicon, are intermediate between conductors and insulators in their ability to conduct electricity.

5. **Pipes and Wires** Just as pipes have a certain *resistance* to the flow of water, so do wires have a certain *resistance* to the flow of electricity. To make the analogy more precise, the water pipes should be filled with fine sand or, better yet, with powder. Then the friction on the water occurs throughout the volume of the pipe, by collisions of the water against the sand. This is like the friction on the electrons that occurs throughout the volume of the wire, by collisions of the electrons against the atomic nuclei. The sand also serves to prevent turbulent flow of the water. Turbulent flow of electricity takes place only in ionized plasma, such as the atmosphere of a star or inside some vacuum tubes.

Here is an important difference between water in pipes and electrons in wires: the electrons are always in the wires, ready to produce an electric current, whereas sometimes the pipes have to fill with water before they can produce a water current.

A quantity of water, in motion becoming a water current, is driven through a pipe from higher pressure to lower pressure. Similarly, a quantity of electricity, in motion becoming an electric current, is driven through a wire from higher voltage to lower voltage, as described by (R.2). The amount of mass current or electric current is proportional to the amount of the pressure *difference* or voltage *difference*, the amount of current being *inversely proportional* to the *resistance* of the pipe or tube. Monitoring the pressure difference (or voltage difference) along the length of a current-carrying pipe (or wire) of uniform constitution and cross-section shows that it varies linearly from the higher pressure (or voltage) to the lower one. Just as, at a given instant of time, different water molecules enter one end of the pipe and exit the other end, so different electrons enter one end of a wire and exit the other end.

6. **Pumps and Batteries of Voltaic Cells** Just as water pumps generate pressure differences between the ends of a pipe, so do voltaic cells generate voltage differences across the ends of a wire. A water pump can drive water current around a *water circuit* that includes a fountain. For example, the water circuit could be: pump-to-pipe#1-to-fountain-to-pipe#2-to-pump-to-pipe#1.... Similarly, a voltaic cell can drive electric current around an *electric circuit* that includes, for example, a lightbulb. The electric circuit would be: voltaic cell to wire#1-to-lightbulb-to-wire#2-to-voltaic cell-to-wire#1.... Further, a capacitor can drive current through a lightbulb, but such current typically lasts only a short time because capacitors usually discharge relatively quickly.

Just as the water company provides water at high pressure, the electric company provides electricity at high voltage. However, whereas the water company provides dc power (at a certain pressure, corresponding to the

In describing water, we do not say that "a high pressure passed across the pipe"; we say that "a high current passed through the pipe, driven by the high pressure difference across its ends." However, in describing electricity, the popular press often employs incorrect usage, such as "he received a shock when a high voltage passed through him." Sometimes *surged* ≡ *passed* might be employed. By Ohm's law, (R.2), much better usage would be "on touching the wire, there was a high voltage difference between his hand and the ground, causing a large electric current to pass through him."

height of the water tower), the electric company provides an oscillating, or ac, power. Recall that dc stands for *direct current*, and ac stands for *alternating current*. Voltage is standardized throughout each nation (otherwise electrical equipment would not operate properly), whereas water pressure can vary considerably from town to town. However, ac power varies from nation to nation: in Europe, it typically has a frequency of 50 Hz and a voltage of 220 V.

The idea of a battery of voltaic cells is due to Volta. Chemical energies limit voltaic cells to emfs on the order of a few volts, so to get higher voltages with voltaic cells, we follow Volta and put them in series. Before Volta, Franklin had employed batteries of Leyden jars, both in series and in parallel.

7. **Breakdown Phenomena**   Just as the walls of a reservoir or a pipe normally are impermeable to the flow of water, so electrical *insulators*, such as air or glass or wood, normally are impermeable to the flow of electricity. In some cases, there is temporary electrical breakdown in an insulator—for example, sparking in air—corresponding to the temporary opening of a relief valve in a high pressure tank. Sometimes, however, there is catastrophic electrical breakdown—such as sparking, or even lightning—corresponding to the bursting of a water tank under too high a pressure. *Sparking* and *lightning are very variable*. A lightning bolt might carry a peak current of 10,000 A and transfer a net negative charge of a few C from a cloud to the earth. (However, lightning sometimes transfers charge the other way; when it does the bolts are exceptionally powerful!) Under ordinary conditions, negative charge flows upward within the atmosphere from the earth, partly due to negative ions flowing upward and partly from positive ions flowing downward. Lightning is a complex and still poorly understood phenomenon.

R.7   Answers to the Practice Exam Questions

With the fluid analogy in mind, let us consider the questions posed earlier.

1. Why does the hair of the band members stand on end?

We first answer this question for the artificial situation where the power is dc, rather than ac. Ordinarily, the platform and the band members on it will be electrically neutral. However, if the platform has become charged up with either an excess or deficit of electrons, then the band members, being good conductors relative to dry wood—animals are composed mostly of salty

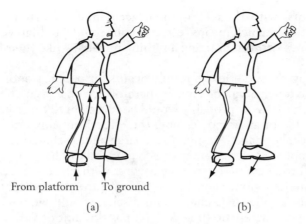

From platform   To ground

(a)   (b)

Figure R.6 Depictions of electric current flow.
(a) Electric shock. (b) Electrical discharge.

water—also will charge up. Because the band members are conductors, this excess charge goes to their surface, which includes their hair. Strands of hair, being relatively light, and charged with the same sign, will repel one another (they will also be repelled by the charge all over the platform); thus their hair stands on end. (Some people think that, somehow, the hair stands up on end because it is discharging into the air. Not so. Hair can stand up on end even when the voltage is too small to cause noticeable discharge into the air.)

2. Should the band members jump off or step off?

Assume that the band members are wearing shoes with soles of leather (a much better electrical conductor than rubber). When they step down from the platform, they will provide a relatively good path for electricity to flow from the platform to ground, and vice versa. The electricity will be supplied constantly by the power source, and thus will flow through them as long as they are in contact with both platform and ground. The electrical path from one foot to the other passes the midriff and will include the region of the heart, so the chance of electrocution will be significant. See Figure R.6(a).

On the other hand, if the band members jump to the ground, they will be much safer. At the instant they jump, they have a small excess of electric charge, which they retain until they hit the ground and receive a small, momentary shock as they discharge. See Figure R.6(b). They then will be at the same voltage as ground, and there will be no more current flow, unlike what would happen if they were touching the platform with one foot, and the ground with the other. In a sense, while in the air they are like a charged capacitor, and when they discharge, they are like a discharging capacitor.

R.7.1 *How Long Does It Take to Charge and Discharge?*

Actually, the power provided by the electric company is ac. Anecdotal evidence indicates that the hair-standing-on-end effect does *not* usually occur for ac voltages: drivers of earth-moving equipment have been known to operate their equipment for hours, with their equipment accidentally shorted ("short" for "short wire connection," which means "directly connected") to high voltage (e.g., 440 V) ac power lines without noticing anything unusual—like their hair standing on end.

The hair-standing-on-end effect certainly occurs for people standing on insulated platforms who touch a Van de Graaf generator as it charges up to thousands of dc volts. Although 440 V dc probably would be sufficient to cause hair to stand on end, for the 440 V of an ac power line, the effect probably does not occur because the time it takes hair to respond to charge and discharge exceeds $1/60$ of a second. In Chapter 8, we consider the charge and discharge of a capacitor of capacitance C through a path, such as a wire, with electrical resistance R. (Do not confuse the italic-font capacitance, C, with the Roman-font coulomb, C.) Technically, this is called RC charge and discharge, and we will show that it takes place in the characteristic time RC. (R has the unit of $\Omega = V/A$, and C has the unit of $F = C/V$, so RC has the unit of $(V/A)(C/V) = C/A = $ second.) Since it takes on the order of a few seconds for someone's hair to stand up when connected to a Van de Graaf generator, for this situation, RC may be as large as a few seconds. This long charging time is an indication that, although Van de Graaf generators can develop a high dc voltage, they cannot provide a large continuous current. For that reason the spark, however uncomfortable, discharges the generator, which then takes a few seconds to recharge. (Batteries produce lower voltages, but more sustained currents.) A fluid analogy to RC discharge would be the discharge of a water tank (of finite fluid capacity) through a water pipe of finite fluid resistance. The larger the tank or the narrower—and thus more resistive—the pipe, the longer the time to charge or discharge.

R.7.2   *How Large Is a Dangerously Large Current?*

The previous example touches on the question of electrical safety. *How large a current can pass through a person without hurting him or her?* Current is at issue here, not voltage. Although voltage drives the current, the current passing through a person is responsible for detrimental effects.

There is no single answer to this question. At least three factors are important: (1) the electrical resistance varies significantly with the moisture on the person's skin (lie detectors use the fact that people under emotional stress tend to sweat, and sweat conducts electricity much better than dry skin; thus sweat decreases the electrical resistance of the skin); (2) the electrical resistance depends on where electrical contact is made; and (3) the nature of the damage depends on where the current passes.

Thus, a person who touches two wires at very different voltages will get a shock that depends on where the two wires contact the person. If the contact is across two fingers on the same hand, the current will flow across the hand. Burns, perhaps serious, may result. However, if one hand touches one wire, and the other hand touches the other wire, the current will pass across the region of the heart. This may disturb the natural electrical self-excitation of the heart, setting off the heart into a disorganized twitching of the heart muscles called *ventricular fibrillation*. Here, little blood is actually pumped, because the chambers of the heart contract out of sequence. This is much like what happens when the ignition of an automobile is far out of tune. A current of only 0.1 A (and sometimes only

0.001 A) may be enough to cause this to occur. In such cases, it is often easier to stop the heart completely by giving it an even larger shock, and then to restart it properly by yet another shock.

R.8   Beyond the Electric Fluid Model

R.8.1   *Electrical Polarization and Electrostatic Induction*

We have yet to discuss some very important electrical phenomena that are not included in the electric fluid model. The term *electrical polarization* describes a rearrangement, with no change in the net charge, of the electric charge on any kind of material, insulator or conductor, caused by an external source of electric charge. One side of the material tends to become more positive and the other side tends to become more negative, in proportion to the amount of external source charge. For insulators, polarization of individual localized atomic or molecular orbitals occurs. For conductors, the term *electrostatic induction* is used to describe polarization. (Electrostatic induction should not be confused with electromagnetic induction, which involves the motion of electricity in nonstatic situations.)

Conductors typically are either electrolytes or metals. In electrolytes, electrostatic induction occurs by the motion of positive ions one way and negative ions the other way. In metals, electrostatic induction occurs by the distortion of delocalized orbitals from one side of the conductor to the other. Electrostatic induction in metals occurs more rapidly than in ionic conductors because electrons, having less inertia, can adjust more rapidly than can ions.

R.8.2   *Humidity and Sparking*

It has been known for centuries that humidity inhibits the production and retention of static electricity. This is because water molecules (H_2O) in their natural state have an excess of positive charge near the H nuclei, and an excess of negative charge near the O nucleus. For that reason, water molecules can be attracted to a source of static electricity, where they either gain or lose charge—in the form of either an ion or, more likely, the relatively light and negatively charged electron. Thus, static electricity experiments do not work well during the summer, or in poorly ventilated rooms filled with large numbers of people. On the other hand, in winter, when the humidity is usually low, it is not uncommon to receive an electrical shock when you walk across a rug in shoes with rubber soles, and then touch a good conductor, such as a doorknob. This process is considerably more complex than described in many accounts, which normally dismiss it with the phrase *static electricity*.

Sparking involves four stages. (1) The shoes *charge up*, with many billions of negatively charged electrons transferred by rubber from the rug to the rubber soles. (2) Positive ions in the body are attracted to, and negative ions in the body (including the hand) are repelled by, the negative charge on the shoes, as in Figure R.7(a). (The charge on the rug cancels the charge on the shoes, the net charge on the person is zero, and the net charge on the doorknob–doorplate assembly is zero.) (3) As the negatively charged finger approaches the doorknob, charge on the doorknob rearranges by electrostatic induction, giving an excess

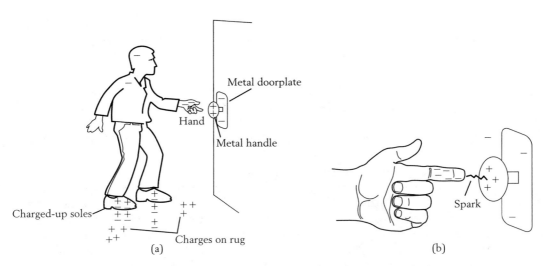

Figure R.7 An example of spark discharge. (a) Electrostatic induction with charge conservation. (b) The sparking process.

of positive charge near the finger. (4) When the person's hand approaches close enough to the doorknob, a spark occurs, and the finger rids itself of excess negative charge, as in Figure R.7(b). *In Figure R.7 there are too many charges to depict (some* 10^{23}, *with nearly equal amounts of positive and negative!), so we draw only uncompensated charge.* The sparking process itself is surprisingly complex; it is initiated not by electrons on the finger or the door handle, but rather by electrons in the air.

R.8.3 *Some Inadequacies of the Electric Fluid Model*

Some deficiencies in the simple electric fluid analogy have already been mentioned. Mass is positive only, whereas electric charge can have either sign. Negative pressures exist, but only down to the value at which bubble formation, or cavitation, occurs. However, there is no such limit on how negative a voltage can be—as long as electrical breakdown, which can occur for both positive and negative voltages, does not occur. Another difference is that charge transfer by friction has no fluid analog. Neither does electrostatic induction or electrical polarization. Thus, *despite the analogy to a fluid under pressure, it is important to study electricity on its own terms.* Nevertheless, if the electric fluid is extended to include these special properties, its usefulness can be increased.

Here is yet another difference between ordinary fluids and the electric fluid. When a pump drives fluid through a pipe, stresses are set up in both the fluid and the pipe, and this drives the fluid through the pipe. In terms of energy flow, the pump provides energy via the fluid to distant regions of the fluid. On the other hand, when a battery is placed in an electric circuit, it causes charge to be distributed over the surface of the wires in the circuit. This surface charge actually drives electricity through the wire. In terms of energy flow, the battery provides energy via the region outside the wire to distant regions of the wire. We will say more about this in the next section.

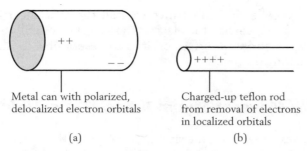

Metal can with polarized,
delocalized electron orbitals

Charged-up teflon rod
from removal of electrons
in localized orbitals

(a)

(b)

Figure R.8 Different interpretations of positive and negative charges. An excess of electrons is represented by a negative; a deficit, by a positive. (a) For conductors, the charges represent the collective effect of some 10^{23} delocalized electron orbitals. (b) For insulators, the charges represent the effect of individual electrons.

R.8.4 *How to Visualize the Particles of Electricity*

Figures often represent electricity by pluses and minuses (see Figure R.8). Although the density of the pluses and minuses represents only the average electric charge, we tend to interpret the figures as if a minus corresponds to an individual electron, and a plus corresponds either to a positively charged nucleus or the absence of an individual electron. For insulators, where the electron orbitals really are localized, this interpretation can be accurate. However, such an interpretation is misleading when applied to electrical conductors, which contain charge carriers that extend throughout the entire conductor. For electronic conductors (such as metals), all the conduction electron orbitals, which overlap one another and extend over the entire conductor, adjust slightly when a charged rod is brought near. There, a minus represents an excess and a plus represents a deficit in the average density of the 10^{23} or so conduction electrons in a mole of electronic conductor. For ionic conductors (such as salt solutions), the ions redistribute slightly when a charged rod is brought near. Thus, for conductors the pluses and minuses represent a more collective effect.

R.8.5 *Resolving an Apparent Contradiction*

When an electron is added to a conductor—for example, a small blob of solder sitting on an insulating surface—the electric fluid model tells us that the excess electric charge due to this electron goes to the surface of the blob of solder. If the electron itself were to go to the surface, then it would go there to occupy a surface orbital (i.e., an orbital localized on the surface). However, unoccupied surface orbitals typically are at relatively high energies compared to unoccupied bulk orbitals. Therefore the added electron goes into a bulk orbital and roams over the entire blob of solder, including the interior. To satisfy the requirement that the excess charge go to the surface, all the conduction electrons—including the electron that was just added—then adjust their orbitals slightly outward in just the right way to place on the surface a net charge equal to that of one electron. Again, by a collective effect involving all the charge carriers, rather than any individual charge-carrier, the electric charge gets redistributed over the electrical conductor. Electrons in surface orbitals are responsible for the behavior of many semiconducting devices.

R.8.6 *How Electricity Moves: Vacuum Tubes, Wires, Lightning*

How electricity moves depends on the situation. Consider the following analogy. Imagine that there is a ballroom, with doors at opposite ends. Let us count people entering and leaving.

If the ballroom is completely empty, then for person Z to leave, he must have first entered. A similar thing happens in a vacuum tube (such as a TV tube), where electrons are emitted at the negatively charged cathode and travel without stopping until they hit a screen, where we can count them by the light that is emitted.

If the ballroom is completely full, then for person Z to leave, he must first have been pushed out by person Y standing next to him, who was pushed by person X, by a process that leads back to person A, who entered at the other end. This is somewhat like what happens in a wire filled with electrons, where charge from electrons enters one end, and charge from other electrons leaves at the other end. (Because the electric force is of long range, the image of pushing only on one's neighbors cannot be taken literally.) A similar pushing process also happens in electrolytes (like salt water) filled with ions, where an ion enters at one end and another ion leaves at the other end. (A more literal pushing process, mediated by molecular collisions, is also the means by which sound gets from its source to your ear; the source pushes adjacent air, that air pushes air a little farther away, and so on, until the air next to your ear gets pushed, and then that air pushes against your ear.)

Now consider that the ballroom is filled with dancers, male and female. Assume that, as long as they are together, they stay together. However, if they come apart, although they try to come together again (and sometimes they succeed, perhaps with different partners), the males tend to go to the left door and the females to the right door. Suddenly a male comes barreling into the room from the right door, traveling leftward, crashing into dancing partners and separating them. This separation of dancers from their partners gives the possibility that many more males will leave to the left than actually entered at the right, and many more females will leave to the right than entered at the left. This resembles what happens in sparking, where a high-energy electron passes through a material, kicking electrons off neutral atoms, and causing an *electron avalanche*. (Note that males and females have nearly the same mass, but electrons are much less massive than atoms. Hence this analogy would be misleading in describing momentum transfer due to collisions.)

Lightning bears similarities both to an electron avalanche and to a wire. (No charges can get from cloud to ground—or vice versa—without collisions, unlike the case of the vacuum tube.) Initially, there is only neutral air. Then there is something like an electron avalanche, filling the air with electrons and ions. Finally, one set of electrons start entering at one end and another set start leaving at the other. We will discuss lightning in somewhat more detail in Chapter 6.

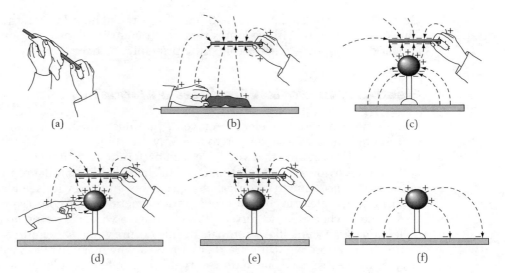

Figure R.9 Field-line and charge representation of a sequence of charging operations. (a) Charging process. (b) Charge separation. (c) Electrostatic induction on conducting sphere. (d) Just before discharge of sphere. (e) After discharge of sphere. (f) Isolated, charged, spherical conductor.

R.8.7 *Visualizing Electric Charge and Its Effect: The Electric Field*

One of the most powerful tools to assist our understanding of electricity was developed in the 19th century by Michael Faraday. It is a visual schema, whereby what are called *electric field lines* are considered to leave positive charges and enter negative charges. The number of field lines is proportional to the amount of charge. In Figure R.9 we illustrate this with a sequence of drawings.

In Figure R.9(a), an initially neutral teflon rod is rubbed with a piece of cloth. When the cloth and the hand holding it are pulled off, as in Figure R.9(b), there are eight negative charges on the rod, and an equal number of field lines entering the rod. In addition, a number of field lines come from positive charges on the cloth and from the hands of the person who did the rubbing, as well as a few that are due to charges that are now so far away that we see only their field lines. In Figure R.9(c), the rod is brought up to a conducting sphere that rests on an insulated stand. The sphere is thereby subject to electrostatic induction but, being neutral, it has as many field lines entering (four) as leaving. In Figure R.9(d), the left-hand finger approaches the sphere, the finger being subject to electrostatic induction as well as the sphere. Note the high concentration of equal and opposite charge between the finger and the sphere. (We have not drawn two of the positive charges on the finger and two of the negative charges on the sphere, nor the arrows of the associated field lines, to avoid cluttering the diagram.) Figure R.9(e) depicts the situation after the finger has touched the sphere, drawn off some charge, and then been removed. Figure R.9(f) depicts the situation when the charged rod is removed, leaving behind the charged sphere. Now the positive charge on the sphere rearranges, and subjects the tabletop to electrostatic induction, so the field lines from the sphere lead to the tabletop. This discussion implicitly assumes that electric charge does not spontaneously appear or disappear—what is called the principle of charge conservation.

R.8.8 *Electricity in Motion Produces a Magnetic Field*

Electricity in motion, or electric current, produces a *magnetic field*. It can be represented in a fashion similar to the way we represent the electric field in Figure R.9. (We assume that you have seen patterns of iron filings around a magnet, which provide a representation of its magnetic field. A similar representation of the *electric field* due to electrical charge can be obtained with seeds suspended in oil.) Because electric current produces a magnetic field, a compass needle will be deflected when electric current flows through a wire. However, no analogous deflection of anything like a compass needle occurs when water current flows in a pipe. In addition, we will learn that a time-varying magnetic field will *induce* a circulating electric field, and a time-varying electric field will *induce* a circulating magnetic field. For these phenomena, there are no analogies in traditional fluid flow. Again, if the electric fluid model is to be used in such situations, it must be extended to include these properties.

R.8.9 *How to Visualize the Flow of Electrical Energy*

The flow of electrical energy provides yet another major difference between the flow of water and the flow of electricity. Consider imaginary circular slabs cut from a cylindrical pipe and a cylindrical wire. For water flowing along the pipe, the energy flow vector exists only within the fluid, and points along the direction of the pipe, and the net flow into the slab occurs because the energy flow vector is greater on the input side than the output side. See Figure R.10(a). However, for electricity flowing along a wire, the energy flow vector exists throughout all of space. It originates at the source of electrical energy (which might be a capacitor or a battery), and it flows toward the electrical resistances within the circuit. Since the wires of the circuit typically have only a small electrical resistance, the energy flow vector (called the Poynting vector) mostly points parallel to the wires, until it reaches the resistors, where it points inward. See Figure R.10(b).

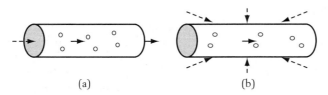

(a)   (b)

Figure R.10 Distinction between matter flow and energy flow. (a) Water flows within a pipe (associated with energy flow). (b) Electrons flow within a wire, but energy associated with the electric field flows even outside the wire (e.g., radio and TV waves).

R.8.10 *Conclusion*

Despite the greater sophistication of modern conceptions of the atomic and electronic nature of matter, the electric fluid model—suitably modified—continues to be a valuable way to represent the electric charge and electric current for an electrical conductor.

In reading the next eight chapters, keep in mind that the electric fluid is characterized by three important properties: electric charge (which is conserved), electric field (a vector that is produced by electric charge), and electric voltage (a scalar that also is produced by electric charge). In Chapter 1, we consider charge conservation. Next, we consider electric force (Chapter 2), electric field (Chapter 3), the relation between charge and field (Chapter 4), and the relations between field and voltage and between charge and voltage (Chapter 5). With the concept of voltage finally defined precisely, we turn our attention to capacitance C in Chapter 6 (where every part of a piece of metal in equilibrium has the same voltage), to resistance R in Chapter 7 (where the voltage varies along a piece of metal that is not in equilibrium), and both in Chapter 8. At the end of Chapter 8, before we begin the study of magnetism, we discuss the limitations of the lumped circuit approach (R's and C's) to the response of electric circuits.

R.9 Review of Vectors in Three Dimensions

In the 1880s, vectors were invented independently by the physicist Gibbs and the self-taught electrical engineer Heaviside. Heaviside's motivation was to rewrite, in compact form, the equations of electricity and magnetism developed by Maxwell. The dashed lines in Figure R.9 represent a vector: the electric field vector $\vec{E}$. There is a corresponding magnetic field vector $\vec{B}$. Using his new vector language, Heaviside first wrote down what are now called Maxwell's equations. Mathematicians later generalized the vector idea, but the generalization loses some of the specificity of meaning. In what follows, you will learn why the rules for manipulating vectors—and, in particular, the definitions of the scalar product and the vector product—have *not* been chosen arbitrarily.

R.9.1 *What Is a Vector?*

The first definition you probably had of a vector is that it is a quantity with *magnitude* and *direction*. A position vector $\vec{r}$ in three dimensions can be represented by its components along the $\hat{x}$-, $\hat{y}$-, $\hat{z}$-axes, which gives the triplet of numbers (x, y, z) or $\langle x, y, z \rangle$. Under rotation, the components of a position vector transform according to a specific rule. That rule (which we're not going to specify because in general it is very complex) must preserve (1) the magnitude of any vector and (2) the angle between any two vectors.

Vectors more general than position vectors can be defined. A vector can be a velocity or an acceleration or a force; by taking higher time derivatives of the acceleration, we can form an infinite number of vectors. A vector can be an electric or magnetic field. Such vectors transform under rotations in the same way that position vectors transform. As a consequence, the magnitude of any vector, and the angle between any two vectors—even vectors of different types, like position and velocity—do not change under rotations.

If a triplet of numbers doesn't transform properly under rotations, then it isn't a vector. For example, a triplet of numbers that doesn't change at all when the coordinate system is rotated is not a vector. Thus the triplet (P_1, P_2, P_3), where P_1, P_2, and P_3 are the phone bills for your first three months in college, is a triplet of numbers, but it's not a vector under rotations. Neither is the triplet $(|x|, |y|, |z|)$; it transforms under rotations, but not in the right way. Our definition of a vector in terms of its behavior under rotation is very restrictive.

Note that each of the entries in a vector $\vec{a} \equiv (a_x, a_y, a_z)$ must have the same dimension; we can't have a_x a distance and a_y a velocity.

Get in the habit of writing arrows over vectors. Something like $F = m\vec{a}$, with a vector on the right-hand side and a scalar on the left-hand side, is mathematical nonsense.

$$\vec{F} = m\vec{a} \rightarrow \text{GOOD}; \qquad F = m\vec{a} \rightarrow \text{BAD}.$$

Note: $\vec{F}_1$ is a vector whose name is the subscript 1. Hence $\vec{F}_x$ is a vector whose name is the subscript x, not the x-component of $\vec{F}$. Also, the x-component of $\vec{F}$ is written as F_x or $(\vec{F})_x$, not as $|\vec{F}_x|$.

R.9.2 *What Is a Scalar?*

By scalar, we mean a number (perhaps one with dimensions, such as a mass) that doesn't change under rotations. The single number, or singlet, $\psi' \equiv a_x$ is *not* a scalar under rotations, because a_x of the vector $\vec{a}$ changes under rotations; however, the magnitude $|\vec{a}|$ *is* a scalar under rotations.

R.9.3 *Addition and Subtraction of Vectors*

The rules that follow are clearly true for position vectors. They are also true for other types of vectors, such as forces or velocities or accelerations.

To add and subtract two vectors $\vec{a}$ and $\vec{b}$, they must have the same dimension (e.g., both position vectors or velocity vectors, but not $\vec{a}$ a position vector and $\vec{b}$ a velocity vector). Then

$$\vec{a} + \vec{b} = \vec{b} + \vec{a}. \tag{R.5}$$

That is, vector addition satisfies a commutative rule. (So does addition of the real numbers.) You should verify (R.5) by (1) in the xy-plane, drawing two vectors (call them $\vec{a}$ and $\vec{b}$), their tails both on the origin; and (2) adding them with the tail of $\vec{a}$ on the origin and the tail of $\vec{b}$ on the tip of $\vec{a}$ (this is $\vec{a} + \vec{b}$), and adding them in the reverse order.

For three vectors with the same dimension, we have the right and left associativity rules that

$$\vec{a} + (\vec{b} + \vec{c}) = (\vec{a} + \vec{b}) + \vec{c} = \vec{a} + \vec{b} + \vec{c}. \tag{R.6}$$

That is, the order in which we add vectors, *no matter how many of them*, doesn't matter (similarly, for addition of the real numbers). Suppose we are to sum 100 vectors, the second 50 of which are the negatives of the first 50. This rule allows us to rearrange them to get zero for each pair of opposites, and then sum to get zero, rather than moronically (e.g., like a computer) adding them all up and then subtracting them all back down to zero.

Subtraction of $\vec{b}$ from $\vec{a}$ is defined as the vector addition of $\vec{a}$ and $-\vec{b}$, $-\vec{b}$ has the same magnitude as $\vec{b}$, but points in the opposite direction. (If we think of the real numbers as having a direction with respect to the origin, then for a real number d, the real number $-d$ has the same magnitude $|d|$ but points in the opposite direction.) Thus we write

$$\vec{a} - \vec{b} \equiv \vec{a} + (-\vec{b}).\tag{R.7}$$

Note that

$$\vec{a} - \vec{b} = -(\vec{b} - \vec{a}).\tag{R.8}$$

That is, vector subtraction satisfies an anticommutative rule (so does subtraction of real numbers).

R.9.4 *Multiplication of Vectors by Scalars*

If λ is a scalar, and $\vec{a}$ is a vector (e.g., an acceleration or a position), the composite quantity $\lambda\vec{a}$ is a vector that is along (opposite) $\vec{a}$ if λ is positive (negative). If $\vec{a}$ has magnitude $|\vec{a}|$, then $\lambda\vec{a}$ has magnitude $|\lambda||\vec{a}|$. The commutative law

$$\lambda\vec{a} = \vec{a}\lambda\tag{R.9}$$

holds, so it doesn't matter which way we multiply vectors by scalars. Note that if λ has dimensions (e.g., mass), then $\lambda\vec{a}$ has different dimensions than $\vec{a}$ (e.g., $\vec{F} = m\vec{a}$ of Newton's second law).

Under multiplication by the scalar λ, vector addition satisfies the right- and left-distributive laws:

$$\lambda(\vec{a} + \vec{b}) = \lambda\vec{a} + \lambda\vec{b}, \qquad (\vec{a} + \vec{b})\lambda = \vec{a}\lambda + \vec{b}\lambda.\tag{R.10}$$

R.9.5 *Vectors in Cartesian Coordinates: Magnitudes and Unit Vectors*

Let the unit vectors $\hat{i}$, $\hat{j}$, and $\hat{k}$ point along the x-, y-, and z-axes, respectively. They are normal (perpendicular) to one another, forming a right-handed triad, as determined by the *vector product right-hand rule*. See Figure R.11(a), where the thumb is along $\hat{k}$, the curled fingers are along $\hat{j}$, and the uncurled fingers would be along $\hat{i}$. This rule is so important that this chapter illustrates it twice.

With the unit vectors defined, let us now consider two vectors $\vec{a}$ and $\vec{b}$, expressed in terms of their cartesian components. That is,

$$\vec{a} = a_x\hat{i} + a_y\hat{j} + a_z\hat{k}, \qquad \vec{b} = b_x\hat{i} + b_y\hat{j} + b_z\hat{k}.\tag{R.11}$$

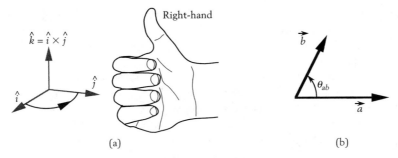

Figure R.11 (a) A unit triad and a curled right hand. (b) Two vectors in their common plane, and the angle between them.

By the Pythagorean theorem, the magnitude $|\vec{a}|$ of $\vec{a}$ can be obtained from

$$\vec{a} \cdot \vec{a} = a_x^2 + a_y^2 + a_z^2 = |\vec{a}|^2. \tag{R.12}$$

Under multiplication of a vector $\vec{a}$ by a scalar λ, each component of $\vec{a}$ is multiplied by λ. This is relevant to unit vectors. We can convert any $\vec{a}$ to a unit vector $\hat{a}$ via

$$\hat{a} = \frac{\vec{a}}{|\vec{a}|}, \tag{R.13}$$

which corresponds to multiplying $\vec{a}$ by $\lambda = |\vec{a}|^{-1}$.

Example R.1   Magnitude of a vector

Let $\vec{a} = (12, 15, -16)$ in units of cm. Find $|\vec{a}|$.

Solution: Equation (R.12) gives $|\vec{a}| = \sqrt{(12)^2 + (15)^2 + (-16)^2} = 25$ cm.

Example R.2   Scalar multiplication and unit vectors

For $\vec{a}$ of the previous example, find $\hat{a}$.

Solution: Because $|\vec{a}|^{-1} = 0.04 \text{ cm}^{-1}$, by (R.13) we find $\hat{a} = (0.04) \cdot (12, 15, -16) = (0.48, 0.6, -0.64)$. You can verify that $|\hat{a}| = 1$.

Example R.3   Rotating a vector

We now put together the different parts of this section. Let $\vec{a} \equiv (3, 4, 0)$, corresponding to a vector that is 5 units long, at an angle of $\tan^{-1}(4/3) = 53.1$ degrees counterclockwise to the x-axis.

(a) Find the vector $\vec{a}'$ that corresponds to a clockwise rotation by 25 degrees.

(b) Show that the magnitudes $|\vec{a}'| = |\vec{a}|$.

Solution: At the outset, note that the rotation should yield a vector that is 5 units long, at an angle of $53.1 - 25 = 28.1$ degrees counterclockwise to the x-axis.

(a) Because a vector is the sum of its components, we can consider how the x- and y-components transform separately. The x-component of $\vec{a}$, or 3, transforms into a part $3\cos(25)$ along x, and a part $-3\sin(25)$ along y. The y-component of $\vec{a}$, or 4, transforms into a part $4\cos(25)$ along y, and a part $4\sin(25)$ along x. The sum has x-component $3\cos(25) + 4\sin(25) = 4.41$, and y-component $-3\sin(25) + 4\cos(25) = 2.36$. Thus $\vec{a}' \equiv (4.41, 2.36, 0)$. This makes an angle of $\tan^{-1}(2.36/4.41) = 28.1$ degrees counterclockwise to the x-axis, as expected.

(b) By (R.12), $|\vec{a}'| = \sqrt{4.41^2 + 2.36^2 + 0^2} = 5.00 = |\vec{a}|$, as expected. Thus $|\vec{a}|$ transforms as a scalar.

R.10   Multiplication of Vectors

R.10.1   *Can We Multiply Two Vectors to Obtain a Third Vector?*

Consider two vectors $\vec{a}$ and $\vec{b}$, as in (R.11). Choose a coordinate system with a plane containing $\vec{a}$ and $\vec{b}$, where the angle between them is called $\theta_{\vec{a},\vec{b}}$ and is less than $180°$. See Figure R.11(b). (There is an infinite number of such coordinate systems because we can rotate them by any angle about their normal.)

Can we form from $\vec{a}$ and $\vec{b}$ a new vector? Like the rule for multiplication of ordinary numbers, the rule for vector multiplication should give a result that is proportional to the magnitudes $|\vec{a}|$ and $|\vec{b}|$ of each of the vectors. (We call this proportionality to both magnitudes *bilinearity*). Also, it should give a vector that transforms under rotations like a vector. We hope it will also have certain desirable properties that make it easy to rearrange terms in an algebraic expression—such as commutivity (or anticommutivity) and distributivity (right and left). Before answering this question—in the affirmative (but we're going to have to replace commutivity by anticommutivity)—we raise a red flag.

Consider the triplet

$$\psi \equiv [|a_y a_z|^{1/2}, |a_z a_x|^{1/2}, |a_x a_y|^{1/2}] \tag{R.14}$$

constructed from the true vector $\vec{a} \equiv (a_x, a_y, a_z)$. For $\vec{a} \equiv (5, 0, 0)$, (R.14) gives $\psi \equiv [0, 0, 0]$. Under a rotation about z by $\tan^{-1} 4/3 \approx 53°$, $\vec{a}$ goes from $(5, 0, 0)$ to $(3, 4, 0)$. Then, by (R.14) ψ goes from $[0, 0, 0]$ to $[0, 0, 12^{1/2}]$. This creature ψ certainly does not behave like a vector under rotations, for it goes from the origin to a point on the z-axis when we rotate about the z-axis, and its magnitude has changed! *The moral is that we can't just make up an arbitrary triplet and expect it to behave like a vector under rotations.*

For that reason, before considering how to construct a third vector from the vectors $\vec{a}$ and $\vec{b}$, let's first consider the simpler problem of how to construct a scalar from $\vec{a}$ and $\vec{b}$.

R.10.2   *Scalar Product and Its Algebraic Properties*

We'll now construct the *scalar product*, so-called because it is a scalar under rotations; we also call it the *dot product* because of the dot symbol ($\cdot$) we use to represent this operation.

The scalar product arises naturally in many physical contexts, such as the definition of work W. If a constant force $\vec{F}$ moves an object by the vector distance $\vec{r}$, the work done is

$$W = |\vec{F}||\vec{r}||\cos\theta_{\vec{F},\vec{r}}|, \tag{R.15}$$

where $\theta_{\vec{F},\vec{r}}$ is the angle between $\vec{F}$ and $\vec{r}$. This quantity is a scalar under rotation because $|\vec{F}|$, $|\vec{r}|$, and $\theta_{\vec{F},\vec{r}}$ don't change under rotation.

More generally, to obtain the scalar product of two vectors $\vec{a}$ and $\vec{b}$, we multiply the magnitude $|\vec{a}|$ of the first vector by the projection $|\vec{b}|\cos\theta_{\vec{a},\vec{b}}$ of the second vector along the first. Thus the scalar product is a measure of the projection of one vector on another. In equation form,

$$\vec{a}\cdot\vec{b} = |\vec{a}||\vec{b}|\cos\theta_{\vec{a},\vec{b}}. \tag{R.16}$$

Because $|\vec{a}|$, $|\vec{b}|$, and $\theta_{\vec{a},\vec{b}}$ are scalars under rotations, the dot product of (R.16) also is a scalar under rotations. Clearly, the scalar product is bilinear in $\vec{a}$ and $\vec{b}$. In scalar multiplication, $\vec{a}$ and $\vec{b}$ need not have the same dimensions.

The corresponding angle $\theta_{\vec{b},\vec{a}}$ from $\vec{b}$ to $\vec{a}$ equals $\theta_{\vec{a},\vec{b}}$. Thus the dot product in the reverse order is

$$\vec{b}\cdot\vec{a} = |\vec{b}||\vec{a}|\cos\theta_{\vec{b},\vec{a}} = |\vec{b}||\vec{a}|\cos\theta_{\vec{a},\vec{b}} = |\vec{a}||\vec{b}|\cos\theta_{\vec{a},\vec{b}}. \tag{R.17}$$

Hence, comparing (R.16) and (R.17), we see that the order of the vectors doesn't matter:

$$\vec{a}\cdot\vec{b} = \vec{b}\cdot\vec{a}. \quad \text{(commutative law)} \tag{R.18}$$

The dot product is said to be *commutative*; the projection of $\vec{b}$ on $\vec{a}$ is the same as the projection of $\vec{a}$ on $\vec{b}$.

The dot product is also said to be *right distributive*, meaning that the projection of $\vec{b}+\vec{c}$ on $\vec{a}$ in $\vec{a}\cdot(\vec{b}+\vec{c})$ is the sum of the projections of $\vec{b}$ on $\vec{a}$ and of $\vec{c}$ on $\vec{a}$. See Figure R.12(a). We use the notation $\vec{c}_{par}$ to indicate the component in the $\vec{a}\vec{b}$-plane of the vector $\vec{c}$. This takes advantage of the fact that we can choose a coordinate system such that $\vec{a}$ is along the x-axis, and $\vec{b}$ is in the xy plane. Although $\vec{c}$ must be given all three components, its component along z doesn't matter for the purposes of projection along x. Thus

$$\vec{a}\cdot(\vec{b}+\vec{c}) = \vec{a}\cdot\vec{b} + \vec{a}\cdot\vec{c}. \quad \text{(right distributive law)} \tag{R.19}$$

The dot product is also *left distributive*, meaning that the projection of $\vec{b}+\vec{c}$ on $\vec{a}$ in $(\vec{b}+\vec{c})\cdot\vec{a}$ is the sum of the projections of $\vec{b}$ on $\vec{a}$ and of $\vec{c}$ on $\vec{a}$. We can also obtain this left distributive law by applying (R.18) (twice) and (R.19) (once):

$$(\vec{a}+\vec{b})\cdot\vec{c} = \vec{c}\cdot(\vec{a}+\vec{b}) = \vec{c}\cdot\vec{a} + \vec{c}\cdot\vec{b} = \vec{a}\cdot\vec{c} + \vec{b}\cdot\vec{c}. \quad \text{(left distributive law)} \tag{R.20}$$

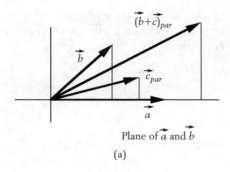

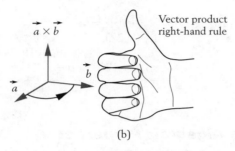

Figure R.12 (a) The distributive law of the scalar product. (b) The vector product right-hand rule.

R.10.3 *Scalar Product in Cartesian Components*

Let's now consider how to evaluate $\vec{a} \cdot \vec{b}$ in terms of the components of $\vec{a}$ and $\vec{b}$. To do this, first we have to work out the scalar product between the unit vectors. By (R.16) the scalar product of two unit vectors yields one or zero, according to whether or not a unit vector is dotted with itself (angle zero and cosine unity) or with a perpendicular unit vector (angle 90° and cosine zero). That is,

$$\hat{i} \cdot \hat{i} = \hat{j} \cdot \hat{j} = \hat{k} \cdot \hat{k} = 1, \qquad \hat{i} \cdot \hat{j} = \hat{j} \cdot \hat{i} = \hat{j} \cdot \hat{k} = \hat{k} \cdot \hat{j} = \hat{k} \cdot \hat{i} = \hat{i} \cdot \hat{k} = 0.$$
(R.21)

We can now apply the two commutative laws and (R.21) to $\vec{a}$ and $\vec{b}$ written out in component form. This yields

$$\vec{a} \cdot \vec{b} = (a_x\hat{i} + a_y\hat{j} + a_z\hat{k}) \cdot (b_x\hat{i} + b_y\hat{j} + b_z\hat{k}) = a_xb_x + a_yb_y + a_zb_z, \qquad \text{(R.22)}$$

where the distributive laws permitted us to rearrange (not shown) the nine dot products of unit vectors, and (R.21) permitted us to reduce those nine dot products to only three non-zero terms.

Note that, if $\vec{b} = \vec{a}$, then $\theta_{\vec{a},\vec{b}} = 0$, and its cosine is one. Thus, by (R.12) the dot product should yield the magnitude squared of the vector $\vec{a}$. Indeed, for this case (R.22) reduces to (R.12).

Comparing (R.16) and (R.22) we find that

$$\vec{a} \cdot \vec{b} = |\vec{a}||\vec{b}| \cos\theta_{\vec{a},\vec{b}} = a_xb_x + a_yb_y + a_zb_z. \qquad \text{(R.23)}$$

Equation (R.23) permits us to quickly, accurately, and easily compute the magnitude of any vector (set $\vec{b} = \vec{a}$ to get $|\vec{a}|$, and the angle between any two vectors). Our scalar product is bilinear, commutative, and distributive.

Example R.4   **Scalar product**

Let

$$\vec{a} \equiv (12, 15, -16), \qquad \vec{b} \equiv (175, -168, 576). \tag{R.24}$$

Find $|\vec{a}|$, $|\vec{b}|$, $\vec{a} \cdot \vec{b}$, and θ.

Solution: By (R.12), $|\vec{a}| = \sqrt{(12)^2 + (15)^2 + (-16)^2} = 25$ and
$|\vec{b}| = \sqrt{(175)^2 + (-168)^2 + (576)^2} = 625$. By (R.22),
$\vec{a} \cdot \vec{b} = (12) \cdot (175) + (15) \cdot (-168) + (-16) \cdot (576) = -9636$. By (R.24),
$\cos \theta_{\vec{a},\vec{b}} = (-9636)/(25 \cdot 625) = -0.6167$, so $\theta = 128°$ or $\theta = 232° \equiv -128°$.

R.10.4   *Vector Product and Its Algebraic Properties*

We'll now construct the *vector product*, so-called because it is a vector under rotations; we also call it the *cross product* because of the cross symbol ($\times$) we use to represent this operation.

The vector product arises in many physical contexts, such as the definition of torque $\vec{\tau}$—sometimes called $\vec{\Gamma}$. (The Greek letter τ is spelled *tau*, the Greek letter γ is spelled *gamma*, and Γ is capital γ.) Let a constant force $\vec{F}$ act on an object by the vector distance $\vec{r}$ with respect to some origin P. Then the torque $\vec{\tau}$ about P, due to $\vec{F}$, has magnitude

$$|\vec{\tau}| = |\vec{r}||\vec{F}| \sin \theta_{\vec{r}, \vec{F}}|, \tag{R.25}$$

where $\theta_{\vec{r}, \vec{F}}$ is the angle between $\vec{r}$ and $\vec{F}$. $|\vec{\tau}|$ is a scalar under rotation because $|\vec{F}|$, $|\vec{r}|$, and $\theta_{\vec{r}, \vec{F}}$ don't change under rotation. Because the torque $\vec{\tau}$ causes the object to spin about an axis given by the vector product right-hand rule (applied to $\vec{r}$ and $\vec{F}$), we identify the direction of $\vec{\tau}$ with this axis.

More generally, to obtain the vector product $\vec{a} \times \vec{b}$ of two vectors $\vec{a}$ and $\vec{b}$, we obtain its magnitude as the area of the parallelogram formed by $\vec{a}$ and $\vec{b}$, so

$$|\vec{a} \times \vec{b}| = |\vec{a}||\vec{b}|| \sin \theta_{\vec{a},\vec{b}}|. \tag{R.26}$$

This magnitude is a scalar under rotation because $|\vec{a}|$, $|\vec{b}|$, and $\theta_{\vec{a},\vec{b}}$ don't change under rotation. The vector product typically has different dimensions than either of the original vectors. Clearly, (R.26) is bilinear in $|\vec{a}|$ and $|\vec{b}|$.

In general, to obtain the direction of $\vec{a} \times \vec{b}$, we employ the vector product right-hand rule that is used to go from the unit vectors $\hat{i}$ and $\hat{j}$ to the perpendicular unit vector $\hat{k}$. More generally, for two vectors $\vec{a}$ and $\vec{b}$, swing your right hand in the plane of $\vec{a}$ and $\vec{b}$ from $\vec{a}$ to $\vec{b}$ through the angle of less than $180°$; your thumb will then point along the direction of $\vec{a} \times \vec{b}$. See Figure R.12(b). (We assume that $\vec{a}$ and $\vec{b}$ are not collinear.)

A "stupidity check" that you're not doing something totally wrong (the sort of "zeroth rule" you'd like to have about anything) is (1) identify the plane of $\vec{a}$ and $\vec{b}$; (2) take $\vec{a} \times \vec{b}$ normal to this plane; (3) if your application of the vector product right-hand rule is not normal to that plane, you know you've done something wrong, and you should start over.

With this rule, the effect of the operation $\times$ on the unit vectors is

$$\hat{i} \times \hat{j} = \hat{k} = -\hat{j} \times \hat{i}, \qquad \hat{j} \times \hat{k} = \hat{i} = -\hat{k} \times \hat{j}, \qquad \hat{k} \times \hat{i} = \hat{j} = -\hat{i} \times \hat{k},$$

$$\hat{i} \times \hat{i} = \vec{0}, \qquad \hat{j} \times \hat{j} = \vec{0}, \qquad \hat{k} \times \hat{k} = \vec{0}, \qquad \text{(R.27)}$$

where $\vec{0} \equiv (0, 0, 0)$ has *all three* components set to zero.

In general, the cross-product rule is *not* commutative: it gives

$$\vec{a} \times \vec{b} = -\vec{b} \times \vec{a}. \qquad \text{(R.28)}$$

Because of the sign change, the cross-product is said to be *anticommutative*.

We can prove the right distributive law

$$\vec{a} \times (\vec{b} + \vec{c}) = \vec{a} \times \vec{b} + \vec{a} \times \vec{c} \qquad \text{(R.29)}$$

by a geometrical construction. First, we project $\vec{b}$ and $\vec{c}$ onto the plane perpendicular to $\vec{a}$. This is done in Figure R.13(a), where we take $\vec{a}$ to be normal to the page, and we employ the subscript *perp* to indicate only the parts of $\vec{b}$, $\vec{c}$, and $\vec{b} + \vec{c}$ that are perpendicular to $\vec{a}$. Clearly, $(\vec{b}_{perp} + \vec{c}_{perp}) = \vec{b}_{perp} + \vec{c}_{perp}$. Next, we consider the cross-products, obtained from the projections on multiplying by $|\vec{a}|$ and rotating counterclockwise by $90°$. This is done in Figure R.13(b), which shows that (R.29) indeed is satisfied.

Using anticommutivity (twice) and right distributivity (once), we have

$$(\vec{a} + \vec{b}) \times \vec{c} = -\vec{c} \times (\vec{a} + \vec{b}) = -\vec{c} \times \vec{a} - \vec{c} \times \vec{b} = \vec{a} \times \vec{c} + \vec{b} \times \vec{c}$$

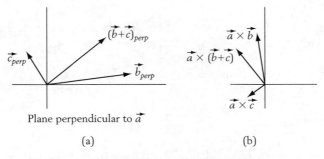

Figure R.13 (a) The distributive law for the addition of the components of two vectors normal to a third. (b) The distributive law of the vector product.

Thus we have derived the left distributive law

$$(\vec{a} + \vec{b}) \times \vec{c} = (\vec{a} \times \vec{c}) + (\vec{b} \times \vec{c}). \tag{R.30}$$

In summary, our vector product is bilinear, anticommutative, and distributive.

With (R.29) and (R.30), we can now evaluate the vector product in component form. We have

$$\vec{a} \times \vec{b} = (a_x\hat{i} + a_y\hat{j} + a_z\hat{k}) \times (b_x\hat{i} + b_y\hat{j} + b_z\hat{k})$$
$$= (a_yb_z - a_zb_y)\hat{i} + (a_zb_x - a_xb_z)\hat{j} + (a_xb_y - a_yb_x)\hat{k}, \tag{R.31}$$

where the two distributive laws permitted us to rearrange (not shown) the nine cross-products of unit vectors, and (R.27) permitted us to reduce those nine cross-products to only six non-zero terms.

The three apparently complex terms in (R.31) are related. Define the vector $\vec{d} \equiv \vec{a} \times \vec{b}$. On making the *cyclic permutation* $(x, y, z) \rightarrow (y, z, x)$, the $d_x = (a_yb_z - a_zb_y)$ term multiplying $\hat{i} \equiv \hat{x}$ becomes the $d_y = (a_zb_x - a_xb_z)$ term multiplying $\hat{j} \equiv \hat{y}$; the $d_y = (a_zb_x - a_xb_z)$ term multiplying $\hat{j} \equiv \hat{y}$ becomes the $d_z = (a_xb_y - a_yb_x)$ term multiplying $\hat{k} \equiv \hat{z}$; and the $d_z = (a_xb_y - a_yb_x)$ term multiplying $\hat{k} \equiv \hat{z}$ becomes the $d_x = (a_yb_z - a_zb_y)$ term multiplying $\hat{i} \equiv \hat{x}$.

Equation (R.31) may be rewritten as a determinant; see Problem R-10.15.

R.10.5 *Testing for Vectorness*

Equation (R.31) has x-, y-, and z-components, which makes it *look* like a vector, and we have constructed it in such a way as to be a vector, but how can we be sure it really is a vector? Does $\vec{a} \times \vec{b}$ of (R.31) transform like a vector under rotations? And is its magnitude $|\vec{a} \times \vec{b}|$ given by (R.26)?

Because of the vector product right-hand rule, $\vec{a} \times \vec{b}$ should be perpendicular to the two vectors $\vec{a}$ and $\vec{b}$, even under rotations. If that is true, then the scalar product of $\vec{a} \times \vec{b}$ with $\vec{a}$ and $\vec{b}$ should be zero. Let's check it out.

In (R.23), take $\vec{a}$ from (R.11) and replace $\vec{b}$ by $\vec{a} \times \vec{b}$ of (R.31); this leads to

$$\vec{a} \cdot (\vec{a} \times \vec{b}) = a_x(a_yb_z - a_zb_y) + a_y(a_zb_x - a_xb_z) + a_z(a_xb_y - a_yb_x) = 0. \tag{R.32}$$

Here $a_x(a_yb_z)$ is canceled by $a_y(-a_xb_z)$, and there are analogous cancellations for the other terms. In a similar way, we can show that $\vec{b} \cdot (\vec{a} \times \vec{b}) = 0$. Thus, as expected, $\vec{a} \times \vec{b}$ is perpendicular to both $\vec{a}$ and $\vec{b}$, even if $\vec{a}$, $\vec{b}$, and $\vec{a} \times \vec{b}$ are all rotated. Therefore the *direction* of $\vec{a} \times \vec{b}$ rotates like a vector.

We still must verify that the magnitude $|\vec{a} \times \vec{b}|$ satisfies (R.26), and thus doesn't change under rotations. With (R.31), (R.12), and (R.23) we obtain (leaving out some algebra)

$$|\vec{a} \times \vec{b}|^2 = (a_yb_z - a_zb_y)^2 + (a_zb_x - a_xb_z)^2 + (a_xb_y - a_yb_x)^2$$
$$= (a_x^2 + a_y^2 + a_z^2)(b_x^2 + b_y^2 + b_z^2) - (a_xb_x + a_yb_y + a_zb_z)^2$$
$$= |\vec{a}|^2|\vec{b}|^2 - |\vec{a}|^2|\vec{b}|^2 \cos^2\theta_{\vec{a},\vec{b}} = |\vec{a}|^2|\vec{b}|^2 \sin^2\theta_{\vec{a},\vec{b}}. \tag{R.33}$$

This is in agreement with (R.26). Note that $|\vec{a} \times \vec{b}|$ doesn't change under rotations, because $|\vec{a}|$, $|\vec{b}|$, and $\sin\theta_{\vec{a},\vec{b}}$ don't change under rotations.

Example R.5 Vector product

For $\vec{a}$ and $\vec{b}$ of (R.24), find $\vec{a} \times \vec{b}$ and θ.

Solution: Equation (R.33) applied to $\vec{a}$ and $\vec{b}$ of (R.24) gives $\vec{a} \times \vec{b} \equiv$ (5952, −9712, −4641), so $|\vec{a} \times \vec{b}|^2 = 1.51288129 \times 10^8$ exactly, and $|\vec{a} \times \vec{b}| = 12{,}299.924$ (to ridiculous accuracy). With $|\vec{a}| = 25$ and $|\vec{b}| = 625$, (R.26) yields $|\sin\theta| = 0.787$, corresponding to $\theta = \pm51.924°$ or $\theta = \pm128°$. This θ is consistent with Example R.4, which used the scalar product.

Problems

R-2.1 Describe how the ground wire serves to protect the drill operator in Figure R.2.

R-2.2 Electric-powered trains obtain their electricity from the so-called third rail, which is at a high ac voltage. Discuss the possibilities for what might happen if we step on the third rail, according to the material of our shoes, what we are touching with our hands or other foot, and so on.

R-2.3 A 5 Ω resistor has end A at 5 V ($V_A = 5$ V) and end B at 5 V ($V_B = 5$ V). (a) Determine the current, the direction of the current flow, and the rate of heating. Repeat for (b) $V_A = 5$ V, $V_B = −5$ V; (c) $V_A = 5$ V, $V_B = 0$ V; (d) $V_A = 0$ V, $V_B = 5$ V; and (e) $V_A = −5$ V, $V_B = 5$ V.

R-2.4 Determine the electrical resistance and the current passing through a lightbulb rated at 60 W when it is connected to a 120 V power source.

R-2.5 Explain why, for purposes of electrical safety, if we are working with a circuit or equipment that could involve dangerously large voltages, it is prudent to wear insulated shoes, and to place one hand behind our back, and limit to the other hand our poking around at the connections.

R-2.6 If the average current in a bolt of lightning is 120,000 A, and it lasts for 2×10^{-4} s, how much charge does the lightning bolt transfer?

R-3.1 Give some examples (beyond those mentioned in the text) of the uses of electrical power.

R-3.2 Six-year-old Charlie is playing in the living room with a battery-powered toy that uses six D cells. The power in the house goes out, and you want to borrow two cells for a flashlight. He complains that you will use up all the power and there won't be any left for his toy. What do you tell him?

R-4.1 A voltaic cell is rated at 2 V. A slow discharge through a resistor involves a nearly constant current of 0.1 A, which suddenly stops after 80 minutes. Find: (a) the resistance of the resistor; (b) the rate of power dissipation; (c) the total power dissipated; and (d) the "charge" on the voltaic cell.

R-4.2 A typical car battery is discharged through a resistor equal to its internal resistance. (a) Find the rate of heating of the car battery and the resistor. (b) Find how long it will take to produce 12 J in the resistor. (One joule equals a watt-second, or J = W-s).

R-5.1 Frankie, a computer newbie, is using his computer at work. Suddenly the lights go out. He looks at his monitor, and it has gone dark. He calls the computer company's help desk. (a) Is this the appropriate response? (b) What would you tell him? (There is a story on the World Wide Web that something like this actually happened. The agent at the help desk was unable to provide assistance, until finally told that the lights also had gone out.)

R-5.2 Give three examples of how electricity is used within a computer.

R-6.1 Joan stands on an insulating platform and touches a disconnected and discharged electrostatic generator. The electrostatic generator is then connected and turned on. Her hair stands on end. Explain.

R-6.2 Laura stands on ground and touches a connected and charged electrostatic generator. (a) Explain what happens next. (b) What difference might it make if she were standing on an insulating platform?

R-6.3 Discuss the following statement: a more accurate terminology than ac and dc would be dv and av (*direct voltage* and *alternating voltage*).

R-6.4 When two identical faucets are wide open, the maximum flow of water is double that for one faucet. Give an electrical analog.

R-6.5 Sometimes we receive an electric shock when leaving a car, especially in dry weather. This can be avoided by getting up from one's seat while touching the outside surface of the car. Explain why.

R-6.6 When two identical hoses are connected to one another, the maximum flow of water is less than for one hose. Explain why, and give an electrical analog.

R-8.1 Consider three people (A,B,C) of equal mass, centered at $x = 1, 2, 3$, which we represent as (1,2,3). Their center of mass is at $x = 2$. Find the center of mass for the following three cases: (a) Shift them all to the right by one, giving (2,3,4). (b) Shift only A, to $x = 4$, giving (4,2,3). (c) Shift only B, to $x = 5$, giving (1,5,3). [Answer: All should be at $x = 3$.] (d) Here's the physics of this problem: describe the shifts as due either to a collective effect or to an individual effect.

R-8.2 Explain the difference between charge transfer as it occurs when there is sparking in air and when there is an electric current in a wire.

R-8.3 For both conductors and insulators, we employ + and − signs to represent charge. Nevertheless, there is a difference in interpretation at a microscopic level. Explain.

R-8.4 Describe the flow of electric current in terms of the behavior of the charge carriers: (a) in a salt solution, and (b) in a piece of copper wire.

R-8.5 Describe the response of the other charge carriers to the insertion of a single charge carrier within (a) a glass beaker containing a salt solution, and (b) a piece of copper wire sitting on an insulating surface.

R-8.6 Describe where you think the charges come from in each part of Figure R.9.

R-8.7 For pure water, the individual water molecules undergo rapid random motion, due to thermal effects. (a) Is there random motion of the Na^+ and Cl^- ions (and the water molecules) in a solution of salt water? For a metal wire, the individual electrons are in large overlapping orbitals that permit electron motion more rapid than due to thermal effects. (b) For pure water, salt water, and a metal wire, discuss how a net fluid flow or net electric current (as appropriate) can come about by superimposing an average velocity on the random motions.

R-8.8 In a low-humidity room, we rub a comb through a piece of fur, and then separate them. Assume that, during the rubbing, the comb gains electrons and the fur loses electrons. Compare the rate at which the comb and the fur seem to lose charge if first we introduce water vapor near the comb, and then if we introduce water vapor near the fur. Think in terms of charge transfer of electrons, and the time it takes for water molecules to get from one place to another.

R-9.1 Professor X believes in giving no partial credit because that way he gives students an incentive to get ideas *completely* correct. What credit (full or none) would Professor X give if, on an exam, he asked students to write down Newton's second law of motion, and he saw the equation: (a) $\vec{F} = ma$? (b) $F_x = ma$? (c) $F_x = ma_x$? (d) The set of equations $F_x = ma_x$, $F_y = ma_y$, $F_z = ma_z$? (e) If a student didn't define F or a, but simply wrote $F = ma$, would Professor X give credit? (f) If a student explicitly defines $F \equiv |\vec{F}|$ and $a \equiv |\vec{a}|$, would Professor X give credit for $F = ma$? *Hint:* What about directional information?

R-9.2 If right distributivity holds for an imaginary operation $\otimes$, and if $\vec{a} \otimes \vec{b} = c\vec{b} \otimes \vec{a}$, where c is a number, find the possible values of c that make left distributivity hold.

R-9.3 Show that (R.9) implies that the two equations in (R.10) are equal so that if right distributivity is true, then left distributivity is true, and vice versa.

R-9.4 Give a specific example of a vector $\vec{a} \equiv (a_x, a_y, a_z)$, and a specific rotation, to show that $\psi' \equiv a_x$ is not a scalar under rotation.

R-9.5 Give a specific example of a vector $\vec{a} \equiv (a_x, a_y, a_z)$, and a specific rotation, to show that $\psi^* \equiv (|a_x|, |a_y|, |a_z|)$ is not a vector under rotation.

R-9.6 (a) Prove or disprove the statement $|a(b + c)| = |ab| + |ac|$ for all real numbers. To disprove this statement, you need to find only a single counterexample; a million examples won't prove it. (b) Is it true for positive numbers only?

R-10.1 Define the $\otimes$ operator to produce a vector $\vec{a} \otimes \vec{b}$ whose direction is given by the right-hand rule, and whose magnitude is given by $|\vec{a} \otimes \vec{b}| = |\vec{a}||\vec{b}|| \sin 2\theta_{\vec{a},\vec{b}}|$. (a) Show that its direction transforms under rotation like a vector. (b) Give an example showing that it is not distributive.

R-10.2 The z-component of $\vec{a} \times \vec{b}$ is $a_x b_y - a_y b_x$. This mathematical combination is antisymmetric. In 3d, the number of true vector components is 3 and the number of such antisymmetric combinations is 3. This equality was necessary to construct the vector product. (a) How many antisymmetric combinations are there in 2d? (b) In 4d? (c) Is it possible to use these antisymmetric combinations to produce a vector product in 2d? (d) In 4d? [Answers: (a) One. (b) Six. (c) No. (d) No.]

R-10.3 Mr.$\odot$-man wants to sell us a scalar product that he claims is bilinear, commutative, and distributive. He tells us that the rule is $\vec{a} \odot \vec{b} \equiv |\vec{a}||\vec{b}| \cos(2\theta_{\vec{a},\vec{b}})$. (a) Show that it is bilinear and commutative, and a scalar under rotations. However, it is not clear that this rule satisfies the right or left distributive law. (b) Using the set of vectors $\vec{a} = \hat{i}$, $\vec{b} = \hat{i} + \hat{j}$, and $\vec{c} = -\hat{i} + \hat{j}$, test whether or not $\vec{a} \odot (\vec{b} + \vec{c}) = \vec{a} \odot \vec{b} + \vec{a} \odot \vec{c}$. [Answer: $\vec{a} \odot (\vec{b} + \vec{c}) = -2$, $\vec{a} \odot \vec{b} = 0$, and $\vec{a} \odot \vec{c} = 0$. Clearly, $\vec{a} \odot (\vec{b} + \vec{c}) \neq (\vec{a} \odot \vec{b}) + (\vec{a} \odot \vec{c})$. Sorry, Mr.$\odot$-man, your scalar product is not distributive! We're buying our scalar product from Mr.dot-man.]

R-10.4 (a) From the scalar product in two dimensions, and with the unit vectors $\hat{a}_{1,2}$ making angles $\theta_{1,2}$ to the x-axis, show that $\cos(\theta_2 - \theta_1) = \cos \theta_2 \cos \theta_1 + \sin \theta_2 \sin \theta_1$. (b) Derive the cosine addition law by letting $\theta_1 \to -\theta_1$. (c) Derive the sine addition law by letting $\theta_1 \to (\pi/2 - \theta_1)$ and using $\cos(\pi/2 - \theta) = \sin \theta$.

R-10.5 (a) From the vector product applied to the unit vectors $\vec{a}_{1,2}$ at angles $\theta_{1,2}$ to the x-axis, show that $\sin(\theta_2 - \theta_1) = \sin \theta_2 \cos \theta_1 - \cos \theta_2 \sin \theta_1$. (b) Derive the sine addition law by letting $\theta_1 \to -\theta_1$. (c) Derive the cosine addition law by letting $\theta_1 \to (\pi/2 - \theta_1)$ and using $\sin(\pi/2 - \theta) = \cos \theta$.

R-10.6 Derive the sine addition law by drawing two right triangles of angles α and β, with α measured counterclockwise from the x-axis and β added counterclockwise to α. See Figure R.14, where OP, of unit length, defines the spacial scale. *Hint:* First show that the vertical projections (AB and BC) are $\sin \alpha \cos \beta$ and $\cos \alpha \sin \beta$.

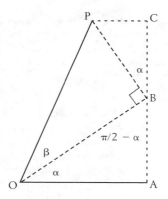

Figure R.14 Line OP has unit length (Problem R-10.6).

R-10.7 Derive the cosine addition law from the diagram in the previous problem. See Figure R.14. *Hint:* First show that the horizontal projections (OA and PC) are $\cos \alpha \cos \beta$ and $\sin \alpha \sin \beta$.

R-10.8 Let $\vec{a} \equiv (3, -4, 2)$ and $\vec{b} \equiv (2, 6, -1)$. (a) Find $|\vec{a}|$, $|\vec{b}|$, $\vec{a} \cdot \vec{b}$, $\vec{a} \times \vec{b}$. (b) Find the unit vector along $\vec{a}$, $\vec{b}$, and $\vec{a} \times \vec{b}$. (c) Find the angle between $\vec{a}$ and $\vec{b}$ from the scalar product and from the vector product.

R-10.9 Rotate the vectors in the previous example about the z-axis so that $\vec{a} \to \vec{a}' \equiv (5, 0, 2)$. (a) Find the angle of rotation. (Note: Using $\vec{a} \cdot \vec{a}'$ gives the angle of rotation about the direction of $\vec{a} \times \vec{a}'$, not the angle of rotation about z.) (b) Find the new vector $\vec{b}'$ by applying this angle of rotation to $\vec{b}$. (c) Find the new vector $(\vec{a} \times \vec{b})'$ by applying this angle of rotation to $\vec{a} \times \vec{b}$. (d) Find $\vec{a}' \times \vec{b}'$. (e) Compare $(\vec{a} \times \vec{b})'$ and $\vec{a}' \times \vec{b}'$. (f) Find $|\vec{a}'|$, $|\vec{b}'|$, $|\vec{a}' \times \vec{b}'|$. (g) Find the angle between $\vec{a}'$ and $\vec{b}'$. (h) Do the vector lengths and angles transform as you expect them to?

R-10.10 Five versions of the vector product appear in this text. Here is one. The torque $\vec{\tau}$ on an electric dipole $\vec{p}$ in an electric field $\vec{E}$ is given by $\vec{\tau} = \vec{p} \times \vec{E}$. If $\vec{p} \equiv (0, 3, 0)$ C-m, and $\vec{E} \equiv (2, -4, 1)$ N/C, find $\vec{\tau}$.

R-10.11 The torque $\vec{\tau}$ on a magnetic dipole $\vec{\mu}$ in a magnetic field $\vec{B}$ is given by $\vec{\tau} = \vec{\mu} \times \vec{B}$. If $\vec{\mu} \equiv (-1, 2, 0)$ A-m^2, and $\vec{B} \equiv (2, -1, 5)$ N/A-m, find $\vec{\tau}$.

R-10.12 The force $\vec{F}$ on a charge q moving at velocity $\vec{v}$ in a magnetic field $\vec{B}$ is given by $\vec{F} = q\vec{v} \times \vec{B}$. If $q = 4.5 \times 10^{-17}$ C, $\vec{v} \equiv (6, 3, 1)$ m/s, and $\vec{B} \equiv (2, -1, 5)$ N/A-m, find $\vec{F}$.

R-10.13 The force on a length ds of a current-carrying wire in a magnetic field $\vec{B}$ is given by $d\vec{F} = I d\vec{s} \times \vec{B}$, where I is the current and the directed length $d\vec{s}$ points along I. If $I = 2.4$ A, $d\vec{s} \equiv (0, 0, 0.002)$ m, and $\vec{B} \equiv (2, -1, 5)$ N/A-m, find $d\vec{F}$.

R-10.14 The magnetic field $d\vec{B}$ produced by a length ds (at source position $\vec{r}\,'$) of a current-carrying wire at the observer position $\vec{r}$ is given by $d\vec{B} = k_m I d\vec{s} \times \hat{R}/R^2 = k_m I d\vec{s} \times \vec{R}/R^3$, where $k_m = 1.0 \times 10^{-7}$ N/A^2, I is the current, the directed length $d\vec{s}$ points along I, and $\vec{R} = \vec{r} - \vec{r}\,'$ points from the source $I d\vec{s}$ at $\vec{r}\,'$ to the observer at $\vec{r}$. If $I = 2.4$ A, $d\vec{s} \equiv (0, 0, 0.002)$ m, $\vec{r} \equiv (0.5, -1.2, 1.4)$ m, and $\vec{r}\,' \equiv (-0.6, 0.8, -1.3)$ m, find $d\vec{B}$.

R-10.15 Show that (R.31) is obtained by finding the determinant of a "matrix" where the first row is the set of unit vectors $\hat{i}$, $\hat{j}$, and $\hat{k}$; the second row is the components a_x, a_y, and a_z; and the third row is the components b_x, b_y, and b_z.

R-10.16 Show, by explicit computation using (R.31), that if $\vec{a}$ and $\vec{b}$ are parallel, then each component of their cross-product is zero.

R-10.17 A plane is a set of points $\vec{r}$ normal to some direction $\hat{n}$ and passing through a specific point $\vec{r}_0$. Show that $(\vec{r} - \vec{r}_0) \cdot \hat{n} = 0$.

R-10.18 Find the distance s from a point $\vec{r}\,' \equiv (x', y', z')$ to the plane of points $\vec{r}$ satisfying $(\vec{r} - \vec{r}_0) \cdot \hat{n} = 0$, where $\vec{r}_0 \equiv (x_0, y_0, z_0)$ is a specific point in the plane, and $\hat{n} \equiv (n_x, n_y, n_z)$ is a unit vector normal to the plane. *Hint:* $\vec{r} - \vec{r}_0$ has a component along $\hat{n}$ whose value is s.

R-10.19 Given the plane $ax + by + cz + d = 0$, find an $\vec{r}_0$ (it is not unique) and a direction $\hat{n}$ (unique up to a sign) that makes points $\vec{r}$ in this plane satisfy $(\vec{r} - \vec{r}_0) \cdot \hat{n} = 0$. *Hint:* Set $x_0 = y_0 = 0$, and show that $\hat{n}$ is proportional to (a, b, c).

R-10.20 Let $\vec{r} \equiv (3, -2, 4)$ and $\vec{r}\,' \equiv (3.2, -2.1, 3.8)$. (a) Determine $d\vec{s} \equiv \vec{r}\,' - \vec{r}$. (b) With $d\vec{s} \equiv \hat{s} ds$, where $ds = |d\vec{s}|$, determine ds and $\hat{s}$. (c) At $\vec{r}$, let the voltage be $V = 4.2$ volts, and let the electric field be $\vec{E} \equiv (34, -15, 56)$ volt/m. If $dV = -\vec{E} \cdot d\vec{s}$ gives the voltage change on moving by $d\vec{s}$ to $\vec{r}\,'$, estimate dV. (d) Estimate V at $\vec{r}\,'$.

R-10.21 For a surface element centered at $\vec{r} \equiv (-5, -2, 6)$ m, with normal $\hat{n}$ and area dA, the "electric flux" $d\Phi_E$ passing through it is given by $d\Phi_E = (d\Phi_E/dA) dA$, where $d\Phi_E/dA \equiv \vec{E} \cdot \hat{n}$. Let $\vec{E} \equiv (13, 27, -18)$ volt/m. Let $\hat{n}$ point along $(2, -3, 7)$, and $dA = 0.52$ mm^2. (a) Determine $\hat{n}$. (b) Evaluate $d\Phi_E/dA$. (c) Evaluate $d\Phi_E$.

"It is of great advantage to the student of any subject to read the original memoirs on that subject, for science is always most completely assimilated when it is in the nascent state."

—James Clerk Maxwell,
Preface to *A Treatise on Electricity and Magnetism*, 1873

"A penny saved is a penny earned."

—Benjamin Franklin,
stating a conservation law in *Poor Richard's Almanac*

Chapter 1

A History of Electricity and Magnetism, to Conservation of Charge

Chapter Overview

Section 1.1 provides a brief introduction. Section 1.2 presents some of the early history of electricity and magnetism and an explanation of the *amber effect*, by which electricity was discovered. Sections 1.3 and 1.4 present the history of electricity up to the point where scientists became aware of the concept of *charge conservation*. Section 1.5 discusses charge conservation itself. Section 1.6 returns to a discussion of history, in the context of the discovery and understanding of *electrostatic induction*. Section 1.7 discusses modern views of electric charge, and in Section 1.8 we discuss *charge quantization*, whereby electric charge seems to come only in integral units of the proton charge e. Section 1.9 discusses how to employ integral calculus to evaluate the charge on an object when the charge is distributed over a line, over an area, or throughout a volume. It provides a review of some important geometrical facts, and it concludes with a brief discussion of the dimensionality of length, area, and volume. Section 1.10 contains some optional home experiments based on the ideas of this chapter. It considers only phenomena that can be studied relatively easily, with equipment no more sophisticated than comb, paper, tape, aluminum foil, plastic wrap, soft drink can, and a few other commonly available objects. Section 1.11 contains some electrical extras: how electricity produces light, and some historical tidbits. ■

1.1 Introduction

Students have many of the same conceptual difficulties as the pioneers in any area of science. By presenting some of the history of electricity and magnetism, we hope to aid the student in eliminating misconceptions. Rather than the original memoirs, we will present summaries since the primary purpose of this work is to teach physics, rather than history.

This chapter presents the history of electricity and magnetism before any laws were established quantitatively. You will (1) learn of the phenomena that puzzled scientists of the time, (2) identify the conceptual hurdles that had to be surmounted, and (3) learn how we interpret these phenomena now.

1.2 Early History

Both loadstone and amber puzzled the ancients.

Lodestone Magnetism was known to the ancient Greeks in the form of permanently magnetized pieces of iron oxide (the lustrous black mineral *magnetite*, or Fe_3O_4) called lodestone. Both attraction and repulsion between two lodestones was known. The Chinese knew of the direction-finding properties of lodestones some 3000 years ago, and there are reports of Chinese mariners navigating the Indian Ocean with magnetic compasses some 1800 years ago.

The first recorded discussion of magnetism in the West was written by Petrus Peregrinus (also known as Pierre de Maricourt) in 1269. Using a magnetized needle, he plotted lines representing the needle's orientation on the two-dimensional surface of a spherical lodestone, finding a convergence of the lines at two opposing ends of the sphere. By analogy to the earth, he called these two ends *poles*. He also showed how to make a floating and a pivoted compass. (By this time, magnetic compass needles were being used for navigation in many parts of the world.) He recognized that *opposite poles attract*, and considered that to be the fundamental statement about the behavior of magnets.

Amber Electricity ultimately derives its name from "electron," or $\eta' \lambda \epsilon \kappa \tau \rho o \nu$: Greek for amber. Amber is a fossilized resin with over 200 different varieties. It has been used in jewelry since antiquity; the "amber routes" from the Baltic to the Adriatic and the Mediterranean were among the earliest trade routes in human history. One can imagine ancient people shining up their amber, and then finding that it attracted bits of chaff—the *amber effect*. The opaque golden yellow variety of amber is perhaps the origin of its Greek name: "electron" derives from the ancient Greek word for the sun, "elios" or "elector." More commonly found is the transparent dark-honey-colored variety usually employed as a gemstone, and often containing fossilized insects.

> **Home Experiment 1.1** The Amber Effect
>
> As Gilbert showed (see Section 1.3), amber is not alone in producing the amber effect. Rub a comb through your hair and try to attract small pieces of paper, aluminum foil, styrofoam, and the like. Try to attract water from a slowly flowing (or dripping) faucet. If you don't have a comb, try rubbing a styrofoam cup against your clothing, and so on. *Use your imagination.*

1.2.1 How the Amber Effect Works

Consider a neutral piece of paper and a comb charged up by passing it through your hair. See Figure 1.1. (We now know that such a comb possesses an excess

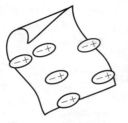
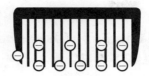

Neutral paper with
polarized molecules

Excess electrons

Figure 1.1 The amber effect, with a comb for amber and paper for chaff. The excess of electrons on the comb polarizes the neutral paper. Because the electrical force falls off with distance, the force on the comb due to the nearer positive charge of polarization exceeds the repulsive force due to the somewhat further negative charge of polarization.

of negatively charged electrons; the electrons have been transferred to the comb from your hair.) To explain the amber effect requires three steps of reasoning.

1. The negative charge on the comb polarizes the paper, with a net positive charge drawn nearer to the comb, and an equal amount of negative charge repelled by the comb. Thus the paper remains neutral when polarized.

2. The paper's positive charge is attracted to the comb, and its more distant negative charge is repelled by the comb.

3. Because the electrical force falls off with distance, the attractive force due to the closer charge dominates the repulsive force due to the further charge. This leads to a net attraction. How the net force falls off with distance depends on the details of how the electrical force falls off with distance.

Food for Thought. Repeat this argument with negative charge on the comb replaced by positive charge. You should conclude that the force is attractive.

If the paper is replaced by an electrical conductor (such as a piece of aluminum foil), the same qualitative argument applies. As discussed in the previous chapter, for conductors the process of polarization is called *electrostatic induction*.

1.3 Seventeenth-Century Electricity and Magnetism

1.3.1 *Gilbert's Systematic Study of the Amber Effect*

In 1600, the Englishman William Gilbert, physician to Queen Elizabeth, published *De Magnete*. This work, written in Latin, was widely read, and was even referred to by Shakespeare. Primarily devoted to magnetism, it pointed out that the earth appears to act like a huge magnet. Gilbert personally confirmed or denied a vast number of claims and reports by others, and showed how to repeat his experiments, encouraging others not to take him at his word. Unlike Peregrinus, who used only spherical magnets, Gilbert also used long thin magnets, and thus was aware that like magnetic poles repel.

The longest chapter in his six short "books" (Chapter II of Book II) systematically studies the amber effect. *Gilbert clearly distinguished amber from lodestone.*

1.3.2 *The Versorium: The First Instrument for Detecting Electricity*

Gilbert employed a device that he called a *versorium*. This small, pivoted needle made of wood or metal is similar to the magnetic compass needle in that it can detect the presence of electricity, just as a compass needle can detect the presence of magnetism. However, unlike the ends of a compass, which have a permanent magnetic polarity, the ends of the versorium possess no permanent electric polarity.

The material of the versorium polarizes only when an electric charge is brought near it, and this polarization disappears when the charge is withdrawn. Either end will be attracted to the amber. See Figure 1.2(a) and (b). Iron filings, which magnetize only when a magnet is brought near them, and lose their magnetization when the magnet is withdrawn, are analogous to the versorium.

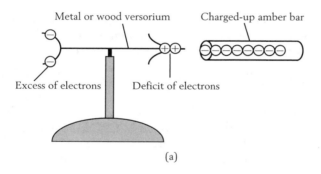

(a)

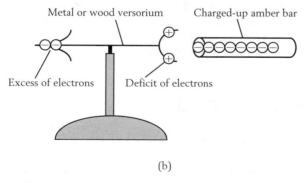

(b)

Figure 1.2 The versorium, which has no permanent polarization, so either end can be attracted (by the amber effect) to either positive or negative charge. (a) Head near charged-up amber. (b) Tail near charged-up amber. Must the versorium be made with asymmetric ends?

Gilbert found that many other substances can be substituted for amber, including diamond, sapphire, glass, rock crystal, artificial gems, sulfur, sealing wax, and hard resin; and that many others can be substituted for chaff, including "all metals, wood, leaves, stones, earths, even water and oil." To categorize the class of materials that display the amber effect, Gilbert modified the Greek word for amber, thus coining the noun *electric* for those substances that, like amber, would attract chaff when rubbed. He used the phrases *electric attraction*, *electrified state*, and *charged body*. (The words *electrical* and *electricity* were first used in 1618 and 1646, respectively.) Gilbert was aware that surface or atmospheric moisture inhibits the attractive strength of amber but does not affect that of lodestone. Materials like the metals, and so on could not be electrified by Gilbert; he thus classified them as *nonelectrics*.

By the early 18th century, better generators of electricity (rubbed glass) and better detectors of electricity (cloth threads and leaf brass) had been discovered.

A compass needle made of lodestone or magnetized iron—both electrical conductors—can serve as a detector of electricity. This is because all electrical conductors—magnetic or not—are susceptible to electrostatic induction. The versorium was a major advance both for its sensitivity and because it would respond only to electricity.

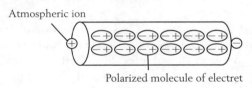
Atmospheric ion

Polarized molecule of electret

Figure 1.3 An electret, which has a permanent polarization. Also shown are some atmospheric ions that have been attracted to the electret.

Electrets—also known as *ferro-electrics*—such as tourmaline or carnuba wax possess permanent electric poles, and thus are the electric analogs of permanent magnets. Electrets can appear to lose their charge because their poles attract atmospheric ions of the opposite sign, thereby canceling some of the effect of each pole. See Figure 1.3.

1.4 Eighteenth-Century Electricity

1.4.1 *Gray: Conductors versus Insulators*

Using a rubbed glass tube about 0.5 m long, an innovation dating to 1708, in 1729 the Englishman Gray noticed that feathers were attracted both to the tube itself and to the cork at its end. See Figure 1.4. He extended the "length" of the tube by sticking into the cork a wooden pole with an ivory ball at its other end. With the glass tube charged, leaf brass was attracted to the distant ivory ball. He next made an even longer object by sticking a long string into the cork, the other end of the string attached to a kettle. With the glass tube charged, and a 52-foot vertical drop of string, the kettle still attracted leaf brass! (Electrostatic induction, due to the charge on the tube acting on successively longer conductors, is responsible for this extended electrical effect.)

In trying to extend the effect even further, Gray next made a major discovery. He could *not* attract leaf brass to the end of a long horizontal string when it was suspended by other strings nailed into wooden ceiling beams. His friend Wheeler, in whose barn these experiments were performed, proposed suspending the string by silk threads, whose fineness might prevent the loss of the "electrical

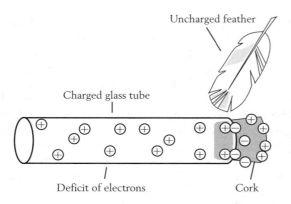

Figure 1.4 Polarization of a piece of cork by a positively charged glass tube, and attraction to the cork of an uncharged feather.

virtue." This worked, but another material of the same fineness—brass wire—chosen for its greater strength, did not. Thus was made the distinction between electrical "receivers" (i.e., conductors, like brass) and "isolators" (i.e., insulators, like silk thread).

1.4.2 *To Retain Electrification, Use an Insulating Stand: Dufay*

In 1730, the French scientist Dufay systematized the study of phosphorescent materials (publication of his recipes caused a dramatic fall in their prices). In 1732, he began a systematic study of electrical materials. He first confirmed Gray's experiments on the communication of electricity by thread, observing that moistening the thread promoted communication of the electrical virtue. He next showed that (1) despite random observations by previous scientists, the color of an object is unrelated to its electrical properties; (2) although a rubbed tube can produce electrical effects that penetrate dry silk curtains, no such penetration occurs for wet curtains; (3) when properly dried, warmed, and placed on an insulating stand, all rubbable objects but metals can be electrified by friction; and (4) all substances, even metals and fluids (but not flame) can be electrified by contact when placed on an insulating stand. Only in the late 1770s were metals electrified (charged) by friction; previous to that they were charged either by contact or by sparking. Following Dufay, it became customary to perform electrical experiments on an insulating stand. Dufay would remove the charged tube, his source of electricity, before studying a given material, so that electrostatic induction was not a factor in his further experiments.

1.4.3 *Two Classes of Electricity, and "Opposites Attract, Likes Repel": Dufay*

In 1733, Dufay made the important discovery that two gold leaves, each electrified by falling on an electrified glass tube and then *repelled* by the tube, also *repelled* each other. (The glass and the gold leaves were all positively charged.) Unlike others who occasionally had observed this repulsion effect, Dufay characteristically subjected it to systematic study. Bringing up a piece of amber (negatively charged), he was surprised to find that the amber *attracted* the gold leaf. Further investigation revealed that there are two classes of materials, *resinous* (amberlike, corresponding to negative charge) and *vitreous* (glasslike, corresponding to positive charge), which repel within each class, but attract between the classes. Dufay found that, for the repulsion to occur, both objects had to be sufficiently electrified; otherwise, there would be a weak attraction (due to the amber effect). Some substances could take on either kind of electricity, depending on how they were rubbed or touched.

1.4.4 *Storage of Electricity: The Leyden Jar ≡ Condensor ≡ Capacitor*

In Germany in the late 1730s, Bose reproduced the experiments of both Hauksbee and Dufay, including one where a person, suspended by silk threads, was charged up. Bose replaced the suspended person by an insulated tin-plated telescope tube, 6 m long and 10 cm in diameter. It was placed about 4 mm from a glove-rubbed rotating glass globe, which served as an electrical generator. Powerful sparks rapidly crossed between the tube and globe. Addition of a bundle of threads to the end of the metal tube, to soften the jostling of the tube against the globe, increased the flow of electricity to the tube. This so-called *prime conductor* was the first clearly defined charge storage device, and made electrical experiments more powerful, reliable, and convenient. With the availability of relatively inexpensive apparatus, amateurs could get into the act; we now describe the important discovery made independently by two of them.

In early 1746, Musschenbroek in Leyden tried to "draw fire" (i.e., a spark) from a beaker of water held in his hand while he stood on an insulated stand. A magistrate, Cunaeus, visited Musschenbroek's laboratory and tried it himself. Not knowing to stand on an insulator, he stood on the ground with the beaker of water in his hand. On touching the electrified bar with his other hand, he then received a powerful shock. (Cunaeus, standing on the ground, was subject to the full voltage difference between the charge source and ground; Musschenbroek, standing on the insulator, was subject to only a small fraction of this voltage difference.) Repeated by others, the shock was formidable enough to cause nosebleed, convulsions, and temporary paralysis. A cleric in Pomerania, Kleist, made the same discovery as Cunaeus. The Leyden jar, as it became known, was the first two-plate *condenser* (as it was called some 35 years later by Volta, because it appears to condense, or concentrate, the electricity). It would today be called a *capacitor* (because it has the capacity to store charge). See Figure 1.5.

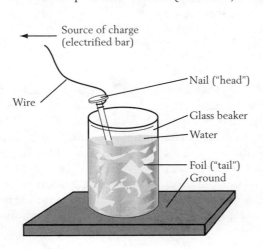

Figure 1.5 Leyden jar.

Labels: Source of charge (electrified bar); Wire; Nail ("head"); Glass beaker; Water; Foil ("tail"); Ground

From France to Japan, chains of sometimes over 100 people hand in hand would discharge Leyden jars. Everyone in the chain was shocked when, with the person at one end holding the tail wire, the person at the other end touched the head wire. (See Figure 1.5.) In 1746, when Benjamin Franklin entered the picture, the Leyden jar defied explanation.

1.4.5 *Franklin and the Electric Fluid Model*

Born in 1706, by 1745 Franklin had enough income from his publishing activities to permit him the leisure for purely intellectual pursuits. That year, a glass tube and a copy of the *Gentleman's Magazine* arrived from England at the Library Company, founded by Franklin in his adopted city of Philadelphia. The magazine contained an article on electricity that described work up to Bose. The article indicated that lightning and electricity were related, and that the electric spark was closely related to fire. The Leyden jar was not mentioned.

Franklin and his friends immediately began performing their own electrical experiments. In a series of letters, Franklin summarized and interpreted their experiments. One letter discussed their discovery of the "power of points" to draw off and and emit "the electrical fire," and described various demonstrations of electrification and sparking. It also described a simple new electrical generator using a rotating glass globe. Perhaps most important, *it also described an experiment that led Franklin to his first version of the electric fluid model.* Franklin called his experimenters *A*, *B*, and *C*; here is our version.

Are You Electrised?

Leonardo and Michaelangelo are standing upon wax, an insulator. They charge up when Leonardo rubs the glass tube himself (thus getting charged by friction). Michaelangelo then touches the other end of the tube. See Figure 1.6. Both Leonardo and Michaelangelo appear to be "electrised" (i.e., charged) to Raphael, on the ground, as determined by the spark between him and Leonardo, and between him and Michaelangelo. On recharging, the spark between Leonardo and Michaelangelo is much stronger than between Raphael and either Leonardo or Michaelangelo. Moreover, after Leonardo and Michaelangelo touch, "neither of them discover any electricity." Franklin considered that Leonardo is "electrised negatively" or "minus," Michaelangelo is "electrised positively" or "plus," and that initially Raphael was not "electrised." Positive charge on the tube subjects Michaelangelo to electrostatic induction, and on contact draws mobile negative charge from him, leaving Michaelangelo with a net positive charge.

This one-fluid model of electricity, with the idea of an excess (plus) and a deficit (minus) of electricity, and implicitly containing the idea of charge conservation, can be made consistent with Dufay's two types of electricities, of which Franklin initially was unaware. In his initial version of the model, Franklin used the ideas that, when rubbed, the glass acted like a pump for the electric fluid, and that the ground was a source of electricity. Thus, a person on the ground rubbing the glass would draw off more charge than if he were on an insulating stand, a fact that had previously been unexplained. Franklin also used the idea that glass is completely impermeable to the actual movement of electricity.

1.4.6 *Franklin and the Leyden Jar: How It Works and Where the Charge Resides*

Franklin soon learned of the Leyden jar and used these ideas to explain its operation. He argued that, when the Leyden jar was grounded, the excess of electrical

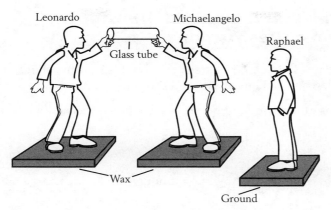

Figure 1.6 Franklin's crucial experiment. Leonardo and Michaelangelo become charged oppositely, while Raphael remains uncharged. The spark between Leonardo and Michaelangelo exceeds that between Raphael and either Leonardo or Michaelangelo.

fire on the inner surface created an "atmosphere" that drove an equal amount of electrical fire from the outer surface to ground. This left a net of zero electrical fire (the sum of the electric charge on the inner and outer surfaces) that was the same after electrification as before. Support for this conclusion of overall neutrality came from an additional experiment. He attached a cork "spider" to an insulating string suspended between metal extensions of the "head" (connected to the jar interior; compare Figure 1.5) and the "tail" (connected to jar exterior). The spider repeatedly swung from head to tail and back again. When the cork stopped, the jar was discharged. This was perhaps the first electrical motor.

Franklin next performed a series of experiments to determine where the electrical fire resides. He first charged a Leyden jar and removed its connections to the inside (the head) and to the outside (the tail). While holding the outside of the jar with one hand, on touching his other hand to the water he received a strong shock. Thinking the charge to reside in the water, he reassembled the jar, charged it, and again removed its head and tail wires and their connections. Pouring what he thought was charged water into a second Leyden jar, he was surprised to find that the second jar was uncharged. However, on pouring fresh water into the original jar and reinstalling its wiring, he found that the original jar once again was charged. This indicated that the *charge resided on the glass inner surface of the original jar.* Further tests directly established this result, and also established that an opposite charge also resided on the glass outer surface. European scientists could not fail to be impressed both by his clear and direct experimental method and by his cogent theoretical reasoning.

At this point, we cut off our history of electricity, before it takes us too far from our goal of actually studying the laws of electricity. We have reached, however, a fundamental quantitative law of electricity—conservation of electric charge. Let us now consider its consequences for a number of situations confronted by experimenters at the time, and for experiments we ourselves can perform.

1.5 Electric Charge Is Conserved: Transfer, but No Creation or Destruction

Franklin's electrical fluid could be transferred but, like other fluids, it could not be created or destroyed. Thus, it satisfied the conservation law

$$\text{amount of electric fluid before} = \text{amount of electric fluid after} \qquad (1.1)$$

In modern terms, we write this as

$$\text{amount of electric charge before} = \text{amount of electric charge after} \qquad (1.2)$$

Like the conservation laws of mechanics (energy, momentum, and angular momentum), it is both simple and profound. It may not predict the details of a process, but it does make an overall statement relating before and after. We now show what it has to say about a number of methods by which electric charge can be transferred. In almost all cases, when two objects are rubbed, negatively charged electrons are transferred from the least well-bound orbitals of the materials involved. Electrons are much lighter and more mobile than the much heavier, positively charged nuclei.

1.5.1 Charging by Friction

Let two initally uncharged insulators A and B be rubbed against each other. Charge is transferred, so by charge conservation they develop equal and opposite amounts of electricity: since $0 = Q_{before} = Q_{after} = Q_A + Q_B$, we have $Q_B = -Q_A$. See Figure 1.7. This situation also can be produced when a piece of sticky tape is placed on a tabletop and then quickly removed, the tape getting charge Q_A and the tabletop $-Q_A$. The tape is an insulator, so the charge on it remains in place. The table typically will be a conductor, so the charge on it quickly rushes off to ground.

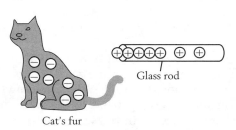

Glass rod

Cat's fur

Figure 1.7 Charge separation by friction. Note the overall neutrality.

Example 1.1   Cat's fur and glass rod

Cat's fur is rubbed against a glass rod to produce a charge on the fur of $Q_A = 10^{-9}$ C. Find the charge Q_B on the rod.

Solution: $Q_B = -Q_A = -10^{-9}$ C.

1.5.2 Charging by Contact

Let conductor A with charge Q_A be brought near a neutral conductor B. Both are mounted on insulated stands. See Figure 1.8(a), where the positives

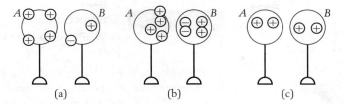

Figure 1.8 Charging by contact between conductors. The conducting balls are supported by insulating rods: initially (a), just before contact (b), and after contact (c).

(negatives) represent a deficit (excess) of electrons, and B is polarized, because of electrostatic induction. In Figure 1.8(b), B is more polarized because A is closer to it, and in response the positives on A are closer to B. On contact, charge is transferred, as in Figure 1.8(c). If the objects are spheres of the same radius, by symmetry they will share the charge equally.

Example 1.2 **Contact between identical conducting spheres**

A and B are two identical conducting spheres. Initially, A has charge $Q_A = 2 \times 10^{-8}$ C, and B is uncharged. The spheres are brought into contact with each other. Find the final charges on the spheres.

Solution: $Q_{total} = Q_A^{initial} + Q_B^{initial} = 2 \times 10^{-8}$ C. The spheres are identical, so $Q_A^{final} = Q_B^{final} = Q_{total}/2 = 2 \times 10^{-8}/2 = 10^{-8}$ C.

If the objects are not spheres of the same radius, the charge typically will not be shared equally. If $Q_A > 0$, then the final charge $Q'_A > 0$ but $Q'_A < Q_A$. Thus, for $Q_A = 2 \times 10^{-8}$ C, the value $Q'_A = 1.5 \times 10^{-8}$ C is allowed (and $Q'_B = 0.5 \times 10^{-8}$ C, by charge conservation). However, neither $Q'_A = 9 \times 10^{-8}$ C (since it exceeds the original charge) nor $Q'_A = -2 \times 10^{-8}$ C (since it is of opposite sign to the original charge) is allowed.

If A is a conductor, then any part of A can be charged so that B can easily take up charge. If A is an insulator, only parts of it might be charged (e.g., in places where it was rubbed). Thus B (insulator or conductor) may pick up charge only when it makes contact near the parts of A that are charged. See Figure 1.9.

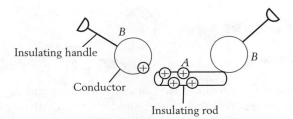

Figure 1.9 Charging of a conductor by contacting a charged insulator.

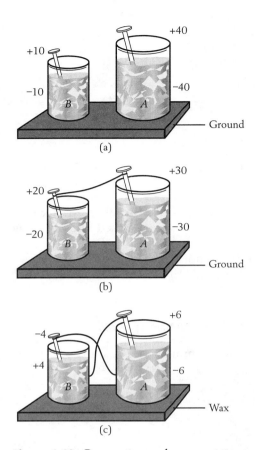

Figure 1.10 Connecting and reconnecting two Leyden jars: the effect of charge-storing capacity. (a) Initial configuration. (b) After connection of the internal wires to each other, and of the exteriors to each other, via ground. (c) After connection of the interior of one to the exterior of the other, and vice versa. Numbers indicate charge, in units of μC.

Example 1.3 **Connecting and reconnecting two Leyden jars**

Consider two Leyden jars, A and B, with A larger than B. Initially, they have charges $Q_A = 40\ \mu$C and $Q_B = 10\ \mu$C on their inner surfaces (with opposite amounts on their outer surfaces). See Figure 1.10(a). When connected, they always share charge in the same ratio. Assume that, in a previous measurement when they were connected inner to inner and outer to outer, the larger one held 60% of the total charge. With insulating gloves, connect them inner to inner, and outer to outer, as in Figure 1.10(b). What is the final charge on the jars?

Solution: Since the total charge available is $40 + 10 = 50\ \mu$C, after connection the larger jar (A) will get 30 μC (i.e., 60% of the total charge) and the smaller jar (B) will get 20 μC. See Figure 1.10(b).

Example 1.4 **Further reconnection of two Leyden jars**

Next, using insulating gloves, disconnect the jars of Example 1.3, and reconnect them with the inner surface of the larger jar connected to the outer surface of the smaller jar, and vice versa, as in Figure 1.10(c). Find the final charge on each jar.

Solution: Now the total charge available to the inner surface of the larger jar and the outer surface of the smaller jar is $30 + (-20) = 10\ \mu$C. When the jars come to equilibrium, the inner surface of the larger jar will hold 60% of 10, or 6 μC, and the outer surface of the smaller jar will hold 4 μC.

1.6 Electrostatic Induction

1.6.1 *Canton's Tin Cylinder*

With the excess electrical fire on the inside driving off electrical fire from the outer surface, despite the impermeable glass, Franklin's model distinguished between conduction of the "electrical virtue" itself and its effect on matter, such as polarization or electrostatic induction. In 1753, Canton, in London, performed

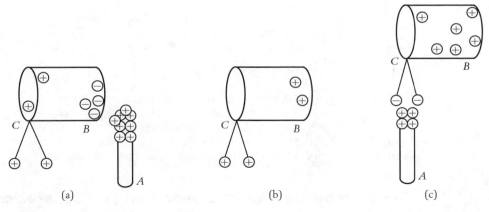

Figure 1.11 Canton's can and electrostatic induction. (a) Charged rod polarizing can, by electrostatic induction. (b) Can, after sparking between the rod and the can has charged the can. (c) Rod brought up to detecting cork, by induction charging them oppositely to how they had previously been charged.

an experiment with a positively charged glass rod and a suspended, insulated, tin cylinder, with two corks hanging at one end as a detector of electricity. Franklin simplified and clarified this work, and wrote an addendum to Canton's paper, modifying his atmospheres hypothesis because it could not explain the repulsion of two negative charges for each other. (An indication of Franklin's willingness to change his views when faced with contradictory evidence is the remark he made about his theory of the origin of storms: "If my hypothesis is not the truth, it is at least as naked. For I have not, with some of our learned moderns, disguised my nonsense in Greek, cloyed it in algebra, or adorned it in fluxions [the calculus].")

A positively charged glass rod A is used as a source of positive electricity (take $Q_A = 8 \times 10^{-8}$ C), and a tin cylinder is suspended on insulating silk threads (not shown). Cork balls in electrical contact with the cylinder via conducting string at end C were used as detectors. When the glass rod approaches end B, as in Figure 1.11(a), it induces a negative charge (take $Q_B = -4 \times 10^{-8}$ C) in the vicinity of end B and a positive charge $Q_C = -Q_B$ in the vicinity of end C. Some of Q_C travels along the strings to the cork balls, causing them to repel one another. When the positively charged rod is brought close enough to the cylinder, a spark occurs, transferring positive charge (take it to equal half the original charge Q_A) to the cylinder. Thus the glass rod now has $Q'_A = 4 \times 10^{-8}$ C. When the charged glass rod is withdrawn, so that electrostatic induction no longer occurs, this positive charge distributes itself over the can, as in Figure 1.11(b). The cork balls still repel one another, but by slightly less than in Figure 1.11(a). When the positively charged rod is brought back and placed near the cork balls on end C, it induces in the cork balls a negative charge in addition to the positive charge already on them. Hence the magnitude of the charge on the cork balls decreases, and they repel less strongly. Eventually, the rod is close enough that the cork balls do not repel at all. Bringing the rod even closer causes the cork balls to have a net charge that is negative, and once again they repel. See Figure 1.11(c). In Figure 1.11(a) the induction on the can at B, due to rod A, is similar to the induction on the cork balls in Figure 1.11(c) at C, due to rod A.

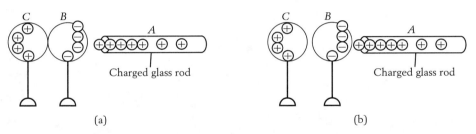

Figure 1.12 Charging by electrostatic induction on conducting spheres.
(a) Induction with contact. (b) Induction with contact and separation.

1.6.2 *Charging by Electrostatic Induction and Separation*

Consider the insulating glass rod A with a charge $Q_A = 10^{-9}$ C, and two initially uncharged conductors, B and C, on insulating pedestals. B and C are brought in contact with one another while B is closer to A than to C. See Figure 1.12(a). The combination of B and C is subject to electrostatic induction, with $Q_B < 0$. By charge conservation, $Q_C = -Q_B > 0$. Clearly, when A is very far away, $|Q_B|$ is very small, and as A approaches, $|Q_B|$ increases. At any time we can slightly separate C from B, thus *charging by induction*. See Figure 1.12(b). The largest possible value for $|Q_B|$ occurs when A is both very small compared with B, and very close to (or enclosed by) B. In this case $|Q_B| = Q_A = 10^{-9}$ C. If B and C are very mismatched in size, $|Q_B|$ and $|Q_C|$ may be very small because nearly all the electrostatic induction takes place on the larger of them.

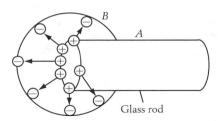

Figure 1.13 Close-up of interior of a conducting sphere with a hole through which a charged rod can pass.

In Review/Preview we mentioned Faraday's concept of electric field lines, or lines of force, which originate on positive charges and terminate on negative ones. If A is enclosed by B, then all the electric field lines that originate on A terminate on the inner surface of B, so that $Q_B = -Q_A = -10^{-9}$ C. For a close-up of Figure 1.12(b) when B has a hole by which A can enter, see Figure 1.13. The arrows are the field lines.

1.6.3 *Charging by Electrostatic Induction and Contact (or Sparking)*

Again let a glass rod A with $Q_A = 10^{-9}$ C be brought near neutral conductor B, subjecting the latter to more electrostatic induction than the more distant C. See Figure 1.14(a). The charge separation in this case cannot exceed Q_A. If B is then touched briefly to conductor C, as much charge is transferred (4 units of negative charge from C to B) as if they originally had been in contact (as in the previous example). See Figure 1.14(b). (If B were brought close enough for sparking without touching, again charge would be transferred, but perhaps not as much as with contact.) As in the previous examples, $Q_C = -Q_B$ from charge

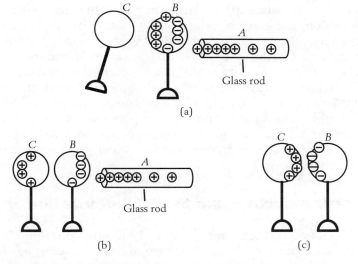

Figure 1.14 Charging by electrostatic induction and contact (or sparking): before contact (a), after contact (b), and after removal of glass rod (c).

conservation, and $Q_C < Q_A$ because the amount of induction is limited. Thus $Q_C = 2 \times 10^{-9}$ C is not a possible charge transfer. Figure 1.14(c) depicts the situation when the charged rod A is removed.

1.6.4   *Volta and the Electrophorous*

Volta is now known primarily for his development of the *voltaic cell* (to be discussed in Chapter 7), but he performed a number of other valuable electrical investigations. These included his studies of condensors (capacitors), and his invention of the *electrofore perpetuo*, or *electrophorus* (1775). By using electrostatic induction on external conductors, this charged, insulating device could be used repeatedly to charge other objects, without itself losing any charge. The electrophorus could be used in place of the combination of a frictional electricity device (for charge generation) and the Leyden jar (for charge storage).

The electrophorus consists of an insulating resinous "cake" that is charged up by rubbing (we assume, negatively, as with resin). See Figure 1.15 for a modern version that employs a styrofoam "cake." A metal plate with an insulating

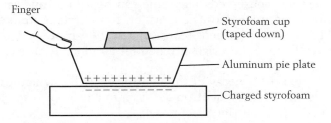

Figure 1.15 A simple electrophorous. The charged styrofoam, an insulator, touches the aluminum pie plate at only a relatively few points.

handle (here, a styrofoam cup) is placed on top of the cake. Because of electrostatic induction, the bottom surface of the metal plate (touching the negatively charged styrofoam cake) becomes positively charged and its rim becomes negatively charged. (A small amount of charge can be transferred from the insulating cake to the metal plate at the places where they make contact.) The rim is now touched by the experimenter (assumed to be connected to ground), thus drawing off negative charge, leaving the plate positively charged. The plate is then lifted off by its insulating styrofoam cup handle, and used as a source of positive charge. In the process, the electrophorus does not lose its charge. Hence the process can be repeated with other metal plates or with the same metal plate after it has been discharged.

1.6.5 *Electrostatic Induction and the "Depolarization Spark"*

We close this section by mentioning a curious and often puzzling effect associated with a conductor near a powerful source of charge (such as an electrostatic generator or a stormcloud), so the conductor is subject to electrostatic induction. If the source of charge suddenly discharges, the conductor is no longer subject to electrostatic induction, and thus it *depolarizes*. This has the same effect as an electric shock. In this way stormclouds cause static on phone lines, sometimes great enough to "blow" poorly protected and sensitive electrical equipment connected to phone lines. Probably numerous people who claim to have been hit by lightning have "only" been suddenly depolarized when a nearby stormcloud discharged elsewhere. (You can also be shocked if you an in the path of current flowing along the ground, away from a lightning hit.)

1.7 Modern Views of Charge Conservation

Since Thomson's 1897 identification of cathode rays as a "corpuscle," or particle—the electron—our knowledge of the nature of matter has increased significantly. We now know that neither Franklin's one-fluid model, nor even a competing two-fluid model (one positive, one negative), is appropriate. If we were to talk in terms of types of electric fluids, there would now be hundreds of them, one for each of the charged elementary particles that have been discovered in the 20th century.

Although chemical reactions change the nature of molecules and atoms by rearranging electric charge, nevertheless they do not violate the fundamental principle that charge is conserved. The same can be said of nuclear reactions. For example, alpha decay involves the emission from certain atomic nuclei of an alpha particle (a helium nucleus, consisting of two protons and two neutrons). This leaves behind the originally neutral atom with an excess of two electrons.

Another example is beta decay of the neutron n (either when it is within a nucleus or outside one). Here a neutron decays into a proton p, an electron e^-, and what is called an antineutrino $\bar{\nu}$:

$$n \rightarrow p + e^- + \bar{\nu}.$$

The neutron in free space is unstable, with a half-life of about 10.4 minutes.

Conservation of individual particle number has been given up, but electric charge continues to be conserved.

Charge is conserved *locally*. A neutron in Houston, Texas, cannot decay into a proton in Los Angeles, California, and an electron and an antineutrino in Tokyo, Japan.

1.8 Charge Quantization

Electric charge is quantized in units of

$$e = 1.60217733(\pm 49) \times 10^{-19} \text{ C}. \tag{1.3}$$

For our purposes, $e = 1.60 \times 10^{-19}$ C will be good enough. It is not yet clear why such quantization occurs; it is still a mysterious fact of nature.

When you rub a comb through your hair, a charge perhaps as large as 10^{-9} C might be transferred. (In the next chapter, we will study an example that yields about this much charge.) This corresponds to on the order of 10^{10} electrons! The fact that bulk matter is ordinarily neutral has enabled us to argue that positive charges (protons) and negative charges (electrons) have equal and opposite amounts of electricity. By letting nominally neutral compressed H_2 gas escape from a closed, conducting tank, and measuring the change in charge of the tank, it has been established that the magnitudes of the proton and electron charges are the same to about one part in 10^{20}. Semiconductor devices are now made that depend for their proper operation on the fact that charge is quantized.

At the subatomic level, the nucleons can be thought of as composites of objects called *quarks*, which come in six varieties having quantized charges of $\pm(1/3)e$ and $\pm(2/3)e$. Quarks have not been, and are not expected to be, observable in isolation. However, they provide a useful framework for understanding the behavior of subnuclear objects.

Electric charge does not depend on temperature, pressure, gravitational potential, or any of a host of other variables. It is a truly fundamental quantity. Later we will study magnets, which appear to have magnetic charge at their poles. There, the pole strength is temperature dependent, an indication (but not a proof!) that there is no true magnetic charge on magnets; true magnetic charge should be independent of temperature.

1.9 Adding Up Charge

Our discussion of the early history of electricity showed that early scientists had a difficult time distinguishing the "stuff" of electricity from its effects. We conclude this chapter with a quantitative discussion of this stuff—electric charge.

So far we have discussed only discrete amounts of charge, as if charge were localized at points, or distributed as a whole over an object. We now consider the problem of adding up the charge on an object with a charge distribution. We will integrate over charge, which—like mass—is a *scalar* quantity. If you have previously done integrations over mass distributions, this will look very familiar.

When we add charge to ordinary materials, how it distributes depends on the type of material. *If the material is an insulator, the charge tends to remain where it is added. If the material is a conductor, the charge almost always rearranges.* In what follows, we will consider mostly hypothetical examples, which we may consider to involve insulators with fixed charge distributions. The charge distributions for conductors are usually very difficult to obtain, and typically are quite nonuniform. We will consider one case involving a charged conductor.

There are three types of charge distributions: linear (as for a string), areal (as for a piece of paper), and volume (as for a lump of clay). In each case, we will be interested in a small amount of charge dq in a region. We will work out some examples of each.

1.9.1 *Quick Summary of Integral Calculus*

The symbol for integration, $\int$, was deliberatedly crafted by Leibniz (who co-invented calculus, independently of Newton) to look like the letter S, for summation. Summation is the basic idea underlying integration. A numerical integration is called a *quadrature*. Integrals evaluated in equation form are said to be evaluated *analytically*.

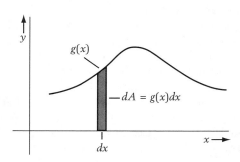

Figure 1.16 Area $dA = g(x)dx$ under the curve $g(x)$, between x and $x + dx$.

Consider a smooth function of the variable x, which we will call $g(x)$. Let us find the area A under the curve defined by $y = g(x)$ between the values $x = a$ and $x = b$. We break up the interval (a, b) into infinitely many tiny lengths dx. (Each dx is said to be *infinitesimally small*.) The area A is the sum of an infinite number of tiny areas $dA = ydx = g(x)dx$, where the corresponding infinite number of dx's go from $x = a$ to $x = b$. See Figure 1.16. Thus $A = \int_a^b g(x)dx$. The question is "How can we evaluate A analytically?"

The fundamental theorem of the calculus says that if $g(x)$ is the derivative of some function $f(x)$, or $g = df/dx$, then $A = f(b) - f(a)$. f is called the *antiderivative* of g, and A is the *definite integral* of $g(x)$ from a to b. Putting it all together, we have

$$f(b) - f(a) = \int_a^b \frac{df}{dx}dx. \tag{1.4}$$

Sometimes it is useful to write $df \equiv (df/dx)dx$, in which case (1.4) becomes

$$f(b) - f(a) = \int_a^b \frac{df}{dx}dx = \int_{f(a)}^{f(b)} df. \tag{1.5}$$

Since dx is an infinitesimal, and df/dx (the slope of f versus x) is assumed to be finite, df also is an infinitesimal.

Usually it is not simple to find the antiderivative of a function. However, consider that we know a function (e.g., $f(x) = 2x$) and its derivative (e.g., $df/dx = 2$). The function itself is the antiderivative of the derivative of the function. Actually, this is not quite correct: the antiderivative $f(x)$ of any function

$g(x) = df/dx$ is uniquely defined only up to a constant. Thus Molly might say that $2x$ is the antiderivative of 2, and Moe might say that $2x + 1$ is the antiderivative of 2—and both of them would be right! One way around this nonuniqueness of the antiderivative is to write

$$f(x) = \int g(x)dx + \text{constant}. \tag{1.6}$$

However, if $f(x)$ is known at a single point—such as $x = a$, where the integration begins—then the constant in (1.6) can be determined.

Here is one function and its derivative that surely you know:

$$f(x) = x^n, \qquad \frac{df}{dx} = nx^{n-1}. \tag{1.7}$$

Hence, by the fundamental theorem of the calculus,

$$\int nx^{n-1}dx = x^n + \text{constant}. \tag{1.8}$$

This is true only for $n \neq 0$; as you know, the integral for $n = 0$ gives $\ln(x)$. Letting $n \to n + 1$ in (1.8) gives

$$\int (n + 1)x^n dx = x^{n+1} + \text{constant}. \tag{1.9}$$

Then, dividing both sides of (1.9) by $n + 1$, and renaming the constant, we have

$$\int x^n dx = \frac{1}{n + 1}x^{n+1} + \text{constant}'. \tag{1.10}$$

In the next section, we will repeatedly employ (1.10) to perform various integrals. Even for the one case where the integral can't be done directly by using (1.10), by a change of variables (1.10) will still do the job! For more on calculus, see Appendix A.

1. **Linear charge distribution** λ: $dq = \lambda ds$ gives dq distributed over the length ds.

See Figure 1.17(a). Here we write

$$q = \int dq = \int \frac{dq}{ds}ds = \int \lambda ds, \qquad \lambda \equiv \frac{dq}{ds}, \tag{1.11}$$

where $ds > 0$ is the small length over which dq is distributed, and $\lambda \equiv dq/ds$ is the *charge per unit length*. We must take ds so small that λ really is constant over ds; in practice, we must take ds to be infinitesimal. However, λ can vary over the length of the line. We can think of this as input and output. The charge density λ and the region of integration (defined by all the ds's) are the input, and the charge q is the output.

(a) Uniformly charged string. Consider a line of arbitrary shape (perhaps a loose piece of string) with a uniform charge distribution (here, charge

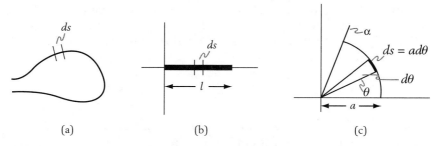

Figure 1.17 Examples of line elements. (a) Arbitrary line and line element ds. (b) Straight line on x-axis. (c) Arc of a circle.

per unit length) λ and total length l. See Figure 1.17(a). By "uniform" we mean that λ is constant over the line.

By total length l, we mean that by adding up the lengths ds we get $l: l = \int ds$. (We sometimes write s for l.) This can be thought of as a quadrature, where we literally add up the lengths. For example, by mentally breaking up the string into 100 small pieces, we can measure each ds, and add them all up. Or we can enter each ds into a column of a spreadsheet—more on spreadsheets in Appendix C—and have the spreadsheet add them up. Alternatively, by stretching out the string against a ruler with the ends at specific coordinates, we can do the integral (i.e., obtain the length) by subtracting the end readings, via (1.4).

By (1.11), the string has a total charge

$$q = \int dq = \int \lambda ds = \lambda \int ds = \lambda l. \tag{1.12}$$

The *average* charge per unit length $\bar{\lambda}$ is

$$\bar{\lambda} \equiv \frac{q}{l} = \frac{\lambda l}{l} = \lambda, \tag{1.13}$$

as expected for a uniformly charged object.

(b) Straight line, nonuniformly charged.   Consider a string of length l that lies on the x-axis from the origin to $x = l$, and has a charge distribution $\lambda = cx$, where c is some constant with units of charge per length squared. See Figure 1.17(b). If we integrate from $x = 0$ to $x = l$, so $dx > 0$, the (positive) length element is $ds = dx$. The length is $\int ds = \int_0^l dx = x|_0^l = l - 0 = l$, as expected. The total charge q is given by

$$q = \int dq = \int \lambda ds = \int_0^l (cx) dx = \frac{1}{2} cx^2 \Big|_0^l = \frac{1}{2} cl^2. \tag{1.14}$$

From (1.14), we find that $c = 2q/l^2$, which has the expected units. The *average* charge per unit length $\bar{\lambda}$ is

$$\bar{\lambda} \equiv \frac{q}{l} = \frac{cl}{2}. \tag{1.15}$$

$NB \equiv$ *note bene* $\equiv$ *note well*: The answer shouldn't matter which way you integrate (i.e., add up) the charge. If you accidentally (or perversely) chose to integrate the "wrong way"—from $x = l$ to $x = 0$, for which $dx < 0$—you would still find the same charge q. Here's how. The (positive) length element would now be $ds = -dx > 0$, so $\int_l^0 \lambda ds = \int_l^0 \lambda(-dx) = \int_0^l (cx)dx = q$, as before.

(c) Arc of a circle of radius a and angle α. See Figure 1.17(c). Here $ds = ad\theta$, where angles are measured in radians. The total arc length $l = \int ds = \int_0^\alpha ad\theta = a\alpha$. As expected, if $\alpha = 2\pi$ (a full circle), then $l = 2\pi a$. Let us take $\lambda = c\theta^2$, where c is now a constant that must have units of charge per unit length. Then

$$q = \int dq = \int \lambda ds = \int_0^\alpha (c\theta^2)ad\theta = ca\frac{1}{3}\theta^3\Big|_0^\alpha = \frac{1}{3}ca\alpha^3. \quad (1.16)$$

From (1.16) we find that $c = (3q/a\alpha^3)$, which has the expected units. The *average* charge per unit length $\bar{\lambda}$ is

$$\bar{\lambda} \equiv \frac{q}{l} = \frac{(1/3)ca\alpha^3}{a\alpha} = \frac{1}{3}c\alpha^2. \quad (1.17)$$

2. Areal charge distribution σ: $dq = \sigma\, dA$ gives dq distributed over the area dA.

See Figure 1.18(a). Here we write

$$q = \int dq = \int \frac{dq}{dA}dA = \int \sigma dA, \qquad \sigma \equiv \frac{dq}{dA}, \quad (1.18)$$

where $dA > 0$ is the small area over which dq is distributed, and $\sigma \equiv dq/dA$ is the charge per unit area. We must take dA so small that σ really is constant over dA; in practice, we must take dA to be infinitesimal. (Sometimes area integrals are written as $\iint dA$, to remind us that areas involve two coordinates.)

(a) Uniformly charged area. Consider an area of arbitrary shape (perhaps a cut up piece of paper) with a uniform charge distribution (here, charge

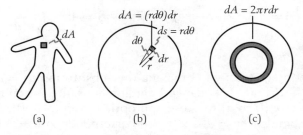

Figure 1.18 Examples of areal elements on a plane. (a) Arbitrary area element dA. (b) Area element in two-dimensional radial coordinates. (c) Area element for an annulus.

per unit area) σ and total area A. See Figure 1.18(a). We can obtain this area by quadrature: we break up the figure into many small areas dA, and add them up, obtaining a value that we call A. (In practice, adding them up would give some number, but we want to work with symbols here.) Alternatively, if the shape had been cut from a sheet of paper of known dimensions—so that we could obtain the area—and if we could weigh both the full sheet and the shape, then we could deduce the area of the shape. The shape has a total charge

$$q = \int dq = \int \sigma dA = \sigma \int dA = \sigma A. \qquad (1.19)$$

The average charge per unit area $\bar{\sigma}$ is

$$\bar{\sigma} \equiv \frac{q}{A} = \frac{\sigma A}{A} = \sigma, \qquad (1.20)$$

as expected for a uniformly charged object.

(b) Conducting disk of radius *a*.   This is the only conductor we will consider. The charge per unit area on each side of the disk depends on the local radius r and is given by $\sigma = \alpha(a^2 - r^2)^{-1/2}$, where α now is a constant with units of charge per unit length. Note how σ grows as r increases, approaching infinity as r approaches the edge, at $r = a$. (The difficult problem of how charge is distributed over an isolated conducting disk was first solved by the 26-year-old William Thomson—later to be known as Lord Kelvin.)

Although the area element for a circular geometry is $dA = (rd\theta)dr$ (see Figure 1.18b), we can integrate over θ from 0 to 2π, to obtain the partially integrated form $dA = 2\pi rdr$ (see Figure 1.18c). This area is the well-known product of the perimeter $2\pi r$ of a circle and the thickness dr. As expected, the total area is $A = \int dA = \int_0^a 2\pi rdr = \pi a^2$. Including both sides, the charge on the disk is

$$q = 2 \int \sigma dA = 2 \int_0^a \alpha(a^2 - r^2)^{-1/2}(2\pi r\, dr) = 4\pi\alpha \int_0^a \frac{rdr}{(a^2 - r^2)^{1/2}}. \qquad (1.21)$$

This can be integrated by (1) making the substitution $u^2 = a^2 - r^2$, so that $udu = -rdr$; and (2) changing the limits of integration from those appropriate to r (0 to a) to those appropriate to u (a to 0). (Note that $r = 0$ corresponds to $u = a$, and $r = a$ corresponds to $u = 0$.) Then, on switching the limits of integration, (1.21) becomes

$$q = 4\pi\alpha \int_a^0 \frac{(-udu)}{u} = 4\pi\alpha \int_0^a du = 4\pi\alpha a. \qquad (1.22)$$

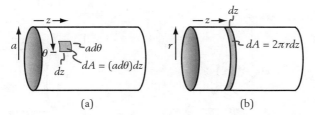

Figure 1.19 Examples of areal elements on a cylinder. (a) Area element defined by angular extent and extension along axis. (b) Area element of a ring.

From (1.22), we find that $\alpha = q/(4\pi a)$, which is indeed a charge per unit length. The average charge per unit area for each side $\bar{\sigma}$ is

$$\bar{\sigma} \equiv \frac{q}{2A} = \frac{q}{2\pi a^2} = \frac{2\alpha}{a}. \tag{1.23}$$

Even though the charge per unit area approaches infinity as $r \to a$, the total amount of charge is finite. Isn't mathematics amazing?

(c) Cylindrical surface of radius *a* and length *l*.   Let us take $\sigma = \alpha z$ so that the charge density varies only along the z-axis. The constant α must have units of charge per unit volume. In this case, although the area element for this geometry is $dA = (ad\theta)dz$ [see Figure 1.19(a)], we can integrate over θ from 0 to 2π, to obtain the partially integrated form $dA = 2\pi a\,dz$ [see Figure 1.19(b)]. This area is the well-known product of the perimeter $2\pi a$ of a circle and the thickness dz. The total area is $A = \int dA = \int_0^l 2\pi a\,dz = 2\pi al$. The total charge is

$$q = \int \sigma\,dA = \int_0^l (\alpha z)(2\pi a\,dz) = 2\pi a\alpha \int_0^l z\,dz = \pi\alpha al^2. \tag{1.24}$$

From (1.22), we find that $\alpha = q/(\pi al^2)$, which is indeed a charge per unit volume. The average charge per unit area $\bar{\sigma}$ is

$$\bar{\sigma} \equiv \frac{q}{A} = \frac{q}{2\pi al} = \frac{\alpha l}{2}. \tag{1.25}$$

3. Volume charge distribution ρ: $dq = \rho\,dV$ gives dq distributed over the volume dV.

See Figure 1.20(a). Here we write

$$q = \int dq = \int \frac{dq}{dV}dV = \int \rho dV, \qquad \rho \equiv \frac{dq}{dV}, \tag{1.26}$$

where $dV > 0$ is the small volume over which dq is distributed, and $\rho \equiv dq/dV$ is the charge per unit volume. We must take dV so small that ρ really is constant over dV; in practice, we must take dV to be infinitesimal. (Sometimes volume integrals are written as $\iiint dV$ to remind us that areas involve three coordinates.)

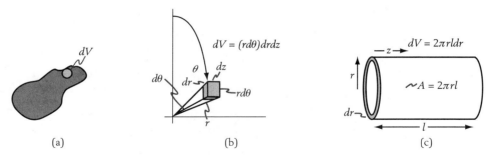

Figure 1.20 Examples of volume elements. (a) arbitrary volume element. (b) volume element in three-dimensional cyclindrical coordinates. (c) volume element of a thin-walled cylinder.

(a) **Uniformly charged volume.** Consider a volume of arbitrary shape (perhaps a lump of clay) with a uniform charge distribution (here, charge per unit volume) ρ and total volume V. See Figure 1.20(a). We can obtain this volume by quadrature: we break up the shape into many small volumes dV, and add them up, obtaining a value that we call V. (As with the area example, we want to work with symbols.) Alternatively, if the shape had been sculpted out of a block of wood of known dimensions—so that we could obtain the volume of the block—and if we could weigh both the complete block and the sculpted shape, then we could deduce the volume of the shape. The shape has a total charge

> **Volumes That You Should Know by Heart**
>
> The volume of a right-circular cylinder of radius r and length l ($\pi r^2 l$), and the volume of a sphere of radius r ($4\pi r^3/3$).

$$q = \int dq = \int \rho \, dV = \rho \int dV = \rho V. \qquad (1.27)$$

The average charge per unit volume $\overline{\rho}$ is

$$\overline{\rho} \equiv \frac{q}{V} = \frac{\rho V}{V} = \rho, \qquad (1.28)$$

as expected for a uniformly charged object. In the next few examples, we'll use the same symbol, α, to refer to totally different quantities.

(b) **Cylinder of radius a and length l, $\rho = \alpha r^3$.** Here the volume element starts out as $dV = (rd\theta)dr\,dz$ (see Figure 1.20b). However, because ρ has no dependence on θ or z, we can integrate over these to obtain the partially integrated form $dV = 2\pi r l \, dr$ (see Figure 1.20c). This volume is the well-known product of the surface area $2\pi r l$ of a cylinder and the thickness dr. The total volume is $V = \int dV = \int_0^a 2\pi l r \, dr = \pi a^2 l$. The total charge is

$$q = \int \rho dV = \int_0^a (\alpha r^3)(2\pi lr\,dr) = (2\pi \alpha l)\int_0^a r^4 dr = \frac{2\pi \alpha la^5}{5}.$$

$$(1.29)$$

From (1.29), we find that $\alpha = (5q/2\pi a^5 l)$, which has units of charge divided by distance to the sixth power. The average charge per unit volume $\overline{\rho}$ thus is

$$\overline{\rho} \equiv \frac{q}{V} = \frac{2\pi \alpha la^5/5}{\pi a^2 l} = \frac{2a^3\alpha}{5}. \qquad (1.30)$$

(c) Cylinder of radius a and length l, $\rho = \alpha z^4$. As in the previous example, the volume element starts out as $dV = (rd\theta)dr\,dz$. However, because ρ has no dependence on θ or r, we can integrate over these to obtain the partially integrated form $dV = \pi a^2 dz$. This corresponds to the volume of a pancake of radius a and thickness dz, that is, to the well-known product of the area πa^2 of a circle and the thickness dz. The total volume is, as in the previous example, $V = \pi a^2 l$. The total charge is

$$q = \int \rho dV = \int_0^l (\alpha z^4)(\pi a^2 dz) = (\pi \alpha a^2)\int_0^l z^4 dz = \frac{\pi \alpha a^2 l^5}{5}. \quad (1.31)$$

From (1.31), we find that $\alpha = (5q/2\pi a^2 l^5)$, which has units of charge divided by distance to the seventh power. The average charge per unit volume $\overline{\rho}$ is

$$\overline{\rho} \equiv \frac{q}{V} = \frac{\pi \alpha a^2 l^5/5}{\pi a^2 l} = \frac{\alpha l^4}{5}. \qquad (1.32)$$

(d) Sphere of radius a, $\rho = \alpha(a^2 - r^2)$. Here, the volume element starts out as $dV = (dA)dr = (r\sin\theta d\phi)(rd\theta)dr$, where Figure 1.21(a) depicts the area dA. Because ρ has no ϕ-dependence, the integral on ϕ from 0 to 2π can be performed, yielding the partially integrated form $dV = (dA)dr = 2\pi r^2 \sin\theta d\theta dr$, where Figure 1.21(b) depicts the new

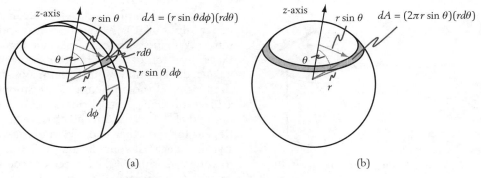

Figure 1.21 Area elements on a sphere. (a) Area element in spherical coördinates. (b) Area of ring. For corresponding volumes, multiply by dr.

area dA. Next, since ρ has no θ-dependence, the integral on θ can be performed, yielding the partially integrated form $dV = 4\pi r^2 dr$. This volume is the well-known product of the surface area of a sphere $4\pi r^2$ with its thickness dr. The total volume is $V = \int dV = \int_0^a 4\pi r^2 dr = 4\pi a^3/3$. The total charge is

$$q = \int \rho \, dV = \int_0^a \alpha(a^2 - r^2)(4\pi r^2 dr) = 4\pi\alpha \int_0^a (a^2 r^2 - r^4)dr$$

$$= 4\pi\alpha \left(a^2 \cdot \frac{a^3}{3} - \frac{a^5}{5} \right) = \frac{8\pi\alpha a^5}{15}. \tag{1.33}$$

From (1.33), we find that $\alpha = (15q/8\pi a^5)$, which has units of charge divided by distance to the fifth power. The average charge per unit volume $\overline{\rho}$ is

$$\overline{\rho} \equiv \frac{q}{V} = \frac{8\pi\alpha a^5/15}{4\pi a^3/3} = \frac{2a^2\alpha}{5}. \tag{1.34}$$

1.9.2 On Dimensions

Don't forget that perimeters have dimension of length, areas have dimension of (length)^2, and volumes have dimension of (length)^3. Consider a television announcer who tells you that from the characteristic lengths a, b, and c, and the equation $w = abc$, you can find the width w of his foot. He would be ill-informed about matters of dimension. What about $w = (abc)^{1/3}$? $w = a^2 + b^2 + c^2$? $w = ab/c$? $w = a^2bc$? $w = ab/c + a^2 + b^2 + c^2$? Which of these can be areas? volumes? Which make no sense either as a length, area, or volume? [Areas are $w = a^2 + b^2 + c^2$ and $w = ab/c$; volume is $w = abc$; length is $w = ab/c$ and $w = (abc)^{1/3}$; nonsense is $w = ab/c + a^2 + b^2 + c^2$; and $w = a^2bc$ has the dimension of length to the fourth power.]

1.10 Home Experiments

Optional

Home Experiment 1.2   A simple version of Gilbert's versorium

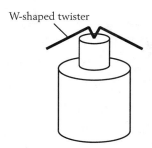

W-shaped twister

Figure 1.22 Home-made versorium. Twister and small, smooth-topped bottle.

Get a twister (such as those used to keep fruit and vegetables in plastic bags—you can also use a paper clip, but it is harder to bend) and bend it into a W-shape with long arms (see Figure 1.22). Place it on top of a small bottle so that it is well balanced. (This may require minor adjustments of the twister. If the bottle top is not smooth, place a piece of tape on it.) Bring a rubbed comb to one end of the versorium and observe its response. Repeat for the other end. Try rubbing a piece of styrofoam against your clothing and bringing it up to your versorium. Take a piece of tape and stick an edge to a tabletop. Then quickly pull it off the tabletop. Determine the response of both ends of the versorium to the tape.

Home Experiment 1.3   Compass needle and comb

This experiment requires a compass needle on a simple pivot. [Note: There should be no metallic case surrounding the compass needle—the metallic case will cause electrical "screening" (due to electrostatic induction), thus weakening the response of the compass needle to the comb.] Charge up a comb and bring it near both ends of the compass needle. Both ends of the needle should be attracted to the comb.

Home Experiment 1.4   Electrical "screening"

Place a piece of writing paper between the versorium and a distant charged comb. Bring the comb nearer. Observe how the versorium responds. See Figure 1.23. Remove the paper and observe how the versorium responds. Repeat using a paper napkin or toilet paper instead of writing paper. Repeat with a double thickness. Repeat with wet paper. Repeat with a piece of aluminum foil, and with a plastic bag. At least one of these materials should *not* screen. In fact, use of the term *screening* is misleading. In response to the charge on the comb, a redistribution of charge occurs on the screening material (for a conductor, by electrostatic induction; for an insulator, by polarization). This tends to cancel the effect of the charge on the comb.

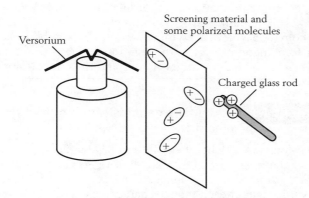

Figure 1.23 Electrical "screening." The electric field of the charge on the rod is screened out by charge induced on the screening material, a conductor or a highly polarizable nonconductor.

Home Experiment 1.5   Suspension demonstration of mutuality of the electrical force

Rub your comb, and then wrap a string around a few of its tines (i.e., its teeth), and tape them together. (Or just use comb and tape.) Suspend the string from a doorway, so the comb can move horizontally. Determine the response of the comb when you bring your finger up to it. See Figure 1.24.

Home Experiment 1.6   Electrifying a metal by friction

Cut a 2-inch piece of tape, touching it as little as possible (to minimize charging). Ensure that it is uncharged by waiting a few minutes, or by moistening it and then gently blowing on it to speed up the drying process. Test it for no charge with the versorium. Then stick the tape to one side of a small coin (the

Figure 1.24 Suspension demonstration of the mutuality of the electrical force. This is the amber effect again, now with the amber moving to the polarized material.

Door frame

smaller the coin the better, to have the tape cover its edges). Rest the tape side of the tape–coin combination on an insulating surface. With one hand, hold the edges of the tape in place (don't touch the coin). With your other hand, rub the coin with a piece of styrofoam or plastic or bubble pack. Lift up the tape–coin combination, and test the coin for charge with the versorium. If you touch the coin, you draw off its charge, and the effect is not seen.

Home Experiment 1.7 Sticky tape demonstration of two types of electricity

Cut two 6-inch-long tapes (label them T1 and B1), and place the sticky side of T1 on top of the nonsticky side of B1. Pull them apart. Do they attract or repel? Is each attracted to your finger? (This is explained by electrostatic induction on your finger, if each tape is charged.) Stick the tape ends to the edge of a table. Prepare another set, labeled T2 and B2. Record how the six combinations of pairs of tapes interact (attraction or repulsion). Stick the ends of the tapes to the edge of a table, a few inches apart. Rub a comb through your hair, and determine how the comb interacts with each tape. Rub the side of a styrofoam cup (or a styrofoam packaging "peanut") against your clothes and determine how it interacts with each tape. Determine the sign of T1, and so on under the assumption that the comb is resinous (negative).

Home Experiment 1.8 An electrical "motor"

Figure 1.25 shows an electrical "motor" similar to that of Franklin.

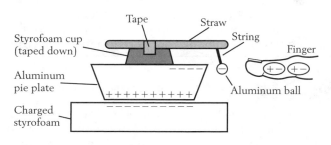

Tape   Straw

Styrofoam cup (taped down)

String

Finger

Aluminum pie plate

Aluminum ball

Charged styrofoam

Figure 1.25 An electrical "motor." The string charges by contact with the pie plate, and then is attracted to the finger (as in the previous figure), causing the string to discharge. The string then returns to the pie plate, and the process repeats.

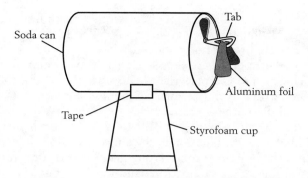

Figure 1.26 A homemade electroscope. Separation of the pieces of aluminum foil serves to indicate the presence of charge.

Home Experiment 1.9   **Electrostatic induction studied with an electrophorus and an electroscope**

A styrofoam plate can be used as an electrophorus, when rubbed with a cloth or, more effectively, with a plastic oven bag. An aluminum pie plate with a taped-on styrofoam cup as handle can be used as the metal plate. To help prevent charge from coming off the styrofoam, a sheet of plastic transparency or plastic wrap may be placed upon it. Next construct a simple electroscope: place a metal soft-drink can horizontally upon an insulating base made of a styrofoam cup. Cut a thin strip of aluminum and bend it to hang over the tab of the can. See Figure 1.26. Then, repeatedly use the electrophorus and the pie plate to charge up the soda can; the "threads" of aluminum foil should get farther apart as the can becomes more charged.

1.11   Electrical Extras

Optional

Here are a few electrical extras that you may find of interest.

1.11.1   *Light and Electricity*

Light can be emitted by a large number of processes that involve electrically charged particles, including but not exclusive to (1) an electron combining with an atomic or molecular ion; (2) an electron combining with an atom or molecule in an excited state; (3) a positively charged ion combining with a negatively charged ion. The nature of the light observed depends upon the ions, atoms, and molecules involved. Typically, when light is observed, it originates in the air, but it can also originate at the surface of a charged material.

The air contains a small but significant fraction of charged particles, caused by collisions of high-energy particles (mostly protons) from outer space, called cosmic rays. These collide with and ionize atoms and molecules in the air, producing positive ions and free electrons. Most of the electrons rapidly attach to uncharged molecules (the larger the molecule, the easier this attachment) to produce negative ions, but a small fraction of electrons remain free. The average distance traveled by a particle between collisions is called its *mean free path*.

Let us consider how light can be produced when an object is charged up. The following electrode properties are significant: (1) the amount of the charge; (2) the sign of the charge; (3) the material of the surface; (4) the shape of the surface. Typically, the farther from the surface, the less significant its properties in influencing electrical discharge and any associated light.

If the surface charge density is negligible, the only processes that can produce light are those that involve recombination in the air. As we know, it is dark within a windowless room with no light source. From this we infer that the intensity from recombination of background electrons and ions (cosmic rays freely pass through walls) is too low to be detected by the naked eye.

As the surface charge density grows, recombination processes at the surface become possible. (1) For negatively charged surfaces, positive ions are attracted to the surface, where they can pick up an electron and perhaps radiate light. (2) For positively charged surfaces, negatively charged electrons and negatively charged ions are attracted to the surface, where an electron can be captured and perhaps radiate light. Nevertheless, for small surface charge densities the rate at which such processes occur is too low to be observed by the naked eye.

For yet higher surface charge densities, electrons within a few mean free paths of the surface can gain enough energy, on colliding with an atom or molecule in the air, or on the surface, to kick off a second (or *secondary*) electron. (1) For negatively charged surfaces, the electrons tend to move away from the surface, into regions where the electric force due to the surface charge, is weaker. (2) For positively charged surfaces, the electrons tend to move toward the surface, into regions where the electric force due to the surface charge, is stronger. In both cases, the electron and ion densities near the surface increase well above the background level due to cosmic rays, and the resultant increase in the number of recombination processes yields enough light to be visible. Nevertheless, there are differences between the two cases, especially if the electrode is a sharp point, where the influence of the tip falls off rapidly with distance from the tip.

For large enough charge densities near the electrode, the electrode, effectively, grows in size. Positive ion densities are large near the positive electrode, and electron densities are large near the negative electrode. This is called *space charge*.

In what follows, we will mention some of the visual effects of electricity that were observed under less than precise conditions and with less than ideal descriptions of how they were observed. Therefore do not be upset if explanations of the phenomena do not leap forward.

1.11.2 *Some Historical Tidbits*

In Lyon, around 1550, a book was published by Fracastoro (better known for his works on medicine, especially epidemiology—he gave the venereal disease *syphilus* its name). Among many topics, he described some measurements made using a plumbline (*perpendiculo*) of the sort contained in a navigator's box (*navigatori á pyxide*). No details are given of this device, but presumably it was a filament suspension (i.e., a string), a simple example of which is given in Figure 1.27. Measurements could then be made: *"and we then saw clearly how a magnet attracted a magnet, iron attracted iron, then a magnet attracted iron, and iron a magnet . . . moreover, amber snatched up little crumbs of amber, silver attracted silver and, what most amazed us, we saw a magnet attract silver."* Later, Gilbert argued that the "silver" that Fracastoro had seen attracted by lodestone must have been debased by the addition of iron.

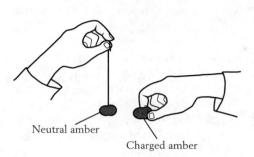

Neutral amber

Charged amber

Figure 1.27 A plumbline, whose deviation from vertical gives a measure of the interaction between two objects.

Also published in Lyon, around 1550, was a book by Cardano (more well known for his work in solving cubic and quartic equations). It noted some differences between rubbed amber and lodestone, not all of which were true. He made the following statements: (1) Rubbed amber attracts many types of objects; whereas lodestone attracts only iron. (2) Rubbed amber attracts only rather light objects, without itself being attracted; whereas lodestone is pulled as it attracts iron. (3) Rubbed amber cannot attract beyond "screens" of metal or certain fabrics; whereas lodestone can attract beyond these screens. (4) Rubbed amber seems to attract from all parts; whereas lodestone attracts iron only to its poles. (5) Amber attracts more effectively after moderate warming; whereas moderate warming does not affect lodestone.

Around 1660, the French Jesuit Honoré Fabri (an accomplished mathematician and discoverer of the Andromeda nebula), using amber on a pivot, demonstrated *mutuality*; just as "electrics" attract all sorts of objects, so too are electrics attracted toward them. The Accademia del Cimento (1657–1667), of which Fabri was a member, in their *Saggi* published another demonstration of the mutuality of the amber effect using suspension, rather than a pivot. See Figure 1.24.

Robert Boyle, better known for his discovery of Boyle's gas law around 1660, read the *Saggi*. Finding that one piece of amber attracted another (presumably one was charged and the other was neutral), he suspended a piece of amber from a silk string and found that, on rubbing it with a pin cushion, the amber moved toward the pin cushion. Boyle interpreted his result as if only the amber were electrified. Boyle further noticed that rubbed diamonds emit a faint glow. Boyle also produced a good vacuum (a pressure of 1/300 of an atmosphere) and found that in vacuum a feather or other light object would be attracted by an electrified body just as well as in air.

In 1660, Guericke, known also for his invention of the vacuum pump, developed a new source of electricity. Molten sulfur and other minerals were poured into a glass globe, and the globe was broken after the sulfur had solidified; the glass globe alone would have served. A handle then was pushed through the sulfur globe, and the globe was placed in a frame from which it was rubbed by turning the handle. See Figure 1.28. Larger and more readily charged than previous electrical devices, Guericke used it to make many interesting observations, which he considered to be special properties of the material of the globe. (1) He could levitate a feather above the charged globe, and he could then walk around, globe in hand, and make the feather move with him. (2) The feather, when brought near objects, would tend to go to the sharpest edges. (3) A linen thread attached at one end to the globe would attract chaff at its other end.

(4) The globe would sometimes crackle and, when placed in a dark room, would glow like pounded sugar (or mint-flavored Lifesaver candies).

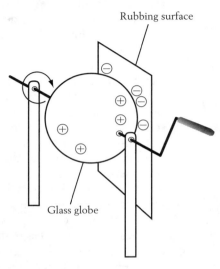

Figure 1.28 Guericke's globe was charged by rotating it by an attached handle, while rubbing it.

In 1706, Francis Hauksbee, instrument maker for the British Royal Society, studied the glow observed in the vacuum above the mercury of a shaken barometer. He found this glow—called the "mercurial phosphorus"—to be associated with mercury droplets crashing against the glass. (This sparking is caused by friction, just as flint can produce a spark when rubbed against a hard object. Such sparking is electrical in nature.) He next studied the light emitted when a small spinning glass globe rubbed against wool, both within an evacuated larger globe. The evacuated larger globe alone, when rubbed while rotated on a wheel, gave off a glow bright enough to read by. More important for posterity, Hauksbee found both a more portable source of static electricity—a glass tube about 30 inches long and 1 inch in diameter—and more sensitive detectors of electricity than straw and paper—leaf brass and lampblack.

In 1708, Stephen Gray noticed a conical glow extending from his finger as it approached a charged globe, and he rediscovered Guericke's repulsion-of-the-feather effect. In the late 1720s, Gray and Wheeler identified a number of other substances that would serve either as supports or as good "receivers" of electricity. Gray found that boys suspended by silk threads and touched by a charged glass tube could attract chaff and brass leaf. See Figure 1.29. The

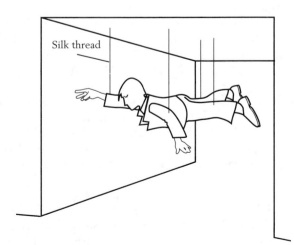

Figure 1.29 A "Gray boy" suspended by silk strings and electrically charged would himself become an electrical storage device.

boys were, unknowingly, serving as electrical storage devices. By mounting metal objects on insulators, Gray electrified metals by contact.

In Paris, around 1732, while suspended on silk threads and charged up, á la Gray, Charles Dufay touched another person whose feet were on the ground, and suddenly felt a shock, accompanied by a spark and a snapping sound. Similar effects occurred on touching a grounded metallic object, but sparking did not occur on touching a grounded "electric," such as amber. The sparking was much more spectacular than Gray's earlier observation of continuous conical glow discharge.

Gray then extended Dufay's observations about sparking by noting that the same snap and spark occurred if a person were replaced by a blunt metal object, but that a continuous conical glow occurred if a sharp metal object were employed. Both Gray and Dufay emphasized that sparking transfers electricity. Gray further noted that if a charged person suspended on silk threads sparks a second person standing on a wax cake, the first one loses electricity and the second one gains electricity. Nevertheless, the *transfer* of an amount of electricity (as in a spark) was still not clearly distinguished from the *effect* of electricity (as in the forces that rotate a versorium).

Franklin's study of the "power of points" to discharge an object led him to study the "electric wind" (Problem 1-11.4) that blows a candle flame from a positive point toward a negative point, and to make the daring suggestion that it is possible to capture lightning: build a sentry box with an insulating stand and a 25-foot iron rod to capture the lightning, which the sentry could draw off in sparks (he later proposed that the rod be grounded). Most importantly, his study of the power of points led him to conceive an important practical invention— the *lightning rod*—to prevent the catastrophic effects associated with lightning strikes to tall buildings. Ringers of church bells during storms had a particularly high mortality rate.

Franklin's letters were published in England in 1751 and quickly were translated into French. His experiments were shortly demonstrated before the French king, Louis XV, who directed that a letter of thanks be written to Franklin for his suggestion of the lightning rod. As lightning rods began to appear in the English Colonies and all over Europe, it was noticed that blunt rods worked about as well as pointed rods. In a related development, the bases of long rods were found to collect electric charge, even in the absence of lightning. See Figure 1.30. Thus began the study of atmospheric electricity.

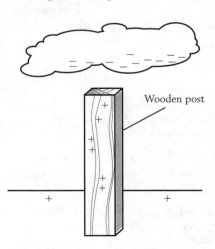

Figure 1.30 Induction of charge in a vertical rod (e.g., iron or fresh-cut wood). Negative charge at the cloud base attracts positive charge to the top of the rod.

In Paris, around 1753, Le Roy observed an asymmetry in the discharge involving conductors with points. Brushes (or cones) came from positive points (where the space charge is due to positive ions), and stars (or spheres) came from negative points (where the space charge is due to electrons).

In 1753, in St. Petersburg, Richmann, known for developing accurate electrometers, was killed by a lightning bolt that traveled down his chimney to where he was adjusting his apparatus to study atmospheric electricity. The eyewitness account, retold in Priestley's 1755 *History of Electricity*, makes it appear that the lightning bolt traveled down the chimney to the apparatus and there produced what is now called *ball lightning*. A glowing blue ball emerged from a rod connected to the apparatus and immediately traveled the foot or so to Richmann's head, where it discharged and killed him. Franklin was aware of such potential perils. In his famous discussion of flying a kite during a thunderstorm, he specified both that the kite must be held in the hand by silk, not by the twine that conducts the electricity from the kite to the ground, and that the experimenter should be under a window or door frame. Franklin's key was attached to the twine at the same junction as the silk. See Figure 1.31.

In 1759, Aepinus provided a mathematical demonstration that Franklin's ordinary matter (i.e., the part that was not electric fluid) had to be self-repulsive in order that two unelectrified objects not attract each other. Problem 1-11.19 gives an explicit example of this idea. (Aepinus also provided a semiquantitative mathematical analysis of the amber effect—a mathematically adorned version of the explanation given near the beginning of the chapter.)

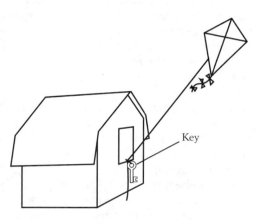
Figure 1.31 Franklin's kite experiment.

Key

Problems

1-2.1 Explain how a positively charged glass rod can attract a neutral object, and vice versa. You must use the idea that all objects are to some extent polarizable, so that a neutral object can have zero net charge, but there can be more positive charge in one region and more negative charge in another region. You must also use the fact that the force between electric charges falls off with distance.

1-2.2 In the 1730s, Schilling showed that a hollow glass ball would move about in water in response to an electrified glass tube. Why might this occur?

1-2.3 Discuss how two sticky tapes, both positively charged, may attract each other for one relative orientation, but repel for another orientation.

1-2.4 Discuss how the attraction of charged amber for a neutral piece of paper should vary if the charge on the amber is doubled. Consider both the amber and the polarization it induces in the paper.

1-3.1 How would the response of the versorium of Figure 1.2 change if the negatively charged amber were replaced by positively charged glass?

1-3.2 A compass needle turns when an object is brought toward it. Does this indicate that the object is magnetic?

1-3.3 An object is attracted to both ends of a compass needle. Suggest how this might come about.

1-4.1 Under the conditions of the experiments done before Gray, metals and water were considered to be nonelectrics. (a) What does this suggest about how the experiments were done? (b) How would you get metal or water to retain an electric charge?

1-4.2 With the discovery of the distinction between insulators and conductors began a more systematic study of the phenomena of electricity. In 1738, Desaguliers introduced the modern terminology of *conductor* and *insulator*, in the qualitative sense of materials used to communicate or prevent the communication of electricity. Using metals and water as examples, explain why this choice of terminology is an improvement on the terminology *electric* and *nonelectric*.

1-4.3 Classify as either conductors or insulators your body, dry clothing, plastic, your comb, a styrofoam cup, and a penny.

1-4.4 Identify Gray's string, silk thread, and brass wire as conductor or insulator, and state why you come to these conclusions.

1-4.5 Explain why, in the Leyden jar experiments of Musschenbroek–Cunaeus and of Kleist, it was necessary for them to stand on ground.

1-4.6 In Franklin's experiment (Figure 1.6), explain why the spark between the two who held the tube when it was rubbed is greater than between either of them and the person on the ground.

1-4.7 In another version of the electric fluid model (by Watson, a contemporary of Franklin), when applied to the Leyden jar, the charge Q_{out} is not driven from the outer conductor. Hence the outer conductor would not develop charge $-Q_{out}$. Now consider a jar that is charged in two different ways: (1) the person holding the jar stands on ground during charging; (2) the person holding the jar is standing on wax during charging. Compare the shocks that would be predicted by Watson and by Franklin.

1-4.8 (a) Explain why everyone in the human chain was shocked when the Leyden jar was discharged. *Hint:* Draw a schematic diagram of what happens when a Leyden jar discharges. (b) Given that people are better conductors than the earth, would any charge go into the ground?

1-4.9 Objects A and B, sitting on wax, are both known to be charged, because a neutral string is attracted to them. If you touch your finger to object A, it gives a big spark once, after which the string is no longer attracted to A. If you touch your finger to object B, it gives a small spark, after which the string is attracted to B, but not quite as strongly as before. Characterize A and B as conductors or insulators.

1-5.1 (a) In economic transactions, money is exchanged, but neither created nor destroyed. Discuss Franklin's maxim "A penny saved is a penny earned," as an example of a conservation law. (b) Consider the effect of thermal insulation on the power consumption of a building in the context of this maxim.

1-5.2 A positively charged plastic rod is touched to an electrically isolated conductor, and then is removed. If the conductor was initially neutral, what is its new state of charge?

1-5.3 Leyden jars A and B of Figure 1.10 are each charged up, 6 units of charge on A, and 2 units on B. They are now connected, head to head and tail to tail. They are then reconnected with the head wire of A connected to the tail wire of B, and vice versa. How is the charge distributed in these cases?

1-5.4 Let the outside of a Leyden jar be placed in contact with ground. Let the inside be placed in contact with a source of charge (or prime conductor), of charge $Q_{PC} = 10^{-6}$ C, and then disconnected. See Figure 1.5. Let a charge $Q_{in} = 10^{-7}$ C be transferred from the prime conductor to the inside of the Leyden jar. (a) How much charge Q_{PC} now resides on the prime conductor? In addition, charge is driven from the outside of the jar to ground, leaving behind a charge Q_{out}. If the glass of the jar is very thin, then $Q_{in} \approx -Q_{out}$. (b) How much charge is on the outside of the Leyden jar? (c) How much charge has been driven to ground? Now place the jar on an insulating stand, and remove the contact with the prime conductor. Someone placing one hand on the inner conductor and the other on the outer conductor will receive a shock, as charge transfers between the inner and outer conductors, in order to neutralize. (d) How much charge will transfer?

1-5.5 Explain each of the following properties of the Leyden jar: (a) when insulated (i.e., removed from the electrified bar) it retained its "charge" for hours or even days; (b) when the bottom wire (connected to its outside) was grounded, it retained its power so long as the top wire (connected to the inside) was not touched; (c) when the bottom (outside) wire was insulated, only a weak spark could be obtained from the top (inside) wire; (d) after performing (c) and then connecting the inside to ground, the jar would regain its strength.

1-6.1 In his experiments, Dufay would remove the charged tube before studying a given material. Comparing Dufay with an experimenter who kept the charged tube in place, whose experiments would be more reproducible, and why?

1-6.2 During the early 1750s in France, Delor and Dalibard studied atmospheric electricity using a long vertical conducting rod with an electrometer at its base. (a) Using the concept of electrostatic induction, explain how this might work. The atmospheric charge more often was found to be negative than positive contrary to Franklin's expectations. (b) If the earth and its atmosphere together are neutral, what does this say about the charge on the earth? (c) Blunt rods seemed to work as well as pointed ones. Using the idea that both blunt and pointed rods are small and far away from clouds, try to explain why they were equally effective.

1-6.3 A person near a charged object that suddenly discharges (such as a Van de Graaff generator) might feel a shock. Why? In principle, you can be electrocuted *without* contacting a high-voltage source.

1-6.4 We can simulate Volta's "cake" by using a piece of styrofoam that has been rubbed. In this case, let a total charge $Q_A = -10^{-8}$ C be on the styrofoam, distributed over the surface. A metal sheet is placed upon the cake and then is touched by the experimenter. (a) About how much charge Q_B is now on the metal sheet, and how much charge Q_C was drawn off by the experimenter? (Eventually, the charge Q_C goes to ground.) (b) How is the cake affected by the charging process?

1-6.5 An electroscope is a detector of electricity. Consider an electroscope that uses hanging gold leaves connected to one another and to a conducting support. When either a positively or negatively charged object is brought near, the leaves repel, similarly to what happens in Figure 1.11. Explain.

1-6.6 Boyle learned from a "Fair Lady" that, when combed, "false locks of hair" often stuck to the cheeks of their wearer. He concluded that this was related to the amber effect. However, he was puzzled that the most powerful electric—amber—could barely attract the (uncharged) hair, whereas when combed, the (electrified) hair could be attracted to the seemingly unelectrified cheek. Use the ideas of polarization and electrostatic induction to explain (a) the strong attraction of newly combed "false locks of hair" to a "Fair Lady's cheek" and (b) the weaker attraction of the good "electric" amber for uncombed hair.

1-6.7 At birthday parties, children often rub balloons and then bring them to a wall, where initially they stick, but later they fall off. (a) Explain these phenomena. (In the 1750s, it was a novelty to stick charged stockings to walls.) (b) Does it make a difference if the wall is a conductor or an insulator? (c) Does it make a difference if the room is humid or dry?

1-6.8 A source charge is brought up to (but does not touch) a conducting sphere resting on an insulator. It is represented by 10 minuses. If electrostatic induction is represented by plus and minus charges on the sphere, which of the following are possible (and if not possible, why not)? (1) 8 minuses near the source charge and 8 pluses on the opposite side; (2) 8 pluses near the source charge and 2 minuses on the opposite side; (3) 5 pluses near the source charge and 5 minuses on the opposite side; (4) 13 minuses near the source charge and 13 pluses on the opposite side.

1-6.9 A negatively charged plastic rod is brought near (but does not touch) a grounded metal sphere. Next, the ground connection is removed. Finally, the negative rod is removed. Discuss the charge transfer and electrostatic induction at each step.

1-6.10 The relative amount of charge shared by two conductors in contact depends upon how they are brought together: a sphere with a tip gains less charge when touched by a perfect sphere on the side opposite the tip than when touched at the tip because charge tends to go to the tip. See Figure 1.32. Discuss.

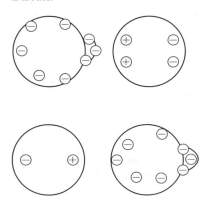

Figure 1.32 Problem 1-6.10.

1-7.1 Alpha particles are nucleii of helium atoms. They can be obtained by ejecting two electrons from a neutral helium atom. What is the charge of an alpha particle?

1-7.2 If a particle decays, and two protons are ejected (in addition to other, uncharged, particles), what can you say about the charge state of the initial particle?

1-7.3 For a particle denoted by X there is an antiparticle denoted by $\overline{X}$ with the same mass but opposite charge. However, for historical reasons, we write e^- for the electron and e^+ for its antiparticle, called the *positron*. (Some, but not all, uncharged particles can be their own antiparticles.) Which of the following reactions are allowed by conservation of charge?

(a) $e^- + e^+ \rightarrow p + \overline{p}$

(b) $e^- + e^+ \rightarrow p + p + 2n$

(c) $e^- + n \rightarrow \overline{p} + 2n$

1-7.4 What are the charge carriers when ordinary salt NaCl is dissolved in water? In a dilute solution of H_2SO_4? In a concentrated solution of H_2SO_4?

1-7.5 Rubbing a plastic rod with a certain cloth gives the rod a net negative charge. Discuss this in terms of the transfer of either electrons or protons or both.

1-7.6 In a dry room, an oven bag is rubbed against styrofoam, giving the styrofoam a negative charge. Each is quickly placed in its own dry enclosure, the enclosures connected by a long tube of insulating material. An atomizer sprays water in the styrofoam enclosure. Assume that water molecules *pick up* an electron from the styrofoam much more easily than they give up one. (a) Compare the rate at which the styrofoam and the oven bag discharge. (b) If the charges on the styrofoam and the oven bag initially sum to zero, will this be true at all times? (c) What would happen if the atomizer sprays water in the oven bag enclosure? See Figure 1.33.

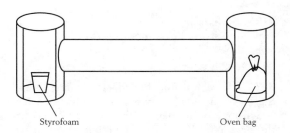

Figure 1.33 Problem 1-7.6.

1-7.7 Repeat Problem 1-7.6 if we assume the water molecules *give up* an electron to the styrofoam much more easily than they pick up one.

1-7.8 P. Heller has found that, unless they are heated before they are charged, certain porous materials lose their charge relatively rapidly. (a) If water can get in the pores, how might you explain this phenomenon? (b) Would positive and negative charge necessarily be lost at the same rate?

1-8.1 (a) How does the statement "I feel half-dead" violate a rule about people quantization? (b) How do birth and death violate a rule about people conservation?

1-8.2 A glass rod has a charge of $(3.2 \pm 0.1) \times 10^{-10}$ C. (a) To an excess or deficit of how many electrons does this correspond? (b) Estimate the uncertainty in the number of electrons.

1-8.3 A lightning flash may transfer charge either from the ground to the clouds or vice versa, but it usually goes from cloud to ground. In a characteristic lightning flash, about 10 C are transferred. To how many electrons does this correspond?

1-8.4 Two undergraduates perform Millikan's famous experiment to determine the charge on oil drops. One finds a cluster of points around the value -0.25×10^{-19} C, with an estimated error of $\pm 25\%$. Another finds a cluster of points around the value 3.5×10^{-19} C, with an estimated error of $\pm 25\%$. Which of these results is more likely to be correct, and why?

1-8.5 A plastic rod has a charge of -5×10^{-10} C. How many more electrons than protons does it possess?

1-8.6 It is believed that protons and neutrons each consist of three objects that have been named *quarks*. Quarks have either charge $-e/3$ or charge $2e/3$. How many quarks of each charge state reside within protons and neutrons?

1-9.1 A rod has charge per unit length $\lambda = Ax^2$, and extends from $x = a$ to $x = 3a$. What units must A have? (a) Show that the charge on the rod from $x = a$ to $x = 2a$ is $(7/3)Aa^3$, and that from $2a$ to $3a$ the charge is $(19/3)Aa^3$. (b) Show that the total charge is $(26/3)Aa^3$, and that the average charge per unit length for the rod is $(13/3)Aa^2$.

1-9.2 An arc on a circle of radius R goes from $\theta = 0$ to $\theta = \alpha$. If the total charge is Q, show that the average charge per unit length is $\bar{\lambda} = (Q/\alpha R)$.

1-9.3 Two rods, of lengths l_1 and l_2, have charges q_1 and q_2. (a) Find the charges per unit length for each rod, individually. (b) Find the charge per unit length, averaged over both rods. (c) Check your result for $l_1 \to 0$. (d) Check your result for $l_1 = l_2$.

1-9.4 A disk of radius a has a charge per unit area $\sigma = Br^2$. (a) What units must B have? (b) Find the charge dQ in an annulus of radius r and thickness dr. (c) Find the total charge Q on the disk as a whole. (d) Find the average charge per unit area $\bar{\sigma}$.

1-9.5 A sphere of radius b has a charge per unit volume $\rho = C + Br$. (a) What units must C and B have? (b) Find the charge dq in a spherical shell of radius r and thickness dr. (c) Find the total charge Q on the sphere as a whole. (d) Find the average charge per unit volume $\bar{\rho}$.

1-9.6 A cylinder of radius a and length l has a charge per unit volume $\rho = C + Br^2$. (a) What units must C and B have? (b) Find the charge dq in a cylindrical shell of radius r and thickness dr. (c) Find the total charge Q on the cylinder as a whole. (d) Find the average charge per unit volume $\bar{\rho}$.

1-9.7 A cylinder of radius a and length l (from $z = 0$ to $z = l$) has a charge per unit volume $\rho = C + Bz^2$. (a) What units must C and B have? (b) Find the charge dq for a circular slice of height z and thickness dz. (c) Find the total charge Q on the cylinder as a whole. (d) Find the average charge per unit volume $\bar{\rho}$.

1-9.8 A cylinder surface of radius a and length l (from $z = 0$ to $z = l$) has a charge per unit area $\rho = C + Bz^2$. (a) What units must C and B have? (b) Find the charge dq for a circular slice of height z and thickness dz. (c) Find the total charge Q on the cylinder as a whole. (d) Find the average charge per unit volume $\bar{\rho}$.

1-9.9 (a) Let $y = a \tan \theta$, where a is a constant. Show that $dy = a \sec^2 \theta\, d\theta$. (b) Let $y^2 = a \sec \theta$, where a is a constant. Show that $2y\,dy = a \sec \theta \tan \theta\, d\theta$. (c) Let $r^2 = a^2 + x^2$, where a is a constant. Show that $2r\,dr = 2x\,dx$.

1-10.1 In the experiment where the tapes are pulled apart, explain how each could be attracted to your finger.

1-10.2 Consider the following information. When A and B are rubbed together, A becomes negatively charged. When A and C are rubbed together, A becomes negatively charged. (a) Do you have enough information to tell whether B will become positively charged when rubbed against C? (b) What does this say about your ability to predict the sign of the charge that one material will have after rubbing against any other material?

1-10.3 In one of the experiments you were told to slowly peel a piece of tape off the table. It is expected that this way it should get less charge than if it were pulled off quickly. How could you test this hypothesis?

1-10.4 (a) Define electrostatic "screening." (b) Use this idea to explain why Dufay found that, although a rubbed tube could produce effects that penetrated dry silk curtains, no such penetration occurred for wet silk curtains.

1-10.5 Let charges $+10$ units and -10 units be held 1 cm above and 1 cm below a table. The force between them is one unit. Now let a thin 20-cm-by-20-cm sheet of aluminum be placed on the table. (a) Sketch the charge distribution on the aluminum sheet. (b) Is the total force on the $+10$ units of charge larger, smaller, or the same as before the sheet was added? (c) Is the force between the $+10$ and -10 units of charge larger, smaller, or the same as before the sheet was added?

1-10.6 (a) Does the fact that a compass needle can be attracted to a charged comb indicate that the comb is magnetic? (b) Cite a fact about this effect to indicate that the needle's response to the comb differs from its response to a permanent magnet.

1-11.1 Huygens performed an experiment in which moist wool—but not dry wool—was driven away after contacting a charged sphere. Propose an explanation.

1-11.2 Around 1758, in England, Symmer found that when he placed two stockings on one leg, and he removed them both, they displayed little electricity. However, on pulling them apart, each attracted small objects from a distance of 5 or 6 feet, and they attracted each other strongly. Once together, they had only a weakened electrical effect, that would extend only a few inches. Explain, and compare with the appropriate sticky tape experiment of Section 1.10.

1-11.3 In 1739, Wheeler presented—but did not publish—a paper based on experiments performed in 1732, in which he showed that a silk thread will (1) be attracted and then repelled by amber if the silk is unelectrified and insulated, (2) be repelled only if the silk is electrified and insulated, or (3) be attracted only if the silk is grounded. Explain Wheeler's observations; you must first determine whether the silk was acting as a conductor or as an insulator. (Wheeler also commented that electricity attracts objects that themselves do not attract, and repels other objects that do attract, apparently unaware of Dufay's work showing that there are two classes of electrical charge.)

1-11.4 In the "electric wind," a flame bends away from a positively charged point and toward a negatively charged point. The heat of a flame ionizes many of the atoms, kicking off their electrons, so the flame is a *plasma* of heavy positive ions and light negative electrons. Consider two effects—the electrical force directly acting on the positive ions and on the (negative) electrons, and the collisions of positive ions and of electrons with neutral atoms. Use them to explain the direction of the electric wind. Hint: The positive ions are of about the same mass as the neutral molecules, but the electrons are much less massive. The poorer the matching of mass, the less momentum is transferred in collisions. This is an example of impedance matching (Section R.2).

1-11.5 Around 1770, Cavendish argued that a conducting disk has a total charge distribution that is approximately the sum of a uniform part and a part associated with its perimeter. (a) Explain how this would make electric effects greater at the perimeter. (b) For a charged needle, to what might the total charge distribution be due? (c) How might this explain the "power of points?"

1-11.6 Gray found that two oak cubes, one solid and the other hollow, of the same exterior dimensions, received electricity in equal amounts when connected by a pack thread (a conductor) in contact with an electrified tube. What does this suggest about where the charge resides on an electrical conductor? (Assume that the oak was not thoroughly dry, and therefore was a conductor.)

1-11.7 Franklin observed that an insulated cork lowered into a charged metal cup is not attracted to its sides, and does not gain electricity upon touching the charged cup from within. What does this suggest about where the charge resides on an electrical conductor?

1-11.8 In 1675, Newton observed that when the top of a lens was rubbed (reproducibility required vigorous brushing with hog bristles), bits of paper just below the lens would be attracted to the lens. Is electrical screening occuring here? Interpret in terms of action at a distance.

1-11.9 Consider Fracastoro's experiments. Give possible explanations for (a) how amber could attract little crumbs of amber; (b) how silver could attract silver (give two mechanisms, one electric and one magnetic); (c) how a magnet could attract silver (give two mechanisms, one electric and one magnetic).

1-11.10 To the best of your knowledge, discuss the correctness of all five of Cardano's statements.

1-11.11 In Fabri's experiments, relate the idea of mutuality to what you know about Newton's laws of motion.

1-11.12 (a) Give an alternative explanation of Boyle's amber and pin cushion experiment. (b) Discuss Boyle's observation that rubbed diamonds emit a faint glow. (c) What does Boyle's vacuum experiment say about air-flow-dependent mechanisms to explain electrical forces?

1-11.13 Suggest explanations for each of the four points discussed in the text relating to Guericke.

1-11.14 (a) In Gray and Wheeler's experiments, explain the role of the supporting string (besides supporting the boys)? (b) Explain how, by mounting metal objects on insulators, Gray could electrify metals by contact.

1-11.15 Explain why sparking occurred in the experiments of Dufay on suspended, electrified people.

1-11.16 (a) Suggest why church bell ringers had a high mortality rate during lightning storms. (b) Explain how lightning rods protect tall buildings.

1-11.17 On the basis of Le Roy's observations, suggest how Gray's 1708 conical discharge may have been obtained.

1-11.18 (a) Explain why Franklin specified silk, not twine, to be held by the experimenter.

(b) Explain why he specified under a window or door frame. (c) What is the purpose of the key in Franklin's experiment?

1-11.19 Aepinus knew that the electric fluid is self-repulsive (charge spreads out on a conductor), and that the electric fluid is attracted to ordinary matter. Now consider two unelectrified objects, AB and A′B′, where A and A′ are electric fluid, and B and B′ are ordinary matter. Assume that the repulsion between A and A′ (to be specific, take it to be 1 N) equals the attraction between A and B′, taken to be the same as the attraction between A′ and B. (a) If B and B′ have no interaction, show that objects AB and A′B′ attract each other with a net force of 1 N. (b) If B and B′ repel as strongly as do A and A′, show that AB and A′B′ do not attract each other (e.g., with zero net force).

1-G.1 Why do static electricity experiments typically work better in the winter than in the summer?

1-G.2 A metal sphere of unknown charge hangs vertically from an insulating string. A negatively charged plastic rod attracts the sphere. (a) Can the metal sphere be positively charged? (b) Neutral? (c) Weakly charged negatively? (d) Strongly charged negatively?

1-G.3 You are given a metallic sphere of charge Q, which is enclosed in a thin glass case. You are also given three identical conducting spheres on insulating stands that can be moved about and that can be touched (e.g., grounded). With $q < Q$, how would you give the three identical spheres charges q, $-q/2$, and $-q/2$?

1-G.4 A lit match placed below a charged object can cause the object to discharge. (a) Suggest why. (b) Discuss whether or not there should be any difference in discharge rate, according to the sign of the charge.

1-G.5 (a) Owners of personal computers (PCs) who wish to add memory (random access memory, or RAM for short) or to replace a hard drive (HD) are warned to turn off the power before opening the case. Explain this warning. (b) Owners of PCs also are warned, that after opening the computer case, they should touch the metal frame of the power supply before making their installation. Explain this warning.

1-G.6 Unless two hands are used, it is often difficult to place with precision a piece of transparent tape, freshly pulled off the roll. Explain why.

1-G.7 You are given two identical metal spheres on insulating stands. (a) Using a negatively charged rod, how would you give them equal and opposite charges? (b) What would happen if you followed the procedure of part (a), but the spheres were of different radii?

1-G.8 You are given two identical metal spheres on insulating stands. (a) Using a negatively charged rod, how would you give them equal charges? (b) What would happen if you followed the procedure of part (a), but the spheres were of different radii?

1-G.9 Explain why, when you remove a T-shirt or a sock in a darkened room, you sometimes see or hear sparks. *Note:* Polyester is usually more effective than cotton at producing this effect.

1-G.10 A person on an insulated stand touches a Van de Graaff generator. She has no apparent ill effects (e.g., she is not shocked) although her hair stands on end. Would the same be true of a person standing on the ground who then touches the Van de Graaff?

1-G.11 Object A attracts objects B and C, but B and C repel. B and C are repelled by a negatively charged rod. What sign of charge can A have?

1-G.12 Some materials, such as graphite-impregnated paper, appear to be conductors on long time scales, but insulators on short time scales. Consider a piece of graphite-impregnated paper resting on a metal desk. If a positively charged object is touched briefly to the paper, discuss the state of charge of the paper both for short times and long times.

1-G.13 Objects A, B, and C are each held by insulating handles and are each uncharged. A is rubbed against B. A, a conductor, is then discharged. Next, A is rubbed against C. (a) If B and C now repel, classify as best you can the charge on A, B, and C. (b) If B and C now attract, classify as best you can the charge on A, B, and C.

1-G.14 When the uncharged insulators A and B are rubbed against each other, A develops a negative charge. If a negatively charged object C is placed nearer A than B while the rubbing takes place, indicate how this can affect the sign of the charge on A.

1-G.15 Imagine a universe with two classes of objects, A and B, where unlikes attract by lowering their energy by S, and likes repel by increasing their energy by S. Find the energies of the "molecules" AA, AB, and BB, and discuss their relative stability.

1-G.16 Imagine a universe with three classes of objects, A, B, and C, where unlikes attract by lowering their energy by $2S$, and likes repel by increasing their energy by S. Find the energies of the "molecules" AB, ABC, AAB, AABB, AAAB, AAAAB, AAAAAB, AAAAAAB. Discuss their relative stability.

Chapter 2

Coulomb's Law for Static Electricity, Principle of Superposition

Chapter Overview

Section 2.2 discusses the discovery of the inverse square law, and Section 2.3 presents the law explicitly. In Section 2.4, we estimate the characteristic force on an electron in an atom. Since the electrical force holds atoms, molecules, cells, and tissues together, from the size of this atomic force we can estimate the strength of materials. Section 2.5 introduces and applies the principle of superposition, which holds for the addition of forces of any origin, both electrical and nonelectrical. Section 2.6 shows two ways in which symmetry considerations can be used to simplify calculations. Section 2.7 considers the force on a point charge due to a charged rod, both using numerical integration (quadrature) and the analytical methods of integral calculus. Section 2.8 discusses problem solving and study strategy.

2.1 Introduction

The previous chapter summarized what was known, both qualitatively and quantitatively, up to about 1760. It included the law of conservation of charge. The present chapter presents another quantitative law: Coulomb's law for the force on one point charge due to another, which varies as the inverse second power of their separation—a so-called *inverse square* law. Chapters 1 through 6 all deal with what is called static electricity, as produced, for example, by rubbing a comb through your hair.

2.2 Discovering the Laws of Static Electricity

Optional

Before the spacial dependence of the electrical force had been established, the following results were already known:

1. There are two classes of electric charge; those in the same class repel, and those in different classes attract. This is summarized by the statement that "opposites attract and likes repel" (Dufay).

2. The electrical force lies along the line of centers between the two charges.

3. The force is proportional to the amount of charge on each of the objects. This proportionality seems to have been known by Aepinus, who, in 1759, knew everything about the force law except its specific dependence on distance.

There were also a number of observations that pointed, either directly or indirectly, to the electrical force satisfying an inverse square law.

Gray. Around 1731, Gray had found that two oak cubes, one solid and the other hollow, of the same exterior dimensions, received electricity in equal amounts when connected by a slightly conducting "pack thread" that was touched in the middle by an electrified tube. (If oak is not dry enough, it serves as an electrical conductor.) This fact implies that for both solid and hollow cubes, the charge resides on the outer surfaces. At the time no one realized that such surface charging implies an inverse square law for the electrical force.

Franklin and Priestley. Around 1755, Franklin noticed that an uncharged, insulated cork lowered into a charged metal cup is neither attracted to the cup's interior surface nor gains electricity on contacting that surface. He encouraged his friend Priestley to investigate this phenomenon further. The latter concluded, in 1767, by an analogy to gravity, that this implied the force law was an inverse square. Newton previously had showed that there is no gravitational force on a mass within a shell of uniform mass per unit area.

Robison. Robison performed electrical measurements using a balance that countered the torque due to electrical repulsion by the torque due to gravitation, finding in favor of an inverse square law. See Figure 2.1, where the plane of the holder assembly (including the charged spheres) is normal to the axis of the handle. Robison did not publish his results of 1769 until 1801, in a supplement to the *Encyclopedia Brittanica*.

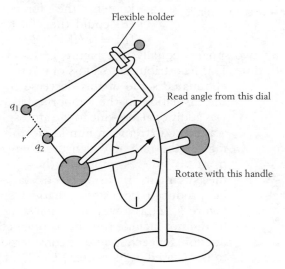

Flexible holder

q_1

r

q_2

Read angle from this dial

Rotate with this handle

Figure 2.1 Robison's gravity balance. Spheres are given charges of the same sign, and they repel until the electrical repulsion balances the pull of the earth's gravity on the upper sphere. For a given orientation angle, the separation is measured.

Cavendish. Cavendish—wealthy, eccentric, and reclusive—performed many important scientific studies, including torsion balance studies of the gravitational force between two spheres, and what are probably the first measurements of the relative electrical conductivities of different materials, obtained by comparing the shocks he received on discharging a Leyden jar through different wires. Probably inspired by Priestley's work, and thus thinking in terms of an inverse square law, in 1770 Cavendish measured the charge on the inner of two concentric metallic shells connected by a fine wire. He found it to be zero, within experimental accuracy. This indirect method established that, if the electrical interaction satisfies a power law, then within a few percent it is an inverse square law. Not until Maxwell read Cavendish's notebooks, nearly 100 years later, was it appreciated how much Cavendish had done and understood.

Coulomb. Coulomb, a military engineer, performed numerous first-rate studies in physics, including friction and the elasticity of thin wires. The latter work led him to invent the torsion balance (independently of Cavendish), which he used to study static electricity and permanent magnets. See Figure 2.2. For springs, Hooke (around 1650) found that the force F opposing a length change by x is proportional to x: $F = -Kx$, where K is a measureable *spring constant*. Coulomb found a similar relation for torsion fibers: the torque τ opposing an angular twist by ϕ (in radians) is proportional to ϕ: $\tau = -\kappa\phi$, where κ is a measureable *torsion constant*. The torque could thus be determined from the angular displacement. Since the moment arm l was known, the force magnitude $|F|$ could then be deduced, via $|F| = |\tau/l| = |\kappa\phi/l|$.

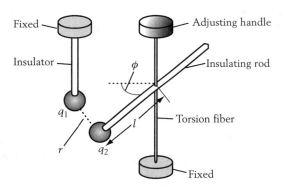

Figure 2.2 Coulomb's torsion balance. Spheres are given charges of either sign, and they rotate until the torque from the electrical force balances the torque from the torsion fiber. For a given orientation angle, the separation is measured.

Coulomb charged up two spheres equally (see Figure 1.8) and found that the force decreased with time. This he attributed to a loss of electric charge. He eliminated some of this decrease by improving the insulation in the supports. However, there was additional loss, due to the atmosphere, which was more extreme in humid weather. Accounting for the rate at which the charge decreased from his spheres improved the accuracy of his measurements. By using electrostatic induction, he produced oppositely charged spheres (see Figure 1.12). (However, in analyzing his results, he did not include the effects of electrostatic induction: for each sphere he considered the charge to be located at its center.) Coulomb published

Effect of Cosmic Rays

Even in dry weather such loss of charge occurs due to stray positive and negative ions in the air; the sphere attracts ions of charge opposite to its own. Only around 1910 was it discovered that such ions are produced by high-energy particles from outer space, called *cosmic rays*. Most cosmic rays are protons.

his work, which was well known in his native France, but 20 years passed before it was to be appreciated elsewhere.

2.3 The Inverse Square Law of Electricity: Coulomb's Law

Robison, Cavendish, and Coulomb all concluded that the electric force between two distinct point objects with charges q and Q varies as the inverse square of the separation. Thus

$$|\vec{F}| = \frac{k|q\,Q|}{r^2}, \qquad \text{(force between two charges)} \qquad (2.1)$$

where the constant k depends upon the units for force, distance, and charge.

When charge is measured in SI units of the coulomb, distance is measured in terms of meters, and force in terms of newtons, the constant k can be determined. It takes on the value

$$k = 8.9875513 \times 10^9 \frac{\text{N-m}^2}{\text{C}^2}, \qquad (2.2)$$

which usually will be taken to be $k = 9.0 \times 10^9$ N-m^2/C^2. In later chapters, it also will be useful to use the quantity ϵ_0, called the *permittivity constant*, or the *permittivity of free space*, given by

$$k = \frac{1}{4\pi\epsilon_0}, \qquad \epsilon_0 = \frac{1}{4\pi k} = 8.85418781762 \times 10^{-12} \frac{\text{C}^2}{\text{N-m}^2}. \qquad (2.3)$$

SI units were not available to Coulomb, but that was not necessary in order to establish the inverse square law. Recall that the coulomb is defined in terms of the unit of electric current, which is the ampere. Thus a coulomb is the amount of charge that passes when an ampere of current flows for one second, or $C = A\text{-s}$.

The results (1–3) in Section 2.2, and (2.1), can together be expressed as a single vector equation for the force $\vec{F}$ on a charge q due to a charge Q. *When specifying a vector, we will employ an arrow or—if it is a unit vector—a hat above it.* Using the notation that the unit vector $\hat{r}$ points toward the observation charge q, from the source charge Q, we have

$$\vec{F} = \frac{kq\,Q}{r^2}\hat{r}. \qquad \text{(force } \vec{F} \text{ on charge } q, \hat{r} \text{ toward } q) \qquad (2.4)$$

The bare geometry in the problem statement (i.e., what the problem provides) is given in Figure 2.3(a). No matter what the charges q and Q, in finding the force on q, $\hat{r}$ points to q from Q.

If q and Q are like charges, the force on q also points to q from Q. The corresponding geometry solving the problem (i.e., what the student must provide) is

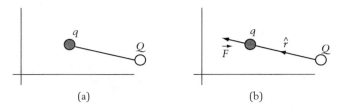

Figure 2.3 Geometry of interacting charges of the same sign. (a) The charges alone. (b) The geometry associated with the force $\vec{F}$ acting on q, with unit vector $\hat{r}$ pointing to q.

given in Figure 2.3(b). The tail of the force on q is placed on q. In this chapter, we will sometimes give both the bare geometry and the solution geometry that students must learn to provide; in later chapters, we will give only the latter.

To obtain the force on Q due to q, we use the unit vector to Q from q, which is opposite the unit vector to q from Q. Hence the force on Q due to q is opposite the force on q due to Q, and action and reaction is satisfied. A figure of the force on Q would place the tail of the force on Q.

2.4 Simple Applications of Coulomb's Law

As mentioned repeatedly, the electrical force, or C force, holds together atoms, molecules, and solids, and indeed holds together our very bodies. For that reason, it is important to get a feeling for how large a force it provides, both within atoms and within nuclei. We will not pursue these questions in great detail because classical mechanics (i.e., Newton's laws) cannot be applied *literally* at such small distances. In that case, *quantum mechanics*, an advanced topic, provides an accurate description.

2.4.1 *Coulomb's Law, Atoms, and the Strength of Materials*

A good rule of thumb is that atoms have a characteristic dimension of about 10^{-10} m, a unit that has been named the *angstrom*, or Å. Some atoms are larger, and some are smaller, but that is a good starting point. (Remember, it is more important to get the exponent correctly than to get the prefactor, although both are needed for precision work.) Therefore, consider the force on an electron in a hydrogen atom, using a separation of $r = 10^{-10}$ m. (Actually, for the hydrogen atom, the appropriate distance is about half that.) Coulomb's law, with both the electron and proton having the same magnitude $|q| = |Q| = e = 1.6 \times 10^{-19}$ C for the charge, yields for the electrical force between the electron and proton $F^{el}_{e,p} = ke^2/r^2 = 2.3 \times 10^{-8}$ N. This appears to be small, but not in comparison with the force on an atom due to the earth's gravity. Take $m_p = 1.67 \times 10^{-27}$ kg for the atomic mass (essentially, the mass of the proton since the electron is so much less massive). Then, with $g = 9.8$ m/s^2, $F^{grav}_{p,earth} = mg = 1.64 \times 10^{-26}$ N. Thus, comparison of the electrical force within the atom to the earth's gravitational force on the atom shows that the latter is negligibly small. This has profound structural significance for individual atoms and even for large molecules: their structure is indifferent to the local gravitational environment. Only on the scale of larger objects, such as trees and people, does gravity affect structure.

Application 2.1 **Electron–proton gravity within the atom is negligible**

The gravitational force of attraction between the electron and the proton is extraordinarily small. With $G = 6.67 \times 10^{-11}$ N-m^2/kg^2 and $m_e = 9.1 \times 10^{-31}$ kg, we obtain $F_{e,p}^{grav} = Gm_e m_p/r^2 = 1.01 \times 10^{-47}$ N, a force about 10^{39} times smaller than the electric force between them.

Application 2.2 **Material strength is atomic force per atomic area**

Let us take an interatomic force of 10^{-8} N to correspond to the force between nearby atoms in a bulk material. Taking atoms to be typically about 3×10^{-10} m apart, so with a cross-section of an atomic separation squared, or $(3 \times 10^{-10}$ m$)^2 = 9 \times 10^{-20}$ m$^2 \approx 10^{-19}$ m^2, this gives a force per unit area of on the order of 10^{11} N/m^2. A commonly measured property of materials is the force per unit area needed to produce a given fractional change in the atomic separation. This is known as the *elastic constant*. For real materials, the elastic constants are also on the order of 10^{11} N/m^2. This agreement indicates (but does not prove) that electrical interactions are responsible for the elastic properties of materials. Assuming that breakage occurs when the fractional change in atomic separation is on the order of 0.1 gives a tensile stress on the order of 10^{10} N/m^2, much higher than for real materials: the tensile stress of iron is on the order of 10^9 N/m^2, and for string it is on the order of 10^7 N/m^2. This indicates that something else determines when a material breaks. In the 1930s, it was discovered that details of atomic positioning, and slippage at the atomic level via what are called *dislocations*, are responsible for the relatively low tensile stress of most materials.

Estimate of adhesive strength. Let us take a modest interatomic force of 10^{-10} N to correspond to what might occur for an adhesive. Let us also take there to be one such interatomic force per 10^{-7} m in each direction along the surface. (This corresponds to about 1 every 1000 atoms.) For a 1 cm$^2 = 10^{-4}$ m^2 area, there are 10^{-4} m$^2/10^{-14}$ m$^2 = 10^{10}$ such interatomic forces, leading to a net force of 1 N, an appreciable value. Clearly, the Coulomb force is strong enough to explain the behavior of adhesives—and the adhesion between living cells.

2.4.2 *Nuclei and the Need for an Attractive Nuclear Force*

The Coulomb force also acts within atomic nucleii, whose characteristic dimension is 10^{-15} m, which is called a *fermi*. There are two protons in a He nucleus, which repel each other because of the Coulomb force. We could compute this force from (2.1), but it is easiest to obtain it by noting that in this case the charges have the same magnitude as for the electron and proton of the previous section, but the distances are smaller by a factor of about 10^5. Since the Coulomb force goes as the inverse square, the force of repulsion between two protons in a helium nucleus is larger by about 10^{10} relative to the electron–proton force in an atom. Thus $F^{el}_{p,p} = 10^{10} F^{el}_{e,p} = 2.3 \times 10^2$ N, which would support a mass exceeding 20 kg under the earth's gravity. This is enormous for such a small object. What keeps the nucleus from blowing apart is an attractive *nuclear force* between the nucleons (protons and neutrons). This force has a very short characteristic range, on the order of a fermi, but it is very strong so that within its range it can dominate the Coulomb repulsion.

Coulomb Repulsion and the Types of Helium Nuclei

Consider the possible types of He nuclei. Because of the Coulomb repulsion, too many protons relative to neutrons is bad for nuclear stability. That is why there is no such thing as a stable ^{2}He nucleus: only one pair of attractive nuclear interactions (between the two protons) is insufficient to overcome the Coulomb repulsion. On the other hand, ^{3}He has two protons and one neutron, with three pairs of attractive nuclear interactions (proton-proton, and two proton-neutron) that overcome the Coulomb repulsion (proton-proton). The isotope ^{4}He, with two protons and two neutrons, has six pairs of attractive nuclear interactions, and is yet more stable than ^{3}He. Additional neutrons must reside in nuclear orbitals that are far from the center of mass of the nucleus, and thus do not participate fully in the attractive interaction of the other nucleons. The isotopes ^{5}He and ^{6}He, although observed experimentally, are unstable, and helium nuclei with larger numbers of neutrons have not been observed at all. Neutron stars exist only because they are so massive that the gravitational attraction is large enough to keep them together.

2.4.3 *A Simple Charge Electrometer: Measuring the Charge Produced by Static Electricity*

Charge electroscopes (such as gold-leaf electroscopes, or the aluminum-foil electroscope of Figure 1.25) and *charge electrometers* are devices for measuring the charge on an object, the electrometer being more quantitative. They use the repulsive force between like charges. Figure 2.4 depicts an experiment to determine how much electricity can be produced by rubbing. Hanging from a common point are two threads of length l and two identical small conducting spheres of mass m, which have been given the same charge q by the charge-sharing process described in Section 1.5.2. Let us find the relationship between the angle θ and the charge q; clearly, the larger the charge, the larger the angle of separation.

This is a problem in statics, where each ball has three forces (each with a different origin) acting on it: gravity (downward), electricity (horizontally, away from the other ball), and the string tension T (along the string, from the ball to the point where the

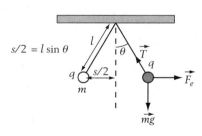

Figure 2.4 A simple electrometer. The two spheres are of equal mass m and equal charge q. By measuring the separation s or the angle θ (which are related), the electric force and the charge can be determined.

string is taped to the stick). Neither the electrical force nor the tension is known. We first discuss some geometry: from the separation s we can deduce the angle that the strings make to the normal:

$$\sin\theta = \frac{s/2}{l} = \frac{s}{2l}. \tag{2.5}$$

From the equations of static equilibrium applied to either ball (to be explicit, we'll consider the ball on the right), we can obtain two conditions. These can be used to eliminate the tension and to relate the electrical force to the gravitational force and to the geometry of the problem.

First, the ball on the right (which we will consider to be the *observer*) feels a downward force of mg from gravity, and an upward force component $T\cos\theta$ from the string tension. In equilibrium, since the sum of the vertical forces is zero, or $\sum F_y = 0$, we have $mg = T\cos\theta$. This can be rewritten as

$$T = \frac{mg}{\cos\theta}. \tag{2.6}$$

In addition, the ball feels a rightward electrical force $F_e = kq^2/s^2$ and a leftward force component $T\sin\theta$ from the string tension. In equilibrium, the sum of the horizontal forces is zero, or $\sum F_x = 0$, so

$$F_e = \frac{kq^2}{s^2} = T\sin\theta = mg\tan\theta, \tag{2.7}$$

where we have eliminated T by using (2.6). Solving for q, we obtain

$$q = s\sqrt{\frac{mg\tan\theta}{k}}. \tag{2.8}$$

To be specific, take length $l = 10$ cm $= 0.1$ m, mass $m = 4$ g $= 0.004$ kg, and separation $s = 2.5$ cm $= 0.025$ m. Then, by (2.5), $\sin\theta = 0.125$, so $\cos\theta = \sqrt{1 - (0.125)^2} = 0.992$ and $\tan\theta = 0.125/0.992 = 0.126$. Using m, s, and θ from the statement of the problem, and $g = 9.8$ m/s^2, (2.8) gives $q = 1.85 \times 10^{-8}$ C as a typical amount of charge that can be obtained by rubbing a comb through one's hair. By analyzing Figure 2.4 quantitatively, we have turned a qualitative electroscope into a quantitative electrometer!

At small θ, where $\sin\theta$ and $\tan\theta$ both vary as θ, s varies as θ and thus q varies as $\theta^{\frac{3}{2}}$. Correspondingly, θ varies as $q^{\frac{2}{3}}$. This rises very quickly for small q, so a measurement of angle is quite sensitive for very small charge. However, for very large charge, all angles will be near $90°$, so a measurement of angle is insensitive for very large charge.

A complete solution would require the tension T of the string, given by (2.6). This quantity becomes relevant if we have a weak string that easily can be broken. In most mechanics problems, physicists don't worry about such questions, but mechanical and civil engineers make a living out of them.

This problem has gravity, strings, and electricity, and at first it seems like apples and oranges and bananas. Just as you can add the scalars representing the masses or calorie contents of different types of fruit, so you can add the vectors representing different types of force.

2.5   Vectors and the Principle of Superposition

2.5.1   *What We Mean by a Vector: Its Properties under Rotation*

Sections R.9 and R.10 discuss vectors in detail. If you aren't yet comfortable with vectors, and you haven't already read those sections, read them now.

It is so important to drive this message home that we'll repeat what you already know. Vectors are characterized by magnitude and direction—*and by their properties under rotation.* Quantities like force, position, velocity, and acceleration are vectors. Their magnitudes do not change under rotation, and the orientation between two such quantities does not change under rotation. If a position vector and a force vector are at $40°$ to each other, then after any rotation they remain at $40°$ to each other.

In contrast, consider pressure P, temperature T, and energy E. None of these three quantities change under a rotation in space; they are scalars. Hence the three-component object (T, P, E) does not transform as would a vector under rotations in space. Merely having three components doesn't assure "vectorness."

2.5.2   *The Principle of Superposition: Add 'Em Up*

Because forces are vectors, when there are individual forces acting on a single object, the net force is obtained by performing vector addition on all the forces. This is called the *principle of superposition.* We used this principle in the electrometer example, where the three forces each had a different source. In what follows, we will use the principle of superposition to add up many forces of electrical origin. A force $\vec{F}$ may have components along the x-, y-, and z-directions. We specify these directions as the set of unit vectors $(\hat{x}, \hat{y}, \hat{z})$, or $(\hat{i}, \hat{j}, \hat{k})$. Indeed, we will sometimes write $\rightarrow$ for $\hat{i}$ and $\leftarrow$ for $-\hat{i}$, $\uparrow$ for $\hat{j}$ and $\downarrow$ for $-\hat{j}$, and $\hat{\odot}$ for $\hat{k}$ (out of page) and $\hat{\otimes}$ for $-\hat{k}$ (into page). Thus $(\hat{x}, \hat{y}, \hat{z}) \equiv (\hat{i}, \hat{j}, \hat{k}) \equiv (\rightarrow, \uparrow, \hat{\odot})$. You should already be accustomed to seeing numerous ways to write the same thing. For certain problems involving gravity, it is often convenient to let *down* correspond to $\hat{y}$. The ways in which we write the laws of physics all depend upon conventions (e.g., we use right-handed, not left-handed, coordinate systems). However, the laws themselves do not depend on these conventions.

Explicitly, we write

$$\vec{F} = F_x\hat{x} + F_y\hat{y} + F_z\hat{z}, \tag{2.9}$$

where F_x is the x-component of $\vec{F}$, and so forth. By the rules of the scalar product, where $\hat{x} \cdot \hat{x} = 1$, $\hat{x} \cdot \hat{y} = 0$, and so forth, we have

$$\vec{F} \cdot \hat{x} = F_x. \tag{2.10}$$

Similarly, we can determine F_y and F_z. Thus, if $\vec{F}$ is specified in terms of its magnitude $F = |\vec{F}|$ and its direction, it also can be written in terms of its components.

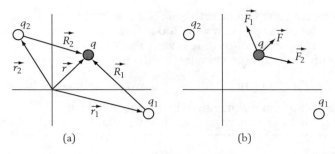

Figure 2.5 (a) Geometry for the force on q due to q_1 and q_2. Lowercase vectors refer to distances from the origin, and uppercase vectors refer to relative distances. (b) Force on q, due to q_1 and q_2.

Likewise, if $\vec{F}$ is specified in terms of its components, then its magnitude can be computed from

$$|\vec{F}| = \sqrt{F_x^2 + F_y^2 + F_z^2}. \tag{2.11}$$

When, in addition to the charge q at $\vec{r}$, there are many other charges, by the principle of superposition the total force is the vector sum of all the forces acting on q. Consider a situation with many charges q_i at respective positions $\vec{r}_i$. Figure 2.5(a) depicts only q_1 and q_2, in addition to q. For the pair q and q_1, we have $\vec{r} = \vec{R}_1 + \vec{r}_1$, so $\vec{R}_1 = \vec{r} - \vec{r}_1$. More generally, $\vec{R}_i = \vec{r} - \vec{r}_i$. With $\hat{R}_i = \vec{R}_i / |\vec{R}_i|$, we see that $\hat{R}_i$ *points to the observation charge q at $\vec{r}$*. For example, if in Figure 2.5(a) $\vec{r} \equiv (4, 4, 0)$ and $\vec{r}_1 \equiv (16, -3, 0)$, then $\vec{R}_1 \equiv (-12, 7, 0)$, $|\vec{R}_1| = \sqrt{193} = 13.89$, and $\hat{R}_1 \equiv (-0.864, 0.504, 0)$.

We generalize (2.4) to obtain the force on q due to q_i as $\vec{F}_i = (kqq_i/R_i^2)\hat{R}_i$. Summing over all $\vec{F}_i$ yields the total force on q. Explicitly, it is

$$\vec{F} = \sum_i \vec{F}_i = \sum_i \frac{kqq_i}{R_i^2} \hat{R}_i.$$

(force $\vec{F}$ on charge q at $\vec{r}$, $\hat{R}_i$ toward q, $\vec{R}_i = \vec{r} - \vec{r}_i$) (2.12)

Let's talk in terms of input and output. The input consists of the charge q and its position $\vec{r}$, and the charges q_i at the positions $\vec{r}_i$. For two source charges, this is given in Figure 2.5(a). The output consists of the individual forces $\vec{F}_i$ and their vector sum $\vec{F}$. This is depicted in Figure 2.5(b) for the cases where the charges q, q_1, and q_2 are all positive or all negative. The relative lengths of $\vec{F}_1$ and $\vec{F}_2$ can only be determined when actual values for q, q_1, and q_2 are given; therefore, Figure 2.5(b) is only a schematic. Each force on q is drawn with its tails on q, as if a person were pulling on a string attached to q.

Example 2.1 **Adding force magnitudes almost always produces garbage**

Adding vector magnitudes to obtain the magnitude of a sum of vectors only works when the vectors have the same direction. Thus, adding vector magnitudes usually gives incorrect answers. For example, if two horses pull on opposite sides of a rope, each with 200 N force, the resultant vector force on the rope is zero, so its magnitude is zero, not 400 N.

Let's dignify this important result with the unnumbered equation

$$\vec{F} = \vec{F}_1 + \vec{F}_2, \text{ but } |\vec{F}| \le |\vec{F}_1| + |\vec{F}_2|, \text{ when adding two forces.}$$

Another example of this result is given by the forces in Figure 2.5(b). This sort of equation (called a constraint) holds when we add two vectors of any type.

Example 2.2   Cancellation of two collinear forces

Consider that, for two charges q_1 and q_2 whose positions are known, we would like to know where to place a third charge q so that it feels no net force due to q_1 and q_2. We are free to choose a geometry where the charges are along the x-axis, with q_1 at the origin, and q_2 a distance l to its right. Thus our specific question is: where should q be placed in order to feel no net force? (Our answer will be independent of q, a fact that is related to the concept of the electric field, which will be introduced in the next chapter.) Before performing any calculations, note that *for there to be zero net force, the two forces on q must have equal magnitude but opposite direction.* The latter condition can only hold if the third charge is placed on the line determined by q_1 and q_2. There are two possibilities.

(a) q_1 **and** q_2 **have the same sign.** In this case, q should be placed between q_1 and q_2, at some distance s from q_1 (to be determined) where q will feel canceling attractions (or repulsions) $\vec{F}_1 = -\vec{F}_2$ from q_1 and q_2. To be specific, let all charges be positive. See Figure 2.6(a). Then $\vec{F}_1$ points away from q_1 and $\vec{F}_2$ points away from q_2, by "likes repel." As in Figure 2.5(b), the forces on q are drawn with their tails on q. Note that the position s where the forces $\vec{F}_1$ and $\vec{F}_2$ cancel will not change if the signs of all the charges are reversed, because again "likes repel." Moreover, the position s where the forces $\vec{F}_1$ and $\vec{F}_2$ cancel will not change if the sign of q is reversed, or if the signs of both q_1 and q_2 are reversed, because then, although the forces change direction, they still cancel.

Let the magnitudes $F_{1,2} = |\vec{F}_{1,2}|$. By (2.1) or (2.4), the equilibrium condition that $F_1 = F_2$ gives

$$\frac{kqq_1}{s^2} = \frac{kqq_2}{(l-s)^2}. \tag{2.13}$$

Canceling kq, taking the positive square root (remember, we have already

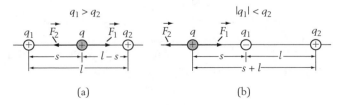

(a)   (b)

Figure 2.6 Locating the zero-force position. (a) When the two source charges have the same sign. (b) When the two source charges have opposite sign. How would this figure look for $q < 0$?

determined that $0 < s < l$), and inverting each side, we obtain

$$\frac{s}{\sqrt{q_1}} = \frac{l-s}{\sqrt{q_2}}.$$ (2.14)

Solving for s, we obtain

$$s = \frac{l}{\sqrt{q_2/q_1} + 1}.$$ (2.15)

Dimensionless Ratios

In taking limits where some quantity goes to zero or to infinity, it is often convenient to make that quantity a dimensionless ratio. For the test charge to approach the weaker charge, l is fixed and s gets smaller. We could obtain the same limit by keeping s fixed and letting l get larger. Using the dimensionless ratio s/l takes care of both possibilities.

This value of s is independent of the value of q; doubling q doubles each force, so they continue to cancel at the same position. As checks, note that (1) $s = l/2$ for $q_1 = q_2$ (i.e., the test charge is equidistant between two equal charges); (2) $s/l \to 0$ as $q_2/q_1 \to \infty$ (i.e., the test charge approaches the weaker charge, here q_1); and (3) $s/l \to 1$ as $q_2/q_1 \to 0$ (i.e., again the test charge approaches the weaker charge, now q_2).

(b) q_1 and q_2 **have opposite sign.** It will be sufficient to consider the case $q_1(<0)$, $q_2(>0)$, and $q(>0)$, because the other cases of this type are related to this one simply by changes in direction of both forces. To be specific, let $|q_1| < |q_2|$. Then q should be placed on the same line, but to the left of the weaker charge q_1, so that proximity can compensate for weakness. See Figure 2.6(b). The two forces will cancel when q is placed a distance s (to be determined) to the left of q_1. By (2.1), the equilibrium condition that $F_1 = F_2$ gives

$$\frac{kq|q_1|}{s^2} = \frac{kq|q_2|}{(l+s)^2}.$$ (2.16)

Solving for s as before, we now obtain

$$s = \frac{l}{\sqrt{|q_2/q_1|} - 1}, \qquad (|q_2/q_1| > 1).$$ (2.17)

As checks, note that (1) $s \to \infty$ as $|q_2/q_1| \to 1$ (i.e., for equal strength source charges, the only way their forces cancel out, if one is nearer q than the other, is if q is at infinity); and (2) $s \to 0$ as $|q_1/q_2| \to 0$ (i.e., the test charge approaches the weaker charge, here q_1).

One Size Doesn't Always Fit All

We are often tempted to develop a totally general equation that will include all cases—"one size fits all." However, if we generalize too soon, we might not notice some fundamental distinctions. This often occurs in computer programming, where an algorithm developed for one situation does not work properly when applied to another. The related cases of like and unlike charges really should be treated separately.

Example 2.3   Addition of two noncollinear forces

Now consider the force on q due to q_1 and q_2 when the three charges do not lie along a line. In principle, the charges can be anywhere with respect to a fixed coordinate system, each of them requiring three numbers (x, y, z) to specify its position. However, because there are only three charges, and three noncollinear points define a plane, which we may choose to be the xy-plane, only two numbers (x, y) per charge will be required to specify each position. Further, we can place q at the origin, so $\vec{r} = \vec{0}$, leaving only two numbers (x, y) per charge to specify the positions of q_1 and q_2. The last allowed simplification is to take q_1 to lie along a specific direction, such as $\hat{x}$. The third charge, q_3, can lie anywhere in the xy-plane. Thus, the geometry is specified with a total of three numbers, which we take to be the distances R_1 and R_2 of q_1 and q_2 to the origin, and the angle θ_2 that R_2 makes with respect to the x-axis (we have already taken $\theta_1 = 0$). The problem is to find the net force on q. To be specific, let $q = 2.0 \times 10^{-9}$ C, $q_1 = -4.0 \times 10^{-9}$ C and $q_2 = 6.0 \times 10^{-9}$ C, $R_1 = 0.2$ m, $R_2 = 0.3$ m, and $\theta_2 = 55°$. See Figure 2.7(a). Find the net force on q.

Solution: There are at least two ways in which we can proceed to solve this problem. We shall call one the *common sense* method, which is particularly appropriate when there are only a few forces involved. (We call it the common sense method because it has been said, with much truth, that science is simply common sense, but more refined.) We shall call the other method the *formal* method because it is a bit akin to the higher mathematics to which J. J. Thomson was referring in the quote at the end of Section R.5. When we have a choice, the first method is preferable, but there are times when the only practical method is the second one. We will use both in the present case.

If you don't know at least two ways to solve a problem, you may not really understand the problem in a deeper sense. Competent practitioners in any area can solve a given problem in multiple ways.

The common sense method first finds the magnitude of each force acting on q (the observer) by using (2.1), then gets the direction of each force by "opposites attract, likes repel," and finally performs the vector addition. That is more or less what we did in the previous example. The formal method uses (2.12) to compute each force in terms of its vector components (so we actually compute the magnitudes of the individual forces), and adds up the components.

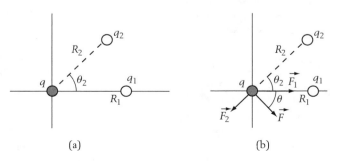

(a)   (b)

Figure 2.7 Force on q at origin, due to q_1 and q_2.
(a) Geometry of the problem. (b) Solution of the problem in terms of individual forces and the total force.

Common sense method. First, we find the magnitude $F_1 = |\vec{F}_1|$ of the force acting on q due to q_1. By Coulomb's law, as in (2.1),

$$F_1 = \frac{k|qq_1|}{R_1^2} = \frac{(9 \times 10^9 \text{ N-m}^2 / \text{C}^2)|(2 \times 10^{-9} \text{ C})(-4 \times 10^{-9} \text{ C})|}{(0.2 \text{ m})^2}$$

$$= 1.8 \times 10^{-6} \text{ N}.$$

Note that F_1 must be positive because it is the magnitude, or absolute value, of $\vec{F}_1$. Similarly,

$$F_2 = \frac{k|qq_2|}{R_2^2} = \frac{(9 \times 10^9 \text{ N-m}^2 / \text{C}^2)|(2 \times 10^{-9} \text{ C})(6 \times 10^{-9} \text{ C})|}{(0.3 \text{ m})^2}$$

$$= 1.2 \times 10^{-6} \text{ N}.$$

Using "opposites attract, likes repel," we can now draw the force diagram (compare Figure 2.7b). This leads to the force components

$$F_x = F_1 - F_2 \cos\theta_2 = 1.8 \times 10^{-6} \text{ N} - 1.2 \times 10^{-6} \text{ N} \cos 55°$$

$$= 1.112 \times 10^{-6} \text{ N},$$

$$F_y = -F_2 \sin\theta_2 = -0.983 \times 10^{-6} \text{ N}.$$

Thus, as in Figure 2.7, $\vec{F}$ lies in the fourth quadrant. We also have

$$F = |\vec{F}| = \sqrt{F_x^2 + F_y^2} = 1.484 \times 10^{-6} \text{ N},$$

and $\vec{F}$ makes an angle with slope

$$\tan\theta = \frac{F_y}{F_x} = -0.884,$$

corresponding to an angle θ in the fourth quadrant, with $\theta = -41.5°$, or -0.724 radians. If $\vec{F}$ had been in the second quadrant, its tangent would also have been negative, so to obtain the correct angle we would have had to add $180°$ (or π radians) to the inverse tangent of $\tan\theta$.

Finding $\vec{F}$ for Rotated Source Charges

If the positions of both q_1 and q_2 are rotated by 24° clockwise, $\vec{F}$ too is rotated by 24° clockwise, to an angle of $-41.5 - 24 = -65.5°$ relative to the x-axis. Its magnitude is unchanged. Thus, the value of F_x for the rotated charges would be $|\vec{F}|\cos(-65.5) = 0.613 \times 10^{-6}$ N. This way to calculate the rotated F_x is simpler than recomputing and adding the x-components of the rotated versions of the individual forces.

Formal method. Even with the formal method, there are at least two ways to proceed. We may rewrite (2.12) as

$$\vec{F} = \sum_i F_i^* \hat{R}_i, \qquad F_i^* \equiv \frac{kqq_i}{R_i^2}, \qquad (2.12')$$

or as

$$\vec{F} = \sum_i \frac{kqq_i}{R_i^3} \vec{R}_i, \qquad \vec{R}_i = \vec{r} - \vec{r}_i. \qquad (2.12'')$$

Using (2.12'), we must first compute the *signed* force F_i^* due to the i charge, and then the unit vector $\hat{R}_i$. Thus, as an intermediate step, we can readily

Table 2.1 Force Calculation

1	2	3	4	5	6	7	8	9	10	11
x_1	y_1	z_1	X_1	Y_1	Z_1	R_1	F_1^*	$F_1^*\left(\frac{X_1}{R_1}\right)$	$F_1^*\left(\frac{Y_1}{R_1}\right)$	$F_1^*\left(\frac{Z_1}{R_1}\right)$

determine the magnitude of each of the individual forces, via $F_1 = |\vec{F}_i| = |F_i^*|$, and we can also readily determine the direction of the force. Use of (2.12′), although part of the formal approach, is closely related to the commonsense method. In Appendix B we use (2.12′) to solve the last example using a spreadsheet.

Equation (2.12″) requires fewer operations than (2.12′). This is because (2.12″) does not require the intermediate computation of $\hat{R}_i = \vec{R}_i/|\vec{R}_i|$. However, with (2.12″) we do not obtain the magnitude and direction of each individual force. Both procedures work. (Different strokes for different folks.) Table 2.1 indicates specifically what computations would have to be made using (2.12″) for the force $\vec{F}_1$ on q at (x, y, z) due to q_1 at (x_1, y_1, z_1). The table uses $X_1 = x - x_1$, and so on, and F_i^* of (2.12′). Columns 9, 10, and 11 contain the x-, y-, and z-components of $\vec{F}_1$.

2.6 Use of Symmetry

When the source charge is symmetrically placed, and the observation charge is at a position where this symmetry is evident, certain simplifications can be made. We will discuss two ways to use symmetry. First, in doing a computation we notice that certain terms must add or cancel. Second, by some general principle or principles, and the fact that force is a vector (so that it rotates when the source charges rotate), we learn that certain possibilities are disallowed.

Example 2.4 Force due to two equal charges

Let there be two equal source charges Q symmetrically placed on the x-axis, and the observation charge q be along the y-axis. In what direction does the net force on q point?

Solution: See Figure 2.8(a), where the individual forces $\vec{F}_1$ and $\vec{F}_2$, and their resultant $\vec{F}$ have been drawn. By symmetry, $\vec{F}$ must be along the y-axis because the x-components of the individual forces cancel.

Example 2.5 Force due to two equal and opposite charges

In Example 2.4, let one source charge be replaced by $-Q$, as in Figure 2.8(b). In what direction does the net force on q point?

Solution: By symmetry, the net force must be along the x-axis because now the y-components of the individual forces cancel.

The above arguments are computational in nature. Here is a noncomputational symmetry argument. Consider first the two equal source charges Q in Figure 2.8(a). Rotating them by 180° about the y-axis gives an equivalent

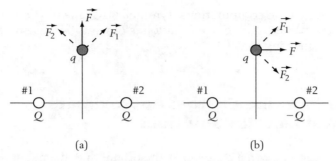

(a)　　　　　　　(b)

Figure 2.8 Force on q along perpendicular bisector between two charges. (a) Equal charges Q. (b) Equal and opposite charges $\pm Q$.

configuration, and therefore the net force on q does not change. Specifically, F_x does not change under this interchange of charges. However, for any vector (including the net force), a rotation about y must change the sign of its x-component. The only way for F_x to both change sign and not change sign is if $F_x = 0$. Consider now the equal and opposite source charges $\pm Q$ of Figure 2.8(b). Rotating by $180°$ about the y-axis, which interchanges the charges, must preserve the y-components of the forces. But interchanging the charges reverses the directions of the individual forces, including the y-components. The only way for F_y to both change sign and not change sign is if $F_y = 0$.

Example 2.6　**Force due to three equal charges**

Let there be three equal source charges $q_1 = q_2 = q_3 = Q$ placed at the corners of an equilateral triangle. Let an observation charge q be placed at the very center of the triangle. What is the net force on the observation charge, q?

Solution: See Figure 2.9, where the individual forces on q are given. The net force on q is zero, as can be established by algebraic computation, by numerical computation using particular values for the side of the triangle and the charges, and by the following argument. The forces on q due to each of these charges all have the same magnitudes, but are rotated by $\pm 120°$ relative to one another. The forces thus form the arms of an equilateral triangle that, under vector addition, sum to zero.

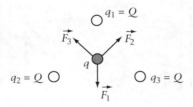

Figure 2.9 Individual forces on q at the center of three equal charges $q_1 = q_2 = q_3 = Q$ at the vertices of an equilateral triangle.

A noncomputational symmetry argument for Example 2.6 begins by observing that the net force must lie in the xy-plane. Because force is a vector, if the source charges are rotated by $120°$, either clockwise or counterclockwise, the force must rotate in the same way. However, since $q_1 = q_2 = q_3 = Q$, this rotation produces the original charge distribution, and therefore the same force. Hence the force must be zero. For a large number (e.g., 47) of equal source charges at equal radii and angles, you can generalize this argument to show that the force on a charge at their center is also zero. For this more complex case, there is no computational symmetry argument. (Of course, for an even number of charges, the net cancellation is obvious because opposite pairs would produce canceling forces.)

2.7 Force Due to a Line Charge: Approximate and Integral Calculus Solutions

Let us obtain the force $\vec{F}$ on a charge q at the origin, due to a net charge Q that is uniformly distributed over the line segment from $(a, -l/2, 0)$ to $(a, l/2, 0)$. See Figure 2.10(a).

Note that $F_y = F_z = 0$, by symmetry. In principle, to find F_x requires calculus, where we break up the continuous line charge into an infinite number of infinitesimal point charges dQ, and add up their (vector) forces $d\vec{F}$ on q. See Figure 2.10(b). However, we will first consider what happens when we approximate the line charge by a finite number of point charges, and we add up their vector forces, using a spreadsheet.

Approximate approach. Spreadsheets can calculate numbers from algebraic formulas, but cannot perform algebra. (See Appendix B for an introduction to spreadsheets, in case you are not already familiar with them.) Therefore we will have to use specific values for q, Q, a, and l. We can employ our earlier problem, merely extending the number of rows to accommodate the number of charges in our approximation. Let us take $q = 10^{-9}$ C, $Q = 5 \times 10^{-9}$ C, $a = 1$ m, and $l = 7$ m. Now consider various approximations to the line charge Q.

1. Approximating Q by a single charge may be done by putting all of Q at the midpoint $(1, 0)$ of the line. In this case, there is no y-component, so all the force is along the x-direction. This gives a force of 45×10^{-9} N.

2. Approximating Q by two subcharges may be done by breaking the line into two equal segments of length $7/2$, and placing $Q/2$ at the segment midpoints, given by $(1, 7/4)$ and $(1, -7/4)$. This gives a force of 5.5×10^{-9} N.

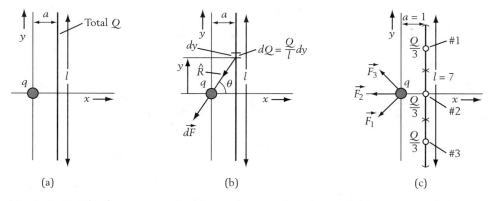

(a)   (b)   (c)

Figure 2.10 The force on a point charge due to a line charge. (a) Statement of the problem. (b) Force due to an element dq. (c) Force due to a discretization of the line into three equal charges.

Table 2.2 Force as a function of the number n of subcharges

n	1	2	3	4	5	6	7	8	9	10	11	12
F_n	45	5.50	16.83	10.60	13.22	12.01	12.53	12.30	12.40	12.3540	12.3714	12.3631

3. Approximating Q by three subcharges may be done by breaking the line into three equal segments of length $7/3$, and placing $Q/3$ at the segment midpoints, given by $(1, 7/2 - (1/2) \cdot 7/3) = (1, 7/3)$, $(1, 7/2 - (3/2) \cdot 7/3) = (1, 0)$, and $(1, 7/2 - (5/2) \cdot 7/3) = (1, -7/3)$. See Figure 2.10(c). This gives a force of 16.83×10^{-9} N. [These calculations are worth doing yourself, either with pencil and paper or with a spreadsheet, to verify that you really understand how to use (2.12).]

More generally, we have Q/n on n subcharges, at the positions $(1, 7/2 - (1/2) \cdot 7/n)$, $(1, 7/2 - (3/2) \cdot 7/n)$, and so on. By using the calculation capability of the spreadsheet, we can compute F_n for various n. By the symmetry of the problem, only the x-components of the individual forces need be calculated. Moreover, the net force $\vec{F}$ points to the left, so F_x has a negative sign. Table 2.2 shows the *magnitude* of the sum of the x-components for n segments. It is given as F_n, and is in units of 10^{-9} N for n up to 12. The sums are converging.

Calculus approach. First set up the problem, which means picking out a typical piece of charge. Figure 2.10(b) shows one dQ in the interval dy centered at y. The line is divided into infinitely many dQ's spanning the range from $y = -l/2$ to $y = l/2$. Corresponding to this y is an angle that we call θ, and a direction $\hat{R}$ to the observation charge q. The dQ acts with a force $d\vec{F}$ on q. Our goal is to add up the $d\vec{F}$'s for all the dQ's that constitute the line charge, to obtain the total force $\vec{F}$ acting on q.

There is an order in which you must perform ordinary mathematical operations: multiplication and division are performed before addition and subtraction. Similarly, there is an order in which you must perform the mathematical operations associated with integral calculus: *identify the type of differential you are adding up; find its vector components (if it is a vector); then separately add up each set of components.* In the present case, we add up the x-components dF_x to obtain F_x, and so on.

As in the previous subsection, to find the force on q we need only consider the x-component of the force, or F_x. (By the symmetry of the problem, $F_y = 0$.) Now, instead of n segments of length $7/n$ and charge $(Q/7)(7/n) = (Q/n)$, we have an infinite number of segments of infinitesimal length $ds = dy > 0$. The charge per unit length is $\lambda = dQ/ds = (Q/l)$ because the charge Q is distributed uniformly over l. Thus dy has charge $dQ = \lambda dy = (Q/l)dy$. (Of course, if we add up all the dQ's, we will obtain Q.) We now apply the commonsense method to find dF_x due to dQ. For generality, instead of $a = 1$ and $l = 7$, we will employ the symbols a and l.

From (2.1), the magnitude $dF = |d\vec{F}|$ of the force $d\vec{F}$ between q and dQ, separated by $R = \sqrt{a^2 + y^2}$ is $dF = kq(dQ)/R^2$. We now find the component dF_x of the force $d\vec{F}$ along x. From $d\vec{F}$ in Figure 2.10(b), this component is

---

> **Notational Choices**
>
> There are many notational choices we can employ to solve a problem, especially in choosing intermediate variables. However, the final answer must be independent of intermediate notation.

$dF_x = -dF \cos\theta$. Then, with $dQ = (Q/l)dy$, we obtain

$$dF_x = -dF \cos\theta = -\frac{kq(dQ)\cos\theta}{(a^2 + y^2)} = -\frac{kq\,Q\,\cos\theta\,dy}{l(a^2 + y^2)}. \qquad (2.18)$$

We've got to make a decision now: to eliminate y in favor of θ, or vice versa. We choose to eliminate y. Figure 2.10(b) shows that $y = a\tan\theta$. Thus

$$dy = (dy/d\theta)d\theta = a(d\tan\theta/d\theta)d\theta = a\sec^2\theta\,d\theta$$

and

$$a^2 + y^2 = a^2(1 + \tan^2\theta) = a^2\sec^2\theta,$$

so (2.18) becomes

$$dF_x = -\frac{kq\,Q\,\cos\theta(a\sec^2\theta\,d\theta)}{la^3\sec^2\theta} = -\frac{kq\,Q\,\cos\theta}{al}d\theta. \qquad (2.19)$$

We're now ready to actually employ the integral calculus (so far we've only used differential calculus to express dF_x). Before doing so, let's note that for two reasons this problem is more complex than integrating to add up the total charge Q on an object (as in Section 1.9). First, force is a vector, whereas charge is a scalar, so we have to determine vector components in this case. Second, the value of the force on q depends on its position, whereas the amount of charge on an object doesn't depend on the position of whoever is adding up that charge. In Chapter 5, where we discuss electrical potential, we will perform integrals of intermediate complexity: they are scalars (as for charge), but they depend on the position of the observer (as for force).

Now we do the integral to obtain F_x. We have

$$F_x = \int dF_x = -\frac{kq\,Q}{al}\int_{\theta_-}^{\theta_+} \cos\theta\,d\theta = -\frac{kq\,Q}{al}\sin\theta\Big|_{\theta_-}^{\theta_+}. \qquad (2.20)$$

From Figure 2.10(b), with $a = 1$ and $l = 7$, the maximum and minimum angles θ_+ and θ_- satisfy $\sin\theta_+ = (l/2)/\sqrt{a^2 + (l/2)^2} = -\sin\theta_-$. Thus

$$F_x = -\frac{kq\,Q}{a\sqrt{a^2 + (l/2)^2}}. \qquad (2.21)$$

For $a = 1$ m, $l = 7$ m, $q = 10^{-9}$ C, and $Q = 5 \times 10^{-9}$ C, (2.21) yields $F_x = -12.362450 \times 10^{-9}$ N. It is very close to the value obtained numerically from Table 2.2. For a different set of inputs—$a = 3$ m, $l = 4$ m, $q = 4 \times 10^{-9}$ C, and $Q = -2 \times 10^{-9}$ C—(2.21) yields $F_x = -6.656 \times 10^{-9}$ N. The advantage of the general analytical expression (2.21) over brute force summation should be apparent. On the other hand, a computer can reevaluate a spreadsheet sum very quickly when the inputs change.

Checks are very important, in both numerical and analytical work. If you derive a result that is very general, it should work for specific cases whose answer you already know, and therefore you should look for specific cases that can serve

Look Again!

> Because of what it will teach you about calculus and vectors, the derivation of (2.21) is in some ways the most important example in the book. Reread the statement of the problem and its solution until you understand it well enough to explain every step to someone else. Then try to reproduce the solution on your own. Understand it, don't memorize it. Learning how to solve this type of problem, involving integral calculus, represents a major level of intellectual achievement.

as checks. Without an analytic check that the spreadsheet or computer gives the correct result in at least one case, we cannot be sure that it gives the correct result in *any* case. With this in mind, comparison of our analytic work of (2.21) with Table 2.2 produced from the spreadsheet shows excellent agreement. Another check that can be made is to take the limit where the length l goes to zero; the line charge contracts to a point charge. Then (2.21) yields $F_x = -kq\,Q/a^2$, which is what we expect, by (2.1).

For an infinite wire, it is convenient to employ the charge per unit length λ, rather than the charge Q, which becomes infinite. In the present case, we have $\lambda = Q/l$, and now we let $l/a \to \infty$. Thus we can neglect a in $\sqrt{a^2 + (l/2)^2}$. Thus $Q/\sqrt{a^2 + (l/2)^2} \to Q/(l/2) = (2Q/l) = 2\lambda$. Hence (2.21) yields

$$F_x = -\frac{2kq\lambda}{r}, \qquad \lambda = Q/l, \qquad l/a \to \infty. \tag{2.22}$$

Here is a way to check this r^{-1} result. Consider two identical infinite, parallel rods, and an observation charge q in their plane at a position that makes one of the rods twice as far away. See Figure 2.11.

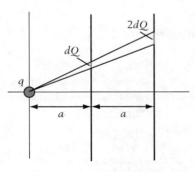

Let a pair of closely spaced lines originate radially at the observer and intersect both rods. The distance r' to the intersection of the radial lines is twice as big for the further rod as for the nearer rod. If the charge intersected by the radial lines is dQ for the nearer rod, then the charge intersected is $2dQ$ for the further rod. By (2.4), with its inverse square dependence on distance and its linear dependence on charge, for each corresponding element intersected by the radial lines the force due to the further rod is $(1/2)^2 2 = 1/2$ that due to the nearer rod. Hence, if both rods are infinite, adding up the effects for all elements of charge will give a total force due to the further rod that is one-half that due to the nearer rod. This is in agreement with the r^{-1} result

Figure 2.11 Charge q and elements of charge dq and $2dq$ from two rods of equal charge density. The magnitudes of the forces on q are the same in each case.

of (2.22). Only by doing the integral, however, can we obtain the coefficient of proportionality.

In contrast to the spreadsheet result, the integral calculus result is exact. If we set up the spreadsheet calculation with an arbitrary position for q, so we also compute F_y, then by changing the position of q, the spreadsheet will nearly instantly calculate the force at *any* position. However, numerous changes have to be made to obtain the integral calculus result. Although this problem—to find the force on q if it is placed at *any* position in the plane—can be done in a closed form, there are many problems where even a slight change in the observation point will cause the resulting integrals to be vastly more complicated or even unsolvable. It is straightforward to use calculus to obtain the force on a charge q placed at the center of a uniformly charged half-circle. However, if q is moved slightly off-center, the problem cannot be solved by elementary methods of calculus. The numerical approach will work equally well for q both on- and off-center.

Don't think that the numerical approach is always applicable. Try evaluating

$$\frac{(1+x)^{1/2} - 1}{x}$$

for $x = 10^{-30}$. To 30 decimal places, the answer is nearly one-half, but your calculator will give you zero because it doesn't keep numbers to 30 places. However, by making a straight line approximation to $(1+x)^{1/2}$ near $x = 0$, with slope at $x = 0$ given by $(d/dx)(1+x)^{1/2}|_{x=0} = \frac{1}{2}(1+x)^{-1/2}|_{x=0} = \frac{1}{2}$, we can obtain the desired result.

2.8 Study and Problem Solving Strategy

As you have seen, the subject of electricity and magnetism, or E&M to the cognoscenti, requires a mathematical background in algebra, geometry, and trigonometry. Chapter 1 required integral calculus, and the current chapter requires vectors and integral calculus.

2.8.1 *Some Advice on How to Succeed in E&M*

This chapter introduces more difficult material, involving both vectors and calculus. *In performing integrals over vectors, first obtain the small vector you are adding up, and then find its components.* Only after this should you consider the integral calculus aspect (which involves, after all, just a method to perform summation). Many students worry so much about getting the calculus right that they miss the vector aspects of a problem.

The biggest hindrance to student understanding is an inability to see common-sense simplicity. There is a natural and understandable reason for this; many students are so involved in learning how to perform technical details that they don't see the forest for the trees. If you can't efficiently and correctly deal with the details, then you won't have the leisure to sit back and think about the overview; you'll be exhausted simply by the task of getting the details right. Nevertheless, you have not completed a problem until you have looked back on it and asked the common sense questions that your grandmother might ask: for example, "are there any comparisons that can be made to related problems?" In the next chapter, we will study the electric field along the axis of a uniform disk of charge. Far away from the disk, the field should look like that for a point charge; up close it should look like that for an infinite sheet. Both of these extreme cases provide common sense checks, and are questions that could be asked by someone who has not even studied E&M!

Finally, a word about *proportionality and scaling*. In Section 2.4, we determined the force between two protons in a helium nucleus not by computing it directly, but by comparing it with the already known force between a proton and an electron in a hydrogen atom. We used the fact that the magnitudes of the charges on the electron and proton are the same, but that the separation of two protons in the nucleus is a factor of 10^{-5} smaller than the electron–proton separation in the atom. Then, applying the inverse square law for electricity, we deduced that the Coulomb repulsion between two protons in the nucleus is 10^{10} bigger than the Coulomb attraction between an electron and proton in a hydrogen atom. This type of *proportional reasoning* is essential in scientific and engineering problems. Scientists and engineers repeatedly must consider how certain effects scale as various dimensions or velocities change. They don't recompute or remeasure everything—when appropriate, they scale the results. This is the basis of wind tunnels, for example, as used in aircraft design. Successful science and engineering students know how to employ this method of reasoning.

2.8.2 *Some Comments about Problem Solving*

Asking your own questions. Problems do not come out of nowhere. Someone has to think them up. For this book, the author had to think them up—with the assistance of a vast array of problems available from other books on this subject. Here is a secret. Not only can problems be solved, *they can also be made up*. You can do it yourself (it is an example of what has been called *active learning*). For each chapter, spend a minute or two thinking about how to make up an interesting variant on at least one problem. Here are some possibilities.

1. If you don't know what an equation means, try putting in numbers or, if appropriate, try drawing a graph. This is the most important rule of all! It's how scientists and engineers get started when they confront an equation whose meaning they don't understand.

2. Think about how to turn a doable problem to an undoable one.

3. Think about how to turn an undoable problem to a doable one—and do it.

4. Think about how to design an experiment. For example, in the electrometer problem, the string might have a certain breaking strength, and we want to know how much charge we can put on an object before the tension exceeds this breaking strength.

5. Make up a problem in which numbers, graphs, and equations are relevant; in the problems on the cancellation of two electrical forces, a sketch of the strengths of each force as a function of position is very revealing.

6. Think up what-if questions.

Although the bread and butter of physics is its ability to give precise answers to difficult but well defined questions, you should avoid the tendency to think *exclusively* in terms of stylized, closed-form mathematics problems. Often a simple qualitative question, whose answer can be given with a simple yes or no, or a direction, or greater than or less than, can teach a concept more efficiently

"You can't learn to swim if you don't jump in the water."
　　　　　　　　　　　　　　　　　　　　　　　　—Anonymous

"Tourist to passerby in Manhattan: How do you get to Carnegie Hall?"
Violinist Jascha Heifitz, without breaking stride: "Practice, practice, practice!"

"To learn to play the blues, first you have to learn to play one song really well."
　　　　　　　　　　　—Mance Lipscomb (1971), musician from Navasota, Texas

"The secret to success is being able to find more than one way to get the job done."
　　　　　　　　　　　　　　　　　　　　　　　　—Anonymous

and effectively than a full-blown problem that requires an enormous amount of calculation. In any situation, even outside physics, one must be careful not to lose the overview in a confusion of details.

Thinking clearly. When you begin a problem, draw a figure in which the variables are clearly defined. When you complete a problem, ask, "What have I learned?" Think about how to modify the problem, and ask, "What changes does the modification cause?" Learn how to recognize problems that you have seen before, even when in disguise: if an automobile mechanic knows how to change a tire on a Ford, he cannot not plead ignorance when someone brings him a Chevrolet.

Styles of studying. There are different styles of studying. Many students work by themselves and don't spend much time giving explanations to others. In so doing, they lose the opportunity to learn while explaining. On the other hand, those who work only in groups are missing the opportunity to build their intellectual muscles; you don't go to the gym to watch others get in shape. Everyone should be doing some of both.

Problems

2-2.1 (a) If, in Robison's experiment (see Figure 2.1), q_1 (with mass m_1) is directly above q_2, show that the equilibrium condition is $kq_1q_2/r^2 = m_1g$. Neglect the mass of the holder. (b) If m_1 doubles, how does r change? Does the dependence of r on m_1 make sense qualitatively? (c) For $q_1 = q_2 = q$, $m_1 = 75$ g, and $r = 4.5$ cm, find q. (d) How would the equilibrium condition change if the top arm (held at a loose pivot) had mass M and length l?

2-2.2 (a) In a Coulomb's law experiment, as in Figure 2.2, a torque of 1.73×10^{-4} N-m is measured for a $2°$ twist. Find the torsion constant κ. (b) $q_1 = q_2 = 2.4 \times 10^{-8}$ C causes a twist of $5°$. Find r.

2-3.1 Joan and Laura are separated by 15 m. Joan has a charge of 4.50×10^{-8} C, and Laura has a charge of -2.65×10^{-6} C. Find the force between them, and indicate whether it is attractive or repulsive.

2-3.2 Two regions of a thundercloud have charges of ± 5 C. Treating them as point charges a distance 3 km apart, determine the force between these regions of charge, and indicate whether it is attractive or repulsive.

2-3.3 The basketball player Michael Jordan is about 2 m tall, and weighs about 90 kg. What equal charges would have to be placed at his feet and his head to produce an electrical repulsion of the same magnitude as his weight?

2-3.4 Two point charges are separated by 2.8 cm. The force between them is 8.4 mN, and the sum of their charges is zero. Find their individual charges, and indicate whether the force is attractive or repulsive.

2-3.5 Show that, at fixed separation a, the maximum repulsion between two point charges of total charge Q occurs when each charge equals $Q/2$.

2-3.6 Two point charges sum to -5 μC. At a separation of 2 cm, they exert a force of 80 N on each other. Find the two charges for the cases when (a) the forces are attractive, and (b) when they are repulsive. [Answer: (a) $q_1 = -5.63$ μC, $q_2 = 0.63$ μC; (b) $q_1 = -4.14$ μC, $q_2 = -0.858$ μC.]

2-3.7 Consider two space ships of mass $M = 1000$ kg in outer space. What equal and opposite charges would have to be given to them so that once an earth day they make circular orbits about their center at a separation of 200 m? Neglect their gravitational interaction, which at that distance would cause them to orbit only every 563 days.

2-3.8 Two point charges sum to -0.5 μC. At a separation of 2 cm, they exert a force of 80 N on each other. Find the two charges for the cases when (a) the forces are attractive, and (b) when they are repulsive.

2-3.9 In esu-cgs units, where length is measured in cm, time in s, and mass in g, we take $k_{esu} = 1$. (a) Find the esu unit of charge, called the *statcoulomb* (sC) in terms of the SI unit of charge (C). (b) Find the charge of an electron in esu units. [Answer: 1 C$=3 \times 10^9$ sC, $e_{esu} = 4.8 \times 10^{-10}$ sC.]

2-4.1 Refer to the electrometer example of Figure 2.4. Let the tension at breaking be $T_{max} = 2mg$. Find the value of θ_{max} and an algebraic expression for q_{max}.

2-4.2 In the electrometer example, find the angles and tensions if $q_L = 0.925 \times 10^{-8}$ C and $q_R = 3.7 \times 10^{-8}$ C? (*Hint:* No complex calculation is necessary. You can use results already obtained. If you are stumped, have a look at Problem 2.4.4a.)

2-4.3 In the electrometer example of Figure 2.4, if the angles are θ for $q_L = q_R = q$, what are they for $q_L = 2q$, $q_R = \frac{1}{2}q$? *Hint:* If you are stumped, have a look at Problem 2-4.4(a).

2-4.4 Two identical masses are suspended by identical strings. If the charges are the same, the strings make equal angles to the vertical. (a) If the charges are different, are the angles different? (b) If the masses are different, are the angles different? (*Hint:* Make the charges or masses very different.)

2-4.5 A positively charged bead q can slide without friction around a vertical hoop of radius R. A fixed positive charge Q is at the bottom of the hoop. Counterclockwise angular displacements of q relative to Q correspond to $\theta > 0$. (a) Find the electrical force on q, as a function of θ. (b) Find the component of this force along the hoop. See Figure 2.12.

Figure 2.12 Problem 2-4.5.

2-5.1 Let $Q_1 = 5 \times 10^{-8}$ C be at $(0, 0)$, and $Q_2 = -4 \times 10^{-8}$ C be at $(3, 0)$, in m. A charge Q_3 is placed somewhere on the x-axis where the force on Q_3 is zero. (a) If the value of Q_3 is adjusted so that the force on Q_1 is zero, find the force on Q_2. (b) Find where Q_3 should be placed to feel no net force. (c) Find the value of Q_3 that will make Q_1 feel zero net force.

2-5.2 Consider two charges, $Q_1 = 5 \times 10^{-8}$ C at $(1 \text{ cm}, 0)$, and $Q_2 = -4 \times 10^{-8}$ C at $(-2 \text{ cm}, 4 \text{ cm})$. We want to find the position where a third charge Q should be placed for it to feel zero net force. (a) Present two methods for doing this. (b) Solve the problem by either method.

2-5.3 Three charges are at the corners of an equilateral triangle of side $l = 10$ cm. See Figure 2.13. If 2 μC is at the origin, and -3 μC is at $(l, 0)$, find the force on 4 μC, at $(l/2, \sqrt{3}l/2)$.

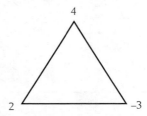

Figure 2.13 Problem 2-5.3.

2-5.4 Three charges Q are placed at the corners $(0, 0)$, $(a, 0)$, and $(0, a)$ of a square. See Figure 2.14. (a) Find the force on q placed at (a, a). (b) Find the force on q placed at $(a/2, -a/2)$. (c) Find the force on q placed at $(-a, -a)$.

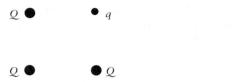

Figure 2.14 Problem 2-5.4.

2-5.5 Repeat the force calculation for the example of the addition of two noncollinear forces on a charge q. However, now let q_1 be at the origin, and rotate about the z-axis to make q_2 along the new x-axis. See Figure 2.15. To produce this requires, in Figure 2.7, rotating the line connecting q_1 and q_2 by an angle ϕ, where $m = \tan\phi = (r_2 \sin\theta_2 - 0)/((r_2 \cos\theta_2 - r_1)) = -8.800$. Thus, in radian measure, $\phi = -83.52°(\pi/180) + \pi = 1.683$ radians. (We add π because we know, by Figure 2.7, that ϕ is in the first or second quadrant, whereas the calculator returns only a value in the first or fourth quadrant.) In degrees, $\phi = 96.5°$. By putting q_1 at the origin, and rotating clockwise by ϕ, we put q_2 along the x-axis, at a distance $r_{12} = \sqrt{(r_2 \sin\theta_2 - 0)^2 + ((r_2 \cos\theta_2 - r_1)^2} = 0.247m$, and we put q a distance r_1 away from the origin, at an angle of $\pi - \phi = 1.458$ radians, or $83.52°$ to the x-axis. Verify that $|\vec{F}|$ is the same as before, and that it is rotated clockwise by $96.5°$ relative to its previous value. This property, that the force rotates by the same angle as the coordinate system, is what we mean when we say that force is a vector. Thus Figure 2.15 was obtained by rotating Figure 2.7.

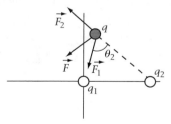

Figure 2.15 Problem 2-5.5.

2-6.1 Use the general symmetry argument to show that, along the perpendicular bisector of the uniformly charged rod discussed in Section 2.7, $F_y = 0$. *Hint:* Assume that $F_y \neq 0$, and then consider how the rod and F_y transform under rotations of $180°$ about the x-axis. Would the argu-

ment work if you considered a reflection $(x, y, z) \rightarrow (x, -y, z)$ of the rod and F_y?

2-6.2 An octagon has charges $-q$ at each of its eight vertices, and charge Q at its center. See Figure 2.16. (a) Use a general symmetry argument to show that the force on Q is zero. (b) Let the charges $-q$ be replaced by infinitely long line charges $-\lambda$ normal to the page. Use a general symmetry argument to show that the force on Q is zero. (To prove this, we don't even need to know the force between Q and a line charge!)

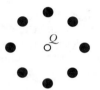

Figure 2.16 Problem 2-6.2.

2-7.1 A line charge with total charge $Q > 0$ uniformly distributed over its length l extends from $(0, 0)$ to $(l, 0)$. See Figure 2.17. (a) Find the force on a charge $q > 0$ placed a distance a to its right, at $(l + a, 0)$. (b) Verify that this force has the expected inverse square behavior for large a.

Figure 2.17 Problem 2-7.1.

2-7.2 A long rod of charge per unit length $\lambda > 0$ is normal to the xy-plane and passes through the origin. In addition, a charge $Q > 0$ is located at $(0, a, 0)$. See Figure 2.18. Find the position where a third charge q will feel zero force.

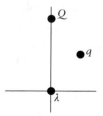

Figure 2.18 Problem 2-7.2.

2-7.3 Use the computational symmetry argument to show that $F_y = 0$ along the perpendicular bisector of the uniformly charged rod of Section 2.7.

2-7.4 A line charge with total charge $Q > 0$ uniformly distributed over its length l extends from $(0, 0)$ to $(l, 0)$. Find the force on a charge q that is placed anywhere in the xy-plane.

2-7.5 Find the force on a charge q at (a, a) due to a charge Q uniformly distributed over a rod of length L with one end at $(0, 0)$ and the other end at $(0, L)$. See Figure 2.19.

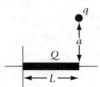

Figure 2.19 Problem 2-7.5.

2-7.6 A rod of length a whose ends are at $(0, 0)$ and $(a, 0)$ has a charge density $\lambda = (Q_0/a^2)x$. See Figure 2.20. (a) Find the total charge Q on the rod. (b) Find the force on a charge q at $(-b, 0)$. (c) Verify that the force has the correct limit as $b \to \infty$.

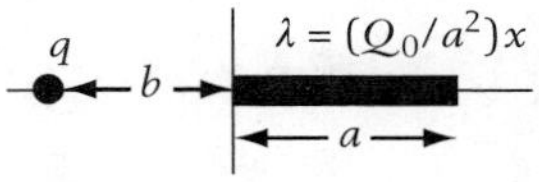

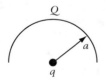

Figure 2.20 Problem 2-7.6.

2-7.7 A charge Q is uniformly distributed over the upper half of a circle of radius a, centered at the origin. See Figure 2.21. Find the force on a charge q at the origin.

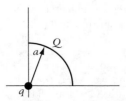

Figure 2.21 Problem 2-7.7.

2-7.8 A charge Q is uniformly distributed over the first quadrant of a circle of radius a. See Figure 2.22. Find the force on a charge q at the center of the semicircle.

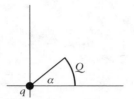

Figure 2.22 Problem 2-7.8.

2-7.9 A charge Q is uniformly distributed over an arc of radius a and angle α that extends from the x-axis counterclockwise. See Figure 2.23. Find the force on a charge q at the center of the semicircle.

Figure 2.23 Problem 2-7.9.

2-G.1 Many electroscopes have a circular conducting base that is both above and connected to a charge detector device (often a needle that can rotate relative to its mount, or flexible foil). They work by a combination of electrostatic induction at the base (due to the source charge), and repulsion of like charges at the detector (needle or foil). See Figure 2.24. (a) If positive charge is brought near the base of an electroscope, what is the sign of the charge attracted to the base? (b) What kind of charge must be at the needle or foil?

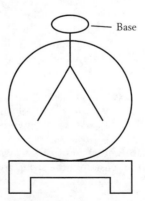

Figure 2.24 Problem 2-G.1.

2-G.2 Show that, for $|nx| \ll 1$, $(1 + x)^n \approx 1 + nx$. This gives the first two terms, for small x, in what is known as the MacLaurin expansion.

2-G.3 Determine the behavior of $(x^2 + 1)^{\frac{1}{2}} - x$ for large x; it approaches zero, but *how* does it approach zero?

2-G.4 In the spreadsheet calculation of Table 2.2, the sums for odd n seem to provide an upper limit for the integral. (a) Explain why. The sums for $n = 2, 4, 6, 8, 10$ seem to be a lower limit for the integral. However, for $n = 12$ and larger, the approximate value is *less* than the exact integral. (b) Explain why. *Hint:* For odd n, all the charge is approximated by charge that is nearer than it really is, but that is not the case for even n.

2-G.5 Two charges $-Q$ are fixed at $(0, a)$ and $(0, -a)$. A third charge, q, is constrained to move along the x-axis. See Figure 2.25. (a) Find the force on q for any value of $|x| < a$; (b) convince yourself that, for $|x| \ll a$, this force is just like that for a harmonic oscillator, and obtain the effective spring constant K; (c) if q has mass m, find the frequency of oscillation of q about the origin. *Hint:* Expand $(a \pm x)^{-2}$ for small x.

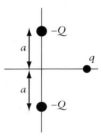

Figure 2.25 Problem 2-G.5.

2-G.6 In the previous problem, the motion of q was stable near the origin for motion along x. Discuss the stability of the motion if small displacements along the y-axis are allowed. Consider $x = 0$ and $|y| < a$.

2-G.7 A long rod of charge per unit length λ is held vertical. At its midpoint, one end of a massless string of finite length l is attached. At the other end of the string is a charge q, with mass m. See Figure 2.26. Find the equilibrium angle of the string to the vertical, and find the tension in the string at that angle.

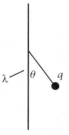

Figure 2.26 Problem 2-G.7.

2-G.8 A square of side a, of uniform charge per unit area σ, is centered about the origin of the xy-plane with sides parallel to the x- and y-axes. A charge q lies a distance l along its perpendicular bisector. See Figure 2.27. (a) By building up the square from lines in the xy-plane of area $l\,dz$, show that the force on q satisfies $|\vec{F}| = 4kq\sigma \sin^{-1}[a^2/(a^2 + 4l^2)]$. (b) Show that, as $l/a \to \infty$, $|\vec{F}| \to (kQq/l^2)$. (c) Show that, as $l/a \to 0$, $|\vec{F}| \to (2\pi k\sigma)q$. *Hint:* Modify (3.21) to $|d\vec{F}| = kqd\,Q/l\sqrt{l^2 + a^2/4}$, use the direction cosine factor $l\sqrt{l^2 + a^2/4}$, and then integrate over dF_x to get F_x. A trig substitution like $z = a\tan\theta$ may be helpful.

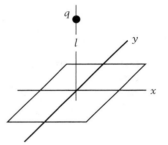

Figure 2.27 Problem 2-G.8.

2-G.9 Consider a ring of radius a, centered at the origin of the xy-plane. It is of uniform charge density and has total charge Q. A charge q lies on the x-axis a distance l from the origin. See Figure 2.28. (a) Find the force on q due to the circle, for $l > a$. (b) Find the force on q due to the ring, for $l < a$. This example shows that the charge on the ring does not behave as if it were centered at its geometrical center (except in the limit where $l/a \to \infty$).

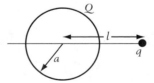

Figure 2.28 Problem 2-G.9.

2-G.10 Conducting spheres are subject to the amber effect. Hence, as two equally charged conducting spheres of radius a approach each other, in addition to the inverse square law of repulsion, there should be an effect due to the charge on one polarizing the other, and vice versa (i.e., electrostatic induction). Using advanced methods, it is possible to determine this effect exactly, but we already know enough to determine the most important contribution. When the separation r is only a few times a, this polarization effect can cause deviations from pure inverse square on the order of a few percent. See Figure 2.29. (a) Does polarization make the net force appear weaker or stronger? (b) What dependence on distance do you expect the most important correction to take? (c) How would you determine, from experimental data of force F versus separation r, how large a coefficient it has? *Hint*: For large r, plot $r^2 F$ *vs.* r^{-3}.

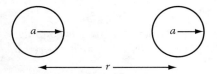

Figure 2.29 Problem 2-G.10.

2-G.11 Repeat the considerations given in Problem 2.6.10 for two conducting spheres of equal and opposite charge.

"That one body may act upon another at a distance through a vacuum without the mediation of anything else . . . is to me so great an absurdity, that I believe no man, who has in philosophical matters a competent faculty of thinking, can ever fall into it."

—Sir Isaac Newton (1692)

"By magnetic curves, I mean lines of magnetic force . . . , which could be depicted by iron filings; or those to which a very small magnetic needle would form a tangent."

—Michael Faraday, *Experimental Researches* (1831)

Chapter 3

The Electric Field

Chapter Overview

Section 3.2 shows how to measure the electric field, and Section 3.3 shows how to determine it by calculation. Section 3.4 discusses how to draw field lines—Faraday's way of looking at electricity. Section 3.5 employs the *principle of superposition* to add up the electric fields due to more than one point charge, and Section 3.6 does the same for continuous distributions of charge. Section 3.7 discusses field line drawing in more detail. Section 3.8 considers the torque on and the energy of an electric dipole in a uniform electric field. Section 3.9 discusses the force on a dipole in a nonuniform electric field (this is related to the amber effect). Section 3.10 considers the deflection of electric charges by a uniform electric field, as in older TV tubes and computer monitors. (Modern TV tubes and monitors use magnetic deflection, as discussed in Chapter 10.) Section 3.11 discusses why electric forces alone are not enough to stabilize a world governed by Newton's laws of motion. ∎

3.1 Introduction

3.1.1 *Development of the Electric Field Concept*

In electrostatics, the electric force between two charges can be thought of as instantaneous action at a distance, no matter the separation between the charges. When Michael Faraday, beginning in the 1830s, espoused an alternative view based on *lines of force* that exist everywhere in space, most other experts in electricity thought it superfluous. Nevertheless, over a century earlier Isaac Newton, despite his quantitative success describing gravity via action at a distance, felt that an instantaneous response is untenable: when a distant star moves, by action at a distance a mass light-years away would have to feel a changed gravitational force instantaneously.

Similarly, Faraday believed that an instantaneous electrical response was untenable. Repeatedly throughout his long career, Faraday tried to determine the speed at which changes in electric forces propagate. Because, as we will discuss in Chapter 15, such changes propagate at a very high speed (that of light, about

3×10^8 m/s), Faraday was unable to measure this speed with the methods of his time.

By thorough and self-critical experimentation, Faraday tested his ideas, rejecting some and refining others, ultimately employing the concept of magnetic lines of force in completely new ways, and extending this concept from magnetism to electricity. Faraday's concepts were first given mathematical form in 1845 by William Thomson. (Thomson was later made Lord Kelvin, for supervising the laying of the first effective trans-Atlantic telegraph cable, in 1865. In this project, he made great practical application of his knowledge of electricity.)

James Clerk Maxwell, in 1855, began his own program to develop Faraday's ideas mathematically. He developed the concept of the *electric field*, the *magnetic field*, and (when time-dependent phenomena were included) the *electromagnetic field*. Because of Maxwell, lines of force are now called *field lines*. In 1865, he found that the resulting equations unified electricity, magnetism, and light. His prediction of electromagnetic radiation, propagating at the speed of light, is one of the greatest of any scientific achievements: radio, TV, and microwave communications are all practical consequences of that work. Thus Faraday was correct about electricity: electric forces *do* propagate at a finite speed. In 1916, Newton was shown to be correct about gravity, when Albert Einstein developed a theory—since verified experimentally—in which the gravitational field propagates with a finite speed that is the same as the speed of light. The concept of the gravitational field did not arise until after the field concept had entered the area of electromagnetism.

The idea of the electric field (sometimes called the *electric force field*) is simple: one electric charge produces an electric field, and another electric charge feels a force due to that field.

An example of a field, and of interactions via a field, can be seen in the interaction of two water striders on a water surface. The weight of each depresses the water's surface, and the depressions produced by each can be felt by the other. See Figure 3.1. The field, in that case, is the distortion of the water's surface. The two water striders interact through that field. When one moves, the distortion changes locally, and there must be a time delay before the change makes its presence felt at the other. Similarly, electric charge produces an electric field, and when one charge moves there must be a time delay before the signal reaches the other charge.

Indispensible for describing dynamic phenomena, the electric field concept also provides insights into static phenomena. Electric field lines for a configuration of electric charges give the pattern of the electric field with only minimal

Figure 3.1 Two water striders on a water surface. One water strider can detect the presence of the other by the latter's deflection of the water surface. The total deflection of the surface is the sum of the deflections due to each.

computation. The electric field concept has two additional advantages, even in electrostatics. First, because electric charge is the source of the electric field, there is a deep relationship between field lines and electric charge (recall Figure R.9). We will develop this in Chapter 4. Second, electrical potential energy can be expressed in terms of the electric field. We will develop this in Chapter 5.

3.2 Obtaining the Electric Field: Experiment

The magnetic field near a magnet can be visualized with iron filings. Similarly, the electric field near an electrically charged body can be visualized with grass seeds. See Figure 3.2(a).

Neither iron filings nor grass seeds have permanent electric or magnetic properties, but they are more magnetizable and polarizable along their long axes. Because the axes of both iron filings and grass seeds have no preferred sense, the field *direction* is ambiguous. Compare Figure 3.2(a) and Figure 3.2(b).

An analogy to gravity leads to a precise definition of the electric field. The gravitational field $\vec{g}$ is defined as the ratio of the gravitational force $\vec{F}_m$ on a test body to its gravitational mass m:

$$\vec{g} = \frac{\vec{F}_m}{m}.$$ (3.1)

Neil Armstrong, the first person to set foot on the moon, can claim to have served as a test mass on both the earth and the moon. He found that $|\vec{g}|_{moon} \approx (1/6)|\vec{g}|_{earth}$. Any mass m on the surface of the moon is subject to a $\vec{g}$ of this magnitude and feels a gravitational force $m\vec{g}$. A directional check on (3.1) is that, since on the earth's surface $\vec{F}_m$ points to the earth's center, (3.1) says that $\vec{g}$ also points to the earth's center, as expected.

By analogy, the electric field $\vec{E}$ at a given point P is defined as the ratio of the electrical force $\vec{F}_q$ on a test body at P to its electrical charge q:

$$\vec{E} = \frac{\vec{F}_q}{q}. \qquad \text{(definition of field)}$$ (3.2)

Experimentally, any value for the test charge q yields the same value for $\vec{E}$. That

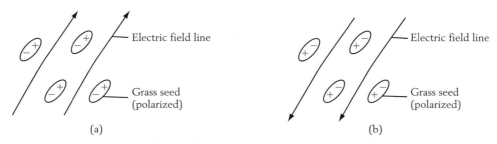

(a)   (b)

Figure 3.2 The orientation of grass seeds in response to an applied electric field is the same if the electric field is reversed. (a) Field in one direction. (b) Reversed field.

is why the field idea works: it is a property of the point in space, not of the test charge. The test charge does not feel a force due to itself; the electric field is due to charge other than the test charge. The unit of the electric field is N/C. $\vec{E}$ is called a field because it is defined at all points $\vec{r}$ in space. To represent $\vec{E}$ at the position $\vec{r}$, draw the vector representing $\vec{E}$ with its tail at the position $\vec{r}$.

Example 3.1 **Electric field of a proton**

Find the electric field acting on the electron due to the proton, in Figure 3.3(a). Take the separation to be 1.0×10^{-10} m.

Solution: First consider the direction of $\vec{E}$. The force on the electron is, by "opposites attract, likes repel," *toward* the proton, which is considered to be the *source* of the electric field acting on the electron. See Figure 3.3(a). From (3.2), because the sign of the charge on the electron is negative ($q = -e$), the direction of the electric field on the electron is *opposite* that of the force on it. Hence the electric field at the site of the electron is *away* from the positively charged proton. (Note that $\vec{E}$ has its tail at the point where it is measured—the electron.) Because the electric field is a property of space, we conclude quite generally that

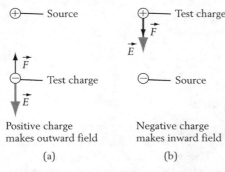

Figure 3.3 Measuring the electric field $\vec{E}$, whose tail is at the point where it is measured—the test charge. (a) Positive source and negative test charge. (b) Negative source and positive test charge.

positive charge produces an electric field that points outward from the charge.

Similarly, as shown in Figure 3.3(b), for the electron as the source,

negative charge produces an electric field that points inward to the charge.

Now consider the field magnitude $|\vec{E}|$. Section 2.4.1 finds that for an electron and proton separated by 1.0×10^{-10} m, the force on the electron has magnitude $|\vec{F}| = 2.3 \times 10^{-8}$ N. With $q = -1.6 \times 10^{-19}$ C, (3.2) yields, for the magnitude of the electric field due to the proton, at the site of the electron,

$$|\vec{E}| = \left| \frac{\vec{F}}{q} \right| = \left| \frac{2.3 \times 10^{-8}\text{ N}}{(-1.6) \times 10^{-19}\text{ C}} \right| = 1.44 \times 10^{11} \frac{\text{N}}{\text{C}}.$$

Any charge q at this position is subject to an $\vec{E}$ of this magnitude and feels an electrical force $q\vec{E}$.

The electric field due to the electron, acting on the proton, has the same magnitude, because the magnitudes of the force on the proton due to the electron, and the magnitude of the proton charge, are the same as in the previous case.

Example 3.2 **Electric field from the force on a test charge**

Section 2.5 (see Figure 2.5) found the force on a charge $q = 2.0 \times 10^{-9}$ C, due to two other charges $q_1 = -4.0 \times 10^{-9}$ C and $q_2 = 6.0 \times 10^{-9}$ C: $F_x = 1.112 \times 10^{-6}$ N, $F_y = -0.983 \times 10^{-6}$ N, and $|\vec{F}| = 1.484 \times 10^{-6}$ N. Find the electric field at the position of q.

Solution: Use of (3.2) gives

$$E_x = \frac{F_x}{q} = 5.56 \times 10^2 \frac{N}{C}, \qquad E_y = \frac{F_y}{q} = -4.92 \times 10^2 \frac{N}{C},$$

$$|\vec{E}| = \left|\frac{\vec{F}}{q}\right| = \frac{|\vec{F}|}{|q|} = 7.42 \times 10^2 \frac{N}{C}.$$

Because q is positive, by (3.2) the direction of $\vec{E}$ is along $\vec{F}$. If the test charge q were negative, then $\vec{F}$ would be in the opposite direction, but $\vec{E}$ would be unaffected.

3.2.1 *Experimental Caution*

The test charge q can produce electrostatic induction or polarization in nearby materials, in proportion to q. These induced charges can then contribute to the electric field. To eliminate this effect the test charge q must be very small:

$$\vec{E} = \lim_{q \to 0} \frac{\vec{F}_q}{q}. \tag{3.3}$$

To see how electrostatic induction can change the electric field, consider an everywhere neutral infinite sheet of aluminum foil. It produces no electric field: $\vec{E} = \vec{0}$. A positive charge q, brought up to the foil, alters the distribution of conduction electrons, shifting their orbitals toward the positive charge. This produces an electric field that, at the site of the positive charge, points toward the foil, with strength proportional to q. Figure 3.4(a) depicts the force $\vec{F}$ acting on q (dark arrow), the field $\vec{E}$ at the site of q (shaded arrow), and the humplike

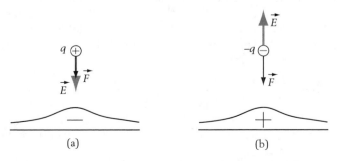

(a)   (b)

Figure 3.4 Force on test charges above a neutral, infinite conducting sheet. (a) $+q$, (b) $-q$.

shape of the charge distribution induced on the foil. (This part of the figure is schematic; the induced charge density does not actually rise up near q, but rather it is largest near q.) If the sign of the charge q is reversed, the sign of the induced surface charge will reverse, which then reverses the direction of the induced electric field: it will now point away from the foil. See Figure 3.4(b). In both cases, there is a force on q, of magnitude proportional to q^2. Therefore, as $q \to 0$, (3.3) yields $\vec{E} = \vec{0}$. (A similar effect occurs when a charge q is brought up to a neutral piece of paper, as in the amber effect.)

The test charge must be of small physical dimension, both to define the observation position of the field measurement and to minimize polarization or induction effects on the test charge itself.

3.3 Obtaining the Electric Field: Theory

Imagine that you have calculated the force $\vec{F}_q$ on a charge q at a specific position, due to q_1 and q_2, as in Figure 3.5(a). Let's say it is 2 N, pointing along $\hat{\imath}$. (In principle, there could be many other charges also contributing to the force on q.) Now mentally replace q by Q and consider how to obtain the force $\vec{F}_Q$ on Q. It is *not* most easily obtained by recalculating the individual forces and adding them up. A simpler approach is to take the ratio of Q to q, and then multiply by the force $\vec{F}_q$ on q. Thus, if $Q = -2q$, then the force on Q has twice the magnitude of the force on q, or 4 N, but is in the opposite direction, or $-\hat{\imath}$. Another simpler approach is to determine $\vec{E}$ by considering q to be a test charge, and then to use $\vec{F}_Q = Q\vec{E}$. Let us develop these ideas.

Consider the force $\vec{F}_q$ on q at $\vec{r}$, due to charges q_i at $\vec{r}_i$. With $\vec{R}_i = \vec{r} - \vec{r}_i$, (2.8) gives

$$\vec{F}_q = \Sigma_i \frac{kqq_i}{R_i^2}\hat{R}_i = q\Sigma_i \frac{kq_i}{R_i^2}\hat{R}_i. \qquad (\hat{R}_i \text{ points to observation charge } q) \quad (3.4)$$

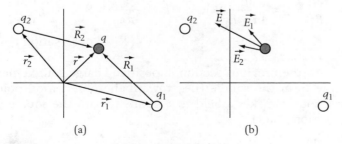

(a) (b)

Figure 3.5 (a) Geometry for the force on q due to q_1 and q_2. Lowercase vectors refer to distances from the origin, and uppercase vectors refer to relative distances. (b) Field at site of q, due to q_1 and q_2.

Then the force $\vec{F}_Q$ on Q is given by the ratio

$$\vec{F}_Q = \frac{Q}{q}\vec{F}_q = Q\Sigma_i \frac{kq_i}{R_i^2}\hat{R}_i. \tag{3.5}$$

Proceeding more systematically, (3.4) may be rewritten, for any charge q at the observation point, as

$$\vec{F}_q = q\vec{E}, \qquad \text{(force on } q\text{)} \tag{3.6}$$

where

$$\vec{E} \equiv \Sigma_i \frac{kq_i}{R_i^2}\hat{R}_i. \quad (\hat{R}_i \text{ to observation point } \vec{r}_i) \qquad (\vec{E} \text{ for point charges}) \tag{3.7}$$

In principle, we should write $\vec{E}(\vec{r})$ because the electric field depends upon position $\vec{r}$.

Let's discuss (3.7) in terms of input and output. The input is the observation position $\vec{r}$ and the individual source charges q_i at the positions $\vec{r}_i$. For two source charges, this is given in Figure 3.5(a). The output is the set of individual electric fields $\vec{E}_i$ and the total electric field $\vec{E}$ at $\vec{r}$. Assuming that $q_1 > 0$ and $q_2 < 0$, Figure 3.5(b) depicts the directions for the fields $\vec{E}_1$ and $\vec{E}_2$. It also depicts their sum $\vec{E}$. The relative lengths of $\vec{E}_1$ and $\vec{E}_2$ can only be determined when actual values for q_1 and q_2 are given; therefore, Figure 3.5(b) is only a schematic. The fields at $\vec{r}$ are drawn with their tails at $\vec{r}$.

A more explicit expression for $\vec{E}$, which is useful for numerical calculations, is obtained by using $\vec{R}_i/|\vec{R}_i|$ rather than $\hat{R}_i$. Then (3.7) becomes

$$\vec{E} = \Sigma_i \frac{kq_i}{R_i^3}\vec{R}_i = \Sigma_i \frac{kq_i(\vec{r}-\vec{r}_i)}{|\vec{r}-\vec{r}_i|^3}. \quad (\vec{R}_i = \vec{r}-\vec{r}_i \text{ to observation point } \vec{r}_i) \tag{3.8}$$

From (3.7), the electric field that a single charge q_1 produces at the site of q is given by

$$\vec{E}_1 = \frac{kq_1}{R_1^2}\hat{R}_1 = \frac{q_1}{4\pi\epsilon_0 R_1^2}\hat{R}_1, \qquad k = \frac{1}{4\pi\epsilon_0} \approx 9\times10^9 \frac{\text{N-m}^2}{\text{C}^2}.$$
$$(\hat{R}_1 \text{ to observation point}) \tag{3.9}$$

In agreement with the previous section, a positive charge source makes a field that points away from the source (i.e., toward the observation point), and a negative charge source makes a field that points toward the source (i.e., away from the observation point).

Example 3.3 Field E_d needed to cause sparking in air

For a charge $q_1 = 10^{-9}$ C (typical of static electricity) at a distance of 1 cm, find the electric field. Repeat for a distance of 1 mm. In both cases, compare with the electric field above which sparking (electrical breakdown) occurs, called the *dielectric strength* E_d. In air at atmospheric pressure, E_d is about 3×10^6 N/C.

Solution: By (3.9), for $q_1 = 10^{-9}$ C and a distance of 1 cm,

$$|\vec{E}_1| = (9 \times 10^9 \text{ N-m}^2/\text{C}^2)(10^{-9} \text{ C})/(10^{-2} \text{ m})^2 = 9 \times 10^4 \text{ N/C};$$

this is less than E_d, so no sparking would occur. For the same $q_1 = 10^{-9}$ C and a distance of 1 mm, $|\vec{E}_1| = 9 \times 10^6$ N/C; this exceeds E_d, so sparking would occur.

Observe that a calculation of $\vec{E}$ at $\vec{r}$, using (3.7), differs from a calculation of $\vec{F}$ on q at $\vec{r}$, using (2.12), only in that the factor q of (2.12) is omitted from (3.7). Before getting into any detailed calculations, we present a more geometric view of electric fields.

3.4 Visualizing the Electric Field: Part 1

3.4.1 *Rules for Drawing Electric Field Lines*

Field lines, or *lines of force*, are used to represent pictorially the electric field $\vec{E}$. To be an accurate representation, they must have the following properties:

1. Field lines point in the direction of the electric field $\vec{E}$. Field lines cannot cross. If two lines did cross, then the force on a charge would have two directions, which is impossible.

2. The *areal density of field lines* (the number of lines per unit area in the plane perpendicular to the field line) is proportional to the magnitude $|\vec{E}|$ of the electric field $\vec{E}$. Thus, the larger the field, the higher the density of the field lines, and vice versa.

By definition, rules 1 and 2 hold for *any* vector field, including the magnetic field.

In addition, for field lines due to electric charges at rest (electrostatics), the following rules apply:

3. Field lines originate on positive charges and terminate on negative charges, the number of field lines being proportional to the charge. (This prescription is not unique because different people, or the same person under differing circumstances, might choose to use different numbers of field lines for the same charge.)

4. Field lines do not close on themselves.

Rule 3 is made more precise in Chapter 4 (which eliminates the ambiguity about the number of lines per unit charge). Rule 4 is derived in Chapter 5. These rules are simple, but they are not obvious. They hold only for electric fields due to electric charges at rest. Magnetic field lines do not satisfy rules 3 and 4, but rather have their own set of rules, which we study in Chapter 11.

These rules yield simple pictures only outside of charge distributions. Everywhere within a ball of charge, field lines are originating or terminating. This causes complications that we need not consider here.

3.4.2 *Applications to Simple Geometries*

Application 3.1 **Positive point charge *q***

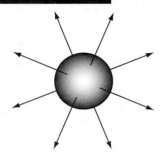

Figure 3.6 Representation of the three-dimensional field lines due to a point charge.

Consider that q produces N field lines. Then the field lines point outward and are uniformly distributed, with one field line for each of the corresponding parts of the total solid angle on the surface of a sphere. See Figure 3.6. It is difficult to reproduce this on a sheet of paper, so the figure must be considered to be schematic. The number of field lines N is fixed, and the area $4\pi r^2$ of a concentric sphere of radius r (through which the lines pass) varies as r^2. Hence the density of field lines (and thus the electric field magnitude $|\vec{E}|$, or intensity) varies as $N/4\pi r^2$, so as expected for a point charge, $|\vec{E}|$ falls off as r^{-2}.

Example 3.4 **A charged sphere *q***

A point charge $q = 10^{-8}$ C produces $N = 4$ field lines. (a) How many field lines are produced by a small charged sphere of $Q = -2 \times 10^{-8}$ C? (b) At a fixed position outside the sphere, how does the magnitude $|\vec{E}|$ of the electric field change if the radius of the charged sphere doubles? (c) How does the magnitude $|\vec{E}|$ of the electric field change if the observer doubles her distance from the charge? Assume that, since the sphere is small, it remains small even when its radius is doubled.

Solution: (a) By symmetry, the field lines must point radially. Since $q = 10^{-8}$ C produces four field lines *outward*, $Q = -2q = -2 \times 10^{-8}$ C produces eight field lines pointing *inward*. (b) Doubling the sphere's radius doesn't change the number of field lines, or their density, so $|\vec{E}|$ doesn't change. (c) Since $|\vec{E}|$ falls off as r^{-2}, when the observer doubles her distance, $|\vec{E}|$ decreases by a factor of 4.

Application 3.2 **Infinite line of negative charge**

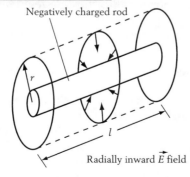

Negatively charged rod

Radially inward $\vec{E}$ field

Figure 3.7 Representation of the three-dimensional field lines due to a line charge.

Unlike the electric field of a point charge, the electric fields of a line charge can be accurately represented on a plane. For a negative line charge of uniform charge per unit length λ, let N lines terminate uniformly in angle over a length l (doubling N doubles l). See Figure 3.7. The number of field lines N is fixed, and the area of a concentric cylinder of radius r and length l (through which the lines pass) is $2\pi rl$. Hence the density of field lines (and thus the field intensity $|\vec{E}|$) varies as $N/2\pi rl$, so $|\vec{E}|$ falls off as r^{-1}.

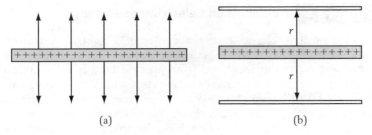

Figure 3.8 Positive sheet of charge intersecting the page. (a) Field lines above and below the sheet. (b) Sheet and two planes, equidistant from the sheet.

Application 3.3 **Infinite sheet of positive charge**

Let the charge per unit area be positive, of magnitude σ, and consider that it produces N field lines per unit area. Then $N/2$ field lines per unit area will point outward in each direction normal to the sheet. (Why outward?) See Figure 3.8(a), which depicts a side view of a sheet that is normal to the y-axis. The locus of points a distance r from the sheet is a pair of parallel planes. See Figure 3.8(b.) Their area does not depend on r. Hence the density of field lines, and thus the field intensity $|\vec{E}|$, does not depend on r, so $|\vec{E}|$ is constant in space; however, $\vec{E}$ changes direction on crossing the sheet, as seen in Figure 3.8(a).

Application 3.4 **Field lines for a uniformly charged disk, near and far**

See Figure 3.9(a), which is only intended to be qualitatively accurate. This disk is taken to be an insulator, so the charge will stay in place. (If it were a conductor, we would have to imagine some extra force holding the charge in place.) Near the disk the field lines are nearly uniformly spaced to correspond to a planar geometry. Far from the disk, the field lines correspond to those of a point charge. Clearly, a great deal can be learned simply by sketching the field lines. Typically, the field lines are not along the normal as they leave the surface.

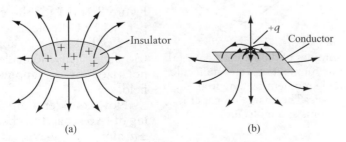

Figure 3.9 Field lines for two-dimensional distributions of charge. (a) A uniformly charged insulator. (b) A neutral conductor in the presence of a point charge.

Field lines and electrical conductors

Consider a positive charge near the center of a neutral, finite conducting sheet. See Figure 3.9(b), which is only intended to be qualitatively accurate. It was drawn using the fact that, if an electric field line has a component along the surface of an electrical conductor, then electric charge will move until the field lines become normal to the surface. Seven field lines enter and seven leave, each normal to the sheet. This example illustrates how much information can be obtained by sketching the field lines, without performing a single calculation.

3.5 Finding $\vec{E}$: Principle of Superposition for Discrete Charges

Having completed our geometrical detour, let's calculate some electric fields, using (3.7) or (3.8).

3.5.1 *The Electric Dipole*

Consider a set of charges that sums to zero, for which the center of the negative charge (of total $-q$) is a distance l from the center of the positive charge (of total q). Its *electric dipole moment* $\vec{p}$ points from the center of the negative charge to the center of the positive charge and has magnitude

$$p = ql. \qquad \text{(dipole moment magnitude)} \qquad (3.10)$$

Like mass, the dipole moment $\vec{p}$ is a quantity associated with a given object.

The set of charges is called an *electric dipole*. See Figure 3.10(a) for the case of a water molecule, said to be a *polar molecule* because it has a permanent electric dipole moment. Polar molecules figure prominently in physics, chemistry, and biology. Without them, organisms would not be able to form cells. Water is a good solvent because of its permanent dipole moment. See Figure 3.10(b) for the case of paper, with a dipole moment only when polarized by an applied electric field. All molecules can be polarized by an applied electric field.

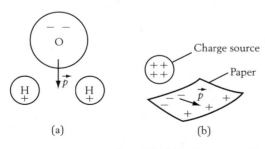

(a)   (b)

Figure 3.10 Examples of dipole moments. (a) Permanent dipole moment of water molecule. (b) Induced dipole moment on a neutral piece of paper. *Note:* The charge on paper is from polarized atoms and molecules.

Now consider a dipole consisting of two separated charges $\pm q$ at $\pm a$ along the y-axis, as in Figure 3.11(a). Since the charges are separated by $l = 2a$, the dipole moment magnitude is given by $p = ql = q(2a)$.

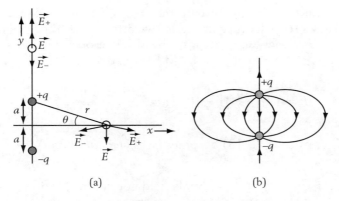

Figure 3.11 A pair of equal and opposite charges.
(a) Individual contributions and the total field along the
x- and y-axes. (b) Field lines from $+q$ to $-q$.

The field-line pattern is sketched in Figure 3.11(b). We now show that electric dipoles have an electric field that falls off at large distances as the inverse cube of the distance.

Field along Dipole Axis. Let the observation point be along the positive y-axis. Since the positive charge is nearer to the observation point, its upward electric field dominates, in agreement with the field direction at the corresponding point in Figure 3.11(b). The unit vector to the observer is $\hat{j}$ for both $\pm q$. Then by (3.7) with $p = q(2a)$,

$$\vec{E} = \frac{kq}{(y-a)^2}\hat{j} + \frac{k(-q)}{(y+a)^2}\hat{j} = \frac{kq(y+a)^2 - kq(y-a)^2}{(y^2-a^2)^2}\hat{j}$$

$$= \frac{4kqay}{(y^2-a^2)^2}\hat{j} = \frac{2kpy}{(y^2-a^2)^2}\hat{j}. \tag{3.11}$$

In the large distance limit, where $y \gg a$, we may neglect the a^2 term in the denominator, so (3.11) becomes

$$\vec{E} = \frac{2kp}{y^3}\hat{j}, \quad y \gg a. \qquad \text{(dipole field along axis of dipole)} \tag{3.12}$$

For a water molecule, as in Figure 3.10(a), $p \approx 6.0 \times 10^{-30}$ C-m. Thus, at 100 nm (nm $= 10^{-9}$ m), a distance of about 1000 times the size of a water molecule, (3.12) gives $|\vec{E}| = 108$ N/C. This is comparable to the field in the earth's atmosphere.

Field Normal to Dipole Axis. Now let the observation point be along the x-axis. It is a distance $r = \sqrt{a^2 + x^2}$ from each charge, so the fields due to each charge have the same magnitude kq/r^2. See Figure 3.11(a). However, the unit vectors to the observation point differ. The horizontal components of the two fields cancel, and the vertical components add. Thus the net electric field points vertically downward, in agreement with the field direction at the corresponding point in Figure 3.11(b). Its magnitude is twice the vertical component of either $\vec{E}_+$ or $\vec{E}_-$. Since each unit vector to the observation point contains a factor $\sin\theta = a/r = a/\sqrt{x^2 + a^2}$ along $-\hat{j}$, by (3.7) the total field is

$$\vec{E} = (2)\left(\frac{kq}{r^2}\right)\left(\frac{a}{r}\right)(-\hat{j}) = -\frac{2kqa}{(x^2 + a^2)^{3/2}}\,\hat{j}. \tag{3.13}$$

In the limit where $x \gg a$, (3.13) becomes (with $p = ql = 2qa$)

$$\vec{E} = -\frac{kp}{x^3}\,\hat{j}, \qquad x \gg a. \qquad \text{(dipole field normal to axis of dipole)} \tag{3.14}$$

3.5.2   *Interaction of a Charge Q with a Dipole p*

From (3.12) and (3.13), at large distances the magnitude of the field $\vec{E}_p$ of the dipole p varies as the inverse cube of the distance. This inverse cube dependence at large distances holds quite generally, with a coefficient that depends on the orientation with respect to the axis of the dipole: $|\vec{E}_p| \sim kp/r^3$. The magnitude of the force $\vec{F}_{Q,p}$ on a charge Q along the axis of a dipole p satisfies, by (3.12) and (3.6),

$$|\vec{F}_{Q,p}| = |Q\vec{E}_p| = \frac{2kQp}{r^3}. \tag{3.15}$$

By Newton's third law (action and reaction), the force $\vec{F}_{p,Q}$ on the dipole p due to the point charge Q must satisfy (3.15). In Section 3.9, we will present an expression for the force on a dipole $\vec{p}$ in the presence of an arbitrary electric field $\vec{E}$.

3.5.3   *Molecular Polarizability and the Amber Effect*

In the amber effect, a charge Q induces a dipole moment $\vec{p}$ on an object, in proportion to the electric field produced by Q. The proportionality constant, α, is known as the *polarizability*:

$$\vec{p} = \alpha\vec{E}. \qquad \text{(field-induced dipole moment)} \tag{3.16}$$

(For asymmetric molecules, α can vary with the molecular axis, but we neglect such complications. Grass seeds are more polarizable along their long axis.)

The dipole moment interacts with the original charge Q, and vice versa. Since $|\vec{E}| = kQ/r^2$ due to Q, (3.16) gives $p = |\vec{p}\,| = \alpha kQ/r^2$. Using (3.15), the force on the induced dipole moment $\vec{p}$ thus satisfies

$$|\vec{F}_{Q,p}| = \frac{2kQp}{r^3} = \frac{2k^2\alpha Q^2}{r^5}. \tag{3.17}$$

The value of α depends upon the atom or molecule. It can be written as $\alpha = V_\alpha/k$, where V_α is a "volume." (A characteristic atomic volume is $(10^{-10}\text{ m})^3 = 10^{-30}\text{ m}^3$.) A scan of the polarizabilities of the atoms in the periodic table reveals that those with the largest and smallest "volumes" are, not surprisingly, the very reactive alkali metal cesium (with $V_\alpha^{Cs} = 59.6 \times 10^{-30}\text{ m}^3$) and the very inert noble gas helium (with $V_\alpha^{He} = 0.205 \times 10^{-30}\text{ m}^3$). The $1/r^5$ dependence of (3.17) increases rapidly as r decreases; perhaps you have observed, on rubbing a comb through your hair and bringing it closer to a small piece of paper, that at some point the paper suddenly "jumps" up to the comb. Consider an atom of carbon, with $\alpha = 1.76 \times 10^{-30}\text{ m}^3$ and $m = 20.0 \times 10^{-27}$ kg. Let it be acted on by a charge $Q = 10^{-9}$ C. The jumping point for this carbon atom is found, approximately, by equating (3.17) to mg. This leads to a distance $r = 1.077$ m at which the carbon atom jumps up to the charge!

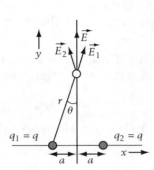

Figure 3.12 A pair of equal charges. Individual contributions and the total field along the y-axis.

Example 3.5   **Field due to two equal charges**

Find the field at a distance y on the y-axis, due to two equal charges $q_1 = q_2 = q$ placed on the x-axis at $\pm a$. See Figure 3.12.

Solution: By symmetry the x-components of $\vec{E}_1$ and $\vec{E}_2$ cancel, and the y-components add. Thus the total field points along $\hat{j}$. By (3.7), it has magnitude $2|\vec{E}_1|\cos\theta = 2[kq/(a^2+y^2)]\cos\theta$, where $\cos\theta = y/(a^2+y^2)^{1/2}$. Thus

$$E_y = \frac{2kqy}{(a^2+y^2)^{3/2}}. \tag{3.18}$$

As expected, the field is zero at the origin ($y = 0$), where the fields of the individual charges should cancel.

3.6 Finding $\vec{E}$: Principle of Superposition for Continuous Charge Distributions

Generalization of (3.7) to the electric field $\vec{E}$ at observation position $\vec{r}$ due to a continuous distribution of source charge dq at position $\vec{r}\,'$ gives

$$\vec{E}(\vec{r}) = k\int dq\,\frac{\hat{R}}{R^2}, \qquad k = \frac{1}{4\pi\epsilon_0}. \quad (\vec{R} = \vec{r} - \vec{r}\,'\text{ to observation point}) \tag{3.19}$$

Here dq is the charge at the source point $\vec{r}\,'$, $\vec{R} = \vec{r} - \vec{r}\,'$ is the vector from the observation point $\vec{r}$ to the source point, and $R = |\vec{R}|$.

As discussed in detail in Chapter 1, there are three types of continuous distributions of charge:

1. $dq = (dq/ds)\,ds = \lambda ds$ for line charge density $\lambda = dq/ds$. Thus the charge on a line segment is $q = \int (dq/ds)\,ds = \int \lambda\,ds$, where the integral extends over the line segment.

2. $dq = (dq/dA)\,dA = \sigma\,dA$ for surface charge density $\sigma = dq/dA$. Thus the charge on an area is $q = \int (dq/dA)\,dA = \int \sigma\,dA$, where the integral extends over the area.

3. $dq = (dq/dV)dV = \rho\,dV$ for volume charge density $\rho = dq/dV$. Thus the charge within a volume is $q = \int (dq/dV)\,dV = \int \rho\,dV$, where the integral extends over the volume.

We will work out examples of the electric fields due to line and surface charge densities. For simplicity, in our previous examples with discrete charges we considered very symmetrical situations. Here we will first consider a very numerical approach. It is adapted to the case where the charge distribution is completely arbitrary (and potentially complicated).

3.6.1 *Numerical Analysis*

Because the electric field is a vector, and vectors in three-space have three components, and because (3.19) involves an integral, students often concentrate on the integral part, and neglect or oversimplify the vector part. To concentrate on the vector aspects, we first discuss the integral as a sum.

From the viewpoint of a spreadsheet analysis, imagine that some elves approximate the source by many tiny elements of source charge dq_i, for example, $N = 1000$. They determine the dq_i and their midpoints $\vec{r}_i$. They also determine the observation position $\vec{r}$. Let $\vec{R}_i = \vec{r} - \vec{r}_i$ be given by its components $X_i = x - x_i$, and so on, and let, by analogy to (2.12'),

$$dE_i^* \equiv \frac{kdq_i}{R_i^2}.$$

(dE_i^* can be negative, so it is not the same as $|d\vec{E}_i|$.) For $i = 1$, Table 3.1 gives a set of quantities leading up to the three components of $d\vec{E}_i$ in columns 10 to 12. Summing columns 10 to 12 would then yield E_x, E_y, and E_z. (Note that the sum of column 9 is meaningless.) From this $\vec{E}$, the force on a charge q would be given as $q\vec{E}$.

The above procedure is well defined and, for complex numerical problems, necessary. Once the elves make the source entries in columns one to four, and make the observer entries elsewhere, all the other results are calculated automatically by the spreadsheet. If the observation point is changed, when the observer-position is changed, the spreadsheet automatically redoes all calculations.

Table 3.1 Spreadsheet entries for field calculation

1	2	3	4	5	6	7	8	9	10	11	12
dq_1	x_1	y_1	z_1	X_1	Y_1	Z_1	R_1	dE_1^*	$dE_1^*(\frac{X_1}{R_1})$	$dE_1^*(\frac{Y_1}{R_1})$	$dE_1^*(\frac{Z_1}{R_1})$

A very similar calculation has already been performed: the spreadsheet example in Chapter 2, for the force on a charge q at the origin, due to a line charge λ. For $q = 10^{-9}$ C, performing the sum for $N = 12$ yielded a force of magnitude 12.36×10^{-9} N. Use of (3.2) then yields $|\vec{E}| = |\vec{F}/q| = 12.36$ N/C. If we were to move the observer position $\vec{r}$ from the origin, the spreadsheet would rapidly compute the new $\vec{F}$ and $\vec{E}$.

3.6.2 *Calculus Analysis*

These examples, important in themselves, also can be used as tests that a numerical calculation gives the correct result.

Example 3.6 Field due to a uniform line charge density λ

Find the electric field at the origin due to a uniform line charge density λ that is parallel to, and a distance a from, the y-axis. See Figure 3.13.

Solution: A previous discussion of this case concluded that $|\vec{E}| \sim r^{-1}$. The previous chapter considered the related problem of the force $\vec{F}$ on a charge q at the origin, due to a charge Q uniformly distributed over a rod of length l, at a distance a along the x-axis. (Hence the charge per unit length is $\lambda = Q/l$.) By (3.2), dividing $\vec{F}$ by q yields $\vec{E}$. Taking the limit where $l \to \infty$ then yields the field due to an infinite line charge. Specifically, Figure 3.13 depicts the force $d\vec{F}$ on q due to the charge $dQ = \lambda\, dy$ in an element of length dy.

The result of summing the $d\vec{F}$'s due to charges dQ on each of the length elements dy is that $F_y = 0$, and F_x is given by (2.21), reproduced here as

$$F_x = -\frac{kq\,Q}{a\sqrt{a^2 + (l/2)^2}}. \tag{3.20}$$

To obtain E_x, by (3.2), we divide (3.20) by q, which leads to

$$E_x = -\frac{kQ}{a\sqrt{a^2 + (l/2)^2}}. \tag{3.21}$$

As a check, note that, as $l/a \to 0$, (3.21) goes to the result for a point charge Q at a distance $r = a$.

Now take the limit where $l/r \to \infty$. Using $E = |\vec{E}| = |E_x|$, and r in place of a, (3.21) yields $E \to kQ/[r(l/2)] = 2kQ/lr$. Thus

$$E = \frac{2k\lambda}{r} = \frac{\lambda}{2\pi\epsilon_0 r}, \quad k = \frac{1}{4\pi\epsilon_0}, \quad \text{(field of infinite line charge)} \tag{3.22}$$

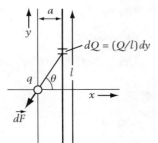

Figure 3.13 Finding the field at the origin due to a line charge: force $d\vec{F}$ on test charge q at origin due to an element dQ.

where $\lambda = Q/l$ is the charge per unit length. As indicated earlier, this varies inversely with r. To obtain a sparking field in air $(E_d = |\vec{E}| = 3 \times 10^6$ N/C) at a distance of 1 mm, (3.22) gives $\lambda = 1.67 \times 10^{-7}$ C/m.

This important geometry is used in particle detectors, such as Geiger counters. When there is ionizing radiation, a large enough field causes electrons to avalanche toward a positively charged wire. The field is largest at the wire.

Example 3.7 Field due to an infinite sheet of uniform surface charge density σ

Find the electric field at the point $P = (0, a, 0)$, due to an infinite sheet of uniform surface charge σ in the $y = 0$ plane. See Figure 3.14.

Solution: This problem could be solved using $dq = \sigma dA$, where $dA = dx dz$, and integrating over both dx and dz. However, the integral over a uniformly charged line (here involving dz rather than dy of Figure 3.13) has already been done in (3.22). Hence the area can be built up as an infinite number of thin strips, each of which can be treated as a line charge. Note that the directions of the fields produced by each line are different.

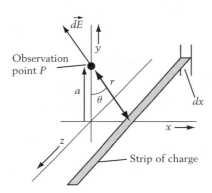

Figure 3.14 Finding the field at the origin due to a sheet of charge: field $d\vec{E}$ due to a strip of charge.

The strip in Figure 3.14, with thickness dx, is parallel to the z-axis, at a distance $r = \sqrt{a^2 + x^2}$ from the point P. As a strip of charge (with a very large length l), it has area $dA = l dx$, charge $dq = (dq/dA)dA = \sigma(l dx)$, and charge per length $d\lambda = dq/l = \sigma dx$. Since $d\lambda$ is independent of l, we may let $l \to \infty$. In (3.22), let us replace E by dE and λ by $d\lambda$. This leads to

$$dE = |d\vec{E}| = \frac{2k d\lambda}{r} = \frac{2k\sigma dx}{r}, \qquad (3.23)$$

which points along the normal from the strip to the observation point. By symmetry, only E_y will integrate to a nonzero value, which requires the direction cosine $\cos\theta = a/r$. Thus

$$dE_y = dE \cos\theta = \frac{2k\sigma dx}{r} \cos\theta. \qquad (3.24)$$

This can be integrated most easily by eliminating x in favor of θ. Since $x = a\tan\theta$, we have $dx = a\sec^2\theta d\theta$. Also, $a = r\cos\theta$, so $r = a\sec\theta$. Then

$$dE_y = \frac{2k\sigma(a \sec^2\theta d\theta)}{a \sec\theta} \cos\theta = 2k\sigma d\theta. \qquad (3.25)$$

Since $\int_{-\pi/2}^{\pi/2} d\theta = \pi$, (3.25) integrates to $E_y = 2k\sigma\pi$. Thus, with E in place of E_y, we have

$$E = 2\pi k\sigma = \frac{\sigma}{2\epsilon_0}, \qquad k = \frac{1}{4\pi\epsilon_0}. \quad \text{(field of flat infinite sheet of charge)} \quad (3.26)$$

As indicated earlier, this is independent of position. It is normal to the sheet and changes direction on crossing from one side of the sheet to the other. To obtain a sparking field in air ($E_d = 3 \times 10^6$ N/C), (3.26) gives $\sigma = 5.31 \times 10^{-5}$ C/m^2.

This important geometry was employed inside older TV tubes to focus the electrons. It is also used in electrostatic precipitators, where it is used to draw off particles of smoke that have been electrically charged.

The present method can be applied to compute the field due to a sheet that is finite along x, by changing the limits of integration. In that case, the field may also have an x-component. To an observer near the center of such a finite sheet, the field will appear to be due to an infinite sheet.

Example 3.8 Field on axis of a ring of uniform linear charge density λ and radius a

Find the electric field a distance z along the axis of a circular ring of uniform linear charge density λ and radius a.

Solution: Let the circle lie at the center of the coordinate system, with its normal along z. For an observation point on the axis, all charge is at a distance $r = \sqrt{a^2 + z^2}$. See Figure 3.15.

By symmetry, only E_z is nonzero. An element of charge dq produces a field of magnitude $dE = |d\vec{E}| = k\,dq/r^2$, and the direction cosine along the z-axis is $\cos\theta = z/r$ for each dq. Thus

$$dE_z = |d\vec{E}|\cos\theta = \frac{k\,dq}{r^2}\frac{z}{r} = \frac{kz}{r^3}\,dq. \tag{3.27}$$

Since kz/r^3 is a constant on the axis, and since $q = \int dq = \int (dq/ds)ds = \int \lambda\,ds = \lambda \int ds = \lambda(2\pi a)$, integration over dq yields

$$E_z = \frac{kz}{r^3}\int dq = \frac{kz\lambda(2\pi a)}{(a^2 + z^2)^{3/2}}. \tag{3.28}$$

Calculus was unnecessary to derive this result: $q = \int dq = \int \lambda\,ds = \lambda(2\pi a)$ is straightforward.

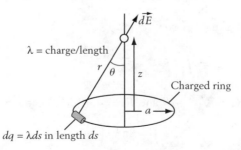

λ = charge/length

$dq = \lambda ds$ in length ds

Figure 3.15 Field $d\vec{E}$ due to part of a ring of charge.

Here are two simple checks. (Think of the *Sesame Street* character Grover, who likes to illustrate *near* and *far*.) Near the ring (here, at its center, $z = 0$), by symmetry the field must be zero; indeed, setting $z = 0$ in (3.28) gives $E_z = 0$. Far from the ring ($z \gg a$), the field should be the inverse square law of a point charge; indeed, using $q = \lambda(2\pi a)$, (3.28) gives $E_z \to kq/z^2$.

For an off-axis observer, the problem would be very difficult because r and $\hat{r}$ would be different for each dq, and there would be three components of $\vec{E}$ to compute. Without advanced techniques, off-axis it would be easier to approximate the integral as a sum on a spreadsheet.

Example 3.9 Field due to a disk of uniform charge density σ and radius a

Find the electric field at a point z on the axis of a disk of uniform surface charge density σ and radius a. See Figure 3.16. The field lines have already been sketched in Figure 3.9(a).

Solution: Let us build up the disk out of rings to take advantage of the results of the previous example. Since only the z-component will be nonzero, we add up the dE_z's due to each of the rings that make up the disk. Thus, consider a typical ring, of radius u and annular thickness du, centered at the origin with normal along the z-axis. Take an observation point that is a distance z along the z-axis. See Figure 3.16. Then the observer is a distance $R = \sqrt{u^2 + z^2}$ from all the charge on the ring. Moreover, the ring has charge $dQ = \sigma dA = \sigma(2\pi u du)$.

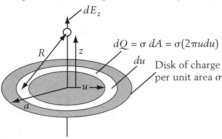

Figure 3.16 Field $d\vec{E}$ due to an annulus of charge.

To apply the results of (3.28), we let

$$E_z \to dE_z, \qquad q = \lambda(2\pi a) \to dQ = \sigma(2\pi u du), \qquad r = \sqrt{a^2 + z^2} \to R.$$

Thus $2\pi \lambda a \to 2\pi \sigma u du$, so (3.28) becomes

$$dE_z = \frac{kz}{R^3}(2\pi \sigma u du). \tag{3.29}$$

Note that $d\vec{E} = dE_z \hat{z}$ for the total field on the axis of the annulus, because $dE_x = dE_y = 0$, by symmetry.

Now change to the variable R, so $R^2 = u^2 + z^2$. Note that $d(R^2) = [\frac{d}{dR}(R^2)]dR = 2R dR$, and similarly $d(u^2) = 2u du$. Since z is constant, we have $2R dR = 2u du$. Then (3.29) becomes

$$dE_z = \frac{kz}{R^3}(2\pi \sigma R dR) = \frac{2\pi k\sigma z dR}{R^2}. \tag{3.30}$$

The total E_z is due to all the rings, so

$$E_z = \int dE_z = -\frac{2\pi k\sigma z}{R}\bigg|_{R_-}^{R_+}. \tag{3.31}$$

Since, by Figure 3.16, $R_+ = \sqrt{a^2 + z^2}$ and $R_- = z$, (3.31) becomes

$$E_z = 2\pi k\sigma z\left(\frac{1}{z} - \frac{1}{\sqrt{a^2 + z^2}}\right) = 2\pi k\sigma\left(1 - \frac{z}{\sqrt{a^2 + z^2}}\right). \tag{3.32}$$

The integrals cannot be done with elementary methods for an observation point off the axis.

Here are two simple checks. Near the disk ($a/z \to \infty$), the field should look like that of an infinite sheet. Indeed, for $a/z \to \infty$, (3.32) goes to $E_z = 2\pi k\sigma$. This agrees with (3.26) for an infinite sheet built up out of lines, rather than rings. Far from the disk ($a/z \to 0$), the field should look like that of a point charge, kQ/z^2, where $Q = \sigma \pi a^2$. To show this, consider the parentheses in (3.32), and divide numerator and denominator by z. That gives

$$1 - \frac{z}{\sqrt{a^2 + z^2}} = 1 - \frac{1}{\sqrt{(\frac{a}{z})^2 + 1}} = 1 - \left(1 + \left(\frac{a}{z}\right)^2\right)^{-\frac{1}{2}}. \tag{3.33}$$

As $a/z \to 0$, this goes to zero. But does this approach zero as $1/z^2$? To see this, set $x = (a/z)^2$ and for $n = -\frac{1}{2}$ apply the result

$$(1 + x)^n \approx 1 + nx, \quad |nx| \ll 1. \tag{3.34}$$

Equation (3.34) follows from the straight-line (or linear) approximation $y = mx + b$ for small x, applied to $y = f(x) = (1 + x)^n$, where $b = f(0) = 1$. To obtain the slope $m = df/dx$ at $x = 0$, take the derivative of $(1 + x)^n$, and then substitute $x = 0$. This gives $df/dx = n(1 + x)^{n-1}$ at any x, so $m = df/dx|_{x=0} = n$.

Applying (3.34) to (3.33) yields

$$1 - \frac{z}{\sqrt{a^2 + z^2}} = 1 - \left(1 + \left(\frac{a}{z}\right)^2\right)^{-\frac{1}{2}}$$

$$\approx 1 - \left(1 - \frac{1}{2}\left(\frac{a}{z}\right)^2 + \ldots\right) \approx \frac{1}{2}\left(\frac{a}{z}\right)^2, \quad \left(\frac{a}{z} \ll 1\right). \tag{3.35}$$

Use of (3.35) in (3.32) then yields that, far from the disk, $E_z \to (2\pi k\sigma)\frac{1}{2}(a/z)^2 = k\sigma(\pi a^2)/z^2 = kQ/z^2$, as expected.

For a position off-axis, the field also has a radial component, which typically is very difficult to calculate.

Example 3.10 **Field due to two parallel sheets of uniform charge densities $\pm\sigma$**

Find the electric field due to two parallel sheets of uniform charge densities $\pm\sigma$, both between the sheets and outside the sheets.

Solution: We again use the superposition principle, here adding up the effect of each sheet, given by (3.26). See Figure 3.17. Outside the region of the sheets, either to the far right or to the far left, the net field is zero because the individual fields $\vec{E}_1 = \vec{E}_+$ and $\vec{E}_2 = \vec{E}_-$, each of magnitude $2\pi k\sigma$, cancel. Within the sheets, however, these fields add, giving a net field $\vec{E}$ pointing from $+\sigma$ to $-\sigma$, of magnitude

$$|\vec{E}| = 4\pi k\sigma. \tag{3.36}$$

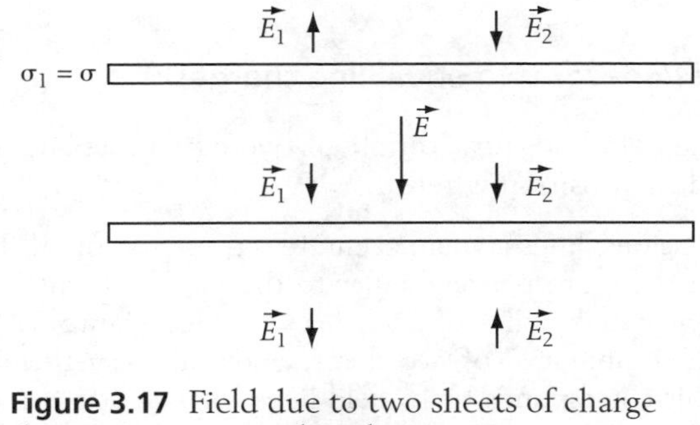

Figure 3.17 Field due to two sheets of charge that intersect normal to the page.

This case is important because parallel plate capacitors have this geometry. Note that $\vec{E}_1$ and $\vec{E}_2$ add vectorially in each region by the principle of superposition; the field due to one sheet is independent of the presence of the other sheet.

3.7 Visualizing the Electric Field: Part 2

3.7.1 *More on Faraday's and Maxwell's Ways of Thinking*

Faraday considered that the source of the lines of force (the charge) is subject to a net force due to the force on all its field lines, using the rules that

1. Each line of force is under tension. Therefore two objects at opposite ends of a line of force (and thus having opposite charges) attract. See Figure 3.11(b).

2. Adjacent lines of force repel one another. Therefore two adjacent objects with the same charge are the source for adjacent repelling lines of force, and hence the objects repel.

Maxwell showed that these ideas are valid, and related the tension and pressure of the lines of force—or field lines—to the electric field strength. He developed an analogy where the field lines leading from a positive charge to a negative charge, as in Figure 3.11(b), are analogous to the fluid flow (*flux*) from a source (e.g., a spigot) to a sink (e.g., a drain). This led Maxwell to think in terms of *flux tubes*, whose sides are parallel to the electric field, with flow direction along $\vec{E}$. The field lines determine the flux tubes, and vice versa.

In the fluid case, equal flux tubes carry equal amounts of fluid per unit time, and the unit flux tube may be defined as one that transports a m^3 of fluid per second. In the electrical case, the unit flux tube is defined so that if there are 300 unit flux tubes per unit area (m^2), then the electric field magnitude is 300 N/C.

It is not possible to draw all the unit flux tubes for a given problem; thus, the flux tubes often are not taken to be unit tubes. Doubling the charge can either double the number of tubes or double the field strength associated with each tube. Thus, the product of the number of tubes and the field strength is related to the electric charge that produces the tube. In general, it is difficult to represent either flux tubes or field lines for three-dimensional geometries. We must often settle for a schematic representation. The examples that follow are simple enough that their field lines can be drawn exactly.

Example 3.11 Field lines for two equal line charges

Discuss and sketch the field lines for a dipole of two equal line charges. Take eight lines per λ. Find the position where $\vec{E} = \vec{0}$.

Solution: In two dimensions, lines give an accurate representation of flux tubes (the other sides of the tubes are perpendicular to the paper). Eight lines leave each positive charge (at equal angles of $360/8 = 45°$). See Figure 3.18(a). Far from these charges, this combination of two charges looks like a single line charge with net charge 2λ. This corresponds to 16 field lines leaving, at equal angles of $360/16 = 22.5°$. See Figure 3.18(a). The boxed point is a position where $\vec{E} = \vec{0}$. It

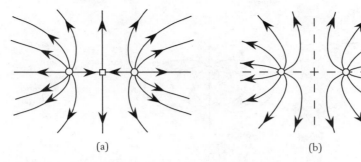

Figure 3.18 Representations of the field lines due to two equal positively charged rods normal to the page. (a) Representation with field lines going from one rod toward the other. These field lines appear to end at the origin (surrounded by a box), with two lines leaving vertically. Thus the net number of field lines leaving the origin is zero. (b) Representation with field lines that do not go from one rod toward the other.

has two lines entering and two lines leaving so that it encloses no net charge. Field lines don't really cross where $\vec{E} = \vec{0}$ since there is no field there. Figure 3.18(b) has been drawn with the field lines at another angle. Faraday would imagine the charges repelling because of the parallel field lines pushing away from each other.

Example 3.12 **Field lines for two line charges on the *x*-axis**

Discuss and sketch the field lines for 2λ at the origin and $-\lambda$ at a distance l to the right. Take eight lines per λ. The lines are oriented normal to the page, as in Figure 3.19.

Solution: Sixteen lines leave the positive charge 2λ (at equal angles of $360/16 = 22.5°$) and eight lines enter the negative charge $-\lambda$ (at equal angles of $360/8 = 55°$). Far away, this combination of two charges looks like a single line charge with net charge $2\lambda + (-\lambda) = \lambda$. This corresponds to eight field lines leaving at equal angles of $360/8 = 45°$. The boxed point on the *x*-axis, where $\vec{E} = \vec{0}$, is a distance s to the right of $-\lambda$, and $l + s$ from 2λ. By (3.23), the fields due to each charge cancel when $2k(2\lambda)/(l+s) = 2k\lambda/s$, giving $s = l$. Two field lines leave this point to right and left. In order that the net number of lines entering this point be zero, two field lines enter it from top and bottom.

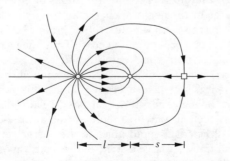

Figure 3.19 Field lines for two rods of charge 2λ and $-\lambda$; the rods are normal to the page. The box denotes the point where the field is zero.

Faraday would imagine the charges attracting because the field lines try to contract.

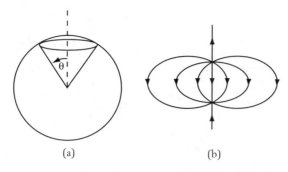

(a)   (b)

Figure 3.20 Geometry associated with dipole field lines. (a) Sphere defining the polar angle θ relative to the dipole axis. (b) Field lines for an electric dipole.

Optional   **Example 3.13**   **Flux tubes and solid angle for a point charge**

Find the solid angle subtended by a cone formed by rotating a line at an angle θ relative to the polar axis, as in Figure 3.20(a). Relate this to flux tubes.

Solution: Your calculus course may have shown that the area A projected on a sphere of radius r of a cone of angle θ is $A = 2\pi r^2 (1 - \cos\theta)$. By definition, the solid angle Ω subtended by this cone is given by $\Omega = A/r^2$ so that

$$\Omega = 2\pi(1 - \cos\theta). \tag{3.37}$$

The larger the solid angle, the larger the flux of the associated flux tube.

Optional   **Example 3.14**   **Field lines and flux tubes for a dipole**

Discuss and sketch the field lines and flux tubes for a dipole of equal and opposite point charges.

Solution: In Figure 3.20(b), the lines represent flux tubes of equal solid angle going from the positive charge to the negative charge. In three dimensions, they are rotated about the polar axis to form the flux tubes. From (3.37), the angles leaving the positive charge constant separations in $\cos\theta$. Thus, measuring with respect to the polar axis, the angles of the tubes entering and leaving the charges are at 0, 60, 90, 120, 180, and so forth. The objects enclosing equal flux are the conical shells that are the differences between successive conical flux tubes. For that reason, they are farther apart for angles near the polar axis. (Related distortions are seen in maps of the world, which also involve representing a three-dimensional situation in two dimensions.) Faraday would imagine the charges attracting because the field lines try to contract. At large distances, the fields vary as (3.12) on the dipole axis, and as (3.14) normal to that axis.

3.8 Force, Torque, and Energy of a Dipole in a Uniform Field

Consider an electric dipole in a completely uniform electric field, as in Figure 3.21. The force $q\vec{E}$ on the positive charge is equal and opposite to the force $-q\vec{E}$ on the negative charge, so there is no *net* force on the dipole. A well-known magnetic analog is an ordinary compass needle in the earth's magnetic field: the needle rotates, but the compass is subject to no net force. This is because the earth's magnetic field is *nearly uniform* in the vicinity of the compass needle, so its north and south poles feel equal and opposite forces.

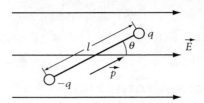

Figure 3.21 Electric dipole of moment $\vec{p}$ in a uniform applied electric field $\vec{E}$. Two equal and opposite charges are separated by l.

Nevertheless, a magnet placed in the nearly uniform magnetic field of the earth feels a torque $\vec{\tau}$ that tends to make it point toward the north pole. We will later describe the magnet as a magnetic dipole in a uniform magnetic field, so we will say that *the magnetic dipole feels a torque that tends to align it with the magnetic field*.

Now consider the analogous case of a permanent electric dipole $\vec{p}$ in a uniform electric field $\vec{E}$ (The dipole could be the water molecule in Figure 3.10a). *The electric dipole $\vec{p}$ feels a torque that tends to align it with the electric field $\vec{E}$.* To determine the value of the torque $\vec{\tau}$, consider that there are two charges $\pm q$ connected by a rod of length l, the charges at positions $\pm \vec{l}/2$. (Thus $\vec{p} = q\vec{l}$.) See Figure 3.21. Let there be a uniform field $\vec{E}$ that points to the right. The force on each of the charges is $\pm q\vec{E}$. Hence the torque $\vec{\tau}$, measured from their midpoint, is

$$\vec{\tau} = \left(\frac{1}{2}\vec{l}\right) \times (q\vec{E}) + \left(-\frac{1}{2}\vec{l}\right) \times (-q\vec{E}) = (q\vec{l}) \times \vec{E}.$$

With $\vec{p} = q\vec{l}$, $\vec{\tau}$ becomes

$$\vec{\tau} = \vec{p} \times \vec{E}. \quad \text{(torque on dipole)} \tag{3.38}$$

For the dipole and field in Figure 3.21, use of the vector product right-hand rule yields a torque that is into the page, which tends to cause a clockwise rotation of the dipole.

We can obtain the energy by using an analogy to gravity. There, a single mass m at position $\vec{r}$ in a uniform gravitational field $\vec{g}$ has gravitational potential energy $U_{grav} = mgy$ or, with $\vec{g} = -g\hat{y}$,

$$U_{grav} = -m(\vec{g} \cdot \vec{r}). \tag{3.39}$$

By analogy, a single charge q at position $\vec{r}$ in a uniform electric field $\vec{E}$ has electrical potential energy

$$U_{el} = -q(\vec{E} \cdot \vec{r}). \tag{3.40}$$

If we now apply this to our two charges, on summing over both we have

$$U_{dip} = -\left[q\left(\vec{E} \cdot \frac{\vec{l}}{2}\right)\right] - \left[(-q)\left(\vec{E} \cdot \left(-\frac{\vec{l}}{2}\right)\right)\right] = -q\vec{E} \cdot \vec{l}, \tag{3.41}$$

or, with $\vec{p} = q\vec{l}$,

$$U_{dip} = -\vec{p} \cdot \vec{E}. \qquad \text{(orientation energy of dipole)} \qquad (3.42)$$

When $\vec{p}$ and $\vec{E}$ are aligned, so $\vec{p} \cdot \vec{E} = |\vec{p}||\vec{E}|$, the energy is minimized, as expected.

3.9 Force on a Dipole in a Nonuniform Field

In the amber effect, there is a force on an induced dipole in a nonuniform field. Equation (3.17) applies only when the electric field is due to a point charge. More generally, there can be a permanent dipole (e.g., due to a water molecule) in an electric field due to many charges. Before getting into mathematical detail, we present some general considerations.

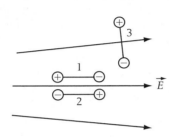

Figure 3.22 Electric dipole moments in a nonuniform applied electric field $\vec{E}$.

Since the force $\vec{F}$ on a dipole $\vec{p}$ in a uniform field is zero, in general the force on the dipole must depend upon the strength of the dipole moment and on *how $\vec{E}$ varies in space*. Consider Figure 3.22, depicting a nonuniform electric field $\vec{E}$ and three dipoles in different orientations. From our earlier considerations about field lines, the electric field is stronger where the field lines are closer. Further, the force on the end of the dipole in the larger field dominates. Thus dipole 1 feels a net force to the right, dipole 2 feels a net force to the left, and dipole 3 feels no net force.

3.9.1 *Quantitative Considerations*

Consider a dipole with q at $\vec{r} + \vec{l}$ and $-q$ at $\vec{r}$. By (3.6), the force on the combination is given by

$$\vec{F} = q\vec{E}\left(\vec{r} + \frac{1}{2}\vec{l}\right) - q\vec{E}\left(\vec{r} - \frac{1}{2}\vec{l}\right), \qquad (3.43)$$

where $\vec{E}$ is evaluated at $\vec{r} \pm \frac{1}{2}\vec{l}$. We wish to evaluate this when $\vec{E}$ varies slowly in space.

First consider the straight-line approximation applied to a function only of x. If a is small, then $f(x+a) - f(x)$ is nearly given by the slope $m = df/dx$ times the coordinate difference a: $f(x+a) - f(x) \approx (df/dx)a$. Now include the coordinate differences in all three directions, and let $f = qE_x$, so $f(x) \to f(x, y, z) = qE_x(x, y, z)$. Then $f(x+a)$ generalizes to $qE_x(x+l_x, y+l_y, z+l_z)$, so $f(x+a) - f(x)$ generalizes to

$$F_x = ql_x\frac{dE_x}{dx} + ql_y\frac{dE_x}{dy} + ql_z\frac{dE_x}{dz} = (\vec{p} \cdot \vec{\nabla})E_x, \quad \vec{p} = q\vec{l},$$

$$\vec{\nabla} \equiv \hat{i}\frac{d}{dx} + \hat{j}\frac{d}{dy} + \hat{k}\frac{d}{dz} \qquad (3.44)$$

with similar equations for F_y and F_z. (The quantity $\vec{\nabla}$ is called the *gradient operator.*) For a dipole $\vec{p}$ aligned with $\vec{E}$, this says that the dipole is attracted to regions of larger $\vec{E}$. Magnetic dipoles have analogous behavior. Verify that this equation agrees with the qualitative considerations for the three dipoles in Figure 3.22.

It is not important that you remember this equation—it is better that you don't! What is important is that, just as we argued above, *this force is proportional to the dipole moment and to how the electric field varies in space.*

3.10 Motion of Charges

Figure 3.23 depicts a positively charged particle (e.g., an alpha particle, which is a helium nucleus, with charge $q_1 = 2e$) moving past another positively charged particle (e.g., the nucleus of an atom, with charge $q_2 = Ze$, where Z is usually much bigger than 2) that is fixed in place. The alpha particle bends away from the nucleus because it is repelled, by "likes repel." Finding the orbit of the alpha particle is a solvable but mathematically complicated problem. It is not necessary to solve such a complex problem to learn how electric fields cause charged particles to be deflected. However, this problem led to the discovery of the atomic nucleus.

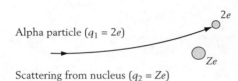

Alpha particle ($q_1 = 2e$)

Scattering from nucleus ($q_2 = Ze$)

Figure 3.23 An alpha particle (a helium nucleus) scattering off the nucleus of an atom of nuclear charge Ze. Not depicted are the very light electrons, which cannot effectively scatter the alpha particle because of the mass mismatch.

Just as the motion of a particle in a uniform gravitational field is relatively straightforward to obtain, so is the motion of a particle in a uniform electric field. The motion of an object depends upon its acceleration, no matter what forces cause that acceleration.

Discovery of the Nucleus

In 1910 Rutherford, winner of the 1908 Nobel Prize in Physics for his discovery of alpha particles, had a student measure the alpha particles that are back-scattered by nuclei. The then current "plum-pudding" model of the atom (due to J. J. Thomson) had positive charge, with most of the mass of the atom uniformly distributed throughout the atom. For this model Rutherford expected to find negligible large angle–scattering. When appreciable back-scattering (nearly 180°!) was found, Rutherford developed a model where the positive charge was concentrated at a massive point. Working out the theory of the alpha particle orbit, he obtained agreement with his experimental results.

3.10.1 *Motion along a Uniform Electric Field*

To draw electrons from it, a cathode is both heated and placed in an electric field due to a positively charged grid screen. At a distance, to the electron the screen looks like a plane.

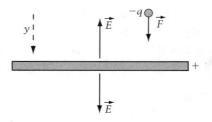

Figure 3.24 Motion of a negative charge $-q$ above a positive, uniformly charged sheet $+\sigma$ that intersects normal to the page.

Consider a uniformly charged sheet in the yz-plane, with positive charge density σ, as in Figure 3.24. Above itself it produces a uniform upward electric field of magnitude $E = 2\pi\sigma$. A negatively charged particle $-q$ would feel a constant force downward; the motion would be very like the motion of a baseball thrown directly upward against gravity. See Figure 3.24. Let us analyze this situation neglecting the force of gravity.

The force is downward, and of magnitude $F = qE$. Let us call this the y-direction. By Newton's second law, this force produces a uniform downward acceleration a_y, where $ma_y = F = qE$. Then

$$\frac{dv_y}{dt} = a_y = \frac{qE}{m}, \tag{3.45}$$

which integrates over dt to yield

$$v_y = \frac{dy}{dt} = v_0 + \frac{qE}{m}t, \tag{3.46}$$

where v_0 is the initial speed. A second integration over dt leads to

$$y = y_0 + v_0 t + \frac{1}{2}\frac{qE}{m}t^2, \tag{3.47}$$

where y_0 is the initial position. This is very like what we have for the gravity problem, except that the acceleration no longer is g, but qE/m.

Example 3.15   **Electron and sheet of positive charge**

Consider in Figure 3.24 an electron ($q = e = 1.6 \times 10^{-19}$ C, $m = 9.1 \times 10^{-31}$ kg) with speed 10^7 m/s at the moment it passes through a tiny hole in a positively charged sheet of $\sigma = 5 \times 10^{-6}$ C/m^2. Find (a) the field, (b) the acceleration, (c) the time for the electron to return to the cathode and (d) the maximum distance of the electron from the sheet.

Solution: (a) By (3.26), $E = 2\pi k\sigma = 2.83 \times 10^5$ N/C. (b) $a_y = qE/m = 4.98 \times 10^{16}$ m/s^2, an acceleration that vastly exceeds that due to gravity. (c) With initial velocity $v_0 = -10^7$ m/s at the hole (negative because it is moving upward), by (3.46) with $v_y = 0$ it takes a time $t = -v_0/a_y = 2.01 \times 10^{-10}$ s for the electron to come to rest, and an equal amount of time for it to return to the cathode. Therefore, $t_{total} = 4.02 \times 10^{-10}$ s. (d) With $y_0 = 0$ and $t = 2.01 \times 10^{-10}$ s, (3.47) yields $y = 2.01 \times 10^{-3}$ m.

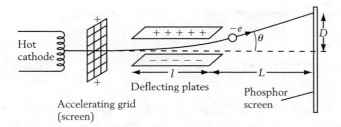

Figure 3.25 Deflection of an electron emitted from a hot cathode. It is attracted to a positively charged grid screen (an anode), overshoots, is deflected by the electric field due to the charge on a pair of parallel plates, and then hits a phosphor screen, where it causes light to be emitted. For many years, television screens used this principle of electrostatic deflection.

3.10.2 *Motion in a Plane, with a Uniform Electric Field*

Older TV tubes and cathode ray tubes, as well as ink-jet printers, use electrostatic deflection to guide the electrons.

Once the electron has passed through the grid and is moving at about the desired velocity, it passes through two pairs of deflecting plates, of length L, to cause the appropriate deflections in the y- and z-directions. See Figure 3.25. (At the TV or oscilloscope screen, the electron collides with a phosphor material, and excites an electron from an atom in the phosphor. The excited electron quickly returns to its ground state, emitting visible light in the process.) Most of the deflection occurs in the region of length D that follows the deflecting plates. However, the deflection angle θ develops while the electron passes through the deflecting plate region. For simplicity, we will not consider motion in the z-direction. See Figure 3.25.

For simplicity, we assume three clear-cut regions: the grid acceleration region (which can be treated as in the previous example), the deflecting plate region, and the post–deflecting plate region.

In the deflecting plate region, the electron has a constant velocity v_x in the x-direction, obtained from the grid region. Thus, measuring from an origin that is at the left edge of the deflecting plates,

$$x = v_x t \tag{3.48}$$

describes the motion in the x-direction, with $x = 0$ at $t = 0$. In the y-direction, there is uniform acceleration starting from rest ($v_y = 0$) at the origin ($y_0 = 0$). With $a_y = F_y/m = (qE)/m$ for the acceleration due to the deflecting plates L, the motion along y is described by

$$v_y = a_y t, \qquad y = v_0 t + \frac{1}{2} a_y t^2, \qquad a_y = \frac{F_y}{m} = \frac{qE}{m}. \tag{3.49}$$

It takes a time $T = L/v_x$ to cross the deflecting plates.

In the post–deflecting plate region, the electron moves in a straight line, at an angle θ determined by the velocity components at time T:

$$\tan\theta = \left.\frac{v_y}{v_x}\right|_T = \frac{a_y(L/v_x)}{v_x} = \frac{a_y L}{v_x^2}. \tag{3.50}$$

By suitably adjusting the charge densities $\pm\sigma$ on the deflecting plates, which by (3.36) produce $E = 4\pi k\sigma$, we can adjust $a_y = qE/m = (e/m)(4\pi k\sigma)$ and therefore the deflection of the electron.

3.11   The Classical World Is Unstable for Electrical

Optional   **Forces Alone**

From the properties of field lines, we can prove that, when Newton's laws of motion (defining *classical mechanics*) apply, a set of electric charges Q_i alone cannot produce any position where a test charge q can be in *stable* equilibrium. Applied to ordinary matter, where the electrical interaction dominates, this means that stability is impossible. This is known as *Earnshaw's theorem*.

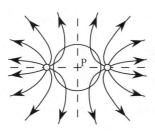

Figure 3.26
Representation of the field lines and an imaginary spherical surface centered along the line between two equal point charges.

Here is a graphical proof. Consider a point P at which $\vec{E} = \vec{0}$. (Note that $\vec{0}$ means that all three components are zero.) Surround P by a sphere of radius so small that it contains no charge. See Figure 3.26, which is based on Figure 3.18(b). If there is an excess of inward (or outward) field lines, then the sphere must contain negative (or positive) source charge. Since there is no charge within the sphere, if four field lines enter then four field lines must leave. Now place a test charge q at P. It feels no force, by $\vec{F} = q\vec{E} = \vec{0}$. However, if it is displaced slightly, in some directions it will feel a force pushing it back to P (stability), and for others it will feel a force pulling it from P (instability). Thus, there is always a direction for which displacements from equilibrium will be unstable.

Since matter is stable, Earnshaw's theorem is an indication that Newton's laws of motion are inadequate at the atomic scale. At the atomic level, classical mechanics (i.e., Newton's laws of motion) must be replaced by *quantum mechanics*. Quantum mechanics eliminates, at the microscopic level, the instability of Earnshaw's theorem.

Problems

3-2.1 In fair weather, at the surface of the earth normally there is a downward directed electric field. (a) Assuming it to have magnitude 125 N/C, what is the force on a water molecule to which an electron has become attached? (b) Compare this with the force of gravity on the water molecule.

3-2.2 (a) A molecule with a single excess electron would need what molecular weight to have a force due to gravity that cancels the electrical force in a field of 2 N/C? (b) In what direction should the field point for such cancellation?

3-2.3 (a) Consider a dust particle of mass 5×10^{-8} kg and an excess of five electrons. How large an electric field would be needed to produce a force of the same magnitude as the gravitational force on the surface of the earth? (b) For an electron, how large a gravitational field would be needed to produce a force of the same magnitude as an electric field of magnitude 100 N/C?

3-2.4 Let the electric field be $\vec{E}$ at a point P near a conducting foil (or any kind of conductor). Because of electrostatic induction, the force on a charge q at P will not yield $\vec{E}$. Show, however, that the average of the fields for two equal and opposite test charges $q > 0$ and $-q$ gives the correct value: $\vec{E} = (\vec{F}_q - \vec{F}_{-q})/2q$.

3-2.5 A string of length l hangs from the ceiling. At its bottom end is a small sphere of mass m and charge $Q < 0$. A uniform horizontal field of magnitude E points rightward. (a) Find the angle θ that the string deflects from the vertical (and which way it deflects). (b) Find the tension T in the string. (c) Evaluate θ and T for $m = 42$ g, $l = 12$ cm, $Q = -3.9$ nC, and $E = 875$ N/C.

3-2.6 A fellow student gives you the following information to represent the force and electric field acting on a charge q: $F_x = 6 \times 10^{-9}$ N and $E_x = 2$ N/C, $F_y = -4 \times 10^{-9}$ N and $E_y = -5$ N/C, $F_z = 4 \times 10^{-9}$ N, $E_z = 0$. (a) Does this data make any sense? (b) If it does, deduce q. If it doesn't, explain why not.

3-2.7 Is $\vec{F} \equiv (1, 2, 3)$ N a possible force for a charge subject to a field $\vec{E} \equiv (2, 4, -1)$ N/C?

3-2.8 A charge -2 μC at the origin is subject to a force of 1.8×10^{-6} N along the $+y$-axis, due to an applied electric field $\vec{E}$. (a) Determine $\vec{E}$ at the origin. (b) Can you use this information to obtain the electric field anywhere else?

3-2.9 A charge Q is known to produce a 5 μN force on a charge 2 nC at a distance of 2 cm. (a) Find the magnitude of the electric field due to Q. (b) Find $|Q|$.

3-2.10 Photocopying uses negatively charged toner particles and a positively charged imaging drum. If the field of the imaging drum has magnitude 4×10^5 N/C, find the force acting on a toner particle that has an excess of 150 electron charges.

3-2.11 One property that makes the electric field $\vec{E}$ a *field* is that it is well defined everywhere in space. Within air, the quantities pressure, density, flow velocity, and temperature are everywhere well defined, and therefore they too are fields. Classify them as *scalar* or *vector* fields.

3-2.12 The intensity of light from a point source (e.g., a small light bulb) falls off as r^{-2}, just as for electric fields. Consider a light source at $(-a, 0, 0)$ to also have charge q. If it produces intensity I_0 and electric field $\vec{E}_0$ at the origin, find the total intensity and the total electric field at the origin if an identical light source with charge q is at $(a, 0, 0)$.

3-2.13 Consider gravity between two point masses m_1 and m_2. (a) Show that the gravitational field $\vec{g}$ due to m_2 points toward m_2. This is opposite the direction of $\vec{E}$ for a positive charge q_2. (b) Write down the gravitational analog of (3.7), making sure that you get the direction correctly.

3-3.1 A charge $Q = 8.5$ nC is at the origin. (a) Find the magnitude and direction of the electric field it produces at $(0, l)$, where $l = 3$ cm. (b) Repeat for $(0, -l)$.

3-3.2 At the origin, an electric field points along the $+y$-axis, with magnitude 560 N/C. (a) Where would you put a charge $Q = 42$ nC to produce such a field? (b) Repeat for $Q = -42$ nC. (c) What charge, placed at $(0, l)$ for $l = 16$ cm, would produce such a field?

3-4.1 For drag-dominated motion, as in wires or electrolytes, $\vec{v}$ points along $\vec{E}$. In the absence of drag, $d\vec{v}/dt$ points along $\vec{E}$. Which of these systems, drag-dominated or dragless, would be more useful for relating measured flow patterns by charged particles and electric field lines?

3-4.2 In considering the field lines for an infinitely long line charge with uniform density, we only considered the field line density along the angular direction. Discuss how the field line density varies along the axis of the line.

3-4.3 In addition to drawing continuous field lines, there is another way to represent a *vector field* like $\vec{E}$, which is the *vector* defined at all points in real space (the *field*): for a representative sample of points in space, draw an arrow pointing along $\vec{E}$, with length proportional to $|\vec{E}|$. (a) What advantages does this method have when there is a sphere containing a uniform volume distribution of charge? (b) What disadvantages does this method have? (c) How do both methods compare relative to the information gained from grass seed alignment?

3-4.4 At point P, Faraday represents the electric field, of magnitude 20 N/C, with a density of eight lines per unit area. At point P′, the density of field lines is 20 lines per unit area. (a) Find the field magnitude at P′. (b) If the strength of the charges producing the field now suddenly quadruples, find the density of field lines at P and at P′.

3-5.1 A charge 4 nC is at the origin and a charge −8 nC is on the y-axis at $y = 0.8$ m. Find the electric field ($\vec{E}$ is a vector) at (a) $y = -0.4$ m, (b) $y = 0.2$ m, (c) $y = 1.2$ m.

3-5.2 Charges 4 nC and −8 nC are placed at the upper right-hand and lower left-hand corners of a square 4 cm on each side. See Figure 3.27. (a) Find the magnitude and direction of the resultant electric field at the lower right-hand corner. (b) Using a symmetry argument, find the resultant electric field at the upper left-hand corner.

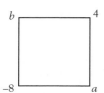

Figure 3.27 Problem 3-5.2.

3-5.3 A circular loop of radius a, centered at the origin, has seven charges equally spaced around it. The middle of the three adjacent $-Q$ charges is uppermost, at $(0, a)$. The other four are q. See Figure 3.28. (a) With $Q, q > 0$, find the direction of the electric field at the center of the loop. (b) Find the magnitude of the electric field.

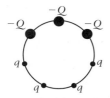

Figure 3.28 Problem 3-5.3.

3-5.4 A square centered at the origin has corners at $(-a, -a)$, $(a, -a)$, (a, a), and $(-a, a)$. If charges Q are at the top left and top right, find the electric field at the other two corners and at the origin. See Figure 3.29.

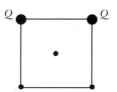

Figure 3.29 Problem 3-5.4

3-5.5 Physics recitation is held in a tower that, viewed from above, rotates clockwise once every two minutes. Each of 23 students is given an electric charge. The total electric field at the front of the class is then calculated, and at 2:01:08 pm is found to have components $E_x = 47$ N/C and $E_y = -36$ N/C as seen by an elevated external observer. (a) If the class members remain in their seats, find the field components 35 seconds later, as seen by the external observer. (b) Compare the calculation done in part (a) with the calculation that would have to be done by adding up the individual fields a second time.

3-5.6 A rod of length 10 cm points along the x-axis. It contains 2000 identical dipoles of moment 0.25 nC-m each, equally spaced and aligned with their dipole moments along the x-axis. Estimate the electric field along the x-axis at 2 m. (*Hint:* For estimation purposes, you may assume that the separation s between dipoles equals the charge separation l within each dipole.)

3-5.7 Consider the charges Q at $(-a, 0)$, $-2Q$ at $(0, 0)$, and Q at $(a, 0)$. Such a combination of charges, with zero net charge and with zero net dipole moment, is called an electric *quadrupole*. (a) Find the electric field along the x axis, for $x > a$. (b) Show that, for $x \gg a$, the electric field varies as x^{-4}. Find the coefficient.

3-5.8 Consider an equilateral triangle of equal charges q, where two charges are at $x = \pm a$ and the third charge is along the positive y-axis a distance $\sqrt{3}a$ above the origin. See Figure 3.30. By computation show that at the center, which corresponds to $s = a/\sqrt{3}$, the field is zero.

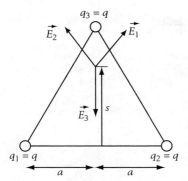

Figure 3.30 Problem 3-5.8.

3-5.9 A circular loop has seven equal charges q equally spaced around it. Use a symmetry argument to show that the electric field is zero at the center of the loop.

3-5.10 (a) For the charges in Figure 3.30, find an expression for the magnitude of the field at any point s on the vertical line through q_3 within the triangle. (b) Show that where the field is zero the distance s satisfies

$$\frac{2s}{(a^2 + s^2)^{3/2}} = \frac{1}{(\sqrt{3}a - s)^2}.$$

(c) Show that, in addition to the root at $s = a/\sqrt{3}$, there is also a root near $s \approx 0.24858a$. (Due to S. Baker.)

3-6.1 A rod in the xy-plane has its ends at $(0, 0)$ and $(L, 0)$. It has uniform charge per unit length λ. (a) Find the total charge Q on the rod. (b) Find the electric field on the x-axis for $x > L$, and verify that it has the expected result for $x \to \infty$. (c) Find the electric field on the x-axis for $0 < x < L$, and verify that it has the expected result for $x = L/2$.

3-6.2 A rod with ends at $(0, 0)$ and $(L, 0)$ has charge per unit length $\lambda = \alpha x$, where α is a constant. (a) Give the units of α and find the total charge Q on the rod. (b) Find the electric field on the x-axis for $x > L$. (c) Verify that the field has the expected result for $x \to \infty$.

3-6.3 Consider a uniformly charged rod of two pieces, from (a, a) to $(3a, a)$ and from $(-3a, a)$ to $(-a, a)$. See Figure 3.31. Each piece has charge $Q/2$. Find the field at the origin. *Hint:* Use superposition.

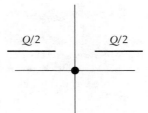

Figure 3.31 Problem 3-6.3.

3-6.4 Consider a uniformly charged arc of radius a and angle α in the xy-plane, centered about the x-axis. See Figure 3.32. If it has charge per unit length λ, find the magnitude and direction of the electric field at the origin.

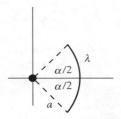

Figure 3.32 Problem 3-6.4.

3-6.5 Consider a ring of radius a in the xy-plane, centered at the origin. Its upper half has charge per unit length $+\lambda$, and its lower half has charge per unit length $-\lambda$. See Figure 3.33. (a) From symmetry considerations, determine the direction of the electric field at its center. (b) Find the magnitude of the electric field at this point.

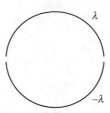

Figure 3.33 Problem 3-6.5.

3-6.6 Consider a ring of radius R, in the xy-plane, centered at the origin. Its lower left-hand quadrant has charge $-Q$, and there is no charge anywhere else. (a) From symmetry considerations, determine the direction of the electric field at the origin. (b) Find the magnitude and direction of the electric field at the origin.

3-6.7 A ring of radius R is centered at the origin in the xy-plane. It is uncharged in the lower half-plane. See Figure 3.34. With θ measured counterclockwise from the x-axis, in the upper half-plane its charge per unit length is $\lambda = \alpha \cos\theta$, where α is a constant. (a) Find the net charge. (b) Find the electric dipole moment, defined as a generalization of (3.10) to be $\vec{p} = \int \vec{r} dq$. Measure $\vec{r}$ relative to the origin. (When an object has zero net charge, it doesn't matter where $\vec{r}$ is measured from.) (c) Find the electric field at the origin.

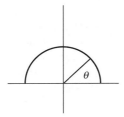

Figure 3.34 Problems 3-6.7 and 3-6.8.

3-6.8 A ring of radius R is centered at the origin in the xy-plane. It is uncharged in the lower half-plane. See Figure 3.34. With θ measured counterclockwise from the x-axis, in the upper half-plane its charge per unit length is $\lambda = \alpha \sin\theta$, where α is a constant. (a) Find the net charge. (b) Find the electric dipole moment. See the previous problem for its definition. (c) Find the electric field at the origin.

3-6.9 Consider a spherical cap of angle α, radius R, and uniform charge per unit area σ. Find the electric field at the center of the cap's sphere. (Figure 3.20a shows a spherical cap of angle θ.)

3-6.10 Consider a slab of uniform charge density ρ within the region $0 < x < L$. Use the principle of superposition on slabs of thickness dx' to find the field due to this charge. Specifically, show (a) for $x > L$ that $E_x = 4\pi k\rho L$; (b) for $x < 0$ that $E_x = -4\pi k\rho L$; and (c) for $0 < x < L$ that $E_x = 2\pi k\rho(2x - L)$. (d) Verify that E_x is continous at $x = 0$ and $x = L$.

3-6.11 Two infinitely long rods are normal to the page, intersecting the page at $(-a, 0)$ and $(a, 0)$. See Figure 3.35. (a) Find the electric field at $(0, y)$ if the

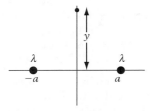

Figure 3.35 Problem 3-6.11.

charge densities are both λ. (b) Repeat if the leftmost charge density is $-\lambda$.

3-6.12 A square of side $2a$ is centered about the origin in the xy-plane. Its border has a uniform charge per unit length λ. (a) Find the electric field at $(0, 0, z)$. (b) Find its total charge. (c) Verify that the electric field goes to the expected result as $z \to \infty$.

3-6.13 You are to produce at the origin a field $|\vec{E}| = 250$ N/C pointing along the $+x$-axis. How would you do this with each of the following sources: (a) a point charge $q = -6\ \mu$C; (b) an infinite line charge $\lambda = 4\ \mu$C/m; (c) an infinite sheet of charge $\sigma = 4\ \mu$C/m^2?

3-6.14 A uniformly charged semicircular arc of radius a and total charge Q lies in the upper half xy-plane, with its center at the origin. See Figure 3.36. Find all three components of $\vec{E}$ for a point a distance z along the z-axis.

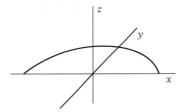

Figure 3.36 Problem 3-6.14.

3-6.15 The field above two uniformly charged, infinite sheets normal to $\hat{y}$ is 20 N/C upward, and the field between them is 30 N/C downward. (a) Find the field below the sheets. (b) Find the charge density on each sheet.

3-6.16 Consider two co-axial disks, each of radius a, and common axis z. One lies on the $z = d/2$ plane and has uniform charge density σ. The other lies on the $z = -d/2$ plane and has uniform charge density $-\sigma$. (a) Show that, between the disks ($|z| < d/2$), the field on the z-axis is

$$E_z = 2\pi k\sigma\left[\left(1 - \frac{(z-d/2)}{\sqrt{a^2 + (z-d/2)^2}}\right) + \left(1 - \frac{(z+d/2)}{\sqrt{a^2 + (z+d/2)^2}}\right)\right], \quad (|z| < d/2).$$

(b) Show that, outside the disks ($|z| > d/2$), the field on the z-axis is

$$E_z = 2\pi k\sigma\left[-\frac{(z-d/2)}{\sqrt{a^2 + (z-d/2)^2}} + \frac{(z+d/2)}{\sqrt{a^2 + (z+d/2)^2}}\right], \quad (|z| > d/2).$$

(c) Show that, for $a \to \infty$, part (a) goes to $4\pi k\sigma$, and part (b) goes to zero, and show that these results are expected for very large uniformly charged plates. (d) Show that, as $|z| \to \infty$, $E_z \to (2\pi k\sigma)(a^2 d/z^3)$, as expected on the axis of a dipole of moment $p = \sigma(\pi a^2)d$. *Hint:* You may find (3.35) helpful.

3-7.1 As indicated by Figure 3.2, field lines can be visualized with grass seeds. What does the fact that grass seeds tend to align along $\vec{E}$ say about which orientation has the lower energy, *along* or *perpendicular* to $\vec{E}$?

3-7.2 Sketch the electric field lines for two line charges λ and -4λ that are normal to the page, and separated by a, as in Figure 3.37. Take four lines per λ. Find the position where the field is zero.

Figure 3.37 Problem 3-7.2.

3-7.3 Sketch the electric field lines for two line charges λ and 2λ that are normal to the page, and separated by a, as in Figure 3.38. Take eight lines per λ. Find the position where the field is zero.

Figure 3.38 Problem 3-7.3.

3-7.4 Find the position where the field is zero. Two uniformly charged circular plates of radius a are co-axial and a distance b apart. They have equal and opposite charges Q. (a) Sketch the electric field lines between the plates. (b) Sketch the electric field very far from the plates. (Sketch the field lines on a scale large enough to show the field lines outside the plates.)

3-7.5 Within a conductor, electric current is driven by an electric field. Discuss why a uniform volume charge distribution is possible for an insulating sphere in equilibrium, but not for a conducting sphere in equilibrium.

3-7.6 A uniformly charged rod of total charge Q has its ends at $(0, 0)$ and $(0, L)$. Sketch the field lines.

3-8.1 The dipole moment of a water molecule (H_2O) is about 6.0×10^{-30} C-m. (a) Find the energy to reorient it from pointing along a field of 30 kN/C to pointing against the field. (b) Find the angle to the field where the torque is a maximum, and find the maximum torque.

3-8.2 Show that the position where the potential energy is zero is irrelevant to the final answer for the dipole energy of (3.42), although it does affect the energy (3.40) of a single charge. Thus we are allowed to use (3.40) in deriving (3.42).

3-8.3 The dipole moment of an ammonia molecule (NH_3) is approximately 5.0×10^{-30} C-m. (a) Find the torque on the dipole when it is aligned normal to a field of 50 kN/C. (b) Find the torque on the dipole when it is at $28°$ to the field. (c) Find the energy change to go from the first position to the second.

3-8.4 A small irregular object is centered about the origin. Far away from it, measurements indicate that the electric field varies as r^{-3}, where r is the distance from the origin. Also, at fixed $r = 40$ cm, as the orientation varies, $|\vec{E}|$ varies, with a maximum value of 14 N/C. (a) Find the net charge Q. (b) Find the dipole moment.

3-8.5 A charge Q is placed along the axis of a dipole $\vec{p}$, as in Figure 3.11. (a) Find the force on the charge Q. (b) Find the force on the dipole $\vec{p}$. (c) If the force is 8×10^{-15} N at $y = 2$ cm, and $Q = 5\,\mu C$, find $|\vec{p}|$.

3-9.1 A dipole $\vec{p}$ points along the x-axis. It is in a field $\vec{E} = Ax\hat{i} + Ay\hat{j} - 2Az\hat{z}$. (a) Find an expression for the force on the dipole in terms of $|\vec{p}|$ and A. (b) Let $|\vec{p}| = 3 \times 10^{-12}$ C-m and $A = 790$ N/C-m. Evaluate the force on the dipole for position $(2, 0, 0)$ (in cm units). (c) Repeat for position $(4, -3, 1)$ (in cm units).

3-9.2 In Figure 3.22, indicate the direction of the force on each of the dipoles, and explain why it has that direction.

3-10.1 A sheet of uniform charge density 5.6 nC/m^2 lies in the xy-plane. An electron starts from rest at $z = 6$ cm. See Figure 3.39. (a) Determine its velocity when it reaches the sheet. (b) Determine how long it takes to reach the sheet.

Figure 3.39 Problem 3-10.1.

3-10.2 An infinite sheet with uniform charge density $\sigma = 4.2 \times 10^{-10}$ C/m² lies on the $y = 0$ plane. On the y-axis, at 12 cm above the sheet, a Li⁺ ion moves in the xy-plane moving downward and rightward, with initial velocity 5.7×10^5 m/s at an angle of 63 degrees to the y-axis. See Figure 3.40. (a) Find the ion's velocity components when it reaches the sheet. (b) Find how long it takes to reach the sheet. (c) Find where on the x-axis it reaches the sheet.

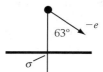

Figure 3.40 Problem 3-10.2.

3-10.3 Consider two horizontal plates with equal and opposite charge densities $\pm\sigma$. An electron enters this region at its midline, with horizontal velocity 4×10^6 m/s. The plates are 5 cm long and 2 cm apart. See Figure 3.41. (a) Find the electric field at which the electron just misses hitting either plate. (b) Find the charge density corresponding to this field. (c) Find the velocity of the electron at this point, and the angle that it makes to the horizontal.

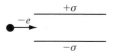

Figure 3.41 Problem 3-10.3.

3-10.4 An electron initially moves along the y-axis with velocity v_0. Within region A, from $y = 0$ to $y = l$, it encounters a rightward electric field of magnitude $\bar{E}_0$. Within region B, from $y = l$ to $y = 2l$, it encounters a leftward electric field of magnitude $\bar{E}_0$. See Figure 3.42. (a) Find how far it deflects in region A. (b) Find how far it deflects in region B. (c) Summarize the changes in electron ve-

locity and position after it has passed through both regions.

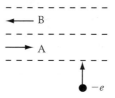

Figure 3.42 Problem 3-10.4.

3-10.5 An ink-jet printer uses an electric field to direct charged ink drops to a sheet of paper. Consider ink drops (mass $m = 8.5 \times 10^{-11}$ kg, charge 1.6 pC) of velocity 15 m/s moving perpendicular to a 120 kN/C electric field. How far will they deflect on passing through a 2 mm region?

3-10.6 An electron moves in a circular orbit about a proton. (a) Relate its velocity v to its radius r. (b) Find its period in terms of r.

3-10.7 An electron moves in a circular orbit of radius r about a wire with positive charge per unit length λ. (a) Relate its velocity v to its radius r. (b) Find its period in terms of r.

3-G.1 (a) Evaluate $[(1 + x)^{-3/2} - 1]/x$ using your calculator for $x = 10^{-14}$. (b) Evaluate it using (3.34).

3-G.2 (a) Evaluate $[(1 + x)^{3/2} - (-1 + x)^{3/2}]/x^{1/2}$ using your calculator for $x = 10^{14}$. (b) Rewrite it so that you can use (3.34), and then evaluate.

3-G.3 Let $n = 100$. (a) Evaluate $(1 + x)^n$ and $1 + nx$ for $x = 0.001$. Compare the results. (b) Repeat for $x = 0.01$ and $x = 0.1$.

3-G.4 (a) Use the principle of superposition to find the electric field midway between an infinitely long line charge $\lambda > 0$ through the origin, along the z-axis, and a charge $q > 0$ that is in the xy-plane at $(a, 0)$. (b) For what value of λ will the electric field at $(a/2, 0)$ be zero? (c) How do these answers change if the line charge is rotated about the x-axis, to coincide with the y-axis?

3-G.5 Consider a conducting disk of radius a and charge Q in the xy-plane and centered at the origin. Its charge density (including both sides) is $\sigma = (2Q/a\sqrt{1 - r^2/a^2})$. (a) Find the electric field along the z-axis. (b) Compare this with the electric field if the disk were uniformly charged. (c) Is the field for the conductor larger or smaller? Why?

(d) For $z \ll a$, what is the electric field? For $z \gg a$? Do these results make sense?

3-G.6 Discuss the difficulties associated with the following measurements of the electric field: (a) using a 2 cm radius insulating sphere coated with net charge q to measure the electric field at 1 cm from a point charge Q; (b) using a 2 cm radius conducting sphere with net charge q to measure the electric field at 1 m from a point charge Q.

3-G.7 Consider a dipole $\vec{p}$ at the origin. For an observer at position $\vec{r}$, decompose $\vec{p}$ into two parts, $\vec{p}_1 = (\vec{p} \cdot \hat{r})\hat{r}$ and $\vec{p}_2 = \vec{p} - (\vec{p} \cdot \hat{r})\hat{r}$. Use (3.12) and (3.14) to show that the dipole field is given by

$$\vec{E} = k\frac{-\vec{p} + 3(\vec{p} \cdot \hat{r})\hat{r}}{r^3}.$$

(Due to Mark Heald.)

3-G.8 Consider a spherical shell of radius a and uniform charge density σ. (a) By doing the integral, show that the electric field inside ($r < a$) is zero. (b) By doing the integral, show that the electric field outside ($r > a$) is the same as if all the charge were at the center of the shell. (This last is an integral that, in the context of gravitation, gave Newton some difficulty. However, you may have learned enough calculus that you can do it. Newton had to *invent* his mathematics as he went along.)

3-G.9 A cylinder of length l and radius a, with charge Q, is centered at the origin, its axis along z. The field at z' is 2 N/C. If the object is now scaled up in size by a factor of two with Q fixed, find the field at $2z'$. *Hint:* Show that this is equivalent to a making a change in the unit of distance.

3-G.10 Consider a fixed dipole $\vec{p}$ at the origin that points along $-z$. A charge $q > 0$ of mass m moves in a circular orbit of radius a about the $+z$-axis, where the plane of the orbit is a distance b from the origin. (a) Show that $a/b = \sqrt{2}$. (b) Show that the charge has velocity $v = \sqrt{3qpk/2^{1/2}m}/r$, where $r = \sqrt{a^2 + b^2}$. (Due to A. Zangwill.)

3-G.11 Consider two line charges $\pm\lambda$ that are both parallel to $\hat{z}$. They intersect the $z = 0$ plane at $(x, y) = (0, 0)$ and $(x, y) = (0, l)$, respectively. See Figure 3.43. Define the dipole moment per unit length to have magnitude $|\vec{\lambda}| = \lambda l$, and to point from negative to positive. Let $l \to 0$, and let λ grow to keep $\vec{\lambda}$ finite. Show that $\vec{E} = 2k[-\vec{\lambda} + 2(\vec{\lambda} \cdot \hat{R})\hat{R}]/R^2$, where R is the nearest

distance from the midpoint of the dipole lines to an observer at (x_0, y_0). This corresponds to a line of dipoles along the z-axis, with dipole moment along $+y$.

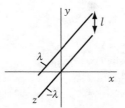

Figure 3.43 Problem 3-G.11.

3-G.12 Consider the semiinfinite sheets defined by $y = 0$ and $x < 0$, and $y = l$ and $x < 0$. Let the upper sheet have uniform charge density σ, and let the lower sheet have uniform charge density $-\sigma$. See Figure 3.44. In the limit as $l \to 0$ this can be thought of as a dipole sheet defined by $y = 0$ and $x < 0$, with finite dipole moment per unit area σl along $+y$. Find its electric field by adding up an infinite number of dipole lines, discussed in the previous problem. For each dipole line, let $\vec{E} \to d\vec{E}$ and $\lambda \to \sigma\, dx$. Then for an observer at (x_0, y_0), show that $\vec{E} = -(2k\sigma/r_0)(\hat{z} \times \hat{r}_0)$, where $r_0 = \sqrt{x_0^2 + y_0^2}$ and $\hat{r}_0 = \hat{y}\cos\theta_0 + \hat{x}\sin\theta_0$.

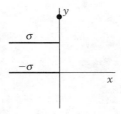

Figure 3.44 Problem 3-G.12.

3-G.13 The previous problem obtained the electric field of a dipole sheet by adding up the fields of many parallel dipole lines. Now obtain the electric field for that same dipole sheet by adding up the electric fields of two nearby parallel planes with equal and opposite charges. To avoid infinities, make the sheets extend from $x = 0$ to $x = -L$, and only after including both sheets take the limit $L \to \infty$.

3-G.14 Consider the amber effect when the neutral object has one axis that is more polarizable than the others. In the dipole system, let $p_{x'} = \alpha_1 E_{x'}$, and $p_{y'} = \alpha_2 E_{y'}$, where $\alpha_2 < \alpha_1$. Let (x', y') be rotated by θ relative to the lab frame (x, y), so $E_{x'} = \cos\theta E_x + \sin\theta E_y$ and $E_{y'} = \cos\theta E_y - \sin\theta E_x$. See Figure 3.45. (a) Show that $p_y = \cos\theta p_{y'} + \sin\theta p_{x'}$ can be rewritten as

$$p_y = (\alpha_1 \sin^2\theta + \alpha_2 \cos^2\theta) E_y$$
$$+ (\alpha_1 - \alpha_2) \sin\theta \cos\theta E_x,$$

and that

$$p_x = (\alpha_1 \cos^2\theta + \alpha_2 \sin^2\theta) E_x$$
$$+ (\alpha_1 - \alpha_2) \sin\theta \cos\theta E_y.$$

(b) For $E_x = 0$ and $E_y = kQ/r^2$, corresponding to a point charge, evaluate p_x and p_y. (c) If $\alpha_1 \ll \alpha_2$ and $\theta \ll 1$, show that the transverse component p_x can dominate, giving a force that is mostly transverse to the applied field.

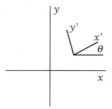

Figure 3.45 Problem 3-G.14.

3-G.15 The 18th-century philosopher Bishop Berkeley (pronounced "Barkley") posed the question "If a tree falls in a forest and there is no one to hear it, does it make a sound?" Show that the answers are different for the two viewpoints of action at a distance and of field (here, the pressure field). Berkeley is also responsible for questioning the mathematical rigor of Newton's approach to calculus. This issue was not resolved until a century later, when mathematicians developed ideas associated with continuity and limits—what students study, perhaps without fondness, using the machinery of epsilon (ϵ) and delta (δ).

*"Now the quantity of electricity in a body is measured in terms, according to Faraday's"
ideas, by the number of lines of force . . . which proceed from it. These lines of force must
all terminate somewhere, either on bodies in the neighborhood, or on the walls and roof of
the room, or on the earth, or on the heavenly bodies, and wherever they terminate there
is a quantity of electricity exactly equal and opposite to that on the part of the body from
which they proceeded.*

—James Clerk Maxwell,
A Treatise on Electricity and Magnetism (1873)

Chapter 4

Gauss's Law: Flux and Charge Are Related

Chapter Overview

Section 4.2 defines electric flux and electric flux density, and obtains the constant of proportionality between electric flux leaving and charge enclosed by a Gaussian surface. Section 4.3 discusses Gauss's law in more detail. Section 4.4 presents a number of examples of the calculation of electric flux, from which the charge enclosed is deduced. Section 4.5 considers the three symmetrical cases where a Gaussian surface S can be found for which the flux per unit area is constant, and uses a knowledge of the charge enclosed to find the electric field. Section 4.6 considers electrical conductors in equilibrium, showing that the electric field is zero inside them (electrostatic "screening"); application of Gauss's law then shows that any charge on the conductor must reside on its surfaces. Section 4.7 shows that, for a conductor in equilibrium, the local value of the surface charge is proportional to the field just outside the surface. Section 4.8 discusses charge measurement using a charge electrometer and Faraday's ice-pail conductor. Section 4.9 proves Gauss's law. Section 4.10 discusses electrostatic screening in more detail, and shows that there can be no analogous gravitational screening. Section 4.11 discusses some properties of electrical conductors that depend on the microscopic nature of the conductor. ∎

4.1 Introduction

Electricity is difficult to comprehend in part because we cannot actually *see* the electric charge that produces specific electrical effects. However, by tracing the path of the field lines due to a set of electric charges, Faraday would have been able to locate the charges, and even determine their magnitudes. This is the basis of field-line drawing rule 3 of Chapter 3.

For more precision, we replace field lines by *electric flux*. (In discussing a fluid, with the electric field replaced by the fluid velocity, the *fluid flux* would be a measure of the rate at which fluid volume leaves.) Gauss's law states that

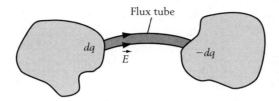

Figure 4.1 Flux tube associated with a charge dq on one conductor, and a charge $-dq$ on another conductor.

the total electric flux leaving a closed surface equals $4\pi k$ times the net electric charge Q_{enc} enclosed within that surface. Because of Gauss's law, *flux tubes* can be drawn, whose sides are parallel to the electric field, and which carry the same amount of flux along any cross-section. This flux can be traced along the tube at one end to a definite amount of positive charge, and at the other end to an equal amount of negative charge. See the flux tube and its associated charge in Figure 4.1, which is drawn for charge on two conductors.

A consequence of Gauss's law is that, if the electric flux leaving the surface of an object is known, either by calculation or by measurement, we can determine the electric charge within that object.

4.2 Motivating Gauss's Law: Defining Electric Flux Φ_E

4.2.1 *The Number of Field Lines N Leaving a Closed Surface*

According to Faraday, the number N of field lines (or lines of force) leaving a closed surface (such as the exterior surface of a football) is proportional to the charge enclosed (Q_{enc}) by that surface. That is,

$$N = \alpha \, Q_{enc}. \tag{4.1a}$$

However, the proportionality constant α depends upon the choice of number of field lines per unit charge: Faraday might choose eight lines per unit charge, whereas Maxwell might choose six lines per unit charge. Our goal is to write a relation like (4.1a) with a new α whose value everyone will agree upon. The new relation is called Gauss's law.

Let's find N for the irregular closed surface of Figure 4.2(a). Break up the full surface into an infinite number of infinitesimal areas dA. Take $\hat{n}$ to point along the outward normal to dA. Measure the number of field lines dN leaving each dA and add them up. Representing this sum as an integral, and using a little circle to denote an integral over a closed surface, we have

$$N = \oint dN = \oint \frac{dN}{dA} dA. \tag{4.1b}$$

Since the number per unit area dN/dA of the field lines is proportional to $|\vec{E}|$, this suggests taking $dN/dA \sim |\vec{E}|$ in (4.1b). That works for a surface dA_0 to which $\vec{E}$ is normal (i.e., $\vec{E}$ points along the normal $\hat{n}_0$ in Fig 4.2b). Then

$$\frac{dN}{dA_0} \sim |\vec{E}|, \qquad \text{for } \vec{E} \text{ along } \hat{n}_0. \tag{4.2a}$$

However, consider a surface dA inclined to dA_0 at an angle θ, where the same number of field lines dN pass through both dA_0 and dA. Now $\vec{E}$ makes

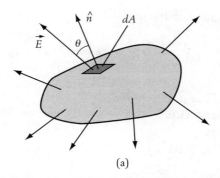
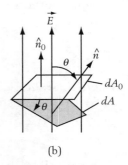

(a)
(b)

Figure 4.2 Defining electric flux Φ_E. (a) Flux leaving an arbitrary closed surface, in terms of the flux through a surface element dA with outward normal $\hat{n}$. Flux per unit area is $d\Phi_E/dA = \vec{E} \cdot \hat{n}$. (b) Reorienting from surface element dA_0 with normal $\hat{n}_0$ along $\vec{E}$ to surface element dA with arbitrary normal $\hat{n}_0$ decreases $d\Phi_E/dA$.

an angle θ relative to $\hat{n}$: $\cos\theta = \hat{E} \cdot \hat{n}$. From Figure 4.2(b), $dA_0 = dA\cos\theta$, so $dA = dA_0/\cos\theta$. With $\vec{E} = |\vec{E}|\hat{E}$, we have

$$\frac{dN}{dA} = \frac{dN}{dA_0/\cos\theta} = \frac{dN}{dA_0}\cos\theta \sim |\vec{E}|\cos\theta = |\vec{E}|\hat{E} \cdot \hat{n} = \vec{E} \cdot \hat{n}. \quad (4.2b)$$

Equation (4.2b) is consistent with the fact that, if $\vec{E}$ is parallel to our surface ($\vec{E} \cdot \hat{n} = 0$), then no field lines cross the surface (so $dN/dA = 0$). Moreover, it also includes information about whether or not the field lines go in or go out: $\vec{E} \cdot \hat{n}$ is positive for field lines leaving, and negative for field lines entering. Indeed, $\vec{E} \cdot \hat{n}$, not $|\vec{E}|$, is the proper measure of dN/dA.

4.2.2 *Defining Electric Flux Φ_E*

Instead of (4.2b) for dN/dA, define

$$\frac{d\Phi_E}{dA} \equiv \vec{E} \cdot \hat{n} = |\vec{E}|\cos\theta. \qquad \text{(flux per unit area)} \qquad (4.3)$$

Then (4.1b) is replaced by

$$\Phi_E = \oint d\Phi_E = \oint \frac{d\Phi_E}{dA}dA = \oint \vec{E} \cdot \hat{n}dA.$$

$$\text{(total flux leaving a closed surface)} \quad (4.4)$$

In (4.4), *there may or may not be a real physical object associated with the closed surface.* Such a surface, because it is intended for use with Gauss's law, is called a *Gaussian surface.*

Some authors use the equivalent definition

$$\Phi_E = \oint \vec{E} \cdot d\vec{A}, \qquad d\vec{A} \equiv \hat{n} dA, \tag{4.5}$$

where $dA = |d\vec{A}|$. The notation $d\vec{S} = d\vec{A}$ is sometimes used.
With (4.4), instead of (4.1b) we write,

$$\Phi_E = \alpha \, Q_{enc}, \tag{4.6}$$

where α will be determined shortly.
Note that

$$\vec{E} \cdot \hat{n} = E_x n_x + E_y n_y + E_z n_z = |\vec{E}| \cos\theta, \tag{4.7}$$

where θ is the angle (of less than or equal to 180°) between $\vec{E}$ and $\hat{n}$. Equations (4.7) and (4.3) give us more than one way to obtain $d\Phi_E/dA$.

Example 4.1 **Surface, field, and flux**

Consider a closed surface and an element of area $dA = 10^{-6}$ m² where the outward normal is $\hat{n} = 0.36\hat{x} + 0.8\hat{y} + 0.48\hat{z}$ and the electric field is $\vec{E} = (2\hat{x} - 3\hat{y} + 4\hat{z})$ N/C at dA. (a) Verify that the normal $\hat{n}$ is a unit vector. Find (b) $|\vec{E}|$, (c) the flux per unit area, (d) the flux, and (e) the angle θ between $\vec{E}$ and $\hat{n}$.

Solution: (a) By (R.11),

$$|\hat{n}| = \sqrt{(0.36)^2 + (0.8)^2 + (0.48)^2} = 1.$$

(b) $|\vec{E}| = \sqrt{(2)^2 + (-3)^2 + (4)^2} = \sqrt{29} = 5.39$ N/C. (c) By (4.3), the flux per unit area is $d\Phi_E/dA = \vec{E} \cdot \hat{n} = (2)(0.36) + (-3)(0.8) + (4)(0.48) = 0.24$ N/C at dA. (d) Since $dA = 10^{-6}$ m², we have $d\Phi_E = (d\Phi_E/dA)dA = 0.24 \times 10^{-6}$ N-m²/C. This is the flux that leaves through dA. (e) By (4.7), $\cos\theta = \vec{E} \cdot \hat{n}/|\vec{E}| = 0.045$, so that $\theta = 87.4°$. Thus, the field line leaves the surface element at near glancing incidence.

Example 4.2 **Flux for rotated surface**

For the same $\vec{E}$ and dA as in Example 4.1, rotate the normal $\hat{n}$ until it makes an angle of 25° to $\vec{E}$. Find the flux through dA.

Solution: By (4.4), the flux $d\Phi_E = (d\Phi_E/dA)dA = |\vec{E}| \cos\theta \, dA = \sqrt{29}\cos(25°) \cdot 10^{-6} = 4.88 \times 10^{-6}$ N-m²/C leaves through dA.

4.2.3 *Relating Electric Flux Φ_E and Charge Enclosed Q_{enc}*

The constant α in (4.6) is determined once and for all by considering a case for which both Φ_E and Q_{enc} can be obtained: a point charge q at the origin.

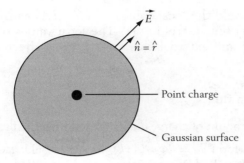

Figure 4.3 Electric flux Φ_E for a concentric Gaussian surface, due to a point charge. Here, the flux per unit area is uniform, and easily computed, so the total flux can be easily computed. A Gaussian surface is a closed (and typically imaginary) surface, used for application with Gauss's law, which relates electric flux to charge enclosed.

It makes a spherically symmetric electric field at $\vec{r}$, given by $\vec{E} = (kq/r^2)\hat{r}$, so $E_r \equiv \vec{E} \cdot \hat{r} = (kq/r^2)$. For a concentric spherical Gaussian surface of radius r, $\hat{n} = \hat{r}$, because the electric field points along the outward normal. See Figure 4.3.

Thus the flux per unit area over this concentric spherical surface of constant radius r takes on the constant value

$$d\Phi_E/dA = \vec{E} \cdot \hat{n} = \vec{E} \cdot \hat{r} = E_r = (kq/r^2).$$

Hence the total flux is

$$\Phi_E = (d\Phi_E/dA)A_{sphere} = (kq/r^2)A_{sphere} = (kq/r^2)(4\pi r^2) = 4\pi kq.$$

Moreover, for this Gaussian surface $Q_{enc} = q$. Placing these Φ_E and Q_{enc} in (4.6) yields $4\pi kq = \alpha q$, so $\alpha = 4\pi k$, where $4\pi k = \epsilon_0^{-1}$.

4.3 Gauss's Law

Using $\alpha = 4\pi k$, (4.6) becomes

$$\Phi_E = \oint \vec{E} \cdot \hat{n} dA = 4\pi k Q_{enc} = \frac{1}{\epsilon_0} Q_{enc}, \qquad \text{(Gauss's law)} \qquad (4.8)$$

or equivalently,

$$Q_{enc} = \frac{1}{4\pi k}\Phi_E = \epsilon_0\Phi_E, \qquad k = 8.9875 \times 10^9 \frac{\text{N-m}^2}{\text{C}^2}. \qquad \text{(Gauss's law)}$$

$$(4.9)$$

Either of these results is known as Gauss's law. They are true for any shape of the Gaussian surface. Coulomb's law looks simpler using k, and Gauss's law looks simpler using ϵ_0.

Equation (4.8) expresses the idea that only Q_{enc} is responsible for the electric flux. It doesn't tell us how the charge is arranged, or how many charges there are; only the net charge. If the charge inside is moved, the flux doesn't change. If the charge outside is moved, the flux doesn't change. Only if charge *enters* or *leaves* the Gaussian surface does the flux change. If we were to move the enormous charge $Q = 10^9$ C from infinity to just *outside* a Gaussian surface, the local values of $d\Phi_E/dA$ on the surface would change considerably, but Φ_E would not change at all. If we were to move this enormous charge to the *inside* of the

Gaussian surface, not only would the local values of $d\Phi_E/dA$ change considerably (including sign changes) but so would Φ_E. The local values of $d\Phi_E/dA$ are affected by both internal and external charge, but Φ_E is affected only by internal charge.

Example 4.3 **Charge from flux**

For Example 4.1, find the amount of charge dq associated with the flux $d\Phi_E$.

Solution: Using (4.9) in the form $dq = d\Phi_E/(4\pi k)$, Faraday would trace from $d\Phi_E$ to dq. With $d\Phi_E = 4.88 \times 10^{-6}$ N-m²/C, this gives $dq = 4.31 \times 10^{-17}$ C. The equation $dq = d\Phi_E/(4\pi k)$ also relates the dq and $d\Phi_E$ of Figure 4.1.

4.3.1 *Numerical Integration of Electric Flux*

For a specific example of electric flux computation for a more complex surface than a sphere concentric with a point charge, consider a potato. See Figure 4.4.

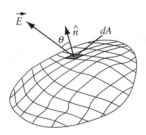

Figure 4.4 Electric flux Φ_E leaving an arbitrary closed surface (a Gaussian surface), with the surface broken into many surface elements.

Imagine that some elves have set up a fine grid on the surface of the potato (with perhaps 1000 surface elements). Into a spreadsheet (see Table 4.1), they enter information about the Gaussian surface, taken to be the outer surface of the potato.

Column A gives the integers i from 1 to 1000. For the i element, the elves measure the area dA_i and determine the three components of the outward normal $\hat{n}_i$. Column B contains the dA_i's, and columns C, D, E contain the three components of the $\hat{n}_i$'s. (So far, the elves have considered only the properties of the surface. These do not change even when the electric field changes.) Now, for each surface element, the elves measure the three components of the electric field $\vec{E}_i$, which are entered in columns F, G, H. The measurements are now complete, and we now can compute the electric flux.

Equation (4.2) gives the flux per unit area for each surface element, and is entered in column I. (In terms of the spreadsheet, the dot product is the sum of the products of columns C and F, D and G, and E and I.) The corresponding flux is the product of the flux per unit area (column I) and the area (column B), and is placed in column J. (In Table 4.1, the entries were made assuming that Example 4.1 corresponds to $i = 1$.) Summing all 1000 entries in column J gives a numerical value for (4.4), the total flux leaving the surface. (If a grid of 1000 points isn't fine enough for an accurate measure, then a finer mesh should be used.

Table 4.1 Spreadsheet calculation of flux

A	B	C	D	E	F	G	H	I	J
i	dA_i	n_{ix}	n_{iy}	n_{iz}	E_{ix}	E_{iy}	E_{iz}	$\vec{E} \cdot \hat{n}$	$\vec{E} \cdot \hat{n} dA$
1	1.0×10^{-6}	0.36	0.8	0.48	2	−3	4	0.24	0.24×10^{-6}

In principle, all numerical integrations should always use at least two meshes, one finer than the other, to be reasonably sure that there isn't an error due to too crude a mesh.) Let's say that the answer, converged to two decimal places, is $\Phi_E = 8.4 \times 10^2$ N-m^2/C. Applying (4.9) to our potato-shaped Gaussian surface, with $\Phi_E = 8.4 \times 10^2$ N-m^2/C, yields that it contains $Q_{enc} = 7.4 \times 10^{-9}$ C, a value characteristic of static electricity.

Gauss's law has let us determine the total, or net, electric charge inside the potato-shaped Gaussian surface, without actually measuring that charge. We measured something else, the electric flux, which by Gauss's law yielded the total charge enclosed. That is something of a miracle. In principle, we could build an extensible surface (like a balloon, or a 1950s comics character called Plastic Man, or the "shape-shifters" of the 1990s television show *Deep Space Nine*) to surround any region, and to determine the electric flux through it. In practice, direct measurements are difficult. As will be discussed, devices using the principle of Faraday's ice pail (with an attached electrometer) remove that difficulty.

4.3.2 *Useful Result for Uniform Flux through Only One Part of the Gaussian Surface*

Consider a Gaussian surface that has been decomposed into partial surfaces. If the flux goes through only one of those parts (a big if), and if the flux is uniform through that part (an even bigger if), then

$$\Phi_E = \frac{d\Phi_E}{dA} A_{flux} = \vec{E} \cdot \hat{n} \, A_{flux}, \qquad \text{(uniform, nonzero flux only through } A_{flux})$$

(4.10)

where A_{flux} is the area of the surface that picks up the flux. There are certain important geometries, which involve either conductors in equilibrium or very symmetrical charge distributions (spherical, cylindrical, and planar), where (4.10) does apply. For these cases, (4.8) and (4.10) combine to yield

$$\frac{d\Phi_E}{dA} = \vec{E} \cdot \hat{n} = \frac{4\pi k Q_{enc}}{A_{flux}}. \qquad \text{(uniform, nonzero flux only through } A_{flux})$$

(4.11)

As a check, note that for a Gaussian surface that is a sphere of radius r concentric with a point charge q, the entire sphere is the part through which the flux is uniform. See Figure 4.3. With $A_{flux} = 4\pi r^2$, (4.11) then yields $\vec{E} \cdot \hat{r} = kq/r^2$, as expected. Equation (4.11) is central to Sections 4.5 and 4.7.

4.4 Computing $\vec{E}$ and Then Using Gauss's Law to Obtain Q_{enc}

Here are some examples of how to compute Φ_E for a closed surface, and then determine the charge enclosed Q_{enc}, by Gauss's law, (4.9).

Example 4.4 **Uniform field and cube**

Consider a Gaussian surface S that encloses a cube of side a, with one corner at the origin, and another corner at (a, a, a). It is in a uniform electric field along the x-axis, so $\vec{E} = E\hat{x}$, where E is a constant. See Figure 4.5. Find the flux through S and the charge enclosed.

Solution: Since $\vec{E}$ is constant, every line that enters S also leaves it, so the net number of lines leaving S is zero. Thus we expect the explicit calculation to yield zero net flux, or $\Phi_E = 0$. Then, by Gauss's law, $Q_{enc} = 0$. Let's now do the actual calculation of Φ_E. The top and bottom of this cube, with $\hat{n} = \pm\hat{z}$, have no flux passing through them because $d\Phi_E/dA = \vec{E} \cdot \hat{n} = 0$ for these cases. There is also no flux through the right and left faces, with $\hat{n} = \pm\hat{y}$. For the back face, where $\hat{n} = -\hat{x}$, $d\Phi_E/dA = \vec{E} \cdot \hat{n} = -E$, it is negative because the field enters the surface. Thus, the flux through the back face is $\Phi_E^{back} = -EA$, where $A = a^2$. For the front face, where $\hat{n} = \hat{x}$, $d\Phi_E/dA = \vec{E} \cdot \hat{n} = E$. It is positive because the field leaves the surface. Thus, the flux through that face

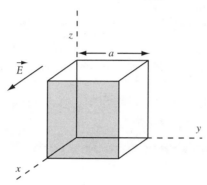

Figure 4.5 Gaussian surface that is a cube.

is $\Phi_E^{front} = EA$. The total flux is $\Phi_E = \Phi_E^{back} + \Phi_E^{front} = -EA + EA = 0$, as expected.

Note: a uniform field can be produced by a sheet of uniform charge density σ, where

$$E = 2\pi k\sigma, \tag{3.26}$$

determines σ. (A field along $+x$ can be produced either by a sheet σ in a constant x-plane with $x < 0$ or by a sheet $-\sigma$ in a constant x-plane with $x > a$. Indeed, there are infinitely many ways to produce a uniform $\vec{E}$ for $0 \le x \le a$.) Since the sheet of charge does not intersect the cube, we expect that $Q_{enc} = 0$. By Gauss's law, (4.8), we then expect that $\Phi_E = 0$, consistent with the discussion above. (If the charged sheet lies in $0 < x < a$, so it intersects S, then the flux is $+EA$ for both front and back faces, for a net flux of $\Phi_E = 2EA$, and $Q_{enc} = (4\pi k)^{-1}\Phi_E = \sigma A$, as expected. We will return to this case.)

Example 4.5 **Uniform field and half-cube**

Now consider that the cube is sliced diagonally, as in Figure 4.6. Find the flux leaving this new Gaussian surface S and the amount of charge enclosed by S when it is in a uniform electric field along the x-axis.

Solution: For this uniform field, every line that enters S also leaves it, so the net number of lines leaving S is zero; again we expect zero net flux, and zero enclosed charge. Let's now do the calculation. As before, there is no flux

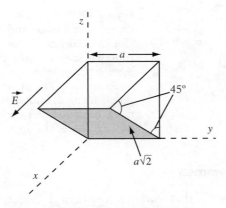

Figure 4.6 Gaussian surface that is half of a cube.

through the top or bottom (there is no bottom), and no flux through the right or left, because the field lines do not pass through any of these surfaces. As before, the flux through the back is $-EA$. For the front, the normal is $\hat{n} = (\hat{x} - \hat{z})/\sqrt{2}$, so $d\Phi_E/dA = \vec{E} \cdot \hat{n} = E/\sqrt{2}$, a decreased flux per unit area. However, for the front face the area has increased to $\sqrt{2}A$, so that $\Phi_E^{front} = (E/\sqrt{2})(\sqrt{2}A) = EA$ does not change. Thus the total flux is $\Phi_E = -EA + EA = 0$, and by (4.9), $Q_{enc} = 0$, as expected.

Example 4.6 **Nonuniform field and box**

Consider a Gaussian surface that encloses a parallelopiped whose front and

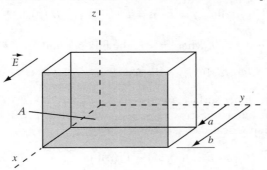

Figure 4.7 Gaussian surface that is a rectangular parallelopiped.

back faces have $x = b$ and $x = a$, and area A. Let the electric field $\vec{E}$ point only along $\hat{x}$, with a component E_x that may depend on x, but not on y or z; $\vec{E} = E_x(x)\hat{x}$. See Figure 4.7. (a) Find the net flux through this Gaussian surface. (b) For $A = 4 \times 10^{-4}$ m^2, $a = 1.5$ m, and $b = 2.5$ m, and $E_x(x) = Cx$, where $C = 2 \times 10^6$ N/C-m, find Φ_E and Q_{enc}.

Solution: (a) Here the field is nonuniform, so more field lines may enter the front than the back, meaning possible net flux and nonzero Q_{enc}. As in the previous examples, there is no flux through the top or bottom, or the right or left. The flux per unit area through the front takes on the uniform value $d\Phi_E/dA = \vec{E} \cdot \hat{n} = \vec{E} \cdot \hat{x} = E_x(b)$; through the back the flux per unit area takes on the different uniform value $d\Phi_E/dA = \vec{E} \cdot \hat{n} = \vec{E} \cdot (-\hat{x}) = -E_x(a)$. The net flux thus is given by

$$\Phi_E = E_x(b)A - E_x(a)A = [E_x(b) - E_x(a)]A. \qquad (4.12)$$

If $E_x(x) > 0$, this corresponds to positive flux for $x = b$ (leaving) and negative flux for $x = a$ (entering). (b) From Gauss's law and (4.12), we can obtain Q_{enc}. Here $E_x(a) = Ca = 3 \times 10^6$ N/C and $E_x(b) = Cb = 5 \times 10^6$ N/C, so by (4.12) $\Phi_E = 8 \times 10^2$ N/C. Hence, by (4.9), $Q_{enc} = 7.07 \times 10^{-9}$ C. If desired, we could also identify charges Q_a and Q_b associated with the fluxes $E_x(a)A$ and $E_x(b)A$ through the front and back faces.

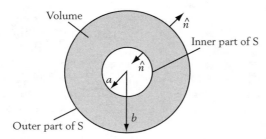

Figure 4.8 Spherically symmetric charge distribution and concentric Gaussian surface S that encloses a spherical shell of finite thickness. Thus S has both an inner surface and an outer surface. The inner (outer) normal points along the outward (inward) radial direction. For each surface, the flux density is uniform.

Example 4.7   Spherical symmetry

Consider a Gaussian surface S that encloses a spherical shell whose inner and outer radii are at $r = a$ and $r = b$, where $a < b$. (Thus S has an inner and an outer surface.) Let there be a nonuniform radial field, so $\vec{E} = \hat{r} E_r(r)$, where $E_r(r)$ means that E_r is a function of r. See Figure 4.8. Find the net flux through S.

Solution: The flux per unit area leaving the inner surface at $r = a$, where $\hat{n} = -\hat{r}$, is

$$(d\Phi_E/dA)|_{inner} = \vec{E} \cdot \hat{n} = \vec{E} \cdot (-\hat{r}) = -E_r(a);$$

for the outer surface at $r = b$, where $\hat{n} = \hat{r}$, the flux per unit area is

$$(d\Phi_E/dA)|_{outer} = \vec{E} \cdot \hat{n} = \vec{E} \cdot \hat{r} = E_r(b).$$

The flux through the inner surface, of area $A_{inner} = 4\pi a^2$, is $-E_r(a)(4\pi a^2)$, and the flux through the outer surface, of area $A_{outer} = 4\pi b^2$, is $E_r(b)(4\pi b^2)$, so the net flux leaving the spherical shell is given by

$$\Phi_E = E_r(b)(4\pi b^2) - E_r(a)(4\pi a^2). \tag{4.13}$$

If $E(a) > 0$, then flux enters at $r = a$, and if $E(b) > 0$, then flux leaves at $r = b$.

Application 4.1   Estimating the charge on the earth!

Assume that a measurement of $|\vec{E}|$ at the surface of the earth ($b = 6.37 \times 10^6$ m) yields $E_r(b) = -130$ N/C, which points toward the earth, indicating that the earth is negatively charged. Taking this field as characteristic of the entire earth and eliminating the inner surface by setting $a = 0$ in (4.13) yields $\Phi_E = -6.63 \times 10^{16}$ N-m²/C. Gauss's law, (4.9), then yields $Q_{earth} = -5.86 \times 10^5$ C. This is quite a lot of charge. It corresponds to a charge per unit area $\sigma_{earth} = Q_{earth}/4\pi b^2 = -1.15 \times 10^{-9}$ C/m². Overall, the earth and its atmosphere are neutral; an equal and opposite amount of positive charge resides in the atmosphere.

Example 4.8   Cylindrical symmetry

Consider a cylindrical geometry with a nonuniform radial field, so $\vec{E} = \hat{r} E_r(r)$, and employ a concentric Gaussian surface of length L that is a cylindrical

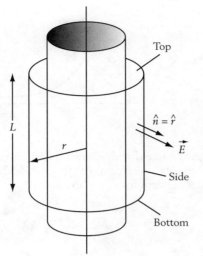

Figure 4.9 Cylindrically symmetric charge distribution and concentric Gaussian surface that is a finite cylinder. Flux passes only through the round side, where the flux density is uniform.

shell of radius r. (Sometimes the symbol $\rho = \sqrt{x^2 + y^2}$ is used for the radius in cylindrical coordinates, but we have already used ρ for the charge per unit volume. There just aren't enough symbols to go around!) The full surface has three parts: the top, bottom, and round parts. See Figure 4.9. (a) Find the net flux through this Gaussian surface. (b) If $L = .2$ m, $r = .02$ m, and $E_r(r) = 4.00 \times 10^4$ N/C, find Q_{enc}.

Solution: (a) The fluxes for the top and bottom are zero because the outward normals are along z; this is perpendicular to $\vec{E}$, which is radial. The flux per unit area leaving the side at r, where $\hat{n} = \hat{r}$, takes on the uniform value $d\Phi_E/dA = \vec{E} \cdot \hat{n} = \vec{E} \cdot \hat{r} = E_r(r)$. The flux through the round surface, of area $A = 2\pi r L$, is $E_r(r)(2\pi r L)$, so the net flux is given by

$$\Phi_E = E_r(r)(2\pi r L). \tag{4.14}$$

(b) For our specific values of L, r, and $E_r(r)$, (4.14) gives $\Phi_E = 1.005 \times 10^3$ N-m²/C. From Gauss's law, (4.9), we then deduce that $Q_{enc} = 8.89 \times 10^{-9}$ C. This positive sign is consistent with the fact that the field points outward.

4.5 Determining $\vec{E}$ by Symmetry: The Three Cases

Charge distributions of three and only three types produce a flux that is uniform or zero for all parts of an appropriately chosen Gaussian surface: centrosymmetric charge distributions with spherical, cylindrical, and planar symmetry. Gauss's law and a knowledge of the charge then enables us to deduce the *magnitude* of the electric field; we simply apply (4.11). (A cube that is concentric with a point charge has some symmetry; each of its faces picks up the same flux, which is one-sixth of the total flux. However, the $\vec{E}$ field is not uniform over each face, so that we can only obtain the average flux for each face, rather than the electric field itself.)

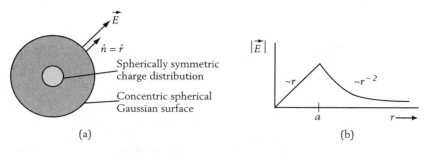

Figure 4.10 (a) Spherically symmetric charge distribution and concentric Gaussian surface for which the flux density is uniform. (b) Electric field magnitude when the charge density is uniform for $r < a$ and zero for $r > a$.

4.5.1 *Spherical Symmetry*

Consider a spherical distribution of charge, which produces an electric field that points radially. Take as Gaussian surface S, a sphere of radius r that is concentric with the charge distribution. See Figure 4.10(a). In general, the charge distribution may extend beyond r.

Here $d\Phi_E/dA$ is uniform over S, and the area that picks up the flux is $A_{flux} = 4\pi r^2$. Thus, by (4.11), with $\vec{E} \cdot \hat{n} = \vec{E} \cdot \hat{r} = E_r$,

$$E_r = \frac{4\pi k Q_{enc}}{4\pi r^2} = \frac{k Q_{enc}}{r^2}. \qquad \text{(spherically symmetric charge distribution)}$$

(4.15)

This looks like Coulomb's law for a point charge, but it applies to *any* spherically symmetric charge distribution. It is deceptively simple. Equation (4.15) says that (1) if charge is within the Gaussian surface of radius r, it contributes as if it were at the center, and (2) if charge is outside the surface, it does not contribute at all. This result is relatively difficult to obtain by direct calculation from $\vec{E} = k \int (dq/r^2)\hat{r}$. Equation (4.15) reproduces the result for a single point charge q, $E = kq/r^2$.

Gauss's Law for Gravitation

Since gravity satisfies an inverse square law, there is a Gauss's law for gravitation, which would have saved Newton a great deal of effort. It was not easy, even for the great Newton, to directly calculate the gravitational field due to a ball of uniform mass density. At a great distance, it is not too bad an approximation to consider the earth to be a point, but for a satellite in near earth orbit, the earth is not a point. By the gravitational analog of Gauss's law, the earth's gravity can indeed be treated as if it is due to all the earth's mass concentrated at the geometrical center of the earth.

Example 4.9 Uniform ball of charge

Consider a spherical ball of radius a and total charge Q that is uniformly distributed over its volume (the next section shows that this cannot be a conductor in equilibrium). Find E_r for all r.

Solution: A concentric spherical Gaussian surface S with $r > a$ has $Q_{enc} = Q$. By (4.15), this implies $E_r = kq/r^2$. (This is the large r part of Figure 4.10b.) For $r < a$, there is a charge per unit volume

$$\rho = \frac{Q}{V_{ball}} = \frac{Q}{\frac{4}{3}\pi a^3}.$$
(4.16)

Then for $r < a$,

$$Q_{enc} = \int dQ = \int \rho dV = \rho \int dV = \rho V = \frac{Q}{\frac{4}{3}\pi a^3}\frac{4}{3}\pi r^3 = Q\left(\frac{r}{a}\right)^3.$$
(4.17)

Placing Q_{enc} in (4.15) then gives $E_r = kQ(r/a)^3/r^2 = kQr/a^3$. This falls to zero at the origin, as expected by symmetry; it also takes the value kQ/a^2 at $r = a$, as expected because then all the charge is enclosed. Of course, a conductor in equilibrium cannot produce such a volume charge distribution; a conductor in equilibrium would have all the charge Q on the outer surface at $r = a$. Figure 4.10(b) plots $|\vec{E}| = |E_r|$ for all r. We can also obtain (4.17) by the ratio of the appropriate volumes: $Q_{enc}/Q = (4\pi\rho r^3/3)/(4\pi\rho a^3/3) = r^3/a^3$.

4.5.2 *Cylindrical Symmetry*

Consider a cylindrically symmetric distribution of charge, infinite in extent. By symmetry, it produces an electric field that points radially. See Figure 4.11(a). As the Gaussian surface S, take a cylinder concentric with the charge distribution, of radius r and finite length L. (E_r should not depend on L.) Here, $d\Phi_E/dA$ is zero over the top and bottom, and is uniform over the cylindrical surface, so the area that picks up the flux is $A_{flux} = 2\pi r L$. Because Q_{enc} is proportional to L,

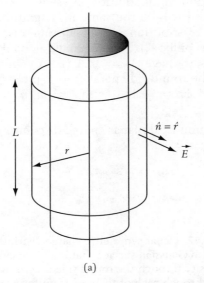
(a)

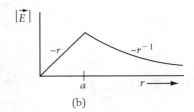
(b)

Figure 4.11 (a) Cylindrically symmetric charge distribution and concentric Gaussian surface that is a finite cylinder. The flux density through the round side is uniform. (b) Electric field magnitude when the charge density is uniform for $r < a$ and zero for $r > a$.

$\lambda_{enc} \equiv Q_{enc}/L$ is independent of L. From (4.11), with $\vec{E} \cdot \hat{n} = \vec{E} \cdot \hat{r} = E_r$,

$$E_r = \frac{4\pi k Q_{enc}}{2\pi r L} = \frac{2k\lambda_{enc}}{r}. \qquad \text{(cylindrical charge distribution)} \quad (4.18)$$

Equation (4.18) implies that, if charge is within the Gaussian surface of radius r, it contributes as if it were on the axis, and if it is outside the surface, it does not contribute at all. For a line charge λ, (4.18) reproduces (3.22), with $\lambda_{enc} = \lambda$: that is, $E_r = 2k\lambda/r$.

Example 4.10 Uniform cylinder of charge

Let the charge per unit volume ρ be a constant for $r < a$, and zero otherwise. Find E_r for all r.

Solution: A concentric cylinder of radius r and length L contains charge $Q_{enc} = \rho\pi r^2 L$, so $\lambda_{enc} = Q_{enc}/L = \rho\pi r^2$. Equation (4.18) then gives $E_r = 2\pi k\rho r$ for $r < a$. As a check, note that, as expected by symmetry, the field is zero at the origin ($r = 0$). For $r > a$, we have $\lambda_{enc} = \rho\pi a^2$. Equation (4.18) then gives $E_r = 2\pi k\rho a^2/r$ for $r > a$. As a check, note that outside the charge distribution the charge behaves like a line charge on the axis. Figure 4.11(b) plots $|\vec{E}| = |E_r|$ for all r.

4.5.3 *Planar Symmetry*

For a planar distribution of charge that is uniform in the yz-plane, the electric field must point along the x-direction. If, in addition, there is a center of symmetry at the origin, the electric field will be the same in magnitude at both x and $-x$. We now enclose the charge distribution with a Gaussian surface S shaped like a tiny right-circular cylinder (a "pillbox"). With normals along $\pm\hat{x}$, and area A for each of the right and left sides, we have $A_{flux} = 2A$. (For the pillbox, which can be made as flat as a pancake, the round side parallel to $\vec{E}$ picks up no flux.) See Figure 4.12. Let $\sigma_{enc} \equiv Q_{enc}/A$. From (4.11), with $E = \vec{E} \cdot \hat{n}$,

$$E = \frac{2\pi k Q_{enc}}{A} = 2\pi k\sigma_{enc}. \qquad \text{(symmetrical planar charge distribution)} \quad (4.19)$$

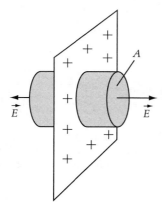

Figure 4.12 Planar symmetric charge distribution and concentric Gaussian surface that is a finite cylinder. The flux density through the round side is zero, and the flux density through each of the caps is uniform.

For a sheet of charge density σ, (4.19) reproduces (3.26) with $\sigma_{enc} = \sigma$; that is, $E = 2\pi k\sigma$.

Example 4.11   **Uniform slab of charge**

Let the charge per unit volume be given by ρ for $-a < x < a$, and zero otherwise. Find E_r for all r.

Solution: Consider a Gaussian surface S that is a symmetric pillbox of area A that extends from $-x$ to $+x$. It contains charge $Q_{enc} = \rho A(2x)$, so $\sigma_{enc} = Q_{enc}/A = \rho(2x)$. Placed in (4.19), this gives $E = 4\pi k\rho x$ for $-a < x < a$. As a check, note that, as expected by symmetry, the field is zero at the origin ($x = 0$). For $x > a$, we have $\sigma_{enc} = \rho(2a)$. Placed in (4.19), this gives $E = 4\pi k\rho a$ for $x > a$. As a check, note that outside the charge distribution the charge behaves like a sheet charge on the axis. More generally, for planar symmetry, if there is a region where $\vec{E}$ varies linearly with distance, that region contains a uniform charge density ρ. For example, if $E_x = cx + d$, then $c = 4\pi k\rho$ relates c and ρ. The constant d could be due to distant sheets or distant slabs of charge. An indication of a uniform charge density is that the slope of the field versus distance is a constant; here $\partial_x E_x = c = 4\pi k\rho$ is constant. An approximately uniform planar charge density can be produced by bombarding a plastic slab with ions.

4.6 Electrical Conductors in Equilibrium

Consider an electrical conductor of arbitrary shape (e.g., a potato, an aluminum-painted rubber duck, or the object in Figure 4.13).

In isolation, the conductor contains a large number (a "sea") of mobile, negatively charged conduction electrons (the "electric fluid"), and a fixed background of positively charged nuclei, with overall neutrality. How does that electric fluid respond when we bring a charged rod near this conductor? When we add some electric charge to the conductor? If the charge is external to the conductor, then (as discussed in Chapter 1) it causes electrostatic induction of the electric fluid (a polarization of the conductor); if the additional charge is placed on the conductor itself (perhaps as additional electrons, or as positively charged ions that stick to the surface), the electric fluid redistributes itself throughout the conductor.

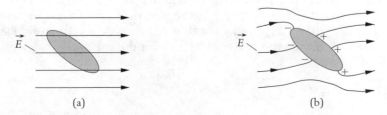

(a)   (b)

Figure 4.13 Conductor in a uniform field. (a) The mobile charge on the conductor not permitted to move (no electrostatic induction). (b) The mobile charge on the conductor permitted to move (full electrostatic induction). The field lines enter negative charge and leave positive charge.

> ### Charge Redistribution on a Conductor is Collective
>
> Recall that the mobile electrons are in orbitals that extend over the entire conductor. When the electric fluid redistributes itself throughout the conductor, it is *not via individual mobile electrons*. Rather, all the mobile electrons on the conductor adjust their orbitals slightly, *collectively* redistributing their total charge in the same way as would the classical electric fluid.

An example of charge redistribution can be seen by comparing Figure 4.13(a) to Figure 4.13(b). Figure 4.13(a) shows a conductor in a uniform electric field *before* the conductor has had time to respond. Figure 4.13(b) shows the conductor and the total electric field *after* the conductor has come to equilibrium. It demonstrates that in equilibrium (a) there are no field lines within the conductor ($\vec{E} = \vec{0}$); (b) excess charge resides on the conductor's surface (no charge lies within the material of the conductor); and (c) just outside the conductor the electric field is normal to the surface. In the remainder of this section, we will derive each of these results, and then discuss what happens when the conductor has a cavity.

4.6.1 *Within Conductors in Equilibrium, $\vec{E} = \vec{0}$*

We now establish that, for an electrical conductor in equilibrium, the electric field is zero within the material of the conductor. This result depends only upon two properties of electrical conductors:

1. In equilibrium, there is no electric current flowing anywhere within an electrical conductor. In contrast, the filament of a flashlight bulb connected to a battery carries an electric current and is *not* in equilibrium. However, after the power has been turned off and the filament has cooled down to room temperature, it *is* in equilibrium. No electric current then flows through it.

2. In an electrical conductor, such as a wire, if there is an electric field, then the charge carriers feel an electrical force, which drives an electric current. Since in equilibrium there is no electric current, $\vec{E} = \vec{0}$ in equilibrium. (In fact, the current is proportional to the electric field. This is a version of Ohm's law.)

Consider a conductor, such as that in Figure 4.13. Let external charge q_{ext} produce a field $\vec{E}_{ext}$, and let electric charge q_{cond} distributed over the conductor itself produce a field $\vec{E}_{cond}$. Then the total field $\vec{E} = \vec{E}_{ext} + \vec{E}_{cond}$. If the conductor is in equilibrium, then by property 1, the electric current must be zero. Further, by property 2, within the conductor (but not outside it), the total electric field $\vec{E} = \vec{0}$. Hence, within the material of the conductor itself, $\vec{E}_{cond}$ must cancel $\vec{E}_{ext}$. That is,

$$\vec{0} = \vec{E} = \vec{E}_{ext} + \vec{E}_{cond}, \text{ so } \vec{E}_{cond} = -\vec{E}_{ext}.$$

(field inside conductor in equilibrium) (4.20)

The field inside the conductor is zero because the field of the conductor *cancels* the external field, *not* because the conductor somehow prevents the external field from entering ("screening"). The result of this subsection is true regardless of the electrical force law. Even if the force law were a radially outward inverse cube (for which Gauss's law does not apply), the electric field would be zero within the material of a conductor in equilibrium.

One might expect the charge associated with the electric fluid to be distributed throughout the conductor. The next section shows that, because of Gauss's law (which follows from Coulomb's law), the charge associated with the electric fluid resides only on the surfaces of the conductor.

4.6.2 *For a Conductor in Equilibrium, Any Net Charge Can Reside Only on Its Surfaces*

Consider a conductor with outer surface S_{outer} (such as in Figure 4.13b) and, if it is hollow, with inner surface S_{inner}. Draw an arbitrary Gaussian surface S that totally passes through only conducting material. Because $\vec{E} = \vec{0}$ within the conductor, the electric flux density $d\Phi_E/dA = \vec{E} \cdot \hat{n}$ is zero for any part of S, so the total flux through S is zero. By Gauss's law, the electric charge within the volume enclosed by S must also be zero. By breaking up the conductor into tiny volumes, we can piecewise eliminate the possibility that there is electric charge in any of those volumes. Thus, *for an electrical conductor in equilibrium, net electric charge can reside only on its surfaces.* This result depends crucially on Gauss's law. The electric field due to the electric fluid arises only from surface charge. In general, this surface charge distribution is complex and difficult to determine.

Concept Quiz 4.1

If an object has a nonzero volume charge distribution (as in Example 4.10) and it is in equilibrium, can the object be a conductor?

Solution: No. It must be an insulator. For example, charged ions might be held in place within an insulator by a nonelectrostatic interaction with the molecules of the object.

Concept Quiz 4.2

If an object has a nonzero volume charge distribution and it is *not* in equilibrium, must the object be a conductor?

Solution: No. It also could be an insulator. For example, charged ions within the insulator might be in the process of rearranging.

Concept Quiz 4.3

If an object has a nonzero surface charge distribution, must the object be a conductor in equilibrium?

Solution: No. The object could be either an insulator or a conductor, and it could be either in or out of equilibrium. For example, a comb rubbed through your hair has a surface charge distribution, but the comb is an insulator. (Moreover, the distribution may or may not be in equilibrium.) Or, a piece of aluminum foil could have a surface charge distribution due to a sudden discharge, but not enough time has passed for the foil to reach equilibrium.

4.6.3 *Just Outside a Uniform Conductor $\vec{E}$ Is Normal to the Surface*

Consider an electrical conductor that is uniform (or locally uniform) along its surface. (Here, *uniform* means that the atomic order is uniform.) If the electric field had a component that is parallel to the surface, an electric current would flow along the surface, contrary to the assumption of equilibrium. Hence, in equilibrium, the electric field is normal to the surface. See Figure 4.14.

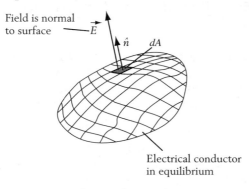

Field is normal to surface — $\vec{E}$

$\hat{n}$ dA

Electrical conductor in equilibrium

Figure 4.14 Electric flux Φ_E leaving an electrical conductor in equilibrium, with the surface broken into many surface elements. The field lines are normal to the surface.

The result that $\vec{E}$ is normal to the surface can also be obtained from Faraday's viewpoint. Imagine freezing in place the mobile charge (the electric fluid) on a neutral conducting object, and then placing it in an external field. See Figure 4.13(a). With the charge on the conductor frozen in place, these field lines, due only to the external field, pass through the conductor unaffected. According to Faraday's view, on unfreezing the mobile charge, the field lines, to whose ends the mobile charge is attached, shorten because of field-line tension (subject to the constraint that the lines not get too dense, because of field-line pressure). The motion stops when the lines are normal to the conductor and have the appropriate separation. See Figure 4.13(b), where three lines terminate on negative charge and three lines originate on positive charge, the conductor having zero net charge. Because of charge rearrangement on the conductor, the *total* electric field is normal to the surface. In Figure 4.13(b), the field due to the external charge is the same as in Figure 4.13(a); the difference in the two cases is solely due to charge rearrangement on the conductor.

Freezing and Unfreezing the Charge Distribution

The idea of freezing and unfreezing the charge distribution isn't theoretical. Certain types of paper can take many seconds to fully respond. For short times only the polarization of localized electrons is noticed, but for longer times ions can move, to give a response more like that of a metal.

4.6.4 $\vec{E} = \vec{0}$ *within a Conductor's Empty Cavity*

Consider an electrical conductor with outer surface S_{outer} that is solid except for a cavity with surface S_{inner}, as in Figure 4.15. Consider what happens if there is external charge or charge on the conductor, but no charge in that cavity.

To study this case, we use an almost trivial but very important result that may be called *the scooping-out theorem*, which states that removal of material from a conductor in equilibrium does not disturb the electric field anywhere.

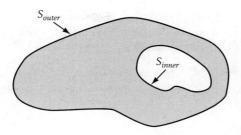

Figure 4.15 An electrical conductor with a cavity.

The proof of the scooping-out theorem is straightforward. There are two ways that scooping-out might produce an electric field: (1) Scooping-out might leave behind volume charge in the interior, which would produce an electric field. However, the interior of the conductor is neutral, so that scooping out the cavity does not remove volume electric charge. (2) If the removed material were polarized, scooping-out might leave behind surface charge, which would produce an electric field. However, the interior of a conductor can only have an induced polarization, proportional to $\vec{E}$. Since $\vec{E} = \vec{0}$, the interior is not polarized, and therefore scooping out the cavity does not leave behind surface charge.

Within the cavity of the conductor in Figure 4.15, since the electric field was zero before there was a cavity, the electric field is zero after scooping out the cavity. There is no charge anywhere on the inner surface of the conductor, since there was no charge there before the cavity was scooped out.

4.7 Just Outside a Uniform Conductor, $\vec{E}_{out} \cdot \hat{n}$ Varies as the Local Surface Charge Density σ_S

Consider a conductor in equilibrium, as in Figure 4.13(b). Since $\vec{E} = \vec{0}$ inside a conductor, all field lines and electric flux entering or leaving part of a conductor can be attributed only to the charge on that part of the conductor. This permits us to determine the surface charge density σ_S.

Very close to an electrical conductor, the surface appears to be flat. (We neglect the atomic nature of matter.) Draw a Gaussian surface S that is a small pillbox of area A enclosing a piece of the surface, oriented with its flat outer face parallel to the surface. See Figure 4.16(a). The electric flux Φ_E comes only from the flat outer face because $\vec{E} = \vec{0}$ inside and because the thin round part of the pillbox may be taken to have negligible area. For a pillbox so small that there is a uniform nonzero flux through only one part of the surface, (4.11) applies with

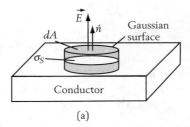

(a)

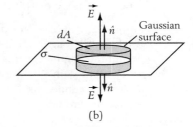

(b)

Figure 4.16 Locally planar charge distributions, and associated pillbox-shaped Gaussian surfaces. (a) At surface of a conductor, only the outside circular face picks up flux. (b) Infinite sheet of charge, the circular surfaces pick up equal amounts of flux.

$A_{flux} = A$. Here $Q_{enc} = \sigma_S A = \sigma_S A_{flux}$, where σ_S is the surface charge density. With $\vec{E}_{out}$ the electric field just outside the conductor, and $\hat{n}$ the outward normal from the conductor (and from the outside of the pillbox), (4.11) yields

$$\vec{E}_{out} \cdot \hat{n} = 4\pi k \sigma_S. \qquad \text{(field just outside any conductor)} \qquad (4.21)$$

This can be rewritten as

$$\sigma_S = \frac{\vec{E}_{out} \cdot \hat{n}}{4\pi k}. \qquad (4.22)$$

If $\vec{E}$ points away from the surface, then σ_S is positive; if $\vec{E}$ points toward the surface, then σ_S is negative. These are as expected.

By measuring $\vec{E}_{out}$, one can deduce σ_S. For example, if $|\vec{E}| = 250$ N/C at P, just outside part of a surface whose local charge density is negative, then by (4.22) $\sigma_S = -2.21 \times 10^{-9}$ C/m^2 near P, and $\vec{E}$ points into the surface. Note the field lines entering the conductor in Figure 4.13(b).

Proof Plane

Coulomb developed a device called a *proof plane*, a small thin conducting disk with an insulating handle, to determine σ_S directly. On part of the conducting surface he would place the proof plane and then lift it off. The amount of charge on the proof plane was proportional to σ_S so that he could obtain relative charge densities. By suitable calibration, absolute measurements can be obtained with a proof plane.

For $\sigma_S = \sigma$, (4.21) for the exterior of a conductor has an extra factor of two relative to the case of the electric field $E_{sheet} = 2\pi k\sigma$ produced by an isolated sheet of charge density σ. This factor of two occurs because, for the conductor, *all* the electric flux leaves one side (the exterior) of the conductor pillbox, whereas for an isolated sheet of charge, only *half* the electric flux leaves each side of the corresponding pillbox, as in Figure 4.16(b). Thus, for the conductor, all the flux gets concentrated on one side. Equivalently, in terms of (4.11), A_{flux} is half as large when applied to a conductor, so the flux is twice as large.

Example 4.12 An infinite conducting sheet

An infinite conducting sheet initially has charge per unit area σ on only one surface. (a) Find the electric field. (b) The system now comes to equilibrium. Find the charge distribution and the electric field outside the conductor.

Solution: (a) On each side of the charged surface (even inside the conductor), the field has magnitude $2\pi k\sigma$ and points away from the surface. Because $\vec{E} \neq 0$ within the conductor, it is not in equilibrium. (b) The charge rearranges so that, by symmetry, in equilibrium each surface has $\sigma_S = \sigma/2$. This gives zero field inside the conductor. By (4.22), the field outside each surface of the conductor has magnitude $4\pi k\sigma_S = 4\pi k(\sigma/2) = 2\pi k\sigma$, the same as before the charge rearranged.

Equations (4.21) and (4.22) hold just outside any conductor in equilibrium. By integrating over (4.22), we obtain the result that the total charge on the surface is

$$Q_S = \int dQ_S = \int \sigma_S dA = \int \frac{\vec{E}_{out} \cdot \hat{n}}{4\pi k} dA = \frac{\Phi_E}{4\pi k} = Q_{enc}, \qquad (4.23)$$

as expected. That is, Q_{enc}, the total charge enclosed by a Gaussian surface enclosing the conductor, equals the total charge Q_S on the outer surface of the conductor, even in the presence of cavities.

Example 4.13 Two parallel charged conducting sheets

Consider two infinite parallel conducting sheets of net charge/area (top and bottom of each sheet) $\sigma_1 > \sigma_2 > 0$, with $\sigma_1 = (3/2)\sigma_2$. Find the equilibrium charge densities on the inside and outside of each sheet. Figure 4.17 presents sheets of finite size; think of these as infinite, so that there are no edges.

Solution: For the planar geometry, unlike the geometry of Figure 4.13, the rearrangement of charge on a conductor has no effect on the electric field outside that conductor. However, as in Figure 4.13 and Example 4.12, the rearrangement of charge does cause the field to be zero within each conductor. In equilibrium, there is charge only on the surfaces of the conducting sheets. Since the field for a sheet doesn't fall off with distance, we can compute the field as the sum over the total charge on each sheet. Superposition applied to (3.26), or $|\vec{E}_{sheet}| = 2\pi k|\sigma|$, yields the fields outside and between the sheets. See Figure 4.13. Including direction, above both sheets

$$\vec{E}_{out} = 2\pi k(\sigma_1 + \sigma_2)\hat{\uparrow} = 5\pi k \sigma_2 \hat{\uparrow}.$$

With $\hat{n} = \hat{\uparrow}$ just above sheet 1, (4.22) then yields $\sigma_1^{top} = \frac{1}{2}(\sigma_1 + \sigma_2) = \frac{5}{4}\sigma_2$. Including direction, between both sheets

$$\vec{E}_{between} = 2\pi k(\sigma_1 - \sigma_2)\hat{\downarrow} = \pi k \sigma_2 \hat{\downarrow}.$$

With $\hat{n} = \hat{\downarrow}$ just below sheet 1, (4.22) then yields $\sigma_1^{bottom} = \frac{1}{4}\sigma_2$. The total charge

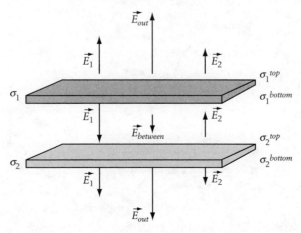

Figure 4.17 Two infinite sheets of charge, and the electric field in the regions above, between, and below them.

density on sheet 1 is $\sigma_1^{top} + \sigma_1^{bottom} = \frac{3}{2}\sigma_2 = \sigma_1$, thus satisfying charge conservation. A similar analysis can be made for sheet 2. The charges on the inner surfaces are equal and opposite ($\sigma_2^{top} = -\sigma_1^{bottom}$) so, as expected, the field lines start on positive charge and end on equal negative charge. The charges on the outer surfaces are equal ($\sigma_1^{top} = \sigma_2^{bottom}$) and produce field lines that extend to infinity so that each outer surface produces the same flux.

4.8   Charge Measurement and Faraday's Ice Pail Experiment

Consider a charge electrometer as discussed in Section 2.4.3, attached to the exterior of a conductor into which objects can be placed. The conductor itself is placed on an insulated stand to prevent the escape of electric charge. With this electrometer we can determine the charge on an object, without tearing apart that object and without even measuring the electric flux passing through the surface of that object. See Figure 4.18. (We assume that the simple electroscope in Figure 4.18 has been calibrated, thereby making it into an electrometer.) The electrometer gives a response that is directly proportional to the local charge density on the outside of the conductor: either it measures the electric field, which by (4.22) is proportional to the surface charge density, or it measures the surface charge density directly.

4.8.1   *Noncontact Experiments*

Faraday hung an electrically charged object A, with charge $+Q$, by an insulating string. On lowering A into the uncharged ice pail, without A touching the sides, the electrometer reading increased, but once A was about 4 inches below the top of the ice pail, the electrometer reading stabilized, even when A was moved about. When A was lifted back out of the ice pail, the electrometer reading

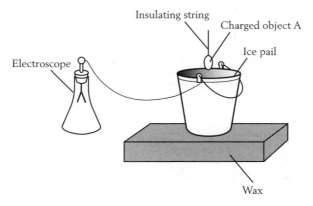

Figure 4.18 Faraday ice pail experiment. The electroscope responds to charge that goes to the outer surface of the ice pail.

returned to zero. Moreover, when A was within the ice pail, the electrometer reading was the same as if all the charge on the object had been placed on the ice pail. This indicated that (1) if the charged object A is far enough within the ice pail, then a charge distribution goes to the outer surface that is the same as if $+Q$ had been placed directly on the ice pail; (2) the charge $+Q$ on A and the charge $-Q$ remaining on the inner surface together produce electric fields that have no effect on the charge distribution on the outer surface of the ice pail. Recall that (4.22) relates the electric field at the surface and the surface charge density.

4.8.2 *Contact Experiments*

On lowering a charged *conducting* object A into the ice pail, the electrometer response increased, as before. Touching A to the inside of the ice pail, temporarily making an ice pail–object combination, yielded no change in the electrometer reading. However, lifting out A, the electrometer reading remained at the *same* value as when A was within the ice pail. Moreover, placing A within another ice pail electrometer gave no response. This indicated that, while A and ice pail were in contact (1) there was no charge on the inner surface of the ice pail–object combination; (2) there was no charge in the volume associated with A, thought of as part of the ice pail–object combination. Hence, touching the charged conducting object to the interior of the ice pail made A transfer its charge to the ice pail.

   Although A originally was attracted to both the interior and the exterior of the neutral ice pail, after contact with the interior it was attracted only to the exterior of the now charged ice pail. (This was noted earlier by Franklin.) This behavior after contact with the interior can be explained using electrostatic induction if now A is uncharged, and the ice pail is charged on the outside but not on the inside.

4.8.3 *Interpretation in Terms of Field Lines*

Once A is far enough inside the ice pail, all the field lines produced by $+Q$ terminate on the inner surface so that a net surface charge $-Q$ has been attracted (by electrostatic induction) to the inner surface. Since the ice pail itself is a conductor in equilibrium, it has zero volume charge density. By charge conservation, the charge $-Q$ on the inner surface must have been attracted from the outer surface, which therefore has a charge $+Q$. Moving $+Q$ around inside the cavity may change the field lines and the distribution of charge density on the inner surface, but it does not change the value $-Q$ of the total surface charge. Moreover, moving $+Q$ around inside the cavity does not change the charge density on the outer surface, nor its integrated value of Q, nor the exterior field lines.

> **Home Experiment 4.1**   **Reproducing Faraday's results**
>
> Many of Faraday's results can be reproduced using only (1) a charged plastic comb as the charge source A; (2) a small food can (or soft drink can) with its top removed, having insulating handles of folded-over sticky tape; (3) a large food can with its top removed (which serves as the ice pail); and (4) a

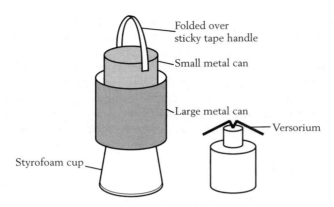

Figure 4.19 A modern, home version of the Faraday ice pail experiment.

plastic-bag-twister versorium as the electrometer. See Figure 4.19. The larger can should be mounted on an insulating surface, such as a styrofoam cup. The versorium should be placed near an edge of the large can, where the electric field is largest. The comb can be charged and used as an insulating charge source within the can. Placing the charged-up comb within the small can, and then touching the small can charges the small can, which then can be used as a conducting charge source within the large can. The small can and the comb are oppositely charged, as shown by the lack of versorium response when both are within the larger can.

By determining how much charge goes to the interior of a hollow conductor, and comparing to the predictions of non-Coulomb's law–based theories, we can set a limit on how well Coulomb's law (and thus Gauss's law) is satisfied. This was the basis of Cavendish's experiments. Measurements made in 1970 show that, if Coulomb's law is assumed to vary as $r^{-(2+\delta)}$, then $\delta < 1.0 \times 10^{-16}$.

4.9 Proof of Gauss's Law

We now prove Gauss's law by using the concept of solid angle. This is perhaps most readily approached from the viewpoint of astronomy. If we look up at the sky and imagine that it is a great sphere of very large radius R, then by definition the solid angle Ω taken up by, say, the constellation Orion, is given by the area A of Orion, projected onto this great sphere, divided by R^2. More generally, for a small projected area $dA_\perp$ perpendicular to the line of sight, we have a small solid angle

$$d\Omega = \frac{dA_\perp}{r^2} = \frac{\hat{R} \cdot \hat{n} \, dA}{R^2}. \tag{4.24}$$

Here $\hat{R}$ is the direction in which the observer is looking, and $\hat{n}$ is the normal to the small area dA. Note that $\hat{R} \cdot \hat{n}$, and thus $d\Omega$, can be either positive or negative. We now develop some properties of the solid angle that are needed for the proof.

For the entire great sphere the area is $4\pi R^2$, so the total solid angle of a sphere is $\Omega = (4\pi R^2)/R^2 = 4\pi$, independent of the (unknown) radius R. Moreover, if we move or deform the great sphere, so long as it contains the observer within it, the total solid angle will remain 4π. Further, if we consider the moon, its total solid angle, front and back, is zero, because the positive solid angle of any part on the bright side is canceled out by the negative solid angle of the corresponding projection onto the dark side. Finally, if we move or deform the moon, the total solid angle will remain zero.

Here is the proof of Gauss's law. Consider an arbitrary Gaussian surface. Let there be a point charge q at the origin, so $\Phi_E = \oint \vec{E} \cdot \hat{n} dA = \oint (kq\hat{R}/R^2) \cdot \hat{n} dA = kq \oint d\Omega = kq\Omega$. That is, $\Phi_E = kq\Omega$, where the solid angle Ω enclosed by the Gaussian surface is measured relative to the position of the charge q. From the discussion of the great sphere and of the moon, either $\Omega = 4\pi$ (where $Q_{enc} = q$) or $\Omega = 0$ (where $Q_{enc} = 0$), so $\Phi_E = 4\pi k Q_{enc}$. By superposition, this result can be established for as many charges as needed, so it is true in general. This establishes Gauss's law.

When charges move, the electric field is no longer given by Coulomb's law. Nevertheless, Gauss's law continues to hold. That field lines are produced by electric charge is a more generally valid idea than that there is action at a distance with an inverse square law.

4.10 Conductors with Cavities: Electrical Screening

Optional

We now extend our earlier discussion of conductors. For our present purposes, we may consider the Faraday ice pail to be equivalent to a closed conductor with a cavity; the large opening of the ice pail serves only to permit us to bring objects in and out.

4.10.1 *Conductors with Cavities*

Consider an electrical conductor with outer surface S_{outer} that is solid except for a cavity with surface S_{inner}, as in Figure 4.15. In the most general case, there is charge *in, on, and outside* the conductor. We will be able to treat this general case by using the principle of superposition applied to three cases:

1. There is a net charge Q_{on} on the material of the conductor. This case was already considered in Section 4.6. Its distribution depends only on the shape of S_{outer} and the amount of Q_{on}.

2. There is a charge Q_{out} outside the conductor, and the conductor has net charge $Q_{on} = -Q_{out}$. In this case, the charge Q_{on} is only on the outer surface, with a distribution that depends on the position and amount of Q_{out}, and on the shape of S_{outer}.

3. There is a charge Q_{in} inside the cavity, and the conductor has net charge $Q_{on} = -Q_{in}$. In this case, the charge Q_{on} is only on the inner surface, with a distribution that depends on the position and amount of Q_{in}, and on the shape of S_{outer}.

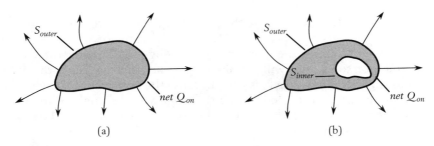

Figure 4.20 An isolated charged conductor in equilibrium, with field lines.
(a) Without any internal cavities. (b) With an internal cavity.

Each of these cases has an electric field that is zero in the bulk and is normal to the surfaces. Therefore we can superimpose linear combinations of these three cases because superposition will not cause any disturbance either in the bulk or on either surface. We will use these cases to analyze the *Faraday cage* and the *Faraday ice pail*.

We now apply the scooping-out theorem to these three cases.

1. **Conductor with net charge Q_{on}**
   First consider an uncharged, solid conductor with (outer) surface S_{outer} that is given a net charge Q_{on}. See Figure 4.20(a).

   Previous considerations show that Q_{on} must be distributed over the surface S_{outer}. Since the interior of the conductor is neutral, by the scooping-out theorem we can scoop out from the conductor a cavity with surface S_{inner} without affecting the electric field anywhere. See Figure 4.20(b). Hence, after scooping out neutral material from the conductor, the electric field in the scooped-out region (the cavity) remains zero. (This argument repeats Section 4.6.3.) Thus the electric field due to Q_{on} has been screened out. Moreover, the charge distribution on the outer surface is unaffected. If Q_{out} is doubled, the charge distribution doubles everywhere on the surface, and the field lines have the same pattern, but are twice as dense.

2. **External charge Q_{out} and conductor with charge $Q_{on} = -Q_{out}$**
   First consider a charge Q_{out} external to a solid conductor with surface S_{outer} and a charge $Q_{on} = -Q_{out}$. See Figure 4.21(a).

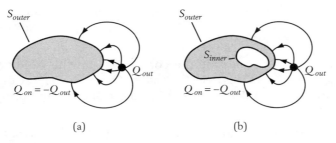

Figure 4.21 A charged conductor near an equal and opposite external charge, with field lines depicted.
(a) Without any internal cavities. (b) With an internal cavity, the conductor is in equilibrium.

By the previous section, in equilibrium the interior of the conductor is neutral, so all the charge on the conductor must reside on the (outer) surface S_{outer}. If Q_{out} is moved to a new position, the surface charge adjusts. There is an attractive force between Q_{out} and the surface charge. If Q_{out} and $Q_{on} = -Q_{out}$ are doubled, the field lines have the same pattern, but become twice as dense.

Since the interior of the conductor is neutral, by the scooping-out theorem we can scoop out a cavity with surface S_{inner} from the conductor without affecting the electric field anywhere. See Figure 4.21(b). Hence, after scooping out neutral material from the conductor, the electric field in the scooped-out region (the cavity) remains zero.

That is, the electric field within the cavity, due to the charge outside the cavity, has been screened out. Moreover, we have not affected the surface charge.

Faraday cage: Superimposing cases 1 and 2, to produce arbitrary charge outside and on a hollow conductor, gives the situation corresponding to the Faraday cage, where there is no electric field within a metallic cage. See Figure 4.22.

Faraday had so much faith in electrical screening that he built an electrified metal cube and lived inside it: *"Using lighted candles, electrometers, and all other tests of electrical states, I could not find the least influence upon them ... though all the time the outside of the cube was very powerfully charged, and large sparks and brushes were darting off from every part of its outer surface."*

For high precision in electrometer readings, a Faraday cage is placed around the electrometer to screen out the effects of uncontrolled external charge, such as that induced on the experimenter by the charge we are trying to measure. Present-day automobiles use electronic circuitry that must be

Faraday Cage

A Faraday cage need not be solid. If the cage is a screen with hole dimension *a*, then approximately a distance *a* from the screen into the cage, the field will be screened out. See Figure 4.22. This is consistent with the Faraday ice pail interior being relatively open, yet still behaving as if it were nearly enclosed. Note that, when sparking occurs, electromagnetic radiation of a wide range of frequencies is emitted, and the system goes out of equilibrium. This does not affect the screening for radiation with wavelengths larger than the hole dimension. For microwave ovens, the screened door prevents the long-wavelength, low-frequency microwave radiation (but not the short-wavelength, high-frequency optical radiation) from escaping. Within a building

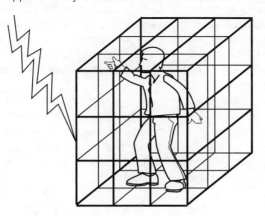

Figure 4.22 A Faraday cage. The person within the cage is unaffected by electrical discharges outside it, or on the cage itself.

or tunnel with steel girders, the longer-wavelength, lower-frequency AM stations are inaudible, but the shorter-wavelength, higher-frequency FM signals can be heard.

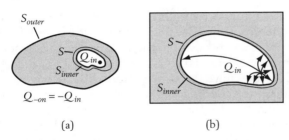

Figure 4.23 Charged conductor with a cavity containing an equal and opposite point charge. (a) View from a distance. There is no external field. (b) Close-up of the vicinity of the cavity. The field lines, which originate on the point charge, also terminate on the inner surface of the cavity.

protected from unwanted electric fields, which can be very intense near radio and television transmitters. What may appear to be spontaneous opening and closing of windows, or activation of cruise control, can occur for a car whose electronics is inadequately screened.

3. Internal charge Q_{in} and conductor with charge $Q_{on} = -Q_{in}$
   Now consider a conductor with charge Q_{in} within a cavity. See Figure 4.23(a). By Section 4.6, if the conductor is in equilibrium, there is no charge within the material of the conductor; any electrostatically induced charge (subject, as usual, to charge conservation) must reside on the surfaces of the conductor. Consider a Gaussian surface S that just barely encloses the actual inner surface S_{inner}, as in Figure 4.23(b). If the system is in equilibrium, the electric field is zero everywhere on S. Hence the flux through S is zero. By Gauss's law, (4.9), $Q_{enc} = 0$ for S_{inner}. Since $Q_{enc} = Q_{in} + Q_{inner}$, we thus conclude that $Q_{inner} = -Q_{in}$: the charge on the inner surface cancels the charge in the cavity. Since there is no charge in the bulk of the conductor, and the total charge on the conductor is $Q_{on} = -Q_{in}$, there is no charge on the outer surface.
   Typically, the charge distribution on the inner surface is nonuniform, being larger in magnitude nearer Q_{in}, and opposite in sign to Q_{in}. There is an attractive force between Q_{in} and the charge on the inner surface. If Q_{in} and $Q_{on} = -Q_{in}$ are doubled, the field lines have the same pattern, but become twice as dense.
   Faraday ice pail: Superimposing case 1 with $Q_{on} = q_1$, and case 3 with $Q_{in} = -Q_{on} = q_1$, so there is zero net charge on the conductor, corresponds to a neutral Faraday ice pail. The charge distribution on the outer surface, being due to case 1, is independent of the position of the charge within the cavity. This explains why the Faraday ice pail works as an electrometer, independent of the position of Q_{in} within S_{inner}.

4.10.2 *Gravitational Screening Cannot Occur*

Many people think that, because gravity and electricity both satisfy an inverse square law, they might be able to find a way to screen out gravity. However, the ability to screen out electric fields is not shared by all materials—electrical insulators cannot do it. What electrical conductors have that electrical insulators do not have, and which permits electrical screening, are mobile electric charges (the electric fluid) of sign opposite that of the immobile background charges, to produce a field that can cancel the applied electric field. For matter to be able to screen out gravity, that matter must have both positive and negative masses, and one of them must be mobile (like the conduction electrons). However, negative gravitational mass does not exist; even a so-called antiparticle has positive mass, equal to the same mass as its corresponding particle. Therefore gravitational screening cannot occur.

4.11 Advanced Topics in Conductors and Screening

4.11.1 *Electrostatic Screening: A Nontrivial Example*

At 21, William Thomson (later to become Lord Kelvin) invented the method of images to solve the problem of the charge density on conductors in the presence of external charges. It works for only a few cases, but that includes the important one of a point charge q and an infinite conducting sheet (e.g., of aluminum foil).

Consider a neutral, infinite sheet of aluminum foil. Freeze the neutral electric charge distribution so that electrostatic induction cannot occur. Along the midplane between two charges q (above) and $-q$ (below), place the sheet of foil. See Figure 4.24(a). Now unfreeze the charge distribution so that electrostatic induction occurs. Because $\vec{E}$ from q and $-q$ already is normal to the sheet, it does not cause electric currents to flow along the sheet. However, it polarizes the sheet by electrostatic induction. The surfaces of each sheet develop equal and opposite charge densities $-\sigma_S$ and σ_S, with net charges $-q$ and q, to terminate and originate the field lines from the point charges. See Figure 4.24(a). The equal and opposite surface charge densities are so close that, outside the sheet, they produce no field; they serve only to produce a field that, inside the foil, cancels the field of the external charge. Outside the sheet the field is due only to q and $-q$.

Below the sheet, the foil may be considered part of a spherical shell of infinitely large radius with center above the sheet. The field lines that originate on q terminate on a surface charge density $-\sigma_S$ on the top surface of the foil (the interior of the infinite spherical shell). Together, q and $-\sigma_S$ produce no effect

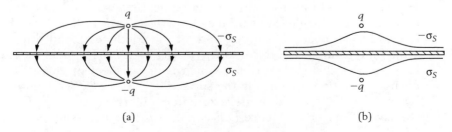

(a)   (b)

Figure 4.24 Electrostatic "screening" of point charges by a thin conducting sheet in equilibrium. (a) Point charges $\pm q$ above and below the sheet, with field lines. (b) Surface charge density, positive on the upper surface and negative on the lower surface.

outside the infinite spherical shell. Therefore, below the sheet, it is as if the only charges are $-q$ below the foil, and σ_S on its lower surface. This converts the original problem to the case of a single charge $-q$ and a sheet of aluminum foil with charge density σ_S.

Equation (4.22) yields σ_S from $\vec{E}_{out} \cdot \hat{n}$, which can be computed using only q and $-q$ (these form a dipole) and finding the field on their midplane. This was considered in Chapter 3, where $2a$ was the separation between the charges. With x the distance normal to the dipole axis, Equation (3.13) yields a field of magnitude $2kqa/(x^2 + a^2)^{3/2}$. Rewriting this result, with r instead of x, yields a field of magnitude $2kqa/(r^2 + a^2)^{3/2}$. Hence, from (4.22), on the $-q$ side, where $\vec{E}$ points along $\hat{n}$,

$$\sigma_S = \frac{\vec{E} \cdot \hat{n}}{4\pi k} = \frac{qa}{2\pi (r^2 + a^2)^{3/2}}. \tag{4.25}$$

Using $u = r^2 + a^2$, so $du = 2r dr$, and $\pi du = 2\pi r dr$, (4.25) yields

$$\int \sigma_S dA = \int \sigma_S (2\pi r) dr = \int_{a^2}^{\infty} \frac{qa}{2\pi u^{3/2}} \pi du$$

$$= \frac{qa}{2}(-2)u^{-1/2}\Big|_{a^2}^{\infty} = -qa\left(0 - \frac{1}{a}\right) = q. \tag{4.26}$$

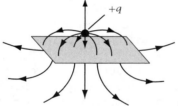

Thus, as indicated, the surface charge is equal and opposite to the charge $-q$, to whose field lines it connects. We plot (4.26) in Figure 4.24(b).

If the sheet is uncharged and finite, this result for σ_S is only approximate and does not hold very well near the edges, where there must be a neutralizing negative charge. Figure 4.25, repeated from Figure 3.9(b), presents the field lines.

Figure 4.25 Field lines for a neutral conductor in the presence of a point charge.

4.11.2 *Outward Electrical Pressure on a Charged Conductor*

There is an outward electrical force acting on the surface charge of conductors, which we may interpret as an electrical pressure. It is due to the force of the external field on the surface charge density. Very near the conductor, either inside or outside, the field may be written as $\vec{E} = \vec{E}_{ext} + \vec{E}_S$. Here $\vec{E}_{ext}$ is due to distant charges (both on and off the conductor), and thus remains the same on crossing the surface, so $\vec{E}_{ext} = \vec{E}_{ext}^{out} = \vec{E}_{ext}^{in}$. The field $\vec{E}_S$ is due to the local surface charge density σ_S, which is like a sheet of charge, and thus $\vec{E}_S$ changes direction on crossing the surface: $\vec{E}_S^{out} = -\vec{E}_S^{in}$. See Figure 4.26. Since by (4.20) the total field inside is zero in equilibrium, or $\vec{E}^{in} = \vec{E}_{ext} + \vec{E}_S^{in} = \vec{0}$, we have $\vec{E}_S^{in} = -\vec{E}_{ext}$. Hence $\vec{E}^{out} = \vec{E}_{ext} + \vec{E}_S^{out} = \vec{E}_{ext} - \vec{E}_S^{in} = 2\vec{E}_{ext}$. Thus

$$P = \left|\frac{d\vec{F}}{dA}\right| = |\sigma_S||\vec{E}_{ext}| = \frac{|\sigma_S||\vec{E}^{out} \cdot \hat{n}|}{2} = |\sigma_S|(2\pi k)|\sigma_S| = (2\pi k)\sigma_S^2 = \frac{|\vec{E}^{out}|^2}{8\pi k}. \tag{4.27}$$

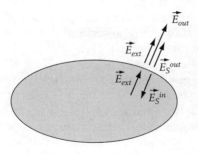

Figure 4.26 Conductor in equilibrium. The field depiction is only for a small region of the upper surface. The fields $\vec{E}_S$ due to the local charge density σ_S, $\vec{E}_{ext}$ due to more distant charge (both on and off the conductor), and the total field $\vec{E}$ are given. $\vec{E}_{ext}$ does not change on going from just inside to just outside the conductor, but $\vec{E}_S$ reverses.

Consider a conducting sphere for which $|\vec{E}_{out}|$ equals the dielectric strength E_d of air, 3.0×10^6 N/C. Then $\sigma_{max}^{air} = E_d/4\pi k = 5.3 \times 10^{-5}$ C/m^2, which gives $P_{max}^{air} = 77$ N/m^2. Since $P_{atm} \approx 10^5$ N/m^2, this outward electrical pressure is insufficient to inflate a balloon against atmospheric pressure. However, in outer space, where there is negligible atmospheric pressure, electrical inflation of an aluminized mylar balloon might be more feasible.

4.11.3 *Microscopic Screening Length*

Our discussion of electrical screening due to surface charge has assumed the surface to be geometrically sharp. This is not literally true because at the microscopic level the surface contains atoms of characteristic dimension 10^{-10} m. Even if deviation from ideal sharpness can be neglected, there is still the issue of the characteristic length over which the screening takes place. Gauss's law applied to the electric fluid model says that to screen out the applied electric field charge literally piles up at the surface. However, for real conductors, the charge must be distributed over a finite distance from the surface, called the *screening length l*.

For metals, where the charge carriers are electrons, this screening length l is on the order of the distance a between atoms. The net electric field, including that of the screening charge due to adjustments in the electron orbitals, both oscillates and decreases in amplitude on moving to the interior. See Figure 4.27(a). (In the mathematics of Fourier series, the *Gibbs phenomenon* is a related effect.) Clearly, the physics of surfaces is complex. Wolfgang Pauli, whose idea it was that only one electron per orbital is allowed, once remarked that "surfaces are the invention of the devil."

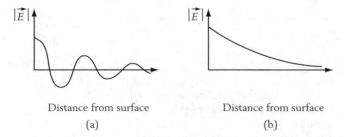

Figure 4.27 Schematic of electric field in vicinity of two types of real conductors in equilibrium. (a) A metal. (b) A semiconductor or an ionic conductor.

In semiconductors, where the conduction electron density can be much lower, the screening length l can be hundreds of times larger, and the net electric field decays exponentially with this length. See Figure 4.27(b).

For ionic conductors (living cells contain such ions as Na^+, K^+, H^+, and Cl^-), the electric field decays exponentially in space, and the characteristic size of the screening length l depends on temperature. It has an order of magnitude given by the product of the distance b between ions with the square root of the small dimensionless ratio of the characteristic thermal and electrical energies. For general information—not because it gives the student a deeper understanding of electricity—this ratio takes the form $l/b = [k_B T/(ke^2/b)]^{1/2}$. (Here, $k_B = 1.38 \times 10^{-23}$ J/K is the Boltzmann constant, K is degrees Kelvin, and the temperature T is given in units of K.) In a human cell, l is on the order of ten atomic dimensions; in the electrolyte of a well-charged car battery, l is on the order of an atomic dimension. Note that ke^2/b^2 has the dimensions of a force, so $ke^2/b = (ke^2/b^2)b$ has the dimensions of an energy. Electrical energy is the subject of the next chapter.

4.11.4 *Depletion Layer in Semiconductors*

Figure 4.27(a) is a good representation of what happens when a weak external electric field is applied to a semiconductor so that the change in density of charge carriers is small compared to the equilibrium charge carrier density. However, when a strong electric field is applied, so many semiconductor charge carriers may be required to screen out the applied field that near the surface they exceed the equilibrium charge carrier density. In this case, the system responds differently. For purposes of discussion, assume that the charge carriers are electrons. There are two cases to consider.

If the applied electric field is directed toward the conductor, that will attract electrons to the vicinity of the surface, at a density higher than the equilibrium density (*accumulation*). See Figure 4.28(a) for a schematic of the charge carrier density within the semiconductor. This is qualitatively similar to what happens for small electric fields.

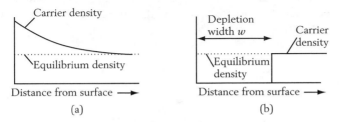

Figure 4.28 Schematic of charge carrier density near the surface of a semiconductor in the presence of an applied electric field. (a) When an excess of charge carriers is attracted to the vicinity of the surface. (b) When charge carriers are repelled from the vicinity of the surface so much that a depletion layer of thickness w is formed.

If the applied electric field is directed away from the conductor, that will repel electrons, causing them to be completely depleted in a region near the surface. The thickness, or width, w of this *depletion layer* is determined by the electric field magnitude $|\vec{E}|$ and the background density n of positive charge that, in equilibrium, cancels the charge of the electrons in charge-carrying states. See Figure 4.28(b). Full depletion leaves behind a positive charge density en. Integration over the depletion width gives a charge per unit area of enw. Applying (4.22) as $|\sigma_S| = |\vec{E}|/4\pi k$ with $\sigma_S = enw$ then gives

$$w = \frac{|\vec{E}|}{4\pi k en}. \qquad \text{(depletion layer width)} \qquad (4.28)$$

Hence, the lower the density of charge carriers, the further the depletion region extends into the semiconductor. Let $n = 10^{18}$ m^{-3}, corresponding to a linear separation between charge carriers of 10^{-6} m. If the linear separation between atoms is 3×10^{-10} m, this corresponds to a linear separation of about one charge carrier per 3300 atoms. If $|\vec{E}| = 2 \times 10^5$ N/C, then (4.28) gives $w = 1.105 \times 10^{-5}$ m, or about ten times the linear separation between charge carriers.

Problems

4-2.1 (a) What happens to the electric flux if we define it with the *inward* normal? (b) Specifically, if $\Phi_E = 34$ N-m^2/C with the usual definition, what is Φ_E with the *inward* normal? (c) How would we have to write Gauss's law using that definition of flux?

4-2.2 A cube of side 12 cm lies in the first octant with one corner at the origin, as in Figure 4.5. A field points along the y-axis, with $E_y = 10$ N/C on the $y = 0$ plane, and $E_y = 20$ N/C on the $y = 12$ cm plane. Take $E_x = E_z = 0$. Find the net flux leaving the cube.

4-2.3 A cube of side 4 cm lies in the first octant with one corner at the origin. It is in a field given by $\vec{E} = (1 - 3x^2)\hat{i}$, where x in m gives $\vec{E}$ in N/C. Find the net flux leaving the cube.

4-2.4 Consider an electric field $\vec{E} \equiv (-2, 4, -5)$ N/C, and a surface element of area 4 mm^2 and normal along $(2.4, -4.5, 6.0)$. (a) Find the unit vector $\hat{n}$ normal to the surface. (b) Find the flux per unit area. (c) Find the flux through this surface element.

4-2.5 A circle of radius R sits on the $z = 0$ plane, its center at the origin. A uniform field $\vec{E}$ points along the z-axis. See Figure 4.29. Find the flux passing through the circle.

Figure 4.29 Problem 4-2.5.

4-2.6 A hemispherical shell of radius R sits above the $z = 0$ plane, its center at the origin. A uniform field $\vec{E}$ points along the +z-axis. See Figure 4.30. Find the flux passing through the shell.

Figure 4.30 Problem 4-2.6.

4-2.7 Calculate the electric flux through a disk of radius R, due to a charge Q a distance d along its axis.

4-2.8 An infinitely long line charge λ is along the z-axis, and passes through the origin. Concentric with it is an infinitely long cylinder of radius R.

Calculate the electric flux per unit length passing through the cylinder.

4-3.1 A Gaussian surface surrounds an elephant, which resides on an insulating platform. The elephant accidentally swallows a glass bead whose interior has a charge $q = 6$ nC. Determine the electric flux through the Gaussian surface (a) before swallowing, (b) while swallowed, and (c) after expulsion of the glass bead through the elephant's trunk.

4-3.2 The center of a sphere of charge Q and radius R is at one corner of a cube of side $a > 2R$. (a) Find the total flux through the cube. (b) In the limit $R/a \to 0$, find the flux through each side.

4-3.3 The center of a sphere of charge Q and radius R is at the center of a cube of side $a > 2R$. (a) Find the total flux through the cube. (b) In the limit $R/a \to 0$, find the flux through each side.

4-3.4 A tiny sphere carrying 650 μC is at the center of a cube of 12 cm per side. (a) Find the flux through one face. The charge now is moved to the center of one face (half inside the cube and half outside it). (b) Approximately how much flux goes through that face? (c) Approximately how much flux goes through the remaining faces?

4-3.5 A surface element of area 2×10^{-4} m^2 has its normal along $(20, 12, -9)$. For this element, $\vec{E} = (6, -2, -5)$ N/C. Find (a) the normal $\hat{n}$; (b) $|\vec{E}|$; (c) $\vec{E} \cdot \hat{n}$; (d) the angle between $\vec{E}$ and $\hat{n}$; (e) the flux $d\Phi_E$ through this element; and (f) the charge dQ associated with this flux.

4-3.6 A uniform field of magnitude 750 N/C points 20° off the normal from the plane of a circular plate of radius 15 cm. See Figure 4.31. Find the flux through the plate, and the charge to which this corresponds.

Figure 4.31 Problem 4-3.6.

4-4.1 A spherical conducting shell of radius 2 cm is uniformly charged. The field just outside its surface is 200 N/C, pointing outward. Find the charge density at the surface.

4-4.2 Measurements of the electric field in a re-

gion reveal that it points along the (vertical) y-axis and is given by $E_y = 2 + 4y^2 - 5y^3$, for E_y in N/C and y in m. Consider a cylinder of 0.02 m^2 cross-section that sits on the $y = 0$ plane and has height $h = 0.05$ cm. Find (a) the net flux leaving the cylinder, and (b) the charge enclosed by the cylinder.

4-4.3 The electric flux through a donkey was initially 150 N-m^2/C, but after eating a potato, it is -275 N-m^2/C. Find the charge within the potato.

4-4.4 A point charge is at the center of a regular tetrahedron. The flux through one of its four sides is 400 N-m^2/C. Find the charge within the tetrahedron.

4-5.1 Outside a spherical distribution of charge, at a radius of 2.5 cm, the electric field has magnitude 28 N/C and points inward. (a) Find the total charge associated with this charge distribution. (b) If the charge were 18 nC, find the electric field at this position.

4-5.2 Outside a cylindrical distribution of charge, at a radius of 4.6 cm, the electric field has magnitude 15.8 N/C and points outward. (a) Find the total charge per unit length associated with this charge distribution. (b) If the charge per unit length were 18 mC/m, find the electric field at this position.

4-5.3 Outside a planar distribution of charge, at a distance of 1.3 cm from its center, the electric field has magnitude 7.3 N/C and points outward. (a) Find the total charge per unit area associated with this charge distribution. (b) If the charge per unit area were 47 mC/m^2, find the electric field at this position.

4-5.4 A spherical shell of inner radius a and outer radius b contains a uniform charge density ρ of total charge Q. (a) Find ρ. (b) Find the electric field for $r < a$, $a < r < b$, and $b < r$. Sketch it.

4-5.5 A cylindrical shell of inner radius a and outer radius b contains a uniform total charge per unit length λ. (a) Find the charge density ρ in terms of λ, a, and b. (b) Find the electric field for $r < a$, $a < r < b$, and $b < r$. Sketch it.

4-5.6 Two concentric conducting shells have radii 2 cm and 4 cm. At $r = 3$ cm, the electric field has magnitude 40 N/C and points inward. At $r = 5$ cm, the electric field has magnitude 40 N/C and points outward. (a) Find the charges on each shell. (b) Find the electric field at $r = 10$ cm.

4-5.7 Two concentric conducting cylinders have radii 1 cm and 3 cm. At $r = 2$ cm, the electric field has magnitude 650 N/C and points inward. At $r = 6$ cm, the electric field has magnitude 40 N/C and points outward. (a) Find the charge per unit length on each shell. (b) Find the electric field at $r = 12$ cm.

4-5.8 Two conducting sheets are normal to the x-axis, at $x = -2$ cm and $x = 2$ cm. At $x = 1$ cm, the electric field has magnitude 650 N/C and points rightward. At $x = 12$ cm, the electric field has magnitude 40 N/C and points leftward. (a) Find the charge per unit area on each sheet. (b) Find the electric field at $x = -8$ cm.

4-5.9 (a) Write down the gravitational analog of Gauss's law. (b) Apply it to find the gravitational field within a ball of radius R and mass M, with uniform mass density. (c) Show that, if a tiny hole is drilled through the center of such a ball, then a tiny mass m, dropped from the surface, will undergo simple harmonic motion, and find its period. (d) Put in parameters appropriate to the earth, and thus determine the period of the motion.

4-5.10 Whenever there is an area for which the flux per unit area is constant, then $(d\Phi_E/dA)A_{flux} = 4\pi k Q_{enc}$. Discuss how this applies to the cases we have studied—in particular, for sheets, cylinders, spheres, and pillboxes outside conductors. Does it apply to a cube-shaped charge, surrounded by a Gaussian surface that is a concentric cube?

4-6.1 Consider two conductors A and B, well-isolated from each other, with surface charge densities σ_A and σ_B. If we can freeze the charge in place, and then bring the conductors near each other, will these charge densities produce electric fields that are normal to each surface? *Hint:* Think in terms of conducting spheres.

4-6.2 Explain why, for purposes of the principle of superposition, a small conducting sphere can be thought of as a point charge. *Hint:* For a charged sphere of small radius, all other charges are very far away relative to the charge on the sphere. What does this say about the amount that the sphere can be polarized, and therefore the effect of this polarization on other charges?

4-6.3 Consider a point charge q outside a neutral conductor, as in Figure 4.32(a). The response

of the conductor can be obtained as a superposition of (1) the equilibrium charge configuration for q itself and a charge $-q$ on the conductor [see Figure 4.32(b)]; and (2) the equilibrium charge configuration for only a charge q on the conductor [see Figure 4.32(c)]. (a) If the net potential of the neutral conductor is $+40$ V, and its potential in the presence of q itself and the conductor with $-q$ is $+15$ V, find its potential if there is only q on the conductor. (b) Discuss the direction of the dipole moment on the conductor, and the direction of the net force on the point charge q.

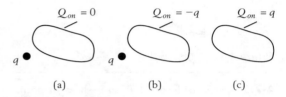

Figure 4.32 Problem 4-6.3.

4-6.4 Consider a neutral dielectric whose molecules are polarized to the right. If molecules are scooped out from the interior, charge is left behind on the surface of the resulting cavity. (a) What is the net charge on the cavity surface? (b) For a spherical cavity, in what direction does the resulting electric field point?

4-6.5 A brass "football" sits on a styrofoam pad. A charge Q_1 is within the cavity of the football, and there is no net charge on the football. (a) Find the charge on the football's inner and outer surfaces. (b) The football is now connected by a wire to ground. Find the charges on its inner and outer surfaces.

4-6.6 Consider two concentric, thin, conducting shells, of radii $r_1 < r_2$ and positive charges Q_1 and Q_2. (a) Find the charge densities on their inner and their outer surfaces. (b) If a charge $Q_3 > Q_2$ is brought up from infinity to just outside r_2, how does this affect the charge densities qualitatively?

4-6.7 P. Heller has developed a "screening" device. It consists of a set of vertical conducting slats a few cm wide and about 20 cm long, close to one another but not in contact, all hung from a circular mount of about 10 cm diameter. When the slats are connected to one another by a wire, they screen the interior of the enclosed cylindrical region, but when disconnected they don't screen. See Figure 4.33, where the connecting wire is the circle. Explain why connecting the slats enables screening to occur.

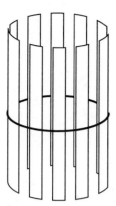

Figure 4.33 Problem 4-6.7.

4-7.1 Just outside point P on a sheet of crumpled aluminum foil, the electric field has magnitude 24 N/C and points inward. Determine the surface charge density at P.

4-7.2 The electric field just outside a uniformly charged conducting sphere of radius 4.7 cm has magnitude 1400 N/C and points outward. (a) Determine the surface charge density on the sphere. (b) Determine the total charge on the sphere.

4-7.3 Explain, in your own words, the factor of two difference between the equations for the electric fields outside a conductor and outside a uniform sheet of charge.

4-7.4 Two thin conducting plates are normal to the x-axis. 1, on the left, has total charge per unit area 2σ, and 2, on the right, has total charge per unit area -4σ, with $\sigma > 0$. See Figure 4.34. (a) Find the electric field to the left, to the right, and between the plates. (b) Find the surface charge density for each surface.

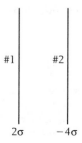

Figure 4.34 Problem 4-7.4.

4-7.5 Three thin conducting plates are normal to the x-axis. 1, on the left, has total charge per unit area 2σ, 2, in the middle, has total charge per unit area -4σ, and 3, on the right, has total charge per unit area -3σ, with $\sigma > 0$. See Figure 4.35. (a) Find the electric field to the left, to the right, and in the two intermediate regions between the plates. (b) Find the charge densities for both sides of each plate.

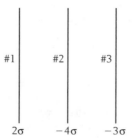

Figure 4.35 Problem 4-7.5.

4-7.6 The field between two parallel, vertically stacked conducting sheets is 200 N/C upward. The field above both of them is 20 N/C downward. Find the electric field below both sheets, and the charge densities on each surface of each sheet.

4-7.7 A conducting spherical shell has a 2 cm radius. Just outside it, the electric field points radially inward and has a magnitude of 900 N/C. (a) Find the surface charge density. (b) Find the total charge on the shell. (c) Find the field just inside the shell. (d) Find the field at 3 cm from the center of the shell.

4-7.8 Let a conducting spherical shell of radius R possess a net charge Q. Let a point charge q be within the shell. (a) Show that, for the shell, $Q_{inner} = -q$ and $Q_{outer} = Q+q$. (b) Show that σ_{outer} is distributed symmetrically, so $\sigma_{outer} = Q_{outer}/4\pi R^2 = (Q+q)/4\pi R^2$. (c) Show that $\sigma_{inner} = Q_{inner}/4\pi R^2 = -q/4\pi R^2$ if q is on-center. (d) Can you find σ_{inner} if q is off-center?

4-7.9 Two thin, conducting, concentric spherical shells have radii a and b, where $a < b$. The total charge on the inner shell is Q, and that on the outer shell is $-2Q$. (a) Find the electric field for $r < a$, $a < r < b$, and $b < r$. (b) Find the charge on the inner and outer surfaces of the inner shell. (c) Repeat for the outer shell.

4-7.10 Three thin, conducting, concentric spherical shells have radii a, b, and c, where $a < b < c$.

The total charges are, respectively, $Q_a = 3Q$, $Q_b = -2Q$, and $Q_c = Q$. (a) Find the electric field for $r < a$, $a < r < b$, $b < r < c$, and $c < r$. (b) Find the charge on the inner and outer surfaces of the inner shell. (c) Repeat for the middle shell. (d) Repeat for the outer shell.

4-7.11 Two thin, conducting, concentric cylindrical shells have radii a and b, where $a < b$. The total charge per unit length on the inner shell is 2λ, and that on the outer shell is -3λ. (a) Find the electric field for $r < a$, $a < r < b$, and $b < r$. (b) Find the charge per unit length on the inner and outer surfaces of the inner shell. (c) Repeat for the outer shell.

4-7.12 Three thin, conducting, concentric cylindrical shells have radii a, b, and c, where $a < b < c$. The total charges per unit length are, respectively, $\lambda_a = 2\lambda$, $\lambda_b = -3\lambda$, and $\lambda_c = 4\lambda$. (a) Find the electric field for $r < a$, $a < r < b$, $b < r < c$, and $c < r$. (b) Find the charge per unit length on the inner and outer surfaces of the inner shell. (c) Repeat for the middle shell. (d) Repeat for the outer shell.

4-8.1 An electrometer is connected to the exterior of an isolated metal cup resting on an insulating surface. The electrometer reads $5\,\mu C$. (a) Where on the cup does the charge reside? (b) A ball on an insulating string now is lowered into the cup. The electrometer now reads $2\,\mu C$. What is the charge on the ball? (c) The ball in the cup is touched briefly to the electrometer. What does the electrometer read? (d) The ball is removed from the cup, and the electrometer reads $2.4\,\mu C$. Find the charge on the ball.

4-8.2 Repeat the previous problem, for parts (a), (b), and (d), giving the number of field lines into or out of each surface, with one field line per $0.2\,\mu C$.

4-9.1 Find the solid angle subtended by a disk of diameter d, at a distance s along its normal.

4-9.2 Find the solid angle subtended by a square of side a, at a distance $a/2$ along its normal. *Hint:* Use symmetry.

4-10.1 Consider a point P on the outer surface of a conductor with a cavity. For $Q_{in} = Q_1 = 2 \times 10^{-6}$ C in the cavity, and $Q_{on} = Q_1$ distributed over the conductor, measurement yields $\sigma_s = 2 \times 10^3$ C/m^2 at P. For $Q_{in} = -2Q_1$, and $Q_{on} = -Q_1$ measurement yields $\sigma_S = -4 \times 10^3$ C/m^2 at P. Deter-

mine σ_S at P when there is only $Q_{in} = 3Q_1$ and $Q_{on} = 0$.

4-10.2 A charge q_1 is outside a conductor, a charge q_2 is inside a cavity within the conductor, and q_3 is on the conductor itself. Discuss the forces on q_1, on q_2, and on the conductor.

4-11.1 A rod of charge per unit length λ is parallel to the z-axis, a distance b from an uncharged infinite conducting sheet in the $y = 0$ plane. (a) Using the method of images, find the electric field as a function of x just outside the surface of the sheet, on the same side as the rod. (b) As a function of x, find the charge per unit area on the surface of the sheet near the rod.

4-11.2 (a) Find the electrical pressure on a cylindrical shell of radius r and charge per unit length λ. (b) Determine the work done against the electrical force to compress it from radius b to radius a.

4-11.3 (a) Find the electrical pressure on a spherical shell of radius r and charge Q. (b) Determine the work done against the electrical force to compress it from radius b to radius a.

4-11.4 A semiconductor layer above a metal has thickness 50 nm and carrier electron density $n = 2.6 \times 10^{18}$/m^3. Find how large an electric field must be applied to produce complete depletion of carrier electrons from the semiconductor.

4-G.1 Why are these equations incorrect?
(a) $\oint \vec{E} \cdot \hat{n} dA = \vec{E} \oint \hat{n} dA$.
(b) $\oint \vec{E} \cdot \hat{n} dA = |\vec{E}| \oint \hat{n} dA$.
(c) $\oint \vec{E} \cdot \hat{n} dA = \oint |\vec{E}| \hat{n} dA$.

4-G.2 Consider a region where there is a uniform field $\vec{E}$ of arbitrary magnitude and direction. (a) Apply Gauss's law to any closed surface in this region to show that $0 = \oint \vec{E} \cdot \hat{n} dA$. (b) Show that $0 = \vec{E} \cdot \oint \hat{n} dA$. (c) Hence deduce that $\vec{0} = \oint \hat{n} dA$ for *any* closed surface. This is a very powerful *mathematical* result.

4-G.3 Consider a charge distribution with planar symmetry $\rho = \rho(x)$. (a) For a Gaussian surface that is a slab of area A and thickness dx, show that $\Phi_E \approx (d/dx)(E_x)(A dx)$. (b) Show that $Q_{enc} \approx \rho(A dx)$. (c) Hence, from Gauss's law, deduce that, for planar symmetry, $(d/dx)(E_x) = 4\pi k\rho$.

4-G.4 Consider a charge distribution with cylindrical symmetry $\rho = \rho(r)$. (a) For a Gaussian surface that is a cylindrical shell of length L, inner radius r, and thickness $dr \ll r$, show that $\Phi_E \approx 2\pi(d/dr)(E_r r)(L dr)$. (b) Show that $Q_{enc} \approx \rho(2\pi r L dr)$. (c) Hence, from Gauss's law, deduce that for cylindrical symmetry, $(d/dr)(E_r r) = 4\pi k r \rho$.

4-G.5 Consider a charge distribution with spherical symmetry $\rho = \rho(r)$. For a Gaussian surface that is a spherical shell of inner radius r and thickness $dr \ll r$, show that $\Phi_E \approx 4\pi(d/dr)(E_r r^2)(dr)$. Show that $Q \approx \rho(4\pi r^2 dr)$. Hence, from Gauss's law, deduce that for spherical symmetry, $(d/dr)(E_r r^2) = 4\pi k r^2 \rho$.

4-G.6 Within a charged region with planar symmetry, the electric field varies as x^3. How does the charge density vary with x? See Problem 4-G.3.

4-G.7 Within a uniformly charged cylindrical region, the electric field points radially and has a constant magnitude. How does the charge density vary with r? See Problem 4-G.4.

4-G.8 Within a charged spherical region, the electric field points radially and varies as r^2. How does the charge density vary with r? See Problem 4-G.5.

4-G.9 Consider an infinite slab of thickness $2R$ with uniform charge density ρ. A smaller parallel slab is scooped out, of thickness $2a < 2R$ and with center a distance b from the center of the original slab, where $a + b < R$. Show that the electric field within the cavity is uniform, and find its magnitude and direction. *Hint:* Superpose two charge distributions.

4-G.10 Consider an infinite cylinder of radius R with uniform charge density ρ. An infinitely long cavity of radius a is scooped out, with center a distance b from the center of the larger cylinder, where $a + b < R$. See Figure 4.36. Show that the electric field within the cavity is uniform, and find its magnitude and direction.

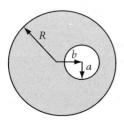

Figure 4.36 Problem 4-G.10.

4-G.11 Consider a sphere of radius R, centered at the origin with uniform charge density ρ. A cavity of radius a is scooped out, with center a at $\vec{b}$, where $a + b < R$. See Figure 4.36. Show that the electric field within the cavity is uniform, and given by $\vec{E} = (2k\rho/3)\vec{b}$.

4-G.12 For a planar distribution of charge $\rho(x)$, consider the charge $Q(x)$ that is enclosed by a rectangular parallelopiped with area A normal to $\hat{x}$, and with sides at 0 and x. (a) Show that $Q(x) = A\int_0^x \rho(x)dx$, and hence that $dQ/dx = A\rho(x)$. (b) For $Q(x) = 2Ax^5$, find $\rho(x)$.

4-G.13 For a cylindrical distribution of charge $\rho(r)$, consider the charge $Q(r)$ that is enclosed by a concentric cylinder of radius r and length L. (a) Show that $Q(r) = L\int_0^r 2\pi\rho(r)dr$, and hence that $dQ/dr = 2\pi r L\rho(r)$. (b) For $Q(r) = 8Lr^4$, find $\rho(r)$.

4-G.14 For a spherical distribution of charge $\rho(r)$, consider the charge $Q(r)$ that is enclosed by a concentric sphere of radius r. (a) Show that $Q(r) = \int_0^r 4\pi r^2\rho(r)\,dr$, and hence that $dQ/dr = 4\pi r^2\rho(r)$. (b) If $Q(r) = 12\exp(-2r)$, find $\rho(r)$.

4-G.15 (a) Consider charges Q at $(a, 0, 0)$ and $(-a, 0, 0)$. Find the field at the origin. (b) Consider small lightbulbs at $(a, 0, 0)$ and $(-a, 0, 0)$. If each produces intensity I_0 at the origin, find the total intensity at the origin. (c) Compare and constrast the addition of electric fields and lightbulb intensities.

4-G.16 Light intensity from a flashlight satisfies something like Gauss's law because intensity from a point source falls off as r^{-2}. However, intensity is always positive. (a) Find how the intensity falls off with distance r for a fluorescent tube of length l and radius a, for $a \ll r \ll l$. (b) Find how the intensity varies with distance r from a uniformly lit square movie screen of side a, for $r \ll a$. (c) At the midpoint between two equal charges q, the electric field is zero. If the intensity of each of two identical flashlights is I_0 at their midpoint, find the total intensity at their midpoint.

4-G.17 A point charge q is at the origin. Consider a square of side a that is normal to $\hat{y}$, at a distance l from q. (a) Show that the electric flux through it is $\Phi_E = 2kq\,\sin^{-1}[a^2/(a^2 + 4l^2)]$. (b) Find the solid angle Ω subtended by the square, as measured from the origin. (c) Verify that Ω has the expected form for $l \to \infty$, $l \to 0$, and $l = a/2$.

4-G.18 A line of charge λ runs along the z-axis. Consider a Gaussian surface that is a cylinder of radius a parallel to the z-axis. Develop a spreadsheet to calculate the flux per unit length along z for this Gaussian surface, due to the line charge. (This requires a numerical value for λ.) Verify that the flux changes discontinuously as the line charge moves from inside to outside the Gaussian surface.

"The whole theory ... of the potential ... belongs essentially to the method [of fields] which I have called that of Faraday. According to the other method [of action at a distance], the potential, if it is considered at all, must be regarded as the result of a summation of the electrified particles divided each by its distance from a given point. Hence many of the mathematical discoveries ... find their proper place ... in terms of conceptions mainly derived from Faraday."

—James Clerk Maxwell,
Preface to *A Treatise of Electricity and Magnetism* (1873)

Chapter 5

Electrical Potential Energy and Electrical Potential

Chapter Overview

Section 5.1 introduces the concepts of electrical potential energy and electrical potential. Section 5.2 briefly reviews the relationship between gravitational force and gravitational energy. Section 5.3 defines electrical potential and its relationship to electrical potential energy. Section 5.4 applies electrical potential to electrical conductors in equilibrium, energy conservation involving electrical potential energy, and the use of a force measurement to deduce voltage differences (thus making voltage differences accessible to mechanical measurements on macroscopic objects). Section 5.5 discusses equipotential surfaces in some detail. Section 5.6 studies the electrical potential and electrical potential energy of point charges to obtain typical atomic and nuclear energies. Section 5.7 shows that the electrical potential of point B relative to point A depends only on their positions, not on the path from A to B. This is a consequence of conservation of energy for the electrostatic force. It implies the non–close on itself field-line rule of Chapter 3.

   Sections 5.8–5.10 are reprises of material touched on lightly in earlier sections. Section 5.8 shows how to calculate the electrical potential from the electric field using the field viewpoint, and Section 5.9 does the same using the action-at-a-distance viewpoint. Section 5.10 shows how to determine the electric field from the spatial variation of the potential. (Example 5.20 of Section 5.10 goes from V to $\vec{E}$ to the charge enclosed by a specific surface, thus relating material from Chapters 1 through 5.) Section 5.11 shows why the electric field on a conductor is often largest where its radius of curvature is smallest, thus explaining Franklin's observation (Section 1.4) of the "power of points" to produce electric fields large enough to cause electrical discharge. Section 5.12 discusses the electrical properties of nonuniform conductors.

   Having a precise definition of electrical potential permits us to discuss the ratings on charge reservoirs, or capacitors, some of the fundamental building blocks for electrical circuits. The capability to store charge, or capacitance, is such an important subject that the next chapter is devoted to it. ∎

5.1 Introduction

In your mechanics course, you studied how the *gravitational force* on a mass *m* led to *gravitational potential energy*. The amount of gravitational potential energy U_g equals the product of *m* and a quantity V_g called the *gravitational potential*.

This chapter shows that the *electrical force* on a charge *q* leads to *electrical potential energy*, denoted by *U*. The amount of electric potential energy equals the product of *q* and a quantity *V* (italicized) called the *electrical potential*, whose unit is the volt (V, not italicized). Just as for gravity, where only differences in the gravitational potential are meaningful, so for electricity, only differences in the electrical potential are meaningful. Sometimes electrical potential energy is called merely *electrical energy*.

There is a simple geometrical relationship between electric field $\vec{E}$—a vector-valued function of position $\vec{r}$—and electrical potential *V*—a scalar-valued function of $\vec{r}$: (1) The scalar $V(x, y, z) = $ constant defines a surface to which the vector $\vec{E}$ is normal. (2) $\vec{E}$ points from higher potential to lower potential. (3) The magnitude of the slope of *V* in the direction of $\vec{E}$ gives $|\vec{E}|$. See Figure 5.1, where the positively charged source where the field lines originate, and the negatively charged source where they terminate, do not lie between the two equipotentials. Compare Figure 5.1 to Figure 4.1, where the charges *dq* and −*dq* that initiate and terminate a flux tube are depicted. In Figure 4.1, the surface on which *dq* resides is at a higher potential than that on which −*dq* resides. If their potential difference is 20 volts, then the flux tube in Figure 4.1 can be decomposed (for example) into 10 sub-tubes across each of which the potential difference is 2 volts. Using the concepts of flux and potential, it is possible to categorize the electric field completely. Note that, because both the electric field and the electric potential have values at all points of space, they are both, in a technical sense, *fields*. Indeed, temperature, like electrical potential, is a scalar field, whereas the electric field is a vector field. (We will often interpret "field" as "electric field" although "potential field" would be a perfectly valid interpretation.)

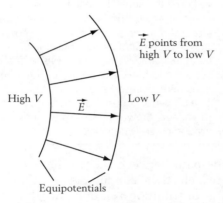

$\overrightarrow{E}$ points from high *V* to low *V*

High *V* $\vec{E}$ Low *V*

Equipotentials

Figure 5.1 Two equipotentials and field lines between them. Each field line points from a region of high potential to a region of low potential.

The quotation from Maxwell at the head of the chapter notes that there are two ways to consider electrical potential. (1) From the action-at-a-distance viewpoint, the electrical potential is obtained as a sum of the contributions from every electric charge, with the potential at infinity conventionally taken to be zero. This requires a complete knowledge of the positions and magnitudes of every charge. To obtain the electrical potential difference between two points thus requires the potential at both points. (2) From the field viewpoint, the electrical potential difference between those two points is obtained as an integral

Voltage and Electrical Potential

Consider a battery that causes two conducting plates to have a specific potential difference (e.g., 12 volts, or 12 V). We do not need to know the details of an electron's motion (e.g., its position and velocity as a function of time) in order to determine its speed on crossing from the negative plate, where it starts at rest, to the positive plate. (See Figure 5.2.) Here, just as for the motion of a stone under gravity, conservation of energy yields the final speed. As indicated in Review/Preview, power companies provide electric current at a specified electrical potential, or *voltage*, just as water companies provide water current at a specified *pressure*. (An important difference is that the electrical potential oscillates in time, but at a rate that, from the viewpoint of an electron, is incredibly slow.) Hence it is sufficient to know the applied voltage and only a few properties of an electrical device to determine the device's performance. We need not know the location of every bit of charge, only the initial and final voltages. Faraday's desire for a visualizable physical picture and Maxwell's desire for a mathematically consistent representation of that picture have led to a powerful conceptual structure, using the idea of voltage, on which modern technology and civilization crucially depends.

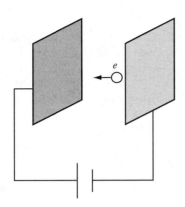

Figure 5.2 An electron moving between two conducting plates held at different electrical potentials by a battery; the large plate is positive, the small plate is negative.

over the electric field along some path between them. The field viewpoint is particularly useful in describing real electrical conductors in equilibrium, where typically the detailed surface charge distribution σ_S is unknown, but the potential over the conductor is known.

Conductors in equilibrium are characterized by a fixed value for the electrical potential, which is the same over the surface and even the volume of the conductor. (Thus they are called *equipotential surfaces*.) You already know something about electrical potential. The terminals of a battery or the prongs of an electrical outlet provide a difference in electrical potential. *Voltage* and *electrical potential* are used interchangeably.

5.2 Gravitational Potential Energy and Gravitational Potential

The idea of electrical potential energy has already been used in Section 3.8, in the context of an electric dipole $\vec{p}$ in a uniform electric field $\vec{E}$. This was done by analogy to gravity. Let us now review some additional results for gravity. Consider an object with mass m subject to a gravitational force $\vec{F}$. The object could be on the moon, on the earth, or in outer space. See Figure 5.3.

$\vec{F} = m\vec{g}$ **Figure 5.3** Gravitational force on a mass m. The gravitational field $\vec{g}$ need not correspond to earth, nor need it be uniform.

The gravitational force $\vec{F}$ on the object equals the product of the gravitational mass m and the local value $\vec{g}$ of the gravitational field:

$$\vec{F} = m\vec{g}. \tag{5.1}$$

(At the surface of the moon, $|\vec{g}| \approx 1.4$ m/s^2, and $\vec{g}$ points toward the center of the moon.) From (5.1), the local value of $\vec{g}$ can be obtained from a knowledge of both $\vec{F}$ and m, via

$$\vec{g} = \frac{\vec{F}}{m}. \tag{5.2}$$

Newton's law of universal gravitation yields, for the force on m due to a point mass M, at a distance r from M,

$$\vec{F} = -\frac{GmM}{r^2}\hat{r}. \qquad (\hat{r} \text{ points to } m) \tag{5.3}$$

From (5.2) and (5.3), we deduce that a point mass M produces

$$\vec{g} = \frac{\vec{F}}{m} = -\frac{GM}{r^2}\hat{r}. \qquad (\hat{r} \text{ points to } m) \tag{5.4}$$

Gravity, like electricity, also satisfies an inverse square law. Unlike electricity, gravity only has the equivalent of "like" charges, and they only *attract*. That is the significance of the minus sign in (5.4).

The gravitational energy U_g is given in terms of the mass m and the gravitational potential, V_g, by

$$U_g = mV_g, \tag{5.5}$$

where $V_g = 0$ at some convenient place.

Example 5.1 V_g for a point mass

In your mechanics course, you may have learned that, if $U_g = V_g = 0$ at infinity, then $U_g = -GMm/r$ gives the gravitational interaction energy between the two masses m and M a distance r apart. What is the gravitational potential, V_g, produced by a point mass, M?

Solution: Equation (5.5) yields $V_g = -GM/r$.

Example 5.2 V_g for a uniform gravitational field

For a uniform downward gravitational field, as on the surface of the earth, $\vec{g} = -g\hat{y}$. With $U_g = 0 = V_g$ for $y = 0$, in your mechanics course you learned that $U_g = mgy$. What is $V_g(y)$?

Solution: From (5.5), we find that $V_g = gy$.

For gravity, energy conservation takes the form

$$\frac{1}{2}mv^2 + mV_g = \text{constant}. \tag{5.6}$$

We cannot overemphasize that the value of V_g itself has no meaning; only differences in V_g have meaning. Moreover, for (5.6) to apply at two positions $\vec{r}$ and $\vec{r}'$, the value of V_g at $\vec{r}$ must be independent of the path from $\vec{r}'$ to $\vec{r}$, and vice versa.

5.3 Electrical Potential Energy and Electrical Potential

We now turn to electricity, and a derivation of results analogous to those for gravity.

5.3.1 *Overview*

The electrical force on a charge q can be written as the product of q with the electric field $\vec{E}$:

$$\vec{F} = q\vec{E}. \tag{5.7}$$

Let us develop a similar relationship between the electrical energy U of a charge q as the product of q with the electrical potential V:

$$U = qV. \quad \text{(electrical energy from electrical potential)} \tag{5.8}$$

From (5.8), the electrical potential V has units of J/C, which defines the volt (V). Just as for gravity, only *differences* in electrical potential energy and electrical potential have meaning. Usually, such differences are measured with respect to "ground," which is taken to be at zero potential. The electrical potential concept has the same advantage as the electric field concept: it is a property of position and is independent of the test charge.

The Tricky Meaning of "Ground"

The origin of the term *ground* lies in early electrostatics experiments, where sometimes the experimenter or the experiment was electrically connected to the ground, and therefore was said to be grounded. In a modern building, lightning rods are connected to ground, or to electrical appliances or outlets that are grounded. However, ground for one building is unlikely to be exactly the same as ground for another building. Even the grounds at two wall plugs in the same laboratory can differ by a millivolt or more; this is because there are often small electric currents flowing from regions with a (temporary) excess of positive charge to regions of a (temporary) excess of negative charge. Therefore it is essential to know explicitly what ground is used in making voltage measurements. Let two electricians, *A* and *B*, use different grounds. If the ground of *A* is 2 V higher than the ground of *B*, then *all* voltages measured by *A* relative to his ground are higher by 2 V than those measured by *B* relative to his ground. However, all voltage *differences* are the same to both *A* and *B*.

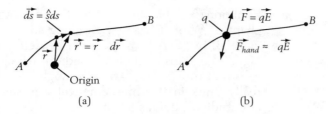

Figure 5.4 (a) A path from point A to point B, needed to compute the difference in electric potential between A and B. (b) To obtain $\vec{E}$, a test charge q is held in equilibrium by an external source of force, considered to be provided by a hand.

5.3.2 *Electrical Potential from Electrical Potential Energy*

Let us obtain an expression for electrical potential energy. Consider moving a charge q from A to B. Move it very slowly so that the work done does not increase the kinetic energy; then all the work W will go into changing the electrical potential energy U. It is a bit like winding up a spring. If our hand does the work, then

$$W_{hand} = \int_A^B \vec{F}_{hand} \cdot d\vec{s}. \tag{5.9}$$

Here, $d\vec{s}$ represents the tiny displacement vector between two nearby points $\vec{r}$ and $\vec{r}\,' = \vec{r} + d\vec{r}$ along the path. See Figure 5.4(a). Clearly, $d\vec{r} = d\vec{s}$, but it is convention to use $d\vec{s}$ rather than $d\vec{r}$. Note that $d\vec{s}$ points along the tangent vector $\hat{s}$ and has length $ds = |d\vec{s}| > 0$, so

$$d\vec{s} = \hat{s}ds. \tag{5.10}$$

Because q is moved ever so slowly, for all practical purposes q is subject to zero net force. See Figure 5.4(b). Thus, in taking this zero net force limit, (5.7) yields

$$\vec{F}_{hand} = -\vec{F} = -q\vec{E}. \tag{5.11}$$

Then, equating W_{hand} to the change in electrical energy yields

$$W_{hand} = U_B - U_A. \tag{5.12}$$

Thus (5.9) and (5.11) yield

$$U_B - U_A = \int_A^B \vec{F}_{hand} \cdot d\vec{s} = -q \int_A^B \vec{E} \cdot d\vec{s}. \tag{5.13}$$

Using the definition (5.8) in (5.13), the electrical potential difference between points A and B is given by

$$V_B - V_A = -\int_A^B \vec{E} \cdot d\vec{s}. \quad \text{(definition of voltage difference)} \tag{5.14}$$

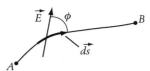

Figure 5.5 Information needed to compute the difference in electric potential between A and B.

Equation (5.14) is the basis for the remainder of this chapter. With $\vec{E}$ known all along some path from A to B, by (5.14) $V_B - V_A$ can be determined. See Figure 5.5. This is the field method, which Maxwell calls that of Faraday. (The self-taught Faraday didn't really think in this mathematical way since he had not studied calculus.) Note that $\vec{E}$, previously given in units of N/C, from (5.14) has the equivalent units of V/m. For (5.14) to be useful, we must establish that the potential difference $V_B - V_A$ is independent of the path between A and B. This is done in Section 5.7, after which Section 5.8 gives a number of examples of how to apply (5.14).

By the fundamental rules of integral calculus, the left-hand side of (5.14) can be rewritten as

$$V_B - V_A = \int_A^B dV = \int_A^B \frac{dV}{ds} ds. \tag{5.15}$$

We obtain dV/ds, the change in potential per unit length, from the right-hand side of (5.14), on using (5.10):

$$V_B - V_A = -\int_A^B \vec{E} \cdot d\vec{s} = -\int_A^B \vec{E} \cdot \hat{s} \, ds. \tag{5.16}$$

Equating the integrands in (5.16) and (5.15) yields

$$\frac{dV}{ds} = -\vec{E} \cdot \hat{s} = -|\vec{E}| \cos \phi,$$

(relationship between electric field and gradients in potential) (5.17)

where ϕ is the angle between $\vec{E}$ and $\hat{s}$. Hence, if the path of $d\vec{s}$ is along the field lines ($\vec{E}$), then $\vec{E} \cdot d\vec{s} > 0$, and the potential *decreases*. In other words, *field lines point from positions of high potential to positions of low potential.* The local direction of $\vec{E}$ defines the normal to a local surface called an *equipotential*; everywhere on an equipotential surface, the potential has the same value. The negative sign in (5.17) is important; it explains *why* $\vec{E}$ points from positions of higher voltage to positions of lower voltage. See Figure 5.1, where the lines that represent surfaces of constant potential are drawn with thicker lines than the electric field lines. Section 5.5 gives a more detailed discussion of equipotential surfaces, and Section 5.9 shows how to use (5.17) to obtain $\vec{E}$ from a knowledge of V. One advantage of the potential V is that it is a scalar from which the vector $\vec{E}$ can be obtained.

Example 5.3 Estimate of $\vec{E}$ from drawn equipotentials

In Figure 5.1, if the equipotentials are at 10 V and 8 V, and are separated by 5 cm at the position of the $\vec{E}$ symbol, estimate $|\vec{E}|$ midway between the equipotentials.

Solution: The two equipotentials have a voltage difference $\Delta V = 2$ V and a separation $\Delta s = 0.05$ m. Then, by (5.17)

$$|\vec{E}| = |dV/ds| \approx |\Delta V/\Delta s| = 2 \text{ V}/0.05 \text{ m} = 40 \text{ V/m}.$$

Example 5.4 Estimate of $\vec{E}$ from the electrical potential

Joan has been given the rule that $V(x) = 2x^3$, where V is in V and x is in m, and she is trying to find $\vec{E}$ at $x = 1$ m. We tell her that we can estimate it from $V(1.1)$ and $V(0.9)$. How do we do this?

Solution: First, we compute that, on the planes $x = 1.1$ and $x = 0.9$, $V(1.1) = 2.662$ V and $V(0.9) = 1.458$ V. Thus, as in the previous example, we have two equipotentials. Since the electric field points from the position where the voltage is high ($x = 1.1$ m) to the position where the voltage is low ($x = 0.9$ m), we immediately can tell her that the electric field points to the left, along $-\hat{x}$. To estimate $|\vec{E}|$ at $x = 1$ m, we use (5.17). First, we take $d\vec{s}$, and thus $\hat{s}$, along $\hat{x}$, corresponding to $\phi = 180°$ in Figure 5.5. Then (5.17) gives us $dV/ds = -|\vec{E}| \cos \phi = |\vec{E}|$. But

$$\left|\frac{dV}{ds}\right| \approx \left|\frac{\Delta V}{\Delta s}\right| = \frac{(2.662 - 1.458)\text{V}}{0.2\text{m}} = 6.02 \text{ V/m}.$$

Hence we tell her that $|\vec{E}| = |dV/ds| \approx 6.02$ V/m at $x = 1$ m. In Section 5.9, we will show how to calculate $\vec{E}$ exactly.

Example 5.5 Estimate of V_B from V_A and $\vec{E}$ in region between nearby points A and B

At point A, let $V_A = 15.2$ mV (mV = millivolts = 10^{-3} V) and let $\vec{E} = (3\hat{x} - 4\hat{y} + 5\hat{z})$ N/C. Estimate V_B, where B is separated from A by a small distance $\vec{r}_B - \vec{r}_A = d\vec{s}$, with $ds = |d\vec{s}| = 1$ mm and $d\vec{s}$ is along $\hat{s} = -0.6\hat{x} + 0.64\hat{y} - 0.48\hat{z}$.

Solution: We solve this by first using (5.17), which gives

$$dV/ds = -\vec{E} \cdot \hat{s} = 6.76 \text{ N/C} = 6.76 \text{ V/m, so } dV \approx (dV/ds)ds = 6.76 \text{ mV}.$$

Then, since points A and B are close to each other, we can rewrite (5.15) as $V_B - V_A = dV$ (we approximate the integral as only one term). Therefore we have $V_B = V_A + dV = 21.96$ mV.

Example 5.6 Numerical integration with a spreadsheet

Indicate how to use a spreadsheet to compute $V_B - V_A$.

Solution: Equation (5.14) requires $\vec{E}$ all along a path from A to B. First break up the path into a large number N of small steps $d\vec{s}_i = \hat{s}_i ds_i$ (e.g., $N = 100$, and $1 \leq i \leq 100$), the first of which begins on A, and the last of which ends on B. These numbers go into column A of the spreadsheet. Then determine (perhaps by employing elves), for each element i of the path, the x-, y-, and z-components of $d\vec{s}_i = \hat{s}_i ds_i$, or ds_i and the three values for $\hat{s}_i$. To be specific, let us do the latter. Then, in column B put ds_i, and in columns C to E put the three components of $\hat{s}_i$. In addition, for each element i, determine the three components of $\vec{E}_i$, which are put into columns F to H. Column I contains $-\vec{E}_i \cdot \hat{s}_i$, and column J contains

Table 5.1 Spreadsheet for calculation of potential difference

A	B	C	D	E	F	G	H	I	J
i	ds_i	s_{ix}	s_{iy}	s_{iz}	E_{ix}	E_{iy}	E_{iz}	$\vec{E} \cdot \hat{s}$	$-\vec{E} \cdot \hat{s}\,ds$
1	1.0×10^{-3}	-0.6	0.64	-0.48	3	-4	5	-0.676	0.676×10^{-3}

$dV_i = -\vec{E}_i \cdot \hat{s}_i ds_i$. The sum over column J then yields $V_B - V_A$. (See Table 5.1, where the entry is taken from Example 5.5.) We should also repeat the calculation with a finer partition of the path length, to check on convergence.

If we know the positions of all the charges, we can compute $\vec{E}$ along the path from A to B, rather than having to measure it.

5.3.3 *Electrical Potential Due to a Uniform Electric Field*

Consider a region of space in which there is a uniform electric field $\vec{E} = -E\hat{y}$. See Figure 5.6.

Such an electric field can be produced in an infinite variety of ways. One is to have a sheet of positive charge above the region; another is to have a sheet of negative charge below the region. In (5.14), with $V_A = 0$ corresponding to $y' = 0$, and $V_B = V$ corresponding to $y' = y$, we have $V_B - V_A \equiv \Delta V = V$. With $d\vec{s} = \hat{y}dy'$, (5.14) becomes

$$\Delta V = V = -\int_0^y (-E\hat{y}) \cdot (\hat{y}dy') = E \int_0^y dy' = Ey. \qquad \text{(uniform electric field)}$$

(5.18)

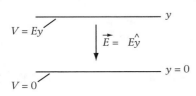

Figure 5.6 Two equipotentials for a region where the electric field is uniform.

This is very like the case of a uniform gravitational field. By (5.18), moving toward the region of higher electrical potential (larger y) is opposite to the electric field. This is true in general, as follows from (5.17). For $E = 10^4$ N/C and $y = .002$ m, (5.18) gives $V = 20$ V. (If we had taken $V_A = -5$ V, then (5.18) would give $V_B = V = 15$ V.) The equipotentials here correspond to constant values of y; they are planes that are perpendicular to the electric field.

5.4 Some Applications of Electrical Potential

5.4.1 *Electrical Conductors Are Equipotentials*

An immediate application of the definition of electrical potential can be made to the properties of electrical conductors in equilibrium. Consider a penny (made of copper, a good conductor) that is in equilibrium, and let the point A on its surface be at 5 V relative to ground. What can be said about the voltage at any other point B of the penny, either on the surface or in the interior? Because $\vec{E} = \vec{0}$ within the material of the conductor, by taking a path that goes through the material from A to B, (5.14) implies that $V_B - V_A = 0$, so $V_B = V_A = 5$ V. That is, the entire volume of the penny (including its surface) has the same potential. Thus the surface of the penny is an equipotential, at 5 V. Indeed, in equilibrium, the entire volume of an electrical conductor has the same potential as the surface. Moreover, if the conductor has a cavity, and there are no charges within the cavity, then the potential within the cavity is the same as at the surface.

5.4.2 *Energy Conservation Using Electrical Potential Energy*

Although potential is an abstract concept, energy is more concrete; hence, if there is a need to make electrical potential less abstract, (5.8) can be used to obtain electrical potential energy.

For a particle of charge q acted on only by electrical forces, conservation of energy takes the form:

$$\text{kinetic energy} + \text{electrical potential energy} = \frac{1}{2}mv^2 + qV = \text{constant,}$$

$$\text{(energy conservation)}   \text{(5.19)}$$

where V is the electrical potential, and qV is the electrical potential energy.

To be specific, consider two conductors of arbitrary shape (e.g., two aluminum, paint-sprayed rubber ducks; see Figure 5.7), connected by a 12-volt battery, so $V_B = V_A + 12$ V. Each of these conducting ducks is an equipotential. An electron initially at rest near the duck connected to the negative terminal (A) will move toward the duck connected to the positive terminal (B). To calculate its velocity on hitting the positive side, employ (5.19) in the form

$$\frac{1}{2}mv_A^2 + qV_A = \frac{1}{2}mv_B^2 + qV_B. \tag{5.20}$$

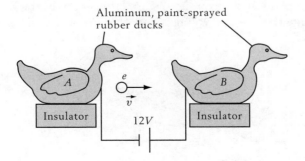

Figure 5.7 An electron moving between two conductors held at different electrical potentials by a 12 V battery; the large plate is positive, the small plate is negative.

By setting $v_A = 0$, $V_A - V_B = -12$ V, $q = -e = -1.6 \times 10^{-19}$ C, $m = 9.1 \times 10^{-31}$ kg, (5.20) yields that $v_B = \sqrt{2e(V_B - V_A)/m} = 2.05 \times 10^6$ m/s. The energy method cannot predict how long it takes to cross from A to B.

Correspondingly, measurement of the change in kinetic energy of a charged particle in going from A to B permits deduction of the change in its electrical potential energy and in its electrical potential.

Example 5.7   Charged dust attracted to a charged duster

A piece of dust with charge $Q = -8 \times 10^{-9}$ C and mass $M = 0.04$ g is at rest at point A. It is attracted to a charged duster, and when it hits the duster at point B, it is moving with a speed of 50 cm/s. Determine the voltage difference $V_A - V_B$.

Solution: Using (5.20), with $v_A = 0$ and $q \to Q$, we find that $QV_A = \frac{1}{2}mv_B^2 + QV_B$. Thus

$$V_A - V_B = \frac{m}{2Q}v_B^2 = \frac{0.04 \times 10^{-3} \text{ kg}}{(2)(-8 \times 10^{-9} \text{ C})}(0.5 \text{ m/sec})^2 = -625 \text{ V}.$$

Note that the negative charge Q should be attracted to a position (B) with a more positive potential than where it starts (A). Indeed, B has a higher potential than A by 625 V.

5.4.3 *A Voltage Electrometer*

Section 2.4 discussed a charge electrometer, where from measuring a repulsive electrical force the charge can be deduced. We now discuss a 19th-century *voltage electrometer*, similar to that used by Kelvin, where from measuring an attractive electrical force the voltage can be deduced. Modern voltage electrometers do not have such a transparent construction.

Consider two conductors initially at voltages V_1 and V_2, where $V_2 > V_1$. Also consider two conducting disks of radius R and area $A = \pi R^2$, concentric with each other at small separation $d \ll R$. On connecting the conductors to the disks, as in Figure 5.8, the disks obtain equal and opposite charges, with positive charge at the higher-voltage plate. If the conductors are large enough, the disks do not draw off much charge. Hence the voltages and charges of the disks will be V_1

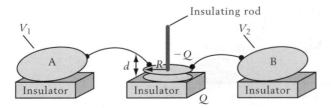

Figure 5.8 A voltage electrometer to measure voltage differences between two conductors A and B. The conducting disks must take up only a small amount of the charge on either conductor, and thus must be relatively small compared to the conductors.

and $-Q$, and V_2 and Q. (The disks in Figure 5.8 are comparable in dimension to the conductors to which they are connected, and thus actually would draw off a relatively large amount of charge. How much charge goes to each disks is related to the subject of *capacitance*, to be taken up in the next chapter.)

By measuring the force of attraction between the disk-shaped plates, and neglecting edge effects (so there will be a uniform electric field of magnitude E between the plates), we can deduce their voltage difference $V_2 - V_1$.

Here's how it is done. In terms of the charge density $\sigma = Q/A$ on the positive plate, the total electric field has magnitude double that from one of the plates. Thus, use of (3.27) gives $E = 2(2\pi k\sigma) = 4\pi kQ/A$. Then, from (3.27) with $y = d$,

$$\Delta V = Ed = \frac{4\pi kd}{A}Q. \tag{5.21}$$

The force magnitude $F = |\vec{F}|$ on the positive plate is the product of its charge Q and the electric field magnitude $E_- = 2\pi k\sigma$ due to the negative plate. Then, using (5.21),

$$F = QE_- = Q(2\pi k\sigma) = \frac{2\pi kQ^2}{A} = \frac{(\Delta V)^2}{8\pi kd^2}A. \tag{5.22}$$

On measuring F and A, use of (5.22) yields ΔV. For $A = 4$ cm^2 and $d = 0.7$ mm, $\Delta V = 150$ V gives $F = 1.444 \times 10^{-8}$ N. The sensitivity can be increased significantly if d can be made very small.

For more general plate geometries $F = \alpha(\Delta V)^2$ applies, but the proportionality constant α typically cannot be calculated easily. On the other hand, if F and ΔV both can be measured by some other means, this proportionality factor can be deduced, and thus the voltage electrometer can be calibrated. For example, if $F = 54$ nN when $\Delta V = 98$ V, then $\alpha = F/(\Delta V)^2 = 5.62 \times 10^{-12}$ N/V^2. A later measurement of F that gives $F = 23$ nN then corresponds to $\Delta V = 63.4$ V.

5.5 Equipotential Surfaces and Electric Fields

The direction of $\vec{E}$ defines the local normal to a surface. By (5.14), for $d\vec{s}$ along that surface (i.e., perpendicular to $\vec{E}$), the potential does not change. Hence, such a surface is an equipotential. See Figure 5.1, which depicts, of the two directions perpendicular to $\vec{E}$, only the one in the plane; the other perpendicular direction is normal to the page.

5.5.1 *Equipotentials from Electric Fields*

For a sheet of charge the equipotentials are planes, since a plane is traced out by moving in the two directions perpendicular to the normal to the sheet. See Figure 5.9(a). For a line charge the equipotentials are cylinders, since a cylinder is traced out by moving in the two directions perpendicular to the radius vector. See Figure 5.9(b). For a point charge the equipotentials are spheres, since a sphere

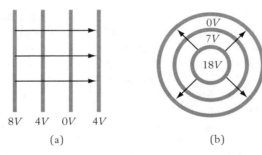

(a)　　　　　　　　　(b)

Figure 5.9 Electrical potentials and field lines: (a) for uniform electric field; (b) for cylindrical symmetry.

is traced out by moving in the two directions perpendicular to the radius vector. In these figures, the equipotentials have been drawn with thick shaded lines, and the field lines have been given arrows.

If there is a preexisting equipotential, and a thin, neutral conductor has the same shape and position of that equipotential, then the thin conductor does not disturb the potential. This is because charge rearranges on the thin conductor so that the field lines enter negative charges on one side and leave equal and opposite positive charges on the other side. Thus any of the equipotentials in Figure 5.9(a) and Figure 5.9(b) could correspond to thin conductors in the shape of a plane or a cylinder, respectively.

Here are two simple examples of field lines and equipotentials for a discrete number of charges: the dipole pair $\pm q$ (whose field lines were given in Figure 3.19a) shown in Figure 5.10(a), and the two equal charges q (whose field lines were given in Figure 3.20) shown in Figure 5.10(b). In order to avoid clutter in these and some figures to follow, not all field lines have been given arrows.

5.5.2　*Electric Field Lines from Equipotentials*

It is also of interest to show the field lines and equipotentials for the interior of a conducting cylinder of square cross-section, whose sides are at different potentials. (Note: In a technical sense, a cylinder can have any cross-section at all, but each cross-section must be the same.) In that way, we can get a feeling for the surface charge density, both in sign and magnitude. See Figure 5.11 for three examples. In each case the field lines are normal to the conducting surfaces,

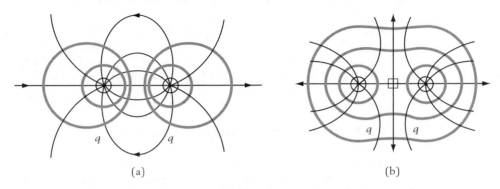

(a)　　　　　　　　　(b)

Figure 5.10 Electrical potentials and field lines. (a) Equal and opposite charges. (b) Equal charges.

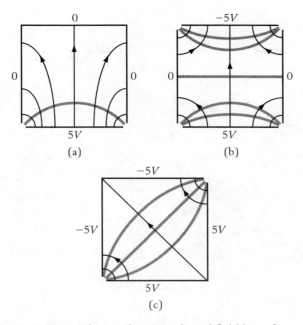

Figure 5.11 Electrical potentials and field lines for geometries with symmetry along the axis normal to the page. The conductors are depicted as the sides of a square, with gaps separating regions of different potential. Field lines originate on positive charge and terminate on negative charge, and point from regions of high potential to regions of low potential.

and point from positions of higher potential to positions of lower potential. A higher density of field lines implies a higher field magnitude $|\vec{E}|$.

Figure 5.12(a), shows the field lines and equipotentials near the edges of two plates with equal and opposite voltages.

Finally, Figures 5.12(b) and 5.12(c) show two sets of field lines and equipotentials for the same geometry. However, the plate voltage differences in Figure 5.12(c) are five times those for Figure 5.12(b). By $|\vec{E}| \approx \Delta V/\Delta s$, the associated electric fields are five times those for Figure 5.12(b). Note that, in Figure 5.12(b), the electric field strength grows on moving from point d to point a because the voltage difference is fixed at 5 V but the plate separations Δs decrease on moving from field line d to field line a.

5.5.3 Maxwell's Trick: Obtaining Surface Charge for Conductors with Special Shapes
Optional

In general, it is extremely difficult to obtain the charge distribution on the surfaces of a given a set of conductors. Maxwell found a useful trick to design conductor shapes for which the charge distribution can be determined.

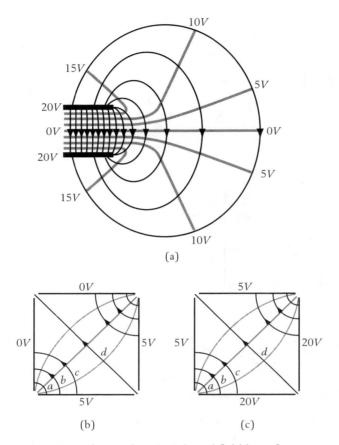

(a)

(b)   (c)

Figure 5.12 Electrical potentials and field lines for geometries with symmetry along the axis normal to the page. (a) Field lines and equipotentials near the edges of two plates with equal and opposite voltages. Parts (b) and (c) are related to Figure 5.11(c).

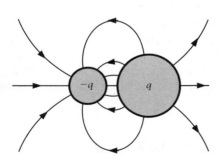

Figure 5.13 Electrical potentials and field lines for conductors with equal and opposite charge. The actual surfaces chosen correspond to equipotentials taken from Figure 5.10(a).

First, choose a set of point charges, and draw a set of equipotentials for those charges, using the theory developed in this chapter. Let us use the dipole equipotentials of Figure 5.10(a) as an example. Next, arbitrarily choose two equipotentials, and then make real conductors in their shapes, giving them charges equal to the sum of the point charges they are to enclose. Now remove the point charges. For example, Figure 5.13 shows two equipotentials, one surrounding the positive and the other surrounding the negative of the dipole. Here is how to find the charge density at any point on the surface of one of the real conductors. From the original point charges find the electric field at that point; this will equal the electric field $\vec{E}_{out}$ just outside the conductor. Then the surface charge is obtained from (4.23) of the previous chapter,

$$\sigma_S = \frac{\vec{E}_{out} \cdot \hat{n}}{4\pi k},$$ (5.23)

where $\hat{n}$ is the outward normal to the conductor.

As described in Section 4.11, this trick was employed by Kelvin for the specific case of two point charges $\pm q$ and their midplane at $V = 0$, but apparently it was first stated in full generality by Maxwell. Note that for an individual point charge the field lines are normal to it. Thus a point charge may be thought of as an infinitesimal conducting sphere.

5.6 Point Charges: Electrical Potential Energy and Electrical Potential

Individual point charges lie at the heart of the action-at-a-distance viewpoint. We will study the case of a single charge, and then use the principle of superposition to obtain the general result.

5.6.1 *Potential Due to a Single Point Charge*

Consider a point charge Q at the origin. See Figure 5.14. Let $V = 0$ at infinity, taken as the point A in (5.14), and let $V_B = V$ at the arbitrary position $\vec{r}$, taken as the point B in (5.14). Then $V_B - V_A = V - 0 = V(\vec{r})$. To avoid integrating from ∞ to r, switch the limits of integration and include a sign change. Then (5.14) becomes $V_B - V_A = \int_B^A \vec{E} \cdot d\vec{r}$, with $\vec{E} = (kQ/r^2)\hat{r}$ and $d\vec{s} = \hat{r} dr$, so

$$V(\vec{r}) = \int_r^\infty \frac{kQ}{r^2}(\hat{r}) \cdot (\hat{r} \, dr) = kQ \int_r^\infty \frac{1}{r^2} \, dr = kQ \cdot \frac{-1}{r} \Big|_r^\infty = \frac{kQ}{r}. \qquad Q \text{ at origin}$$ (5.24)

For Q at $\vec{r}_0$, we have

$$V(\vec{r}) = \frac{kQ}{R_0}, \qquad \vec{R}_0 = \vec{r}_0 - \vec{r}. \qquad Q \text{ at } \vec{r}_0 \qquad \text{(potential of point charge)}$$ (5.25)

The equipotentials are concentric spheres centered about $\vec{r}_0$. For example, if $kQ = 10$ V-m (corresponding to $Q = 1.1 \times 10^{-9}$ C), the concentric sphere with $R_0 = .01$ m is an equipotential at $V = 1000$ V.

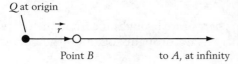

Q at origin

Point B   to A, at infinity

Figure 5.14 Point charge Q at the origin, and points A (at infinity) and B, for use in calculation of the electrical potential due to a point charge.

Caution: Don't Over Generalize

Let $Q > 0$, so $E = |\vec{E}| = kQ/r^2$. Many students notice that if $V = 0$ at infinity ($V_\infty = 0$), then (5.24) yields $V = Er$ for the case of the point charge. This is so similar to (5.18) for the uniform electric field that they are tempted to conclude that such a relationship holds in general. It doesn't. For every other situation we consider, this equation will *not* hold. For example, if there are two charges Q separated by a distance $2a$, the electric field at their midpoint is zero. See Figure 5.15(a). However, the voltage at their midpoint, if $V_\infty = 0$ is

$$V = \frac{kQ}{a} + \frac{kQ}{a} = \frac{2kQ}{a}.$$

Clearly, nothing like $V = Er$ holds in this case. Likewise, if one of the charges is $-Q$, the voltage at their midpoint is

$$V = \frac{kQ}{a} + \frac{k(-Q)}{a} = 0,$$

but the electric field is nonzero. See Figure 5.15(b). Again, nothing like $V = Er$ holds. As we will see, $\vec{E}$ and V are related in all cases, but only for the cases of a sheet charge and a single point charge can the relationship appear as simple as $V = Er$ or $V = Ed$.

$$V \neq 0,\ \vec{E} = \vec{0}$$

(a)

$$V = 0,\ \vec{E} \neq \vec{0}$$

(b)

Figure 5.15 Geometries for calculation of electrical potential. (a) Two equal source charges Q_1. (b) Two equal and opposite source charges $\pm Q$.

5.6.2 *Potential Energy of Two Point Charges*

Except for the absence of a minus sign, (5.24) is much like the corresponding gravitational case $V_g = -GM/r$ for a point source M. Note that a positive charge produces a positive electrical potential, and a negative charge produces a negative electrical potential. The direction toward higher potential (either toward the positive charge or away from the negative charge) is opposite the direction of the electric field (radially outward for the positive charge, and radially inward for the negative charge), as expected from (5.17).

Combining (5.8) and (5.24) for q in the presence of the potential due to Q immediately leads to

$$U = qV = \frac{kqQ}{r}. \qquad Q \text{ at origin} \qquad \text{(electrical energy of two charges)}$$

(5.26)

According to (5.26), if two like charges are brought together, the energy increases ($U > 0$), and if two unlike charges are brought together, the energy decreases ($C < 0$).

Example 5.8   **Energy of an electron in a hydrogen atom**

Let the electron ($q = -e$) and proton ($q = e$) be at a distance of 10^{-10} m, which is a characteristic atomic dimension. Find (a) the potential that the proton produces at the site of the electron, and (b) the electrical potential energy of the electron.

Solution: (a) The potential that the proton produces at the site of the electron is given by $V = ke/r = 14.4$ V. Thus, the volt is a unit of electrical potential that is also characteristic of the electrical potential in the atom. (b) Using this V and (5.8), or by (5.26), the electrical potential energy of the electron (which is the energy required to take the electron infinitely far away from the proton) is $(-e)V = -2.30 \times 10^{-19}$ J $= -14.4$ eV. Here we use the energy unit of the eV $\equiv 1.6 \times 10^{-19}$ C-V $= 1.6 \times 10^{-19}$ J. The eV is a unit of energy that is characteristic of electrical potential energies in the atom. Since the electrical potential energy is zero when the electron and proton are infinitely far apart, the electron and proton "prefer" to be together, where the electrical potential energy is lower, than to be apart.

Example 5.9   **Energy of two protons in a ^{4}He nucleus**

Let the protons be at a distance of 10^{-15} m, which is a characteristic nuclear dimension. Find (a) the potential that one proton produces at the site of the other, and (b) the energy the system would gain if the protons were separated.

Solution: (a) $V = ke/r = 1.44 \times 10^6$ V. This is conveniently expressed in terms of megavolts $= MV = 10^6$ V, a characteristic value associated with nuclei, as 1.44 MV. (b) Using this V and (5.8), or by (5.26), the electrical potential energy (i.e., the energy that the system would gain if they were to be separated) is $eV = 2.30 \times 10^{-13}$ J $= 1.44$ MeV, using the energy unit of the MeV $= 10^6$ eV. The MeV is characteristic of nuclear energies. The positive electrical potential energy means that the protons would "prefer" not to be together, on the basis of the electrical potential energy alone. They are held together by an attractive nuclear interaction that overcomes the positive electrical potential energy.

5.6.3   *Potential Energy of Many Electric Charges*

For two electric charges q_1 and q_2, separated by $r_{12} = |\vec{r}_1 - \vec{r}_2|$, the electrical energy U of (5.26) generalizes to

$$U = \frac{kq_1q_2}{r_{12}}, \qquad r_{12} = |\vec{r}_1 - \vec{r}_2|. \tag{5.27}$$

This is the energy we would have to provide in order to bring the charges from infinity to this separation. The electrical energy computed as the sum over each charge of its charge times the potential due to others gives

$$q_1 \left(\frac{kq_2}{r_{12}} \right) + q_2 \left(\frac{kq_1}{r_{12}} \right) = \frac{2kq_1q_2}{r_{12}},$$

which is *twice* the correct answer. Hence, to obtain the total electrical energy in general, we must perform this sum, and then *divide by two*. This avoids what is called double counting. For the general case, sum over all charges q_i and q_j with separation r_{ij}, making sure not to include the infinite self-interaction corresponding to $i = j$. Thus,

$$U = \frac{1}{2} \sum_{i \neq j} \frac{kq_i q_j}{r_{ij}}, \qquad r_{ij} = |\vec{r}_i - \vec{r}_j|. \tag{5.28}$$

For two charges, where i and j can only take on the values $i = 1$, $j = 2$ and $i = 2$, $j = 1$, (5.28) reproduces (5.27).

<hr>

Application 5.1 | **Energy of a Square of NaCl**

Except in the atomic nucleus, matter is basically held together by electrical energy. Consider ordinary table salt, NaCl, where an atom of Na has 11 electrons and 11 protons, and an atom of Cl has 19 electrons and 19 protons. Let us estimate this electrical energy for a square of NaCl.

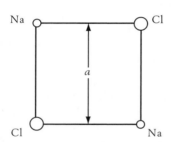

In chemistry, it is well known that atoms are particularly stable if they contain a number of electrons that is one of the closed shell values of 2, 10, 18, 36, 54, and 84. For a crystal of NaCl, Na gives up an electron to Cl so that Na^+ has 10 electrons and Cl^- has 18 electrons, both of these closed shell values. For Na^+ the 10 electrons are outnumbered by the 11 protons of the nucleus, so they compress inward to form a *compact* closed shell ion; hence the Na^+ are drawn as *small spheres* in Figure 5.16. For Cl^- the 18 electrons outnumber the 17 protons of the nucleus, so they expand outward to form an *extended* closed shell ion; hence the Cl^- are drawn as *large*

Figure 5.16 Part of a salt crystal (NaCl). The Na^+ ions, which have lost an electron, have a smaller outer shell of orbitals. The Cl^- ions, which have gained an electron, have a larger outer shell of orbitals.

spheres in Figure 5.16. (From x-ray scattering off NaCl, it has been deduced that the ionic radii are $R_{Na^+} = 0.095$ nm and $R_{Cl^+} = 0.181$ nm.)

The closed shell values come about when one applies the rules of quantum mechanics to study the motion of electrons—subject only to electrical forces—in atoms. Within atoms, although the laws describing electron motion differ from the laws of ordinary mechanics, electricity still provides the interactions that drive electron motion.

<hr>

We will find the electrical potential at the site of one Na^+ ion, and employ it to compute the average electrical energy per ion. Since there are two Cl^- ions with charge $-e$ a distance a away, and one Na^+ ion with charge e a distance $\sqrt{2}a$ away, by summation over (5.24) the electrical potential at the site of the

Na^+ ion is

$$V_{Na^+} = 2\frac{k(-e)}{a} + \frac{ke}{\sqrt{2}a}. \tag{5.29}$$

By (5.8), the electrical potential energy of the Na^+, with charge $q = e$, is thus

$$U_{Na^+} = qV_{Na^+} = -\frac{ke^2}{a}\left(2 - \frac{1}{\sqrt{2}}\right). \tag{5.30}$$

Each Na^+ ion has the same electrical potential energy, which is the same as for each Cl^- ion. To avoid double counting, the energy per ion is obtained by dividing U_{Na^+} by two. Thus,

$$U_{ion} = \frac{1}{2}U_{Na^+} = -\frac{ke^2}{2a}\left(2 - \frac{1}{\sqrt{2}}\right).$$

The total energy U is four times U_{ion}. Applying (5.28) and doing the sum over all 12 terms with $i \neq j$ gives the same $U = 4U_{ion}$.

Taking $a = 10^{-10}$ m and $e = 1.6 \times 10^{-19}$ C yields $U_{ion} \approx -1.49 \times 10^{-18}$ J. In units of the electron-volt, or eV, this is -9.3 eV. A negative potential energy means that the ions "prefer" this configuration to being separated from one another at infinity. A typical value for the binding energy of an ionic crystal is a few eV.

5.7 Electrical Potential Is Path Independent

We now prove that the electrical potential difference between any two points has the same value for all paths between the two points. This implies that the electrical potential is uniquely defined. Thus, on taking a charge q around any closed path, the electrical force does no work on q.

First, consider two paths between points A and B, as in Figure 5.17(a). To be specific, let the source of electric field be a point charge Q, at respective

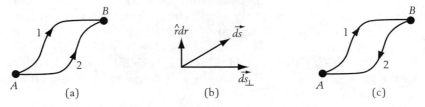

Figure 5.17 Electrical potential between two points A and B is path independent. (a) Two paths from A to B. (b) For the case of a single source charge (not shown), decomposition of displacement $d\vec{s}$ into parts along the radial direction and normal to the radial direction. (c) The voltage change on going from A to B along 1, and then from B to A along 2, is zero. This is equivalent to saying that the electric circulation $\Gamma_E = \oint \vec{E} \cdot d\vec{s}$ along such a closed path is zero.

distances r_A and r_B from A and B. Since Q makes an electric field that points radially outward, by (5.14) displacements $d\vec{s}$ normal to the radial direction cause no change in potential. More explicitly, writing the displacement $d\vec{s}$ in terms of components $\hat{r}dr$ along $\hat{r}$ and $d\vec{s}_\perp$ normal to $\hat{r}$, $\hat{r} \cdot d\vec{s}_\perp = 0$. See Figure 5.17(b). Thus, for either path,

$$V_B - V_A = -\int_{r_A}^{r_B} \frac{kQ}{r^2}(\hat{r}) \cdot (\hat{r}dr + d\vec{s}_\perp)$$

$$= -kQ\int_{r_A}^{r_B} \frac{1}{r^2}dr + 0 = kQ\frac{1}{r}\Big|_{r_A}^{r_B} = \frac{kQ}{r_B} - \frac{kQ}{r_A}. \qquad (5.31)$$

Although each path has a different set of $\hat{r}$ at intermediate steps of the integration, the result is the same for each path. This proves that, for a point charge Q as source, the electrical potential is path independent.

Now we invoke the principle of superposition: since the electrical potential is path independent for one point charge, it must be path independent if it is the sum of the potentials of many point charges. Therefore the electrical potential is path independent in general.

Another way to state this result is to consider what is called the *circulation* Γ_E of $\vec{E}$ around a closed path. It is defined by

$$\Gamma_E \equiv \oint \frac{d\Gamma_E}{ds}ds, \qquad (5.32)$$

where the circulation per unit length $d\Gamma_E/ds$ is defined as

$$\frac{d\Gamma_E}{ds} = \vec{E} \cdot \hat{s}, \qquad (5.33)$$

with $\hat{s}$ the tangent vector. In (5.32), the circle around the integral sign denotes that the path is closed. Comparison of (5.17) with (5.33) shows that

$$\frac{d\Gamma_E}{ds} = -\frac{dV}{ds}. \qquad (5.34)$$

Water swirling down a sink has both fluid circulation and fluid flux. Water swirling about a glass has only a fluid circulation. Water at rest has neither.

In computing the circulation around a path, we may think of the path as beginning at A, reaching B along path 1, and then returning to A along path 2. See Figure 5.17(c). Because of (5.34), (5.14) yields $-(V_B - V_A)_1$ for the first part of the path, and $-(V_A - V_B)_2$ for the second part of the path. Because $V_B - V_A$ is path independent, (5.32) and (5.34) yield

$$\Gamma_E = -(V_B - V_A)_1 - (V_A - V_B)_2 = -(V_B - V_A)_1 + (V_B - V_A)_2 = 0. \quad (5.35)$$

That is, *the circulation of any electric field due to static electric charges is zero.* This explains the last field-line rule in Chapter 3: the field lines due to static electric charges do not close on themselves. If they did, then for a path along the field lines $d\Gamma_E/ds$ would always be positive, and thus Γ_E for this path would be positive.

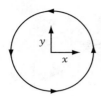

Figure 5.18 Geometry for a closed path, to be used with an electric field with nonzero circulation. Such an electric field cannot be produced by static electric charge.

There are many vectors besides $\vec{E}$ due to electrostatics for which the circulation *can* be nonzero over certain paths: (1) a force $\vec{F}$ that is frictional in nature; (2) the magnetic field (whose symbol is $\vec{B}$) surrounding a wire that carries electric current, if the integral is over a path that encloses the wire; (3) the electric field $\vec{E}$ induced by a time-varying magnetic field (to be discussed in Chapter 12). An example of such an $\vec{E}$ field is given in Figure 5.18. This has a field line that closes on itself, and because electric current flows along field lines, this can cause electric currents to circulate. It has a nonzero circulation and does not correspond to a uniquely defined electrical potential.

Example 5.10 **An $\vec{E}$ that cannot be due to electric charges at rest**

With C a constant having units of V/m², let

$$\vec{E} = C\hat{z} \times (\vec{r}\,) = C\hat{z} \times (x\hat{x} + y\hat{y} + z\hat{z}) = C(x\hat{y} - y\hat{x}), \text{ so } |\vec{E}| = C\sqrt{x^2 + y^2}.$$

Relative to the circle in Figure 5.18, this points along the tangent. Take $d\vec{s}$ to circulate counterclockwise, which is along $\vec{E}$. Compute the circulation for that circle (taking it to have radius a and to be centered at the origin).

Solution: By (5.33),

$$d\Gamma_E/ds = \vec{E} \cdot d\vec{s} = |\vec{E}|ds \cos 0° = C\sqrt{x^2 + y^2}ds = Cads,$$

so

$$\Gamma_E = \oint (d\Gamma_E/ds)ds = C\oint (a)ds = Ca \oint ds = Ca(2\pi a) = C(2\pi a^2).$$

Since the circulation is nonzero, this $\vec{E}$ cannot be due to electric charges at rest.

5.8 Calculating *V* from $\vec{E}$

Here are a few more examples (besides the uniform electric field and the point charge) that use the field viewpoint to determine V from $\vec{E}$ via (5.14). As discussed in Example 5.2, $\vec{E}$ can be obtained by the methods of Chapter 3 (either via measurement or calculation) or, when there is enough symmetry, by the methods of Chapter 4. Thus, this section brings together concepts from more than one chapter.

With this field viewpoint, initially the charges are in place, producing an electric field $\vec{E}$ everywhere, and the observer is at A. The observer then moves to B, integrating over $-\vec{E} \cdot d\vec{s}$ to obtain the change in potential on going from A to B.

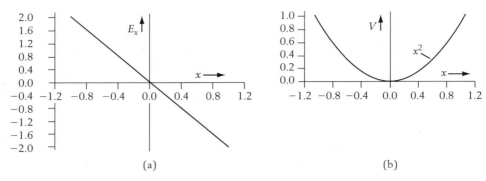

Figure 5.19 (a) $E_x = -2x$. (b) $V(x) = x^2$, with $V(0) = 0$.

Example 5.11 **A problem in one dimension**

Let $\vec{E}$ point only along the x-axis, with $E_x = -2x$. See Figure 5.19(a). (An electric field that varies linearly in a region of space is produced by a slab of constant charge density ρ and is relevant to depletion layers in semiconductors, as discussed in Section 4.11.) Let $V(0) = 0$ at $x = 0$. Find $V(x)$.

Solution: Take B to correspond to x, and A to correspond to $x = 0$. Using $d\vec{s} = \hat{x}dx$, (5.14) yields

$$V(x) - V(0) = -\int_0^x E_x dx = \int_0^x 2x dx = x^2.$$

Since $V(0) = 0$, this gives $V(x) = x^2$. See Figure 5.19(b). The equipotentials are planes normal to the x-axis.

Example 5.12 **Ring of charge**

Consider a charge Q that is uniformly distributed over a ring of radius a that is centered at the origin, its normal along the x-axis. See Figure 5.20. Find V along its axis, using (5.14). (To find V off the axis requires either advanced mathematical methods or numerical methods as discussed earlier.)

Solution: In Chapter 3, $\vec{E}$ was determined for a restated version of this problem. There the axis was z, and the charge was given in terms of the uniform charge per unit length λ as $Q = \lambda(2\pi a)$ along the x-axis. See (3.29). Here is a brief review of the calculation. Recall that a charge dq produces an electric field $d\vec{E} = (kdq/r^2)\hat{r}$ of magnitude $|d\vec{E}| = kdq/r^2$. On the x-axis, the distance $r = (a^2 + x^2)^{1/2}$ is constant, and the x-component of the unit vector along the observer

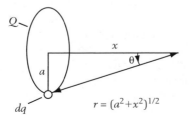

Figure 5.20 Ring of charge, for use in calculating the potential as a sum over the potential dV contributed by each bit of charge dq.

is $(\hat{r})_x = \cos\theta = x/r$, also a constant. Hence, by symmetry, the only nonzero component of the electric field is given by

$$E_x = \int dE_x = \int |d\vec{E}|(\hat{r})_x = \int \frac{kdq}{r^2}\cos\theta = \frac{kx}{r^3}\int dq = \frac{kxQ}{r^3}. \quad (5.36)$$

Because there is a finite amount of charge (unlike the case of a uniformly charged infinite sheet), we may set $V = 0$ at infinity (point A) without running into mathematical difficulties. Then the potential along the x-axis at x (point B) is given by

$$V(x) - V_\infty = V(x) = -\int_\infty^x \vec{E}\cdot d\vec{s}$$

$$= \int_x^\infty \vec{E}\cdot(\hat{x}dx) = \int_x^\infty E_x dx = \int_x^\infty \frac{kQx}{(a^2+x^2)^{3/2}}dx. \quad (5.37)$$

This integral is done by using $r^2 = a^2 + x^2$. Then $2rdr = 2xdx$, or $xdx = rdr$, so (5.37) becomes

$$V(x) = \int_r^\infty \frac{kQr}{r^3}dr = \int_r^\infty \frac{kQ}{r^2}dr = -\frac{kQ}{r}\Big|_r^\infty = \frac{kQ}{r}. \qquad r = \sqrt{a^2+x^2} \quad (5.38)$$

Note that from (5.38) alone we cannot draw surfaces of fixed potential, or equipotentials, because we have V only on the x-axis, not on a surface.

Example 5.13   Concentric cylindrical shells with equal and opposite charge

Consider two infinite, concentric cylindrical shells of radii, $a < b$, with charge per unit length λ on the inner cylinder and charge per unit length $-\lambda$ on the outer cylinder. Find the potential difference between the cylinders.

Solution: To find the electric field between the cylinders, apply Gauss's law with a Gaussian surface that is a concentric cylinder of length L and radius r. See Figure 5.21. For $a < r < b$,

$$E_r A_{flux} = E_r(2\pi r L) = (4\pi k)Q_{enc} = (4\pi k)\lambda L,$$

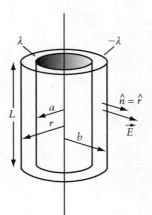

Figure 5.21 Finite section for two concentric cylindrical shells, with charge $\pm\lambda$.

so $E_r = 2k\lambda/r$, and $\vec{E} = E_r\hat{r}$. Since the positively charged inner shell should be at the higher potential, in applying (5.14), take B to correspond to $r = a$ and A to correspond to $r = b$. Then, with $d\vec{s} = \hat{r}dr$,

$$V(a) - V(b) = -\int_b^a \vec{E} \cdot d\vec{s} = \int_a^b (E_r\hat{r}) \cdot \hat{r}dr = \int_a^b \frac{2k\lambda}{r}dr = 2k\lambda\ln r\Big|_a^b = 2k\lambda\ln\frac{b}{a}.$$

$$(5.39)$$

Since $\vec{E}$ is radially outward, the equipotential surfaces $V = const$, which are normal to $\vec{E}$, are concentric cylinders. More generally, $V(a) - V(r) = 2k\lambda\ln(r/a)$.

Note that ground has not been specified. It could be on either cylindrical shell. Since $Q_{enc} = \lambda L - \lambda L = 0$ for a concentric Gaussian surface that is outside the outer shell, the electric field is zero outside the outer shell. Therefore the outer shell is at the same potential as infinity, or $V(b) = V(\infty)$. If there were only a single cylindrical shell, the voltage difference between it and infinity would be infinite, due to the infinite amount of net electric charge.

This example is very comprehensive: it starts with electric charge (Chapters 1 and 2), moves to electric field (Chapter 3, which it evaluates (because of the high degree of symmetry) by Gauss's law (Chapter 4), and then considers and evaluates a potential difference (Chapter 5).

Example 5.14 **Spherical ball**

The previous chapter showed that, for any charge distribution with spherical symmetry, Gauss's law applied to a concentric spherical Gaussian surface yields (4.15), $E_r = kQ_{enc}/r^2$. Consider a ball of radius A, with a uniform volume density of charge Q. See Figure 5.22. Let $V_\infty = 0$. Find $V(r)$ for (a) $r > a$, and (b) $r < a$. (c) Show that $V(0) = 1.5V(a)$.

Solution: (a) Outside the ball ($r > a$), the electric field is the same as for a point charge. On setting $V_\infty = 0$, the result (5.24), $V(r) = kQ/r$, applies. (b) For $r < a$, however, the electric field is less than for a point charge because Q_{enc} is proportional to the volume of a sphere of radius r: $Q_{enc}/Q = r^3/a^3$. Thus $Q_{enc} = Q(r/a)^3$, so $E_r = kQ_{enc}/r^2 = kQr/a^3$. See Figure 5.23(a). Then, with $d\vec{s} = \hat{r}dr$,

$$V(r) - V(a) = -\int_a^r (E_r\hat{r}) \cdot \hat{r}dr = -\int_a^r \frac{kQr}{a^3}dr$$

$$= -\frac{kQ}{a^3}\frac{1}{2}r^2\Big|_a^r = -\frac{kQ}{a^3}\frac{r^2 - a^2}{2}. \qquad (r < a) \qquad (5.40)$$

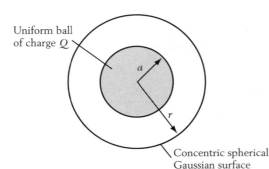

Uniform ball of charge Q

Concentric spherical Gaussian surface

Figure 5.22 Uniform ball of charge.

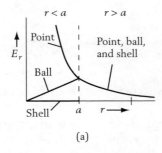
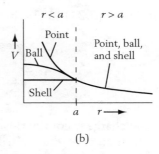

Figure 5.23 Comparison between point charge, uniformly charged ball, and shell of charge. (a) Electric field, which is radial. (b) Electric potential.

Since $V(a) = kQ/a$, this becomes

$$V(r) = V(a) - \frac{kQ}{a^3}\frac{r^2 - a^2}{2} = \frac{3}{2}\frac{kQ}{a} - \frac{kQ}{a}\frac{r^2}{2a^2}. \qquad (r < a) \qquad (5.41)$$

See Figure 5.23(b). The equipotentials are spheres.
   Note that $V(0) = 1.5V(a)$. If $V(a) = 100$ V, then $V(0) = 150$ V.

Example 5.15   **Spherical shell**

Now consider that the charge Q is on a spherical shell of radius a, which is how charge would distribute on a conductor of outer radius a (and arbitrary number and shape for the inner cavities). Let $V_\infty = 0$. Find $V(r)$ for (a) $r > a$, and (b) $r < a$.

Solution: (a) For $r > a$, the electric field is like that of a point charge, just as for the ball of charge. Hence, outside the shell and on the shell the potential is the same as that for a point charge: $V = kQ/r$. (b) For $r < a$, the electric field is zero so that the electric potential does not change on moving from the surface of the shell to the interior. Hence the potential within the shell takes the same value as on the outer surface, kQ/a. See Figure 5.23(a) and Figure 5.23(b).

5.9   *V* as a Sum over Point Charges (Action-at-a-Distance Viewpoint)

The action-at-a-distance viewpoint, as indicated by the quotation from Maxwell at the beginning of the chapter, involves a sum over the effects of individual point charges. This is equivalent to starting with the charges at infinity and the observer at $\vec{r}$. Then the charges are moved in, one by one, from infinity, and the observer adds up each of their contributions to obtain the total potential at $\vec{r}$.

5.9.1   *General Case with a Spreadsheet*

Consider the potential V at observation position $\vec{r}$ due to a set of point charges q_i at positions $\vec{r}_i$. Let $\sum_i q_i$ be finite so that V_∞ may be set to zero. To find $V(\vec{r})$, sum over (5.24) to obtain

$$V(\vec{r}) = \sum_i \frac{kq_i}{R_i}, \qquad \vec{R}_i = \vec{r}_i - \vec{r}.$$

(electrical potential of discrete charge distribution)   (5.42)

If the charge is continuously distributed, with dq at $\vec{r}\,'$, then an integral must be performed over the contributions $dV = kdq/r$ due to each element of charge dq:

$$V(\vec{r}) = \int dV = \int \frac{kdq}{R}, \qquad \vec{R} = \vec{r}\,' - \vec{r}.$$

(electrical potential of continuous charge distribution)   (5.43)

If the integral cannot be performed analytically, it can be performed using a spreadsheet (and perhaps some elves). Just as in the previous chapter, first break up the charge distribution into a large number of tiny pieces. See Figure 5.24. Column A of the spreadsheet in Table 5.2 serves to label the pieces. For each element i, determine dq_i (put this in column B) and the three components of its midpoint $\vec{r}\,' = \vec{r}_i$ (put this in columns C to E). With the three components of the observation position $\vec{r}$ stored elsewhere, compute the three components of the vector $\vec{R}_i = \vec{r}_i - \vec{r}$ (put this in columns F to H). Next compute $R_i = |\vec{R}_i|$, which is put in column I. Finally, compute kdq_i/R_i, which is put in column J. The sum over column J then yields $V(\vec{r})$. As a check on convergence, repeat the calculation with a larger number of points. This procedure is more complex than our earlier scalar calculation for charge (Section 1.9), because the answer now depends on the position of the observer.

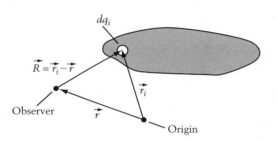

Figure 5.24 Geometry for numerical computation of potential as a sum over dV_i for each charge dq_i.

Example 5.16   **Ring of charge**

See Figure 5.20. For a ring of total charge Q and radius a, use the action-at-a-distance viewpoint to find V along the x-axis.

Table 5.2 Spreadsheet for calculation of potential

A	B	C	D	E	F	G	H	I	J
i	dq_i	$\vec{r}_{i_x}$	$\vec{r}_{i_y}$	$\vec{r}_{i_z}$	$\vec{R}_{i_x}$	$\vec{R}_{i_y}$	$\vec{R}_{i_z}$	R_i	kdq_i/R_i
1	1.0×10^{-3}	-0.6	0.64	-0.48	3	-4	5	-0.676	0.676×10^{-3}

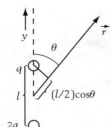

Figure 5.25 Electric dipole, for computation of its electrical potential.

Solution: By (5.43),

$$V(x) = \int dV = \int \frac{k}{r} dq = \frac{k}{r} \int dq = \frac{kQ}{r}. \qquad r = \sqrt{a^2 + x^2} \qquad (5.44)$$

This is the same as (5.38), obtained by the field method, which began with $\vec{E}$, and integrated to obtain V.

Example 5.17   Electric Dipole

Consider an electric dipole consisting of two charges $\pm q$, symmetrically placed on the y-axis so that the origin is at their midpoint, and separated by a distance l, as in Figure 5.25. The charges are located at $\pm(l/2)\hat{y}$, so the dipole moment $\vec{p}$ has magnitude $p = ql$, and points from $-q$ to $+q$, so $\vec{p} = ql\hat{y}$. Let the observer be at $\vec{r}$, which makes an angle θ to the y-axis, so $\hat{r} \cdot \hat{y} = \cos\theta$. Find the potential V.

Solution: The potential for two charges is obtained by simple addition, summing over (5.24) for each point charge with ground taken to be at infinity. With $\vec{R}_\pm = \vec{r} - (\pm)(l/2)\hat{y} = \vec{r} \mp (l/2)\hat{y}$, (5.24) yields

$$V = \frac{kq}{R_+} - \frac{kq}{R_-} = \frac{kq(R_- - R_+)}{R_+ R_-}. \qquad (5.45)$$

From Figure 5.25, if $r \gg l$, then $R_\pm \approx r \mp (l/2)\cos\theta$. Hence

$$R_+ R_- \approx [r - (l/2)\cos\theta][r + (l/2)\cos\theta] \approx r^2 - (l\cos\theta)^2/4 \approx r^2,$$

and $R_- - R_+ \approx l\cos\theta$. We then have

$$V \to \frac{kql\cos\theta}{r^2} = \frac{kql\hat{r} \cdot \hat{y}}{r^2} = \frac{k\vec{p} \cdot \hat{r}}{r^3}. \qquad \left(\frac{l}{r} \to 0\right) \qquad (5.46)$$

Note that $V = 0$ on the midplane, as well as at infinity. By fixing the value of the potential, equipotentials can be obtained for both (5.45) and (5.46). Integrating over $\vec{E}$ along some path to find V, using (5.14), is much too difficult in the present case because there is no path along which $\vec{E}$ is not complex.

5.10 Calculating $\vec{E}$ from *V*

Equation (5.14) shows how to compute the potential difference between two points A and B if the electric field is known along some path between those points. Conversely, if the potential is known, we can find the component of the

Figure 5.26 Electric field in region between two nearby points A and B, for numerical computation of potential difference.

$$dV = V_B - V_A = -\vec{E} \cdot d\vec{s}$$

electric field along this path. To see this, rewrite (5.17) as

$$\vec{E} \cdot \hat{s} = -\frac{dV}{ds}. \qquad (5.47)$$

Thus, to find the component of $\vec{E}$ along the (arbitrary) direction $\hat{s}$, we must know how the potential varies in that direction. See Figure 5.26. The maximum value of $-dV/ds$ is obtained when $\hat{s}$ is along $\vec{E}$.

Example 5.18　**Point charge**

Show that, at a distance r from a point charge q, $E_r = kq/r^2$.

Solution: Here we have $V = kQ/r$, so with $\hat{s} = \hat{r}$, (5.47) gives

$$E_r = \vec{E} \cdot \hat{r} = -\frac{dV}{dr} = -\frac{d}{dr}\frac{kQ}{r} = \frac{kQ}{r^2}, \qquad (5.48)$$

as expected. Also note that a sphere of radius r is an equipotential so that moving along the sphere gives no change in potential, and therefore no component of the electric field in the tangential directions.

Example 5.19　**Ring of charge**

Both (5.38) and (5.44) give

$$V(x) = \frac{kQ}{r}, \quad r = \sqrt{x^2 + a^2} \qquad (5.49)$$

for a ring of uniform charge distribution. Find E_x on the x-axis.

Solution: Employ (5.47) with $\hat{s} = \hat{x}$ and $ds = dx$. Then

$$E_x = \vec{E} \cdot \hat{x} = -\frac{dV}{dx}. \qquad (5.50)$$

Use of (5.49) in (5.50) then yields

$$E_x = -kQ\frac{d}{dx}(x^2 + a^2)^{-1/2} = -kQ\left(-\frac{1}{2}\right)(x^2 + a^2)^{-3/2}(2x) = \frac{kQx}{r^3}. \qquad (5.51)$$

This is the same result as (5.36). Since V is not known off the x-axis, we cannot determine E_y or E_z by this method.

Sometimes, $\vec{E}$ or V can be determined at a point but not in the vicinity of that point. Consider the center of a uniformly charged circular arc, a partial circle. Although $\vec{E}$ can be determined readily at the center of the partial circle, (5.14) cannot be used to compute V at the center because $\vec{E}$ is not known along a path from the center to infinity. Likewise, although V can be determined readily at the center of the partial circle, (5.47) cannot be used to compute $\vec{E}$ at the center because V is not known near the center of the partial circle.

A more formal way to express the electric field in cartesian components is by a generalization of (5.50). When all the components are included, we must replace the derivative by the partial derivative. Then

$$E_x = \vec{E} \cdot \hat{x} = -\frac{\partial V}{\partial x} \equiv -\partial_x V, \quad E_y = \vec{E} \cdot \hat{y} = -\frac{\partial V}{\partial y} \equiv -\partial_y V,$$

$$E_z = \vec{E} \cdot \hat{z} = -\frac{\partial V}{\partial z} \equiv -\partial_z V. \tag{5.52}$$

Another way to write this is

$$\vec{E} = -\vec{\nabla} V = -\frac{\partial V}{\partial x}\hat{x} - \frac{\partial V}{\partial y}\hat{y} - \frac{\partial V}{\partial z}\hat{z}. \tag{5.53}$$

$\vec{\nabla}$ is called the *gradient operator*. To repeat, the minus sign tells us that $\vec{E}$ points from places of higher voltage to places of lower voltage.

Example 5.20   **From V to $\vec{E}$ to Q_{enc} (Chapters 1–5 in a nutshell)**

Now return to Example 5.4, where Joan was given $V(x) = 2x^3$ (where V is in V and x is in m), and she wanted to know $\vec{E}$ at $x = 1$ m. See Figure 5.27(a). We told her that $\vec{E}$ pointed along $-\hat{x}$, and that $|\vec{E}| \approx 6.02$ V/m. Now she wants (a) $\vec{E}$ exactly. Moreover, she wants to know (b) how much electric charge is within (i.e., enclosed by) a rectangular slab of cross-section 1 m² that has opposite faces at 0.9 m and 1.1 m, and area 1 m×1 m×0.2 m.

Solution: (a) By (5.52) or (5.53), since V has no y or z dependence, $E_y = 0 = E_z$ exactly. Moreover, $E_x = -\partial_x(2x^3) = -6x^2$. See Figure 5.27(b). Hence $E_x = -6$ V/m at $x = 1$ m. This points to the left, in agreement with our earlier conclusion from the direction pointing from higher to lower potential. Moreover, it is very close to our earlier estimate of $|\vec{E}| \approx 6.02$ V/m. (b) To find the charge enclosed by the slab, Gauss's law requires the flux through its six sides. The flux through four of the sides is zero because $\vec{E}$ is normal to those sides. For the $x = 0.9$ m side, $E_x = -6x^2$ yields $E_x = -4.86$ V/m, so $\vec{E}$ points along $-\hat{x}$. Since

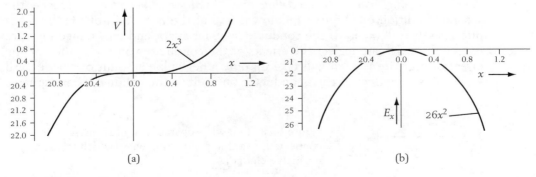

Figure 5.27 $V(x) = 2x^3$ (a); $E_x = -\partial V/\partial x = 6x^2$ (b).

the normal also is along $-\hat{x}$, flux is leaving, of amount

$$|E_x|A_{flux} = (4.86 \text{ V/m})(1 \text{ m}^2) = 4.86 \text{ V}-\text{m}.$$

For the $x = 1.1$ m side, $E_x = -6x^2$ yields $E_x = -7.26$ V/m. Since the normal is along $\hat{x}$, flux is entering, of amount

$$|E_x|A_{flux} = (7.26 \text{ V/m})(1 \text{ m}^2) = 7.26 \text{ V}-\text{m}.$$

The net flux is 2.4 V-m entering, which counts as negative, so $\Phi_E = -2.4 \text{ V}-\text{m} = 4\pi k Q_{enc}$. Then $Q_{enc} = -2.12 \times 10^{-11}$ C.

Hence, from a knowledge of the potential, we have found the charge enclosed by a specific Gaussian surface. This can be done in general although the calculations will be difficult for a complicated surface. This problem is very comprehensive: it starts with voltage (Chapter 5), goes back to electric field (Chapter 3), and then uses Gauss's law (Chapter 4) to determine electric charge (Chapters 1 and 2). You can't get much more comprehensive than this.

5.11 Connecting Two Conductors, Charge Redistribution, and the "Power of Points"

Section 5.4 established that electrical conductors are equipotential surfaces. Thus, if two conductors are connected to each other, charge will flow one way or another, redistributing itself over the surface of each until the two conductors come to the same potential. With this in mind, let us connect two distant conductors by a wire so thin that it does not take up a significant amount of the charge from the two conductors. See Figure 5.28. If the initial charges are $Q_1^{(0)}$ and $Q_2^{(0)}$, and the final charges are Q_1 and Q_2, then by charge conservation, discussed in Chapter 1, the sum of the final charges equals the sum of the initial charges,

$$Q_1^{(0)} + Q_2^{(0)} = Q_1 + Q_2. \tag{5.54}$$

This gives one condition on the two unknowns Q_1 and Q_2. Initially, the two conductors need not have the same potential, so typically $V_1^{(0)} \neq V_2^{(0)}$.

The second condition we would not have known until the present chapter. Here it is: *as long as they are connected, the two conductors (and the wire between them) are a single conductor, and thus they are at the same potential.* Hence, the second condition on the two unknowns is that the potentials of the two conductors satisfy $V_1 = V_2$. If the conductors are far apart, and the connecting wire is very thin, we can neglect the effects of the conductors on each other, or the effect of the wire (which will take up only a negligible amount of charge). All we need is the effect of each conductor on itself. If the conductors are spheres

Figure 5.28 Two well-separated conducting spheres, connected by a thin conducting wire (which takes up negligible charge).

with radii R_1 and R_2, then $V_1 = V_2$ means that

$$\frac{kQ_1}{R_1} = \frac{kQ_2}{R_2}. \tag{5.55}$$

Hence, the larger object has the larger charge. We say that it has a larger *capacitance*. By (5.55), Q_1 (Q_2) is proportional to R_1 (R_2), so the capacitance is proportional to the radius R, not the surface area ($\sim R^2$) or the volume ($\sim R^3$).

Example 5.21   **Connecting two well-separated conducting spheres**

Let two well-separated conducting spheres have $R_1 = 8$ cm and $R_2 = 2$ cm. Before the connection let $Q_1 = 10 \times 10^{-9}$ C and $Q_2 = 0$. Find the final charges and potentials after they have been connected by a fine wire.

Solution: Note that before the connection $V_1 = kQ_1/R_1 = 1125$ V and $V_2 = 0$. By (5.55), after the two spheres make electrical contact, they have $Q_1'/Q_2' = R_1/R_2 = 4$. This gives the four-to-one split of $Q_1' = 8 \times 10^{-9}$ C and $Q_2' = 2 \times 10^{-9}$ C, which sums up to the original total charge. It leads to the common potential of $kQ_1'/R_1 = kQ_2'/R_2 = 900$ V. Notice that the final potential is *not* the average of the initial potentials (although it does lie between the initial potentials). Moreover, the final charge is not equally shared; the conductor with the larger capacitance gets more charge.

   Since $E_1 = kQ_1/R_1^2$, or $E_1 R_1 = kQ_1/R_1$, and similarly $E_2 R_2 = kQ_2/R_2$, we have $E_1 R_1 = E_2 R_2$, or

$$E_1 = \frac{\text{const}}{R_1}. \tag{5.56}$$

Equation (5.56) says that the smaller the radius, the larger the electric field. Hence, the pointier the place on the surface, the larger the electric field at the surface, and the larger the local charge density. If the electric field at the surface exceeds the *dielectric strength* E_d, there can be electrical breakdown in the air (sparking). (That is, the air is "strong enough" to sustain an electric field of magnitude up to E_d without breaking down; more on this in Chapter 6.) *This explains Franklin's observation, discussed in Section 1.4, of the "power of points" to charge and discharge.*

   Caution must be used in generalizing (5.56) to surfaces of arbitrary shape. A charged hollow conductor with a needle sticking out of its inner surface and a smooth outer surface will have zero electric field just outside the needle and a finite electric field just outside the smooth surface.

We are now prepared to begin the study of circuit elements, from which electric circuits are constructed. The first circuit element that we shall study is the *capacitor,* so named because it has the *capacity* to store electric charge. All the ideas from Chapters 1 to 5 will be necessary to study capacitance, and to determine the behavior of capacitors in electric circuits.

5.12 Outside a Nonuniform Conductor, $\vec{E}$ Can Have a Parallel Component

Optional

So far we have only considered conductors whose surfaces are perfectly uniform. Real conductors are not perfectly uniform because they consist of atoms, and the arrangement of atoms along the surface may not be uniform. Many "tin" cans that hold acidic fruits (e.g., pineapple), are coated inside with regions of zinc. These surfaces have a mottled appearance. (Steel poles that support street signs and stop signs are also coated with zinc, and display such patchy regions.) These are regions where different crystal faces are exposed. Thus the surface is not uniform; this has consequences for the electric field outside the crystal.

According to which crystal surface is exposed, the number of surface atoms can vary. In part because of this, the properties of different types of surfaces are different. One of the differing properties is the electrical energy it takes for the highest-energy electron in the crystal to escape from within the crystal to a cm or more outside the crystal. This is called the *work function W*, and it is in part a measure of voltage change on crossing the surface. *W* is related to the ionization energy (the energy it takes to remove an electron from a single atom). For example, atomic cesium (Cs) has a small ionization energy, and in crystalline form Cs has a small work function. Work functions are characteristically on the order of a volt. If a conductor is given a potential that greatly exceeds a volt (e.g., 100 V), then the conductor will be *nearly* an equipotential. In that case, the electric field is nearly normal to the surface of the conductor, as discussed in the previous chapter.

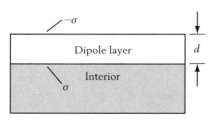

Figure 5.29 Surface of a conductor, where the dipole layer has a thickness d that corresponds to an atomic dimension of on the order of 10^{-10} m. In reality, the charges $\pm\sigma$ are smeared out rather than being located precisely on planes.

To see how the work function can vary along a surface, note that at each surface the atoms make slight adjustments. This causes slightly different charge distributions and dipole moments at the surface. One can think of the region near the surface as two plates with equal and opposite charge densities $\pm\sigma$, with separation d. See Figure 5.29. Applying (5.18) with $V \rightarrow \Delta V$ and $y \rightarrow d$, the voltage difference is then $\Delta V = Ed$, where adding the electric fields of the two sheets gives the interior electric field $E = 2\pi k\sigma + 2\pi k\sigma = 4\pi k\sigma$. Thus, with $p = Qd$ the dipole moment, and A the area,

$$\Delta V = Ed = 4\pi k\sigma d = 4\pi k\frac{Q}{A}d = 4\pi k\frac{p}{A}. \tag{5.57}$$

That is, the voltage difference ΔV across the interface is proportional to the dipole moment per unit area p/A. If ΔV is not the same all along the surface, then there must be a variation in the voltage and in the dipole moment per unit area as we move along the surface. Therefore, by (5.53), there must be a

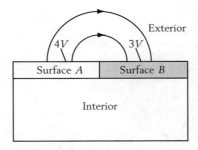

Figure 5.30 Surface of a conductor with a nonuniform surface. The dipole layer voltage across *A* is greater than across *B* so that just outside the surface there must be field lines that point from *A* to *B*. The region labeled "interior" is not precisely defined, nor is the surface region.

component of electric field along the surface. The dipole layer picture should not be taken too literally; the values of d and σ cannot be precisely specified although p/A and ΔV can.

To be more specific, consider that there are adjacent surface structures A and B, with work functions corresponding to voltages of 4 and 3 volts. See Figure 5.30. Starting at a point P, well inside the crystal, and moving to P_A just above A, there is a voltage gain of 4 V; starting at P and moving to P_B just above B, there is a voltage gain of 3 V. Hence there must be an electric field $\vec{E}$ outside the surface and pointing from the region of higher potential (A) to the region of lower potential (B). When there are many "patches" with different work functions, the potential smoothes out when we move away from the surface a distance on the order of a few times the characteristic size of the patches.

5.12.1 *Resolving an Apparent Paradox*

At first sight, we might expect that if, due to a nonuniform work function, there is a component of electric field along the surface, there will also be a component of the electric current along the surface. To understand why this need not be so, we must think in terms of electron orbitals. Orbitals are characterized by their overall energy, not by the local value of the electric field or electrical potential. An orbital's total energy is part electrical and part kinetic, and for that orbital the sum of these two takes the same value everywhere. Thus, when the electrical energy goes up, the kinetic energy goes down. It is possible to have an orbital with no current flow along the surface if an orbital does not correspond to a current-carrying state. This is no less paradoxical than what happens in a hydrogen atom, where the same balancing occurs, and there is no current flow in the radial direction.

In some situations, the electron orbitals are characterized by tangential current flow, with a net angular momentum. Recently, it has been found that, for small rings of conductor, on the order of 10^{-7} m in size, the electron orbitals can be characterized by their angular momentum about the ring. This may have applications for future electronic devices.

5.12.2 *Scanning Tunneling Microscopy*

In the 1980s, a powerful technique became available to examine the surfaces of conducting objects. It is called *scanning tunneling microscopy*, or STM, and it was developed by Gerd Binnig and Heinrich Rohrer, for which they received the 1986 Nobel Prize. In this method, a small conducting tip (perhaps consisting of only an atom or two) is brought near a surface, and a voltage difference between the tip of the probe and the surface causes electrons to "tunnel" across from an orbital on the tip to an orbital on the surface (or vice versa). (Tunneling is a wave phenomenon whereby particles "jump" from one orbital to another.) By

Figure 5.31 Scanning tunneling microscopy (STM) for a GaAs surface with two As vacancies. (Courtesy of M. Weimer, Texas A&M University.)

scanning over the surface, it is possible to make a map of the surface structure on a scale as small as 2×10^{-10} m. Binnig has remarked that STM "has changed our emotional relationship with atoms." See Figure 5.31, which shows two arsenic (As) vacancies on a gallium arsenide (GaAs) surface.

STM provides a microscopically localized probe of the work function of a surface. A less microscopic method to study the work function is *photoemission:* if light of a high enough frequency shines on a surface, then electrons are emitted from that surface; the onset frequency is related to the work function. Photo-emission occurs over a much larger region of the surface than STM, and thus gives an average of the work function for the surface. (The *photoelectric effect* was discovered in 1887 by Hertz in the course of his studies of electromagnetic radiation, which we will discuss in Chapter 15. Einstein received the Nobel Prize in 1921 for his explanation of the photoelectric effect.)

A Scientist and Religion

The famed Nobel Prize–winning physicist, Richard Feynman, was visiting a laboratory where he saw STM scans for the first time. Accompanying him was one of his former students, who assaulted their host with a variety of technical questions about the voltage difference between the STM tip and the surface, the size of the tip, its distance to the surface, and so on. Feynman stopped him short: "Phil, would you please stop asking these questions. Can't you see I'm having a religious experience? Those are atoms I'm looking at."

Problems

5-2.1 Estimate the weight of a 58 kg person on the moon.

5-2.2 An artificial satellite is in earth orbit with nearest and farthest distances to the earth of 1500 km and 4500 km. At the nearest distance from the earth, its velocity is 2.73×10^5 m/s. Find its velocity at its farthest distance from the earth.

5-3.1 (a) How much work must be done to move 5 nC from $V = 5$ V to $V = 9$ V? (b) Repeat for a -5 nC charge from $V = 5$ V to $V = 9$ V. (c) What is the significance of the sign of the work?

5-3.2 You perform 20 mJ of work to move 8 nC from point A, with $V_A = 20$ V, to point B. Find V_B.

5-3.3 For V depending only on x and for $x = 1, 2, 3$ cm the voltages are $V = 24, 26, 28.4$ V. Estimate the electric field at $x = 1.5, 2.0, 2.5$ cm.

5-3.4 A local value of the earth's fair-weather field is 120 V/m, and points downward. What is the voltage difference between the head and toes of a 2 m high person, and which is at the higher voltage?

5-3.5 At $\vec{r}_P \equiv (0.21, 0.43, 0.58)$ m, the electric field is $\vec{E} \equiv (0.25, -0.47, 0.94)$ V/m. At the nearby point $\vec{r}_{P'}' \equiv (0.22, 0.42, 0.56)$ m, the voltage is $V_{P'} = 2.76$ V. Estimate the voltage V_P at P.

5-3.6 Consider two infinite sheets, one on $z = a$ with charge density σ and one on $z = -a$ with charge density $-\sigma$. Let $V = 0$ on the positive plate. Find (a) the voltage on the negative plate, (b) the voltage at $z \to \infty$, (c) the voltage at $z \to -\infty$. (d) Is the voltage at infinity unique in this case?

5-4.1 A proton with energy 2 MeV moves directly toward a gold nucleus ($q = 79e$). If the nucleus is fixed in place, at what distance does the proton stop?

5-4.2 A lightning bolt transfers charge 20 C across a voltage difference of 1.2×10^8 V. (a) How much energy is transfered? (b) How much ice at 273 K would this melt? (c) How much ice at 273 K would this vaporize? (d) How high would this energy propel a 5-ton elephant?

5-4.3 An electron moving along y has an initial velocity of 2.8×10^6 m/s. After traveling 4 cm, its velocity is 4.5×10^6 m/s. (a) Find the potential difference between the starting point and the end point. Which has the higher potential? (b) If the field is uniform, how large is it, and in what direction?

5-4.4 Consider a point charge q near the center of a uniformly charged ring of radius a and charge Q in the xy-plane. See Figure 5.32. It starts from rest slightly off-center and is constrained to move along the ring axis. (a) Using symbols, find its velocity v at large distances. (b) If $q = 8 \times 10^{-14}$ C, $Q = 6$ μC, and $a = 2$ cm, evaluate v numerically.

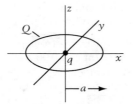

Figure 5.32 Problem 5-4.4.

5-4.5 A particle of charge $q = -2.4 \times 10^{-9}$ C starts from rest in a uniform upward-directed $\vec{E}$ field. See Figure 5.33. After moving 8 cm (in which direction?), it has kinetic energy 4.8×10^{-5} J. Find (a) the work done by the electric force, (b) the change in electrical potential (is the starting point at a higher or lower potential than the end point?), and (c) $|\vec{E}|$.

Figure 5.33 Problem 5-4.5.

5-4.6 A sphere of mass $m = 4.5$ g and charge $q = 2.6 \times 10^{-12}$ C hangs by a 5 cm long thread. It hangs between two conducting sheets with charges σ and $-\sigma$, which produce a horizontal field that causes the thread to make a 20° angle to the vertical. See Figure 5.34. Find (a) $|\vec{E}|$, (b) σ. Relative to the vertical, find (c) the increase in gravitational energy; (d) the decrease in electrical energy; and (e) the decrease in electrical potential.

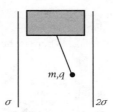

Figure 5.34 Problem 5-4.6.

5-4.7 A 9 V electronics battery maintains two metallic plates at -3 V and 6 V. An electron at rest just outside one plate accelerates and hits the other plate. (a) Which plate does it hit, and how fast is it moving? (b) How much work was done on it? (c) What is its change in electrical potential energy? (d) What is its change in electrical potential?

5-4.8 A Na$^+$ ion at A is subject to an electrical force of 8×10^{-15} N, caused by a negatively charged plate at B that is 2 cm away. (a) Find $V_A - V_B$. (b) How fast will the ion be moving if it hits the plate?

5-4.9 An electron beam of 5×10^{15} electrons/sec, each with energy 40 keV, is incident on a 100 g target of tungsten. If the target's temperature rises at the rate of $5°$K/hr, find the percentage of energy that is absorbed by the target. (The specific heat of tungsten is 134 J/kg-K.)

5-4.10 When special relativity is accounted for, the kinetic energy of a particle of mass m is given by $K = mc^2[(1/\sqrt{1 - (v/c)^2}) - 1]$, where c is the velocity of light. (a) Use K to find the velocity v of an electron accelerated from rest through 2 V and 2000 V. (b) Compare with the v's found using $K = (1/2)mv^2$. The "special" in special relativity means "restricted," because the theory is restricted to inertial frames. In general relativity, the theory applies to any frame of reference, even that of a rock bouncing down the side of a mountain.

5-4.11 Two plates of area 16 cm^2 and separation 2 cm have a 0.14 N force of attraction. Find (a) their voltage difference, (b) their charge, and (c) their charge density.

5-4.12 Two plates of area 45 cm^2 and separation 3 cm have a voltage difference of 1108 V. Find (a) their force of attraction, (b) their charge, and (c) their charge density.

5-5.1 In Figure 5.11(a), change the plates at 0 V to 10 V. Sketch the field lines, and sketch a few more equipotentials. Where is the field the largest? The smallest?

5-5.2 In Figure 5.11(b), change the plates at 5 V and -5 V to 50 V and -50 V. Sketch the field lines, and sketch a few more equipotentials. Where is the field the largest? The smallest?

5-5.3 In Figure 5.11(c), change the plates at 5 V and -5 V to 50 V and -50 V. Sketch the field lines, and sketch a few more equipotentials. Where is the field the largest? The smallest?

5-5.4 Sketch some equipotential surfaces for the vicinity of a living tree (whose water content makes it a good conductor). Model the tree as a vertical rod, and consider the earth to be a good conductor.

5-5.5 Consider a point charge $Q > 0$, in the presence of grounded conductors. (a) Show that the surface charge density σ_s is negative on all conductors. You may think of infinity as a large sphere at zero potential. *Hint:* Field lines cannot connect two

conductors at the same potential. (b) The closer Q is to the conductors, the larger the fraction of its field lines that terminate on them, and therefore the larger the amount of charge on them. Find the magnitude of the maximum total amount of charge on them.

5-5.6 (a) Find the potential at a point that is a distance $1.5a$ above the positive charge in a horizontal dipole pair $\pm q$ separated by $2a$. See Figure 3.11(a). (b) Find the electric field at that point. (c) Use Maxwell's trick of replacing a charge by its surrounding equipotential to find the equipotential at that point, and the normal to the equipotential at that point. (d) Find the necessary charge density at that point.

5-5.7 Consider Maxwell's trick for finding the shapes of conductors that will produce known equipotentials. (a) What equipotential shapes does it give for one charge? (b) Will it work, in principle, for three charges? (c) Assume that, in the presence of two conductors with respective charges q_1 and q_2, you know the equipotentials. If the charges are changed to q_1 and $2q_2$, does this change the shape of the equipotentials? (d) If the charges are changed to $2q_1$ and $2q_2$, does this change the shape of the equipotentials relative to the case of q_1 and q_2?

5-5.8 Consider an infinite conducting sheet that is normal to the page and passes through the x-axis. Bend the sheet about the z-axis so that it becomes a wedge of angle α, where $\alpha = \pi$ corresponds to the original sheet. Now give the sheet a charge. (a) Draw the field lines for $\alpha < \pi$ and for $\alpha > \pi$. (b) Show that $|\vec{E}|$ is small (large) at the corner for wedge angle $\alpha > \pi$ ($\alpha < \pi$).

5-5.9 If two equipotentials cross (this does occur), what does that say about the electric field at the crossing point? *Hint:* Consider two charges $q_1 = q_2 = q$ separated by $2a$, as in Figure 5.10(b). The equipotentials with $V > 2kq/a$ either surround q_1 or q_2. The equipotentials with $V < 2kq/a$ surround both charges. The equipotential for $V = 2kq/a$ is more complex (and not drawn in Figure 5.10b).

5-5.10 Consider a point charge $Q > 0$ and a grounded sphere of radius a at a distance $r > a$. Estimate the charge q on the sphere. Note that $Q + q' > 0$.

5-6.1 A 2.8 nC charge is at the origin and a -3.6 nC charge is on the x-axis at $x = 4$ cm. Find the potential on the y-axis at $y = 5$ cm.

5-6.2 (a) Find the charges needed to bring two conducting balls, of radii 0.5 cm and 0.5 m, to a potential of 2000 V relative to infinity. (b) Contact with which is more likely to be dangerous? Why?

5-6.3 Find the amount of work needed to assemble a square of side a with charges q at the vertices.

5-6.4 Find the amount of work needed to place equal charges q at three corners of a square of side a.

5-6.5 In the quark model, a proton consists of two "up" (u) quarks, each of charge $(2/3)e$, and a "down" (d) quark, of charge $-(1/3)e$. If they form a triangle of side $a = 10^{-15}$ m, find (a) the potential at one of the up quarks due to the other two quarks, (b) the potential at the down quark due to the up quarks, and (3) the total energy of all the quarks.

5-6.6 Sparking can occur when the electric field exceeds the *dielectric strength* E_d, which depends upon the material. For dry air, $E_d = 3 \times 10^6$ V/m. Sparking associated with dust grains that charge up when they collide with one another or with the walls of their container are sometimes responsible for electrical explosions. (a) To what potential must the surface of a spherical dust grain of 0.04 mm radius be raised in order to produce such a field at the surface? (b) To what charge, and how many electrons, does this correspond?

5-6.7 (a) Take $V_\infty = 0$. For two point charges q along the x-axis, separated by a, find the position where $\vec{E} = \vec{0}$ but $V \neq 0$. (b) Can you change V_∞ so that both $\vec{E} = \vec{0}$ and $V = 0$?

5-6.8 (a) Take $V_\infty = 0$. For two charges $\pm q$ along the x-axis, separated by a, find the position where $\vec{E} \neq \vec{0}$ but $V = 0$. (b) Can you change V_∞ so that both $\vec{E} = \vec{0}$ and $V = 0$?

5-6.9 A point charge Q at the origin produces a voltage. The voltage difference between points A and B, at $x = 1$ cm and $x = 2$ cm, is 250 V. Find Q.

5-6.10 A long rod of uniform charge density λ, passing through the origin, normal to the xy-plane, produces a voltage. The voltage difference between points A and B, at $x = 1$ cm and $x = 2$ cm, is 250 V. Find λ.

5-6.11 A large sheet of uniform charge density $\sigma < 0$, passing through the origin, normal to the x-axis, produces a voltage. The voltage difference between points A and B, at $x = 1$ cm and $x = 2$ cm, is 250 V. Find σ.

5-6.12 Two fixed charges Q and $-Q$ are separated by $2a$ along the x-axis. A charge q with mass m is released at rest at the midpoint. (a) Find the change in electrical potential and electrical potential energy when it has moved 3/4 of the way to $-Q$. (b) Find its final speed.

5-6.13 In a NaCl lattice, each Na^+ has six nearest neighbors of Cl^-. If one (small) Na^+ sits at $(0, 0, 0)$, then the (large) Cl^- sit at $(a/2, 0, 0)$, $(-a/2, 0, 0)$, $(0, a/2, 0)$, $(0, -a/2, 0)$, $(0, 0, a/2)$, and $(0, 0, -a/2)$, where $a = 5.63 \times 10^{-10}$ m. The next nearest neighbors are Na^+, four (labeled a) in the $x = 0$ plane at $(0, a/2, a/2)$, $(0, -a/2, -a/2)$, $(0, a/2, -a/2)$, and $(0, -a/2, a/2)$, four (labeled b) in the $y = 0$ plane at $(a/2, 0, a/2,)$, $(-a/2, 0, -a/2)$, $(a/2, 0, -a/2)$, $(-a/2, 0, a/2)$, and four (labeled c) in the $z = 0$ plane at $(a/2, a/2, 0)$, $(-a/2, -a/2, 0)$, $(a/2, -a/2, 0)$, $(-a/2, a/2, 0)$. Find the potential at the site of the Na^+ at the origin, due to these 18 other charges. See Figure 5.35.

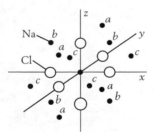

Figure 5.35 Problem 5-6.13.

5-7.1 Let $\vec{E} = (ax + b)\hat{i}$. (a) Find the circulation $\oint \vec{E} \cdot d\vec{s}$ for a circuit that goes clockwise around the square $(0, 0)$, $(0, d)$, (d, d), $(d, 0)$. See Figure 5.36. (b) Is this what is to be expected for electrostatics? (c) If so, what is $V(x, y) - V(0)$?

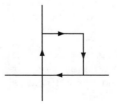

Figure 5.36 Problem 5-7.1.

5-7.2 Let $\vec{E} = (ax + b)\hat{j}$, where a and b are constants. (a) Find the circulation $\oint \vec{E} \cdot d\vec{s}$ for a circuit that goes clockwise around the square $(0, 0)$, $(0, d)$,

(d, d), $(d, 0)$. See Figure 5.36. (b) Is this what is to be expected for electrostatics? (c) If so, what is $V(x, y) - V(0)$?

5-7.3 Let $\vec{E} = \hat{z} \times (\vec{r}) = \hat{z} \times (x\hat{x} + y\hat{y} + z\hat{z}) = x\hat{y} - y\hat{x}$. Consider a square of side b in the first quadrant, with one corner at the origin, called A. Let $B = (b, b, 0)$ be the diagonally opposite corner. See Figure 5.37. Compute $-\int_A^B \vec{E} \cdot d\vec{s}$ for two paths. (a) First take a path that goes along the x-axis to $(b, 0, 0)$ and then goes along the y-axis to B. (b) First take a path that goes along the y-axis to $(0, b, 0)$ and then goes along the x-axis to B. (c) From these two calculations, determine the circulation computed for a clockwise circuit around this square. (d) Can this $\vec{E}$ be due to fixed electric charge?

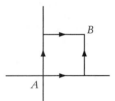

Figure 5.37 Problem 5-7.3.

5-7.4 On a circle of radius a, centered in the xy-plane, let the electric field point tangentially, with $\hat{\theta}$ the counterclockwise tangent. Compute $\oint \vec{E} \cdot d\vec{s}$ going counterclockwise for the following two fields: (a) $\vec{E} = E_0 \hat{\theta} \sin \theta$, (b) $\vec{E} = E_0 \hat{\theta} \sin^2 \theta$.

5-8.1 (a) For a finite conducting cylinder with charge $Q > 0$, in equilibrium, is the potential higher at the surface or the center? (b) Repeat for a finite cylinder with charge $Q > 0$ and a uniform density of charge (this can be a conductor out of equilibrium, or an insulator either in or out of equilibrium).

5-8.2 A conducting sphere of radius 0.5 m is at $V = 2400$ V. Find its charge and the voltage at 2 m from the center of the sphere. Take $V_\infty = 0$.

5-8.3 Two equal and opposite line charges are normal to the page, intersecting it along the x-axis at $x = -10$ cm $(-\lambda)$ and at $x = 10$ cm (λ). $V(5 \text{ cm}, 0) - V(0, 0)$ is 40 V. Find λ.

5-8.4 Let $\vec{E} = Ar^2\hat{r}$ for a situation with spherical symmetry, where A is a constant. Find $V(r)$, with $V(0) = 0$. Describe the equipotential surfaces.

5-8.5 Let $\vec{E} = Ar^2\hat{r}$ for a situation with cylindri-

cal symmetry, where A is a constant. Find $V(r)$, with $V(0) = 0$. Describe the equipotential surfaces.

5-8.6 Let $\vec{E} = Ax^2\hat{x}$ for a situation with planar symmetry, where A is a constant. Find $V(x)$, with $V(0) = 0$. Describe the equipotential surfaces.

5-8.7 Find the potential on the axis of a disk of radius a and uniform charge density σ. Use the electric field from Chapter 3.

5-8.8 Consider two concentric spherical shells. The outer has radius b and charge Q_b, and the inner has radius a and charge Q_a. Take $V(a) = 0$. Find the potential $V(r)$ for $r < a$, $a < r < b$, and $r > b$.

5-9.1 A ring of radius R is centered at the origin of the xy-plane, where we take $\hat{y}$ to be upward. The ring has a uniform distribution of charge $2Q$ over the upper two-thirds of its circumference, and a uniform distribution of charge $-Q$ over the lower one-third of its circumference. See Figure 5.38. (a) Find V at z, along the axis of the ring, if $V_\infty = 0$. (b) Find E_z at z. (c) Do the answers change if the charge on the ring is rearranged? (d) What about E_x and E_y under rearrangement?

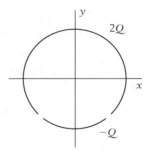

Figure 5.38 Problem 5-9.1.

5-9.2 Sketch some equipotentials and field lines for two parallel line charges λ and -3λ normal to the page, separated by l along the x-axis.

5-9.3 Sketch some equipotentials and field lines for two parallel line charges λ and 2λ normal to the page, separated by l along the x-axis.

5-9.4 An insulating rod lies between $(0, 0, 0)$ and $(-l, 0, 0)$. It has a charge density $\lambda = ax$, where a is a constant. (a) Find the total charge Q. (b) Find the potential $V(x)$ for $x > 0$ if $V_\infty = 0$. (c) Check that the potential approaches that of a point charge Q at large x.

5-9.5 An insulating rod lies between $(-l/2, 0, 0)$

and $(l/2, 0, 0)$. It has a charge density $\lambda = ax$, where a is a constant. (a) Find the total charge Q. (b) Find the potential $V(x, 0, 0)$ for $x > l/2$ if $V_\infty = 0$. (c) Show that for this distribution the dipole moment points along $\hat{x}$ and has magnitude $p = al^3/12$. (Use $\vec{p} = \int \vec{r} dq$, with $dq = \lambda ds$.) (d) Show that, at large x, the potential from part (b) is given by kp/x^2.

5-9.6 Find the potential a distance z along the axis of an annulus (inner radius a, outer radius b) with uniform charge density σ. See Figure 5.39.

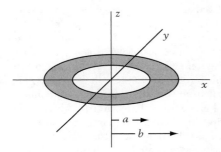

Figure 5.39 Problem 5-9.6.

5-9.7 (a) A spherical shell of radius b has charge $-Q$. If $V = 0$ at infinity, find V within the shell. (b) A smaller shell of radius $a = 0.1b$, and with charge q, is placed concentric within the larger shell. If $q = 0.01Q$, find $V(a)$, and show that it is negative. (c) For what range of values of q will $V(a)$ be positive?

5-9.8 Refer to Example 5.17 and Figure 5.25. Starting from $\vec{R}_+ = \vec{r} - (l/2)\hat{y}$, show that $R_+ \approx r - (l/2)\cos\theta$ for $r \gg l$, where $\cos\theta = \hat{r} \cdot \hat{y}$.

5-10.1 Let $V(x) = 3 + 2x + 5x^2$, where V is in volts for x in m. (a) Find V at $x = 1.9$ m and $x = 2.1$ m. (b) Estimate $\vec{E}$ at $x = 2$ m. (c) Calculate $\vec{E}$ exactly.

5-10.2 Let $V(x) = x^3$, where V is in volts for x in m. (a) Find V at $x = 1.9$ m and $x = 2.1$ m. (b) Estimate $\vec{E}$ at $x = 2$ m. (c) Calculate $\vec{E}$ exactly.

5-10.3 Consider a line of charge Q uniformly distributed from $(-l, 0, 0)$ to $(0, 0, 0)$. (a) Find the potential V at $(x, 0, 0)$, for $x > 0$. (b) From V, find E_x. (c) Compare with the result of a direct calculation of E_x.

5-10.4 Consider a line of charge Q uniformly distributed from $(-l, 0, 0)$ to $(0, 0, 0)$. (a) Find the potential V at $(x, y, 0)$. (b) From V, find E_x and E_y.

5-10.5 From the dipole potential $V = k\vec{p} \cdot \vec{r}/r^3$, show that the electric field due to a dipole at the origin is given by

$$\vec{E} = \frac{k}{r^3}[3(\vec{p} \cdot \hat{r})\hat{r} - \vec{p}]. \qquad (5.58)$$

Hint: Consider E_x, and so on, and then vectorize.

5-10.6 (a) Review (3.42) and use it with the result of the previous problem to show that the electrical energy of interaction between two dipoles $\vec{p}_1$ and $\vec{p}_2$ is

$$U = \frac{k}{r^3}[\vec{p}_1 \cdot \vec{p}_2 - 3(\vec{p}_1 \cdot \hat{r})(\vec{p}_2 \cdot \hat{r})]. \quad (5.59)$$

(b) Evaluate this when the dipoles point parallel to one another, for both the upper and lower pair in Figure 5.40. (c) Which pair has the lower energy? How does this compare with what you would expect from "opposites attract, likes repel"?

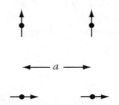

Figure 5.40 Problem 5-10.6(b).

5-10.7 In spherical coordinates, let $V = 5r^4$ for r in m and V in V. (a) Find V at $r = 0.9$ m and $r = 1.1$ m. (b) Estimate $\vec{E}$ at $r = 1$ m. (c) Calculate $\vec{E}$ exactly, at all r and for $r = 1$.

5-10.8 Let $r = \sqrt{x^2 + a^2}$. (a) Show that $dr/dx = x/r$. (b) If $V = V(r)$, use the chain rule of differentiation to show that $E_x = -dV/dx = -(dV/dr)(dr/dx)$. (c) For a ring of charge, with potential $V = kq/r$, show that along the x-axis $E_x = (kQ/r^2)(x/r) = kQx/r^3$, as in (5.51).

5-10.9 Let $V = 2xy - 4x^2 - 5y^2$. Find $\vec{E}$.

5-10.10 Let $V = 2x + 5x^2 - 5x^3$. (a) Find $\vec{E}$. (b) Find the electric flux leaving a unit cube in the first octant. Does this correspond to the presence of electric charge?

5-10.11 The space between the (positive) anode and the (negative) cathode in a vacuum tube diode develops a nonuniform charge distribution with planar symmetry. This causes a voltage $V(x) = Cx^{4/3}$, with $V = 0$ at the cathode $(x = 0)$, and $V = V_a$ at the anode $(x = d)$. See

Figure 5.41, where the darker the region the higher the charge density. (a) Find E_x at any x. (b) If $V_a = 200$ V and $d = 2$ cm, find E_x at $x/d = 0$, 0.25, 0.5, 0.75, and 1.0. (c) Find the charge per unit volume as a function of x. See problem 4-G.3.

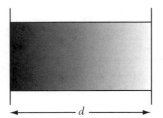

Figure 5.41 Problem 5-10.11.

5-10.12 For charge that is injected into an insulator, the voltage profile is $V(x) = Cx^{3/2}$, where charge is injected at $x = 0$ and collected at $x = d$. (a) Find E_x at any x. (b) If $V(d) = 500$ V and $d = 1.4$ cm, find E_x at $x/d = 0$, 0.25, 0.5, 0.75, and 1.0. (c) Find the charge per unit volume at any x between 0 and d. See problem 4-G.3.

5-11.1 Two metal spheres of radii 2 cm and 4 cm are mounted on insulating stands and are far from each other. They are given charges of 14 nC and −5 nC, respectively. (a) Find their potentials. (b) Find the electric fields at their surfaces. They now are connected by a very thin wire, so that charge can transfer. (c) Find their new charges and potentials. (d) Find the new electric fields at their surfaces.

5-11.2 Let a conductor with charge q be brought from far away to the interior of a Faraday ice pail. On contacting the *interior*, it transfers its charge q. The charge on the exterior of the pail then changes by q. On the other hand, if q were placed directly in contact with the *exterior*, not all the charge would be transferred. What limitation (other than sparking) prevents us from building up an infinite amount of charge on the ice pail by repeatedly bringing up charge to the interior? *Hint:* Think force or energy.

5-11.3 If electrons are transferred to the interior of a conductor, do they go to the outer surface of the conductor? Discuss.

5-11.4 Why does sparking tend to occur at sharp edges of electrical conductors? What about nonconductors?

5-11.5 Two charged metal spheres of radii R and $10R$ are far from each other, but are connected by a very thin wire. Compare their total charge, charge per unit area, electric potential, and electric field.

5-11.6 Two charged metal spheres of radii 2 cm and 20 cm are far from each other, but are connected by a very thin wire. The smaller one is at 2 V. Find their total charge, charge per unit area, electric potential, and electric field.

5-11.7 Consider two distant spheres that are fixed in place. When in electrical contact via a fine wire, they share charge in a 1-to-10 ratio. There is a distant 100 V voltage source. Initially, both spheres are isolated and uncharged. Consider the following sequence of steps: (1) Connect the smaller sphere to the voltage source; (2) disconnect it from the voltage source; (3) connect the two spheres with a fine wire; (4) disconnect the fine wire. (a) Find the potential of the larger sphere. (b) Find the potential of the larger sphere if the sequence is repeated an infinite number of times. (This can be found by a simple physical argument rather than by explicit calculation.) (c) Find the potential on the larger sphere if the sequence is repeated n times and verify that it approaches the correct value for large n.

5-12.1 A conducting surface has a grain boundary separating two surface regions, one with work function 2.2 V and the other with 1.5 V. Estimate the electric field at 200 nm above the boundary. *Hint:* What is the field if the surface regions have the same work function? How can you construct a quantity with units of electric field from the information given?

5-12.2 The electric field near a grain boundary, when the object as a whole is uncharged, is 4300 N/C. What surface charge density could produce such a field?

5-12.3 In an STM experiment, a voltage difference of −0.45 V is measured by an STM tip 0.034 nm away from the surface. Estimate the magnitude of the electric field near the tip.

5-12.4 Water droplets in the atmosphere condense around small aerosol particles that contain molecules that dissolve in the water. Hence the water contains numerous ions, which make it relatively conducting. (a) If a spherical droplet of radius 0.2 mm picks up a charge −50 pC, what is the potential at its surface? (b) At its interior? (c) If two such droplets combine, what is the new surface potential?

5-G.1 A dipole has magnitude $p = 8.0$ nC-m. Consider two points separated by 8 cm, with A at a $30°$ angle to the dipole axis and B on the dipole axis. (a) Find $V_A - V_B$. A frictionless tube connects these two points. (b) If a dust particle with $m = 5$ mg and $q = 6$ nC starts at rest from one point and is constrained to move along the tube, how fast will it be moving when it reaches the other point. (c) At which point does it start?

5-G.2 Within an uncharged conducting spherical shell, a charge q is suspended off-center by an insulating string, which deflects 1 cm to the side. Find the deflection (a) if the shell is placed within an external field, (b) if the shell is given a charge Q.

5-G.3 How would you eliminate stray electric fields from a given region of space?

5-G.4 Consider two conducting co-axial rings of radii a with equal and opposite charges $\pm q$. Their centers are at $\pm h/2$ along the z-axis. h is to be adjusted to make a very uniform field in the vicinity of the midpoint between the rings. See Figure 5.42. Consider V along the z-axis. (a) Show that at $x = 0$ all the even derivatives of $V(0, 0, z)$ are zero. (b) Find the value of h/a that makes the third derivative go to zero.

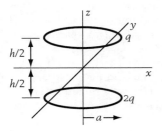

Figure 5.42 Problem 5-G.4.

5-G.5 Discuss the analogy of an equipotential map to a topographic map (height of the earth plotted in two coordinates). How rapidly does the potential change at "cliffs"? On "mesas"?

5-G.6 Consider a two-dimensional situation with $V(x, y) = A\sin qx\exp(-ky)$, where A is in V, and q and k are in m^{-1}. (a) Compute the electric field. (b) Find the electric flux through a cube of length $2\pi q^{-1}$ that is in the first octant and has one corner at the origin. (c) Show that there is no charge within the cube if $q = k$. Equiv-

alently, if the potential V has negative curvature ($\sin qx$) in one direction, and there is no charge in the region described by V, then V has positive curvature in the other direction ($\exp(ky)$ or $\exp(-ky)$). In practice, this usually implies that if a periodicity occurs in one direction, then decay occurs in the other direction. This idea generalizes to three dimensions.

5-G.7 An insulating disk of radius a is centered at the origin with its normal along the x-axis. It is given a charge density $\sigma = Br$. (a) Find its total charge Q. (b) Find its potential $V(x)$. (c) Verify that $V(x)$ has the expected behavior at large x.

5-G.8 In a depletion layer at the surface of a semiconductor, the charge density ρ is finite out to a distance w from the surface, beyond which it goes to zero. Let $V = 0$ at the surface $x = 0$. See Figure 5.43 for the charge density, where the normal is along x. (a) Find $E_x(x)$. (b) Find $V(x)$. (See Section 4.11.4.)

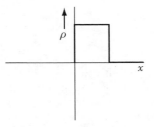

Figure 5.43 Problem 5-G.8.

5-G.9 An insulating Lucite slab of thickness 12 mm is irradiated uniformly by an electron beam of current density 0.12 mA/m^2 for 2 seconds. All the charge remains on the Lucite, trapped with an approximately uniform free charge density over a 2 mm thickness in the middle of the slab. Take $\kappa = 3.2$ for Lucite. (a) Find the free charge density and the induced charge density in the middle of the slab. (b) Find the $\vec{E}$ field and voltage within the charged region. (Take $V = 0$ at the center of the slab.) (c) Find the $\vec{E}$ field and the voltage within the uncharged region. (d) Find the induced charge density on the outer surfaces.

5-G.10 An alpha particle ($q = 2e$) is incident on a gold nucleus ($q = 79e$), and would pass by it at a distance b (the impact parameter) of 10^{-11} m if not for the electrical force. (Recall that $r_{atom} \approx 10^{-10}$ m and $r_{nucleus} \approx 10^{-15}$ m, so $r_{atom} \gg r_{nucleus}$, and thus near the nucleus the rest of the atom can be neglected.) Find the distance of closest approach if the alpha particle has kinetic energy 8 MeV. *Hint:* Use conservation laws.

5-G.11 CO_2, a linear molecule, is neutral and has no permanent electric dipole moment. Let the C, with charge $2q$, be at the origin $(0, 0, 0)$, and the O's, each with charge $-q$, be on the y-axis, a distance a above and below the C, at $(0, \pm a, 0)$. See Figure 5.44. (a) With $V = 0$ at infinity, find V at x along the x-axis. (b) With $q = e$ and $a = 1.2 \times 10^{-10}$ m, evaluate this for $x = 5 \times 10^{-15}$ m, $x = 5 \times 10^{-13}$ m, and $x = 5 \times 10^{-11}$ m. (c) How does V vary with distance at large x?

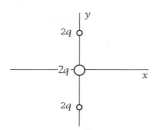

Figure 5.44 Problem 5-G.11.

5-G.12 Equal and opposite charges $\pm Q$ are on the y-axis at $\pm a$ from the origin. (a) Find the potential at x along the x-axis. (b) Find E_x along the x-axis by using $V(x)$. (c) Find E_x along the x-axis by adding up the individual $\vec{E}$ fields.

5-G.13 Equal charges $+Q$ are on the y-axis at $\pm a$ from the origin. (a) Find the potential at x along the x-axis. (b) Find E_x along the x-axis by using $V(x)$. (c) Find E_x along the x-axis by adding up the individual $\vec{E}$ fields.

5-G.14 (a) Show that the energy needed to assemble a spherical shell of radius R and total charge Q by bringing the charge in from infinity bit by bit is $U = kQ^2/2R$. *Hint:* Integrate over dQ' from 0 to Q using $dU = dQ'(V)$, where $V = kQ'/R$. (b) Show that the pressure needed to keep the sphere from expanding is $P = kQ^2/2R^2$. *Hint:* Relate a change in electrical energy to the radial component of the electric force.

5-G.15 Find the potential $V(x)$ on the axis of a uniformly charged disk of radius a and charge density σ, by two methods. Take $V(0) = 0$.

5-G.16 (a) For a quarter-miler who runs in a straight line rightward along the x-axis, compute $\int ds$ and $\int d\vec{s}$. (b) For a quarter-miler who runs around a track, ending at the starting point, compute $\oint ds$ and $\oint d\vec{s}$, and their magnitudes.

5-G.17 Let $V(x, y, z) = V_0 \exp(-x^2/a^2)$. (a) Find $\vec{E}$. (b) Find Q_{enc} for a Gaussian surface that is a parallelopiped extending from $x = 0$ to $x = b$, and of area A normal to x. (c) Find the charge density $\rho(x)$.

5-G.18 Let $V(r) = V_0 \exp(-r/a)$ in cylindrical coordinates. (a) Find E_r. (b) Find Q_{enc} for a Gaussian surface that is a concentric cylinder of radius b and length L. (c) Find the charge density $\rho(r)$.

5-G.19 Let $V(r) = V_0 \exp(-r/a)$ in spherical coordinates. (a) Find E_r. (b) Find Q_{enc} for a Gaussian surface that is a concentric sphere of radius b. (c) Find the charge density $\rho(r)$.

5-G.20 Consider a line (not necessarily straight) going from A to B. The average value of $\vec{E}$ along the tangential direction $\hat{s}$, or $\vec{E} \cdot \hat{s}$, is $\int_A^B \vec{E} \cdot \hat{s} ds / \int_A^B ds$, where $\int_A^B ds$ is the total path length from A to B. For $\vec{E} = -Ax\hat{x} + Ay\hat{y}$ with $A = 10$ V/m^2 and x in m, find $\vec{E} \cdot \hat{s}$ for $(0, 0)$ to $(1, 2)$.

5-G.21 The discussion of NaCl was a bit of a swindle because it didn't discuss the interaction energy U' that keeps the charges from collapsing to a point. (U' represents the repulsion of different atomic shells of electrons, a more advanced topic than electrostatics.) For two charges e and $-e$, consider the energies $U_{el} = -ke^2/r$ and $U' = \alpha k e^2/[a(r/a)^\beta]$, where a is a length and α and β are dimensionless constants. (a) For what range of values for β can $U_{el} + U'$ have a minimum at a finite value of r/a? (b) Find that value, and the total energy for that value.

5-G.22 Superionic conductors are alkali halide-like crystals (Na is an alkali, Cl is a halogen) that are insulators at ordinary temperatures but, when the temperature is raised above a certain value, suddenly become good electrical conductors. X-ray scattering, which can identify the atoms in the crystal, shows that one of the chemical constituents in the lattice melts at that point and can roam relatively freely around the lattice of the remaining atoms. Which is more likely to melt, the alkali or the halogen, and why?

"I wish to inform you of a new but terrible experiment, which I advise you on no account personally to attempt. . . . From the [charged and insulated] gun barrel hung a brass wire, the end of which entered a glass jar, which was partly full of water. . . . This jar I held in my right hand, while with my left I attempted to draw sparks from the gun barrel. Suddenly I received in my right hand a shock of such violence that my whole body was shaken as if by a lightning stroke. . . . In a word, I believed I was done for."

—Peter Van Musschenbroeck (1746)

Chapter 6

Capacitance

Chapter Overview

Section 6.2 discusses *single-plate* capacitors, and Section 6.3 discusses *two-plate* capacitors, with equal and opposite charges, showing why they are so much more effective than single-plate capacitors. Section 6.4 considers capacitors in circuits, both in parallel and in series, as well as in more complicated arrangements. Section 6.5 shows how the capacitance is increased if a polarizable material fills the region between the two capacitor plates. This decreases the magnitude of the electric field within the material, by a factor called the *dielectric constant*. The combination of geometry and dielectric constant is responsible for the large capacitance of the Leyden jar (and thus the dramatic effects observed by Musschenbroek) and of the recently developed electrolytic capacitors. Section 6.6 considers the electrical energy stored by a capacitor, and Section 6.7 relates electrical energy and electrical force. Section 6.8 introduces the coefficients of potential to discuss the general problem where two charged conducting objects do not necessarily have equal and opposite charges. Section 6.9 discusses the dielectric properties and electrical discharge properties of dilute gases, including a discussion of plasma globes and fluorescent tubes. Section 6.10 presents Maxwell's elaboration of Faraday's concept of flux tubes, which are under tension and exert pressure upon one another. ■

6.1 Introduction

The preceding description of the discovery of the Leyden jar illustrates one of the two major uses of *capacitors*: the rapid and powerful discharge of a reservoir of electric charge, such as in a heart defibrillator. The other major use is also as a charge reservoir, but one where a small and controllable discharge at a nearly fixed voltage is desired: the RAM (random access memory) of computers uses the state of storage of microscopic capacitors (charged or discharged) to represent 0's and 1's. Every electronic device, from computers to wristwatches, uses large numbers of capacitors for both of these purposes.

Capacitance is a measure of the capacity of conducting bodies to store charge. Two earlier examples—see Figure 1.10 (where charged Leyden jars are

---

The Leyden Jar as Condensor of Electric Charge

The Leyden jar was the first two-plate capacitor. One plate was the water in the jar, electrically connected to the gun barrel, which served as a reservoir of charge. The other plate was Van Musschenbroeck's right hand, electrically connected to ground, which served as another reservoir of charge. From these reservoirs, charges of opposite sign went to the parts of each plate adjacent to the glass of the jar. On touching his left hand to the gun barrel, there was a pathway through his body for discharge from one plate to the other, and hence the shock. See Figure 6.1. Because the Leyden jar was so effective as a charge storage device, it was considered to be a *condensor* of electric charge. This name is still used to describe the capacitors associated with the ignition systems of gas-powered lawnmowers and pre–electronic ignition automobiles.

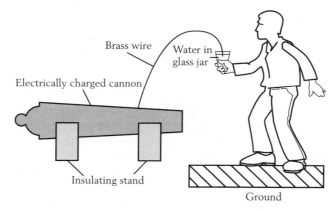

Figure 6.1 The discovery of the Leyden jar: the first two-plate capacitor. A charge source (not shown) brings charge to the cannon, then along the brass wire to the water in the glass jar, and then to the interior surface of the glass jar. Charge is repelled from the exterior surface of the glass jar, through the person's body, and to ground.

---

connected) and Figure 5.26 (where charged spheres are connected)—can be analyzed in terms of capacitance.

6.2 Self-Capacitance of an Isolated Conductor

Let us consider charge storage by an isolated conductor of arbitrary but fixed shape. As discussed in Chapter 4, in equilibrium any net charge on a solid conductor goes to its outer surface. Let the conductor have charge $Q > 0$, and be far from any other objects. Then electric field lines point outward from the object, leading to infinity. (Far from the object, the lines appear to come from a point charge Q.) Since field lines point from higher to lower voltage, the voltage V_∞ at infinity is lower than the voltage V of the conductor. Take $V_\infty = 0$, so $V > 0$.

Let us measure Q with a charge electrometer, as in Section 2.4 (there are other ways, but this is good enough in principle), and let us measure V with a

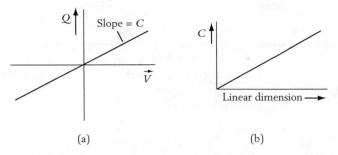

(a)   (b)

Figure 6.2 Charge Q and potential V relative to infinity are measureable quantities. (a) V is proportional to Q. (b) Their ratio, the capacitance $C = Q/V$, is proportional to the characteristic geometric length of the object.

voltage electrometer, as in Section 5.4 (again, there are other ways to measure ΔV). Now plot the curve of Q versus V.

Volta, in the early 1780s did just this, but he employed the angular displacement of a straw electrometer instead of V, and the number of turns of a frictional electrification device instead of Q. The curve he found was a straight line. Its slope is known as the self-capacitance C:

$$C = \frac{Q}{V}. \quad \text{(self-capacitance)} \quad (6.1)$$

The unit of capacitance, with units of C/V, is called the farad (F), after Michael Faraday, who studied material-dependent effects on capacitance (discussed in Section 6.5). The greater the charge Q, for a given voltage V relative to ground, the greater the capacity to store charge. In the form $Q = CV$, it has been called *Volta's law*. See Figure 6.2(a).

Example 6.1   **Measuring self-capacitance**

Consider an isolated object that is too complex for its self-capacitance to be computed: an aluminum paint-sprayed rubber duck. See Figure 6.3(a).

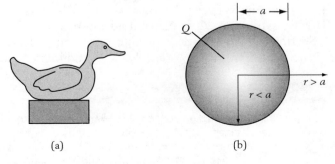

(a)   (b)

Figure 6.3 Capacitance: (a) An isolated conductor resting on an insulating surface. (b) A spherical capacitor: a conducting sphere of radius a and charge Q.

(Rather than having the duck sit on an insulator, close to ground, for better electrical isolation it could be hung from the ceiling by an insulating string.) Let the duck be given a charge $Q = 6 \times 10^{-11}$ C, and let a measurement of its voltage relative to a distant point yield $V = 20$ V. (a) Find its capacitance. (b) If it is now connected to a source of charge (i.e., a charge reservoir) at fixed voltage $V = 8$ V, find the charge Q' on the capacitor, and the change in the amount of charge on the reservoir.

Solution: (a) Equation (6.1) gives $C = Q/V = 6 \times 10^{-11}$C/20V $= 3 \times 10^{-12}$ F, or 3 pF (pF $= 10^{-12}$ F $=$ picofarad). (b) For $V = 20$ V, $Q = CV$ gives $Q' = 2.4 \times 10^{-11}$ C. Since the charge on the capacitor has decreased, by charge conservation it has supplied $Q_{res} = (6 - 2.4) \times 10^{-11}$ C $= 3.6 \times 10^{-11}$ C to the charge reservoir.

Here is another question. If our conductor (e.g., the duck) is scaled up proportionally in all three dimensions by a factor of two, how does that affect C? The answer is it doubles. As will be shown shortly, *C scales linearly with the linear dimension of the object*; not with the surface area (linear dimension squared) nor with the volume (linear dimension cubed). See Figure 6.2(b).

Example 6.2 **Capacitance of a larger-volume duck**

Let the duck of the previous example expand in volume by a factor of five, while still maintaining its same proportions. (a) Find its new capacitance C'. (b) If $V = 8$ V before the increase in volume, and if the duck is electrically isolated during expansion, find the new voltage V'.

Solution: (a) Since the volume (which varies as the cube of the linear dimension) is larger by a factor of five, the linear dimension is larger by $5^{1/3} = 1.71$. Hence the capacitance is now $C' = (1.71)(3 \text{ pF}) = 5.13$ pF. (b) For $V = 8$ V, $Q' = 2.4 \times 10^{-11}$ C, from the previous example. On expansion, $Q' = Q$ but the new voltage is $V' = Q'/C' = 4.68$ V.

We now consider an example showing that Q versus V has a constant slope, and that this slope, C, varies with the linear dimension of the object.

6.2.1 *Theory of Isolated Spherical Capacitor*

Consider a solid conducting sphere, of radius a, which is given a charge Q. See Figure 6.3(b). By symmetry, Q is distributed uniformly over the surface. As discussed in Section 5.8, for $r > a$, it produces the same electric field and potential as a point charge, $V(r) = kQ/r$, and $V_\infty = 0$; and for $r \le a$ its potential

---

Warning of Possibilities for Notational Confusion

In addition to the possibility of confusing voltage V with volts (V), we now have the possibility of confusing capacitance C (italicized) with coulombs C (not italicized). A larger alphabet—such as that used by the Chinese language—might cure this difficulty. Blame written language, which is a product of the human mind, not science, over which we have less control.

> ## Modern Electrometers Measure Charge without Causing Discharge
>
> Multimeters, which can be purchased at electronics stores and discount stores, measure a voltage difference in proportion to the current that it causes. This works if the voltage is maintained by some electrical power source. However, if the voltage is due to a small and nonreplenishable charge source, such as a capacitor, the voltage will fall rapidly as the charge is drawn off in the process of producing the current. This leads to either of two results: if the initial voltage is relatively low, no reading at all; if the voltage is relatively high, a blown fuse because of the large initial current flow through the multimeter. To measure voltage in this case, a modern *electrometer* is used. Unlike the original devices (e.g., calibrated leaf electroscopes), modern electrometers are "active" devices that sense the charge Q on the input side, but use an internal power source to cause, on the output side, current flow in proportion to Q.

takes on the constant value $V(r) = kQ/a$. Thus

$$C = \frac{Q}{kQ/a} = \frac{a}{k}. \qquad \text{(capacitance of sphere)} \qquad (6.2)$$

That is, (6.2) indeed satisfies the rule of Figure 6.2(a), that C, the slope of Q versus V, is independent of V or Q. Moreover, (6.2) satisfies the rule of Figure 6.2(b), that C scales with the linear dimension of the object, in this case its radius. Without saying it, we effectively used the equivalent of (6.2) at the end of the last chapter, where we considered two distant conducting spheres.

Example 6.3   **Two unusual capacitors**

(a) Find the capacitance of an 18th-century cannon, approximated by a sphere of radius 0.5 m. (b) Find the capacitance of a sphere the size of the earth.

Solution: (a) By (6.2), $C = a/k = 5.56 \times 10^{-11}$ F $= 55.6$ pF. (b) The earth has radius 6.37×10^6 m, so (6.2) yields $C = 7.08 \times 10^{-4}$ F.

6.2.2   *Why Q/V Is Independent of Q; Why C Scales with the Linear Dimension*

The potential of a capacitor can be determined by summing up the potential due to all its charges, thought of as point charges. Since, for $V_\infty = 0$, $V = \Sigma_i kQ_i/R_i$, for this sum, doubling each Q_i means doubling V. Hence Q/V is a constant, independent of Q.

Moreover, doubling the linear dimension of the capacitor means doubling each R_i, and thus halving V. As a consequence, $C = Q/V$ doubles. Hence, if an aluminized rubber-duck baby has capacitance C, then an aluminized rubber-duck mother that is doubled in each direction has capacitance $2C$. Equivalently, if equal charges are put on two objects, one being a scaled-up version of the other, the voltage will be higher on the *smaller* object. This scaling with the dimension of the object is a consequence of the fact that the force law for electricity is an inverse square, which leads to $V = kQ/r$ for a point charge Q. Capacitance is a useful quantity because it is a function of the geometry and the material of the capacitor, not of its charge or voltage.

Example 6.4 **Two well-separated spheres**

Two spheres with radii, $r_A = R$ and $r_B = 2R$ are very distant from each other. Each is given the same charge Q. (a) If $V_B = 200$ V, find V_A. (b) If, in addition, $r_B = 4$ cm, find Q and C_B.

Solution: (a) By (6.1), $C = Q/V$ for any isolated conductor at potential V relative to infinity; by (6.2), $C = a/k$ for a sphere of radius a. Combining these two gives $Q/V = a/k$, so $Q = Va/k$. Applied to spheres A and B, with $Q_A = Q_B$, this yields

$$\frac{V_A r_A}{k} = \frac{V_B r_B}{k}, \quad \text{or } V_A r_A = V_B r_B.$$

Hence

$$V_A = V_B \left(\frac{r_B}{r_A}\right) = V_B \left(\frac{2R}{R}\right) = 2V_B = 400 \text{ V}.$$

Even without knowing Q_A or r_A, we have found V_A. (b) $C = a/k$ gives $C_B = 2(.04\text{m})/[9 \times 10^9 \text{ N-m}^2/\text{C}^2] = 4.44 \times 10^{-12}$ F. Then $Q = Q_B = C_B V_B = 8.89 \times 10^{-10}$ C. This is a typical static electric charge.

6.3 Two-Plate Capacitors

For a general two-plate capacitor, let us give to the two conducting plates (which can have arbitrary shape) equal and opposite charges $\pm Q$, and then measure the voltage difference $\Delta V = V_1 - V_2$ between the plates. Here V_1 is associated with Q and V_2 is associated with $-Q$, so that $\Delta V > 0$. See Figure 6.4. We now define the capacitance to be

$$C = \frac{Q}{\Delta V}. \quad \text{(two-plate capacitor)} \quad (6.3)$$

For the same reasons as in the previous section, $C = Q/\Delta V$ is independent of Q or ΔV, and it scales with the linear dimension of the system, by which we mean we must scale both the plates and their separation.

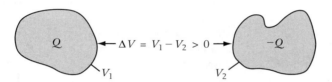

Figure 6.4 Two conductors with equal and opposite charge $\pm Q$ and voltage difference ΔV. In isolation from other conductors and charge, this constitutes a two-plate capacitor.

Capacitor Plates

Keep in mind that capacitor plates are made of conducting material, which in equilibrium have lots of negatively charged conduction electrons (the ones that can move freely about the conductor), and an equal and opposite amount of charge from essentially immobile positive ions. A capacitor plate containing 0.01 of a mole has about 6×10^{21} atoms. If the plate is made of copper, with two conduction electrons per atom, that means $2e = 3.2 \times 10^{-19}$ C of negative charge per atom that can move freely. Multiplying by 6×10^{21}, for the plate as a whole there is -1920 C of charge from the conduction electrons, and 1920 C of charge from the positive ions. Usually, a capacitor will be given no more than a fraction of a coulomb of charge. If the capacitor is given a charge of $Q = 0.001$ C, then the positive plate will have $(0.001 - 1920)$ C of charge from electrons (and $+1920$ C of charge from positive ions) for a net charge of 0.001 C, and the negative plate will have $(-0.001 - 1920)$ C of charge from electrons (and $+1920$ C of charge from positive ions) for a net charge of -0.001 C.

6.3.1 *Parallel-Plate Capacitor*

Consider two parallel plates, each of area A, separated by d, and given equal and opposite charges $\pm Q$. Neglecting end effects, the charge densities on the plates have the uniform values $\pm\sigma = \pm Q/A$. See Figure 6.5. Each plate produces a uniform field of magnitude $2\pi k\sigma$, and since the fields of the two plates add in the region between the plates ($\vec{E}$ points from the positive to the negative plate), $E = 4\pi k\sigma$, as in (3.36). Then the voltage takes the simple form $\Delta V = Ed$, as in (5.18). Hence

$$C = \frac{Q}{\Delta V} = \frac{Q}{Ed} = \frac{\sigma A}{4\pi k\sigma d} = \frac{A}{4\pi kd}. \qquad \text{(parallel plates)} \qquad (6.4)$$

For comparison, let $A = 4\pi a^2$, as for a sphere of radius a. Then (6.4) yields $C = 4\pi a^2 / 4\pi kd = (a/k)(a/d)$, giving a factor (a/d) relative to the capacitance (a/k) [compare Equation (6.2)] of the sphere. This factor of a/d can be made very large. It is one reason that two-plate capacitors can have much larger capacitances than one-plate capacitors.

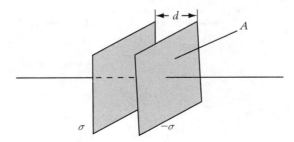

Figure 6.5 Parallel-plate capacitor of plate area A and plate separation d.

Example 6.5 **A two-plate capacitor**

Consider a parallel-plate capacitor with circular plates of radius $a = 1.0$ m, and plate separation $d = 2 \times 10^{-4}$ m. (a) Find its capacitance. (b) What radius would a sphere need to have the same capacitance?

Solution: (a) The plate area is $A = \pi a^2 = 3.14$ m^2. Then, with $d = 2 \times 10^{-4}$ m, (6.4) gives

$$C = \frac{3.14 \text{ m}^2}{[(4\pi)(9 \times 10^9 \text{ N-m}^2/\text{C}^2)(2 \times 10^{-4} \text{ m})]} = 1.39 \times 10^{-7} \text{ F}.$$

(This is smaller than what we can easily obtain with commercial capacitors, but it is much larger than for a sphere of radius $a = 0.5$ m.) (b) To achieve an equivalent capacitance on a spherical capacitor, (6.2) gives $C = a/k$, so $a = kC = (9 \times 10^9 \text{ N-m}^2/\text{C}^2)(1.39 \times 10^{-7} \text{ F}) = 1250$ m!

6.3.2 *Concentric Spherical Two-Plate Capacitor*

Let the outer radius be b and the inner radius be a. Place Q on the inner sphere and $-Q$ on the outer sphere. See Figure 6.6(a).

This situation has spherical symmetry, so $\vec{E} = (kQ_{enc}/r^2)\hat{r}$ of (4.15) applies. For $a < r < b$, $Q_{enc} = Q$, so $\vec{E} = (kQ/r^2)\hat{r}$ for $a < r < b$. Thus

$$\Delta V = V(a) - V(b) = -\int_b^a \vec{E} \cdot d\vec{s}$$

$$= \int_a^b \frac{kQ}{r^2}\hat{r} \cdot \hat{r}\, dr = \int_a^b \frac{kQ}{r^2} dr = -\frac{kQ}{r}\Big|_a^b = -kQ\left(\frac{1}{b} - \frac{1}{a}\right) = kQ\frac{(b-a)}{ab}.$$

Hence

$$C = \frac{Q}{\Delta V} = \frac{ab}{k(b-a)}. \qquad \text{(concentric spherical capacitor)} \qquad (6.5)$$

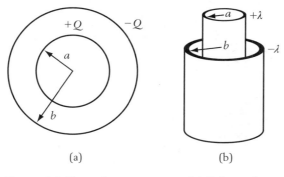

(a)　　　　　　　　(b)

Figure 6.6 Two-plate capacitors: (a) Spherical capacitor with inner radius a and outer radius b. (b) Long cylindrical capacitor with inner radius a and outer radius b.

Applying this to the earth, with $a \approx b = 6.4 \times 10^6$ m and $b - a \approx 5 \times 10^4$ m (about the distance to the ionosphere, which neutralizes the charge of the earth), (6.5) yields $C = 4.53 \times 10^{-2}$ F, much larger than the 7.08×10^{-4} F of Example 6.3(b).

Example 6.6 **Two checks**

(a) Show that, when the separation $b - a = d$ between the two spheres is small compared with a or b, the capacitance is that of two parallel plates of area $A = 4\pi a^2$. (b) Show that, when $b \to \infty$, the capacitance is that of an isolated spherical capacitor of radius a.

Solution: (a) As $a \to b$, with $b - a = d$, $C = ab/k(b-a) \to a^2/kd = A/4\pi kd$, as in (6.4) for a parallel-plate capacitor. (b) As $b \to \infty$, $C = ab/k(b-a) \to ab/kb = a/k$, just as in (6.2) for the single sphere of radius a.

6.3.3 *Concentric Cylindrical Two-Plate Capacitor*

Let the outer radius be b, the inner radius be a, and the length l of each be so long ($l \gg a, b$) that end effects can be neglected. Place charge per unit length λ on $r = a$ and $-\lambda$ on $r = b$. See Figure 6.6(b). We have already obtained the voltage difference between two such cylinders in (5.39) of the previous chapter, but the derivation bears repeating.

This situation has cylindrical symmetry, so $\vec{E} = (2k\lambda_{enc}/r)\hat{r}$ of (4.19) applies. For $a < r < b$, $\lambda_{enc} = \lambda$, so $\vec{E} = (2k\lambda/r)\hat{r}$ for $a < r < b$. Thus

$$\Delta V = V(a) - V(b) = -\int_b^a \vec{E} \cdot d\vec{s} = \int_a^b \left(\frac{2k\lambda}{r}\right)\hat{r} \cdot \hat{r}\, dr$$

$$= \int_a^b \left(\frac{2k\lambda}{r}\right) dr = (2k\lambda)\ln(r)\Big|_a^b = 2k\lambda\ln\frac{b}{a}.$$

Hence

$$C = \frac{Q}{\Delta V} = \frac{l}{2k\ln\left(\frac{b}{a}\right)}. \qquad \text{(concentric cylindrical capacitor)} \qquad (6.6)$$

Taking $b = 3$ mm and $a = 0.2$ mm (approximate values for the co-axial cable of commercial cable companies) yields a capacitance per unit length $C/l = 2.05 \times 10^{-11}$ F/m. A 100 m length of such cable has $C = 2.05 \times 10^{-9}$ F, so to produce $\Delta V = 2$ V requires $Q = C\Delta V = 4.10 \times 10^{-9}$ C.

6.4 Capacitors in Circuits

Equation (6.4) says that doubling the area of a capacitor doubles the capacitance because twice as much charge can be stored at the same voltage. This can be done with two identical capacitors C in parallel, each with the same voltage

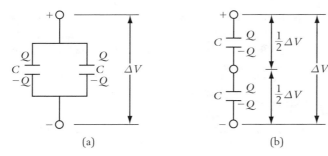

Figure 6.7 Capacitors in circuits. (a) Two identical capacitors in parallel. They have the same potential difference. (b) Two identical capacitors in series. They have the same charge.

difference ΔV. See Figure 6.7(a). Thus two capacitors in parallel should have a larger equivalent capacitance than either by itself.

Equation (6.4) also says that doubling the plate separation d halves the capacitance because the voltage doubles with the same amount of charge. This can be done with two identical capacitors in series, each with the same charge Q. See Figure 6.7(b). Thus two capacitors in series should have a smaller equivalent capacitance than either by itself.

With this background, let us now consider capacitors in series and in parallel. We will later consider more general cases.

6.4.1 *Capacitors in Parallel*

When capacitors are wired together in parallel, they have common upper terminals and common lower terminals. Thus, to place one capacitor in parallel with another, we must make two connections, one for each terminal. See Figure 6.8. Moreover, two new connections are needed to place a second capacitor in parallel with one that is already in a circuit. They are represented schematically like an end view of a parallel-plate capacitor. For purposes of argument, assume a reservoir of charge at a fixed voltage ΔV, to which capacitors C_1 and C_2 can be connected. Our goal is to determine the *equivalent capacitance C*

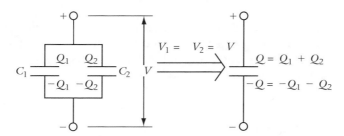

Figure 6.8 Two capacitors in parallel. They have the same potential difference. The equivalent capacitor has the same potential difference and an equivalent charge equal to the sum of the individual charges.

of these capacitors in parallel. (The resulting value of C is independent of how the voltage difference is produced.) To find C, we use two essential principles: (1) *charge conservation*, established in Chapter 1, and (2) the *path independence of the voltage*, established in Chapter 5.

First, because the electrical potential is independent of path, capacitors C_1 and C_2 in parallel with each other each have the same voltage difference as their combined equivalent capacitor: $\Delta V_1 = \Delta V_2 = \Delta V$. Second, by charge conservation the charge Q given to the composite capacitor distributes itself to become charges Q_1 and Q_2 on the individual capacitors: $Q = Q_1 + Q_2$. (For ideal capacitors, the charges on opposite plates are equal and opposite, to produce zero external electric field, thus assuring no current flow in the external system.) Since $Q_1 = C_1\Delta V_1 = C_1\Delta V$ and $Q_2 = C_2\Delta V_2 = C_2\Delta V$,

$$Q = Q_1 + Q_2 = (C_1 + C_2)\Delta V. \tag{6.7}$$

That is, at fixed total charge Q, capacitors in parallel share the charge. By the definition (6.3) of capacitance, $C = Q/\Delta V$, (6.7) then leads to

$$C = C_1 + C_2. \quad \text{(capacitors in parallel)} \tag{6.8}$$

More generally, for many capacitors in parallel, $C = C_1 + C_2 + \ldots$.

Example 6.7 Connecting and cross-connecting two charged capacitors in parallel

Let $C_1 = 3~\mu F$ and $C_2 = 6~\mu F$ initially be isolated with $Q_1^{(0)} = 8~\mu C$ and $Q_2^{(0)} = 10~\mu C$. Label the plates of C_1 as a and a', and those of C_2 as b and b'. Let them first be connected positive to positive (a to b) and negative to negative (a' to b'). After they come to equilibrium, let them next be connected positive to negative (a to b') and negative to positive (b to a'); the left of Figure 6.9 shows this just before the connections are completed. (This cross-connection is like that on going from Figure 1.10(b) to Figure 1.10(c). It is also related to the connecting of two spheres, as in Figure 5.26.) (a) Find the initial voltage differences across each capacitor. (b) Find, after the connection is made, the charge on each capacitor, and the voltage differences across each capacitor. (c) Find, after the cross-connection

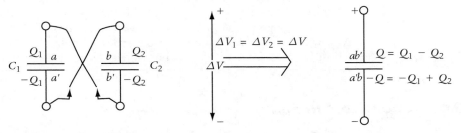

Figure 6.9 Connecting two charged capacitors, positive plate to negative and negative to positive. After equilibration, each pair of plates has the same potential. By charge conservation, the equivalent capacitor has an equivalent charge equal to the sum of the individual charges.

is made, the charge on each capacitor, and the voltage differences across each capacitor.

Solution: (a) From the initial charges and capacitances, the initial voltages are $\Delta V_1^{(0)} = Q_1^{(0)}/C_1 = 2.67$ V and $\Delta V_2^{(0)} = Q_2^{(0)}/C_2 = 1.67$ V. (b) On making the connection, charge will transfer (i.e., charge conservation applies) until the voltage differences become the same. On making the connection, the capacitors are in parallel, with net capacitance given by (6.8), or $C = C_1 + C_2 = 3\ \mu F + 6\ \mu F = 9\ \mu F$. They have total charge $Q = Q_1^{(0)} + Q_2^{(0)} = 8\ \mu C + 10\ \mu C = 18\ \mu C$. Thus, after connecting them, their common voltage difference is $\Delta V_1 = \Delta V_2 = \Delta V = Q/C = 18\ \mu C/9\ \mu F = 2$ V. They then have charges $Q_1 = C_1 \Delta V_1 = (3\ \mu F)(2\ V) = 6\ \mu C$ and $Q_2 = C_1 \Delta V_2 = (6\ \mu F)(2\ V) = 12\ \mu C$, which sums to $18\ \mu C$, as expected. (c) On making the cross-connection, charge will transfer (i.e., charge conservation applies) until the voltage differences become the same. Now, however, the total charge is $Q' = Q_1 - Q_2 = 6\ \mu C - 12\ \mu C = -6\ \mu C$. (The negative means that the plates a and b' are negatively charged.) Again, the capacitors are connected in parallel (even though different plates are now in parallel). Hence, their common voltage difference is $\Delta V_1' = \Delta V_2' = \Delta V' = Q'/C = -6\ \mu C/9\ \mu F = -0.667$ V. They then have charges $Q_1 = C_1 \Delta V_1 = (3\ \mu F)(-0.667\ V) = -2\ \mu C$ and $Q_2 = C_1 \Delta V_2 = (6\ \mu F)(-0.667\ V) = -4\ \mu C$, which sums to $-6\ \mu C$, as expected. Note the importance of charge conservation on both connection and reconnection.

Since the system goes to these states spontaneously after the connections and cross-connections are made, the energy of the final state must be less than the energy of the initial state. We will see this after we study energy storage by capacitors. (Recall that, on releasing a ball, it spontaneously falls under gravity, since it has more gravitational potential energy in its "initial" position than in its "final" position.)

Application 6.1 **On inventing your own rules**

Many students like to invent their own rules for what happens on connecting capacitors—average the charges or average the voltages—but this violates a well-known fact, adapted from gorillas to capacitors. In a cage with 50 bananas, how many bananas does a 500-pound gorilla eat for lunch?

Solution: It depends on whether it is alone, or if there is a 1000-pound gorilla in the same cage. For gorillas, the larger the stomach capacity of the gorilla, the larger the number of bananas it has for lunch; for capacitors, the larger the electrical capacity, the larger the amount of charge it holds for a given voltage difference. For $C_1 \neq C_2$, averaging the charge clearly satisfies charge conservation but gives $\Delta V_1' \neq \Delta V_2'$, which makes the voltage path-dependent. Similarly, averaging over the voltage differences gives $\Delta V_1' = \Delta V_2'$ but violates charge conservation.

6.4.2 *Capacitors in Series*

In contrast to the case of two capacitors in parallel, where two pairs of plates are connected, for two capacitors C_1 and C_2 in series only a single pair of plates is connected. Initially, there is no charge on the plate of either capacitor. See the leftmost part of Figure 6.10. For purposes of argument, assume a reservoir

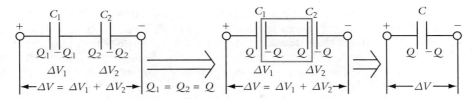

Figure 6.10 Two capacitors in series. They have the same charge. The equivalent capacitor has the same charge and an equivalent voltage equal to the sum of the individual voltages.

of charge at a fixed voltage ΔV, to which one end of each of the capacitors C_1 and C_2 is connected. Our goal is to determine the *equivalent capacitance C* of these capacitors in series. (The resulting value of C is independent of how the voltage difference is produced.) As for capacitors in parallel, we use the two principles of charge conservation and of the path independence of the electrical potential.

On connecting the voltage source, let its positive side put charge Q_1 on the outer plate of C_1, and let its negative side put charge $-Q_2$ on the outer plate of C_2. For ideal capacitors, as we assume here, the plates of each capacitor have equal and opposite charge, to produce zero external electric field, thus assuring no current flow in the external system. Thus, as illustrated in the leftmost part of Figure 6.10, the plates of C_1 and C_2 have charges $\pm Q_1$ and $\pm Q_2$.

Application of charge conservation to the originally neutral region connecting the two capacitors yields $Q_{connect} = 0 = Q_2 - Q_1$. Hence $Q_1 = Q_2$; the charges on adjacent plates must be equal and opposite. Call this common value Q. Then $\Delta V_1 = Q_1/C_1 = Q/C_1$ and $\Delta V_2 = Q_2/C_2 = Q/C_2$. Application of the path independence of the electrical potential, both to the equivalent capacitor subject to ΔV and by summing the voltages across the series of capacitors, yields

$$\Delta V = \Delta V_1 + \Delta V_2 = Q\left(\frac{1}{C_1} + \frac{1}{C_2}\right). \tag{6.9}$$

That is, at fixed total voltage difference ΔV, capacitors in series "share" the voltage difference. By the definition of capacitance, (6.3), written as $1/C = \Delta V/Q$, (6.9) leads to

$$\frac{1}{C} = \frac{1}{C_1} + \frac{1}{C_2}. \quad \text{(capacitors in series)} \tag{6.10}$$

More generally, for many capacitors in series, $C^{-1} = C_1^{-1} + C_2^{-1} + \dots$.

Example 6.8 **"Floating" voltage and the Leyden jar**

After the Leyden jar was discovered, many tried to reproduce the result, but were unsuccessful because they stood on an insulator, such as a cake of wax. (They weren't stupid; it had become customary to perform electrical measurements on an insulator, in order to keep the charge from escaping.) This meant that, instead of their body serving as a wire connected to ground, their feet were serving as a plate of a feet-wax-ground combination (capacitor C_1), their hand was serving as one plate of the Leyden jar (capacitor C_2), and their body was like the wire connecting the two plates of these capacitors in series. Since (6.4) shows that a relatively large separation (as between feet and ground) gives a relatively small capacitance, take the feet-wax-ground capacitance to be $C_1 = 10^{-11}$ F, and the Leyden jar to have the much larger capacitance of $C_2 = 10^{-9}$ F. Further assume that the electrostatic machine produced $\Delta V = 1010$ V on the cannon, relative to ground. Find the total capacitance of the system and the voltage change across each of the capacitors.

Solution: By (6.10), for capacitors in series, $C = 0.99 \times 10^{-11}$ F, so that $Q = C\Delta V = 10^{-8}$ C. Since $Q_1 = Q_2 = Q$, we obtain $\Delta V_1 = Q/C_1 = 1000$ V and $\Delta V_2 = Q/C_2 = 10$ V. They sum to the expected value of 1010 V. In this case, the Leyden jar has only 10 V across it, as opposed to the 1010 V it would have for someone standing on ground. That explains the inability of the other experimenters to charge up the Leyden jar enough to cause a shock. The voltage of the experimenter has "floated up" to 1000 V, very close to the 1010 V of the inside of the Leyden jar.

Note that the capacitor with the larger capacitance requires a larger charge to produce the same voltage as the smaller capacitor. Because capacitors in series have the same charge, this means that the voltage across the larger capacitance (here, the glass of the Leyden jar) is much less than across the smaller capacitance (here, the block of wax between feet and ground).

Distinguishing Capacitors in Series from Capacitors in Parallel

From the left-hand side of Figure 6.8, two capacitors in parallel have two distinct pieces of conductor, having net charges Q and $-Q$. (The two connected conductors on top behave like one piece, and similarly for the two connected conductors on the bottom.) On the other hand, from the middle of Figure 6.10, two capacitors in series have three distinct pieces of conductor, with net charges Q, 0, and $-Q$. If you understand the distinction between parallel and series, then you should be able to convince yourself that three conductors in parallel can be analyzed in terms of only two distinct pieces of conductor, whereas three conductors in series can be analyzed in terms of four distinct pieces of conductor (what are their charges?). (To treat two capacitors in series, with charge on the middle conductor, we must use the methods discussed in Section 6.8.)

6.4.3 *Combinations of Series and Parallel Circuits*

We can produce a wide variety of circuits by combining parallel and series circuits. For example, Figure 6.11 shows a complicated-looking circuit that can be reduced to parallel and series circuits. In this way, the overall capacitance can be obtained. However, a complete description of the circuit involves obtaining the charges and voltages for each capacitor, given, for example, the voltage across the system as a whole.

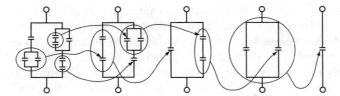

Figure 6.11 Combinations of series and parallel circuits. Parallel and series capacitors are replaced by equivalent capacitance until there is only a single equivalent capacitance. Not all circuits can be analyzed in this way.

Example 6.9   **A combination of three capacitors**

In Figure 6.12(a), there are three capacitors. Let $C_1 = 3$ μF, $C_2 = 6$ μF, and $C_3 = 1$ μF. C_1 and C_2 are in series with each other, and C_3 is in parallel to them. Find their equivalent capacitance, C_{eff}.

Solution: To find the equivalent capacitance of this combination, add C_1 and C_2 in series, to get the equivalent capacitance on the left, C_L, as in Figure 6.12(b). Then add C_L and C_3 in parallel to obtain C, as in Figure 6.12(c). Thus, if $C_1 = 3$ μF and $C_2 = 6$ μF, then by (6.10), $1/C_L = 1/3$ μF + $1/6$ μF = $1/2$ μF, so $C_L = 2$ μF. Then, by (6.8), $C = C_L + C_3 = 2$ μF + 1 μF = 3 μF.

Example 6.10   **Detailed study of Example 6.9**

In the previous example, let the voltage difference across C_1 be 6 V, and let its common end with C_3 be at ground. Find the charge on each capacitor, and the voltage at the other end of C_3.

Solution: Clearly, $Q_1 = C_1 \Delta V_1 = (3$ μF$)(6$ V$) = 18$ μC. Then $Q_L = Q_2 = Q_1 = 18$ μC. Hence $\Delta V_2 = Q_2/C_2 = 3$ V. Thus, the voltage drop across the series combination of C_1 and C_2 is, by (6.9), 6 V + 3 V = 9 V. This must be the same as the voltage drop ΔV_3 across C_3. Thus $Q_3 = C_3 \Delta V_3 = 9$ μC, so $Q_L + Q_3 = 27$ μC, in agreement with $Q = C \Delta V = 27$ μC.

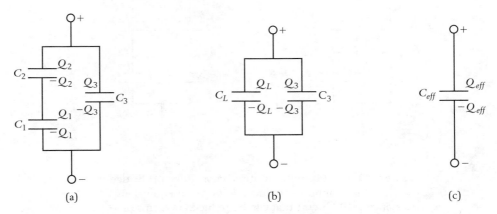

(a)   (b)   (c)

Figure 6.12 Combination of capacitors.

The analysis of Example 6.10 assumes that one capacitor cannot affect another. That is true for ideal two-plate capacitors that produce no external electric fields. However, a wire in a laboratory can charge up, producing an electric field, and that electric field can produce an electrical potential that can affect other wires in the laboratory. Typically, we want to eliminate such nuisance effects. In a more complex analysis, with the wire considered to be a third conductor, from measurements of the coefficients of potential (see Section 6.8) the influence of the wire can be estimated.

Although the response of a conductor to a voltage seems to be reflected only in its net charge, keep in mind that there is a full charge distribution over each conductor, even the connecting wires, and this is an infinitely large set of numbers. Clearly, it can be a very complicated problem to work out the entire charge distribution although in principle it can be measured experimentally.

6.4.4 *More Complex Circuits*

Although we can produce a wide variety of circuits by the preceding procedure of combining parallel and series circuits, there are more general circuits. Pictured in Figure 6.13(a) is a bridge circuit. To obtain the capacitance of this combination, we must assign voltages to each junction and charges to each capacitor, and then repeatedly use $Q = C\Delta V$. Because a similar analysis will be performed for resistors, in Chapter 8, we will not go into such complexity at this time. Consider, however, the case where the voltages at the midpoints, without the bridge capacitor, are the same, so $\Delta V_5 = 0$. Since $Q_5 = C_5 \Delta V_5 = 0$, the bridge capacitor C_5 does not charge up, so it is not an active part of the circuit.

Application 6.2 **The bridge condition**

Disconnecting one end of the bridge capacitor yields the simplified circuit in Figure 6.13(b), which can be analyzed in terms of series and parallel capacitors. The left arm has C_1 and C_3 in series, so $1/C_L = 1/C_1 + 1/C_3$,

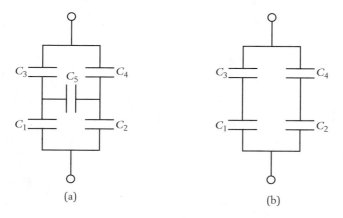

(a) (b)

Figure 6.13 Circuits with multiple capacitors. (a) Bridge circuit, which cannot be reduced in terms of equivalent resistances. (b) Circuit that can be reduced in terms of equivalent resistances.

and the right arm has C_2 and C_4 in series, so $1/C_R = 1/C_2 + 1/C_4$. (Hence the total capacitance, if we wanted it, is $C = C_L + C_R$.) For a total voltage ΔV across the system, with the C_1 and C_2 connection at the low voltage end, the voltages across C_1 and C_2 are

$$\Delta V_1 = Q_L/C_1 = \left(\frac{C_L \Delta V}{C_1}\right) = \Delta V \left[\frac{C_3}{(C_1 + C_3)}\right],$$

$$\Delta V_2 = Q_R/C_2 = \Delta V \left[\frac{C_4}{(C_2 + C_4)}\right].$$

If $\Delta V_5 = 0$, then the two midpoints are at the same potential, or $\Delta V_1 = \Delta V_2$, so $C_3/(C_1 + C_3) = C_4/(C_2 + C_4)$. This leads to the *bridge condition*,

$$\frac{C_1}{C_3} = \frac{C_2}{C_4}. \qquad \text{(bridge condition)} \qquad (6.11)$$

Let C_1 and C_2 be known, let C_4 be unknown, and let C_3 be variable. When C_3 is varied until the voltage difference ΔV_5 across the bridge is zero, the unknown C_4 can be determined from (6.11).

6.5 Dielectrics

One reason that the Leyden jar, a two-plate capacitor, worked so well was that it had a better geometry for storing charge. A second reason was that for a given charge Q the glass between its surfaces increased the capacitance by decreasing the voltage difference ΔV between the plates. The amount by which the voltage difference decreased is a measure of what is called the *dielectric constant*, or *relative permittivity* κ of the glass. The quantity $\epsilon = \kappa \epsilon_0$ is known as the *permittivity*, thus explaining why ϵ_0 is known as the permittivity of free space.

κ is a material property, independent of geometry. See Table 6.1, which gives the dielectric constant κ and the *dielectric strength* E_d, *the field above which sparking—electrical breakdown—occurs*. We first discuss how to measure κ, and then show how it affects the capacitance. In measuring E_d, it should be recognized that breakdown in a capacitor occurs at the position where the field is largest.

6.5.1 Determining the Dielectric Constant κ

Let a parallel-plate capacitor, with area A and separation d, have plates with charge density $\pm\sigma$, so the electric field magnitude $E_0 = 4\pi k\sigma$. (We neglect edge effects.) Now introduce a slab of dielectric (e.g., wood or glass) that just fits into the capacitor. Its molecules become polarized. See Figure 6.14, which includes a representation of some polarized molecules (not drawn to scale).

Table 6.1 Dielectric properties (at standard temperature and pressure)

Material	Air	Water	Glass	Paper	Rubber	Mica	SrTiO₃	Lucite	Wax
Dielectric constant κ	1.0059	80	4–6	3.5	2–3.5	5.5	330	2.8	2.3
Dielectric strength E_d (10^6 V/m)	3	—	9	16	30	150	8	20	10

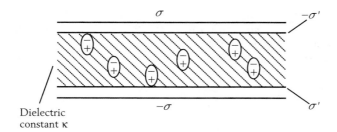

Figure 6.14 Capacitor filled with dielectric, including polarization charges $\pm\sigma'$ on the outer surfaces of the dielectric and source charges $\pm\sigma$ on the inner surfaces of the conducting plates. A gap is drawn between conductor and dielectric, but even if there is none, the polarization charge (intimately associated with molecules of the dielectric) cannot escape to the conductor, or vice versa.

There will be no net charge within the dielectric, but its surfaces develop polarization charges $\mp\sigma'$ opposite the plate charge $\pm\sigma$. Because σ and $-\sigma'$ produce opposing electric fields, the electric field within the dielectric decreases. Within the dielectric the field magnitude is now

$$E = 4\pi k\sigma - 4\pi k\sigma' = E_0\left(1 - \frac{\sigma'}{\sigma}\right), \qquad E_0 = 4\pi k\sigma. \qquad (6.12)$$

Equation (6.12) may be rewritten as

$$E = \frac{E_0}{\kappa}, \qquad \text{(field with dielectric)} \qquad (6.13)$$

where κ is the material-dependent dielectric constant. For real materials, $\kappa > 1$. The more polarizable the material, the larger the value of κ.
    Comparison of (6.13) and (6.12) yields

$$\frac{1}{\kappa} = 1 - \frac{\sigma'}{\sigma}. \qquad (6.14)$$

Here is one way to measure κ directly. If the plates have a fixed charge, then the voltage difference without the dielectric is

$$\Delta V_0 = E_0 d. \qquad (6.15)$$

For the same charge on the plates, but with the dielectric between the plates,

$$\Delta V = E d = \frac{E_0 d}{\kappa}, \qquad (6.16)$$

so that the voltage will decrease by a factor of κ.

Example 6.11 Determining κ by measuring voltage

Let a parallel-plate capacitor in air have voltage difference 20 V. After a dielectric is slid between the plates of the capacitor, the voltage difference is 4 V. Find the dielectric constant κ, and the ratio of the charge induced on the dielectric surfaces to the charge on the plates of the capacitor.

Solution: Equation (6.16) yields $\kappa = \Delta V_0 / \Delta V = 5$. Equation (6.14) yields $\sigma'/\sigma = 0.8$ so that the charge induced on the dielectric surfaces is 80% of charge on the capacitor plates.

6.5.2 *Effect of Dielectric on Capacitance*

What effect does the dielectric have upon the capacitance? By $C = Q/\Delta V$, with fixed Q but ΔV in the denominator *decreased* by the factor κ, the capacitance C *increases* from its value C_0 in air ($\kappa_{air} \approx 1$) by the factor κ. This is true for parallel-plate, cylindrical, and spherical capacitors, provided that the region between the plates is completely filled with a single type of dielectric. Thus

$$C = \kappa C_0. \quad \text{(capacitor filled with dielectric)} \tag{6.17}$$

For $\kappa = 5$, this gives an increase by a factor of five in the capacitance. The geometrical effects associated with two-plate versus one-plate capacitors can easily yield an additional factor of ten. Together, they can lead to an overall capacitance increase by a factor of 50 in going from a one-plate to a two-plate capacitor. This explains the power of the Leyden jar relative to previous methods of charge storage.

Example 6.12 Determining κ by measuring charge

Faraday studied the effects of certain dielectrics by using two identical spherical capacitors, one with an air gap and the other filled with, for example, molten sulfur or paraffin, which cooled and hardened. The plates of the two capacitors were connected, and thus they had the same potential difference. By measuring the charge that flowed to each capacitor, the dielectric constant could be determined. Let the initial charge, for the empty capacitor, be $Q_0 = 1.5 \times 10^{-8}$ C, and let the final charge, where the capacitor is filled with dielectric, be $Q = 6 \times 10^{-8}$ C. Find the dielectric constant.

Solution: $Q = C\Delta V = \kappa C_0 \Delta V$ and $Q_0 = C_0 \Delta V$, so $Q = \kappa Q_0$. For our Q and Q_0, this gives $\kappa = 4$.

Caveat

Two geometrically different capacitors with the same charge and voltage difference are *macroscopically* identical. However, they are not *microscopically* identical when placed in a circuit. For example, the charge distributions on the plates will differ. Moreover, changing the leads to the plates changes the charge distribution on the leads and on the plates.

Electrolytic Capacitors and Supercapacitors

A major advance in capacitor design was the introduction of *electrolytic* capacitors, which have a very thin layer of dielectric separating the two plates of the capacitor. (Thus, unlike Figure 6.14, there would be no air gaps.) There are at least two basic schemes.

In one scheme, the electrodes (in Figure 6.14, the capacitor plates) are asymmetric. The special electrolyte between them normally is conducting, but when a voltage is applied in the proper direction, an insulating oxide layer of very small thickness d (on the order of 10^{-9} m) is deposited. If the electrical leads are connected the wrong way, the oxide layer for such a capacitor is removed by electrolytic decomposition, leading to breakdown of the capacitor. Such capacitors only can be given positive voltage on one side, and for that reason the electrodes are clearly distinguished.

In another scheme, the electrodes are identical. When a voltage is applied, the positive ions from the electrolyte are attracted to the negative electrode. If the voltage is less than a threshold value on the order of a volt (depending upon the electrode and the ions), positive ions press against, but do not react with, the negative electrode. The charge on the electrode and the nearby opposite charge in the electrolyte behave like a capacitor with a very small plate separation. A similar effect with negative ions occurs at the other electrode. Too large a voltage causes a chemical reaction at the electrode–electrolyte interface, thus passing charge through the electrolyte, and making the capacitor ineffective at storing additional charge. By using carbon particles as the electrodes, it has been possible to achieve very large effective areas, thus increasing the capacitance even further.

Electrolytic capacitors with $C = 1$ μF are now commonly available. Even larger capacitances are becoming available. There is talk of capacitors of such large capacitance ("supercapacitors") that, for some applications, they might store enough electrical energy to replace batteries. (A simple application would be to maintain the power in a clock driven by ac power, when the power temporarily goes out.)

Biological Capacitors: The Cell Membrane

Living organisms use electricity in many ways. One use is to provide electrical signals within a cell and between different cells. Even when cells are in their resting state, they are electrically nontrivial because (according to the cell and the organism) the voltage within the cell is lower than outside the cell by some 50 mV to 150 mV.

The *cellular membrane*, or cell wall, separates the exterior and the interior of biological cells. The individual components of this membrane are *amphipathic* molecules, which means that their two ends are very different. The head is a polar molecule, having a permanent electric dipole moment. As a consequence the head group is attracted to water (thus it is called *hydrophilic*, or water loving). The tail group consists of two long fatty acids, called *lipids*, and avoids water (thus it is called *hydrophobic*, or water fearing). See Figure 6.15.

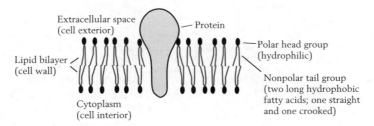

Figure 6.15 Details of a biological cell wall, composed of a lipid bilayer, which separates the cell interior (cytoplasm) from the cell exterior (extracellular space). The wall is composed of two layers of amphipathic molecules, their hydrophic ends in contact; the homophilic ends are in contact with the cytoplasm and extracellular space.

Such molecules spontaneously self-assemble: throw many in water and they form pairs that go tail to tail; then the pairs go side by side to form a *lipid bilayer*—the basis of the cellular membrane. By this means the hydrophilic heads are in contact with the water, and the hydrophobic lipid tails are shielded from the water.

Because cell membranes all have nearly the same thickness d (6–10 nm) and the same dielectric constant ($\kappa \approx 3$, appropriate to fats and oils), cell membranes all have about the same capacitance per unit area, or *specific capacitance* C_m. Since membranes are relatively thin compared to their area, we may use (6.4) with (6.17) to obtain

$$C_m \equiv \frac{C}{A} = \frac{\kappa}{4\pi kd}. \qquad \text{(specific capacitance)} \qquad (6.18)$$

For $\kappa = 3$ and $d = 6$ nm, (6.18) gives $C_m \approx 0.5\ \mu\text{F/cm}^2$, about a factor of two smaller than typical experimental values. The characteristic voltage difference $\Delta V = 100$ mV across a typical cell wall separation of $d = 10$ nm corresponds to an electric field of $E = \Delta V/d = 10^7$ V/m. This exceeds E_d for air! (See Table 6.1.) However, E_d for lipids is much larger than this.

For a nearly spherical cell, from a knowledge of its radius a (giving surface area $A = 4\pi a^2$) and C_m, we can estimate its capacitance $C = C_m A$. (Spherical cell radii can range from $1-100\ \mu$m.) For a nearly cylindrical cell, from a knowledge of its radius a (giving perimeter $p = 2\pi a$) and C_m, we can estimate its capacitance per unit length $C_m p$. (Cylindrical cells, such as nerve axons, the telephone wires of the nervous system, have radii that vary from 1 to 250 μm; the largest is the giant squid axon, whose size has made it a veritable bioelectrical playground.) Figure 6.16 is a schematic of a spherical cell, including negative charges associated with large molecules within the cytoplasm (cell interior), positive ions that cancel the charges on the large molecules, negative ions that go to the inside of the cell wall, and an equal number of positive ions that go to the outside of the cell wall. Not depicted are the vast but equal numbers of positive and negative ions that are dispersed throughout the volume of the cell.

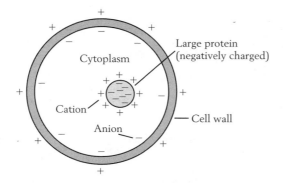

Figure 6.16 Biological cell, with a characteristic negatively charged protein inside, and ions in its immediate vicinity to "screen" its electric field. At the cell wall, there are additional ions on both sides, leading to a voltage difference across the cell wall. This voltage difference is instrumental in the operation of certain large proteins within the cell walls, which serve as ion channels, ion pumps, and ion exchangers, making the human body serve as a huge and complex electrical machine.

The cellular membrane is not a rigid object, for the molecules on either surface (compare Figure 6.15) can diffuse about that surface. Moreover, the cellular membrane also contains proteins that serve as selective ion channels (for passive flow of ions into and out of the cell—think of these as ion-selective resistors), selective ion pumps (for active flow of ions into and out of the cell—think of these as ion-selective voltaic cells), and other structures. (K^+, Na^+, Cl^-, and Ca^{2+} are the most important ions.) The relatively slow ion pumps, and some of the ion channels (over 57 varieties) are always operating. However, some of the ion channels can be switched on suddenly by a change in voltage or ionic concentration. This causes a rapid flow of ions into or out of the cell, leading to effects that make the larger organism (a collection of cells) operate effectively as a whole. Thus your body contains, effectively, a vast number of capacitors that charge and discharge between the cellular and intracellular electrolytes. The beating of your heart is driven by a muscular response set off by this cellular charge and discharge. You are an electrical machine!

6.6 Electrical Energy

6.6.1 *Energy Storage by a Capacitor*

We now consider the energy U it takes to charge up a capacitor. See Figure 6.17. Take $U = 0$ for charge $Q = 0$. Clearly the capacitor would discharge spontaneously if the plates were connected, so we expect that $U > 0$ for $Q \neq 0$. Starting from the configuration where the capacitor is uncharged, successively take small charges dQ' from one plate and put it on the other. At any instant, the charge on the positive plate is Q', so the voltage gain of dQ' is $\Delta V = Q'/C$. Since positive charge dQ' is moved to a plate that already has positive charge Q', this must increase the electrical energy. By (5.8), or $U = qV$,

Figure 6.17 Charge transfer dQ' from one capacitor plate to another, used to determine the energy to charge a capacitor.

with $(U, q, V) \to (dU, dQ', \Delta V)$, the increase dU in electrical energy is

$$dU = dQ'(\Delta V) = \frac{Q' dQ'}{C}. \tag{6.19}$$

This can be integrated immediately, with $U = 0$ for $Q = 0$, to yield $U = Q'^2/2C|_0^Q$, or

$$U = \frac{Q^2}{2C}. \qquad \text{(energy stored by capacitor)} \tag{6.20}$$

Another way to obtain (6.20) is to use $U = Q\overline{\Delta V}$, where the average voltage difference $\overline{\Delta V}$ at which the charge is transferred is $\overline{\Delta V} = (Q/2C)$. With $Q = C\Delta V$, (6.20) may also be written as

$$U = \frac{Q\Delta V}{2}, \qquad U = \frac{C(\Delta V)^2}{2}. \tag{6.21}$$

These results also hold for a single-plate capacitor, where Q is brought in from infinity.

Example 6.13 **A capacitor's energy storage**

For a 20 μF capacitor, let $\Delta V = 240$ V. Find Q and U.

Solution: Equation (6.3) yields $Q = C\Delta V = 4800 \ \mu$C. Equation (6.21) then gives $U = Q\Delta V/2 = 0.576000$ J. If we try to use this energy to lift a live 2 kg duck by the height h, $U = mgh$ yields $h = 14.6$ mm. (Greater height might be attained by discharging the capacitor through the duck, but only if the duck provides energy of its own.)

6.6.2 *Energy Is Minimized by $\Delta V_1 = \Delta V_2$ for Capacitors in Parallel*

Consider two capacitors initially at different potentials. On connecting their upper plates together, and their lower plates together, thus permitting charge transfer, charge is transferred until they reach equilibrium with $\Delta V_1 = \Delta V_2$. See Figure 6.18.

In making this adjustment, the total charge on each set of connected plates (e.g., the connected upper plates) does not change. We expect that such a spontaneously reached situation is a minimum of the energy. Let us see if that is indeed the case. The total electrical energy is given by

$$U = \frac{Q_1^2}{2C_1} + \frac{Q_2^2}{2C_2}. \tag{6.22}$$

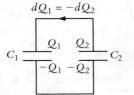

Figure 6.18 Charge transfer from one capacitor to another, used to study energy minimization when two capacitors are brought into electrical contact.

At fixed total charge, $Q = Q_1 + Q_2 = $ constant, so $dQ_1 = -dQ_2$. The differential of U is then

$$dU = \frac{Q_1}{C_1}dQ_1 + \frac{Q_2}{C_2}dQ_2 = \left(\frac{Q_1}{C_1} - \frac{Q_2}{C_2}\right)dQ_1. \qquad (6.23)$$

From (6.23), the energy is minimized when $dU/dQ_1 = 0$. This corresponds to $Q_1/C_1 = Q_2/C_2$, or $\Delta V_1 = \Delta V_2$, as expected.

6.6.3 *Energy Density for a Capacitor*

Let's apply (6.21) to a parallel-plate capacitor containing material of dielectric constant κ. Successively placing (6.4), (6.17), and (6.16) into (6.21) yields

$$U = \frac{1}{2}\frac{\kappa A}{4\pi k d}(\Delta V)^2 = \frac{Ad}{8\pi k}\frac{E_0^2}{\kappa} = \kappa\frac{Ad}{8\pi k}E^2. \qquad (6.24)$$

The electrical energy per unit volume u_E, also known as the electrical energy density, is given by the electrical energy U divided by the volume Ad. Using (6.24), this becomes

$$u_E = \frac{U}{Ad} = \kappa\frac{E^2}{8\pi k}. \qquad \text{(electrical energy density)} \qquad (6.25)$$

This result is true in general. Recall that E is the field within the dielectric itself.

Example 6.14 **Electric field within a capacitor**

Consider a dielectric with $\kappa = 4$. To achieve $u_E = 2 \times 10^8$ J/m^3, a value that is characteristic of energy storage in a voltaic cell, what must E be?

Solution: Equation (6.25) yields $E = (8\pi k u_E/\kappa)^{1/2}$, so here $E = 3.38 \times 10^9$ V/m. This exceeds E_d for most materials. Supercapacitors must withstand such a large E. For air at atmospheric pressure, with $\kappa \approx 1$ and $E_d \approx 3 \times 10^6$ V/m, (6.25) gives only $u_E = 39.8$ J/m^3.

Example 6.15 **Energy calculation of capacitance of spherical capacitor**

We will verify that (6.25) holds more generally, by reconsidering the single-plate spherical capacitor. In this case, $E = kQ/r^2$ for $r > a$, and for a spherical shell of radius r and thickness dr the volume element is $d\mathcal{V} = 4\pi r^2 dr$. (Since V means voltage, we must use another symbol for volume.) How much energy is required to charge the capacitor?

Solution: By (6.25) with $\kappa = 1$ (an air gap), and with $dU = u_E d\mathcal{V}$,

$$U = \int u_E d\mathcal{V} = \int u_E(4\pi r^2)\,dr = \int_a^\infty \frac{(kQ/r^2)^2}{8\pi k}(4\pi r^2)\,dr$$

$$= \frac{kQ^2}{2}\int_a^\infty \frac{dr}{r^2} = \frac{kQ^2}{2a}. \qquad (6.26)$$

Using (6.26) in (6.20) yields $C = Q^2/2U = a/k$, in agreement with (6.2).

6.6.4 *Spontaneous Charge Transfer Causes a Decrease in Electrical Energy*

When two conductors at different potentials are connected, charge flows between them until their potentials are equalized. Since this occurs spontaneously, the total electrical potential energy must decrease in this process. The energy goes into heating up the system, due to drag forces on the charge that passes through the wire connecting the two conductors. Such *resistive heating* will be discussed in detail in the next chapter.

Example 6.16 **Energy loss on connecting two capacitors**

How much energy is lost on connecting and cross-connecting the capacitors in Example 6.7?

Solution: Example 6.7 has $C_1 = 3\ \mu F$ and $C_2 = 6\ \mu F$, and initial charges $Q_1^{(0)} = 8\ \mu C$ and $Q_2^{(0)} = 10\ \mu C$. Thus the initial energy is $(Q_1^{(0)})^2/2C_1 + (Q_2^{(0)})^2/2C_2 = 19\ \mu J$. After connection, $Q_1 = 6\ \mu C$ and $Q_2 = 12\ \mu C$. Thus after connection the energy is $Q_1^2/2C_1 + Q_2^2/2C_2 = 18\ \mu J$. After cross-connection, $Q_1' = 2\ \mu C$ and $Q_2' = 4\ \mu C$. Thus the energy after cross-connection is $Q_1'^2/2C_1 + Q_2'^2/2C_2 = 2\ \mu J$. As expected, the system loses energy at each step.

Example 6.17 **Energy loss on connecting two spheres**

Let two spheres with radii $r_1 = 0.5$ m and $r_2 = 1$ m, and charges $Q_1 = 0$ and $Q_1 = 10^{-9}$ C, be very far apart. What is the change in energy of the system after the spheres are connected, and how much charge is transfered to sphere 1?

Solution: By (6.2), $C_1 = r_1/k = 5.56 \times 10^{-11}$ F and $C_2 = r_2/k = 1.11 \times 10^{-10}$ F. For $Q_1 = 0$ and $Q_2 = 10^{-9}$ C, they have an initial total energy of $U_1 + U_2 = 0 + Q_2^2/2C_2 = 4.5 \times 10^{-9}$ J. When they are connected, charge transfers across them until they have the same voltage, so they may be thought of as being in parallel. Hence $C = C_1 + C_2 = 1.67 \times 10^{-10}$ F. They have a net charge $Q' = Q_1 + Q_2 = 10^{-9}$ C, so after the connection they now have energy $U' = Q'^2/2C = 3.0 \times 10^{-9}$ J. Thus, as expected, the system has lost energy. Note that $\Delta V' = Q'/C = 6$ V, so that $Q_1' = C_1 \Delta V_1' = C_1 \Delta V' = 3.33 \times 10^{-10}$ C, which is the amount of charge transferred by the connection.

6.7 **Force and Energy**

From Chapter 5, if a charge q in an electric field $\vec{E}$ is displaced by $d\vec{s}$, then the change in its electrical energy dU_q is given by the negative of the electrical work dW, or

$$dU_q = -dW = -q\vec{E} \cdot d\vec{s}.$$

Summing over many charges q_i, the total electrical energy change is given by

$$dU = -\vec{F} \cdot d\vec{s}, \qquad \vec{F} = \sum_i q_i \vec{E}_i. \tag{6.27}$$

Here $\vec{F}$ is the total electrical force. We will use (6.27) to study a number of different situations.

6.7.1 Direct Calculation of Attractive Force between Plates of Parallel-Plate Capacitor

The force of attraction between the plates of a parallel-plate capacitor was given in the previous chapter, when we discussed the force electrometer. The result is sufficiently important, of itself and as a check for our calculation by the energy method, that it doesn't hurt to repeat this trivial calculation.

Consider a capacitor of area A whose plate separation d is much less than the characteristic width of the capacitor. For simplicity, consider a capacitor with an air gap, so we take $\kappa = 1$. The force magnitude $F = |\vec{F}|$ on the positive plate is the product of its charge Q with the electric field magnitude $E = 2\pi k\sigma$ of the other, where $\sigma = Q/A$. Thus

$$F = QE = Q(2\pi k\sigma) = \frac{2\pi k Q^2}{A}. \qquad (6.28)$$

6.7.2 Energy Calculation of Attractive Force between Two Capacitor Plates: Fixed Charge

F will now be obtained by calculating the energy change when the upper plate position changes from d to $d + \delta d$. (Since dd is a terrible notation, here we will denote small changes in any quantity, such as d, by δd.) With $C = A/4\pi kd$, this leads to $\delta C = (\delta C/\delta d)\delta d = -(A/4\pi d^2)\delta d$. For a capacitor in isolation, the charge is fixed. Thus, with $U = Q^2/2C$ and (6.4), at fixed Q the differential of U is

$$\delta U = -\left(\frac{Q}{C}\right)^2 \frac{\delta C}{2} = \left(\frac{Q}{C}\right)^2 \frac{A}{4\pi kd^2}\frac{\delta d}{2} = Q^2\frac{2\pi k}{A}\delta d. \quad \text{(fixed charge)}$$
$$(6.29)$$

Thus, at fixed Q, a increase in d causes an increase in U. Clearly, the system will want to spontaneously contract—an attractive electrical force F_y. Explicitly, at fixed Q, and from (6.27) with $d\vec{s} = \hat{y}\delta d$,

$$\delta U|_Q = -F_y\delta d. \qquad (6.30)$$

Comparison of (6.29) and (6.30) yields $F_y = -F$, with the same F as in (6.28). The negative sign in F_y means that the force is downward, corresponding to attraction.

6.7.3 Energy Calculation of Attractive Force between Two Capacitor Plates: Fixed Potential

Let our capacitor be in parallel with a second capacitor of much larger capacitance so that even if the second capacitor transfers charge, its potential remains essentially unchanged. (We can think of the second capacitor as a reservoir at fixed potential.) F will now be obtained by calculating the energy change when the upper plate position changes from d to $d + \delta d$, but we now must include the energy of the second capacitor. See Figure 6.18. Again we employ (6.20) for the energy, but there are now two independent variations. On the one hand, we change the capacitance C by δC on changing the plate separation, as in (6.29); on the other hand we permit charge transfer $\delta Q_1 = -\delta Q_2$, as in (6.23). The differential of U is the sum, or

$$\delta U = -\left(\frac{Q_1}{C_1}\right)^2 \frac{\delta C_1}{2} + \left(\frac{Q_1}{C_1} - \frac{Q_2}{C_2}\right)\delta Q_1. \qquad \text{(fixed voltage)} \quad (6.31)$$

Since we are in equilibrium, the second term is zero because $\Delta V_1 = Q_1/C_1 = \Delta V_2 = Q_2/C_2$. Hence the coefficient of δQ_1 cancels; only the first term survives. However, this is the same as (6.29) for the isolated capacitor, so it leads to the same force as for the isolated capacitor. Thus, as expected, the force of attraction between the plates of a capacitor does not depend on the circuit to which it belongs. Another way of describing the topic of this section is to say that we have calculated the force between the plates at *fixed voltage*.

6.7.4 *Attraction of a Dielectric into a Capacitor*

We now apply (6.27) to find the total electrical force $\vec{F}$ on a slab of dielectric that is partially inserted into the plates of a parallel-plate capacitor. See Figure 6.19.

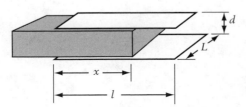

Because this problem is related to the amber effect, we expect the slab to be attracted inward. A direct calculation, involving a sum of forces over all parts of the capacitor, including the fringing field region, would be quite difficult. However, as long as the slab edge isn't near either edge of the capacitor, we can obtain $\vec{F}$ by energy considerations. In that case, the *total* energy change on moving the slab by a small amount involves, effectively, only a transfer of material from the region within the capacitor to the region outside the capacitor. The energy in the difficult-to-calculate-with fringing field region is the same in both cases, so it does not contribute to (6.27).

To be specific, let the capacitor have thickness d and area $A = Ll$, where L is into the page and l is horizontal. Let air fill an area $A_{air} = L(l - x)$, and let

Figure 6.19 Dielectric plate partially inserted into a parallel-plate capacitor. In the interior region, there is a large field but negligible field gradient. In the exterior region, both field and field gradient are negligible. In the fringe region, both field and field gradient are nonnegligible. By the amber effect, forces act only in the fringe region to attract the dielectric into the capacitor. Nevertheless, an energy calculation that neglects the fringe region permits that attractive force to be calculated.

the dielectric fill an area $A_{diel} = Lx$. Thus an increase in x corresponds to the dielectric moving inward. See Figure 6.19. Assume that, within the capacitor, the field lines go directly from the positive to the negative plate. This is a consistant assumption because then, by Gauss's law there is no charge at the air–dielectric boundary, and therefore no charge that tends to bend the field lines.

Treat this as two capacitors in parallel since the common upper and lower plates cause the voltage difference ΔV_{air} across the air gap to be the same as ΔV_{diel} across the dielectric. Taking $\kappa_{air} \approx 1$, (6.4) and (6.17) yield

$$C_{air} = \frac{A_{air}}{4\pi kd} = \frac{L(l-x)}{4\pi kd}, \qquad C_{diel} = \kappa \frac{A_{diel}}{4\pi kd} = \frac{\kappa Lx}{4\pi kd}. \tag{6.32}$$

Since these capacitors are in parallel, (6.8) yields

$$C = C_{air} + C_{diel} = (l - x + \kappa x)\frac{L}{4\pi kd}. \tag{6.33}$$

Clearly, the capacitance increases when the dielectric is pulled into the capacitor, and thus (at fixed charge) the energy decreases because C appears in the denominator of the expression for the energy at fixed charge.

We now determine the force of attraction. In (6.33), increasing x by δx causes a change in C of

$$\delta C = (\kappa - 1)\frac{L}{4\pi kd}\delta x. \tag{6.34}$$

(This corresponds to a shift in area $dA = Ldx$ from air capacitor to dielectric capacitor.) By (6.27) with $d\vec{s} = \hat{x}dx$, by $U = Q^2/2C$ with the charge Q fixed, and by (6.34),

$$\delta U = -F_x\delta x = -\frac{Q^2}{2C^2}\delta C = -\frac{Q^2}{2C^2}(\kappa - 1)\frac{L}{4\pi kd}\delta x. \tag{6.35}$$

Thus

$$F_x = \frac{Q^2}{2C^2}(\kappa - 1)\frac{L}{4\pi kd}. \tag{6.36}$$

Since $\kappa > 1$, the force indeed pulls the slab inward ($F_x > 0$). By (6.36) and (6.33), at fixed Q the force decreases as the slab gets farther into the capacitor.

Example 6.18 **Force pulling slab into capacitor**

Consider a parallel plate capacitor $Q/C = 100$ V, $L = 10$ cm, and $d = 1$ mm. A dielectric slab with $\kappa = 5$ and the same L and d is at the edge. Find the force pulling the slab between the plates. Compare with the force of gravity on a penny, of mass $m = 3.11$ g.

Solution: Equation (6.36) gives $F_x = 1.77 \times 10^{-5}$ N. A penny, of weight $mg = 3.05 \times 10^{-2}$ N cannot be lifted by such a small force.

Example 6.19 **Comparing surface charge densities**

Consider two capacitors with the same geometry, one filled with air and the other filled with dielectric of dielectric constant κ. They are given the same voltage difference. Compare the charge densities in the two cases.

Solution: Because the voltage differences are the same across both air and dielectric, the electric fields are the same across both: $E_{air} = E_{diel}$. Because the charge density induced in the dielectric weakens the field of the charge on the plates, for the dielectric the charge density σ_{diel} on the plates must be larger than the charge density σ_{air} without the dielectric. Specifically, since $E_{air} = 4\pi k\sigma_{air}$ and, by (6.13), $E_{diel} = 4\pi k\sigma_{diel}/\kappa$, we have

$$\sigma_{air} = \frac{\sigma_{diel}}{\kappa}. \tag{6.37}$$

Example 6.20 **Charge densities and net charge as slab goes in**

Let $\Delta V = 100\,\text{V}, \kappa = 5, L = 10\,\text{cm}, l = 10\,\text{cm}$, and $d = 1\,\text{mm}$ in the situation described by Figure 6.19. (a) Find the charge densities for the air region and for the dielectric region. (b) Find the net charge on the plates for the set of distances $x = \{0, 5, 10\}$ cm.

Solution: (a) By $\Delta V = Ed$, $E = \Delta V/d = 1.0 \times 10^5$ V/m, both in the air and in the dielectric. Then $\sigma_{air} = E_{air}/4\pi k = 8.84 \times 10^{-7}$ C/m^2 and by (6.37), $\sigma_{diel} = \kappa\sigma_{air} = 4.42 \times 10^{-6}$ C/m^2. (b) For the set of distances $x = \{0, 5, 10\}$ cm, (6.33) yields the set of capacitances $C = \{8.84 \times 10^{-11}, 2.65 \times 10^{-11}, 4.42 \times 10^{-10}\}$ F. Then $Q = C\Delta V$ gives the set of net charges $Q = \{8.84 \times 10^{-9}, 2.65 \times 10^{-9}, 4.42 \times 10^{-8}\}$ C.

6.8 Coefficients of Potential

So far we have only considered situations where two conductors have equal and opposite charges. We now turn to the more general case.

Consider two conductors of arbitrary shape, each on an insulating stand, with respective net charges Q_1 and Q_2. See Figure 6.20. By the action-at-a-distance approach, there is a contribution $dV = k\,dq/r$ for each bit of charge dq (although the q's on the conductors are unknown). Hence the voltage on any conductor, due to charge on itself or any other conductor, is proportional to the amount of charge on itself and that other conductor. Following a notation introduced by Maxwell, we write

Figure 6.20 Two conductors with arbitrary charges (e.g., ducks) for study of the coefficients of potential.

$$V_1 = p_{11}Q_1 + p_{12}Q_2, \qquad V_2 = p_{21}Q_1 + p_{22}Q_2, \tag{6.38}$$

where the (unknown) coefficients p_{ij} are called *coefficients of potential* and may be determined experimentally. The p_{11} and p_{22} are the coefficients of *self-potential*, and the p_{12} and p_{21} are the coefficients of *mutual potential*. They usually are difficult to calculate; typically, it is more practical to measure them.

Example 6.21 Measuring the coefficients of potential

With one aluminized rubber duck given a charge of $Q_1 = 10^{-9}$ C, and a second, smaller, aluminized rubber duck uncharged ($Q_2 = 0$), the measured voltages are $V_1 = 8$ V, $V_2 = 4$ V. (The second duck's potential has "floated up" toward that of the first, due to the coefficient of mutual potential.) With the second duck given a charge of $Q_2 = 10^{-9}$ C, and the first duck uncharged ($Q_1 = 0$), the measured voltages are $V_1 = 4$ V, $V_2 = 20$ V. Find the coefficients of potential.

Solution: From the first set of information, (6.38) yields $p_{11} = V_1/Q_1 = 8 \times 10^9$ F^{-1} and $p_{21} = V_2/Q_1 = 4 \times 10^9$ F^{-1}. From the second set of information, (6.38) yields $p_{22} = V_2/Q_2 = 20 \times 10^9$ F^{-1} and $p_{12} = V_1/Q_2 = 4 \times 10^9$ F^{-1}. Note that $p_{12} = p_{21}$, a property that holds in general—our example was chosen to satisfy this condition. Here, that means a charge of 10^{-9} C on either duck produces 4 V on the other one. Note that, if both ducks are given the same charge, the smaller duck is at a higher potential because it is closer to that charge.

Example 6.22 Using the coefficients of potential: effect of grounding

Employ the coefficients in the previous example. Let $Q_2 = 10^{-9}$ C, $Q_1 = 0$, and $V_1 = 4$ V, $V_2 = 20$ V, as in the second situation of the previous example. Find (a) the Q_1 that will ground 1, and (b) the new voltage of 2.

Solution: (a) On connecting 1 to ground, V_1 must decrease by 4 V to $V_1 = 0$. For such a decrease, a negative charge Q_1 must flow to conductor 1, obtained from (6.38) by $V_1 = 0 = p_{11}Q_1 + p_{12}Q_2$ with Q_2 unchanged. Solving for Q_1 yields $Q_1 = -Q_2(p_{12}/p_{11}) = -0.5 \times 10^{-9}$ C. (b) Note that the initial voltage difference is $V_2 - V_1 = 16$ V. Because Q_1 is negative, both V_1 and V_2 decrease, but because Q_1 resides on conductor 1, V_1 decreases more than does V_2. Specifically, we have $p_{21}Q_1 = -2$ V, so V_2 is lowered by only 2 V, from 20 V to 18 V. Thus the voltage difference increases to $V_2 - V_1 = 18 - 0 = 18$ V.

Example 6.23 Coefficients of potential for two distant spheres

Consider two spheres separated by a distance R, of radii a and b satisfying $a, b \ll R$. Find the coefficients of potential.

Solution: Even when the spheres are far apart, one can affect the other because it will produce a potential as if it were a point charge. A lesser effect is that, by electrostatic induction, the electric field produced by one will cause the charge on the other to redistribute. Neglecting such charge redistribution,

$$V_a = \frac{kQ_a}{a} + \frac{kQ_b}{R}, \qquad V_b = \frac{kQ_a}{R} + \frac{kQ_b}{b}, \qquad a, b \ll R. \qquad (6.39)$$

Thus

$$p_{11} = \frac{k}{a}, \qquad p_{22} = \frac{k}{b}, \qquad p_{21} = p_{12} = \frac{k}{R}. \qquad \text{(distant spheres)} \quad (6.40)$$

For $a = 0.5$ m, $b = 1$ m, $R = 10$ m, (6.40) yields $p_{11} = 1.8 \times 10^{10}$ F^{-1}, $p_{22} = 0.9 \times 10^{10}$ F^{-1}, $p_{12} = 0.9 \times 10^9$ F^{-1}. If $Q_a = 0$ and $Q_b = 10^{-9}$ C, then (6.38) gives $V_a = 0.9$ V, $V_b = 9$ V. We can also use (6.39) directly to obtain V_a and V_b.

6.8.1 *Properties of the Coefficients of Potential*

The preceding example of two conducting spheres at a distance illustrates a number of general properties of the coefficents of potential p_{ij} for two arbitrary conductors. Many of them are simply common sense restatements of results from Chapter 5.

(a) The p_{ij} are positive because positive charge produces positive potential. Moreover, the coefficients of mutual potential p_{12} and p_{21} approach zero when the conductors get infinitely far apart.

(b) The p_{ij} satisfy $p_{12} = p_{21}$, a result to be established later in this section.

(c) No matter how close 2 comes to 1 (unless 2 is inside 1), $p_{21} < p_{11}$. This is because the charge on 1 is closer to 1 than 2, and thus has a larger effect on 1 than on 2. Similarly, $p_{12} < p_{22}$.

The coefficients of potential have some other properties:

(d) If 2 is inside 1, and it is uncharged, then it is inside the equipotential defined by 1. Having no charge of its own, it does not disturb that equipotential. Hence $V_2 = p_{21}Q_1 = V_1 = p_{11}Q_1$, so $p_{21} = p_{11}$. However, $p_{22} > p_{21}$ continues to hold.

(e) We might think that p_{11} is unaffected by the presence of 2. However, consider the case where 1 is charged positively and 2 is uncharged. By electrostatic induction the induced negative charge on 2 is closer to 1 than the induced positive charge on 2. This decreases the potential of 1. Hence the coefficients of self-potential depend upon the positions (and even orientations) of other conductors. The closer 2 is to 1, the more the decrease. Such effects are neglected in (6.39). They are measureable and are used in devices that detect the approach of a hand to a doorknob, or the movement of a key on a computer keyboard.

6.8.2 *Capacitance in Terms of the Coefficients of Potential*

If the p_{ij}'s have been measured, or can be calculated, then (6.3) can be evaluated using them. With $Q = Q_1 = -Q_2$, and $\Delta V = V_1 - V_2$, (6.38) gives

$$C = \frac{Q_1}{V_1 - V_2} = \frac{Q}{(p_{11}Q - p_{12}Q) - (p_{21}Q - p_{22}Q)} = \frac{1}{p_{11} + p_{22} - p_{12} - p_{21}}.$$
$$(6.41)$$

Hence the capacitance of a two-plate capacitor is a combination of both mutual- and self-potential coefficients. (Don't even *think* about remembering this equation.)

Example 6.24 **Capacitance of a two-duck capacitor**

Find the capacitance of a capacitor where the two plates are the two ducks in our previous two-duck example, where $p_{11} = 8 \times 10^9$ F^{-1}, $p_{22} = 20 \times 10^9$ F^{-1}, and $p_{21} = p_{21} = 4 \times 10^9$ F^{-1}.

Solution: By (6.41), the two conductors, if employed as a two-plate capacitor, have capacitance $C = 5.0 \times 10^{-11}$ F.

For an infinitely wide parallel-plate capacitor, the p_{ij}'s cannot be computed easily because of the infinite amount of charge on each plate. Thus (6.41) cannot be used. Imagine putting charge on one of the plates in order to determine p_{11}. The resulting electric field would fill all space, and the field energy would be proportional to the volume of all space. This makes it difficult to define p_{11}. A similar argument holds for truly infinite cylinders, except that the field energy would vary logarithmically with the distance from the cylinder to infinity.

Example 6.25 **Capacitance of two concentric spheres**

Find the capacitance of two concentric spheres of radii $b > a$. See Figure 6.6(a).

Solution: Let the inner (outer) sphere be 2 (1). If $Q_2 = 0$, then the voltage at b is that of a point charge Q_1 at the origin, or $V_1 = (kQ_1/b)$. By (6.38), $p_{11} = k/b$. Similarly, if $Q_1 = 0$, then $V_2 = (kQ_2/a)$, so $p_{22} = k/a$. Since 2 is within 1, by property (d) of the coefficients of potential, we have $p_{21} = p_{11} = k/b$. Also, by property (b), $p_{12} = p_{21}$. Hence (6.41) leads to

$$C = \frac{1}{(k/b) + (k/a) - (2k/b)} = \frac{ab}{k(b-a)}. \qquad \text{(spherical two-plate capacitor)}$$

This is in agreement with (6.5).

6.8.3 *Energy for Two Conductors: Establishing That* $p_{12} = p_{21}$

Let us consider the energy to charge up two initially uncharged conductors. In the first stage, charge up 1 by bringing in charge dQ_1' from infinity, while holding $Q_2 = 0$. By $dU = dQ(\Delta V)$ and (6.38),

$$dU_1 = dQ_1' \, V_1'|_{Q_2'=0} = dQ_1'(p_{11}Q_1' + p_{12}Q_2')|_{Q_2=0} = p_{11}Q_1'dQ_1'.$$

Hence, integration leads to

$$U_1 = \frac{1}{2}p_{11}(Q_1')^2 \Big|_0^{Q_1} = \frac{1}{2}p_{11}(Q_1)^2.$$

In the second stage, charge up 2 with Q_1 fixed. Then

$$dU_2 = dQ_2' \, V_2'|_{Q_1'=Q_1} = dQ_2'(p_{21}Q_1' + p_{22}Q_2')|_{Q_1'=Q_1} = (p_{21}Q_1 + p_{22}Q_2')dQ_2'.$$

Hence, integration on Q_2' at fixed Q_1 leads to

$$U_2 = p_{21}Q_1Q_2' \Big|_0^{Q_2} + \frac{1}{2}p_{22}(Q_2')^2 \Big|_0^{Q_2} = p_{21}Q_1Q_2 + \frac{1}{2}p_{22}(Q_2)^2.$$

The total energy of charging is thus

$$U = U_1 + U_2 = \frac{1}{2}p_{11}(Q_1)^2 + p_{21}Q_1Q_2 + \frac{1}{2}p_{22}(Q_2)^2. \tag{6.42}$$

Now reverse the order of charging. Mathematically, this corresponds to interchanging 1 and 2 in (6.42), leading to

$$U' = U_1' + U_2' = \frac{1}{2}p_{22}(Q_2)^2 + p_{12}Q_2Q_1 + \frac{1}{2}p_{11}(Q_1)^2. \tag{6.42'}$$

Since the physical states in both cases are the same, the energies U and U' must be the same. Comparing (6.42) and (6.42') yields

$$p_{12} = p_{21}, \tag{6.43}$$

a result stated earlier without proof.

Example 6.26   **Grounding a conductor decreases its electrical energy**

In Example 6.2, we grounded a system of two conductors with $p_{11} = V_1/Q_1 = 8 \times 10^9$ F^{-1}, $p_{22} = V_2/Q_2 = 20 \times 10^9$ F^{-1}, and $p_{21} = V_2/Q_1 = 4 \times 10^9$ F^{-1}, starting with $Q_1 = 10^{-9}$ C, $Q_2 = 0$. Find the initial and final energies.

Solution: The initial energy was, by (6.42) with $Q_2 = 0$, $U_{initial} = p_{11}Q_1^2/2 = 10^{-8}$ J. The final energy, with Q_1 unchanged, but now $Q_2 = -0.5 \times 10^{-9}$ C, is given by the full expression in (6.42). This is $U_{final} = 0.9 \times 10^{-8}$ J. Thus, the system has lost energy. This goes into heating up the system, due to friction on the charge as it passes through the wire connecting C_2 to ground.

6.9 **Material Properties of Dielectrics**

Optional

This section is different from other sections we have studied because it attempts to relate macroscopic phenomena to microscopic quantities.

6.9.1 *Polarizability and Dielectric Constant Are Related*

Chapter 3—see (3.16)—discussed the atomic polarizability α, defined by

$$\vec{p} = \alpha\vec{E}. \tag{6.44}$$

It gives the relation between the induced dipole moment $\vec{p}$ of an atom and the applied electric field $\vec{E}$. Not surprisingly, there is a relationship between the polarizability and the dielectric constant. Although difficult to obtain in general, this relationship can be found for a dilute single-component gas, where we expect that $\kappa \approx 1$. Even knowing about this case is revealing.

Consider the dielectric in Figure 6.14 to be a dilute gas, such as air. Then the induced charge density σ' should be small, and hence in (6.44) the approximation $E \approx E_0$ is valid, where E_0 is due only to the charge $\pm\sigma$ on the capacitor plates. Thus, by (6.12),

$$E \approx 4\pi k \sigma. \tag{6.45}$$

Let the atomic density of the gas be n, and let the parallel-plate capacitor have plate areas A and plate separation d, so it has volume Ad. Then

$$\sigma' = \frac{Q'}{A} = \frac{Q'd}{Ad}. \tag{6.46}$$

Since $Q'd$ is the total dipole moment of the induced charge, and within the capacitor there are $N_{mol} = nAd$ molecules, each of dipole moment p, we have

$$Q'd = N_{mol}p = (nAd)p = (pn)(Ad). \tag{6.47}$$

Then (6.46) becomes, with (6.47) for $Q'd$, with the scalar form of (6.44) for p, and with (6.45) for E,

$$\sigma' = \frac{Q'd}{Ad} = pn = n\alpha E \approx n\alpha(4\pi k\sigma). \tag{6.48}$$

Rearranging (6.14) and then using (6.48) yields

$$\frac{\sigma'}{\sigma} = 1 - \frac{1}{\kappa} = \frac{\kappa - 1}{\kappa} \approx 4\pi k n \alpha. \tag{6.49}$$

Since we assume a dilute gas, in $(\kappa - 1)/\kappa$ we can take $\kappa \approx 1$ in the denominator (but not in the numerator). Then (6.49) yields

$$\kappa \approx 1 + 4\pi k n \alpha. \tag{6.50}$$

Here the relatively small second term of (6.50) contains the polarization effects.

The relationship between κ and α is more complicated for denser systems, like liquids and solids. However, (6.50) captures the intuitively evident physics that *the higher the density and the higher the polarizability, the larger the dielectric constant κ*.

Example 6.27 **From dielectric constant to density**

For a sample of He gas at room temperature, where $\alpha = 0.23 \times 10^{-40}$ C-m^2/V, κ is measured to be 1.000045. Find the density n and the gas pressure P.

Solution: From (6.50), we have $n = (\kappa - 1)/4\pi k\alpha = 1.73 \times 10^{25}/$ m^3. Atmospheric pressure $(P_0 = 1.013 \times 10^5$ N/m^2) and room temperature $T = 293$ K corresponds to $n_0 = 2.47 \times 10^{25}/$m^3. Since, at fixed temperature, by the ideal gas law the pressure P is proportional to the density n, this density corresponds to $P = 0.7$ atm.

We close this section by noting that these ideas work, qualitatively, even for denser materials, as long as the molecules are *nonpolar*—that is, without a permanent electric dipole moment. For *polar* molecules, which have a permanent electric dipole moment, a small electric field can completely align them with the electric field, giving a very large dielectric constant. The extent of their alignment is limited by the thermal energy, which tends to orient them randomly.

6.9.2 *Electrical Breakdown Occurs More Easily at Low Gas Density*

Sparking occurs when a critical value of the electric field, called the *dielectric strength* E_d, is reached. In air, that value is determined by how much energy an electron needs to ionize a molecule of air (either N_2 or O_2). At atmospheric pressure, $E_d = 3 \times 10^6$ V/m.

Consider the motion of an electron through a gas of neutral molecules. (Electrons are always present in gases, due to ionization by cosmic rays.) In a field $\vec{E}$, the characteristic voltage gain ΔV of the electron, on traveling a characteristic mean-free path l from one molecular collision to the next, will be $\Delta V = El$, where $E = |\vec{E}|$. If E is large enough that ΔV is on the order of the ionization voltage V_I of the molecule the electron hits, then another electron is kicked off, so there are now two electrons in the gas. This starts an electron avalanche, leading to a spark. This phenomenon is known as *dielectric breakdown*. At the threshold field E_d for dielectric breakdown

$$V_I \sim E_d l. \tag{6.51}$$

This is not an equality because dielectric breakdown is a statistical process, for which the exact coefficient is difficult to obtain. Note that O_2 has $V_I \approx 12.0$ eV, and N_2 has $V_I \approx 15.5$ eV.

The mean-free path l has the following properties: (1) it increases with decreasing density n of the air molecules (there are fewer to collide with, so it travels farther before it collides); (2) it increases with decreasing scattering cross-section (which is on the order of the square of an atomic dimension a). Hence, l satisfies the proportionality

$$l \sim \frac{1}{na^2}, \tag{6.52}$$

which is also dimensionally correct. Combination of (6.52) with (6.51) leads to

$$E_d \sim \frac{V_I}{l} \sim V_I na^2. \tag{6.53}$$

Thus, the lower the gas density, the lower the dielectric strength, and the easier it is to ionize the gas. In solids and liquids, the dielectric strength is determined by the field needed to ionize the atoms of the material and is much higher than in rarefied gases because the density n is much higher.

Example 6.28   Electrical breakdown in the atmosphere

Consider the atmosphere at $T = 300$ K (near standard temperature of 293 K) and at standard pressure (10^5 Pa, where the pascal is Pa $= $ N/m^2). Assume the ideal gas law in the form $PV = Nk_BT$, where N is the number of molecules and $k_B = 1.38 \times 10^{-23}$ J/K. Taking the collision cross-section to be a^2 with $a = 1.0 \times 10^{-10}$ m, and $V_I = 15$ V, estimate the mean-free path and E_d.

Solution: By the ideal gas law, the gas density $n = N/V = P/k_BT$ is about 2.4×10^{25}/m^3. This corresponds to an average separation $s = n^{-1/3} \approx 3.5 \times 10^{-9}$ m. Equation (6.52) then gives a mean-free path of $l \approx 4.2 \times 10^{-6}$ m. This is much greater than a or s. For $V_I = 15$ V, use of (6.53) as an equality gives $E_d \approx 3.6 \times 10^6$ V/m, very close to the measured value of 3×10^6 V/m.

Example 6.29   Electrical breakdown in plasma globes

What are called *plasma globes* are available at commercial electronics stores. Within the globe, constantly changing filaments of light go from a central electrode to the transparent outer casing. The detailed colors depend upon the gases within the partially evacuated globe. To see the principle involved, consider a partially evacuated chamber of air, with only 10^{-2} of atmospheric pressure. The idea is that, at low gas density, there are few but energetic collisions, and the associated recombination is responsible for the colors of the plasma globe. Estimate the dielectric strength of air at 10^{-2} of atmospheric pressure.

Solution: Since the density of air molecules relative to atmospheric pressure is smaller by a factor of 10^{-2}, the mean-free path l will be 100 times longer (about 4.2×10^{-4} m), and E_d will decrease by a factor of 100, from 3×10^6 V/m to 3×10^4 V/m. Hence, dielectric breakdown is achieved more easily at the lower pressure. Plasma globes have an electrode that produces a high voltage ($\sim$5 kV) at a high frequency ($\sim$30 kHz), and this causes dielectric breakdown in the low-pressure gas within the globe. The light emitted on recombination of electrons with positive ions leads to the observed visual display. If the mean-free path were so long that an electron would cross the globe without hitting any atoms, sparking within the globe would be suppressed. Nevertheless, significant amounts of energy would be deposited at the walls, causing the emission of light or even x-rays if the electrons are energetic enough.

Application 6.3   Electrical breakdown in fluorescent tubes

Fluorescent tubes contain a mixture of argon and nitrogen at atmospheric pressure, including a small drop of vaporized mercury (Hg). Electrons accelerate between the two electrodes at opposite ends of the tube. While crossing the tube, the electrons collide with Hg atoms, losing energy while exciting the Hg atoms. When the Hg atoms deexcite, they emit ultraviolet radiation, which gets absorbed at the walls and is rapidly reemitted as visible light. This rapid reemission at a lower frequency is called *fluorescence*. Absorption and reemission of radiation occurs in all materials. When the reemission process is too fast to be noticed on the human scale of fractions of a second, it is called *phosphorescence*. Recall the quickly disappearing phosphorescent afterglow of a TV or computer screen, and the slowly disappearing fluorescent afterglow of some light switches.

6.9.3 *Electrical Breakdown to the Max: Lightning*

Lightning is not merely an interesting visual display. In the United States alone, each year lightning hits and kills between 75 and 100 people, and starts about 10,000 forest fires, with costs from forest and building fires of about $100 million. There also is economic loss due to currents induced on electrical wiring (e.g., phone lines) connected to sensitive and unprotected electronic equipment (e.g., computer modems). Such currents (in wiring and in people) need not come from a direct hit, but can be caused by sudden depolarization when a cloud overhead suddenly discharges to another cloud or to ground a distance away.

Lightning can be cloud to cloud or cloud to ground, cloud to cloud being about three times more frequent. The most frequent type of cloud-to-ground lightning brings negative charge to the earth. However, about one-tenth of lightning bolts bring positive charge to the earth; such lightning can be particularly destructive.

Because lightning bolts occur quickly, and because electrons have much less inertial mass than ions, the charge transfer is due to electrons. Even when it appears that positive charge is being transferred one way, it is mostly the movement of electrons in the opposite direction.

A typical thundercloud has three regions where electric charge accumulates: (1) closest to ground (about 2 km from ground) might be about $+10$ C, called the p charge; (2) distributed over the middle of the cloud (about 5 km from ground) might be about -40 C, called the N charge; (3) toward the top (about 10 km from ground) might be about $+40$ C, called the P charge. A thundercloud has an electric dipole moment, which can be measured from the ground.

For electrical discharge to occur, an electric field in excess of the breakdown field must develop. In moist air, this is thought to be about 10^5 V/m. Since voltage differences between ground and cloud can be as high as 10^8 V, it is not difficult to see that such fields (i.e., voltage gradients) can develop. Lightning originates in a discharge from the middle of the cloud (the region with negative charge N) to the bottom of the cloud (the region with positive charge p). This negative discharge overshoots, going all the way to the ground in a series of steps, via the *stepped leader*. Once the stepped leader produces an ionized channel, a much larger *return stroke* carries positive charge from ground to cloud; this return stroke is seen as lightning.

(a) The stepped leader proceeds by hundreds of steps, each lasting perhaps a μs (with a rise time of about 0.1 μs) and extending perhaps 50 m, with pause times of perhaps 50 μs. See Figure 6.21(a) and (b), which include a house and a tree. By the time the stepped leader approaches the ground, its channel contains about -5C. (Note: Stepped leaders are quite variable. Steps of 3–200 m, pause times of 30–125 μs, and channel charges of 3–20 C are measured regularly.) During the step, but not during the pause,

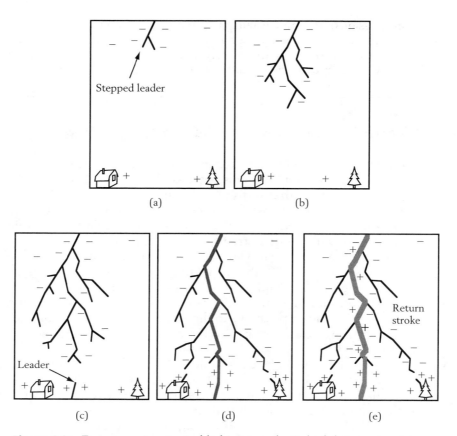

Figure 6.21 Five stages in a typical lightning strike (which brings negative charge to ground). (a) Initiation of stepped leader from cloud. (b) Growth of stepped leader. (c) Return stroke leader from ground rising toward stepped leader. (d) Connection of leader from ground and stepped leader from cloud. (e) Fully developed return stroke from ground constitutes the lightning bolt.

the leader is luminous enough to be visible by special photographic techniques. Luminous stepped leader channels have diameters of 1–10 m.

Starting from a height of 3 km, the total time for lightning to reach ground might be 20 ms. This time gives much information. First, it gives an average downward velocity of 150,000 m/s ($=3$ km/20×10^{-3} s). (Note: average velocities between 1.0×10^5 m/s and 25.0×10^5 m/s are measured regularly.) Second, it indicates that from cloud to ground there might be 400 steps (20 ms total time/50 μs pause time). (Note: This gives a path length of about 400×50 m $= 2$ km, which is less than 3 km; clearly, the *average* pause time must be larger than 20 ms.) Third, it gives an average current (including pauses) of 250 A ($=5$C/20 ms); the peak current during a single step might be 10,000 A.

The leader channel heats to about 30,000 K. Associated with the heating of the channel is a shock wave (high-intensity sound), which propagates outward as thunder.

(b) The return stroke is initiated by its own leader rising up from the ground toward the stepped leader (Figure 6.21c). The two meet some hundreds of feet from the ground (Figure 6.21d), providing a continuous channel of conducting gas from cloud to ground. The return stroke then travels upward (Figure 6.21e) at from one-tenth to one-third of the speed of light, with a total transit time of about 100 μsec, a peak current of about 30 kA, and a rise time of a few ms. The fall time to half the peak current is about 50 μsec, and current on the order of hundreds of amperes can last for a ms or more. It transfers a charge of perhaps -25 C downward to ground, with charges of -3 C to -90 C not uncommon. (For a transfer of 5 C across 2 km, this corresponds to a change in dipole moment of 10^4 C-m.)

Following the return stroke, usually a *dart leader* will pass from cloud to ground, following the path of the stepped leader. This will be followed by *another* return stroke from ground to cloud. There are typically 2 to 3 dart-leader-return-stroke occurrences, but sometimes there are none, and sometimes there are as many as 25. A typical lightning bolt might flash for 0.2 s, but lightning bolts can be as short as 0.01 s or as long as 2 s (there is even a 19th-century report of a 15–20 s flash!).

It is believed that the leader channel contains a highly ionized core of radius 1 mm or so. Assume that all the molecules in a gas column 5 km long and radius 1 mm, at standard temperature and pressure (STP), are singly ionized. This corresponds to a volume of about 16×10^3 cm^3. Since one mole of gas at STP has a volume of 22.4/(1000 cm^3), this corresponds to about 0.7 mole. One mole of ions with charge e contains $6 \times 10^{23} \times 1.6 \times 10^{-19}$ C, or about 96,500 C, known as a *faraday* in the chemical literature. Hence 0.7 mole contains nearly 70,000 C of both positive and negative charge. If only 0.1% of the negative charge is in the form of electrons (rather than negative ions), then there is plenty of charge available to explain how, even though the leader channel has only a net charge of about -5 C, there is a much greater charge transfer by the return stroke.

The area of lightning research is filled with unsolved problems. What atmospheric processes cause the charge separation responsible for lightning? What happens during the pauses of the stepped leader? Exactly where does the charge come from in the return stroke? How does the atmosphere recover after a lightning bolt? Can we predict the brightness of a lightning bolt? Can we predict the intensity of thunder? What role does charge on the earth (a conductor) play? Generations of atmospheric scientists have been at work, trying to answer these questions, and it likely will be many generations more before all the answers have been found.

6.9.4 *More on Electrical Screening*
Optional

1. Ideal conductors: Chapter 4 discussed electrical screening by ideal conductors, which have, implicitly, an infinite density of free charge carriers. In the presence of an applied electric field $\vec{E}_0$, free charge goes to the surface and produces an electric field that, within the conductor, cancels $\vec{E}_0$. For the geometry of Figure 6.14 (but with the dielectric replaced by an ideal conductor), in response to the charge density σ on the capacitor plate free charge from the conductor produces a surface charge density $-\sigma$.

2. **Real conductors:** Section 4.11.3 pointed out that, for real conducting materials, when the finite density of free charge is accounted for there is a *screening length l* over which the total electric field is nonzero within the conductor. This is because the *screening charge* cannot go literally to the surface of the conductor. The higher the density of the free charge carriers, the shorter the l. For the geometry of Figure 6.14 (but with the dielectric replaced by a real conductor), in response to the charge density σ on the capacitor plate free charge from the conductor produces a bulk charge density ρ near the surface. Integrating $\int \rho\, dx$ over a distance a few times l into the conductor gives an effective surface charge density $-\sigma$.

3. **Dielectrics:** The present chapter has discussed the electrical response of dielectrics. In the presence of an applied electric field $\vec{E}_0$ localized charges cause polarization that, over a distance scale on the order of an atomic dimension, puts charge on the surface. This produces an electric field that within the conductor partially cancels $\vec{E}_0$. For the geometry of Figure 6.14, in response to the charge density σ on the capacitor plate, the polarized tips of the atoms and molecules of the material produce a polarization charge density on the surface of $-\sigma(1 - \kappa^{-1})$. [For gases, $\kappa = 1 + 4\pi n\alpha$ of (6.50) shows that the higher the density n of the atoms or molecules of polarizability α, the larger the κ, and thus the more effective the screening.] Hence the total field within the dielectric decreases to $\vec{E}_0/\kappa$. The field within the dielectric does not decrease any further on going deeper into the dielectric because there is polarization charge only on the surface, not in the bulk. The total polarization charge on an object must be zero.

4. **Materials with both free charge and polarization charge:** We can now discuss what happens when there are finite densities of both free charge and of polarization charge. An example is a semiconductor like Si (the basis of most semiconductor electronics) when it has been "doped" with impurities to give it free charges. Here, if the screening length is l due to free charge alone (for which $\kappa = 1$), then it becomes $l\sqrt{\kappa}$ when there is also polarization charge. At the surface the response is the same as for a dielectric: for the geometry of Figure 6.14, in response to the charge density σ on the capacitor plate, the polarized tips of the atoms and molecules of the material produce a polarization charge density on the surface of $-\sigma(1 - \kappa^{-1})$. In addition, there is a bulk charge density ρ near the surface that gives an effective surface charge density $\sigma\kappa^{-1}$ on integrating $\int \rho\, dx$ over a distance a few times the screening length $l\sqrt{\kappa}$ into the conductor. Hence the total charge associated with the material at and near the surface is $-\sigma$, as for the ideal conductor and the real conductor without dielectric. The bulk electrical response of the material is complex. It consists both of free charge and polarization charge, the polarization charge tending to cancel the free charge: (1) the free charge sums to $-\sigma$ per unit area, just what it would be for a conductor; (2) the polarization charge sums to $\sigma(1 - \kappa^{-1})$ per unit area, which completely cancels the polarization charge on the surface. Moreover, the length over which the free charge near the surface extends is larger when there is dielectric, by a factor of

$\sqrt{\kappa}$, because of the partial cancellation of the free charge by the polarization charge.

6.10 Flux Tubes

As he performed more experiments, Faraday changed his views on how to think about electricity. The form that is passed on to us was developed by Maxwell, who made Faraday's field-line ideas quantitative and rigorous by reformulating them in terms of flux tubes. We will develop some of these ideas, following a treatment in J. J. Thomson's *Elements of the Mathematical Theory of Electricity and Magnetism*. Thomson himself edited the third and final edition of Maxwell's classic *Treatise on Electricity and Magnetism*, published in 1891, after Maxwell's untimely death in 1879, at the age of 48.

6.10.1 *Electrical Energy Density $u_E = \frac{E^2}{8\pi k}$*

Consider (5.28) for the electrical energy $U = \frac{1}{2}k\sum'_{ij}q_iq_j/r_{ij}$ of a set of charges q_i and q_j seperated by distances r_{ij}, where the prime means do not include the terms where $i = j$. This can be rewritten, with $V_i = k\sum'_j q_j/r_{ij}$, as $U = \frac{1}{2}\sum_i V_iq_i$. Letting the charges become differentials, both V and U remain finite, with

$$U = \frac{1}{2}\int Vdq, \tag{6.54}$$

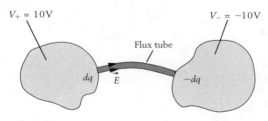

Figure 6.22 Flux tube connecting two conductors in equilibrium. The sides of the flux tube are parallel to the local electric field.

where the factor of $\frac{1}{2}$ is to avoid double counting. We consider both point charges and conductors, in vacuum, treating the point charges as small spherical conductors. The field lines originate on positive charge dq and terminate on negative charge $-dq$ (see Figure 6.22). The integral may be done by considering equal and opposite bits of charge $\pm dq$ via

$$(V_+)(dq) + (V_-)(-dq) = (V_+ - V_-)dq, \tag{6.55}$$

and then integrating only over dq. Because $dq = \sigma_S dA_S$ is at the surface of a conductor, by (4.22) we have

$$dq = \sigma_S dA_S = \frac{E_S dA_S}{4\pi k} = \frac{d\Phi_E}{4\pi k}, \tag{6.56}$$

where $E_S = 4\pi k\sigma_S$ is the magnitude of the field at the surface.

Since the field lines go from higher to lower potential, and $d\vec{s}$ points along $\vec{E}$, we have $\vec{E} \cdot d\vec{s} = Eds$, so

$$V_+ - V_- = \int_+^- Eds, \tag{6.57}$$

where the path of the integral is from dq to $-dq$. Combining (6.56) and (6.57), and changing the order of integration, yields

$$U = \frac{1}{2} \int_{+} (V_+ - V_-) dq = \frac{1}{8\pi k} \int_{+}^{-} E ds \int_{+} E_S dA_S, \qquad (6.58)$$

where $\int_+$ restricts the integration to the surface containing the positive charge.

Now convert from an integral over the surface elements dA_S and the connecting elements ds to an integral over the volume dV of the Faraday flux tube associated with dq. Since, along a flux tube, the flux is constant, we have $d\Phi_E = E_S dA_S = E dA$. Since the flux tubes fill all space, with $dV = dA ds$ (6.58) may be rewritten as

$$U = \frac{1}{8\pi k} \int E ds \int E dA = \frac{1}{8\pi k} \int E^2 dV. \qquad (6.59)$$

From this we deduce that the electrical energy per unit volume u_E is

$$u_E = \frac{dU}{dV} = \frac{E^2}{8\pi k}. \qquad (6.60)$$

This is consistent with (6.25), obtained previously for a strictly uniform field, on setting $\kappa = 1$.

Now consider partial flux tubes all with the same flux $d\Phi_E = E dA$ and the same potential difference $dV = E ds$. Each part contains the energy

$$dU = \frac{E^2}{8\pi k} dA ds = \frac{d\Phi_E dV}{8\pi k}. \qquad (6.61)$$

This is the same for each of the partial flux tubes no matter their volume element $dV = dA ds$.

6.10.2   *Field Line Tension per Unit Area* $T = \frac{E^2}{8\pi k}$

Consider a flux tube connecting one conductor to another. If one of these conductors is compressed so that its surface corresponds to a smaller equipotential surface, then the field lines and flux tubes are not disturbed, except that they lengthen. By (6.61), this costs an extra energy $(E^2/8\pi k) dA ds$. See Figure 6.23. We may think of this as stretching a string by ds against a tension $(E^2/8\pi k) dA$. Hence the tension per unit area dA of the flux tube is

$$T = \frac{E^2}{8\pi k}. \qquad (6.62)$$

The total force exerted by the tension of flux tubes connecting charges $+q$ and $-q$ separated by a distance r has magnitude $|\vec{F}| = kq^2/r^2$. This nontrivial calculation can be done by considering the forces on the midplane separating the charges.

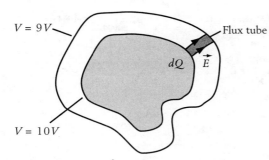

Figure 6.23 Two nearby flux tubes, for calculation of field energy and field-line tension.

6.10.3 *Pressure of Field Lines on One Another is* $P = \frac{E^2}{8\pi k}$: *Radial Field Lines*

The field lines also exert a pressure on one another. To find that pressure, we must examine two cases. First, consider the case where the field varies in strength *along* the flux tube. See Figure 6.24.

Let a tiny length ds of the flux tube have cross-section dA at the top and dA' at the bottom. By construction of the flux tube, the field magnitudes E and E' on the top and bottom are related by

$$d\Phi_E = E\,dA = E'\,dA'. \tag{6.63}$$

From Figure 6.24, there is a vertical force upward, due to the line tension, of

$$F_T = T\,dA - T'\,dA' = \frac{E^2\,dA - E'^2\,dA'}{8\pi k} = \frac{(E - E')E\,dA}{8\pi k}, \tag{6.64}$$

where we have used (6.62) and (6.63).

There is also a downward force F_P from the pressure of lines against one another. See Figure 6.24. We can obtain F_P from considerations of hydrostatics, with a uniform pressure P. An object subject to uniform pressure is in equilibrium, so the force on it from the sides must be equal and opposite to the force on it from the top and bottom. Thus, from Figure 6.24 and some algebraic manipulation,

$$F_P = -P(dA' - dA) = -\frac{P}{E'}(E'\,dA' - E'\,dA)$$

$$= -\frac{P}{E'}(E - E')\,dA \approx -\frac{P}{E}(E - E')\,dA; \tag{6.65}$$

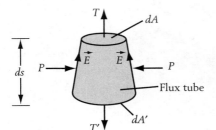

Figure 6.24 Part of a flux tube, for calculation of pressure between neighboring field lines when field is radial.

here we have again used (6.63), and we have set $E' \approx E$ in the denominator. In equilibrium, (6.64) and (6.65) give

$$0 = F_T + F_P = \left(\frac{E}{8\pi k} - \frac{P}{E} \right) (E - E') dA. \tag{6.66}$$

From this, we deduce that

$$P = \frac{E^2}{8\pi k}. \tag{6.67}$$

This is precisely the same in magnitude as the flux tube tension per unit area T.

6.10.4 *Pressure of Field Lines on One Another Is $P = \frac{E^2}{8\pi k}$: Tangential Field Lines*

We now check the consistency of this approach. Consider the case where the field lines vary in strength *perpendicular* to the flux tube. See Figure 6.25.

Let a flux tube of constant area dA have length ds on the left, and length ds' on the right. Take the tube to have width dl, and depth w into the page, so areas on top and bottom satisfy

$$dA = dA' = wdl. \tag{6.68}$$

By construction of the flux tube, the fields on the right and left are related by

$$dV = Eds = E'ds'. \tag{6.69}$$

From Figure 6.25, the net rightward force from pressure is given, from (6.67), (6.68), and (6.69), by

$$F_P = (Pwds - P'wds') = \frac{w(E^2 ds - E'^2 ds')}{8\pi k} = \frac{w(E - E')Eds}{8\pi k}. \tag{6.70}$$

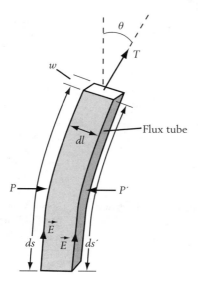

Flux tube

Figure 6.25 Part of a flux tube, for calculation of pressure between neighboring field lines when field is tangential.

Also from Figure 6.25, the net rightward force from tension comes only from the component of the tension force on the top, which is at the small angle θ, where

$$ds = ds' + dl\,\theta. \tag{6.71}$$

Explicitly, the rightward force from tension is, with (6.62) and (6.68),

$$F_T = Td A\theta = \frac{E^2}{8\pi k}wdl\,\theta. \tag{6.72}$$

Use of (6.71) and (6.69) in (6.72) gives

$$F_T = \frac{E^2}{8\pi k}w(ds - ds') = \frac{E^2}{8\pi E'k}w(E'ds - E'ds')$$

$$= \frac{E^2}{8\pi E'k}w(E' - E)\,ds \approx \frac{E}{8\pi k}w(E' - E)\,ds. \tag{6.73}$$

Here we have used $E' \approx E$ to write $E^2/E' \approx E$. Comparing (6.73) with (6.70), we see that

$$F_T = -F_P. \tag{6.74}$$

Hence the force from field-line pressure cancels the force from field-line tension, so the system is in equilibrium, as desired. The concept of pressure of field lines on one another thus is a consistent one, both for field lines that vary in magnitude along the flux tube and for field lines that vary in magnitude perpendicular to the flux tube.

The total force exerted by the pressure of flux tubes connecting charges $+q$ and $+q$ separated by a distance r has magnitude $|\vec{F}| = kq^2/r^2$. This nontrivial calculation can be done by considering the forces on the midplane separating the charges.

Problems

6-2.1 A space cruiser is given a charge 2.5 μC, which raises its voltage by 106 V relative to a distant space station. (a) Determine the capacitance of the space cruiser. (b) Estimate its characteristic size, using a spherical approximation.

6-2.2 A tin can, hung by insulating string from a tall ceiling, has a capacitance of 4.2 nF. A scaled-up version of this can, with a height that is 2.4 times as great, is ejected from a spaceship, having been given a charge of 8.7 nC. Find its voltage relative to a distant point.

6-2.3 The supply of electrons in a material is not limitless; adding or subtracting one electron *per*

atom would surely cause structural changes in the material. Discuss whether excess charge affects the structure of a solid conducting sphere more or less than a conducting shell of the same material and external dimensions.

6-2.4 A grain of Zn (approximated by a sphere of radius 95 nm) is illuminated with ultraviolet radiation, which ejects negatively charged electrons until the Zn voltage is about 1.7 V. Find the charge on the grain, and the number of electrons that have been ejected.

6-2.5 If the capacitance of a spherical aluminum grain is 5 aF (aF $= 10^{-18}$ F), estimate its radius.

6-2.6 Recent advances in technology permit the fabrication of tiny electrical devices where the addition of a single electron to a small piece of metal can produce measureable voltages. (a) Find the capacitance of an aluminum grain about 10 nm in radius. Express your answer in units of 10^{-18} F, or aF, the attofarad. (b) Estimate the voltage change caused by the addition of an electron. (c) Estimate the total number of electrons on the grain, taking there to be three conduction electrons per aluminum atom. Such small grains are called *artificial atoms*, or *quantum dots*, because the electron orbitals on them are unique to very small particles. The density of aluminum is 2.7×10^3 kg/m³, and its atomic weight is 27.0.

6-3.1 Two irregular conducting objects have charges ±5 nC. One is at $V_1 = 20$ V and the other is at $V_2 = -15$ V. They are far from any other conductors. Take $V_\infty = 0$. (a) Determine their capacitance. (b) If these two objects and their separation are now linearly scaled down by a factor of 4, find the charge if the same voltage difference is applied. (If the conductors are spheres, which one is larger?)

6-3.2 A parallel-plate capacitor has circular plates of radius 5.4 cm, and plate separation 2.2 mm. Determine the capacitance, and the voltage difference for charges ±18 nC.

6-3.3 A capacitor consists of two circular plates of radius 5 cm and separation 1.2 mm. (a) Estimate the capacitance. (b) If the field toward the middle of the capacitor is 25 V/m, estimate the charge on the capacitor plates.

6-3.4 A parallel-plate air capacitor C has plate separation d. A piece of metal of thickness $0.8d$ is inserted between the plates. Explain why the new capacitance corresponds to a smaller plate separation, and determine the new capacitance.

6-3.5 The capacitance of a parallel-plate capacitor is proportional to the area A and inversely proportional to the plate separation d. Suggest what might set limits on (a) the minimum d and (b) the maximum A.

6-3.6 Consider two co-axial circular conducting plates of radius a, with equal and opposite charges $\pm q$. They are separated by h. See Figure 6.26. (a) Let $h \gg a$. Describe the surface charge density on each side of the positive plate, and how uniform it is. Is the surface charge density at the edges higher or lower than in the middle? (b) Now let $h \ll a$, so

the plates are near one another. Describe the surface charge density on each side of the positive plate, and how uniform it is. Is the surface charge density at the edges higher or lower than in the middle? (c) Show that the surface charge density at the center of the inner surface of the positive plate is now *four times* (not twice) as large as when the negative plate is far away. *Hint:* See Section 1.9.

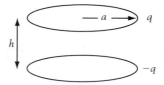

Figure 6.26   Problem 6-3.6.

6-3.7 A parallel-plate air capacitor of capacitance 400 pF has a charge ±500 nC. The plates are 2 mm apart. Find (a) the potential difference, (b) the plate areas, (c) the electric field between the plates, and (d) the surface charge density on the plates.

6-3.8 A spherical capacitor with an air gap has inner and outer radii of 5 cm and 8 cm, at a voltage difference of 250 V. (a) Find its capacitance. (b) Find the charge on the plates. (c) Find the maximum electric field within the capacitor.

6-3.9 A spherical capacitor with a wax-filled gap has an outer radius of 4.8 cm. For a voltage difference of 7600 V, there is breakdown. (a) Find the radius a of the inner plate. (b) Find the capacitance of this capacitor. (c) Find the charge on the capacitor at breakdown. See Table 6.1. *Note:* There are two solutions for a, one corresponding to a small gap and a nearly uniform field, and the other corresponding to a large gap and an inner electrode of very small radius. Consider only the first solution.

6-3.10 A 15 cm long cylindrical capacitor with a wax-filled gap has an outer radius of 0.8 cm. For a voltage difference of 9800 V, thére is breakdown. In what follows, neglect fringing field effects at the ends of the capacitor. (a) Find the radius a of the inner plate. (b) Find the capacitance of this capacitor. (c) Find the charge on the capacitor at breakdown. See Table 6.1. *Note:* There are two solutions for a, one corresponding to a small gap and a nearly uniform field, and the other corresponding to a large gap and an inner electrode of very small radius. Consider only the second solution.

6-3.11 A 24 cm long cylindrical capacitor with an air gap has inner and outer radii of 1.2 cm and

2.4 cm, at a voltage difference of 240 V. In what follows, neglect fringing field effects at the ends of the capacitor. (a) Find its capacitance. (b) Find the charge on the plates. (c) Find the maximum electric field within the capacitor. See Table 6.1.

6-4.1 Frank has many 2 μF capacitors with 100 V breakdown. (a) How should he connect them to obtain a 2 μF capacitor with 300 V breakdown? (b) A 4 μF capacitor with 300 V breakdown?

6-4.2 A capacitor C is rated at a maximum voltage V_m. Explain how, by using four such capacitors, you can get a net capacitance of C, with a maximum voltage of $2V_m$.

6-4.3 Two parallel-plate air capacitors of identical area A have separations d_1 and d_2. Show that, when placed in series, they have the same capacitance as a single capacitor of area A and separation $d_1 + d_2$. Give a physical explanation for this result.

6-4.4 Two parallel-plate air capacitors of identical separations d have areas A_1 and A_2. Show that, when placed in parallel, they have the same capacitance as a single capacitor of area $A = A_1 + A_2$ and separation d. Give a physical explanation for this result.

6-4.5 (a) For capacitors $C_1 = 8\ \mu$F and $C_2 = 6\ \mu$F, find their capacitance in series and in parallel. (b) Find the charge and voltage difference on each capacitor when they are connected in series with a 12 V battery. (c) Find the charge and voltage difference on each capacitor when they are connected in parallel with a 12 V battery.

6-4.6 Terminal a connects C_1, C_2 and C_3. Terminal b connects C_4, C_2 and C_3. See Figure 6.27. Let $C_1 = C_4 = 6.8\ \mu$F and $C_2 = C_3 = 4.8\ \mu$F. Find the voltage and charge associated with each capacitor if $V_a = 50$ V and $V_b = 68$ V. Assume that the plates associated with terminals a and b initially are neutral.

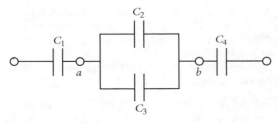

Figure 6.27 Problem 6-4.6.

6-4.7 A 45 nF capacitor C_1 is connected across a 1.5 V battery. The connections to its terminals are then removed from the battery and connected across an unknown, originally uncharged capacitor C_2. If the voltage across C_1 is now 0.34 V, find C_2.

6-4.8 A 6 μF capacitor is in series with a combination of a 3 μF and an unknown capacitor C. The overall capacitance is 4 μF. Find C.

6-4.9 Find the capacitance of a bridge circuit where the bridge capacitor has a capacitance 2 nF, and the other four capacitors have capacitance 4 nF. *Hint:* First use symmetry to determine the voltage difference across the bridge capacitor.

6-4.10 Benjamin Franklin may have been the first person to put capacitors in series, and to charge them by connecting them to a prime conductor (characterized by a fixed voltage, as we now know). He found that the capacitors did not charge as well by this method as when he put them in parallel. Explain.

6-4.11 Figure 6.28 gives a circuit containing a number of capacitors, each of capacitance $C = 6\ \mu$F. Find the capacitance between the terminals.

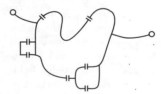

Figure 6.28 Problem 6-4.11.

6-4.12 In equilibrium, two or more conductors in contact with one another take the same potential, so they are equivalent to a single conductor. (a) How many distinct pieces of conductor are there for five capacitors in parallel? (b) For five capacitors in series? (c) For the series case, if the outer conductors have charges $\pm Q$, give the net charges of the internal pieces of conductor.

6-4.13 (a) For the first part of Example 6.7 (connection), work out the wrong "solution" where each capacitor gets the average of the initial charges. (b) Is charge conserved? (c) Are the voltages equalized?

6-4.14 (a) For the first part of Example 6.7 (connection), work out the wrong "solution" where each capacitor gets the average of the initial voltages. (b) Is charge conserved? (c) Are the voltages equalized?

6-5.1 Design a parallel-plate air capacitor with $C = 8$ μF to operate at 400 V in a maximum field that is half E_d in air.

6-5.2 A cylindrical air capacitor of length 4.6 m has a capacitance of 8.6 μF. (a) Find the ratio of the inner and outer radii. (b) If the outer radius is 2 cm, at what voltage is there breakdown?

6-5.3 A parallel-plate capacitor of area 50 cm² and plate separation 0.12 mm has charge ± 3.6 μC. The voltage difference between the plates is 2500 V. Find the dielectric constant.

6-5.4 If you rub a comb through your hair, and then bring it up to a faucet of running water, the water will deflect toward the comb. Transformer oil, with $\kappa = 4.5$, will not deflect nearly as much under the same circumstances. Explain why. *Hint:* Viscosity is not the most relevant property of these fluids.

6-5.5 Despite the large dielectric constant for pure water, it is not normally used inside capacitors. Suggest why.

6-5.6 Let $C_m = 1.1$ μF/m² for a cell wall with $\kappa = 3.6$ that will withstand a maximum voltage difference $\Delta V_m = 250$ V/m before beginning to conduct. (a) Determine the cell wall thickness. (b) Determine the electric field across the cell wall at which conduction begins to occur. (c) Determine the free surface charge density at this field. (d) Determine the corresponding polarization charge density.

6-5.7 A parallel-plate capacitor of $C = 60$ pF is to be filled with Lucite. (a) What is the minimum area it must have to withstand a voltage of 3.5 kV? (b) If the plate area is doubled, what are the new capacitance and the maximum voltage it will withstand?

6-5.8 A cylindrical capacitor of $C = 20$ pF and length 25 cm is to be filled with wax. (a) What is the minimum inner radius it must have to withstand a voltage of 8.2 kV? (b) If the inner and outer radii are doubled, what voltage will it withstand?

6-5.9 A dielectric slab of area A, thickness d_1, and dielectric constant κ_1 is placed in series with a dielectric slab of area A, thickness d_2, and dielectric constant κ_2. This combination is placed within a parallel-plate capacitor of area A and plate separation $d_1 + d_2$. See Figure 6.29. (a) Show that the capacitance is given by $C = (A/4\pi k)/(d_1/\kappa_1 + d_2/\kappa_2)$. (b) Verify that this agrees with the case $\kappa_1 = \kappa_2$. (c) Show that if the charge density is σ on the

positive plate (near 1), then there is a charge density $\sigma(\kappa_1^{-1} - \kappa_2^{-1})$ on the interface between the dielectrics. (Get the sign right by comparing with the charge on the dielectrics near the positive plate.)

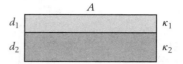

Figure 6.29 Problem 6-5.9.

6-5.10 A dielectric slab of area A_1, thickness d, and dielectric constant κ_1 is placed in parallel with a dielectric slab of area A_2, thickness d, and dielectric constant κ_2. It is placed within a parallel-plate capacitor of area $A_1 + A_2$ and plate separation d. See Figure 6.30. (a) Show that the capacitance is given by $C = (1/4\pi kd)(A_1\kappa_1 + A_2\kappa_2)$. (b) Verify that this agrees with the case $\kappa_1 = \kappa_2$. (c) Show that there is no charge on the interface between the dielectrics.

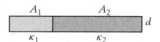

Figure 6.30 Problem 6-5.10.

6-5.11 Consider a sphere of dielectric constant κ, with radius R and total charge Q that is uniformly distributed over its volume. If $V_\infty = 0$, show that the potential at the center of the sphere is $(kQ/R)(1 + (2\kappa)^{-1})$. [Note that (5.41) was derived, implicitly, for $\kappa = 1$.]

6-5.12 A cell wall has $\kappa = 4.3$ and thickness 3.4 nm. (a) Find its capacitance per unit area. (b) Find the capacitance for a spherical cell of radius 8.3 μm. (c) Find the charge associated with a voltage difference of 87 mV.

6-5.13 An aluminum grain has a 1 nm thick insulating oxide layer with $\kappa = 8$ against a thin film of copper. (a) If the contact is a circle of radius 10 nm, estimate the capacitance of the resultant capacitor. (b) Estimate the voltage change if one electron transfers from the aluminum to the copper. [*Note:* In practice, the energy to remove an electron (the *work function*) is different for neutral aluminum and neutral copper. Neglect this effect.]

6-5.14 An aluminum film with a 2.6 nm thick aluminum oxide layer with dielectric constant 8 is against a thin film of copper. If the capacitance across the layer is 8 aF, estimate the contact area.

6-5.15 Consider a spherical shell of inner radius a and outer radius b, its upper half filled with air and its lower half filled with dielectric of dielectric constant κ. See Figure 6.31. With $\kappa_{air} \approx 1$, find its capacitance. *Hint:* Draw the field lines before the dielectric is added. Will adding the dielectric cause the direction of the field lines to change? Will polarization charge appear along the surface that separates the air and the dielectric? The answer to these questions will help you to simplify this problem.

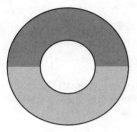

Figure 6.31 Problem 6-5.15.

6-5.16 Find the capacitance for a spherical shell with conducting plates at $r = a$ and $r = c$, and filled with dielectric of dielectric constant κ_1 (κ_2) for $a < r < b$ $(b < r < c)$. See Figure 6.32.

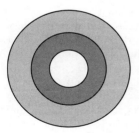

Figure 6.32 Problem 6-5.16.

6-6.1 An air capacitor has an area of 250 cm^2. For a charge of 8 μC, the voltage difference across the plates is 120 V. (a) Find the capacitance and the plate separation. (b) Find the energy stored at this voltage. (c) Find the electric field magnitude within the capacitor. (d) Find the electrical energy density.

6-6.2 (a) What capacitance would store 50 kJ of electrical energy at 860 V? (b) What would the charge be?

6-6.3 A ventricular defibrillator, attached by low-resistance paddles placed above and below the heart, provides an average of 20 A and 10^5 W for 2 ms. If this were provided by a capacitor, estimate its capacitance.

6-6.4 A 50 μF capacitor has 4 kV placed across its terminals. (a) What is the associated charge and energy? (b) How much energy is stored by 2000 such capacitors? (c) How does this compare to the energy stored by a 12 V, 80 A-hr battery? (d) How do the maximum currents compare, if each capacitor can provide 2×10^5 A through a 0.02 ohm resistor, and the battery can provide 1200 A? Note that the capacitors will discharge in about 10^{-6} s, whereas the battery will last for a few minutes at this current.

6-6.5 Capacitors C_A (40 nF) and C_B (20 nF) are connected in series with a 12 V battery. (a) Find the voltages, charges, and energy associated with each. (b) The battery is removed and the plates of like sign (+ to +, and − to −) are connected. Find the voltages, charges, and energy associated with each. (c) Deduce the energy loss on making the connection. (d) Now the plates of opposite sign (+ to −, and − to +) are connected. Find the voltages, charges, and energy associated with each. (e) Deduce the energy loss on making the connection.

6-6.6 (a) By integration over the energy density, find the energy stored by two long concentric cylindrical conductors of radii $a < b$ and length l. Take $+Q$ on the inner cylinder, and neglect edge effects. (b) Determine the capacitance per unit length.

6-6.7 (a) By integrating over the energy density, deduce the energy U of a conducting sphere of radius R and charge Q. (b) Let a sphere with charge Q be slowly compressed from $R + dR$ to R, against the (unknown) electrical force per unit area P_{el}. Find the change in electrical energy and thus deduce P_{el}. Compare with $E^2/8\pi k$.

6-6.8 (a) By integration over the energy density, find the energy stored by two concentric spherical conductors of radii $a < b$. Take $+Q$ on the inner sphere. (b) Deduce the capacitance.

6-6.9 See Figure 6.33. The combination of capacitors is connected to a voltage source across a and c, and then the voltage source is removed. (a) If $V_a - V_b = 60$ V, find the charges and voltages associated with each capacitor. (b) Find the total energy. (c) If b and b' are connected, find the new charges and voltages associated with each capacitor. (d) Find the new total energy. (e) Find the amount and the sign of the charge that must flow from b to b'.

Assume that the plates associated with terminals b and b' initially are neutral.

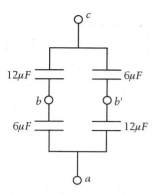

Figure 6.33 Problem 6-6.9.

6-6.10 For an isolated conducting disk of radius a, the charge density is $\sigma = Q/2\pi a\sqrt{a^2 - r^2}$. (a) Show that, if $V_\infty = 0$, then $V = kQ/2a$. (b) Show that the electrical potential energy U of such a conducting disk is $U = kQ^2/4a$. (c) Determine its capacitance.

6-6.11 Consider the earth to be a spherical capacitor with $a = 6.37 \times 10^6$ m and $b = a + d$, where $d = 5 \times 10^4$ m (the distance to the ionosphere). If the fair-weather field is 100 V/m, find the energy stored in the earth's electric field.

6-7.1 A parallel-plate capacitor is charged up by a 6 V battery, which is then removed. A dielectric with dielectric constant $\kappa = 5$ is slid between the plates. Give quantitative answers in terms of the ratio of each quantity before and after. (a) What happens to the electric field within the dielectric? (b) To the voltage difference between the plates? (c) To the charge on the plates? (d) To the capacitance? (e) To the energy stored? Explain how energy is conserved.

6-7.2 A parallel-plate capacitor is charged up by a 6V battery, which is kept in place. A dielectric with dielectric constant $\kappa = 5$ is slid between the plates. Give quantitative answers in terms of the ratio of each quantity before and after. (a) What happens to the electric field? (b) To the voltage? (c) To the charge? (d) To the capacitance? (e) To the energy stored? Explain how energy is conserved.

6-7.3 A slab of dielectric is attracted into a parallel-plate capacitor both when the capacitor

noindent has a fixed charge and when it is attached to a capacitor of such large capacitance that the voltage difference is fixed. Explain why $U = Q^2/2C$ is the relevant energy in the first case but not in the second.

6-7.4 Two capacitors C_1 and C_2 are in series, their net voltage difference maintained by a battery of fixed voltage $\mathcal{E}$. (a) Find the charge and voltage of each capacitor. (b) A dielectric of dielectric constant κ is slipped between the plates of C_1. Find the charge and voltage of each capacitor. (c) What changes are caused by the dielectric?

6-7.5 Consider a capacitor C at a fixed voltage difference V. Show that when the capacitance changes, the energy change is $\delta U_V = [Q^2/(2C^2)]\delta C$, just opposite in sign from (6.35). Hence either $F_x = -dU/dx|_Q$ (corresponding to an isolated capacitor) or $F_x = dU/dx|_V$ (corresponding to a capacitor maintained at a fixed voltage).

6-7.6 (a) Find the energy to charge a spherical capacitor of inner radius a and outer radius b to Q by considering the work it takes to move the charge, using electrical force considerations. Specifically, move dq from $r = b$ symmetrically inward, acting against the force due to the charge q that is already on the inner sphere. (b) Find the capacitance.

6-7.7 Connect a $12\,\mu F$ capacitor to a 6 V battery. Disconnect the battery. Pull the plates apart $(d \to 2d)$. Reconnect to the battery. For the battery, consider that the energy is given by $U_{batt} = \mathcal{E}Q_{batt}$, where $\mathcal{E}$ is the emf of the battery, and Q_{batt} is its charge (this equation is discussed in the next chapter). (a) Analyze the energy transfer, including any work done by or to your hands. (b) Repeat this process if the capacitor is connected to the battery while the plates are pulled apart. Analyze the energy transfer.

6-7.8 Connect an uncharged capacitor $C_1 = 5\,\mu F$ to a capacitor $C_2 = 500\,\mu F$ initially at 6 V. Disconnect the capacitors. Pull the plates of C_1 apart $(d \to 2d)$. Reconnect to C_2. (a) Analyze the energy transfer, including any work done by or to your hands. (b) Repeat this process if C_1 is connected to C_2 while the plates are pulled apart. Analyze the energy transfer.

6-8.1 Consider two conducting spheres of radii a and b, which are very far apart. Specifically, take

$b = 10$ m, and $a = 0.5$ m. Let the large sphere, which serves as a reservoir, initially be at 201 V relative to infinity, and let the small sphere be uncharged. Now connect the spheres with a very fine wire. Find (a) the initial charge on the large sphere, (b) the final potential of the connected spheres, and (c) the final charge on each sphere.

6-8.2 Consider two distant spherical conductors. (a) If 10 nC is added to one sphere, and the other sphere has its voltage increase by 40 mV, estimate the separation r. (b) Does this estimate only apply to spherical conductors?

6-8.3 Recent experiments have shown that small capacitors consisting of two blobs of metal can be influenced by nearby charge on a third blob of metal. Using the p_{ij}'s, show how charge on the third blob affects the voltage difference on the first two. In the experiments, the measured voltage difference was quantized, because the charges on the third blob were due to individual electrons.

6-8.4 Consider two capacitors connected in series. There are three pieces of conductor, so there are six different p_{ij}'s. Let 1 and 3 be on the outside, and let 2 be on the inside. (a) Obtain the overall capacitance in terms of the p_{ij}'s, using $C = Q/(V_3 - V_1)$ and $Q_3 = Q$, $Q_1 = -Q$, and $Q_2 = 0$. (b) What condition must the p_{ij}'s satisfy for $Q_2 \neq 0$ to not influence $V_3 - V_1$?

6-8.5 (a) Explain why, for polarization of a conductor, only the shape is relevant, whereas for polarization of an insulator, both the shape and material are relevant. (b) Consider a conducting sphere of radius a with charge q, and a point charge Q at distance $r \gg a$. Explain why electrostatic induction can be neglected, as a first approximation, in describing their interaction. Hence, far from a small conducting sphere with charge q, the sphere can be thought of as a point charge q.

6-8.6 Consider two distant conducting spheres, 1 and 2. (a) Show that if conductor 2 becomes polarized (e.g., by charge on 1), its polarization charge sets up a dipole moment $\vec{p}_2$ and, by (5.46), a dipole potential $k\vec{p}_2 \cdot \vec{R}/R^3$, where $\vec{p}_2$ is proportional to its volume Ω_2 and to the electric field kQ_1/R^2 at 2 due to 1. (This is related to the amber effect.) (b) Show that this dipole potential *lowers* V_1 in proportion to $kQ_1\Omega_1/R^4$, where Ω_1 is the volume of conductor 1. Thus, p_{11}, which is positive, decreases in proportion to $k\Omega_1/R^4$. Similarly, p_{22} decreases in proportion to

$k\Omega_2/R^4$, where Ω_2 is the volume of conductor 2. (c) From the dependence of p_{11} on the separation R, and the energy of the system when $Q_1 \neq 0$ but $Q_2 = 0$, show that the force of attraction between the two spheres varies as R^{-5}, as expected for the amber effect.

6-9.1 Consider a long organic molecule. (a) Along which axis is it likely to be more polarizable? *Hint:* Assume the extreme situation where the molecule is conducting. Consider the two cases where $\vec{E}$ of a fixed magnitude is along the molecular axis and when it is normal to it. Are the $|\vec{p}|$'s the same for these two cases? (b) Explain why $\vec{p} = \alpha\vec{E}$ is too simple an equation to describe polarization in this case.

6-9.2 Here is a model for the polarizability of an atom. Consider a rigid ball of radius a with uniform charge density and total charge $-e$ in a small external field $\vec{E}$. (a) Show that the equilibrium position $\vec{r}$ of a positive charge $+e$ (relative to the center of the ball of charge) satisfies $\vec{r} = (a^3/ke)\vec{E}$, when $|\vec{E}|$ is small enough that $r < a$. (b) Find the dipole moment $\vec{p}$ of the overall neutral ball–charge combination. (c) Show that the polarizability satisfies $\alpha = a^3/k$. (d) Evaluate α for $a = 10^{-10}$ m.

6-9.3 Consider a spring of constant K that, when extended by x, has dipole moment qx. Show that the polarizability is given by $\alpha = q^2/K$. *Hint:* Minimize the sum of the interaction energy of a dipole (with an external field $\vec{E}$ along x) and the spring's potential energy, as a function of x.

6-9.4 In what follows, take the surface at $x = 0$ and take the electric field to fall off with distance as $\exp[-(x/l\sqrt{\kappa})]$. (a) If the density of localized charge is fixed, so κ is fixed, and the density of localized charge increases, so l decreases, then the screening length $l\sqrt{\kappa}$ decreases. This makes the screening more effective everywhere. For an applied field of 2500 V/m and $\kappa = 5$, at $x = 4$ nm evaluate the field magnitude for $l = 20, 4, 0.8$ nm. (b) A curious effect can occur if the density of delocalized charge is fixed, so l is fixed, and the density of localized charge increases, so κ increases. Although increasing κ causes the field just inside the material to decrease (because it causes polarization charge to literally go to the surface), increasing κ can sometimes cause the field further into the material to become larger (because the screening length increases, by $l \to l\sqrt{\kappa}$). This last effect is most important for large values of κ and relatively large distances into the material, where the electric field is small anyway. For an applied field of 2500 V/m and $l = 2$ nm, at $x = 8$ nm evaluate the field magnitude for $\kappa = 1, 10, 100$.

6-10.1 Use the field-line tension to verify that the attractive force pulling on the midplane of two charges $\pm q$ separated by $2a$ is $kq^2/4a^2$.

6-10.2 Use the field-line pressure to verify that the repulsive force on the midplane of two charges $+q$ separated by $2a$ is $kq^2/4a^2$.

6-10.3 Use the field-line pressure to find the force per unit length between two charged rods $+\lambda$ separated by $2a$ along the x-axis.

6-10.4 Use the field-line tension to find the force per unit length between two charged rods λ and $-\lambda$ separated by $2a$ along the x-axis.

6-G.1 It is not difficult to generate electrical potentials of over a thousand volts by rubbing two insulators against each other. However, on discharge, one can get only a small shock. Discuss why using the concept of capacitance.

6-G.2 A small conducting disk is placed within a parallel-plate capacitor, parallel to the plates. (a) What effect does this have on the field lines? (b) On the capacitance? (c) Repeat if the disk is placed normal to the plates.

6-G.3 If the fringing field of a capacitor is included, how is the capacitance affected? [Answer: It increases because extra charge piles up at the edge, thereby decreasing the field and the voltage drop for other regions.]

6-G.4 Microphones convert the mechanical energy of air motion to an electrical signal. How might a parallel-plate capacitor made of a flexible membrane serve in this role?

6-G.5 A variable air capacitor used in radio tuners has an interleaving of two sets of connected plates. One set are semicircles on the even planes $z = 0, 2d, \ldots, 2nd$. The other set are semicircles on the odd planes $z = d, 3d, \ldots, 2nd + d$. If the semicircles have area A and overlap angle θ (so that $\theta = \pi$ is the maximum overlap), show that the capacitance of this combination is given, approximately, by $C = (2n + 1)A\theta/4\pi^2 kd$.

6-G.6 Show that C for a concentric cylindrical capacitor of radii a and b and length $L \gg$

a, b reduces to that for the parallel-plate capacitor when $b \to a$, on taking $d = b - a$ and $A = 2\pi La \approx 2\pi Lb$.

6-G.7 Here is a method to measure the surface charge on a dielectric. A dielectric slab with an unknown charge density σ on its top is grounded at its bottom. A small distance above the dielectric is a small flat probe of area A. See Figure 6.34. There is capacitance C_d between the top of the dielectric and ground, capacitance C between the probe and the top of the dielectric, and capacitance C_p between the probe and ground. The charge on the surface of the dielectric causes a voltage difference V_p between the probe and ground. Show that $\sigma = (C_d + C_p + C_d C_p/C)(V_p/A)$. *Hint:* The voltage drop is the same along both paths from the top of the dielectric to ground: directly downward through the dielectric and upward through the probe and then to ground.

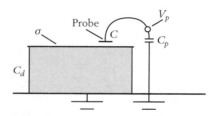

Figure 6.34 Problem 6-G.7.

6-G.8 Consider a large charged conducting sphere, to which an initially uncharged conducting tip is now added. The charge on the sphere now is shared with the tip. How does this change the voltage? The capacitance?

6-G.9 Volta found that a frame of cross-connected wires had a capacitance comparable to that for much larger systems that had no "holes." In other words, it had a relatively large capacitance. How might such a geometry permit a large capacitance? Recall that the electric field is usually larger where the surface is more curved.

6-G.10 The charges on conductors are related to their potentials via

$$Q_1 = q_{11}V_1 + q_{12}V_2, \qquad Q_2 = q_{21}V_1 + q_{22}V_2,$$

where the (unknown) q_{ij} are called *coefficients of induction*. Here potential is measured relative to the potential at infinity. Defining

$$D = p_{11}p_{22} - p_{12}p_{21},$$

where the coefficients of potential p_{ij} are defined in

Section 6.8, show that

$$q_{11} = \frac{p_{22}}{D}, \qquad q_{22} = \frac{p_{11}}{D}, \qquad q_{12} = -\frac{p_{21}}{D},$$

$$q_{21} = -\frac{p_{21}}{D}.$$

6-G.11 Let $p_{11} = 8 \times 10^9$ F^{-1}, $p_{22} = 20 \times 10^9$ F^{-1}, and $p_{21} = p_{21} = 4 \times 10^9$ F^{-1}. (a) From the results of problem 6-G.10, show that $D = 1.44 \times 10^{18}$ F^{-2}, $q_{11} = 5.56 \times 10^{-9}$ F, $q_{22} = 13.9 \times 10^{-9}$ F, and $q_{12} = q_{21} = -2.78 \times 10^{-9}$ F. (b) Show that, if 2 is grounded and a 12 V battery is connected between it and 1, so $V_2 = 0$ and $V_1 = 12$ V, then $Q_1 = q_{11}V_1 = 6.67 \times 10^{-8}$ C and $Q_2 = q_{21}V_1 = -3.33 \times 10^{-8}$ C.

6-G.12 Even a small electric shock can destroy the sensitive circuits of a computer chip. What role (or roles) does the metal coating on the bags have for the computer memory chips (RAM, or random access memory) they contain? What is the coefficient of induction between your hand outside the bag and the chip inside the bag?

6-G.13 If two conductors are connected by a fine wire, they come to the same potential, so they can be considered to behave as a single conductor. (a) Show that

$$C_{self} = \frac{Q_1 + Q_2}{V} = q_{11} + q_{22} + q_{12} + q_{21}.$$

(self-capacitance of connected conductors)

Hence the self-capacitance of two connected conductors is a combination of both the mutual- and self-induction coefficients. (b) Show that, if $q_{12} = 0 = q_{21}$, then $C_1 = Q_1/V_1 = q_{11}$ and $C_2 = q_{22}$, so $C = C_1 + C_2$.

6-G.14 Let $q_{11} = 5.56 \times 10^{-9}$ F, $q_{22} = 13.9 \times 10^{-9}$ F, and $q_{12} = q_{21} = -2.78 \times 10^{-9}$ F. (a) Using the results of problem 6-G.13, find C_{self}. (b) With $Q = 10^{-9}$ C, find the common voltage V, Q_1, and Q_2.

6-G.15 Consider a system of two distant spheres, of radii a and b, separated by $R \gg a, b$. Explain why it is a good approximation to take

$$V_a = \frac{kQ_a}{a} + \frac{kQ_b}{R}, \qquad V_b = \frac{kQ_b}{b} + \frac{kQ_a}{R}.$$

Find the coefficients of induction q_{ij}.

6-G.16 Let two long parallel plates be separated along the x-axis by $2a$, with the midpoint at the origin. Rotate the left plate about

the origin until the angle between the two plates is the small angle α, rather than π. Let there be guard rings (gaps connected only by a thin wire) an additional distance b along the plates. See Figure 6.35. (The guard rings help eliminate the fringing field.) (a) Show that the field magnitude E between the plates is well represented by $E = \Delta V/r\alpha$, where ΔV is the fixed voltage difference between the plates and r is the radial distance. (b) Show that the charge density on the positive upper plate is $\sigma_S = \frac{E}{4\pi k} = \frac{\Delta V}{4\pi kr\alpha}$. (c) Show that the charge associated with the plates, from $r = a$ to $r = b$, and of dimension L into the page, is given by $Q = \int \sigma_S dA = \int \frac{\Delta V}{4\pi kr\alpha} L dr = \frac{\Delta VL}{4\pi k\alpha} \ln\frac{b}{a}$. (d) Show that the capacitance is given by $C = \frac{Q}{\Delta V} = L\frac{\Delta V}{4\pi k\alpha} \ln\frac{b}{a}$.

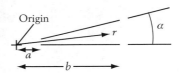

Figure 6.35 Problem 6-G.16.

6-G.17 Consider three blobs of metal, with $q_3 = q$ and $q_1 = -q_2 = Q$. Let q be fixed, but let Q vary. (a) Find the electrical energy U in terms of these charges and the coefficients of potential p_{ij}. (b) Show that U can be written in the form $U = (Q - rq)^2/C + q^2/C_3$, and evaluate C, r, and C_3 in terms of the p_{ij}'s. This form has been used in studying the charging energy when the number of electrons is so small that in writing $Q = ne$ the discreteness of charge is noticeable.

6-G.18 Consider a slab of thickness d and area A with surface charge density $\pm\sigma'$, as in Figure 6.14. Define the direction from $-\sigma'$ to σ' as $\hat{l}$. (a) Within the air gap between the slab and the plates let the field outside the slab be $\vec{E}_{out}$. Let the field within the slab be $\vec{E}_{in}$. Show that $\vec{E}_{out} = \vec{E}_{in} + 4\pi k\sigma'\hat{l}$. (b) Show that the *polarization* $\vec{P}$, defined as the dipole moment per unit volume, is $\vec{P} = (q'd\hat{n})/V = \sigma'\hat{l}$. (c) Show that $\vec{E} + 4\pi k\vec{P}$ is continuous at the interface. Therefore the *electric displacement* $\vec{D} \equiv \epsilon_0(\vec{E} + 4\pi k\vec{P}) = \epsilon_0\vec{E} + \vec{P}$ is continuous at the interface. Since $\vec{P}$ is the polarization of the material, we may think of $\epsilon_0\vec{E}$ as the "polarization of the vacuum." $\vec{D}$ has only *free charge* as its source (such as electrons or ions), not polarization charge (such as from neutral but polarizable atoms or molecules). Maxwell attributed great significance to $\vec{D}$.

6-G.19 Consider an atom with a permanent dipole moment $\vec{p}$ of magnitude p_0. For a strong applied field $\vec{E}$, $\vec{p}$ aligns with $\vec{E}$. For no applied field, $\vec{p}$ points randomly, due to thermal collisions. Assume that, when a weak field $\vec{E}$ is applied, the atoms are aligned with $\vec{E}$ a (small) fraction $p_0 E / k_B T$ of the time, and randomly aligned the rest of the time. (Here k_B is known as the Boltzmann constant, and T is the temperature; $k_B T$ has units of energy, just as does $p_0 E$.) (a) Show that the average dipole moment is proportional to $p_0^2 E / k_B T$. (b) Show that the polarizability is proportional to $p_0^2 / k_B T$.

"The ordinary electrometer indicates tension [voltage]. . . . There was lacking an instrument which would enable us to recognize the presence of the electric current . . . this instrument now exists . . . I think . . . we should give it the name of galvanometer . . . ."

—André Marie Ampère (1822)

"I do not regard my experiments . . . to be complete. . . . My position as a teacher in a grammar school places exceptional obstacles in the way of this fundamental research. . . . Cylindrical conductors of the same substances have the same conducting capacity for different diameters provided the length is proportional to the cross-section."

—Georg Simon Ohm (1826)

Chapter 7

Ohm's Law: Electric Current Is Driven by Emf, and Limited by Electrical Resistance

Chapter Overview

This chapter has three parts. The first part gives a precise definition of electric current and discusses Ohm's law and electrical resistance. The second part discusses sources of emf, whose description can be incorporated in a generalization of Ohm's law. The third part discusses charge carriers within wires and explains the microscopic origin of electrical resistance.

Section 7.1 gives a brief introduction to the chapter. Section 7.2 discusses electric current, for geometries more general than flow along a uniform wire, introducing the concept of the local electric current density $\vec{J}$ (a current per unit area). Section 7.3 states Ohm's law, and for uniform wires introduces the concept of the material's *resistivity* ρ (not to be confused with the ρ used for charge per unit volume). Section 7.4 shows that if the local $\vec{J}$ is proportional to the local electric field $\vec{E}$, thus yielding a local form of Ohm's law, then the usual global form of Ohm's law follows. The proportionality constant relating $\vec{J}$ and $\vec{E}$ is the material's *conductivity* σ (not to be confused with the σ used for charge per unit area), which satisfies $\sigma = \rho^{-1}$. Section 7.5 discusses resistors in series and in parallel, the idealizations usually made in neglecting the resistance of connecting wires, and the real meaning of the expression *path of least resistance*. Section 7.6 shows how additional resistors, in series and in parallel with a galvanometer, can be used to design ammeters and voltmeters.

As a prelude to the discussion of emfs, Section 7.7 discusses some complexities of car batteries. Section 7.8 considers how emf is measured with an electrometer, the internal resistance of a source of emf, and various types of emf. Section 7.9 discusses energy storage by voltaic cells, and energy transfer by voltaic cells in circuits. Section 7.10 considers circuits containing voltaic cells.

Section 7.11 discusses the fundamental process leading to Ohm's law: an average drag force. This idea is first applied to a simple model for parachutists subject to air drag, and then applied to air drag on oil drops, leading to the Millikan oil drop experiment that established *charge quantization.* Finally, drag force is applied to charge carriers in a conductor, subject to collisions with the ionic background. (This model was developed only as recently as 1900, by Drude, following the 1897 discovery of the electron.) In metals and semiconductors, the charge carriers are *electrons,* producing *electronic conduction.* In ionic solids—such as salt—and in ionic fluids (or *electrolytes*)—such as salt water or battery acid—the charge carriers are *ions,* producing *ionic conduction.* Sections 7.12 and 7.13 use the drag model to explain why some materials are good conductors and some are poor conductors (i.e., insulators).

Unassuming as they are, Sections 7.11 and 7.12 are extremely important. Millikan's discovery of charge quantization, coupled with J. J. Thomson's discovery that cathode rays are all negatively charged with the same charge-to-mass ratio for all cathode materials, provides justification for the electric fluid model: it identifies negatively charged electrons as the charge carriers within a conductor. Moreover, it relates the electric current to the average electron flow velocity. In the midnineteenth century, none of this was known. In the absence of such knowledge, James Clerk Maxwell, one of the most important figures in the development of electromagnetic theory, withheld support for the electric fluid model. J. J. Thomson, one of Maxwell's successors at the Cavendish laboratory, became its strong supporter. We believe that so would have Maxwell, had he known Millikan's and Thomson's results. ▪

7.1 Introduction

Electric current—the flow of electric charge—is indispensable to modern society. Toasters use electric current for heating; electric motors use electric current to produce motion; and radio, television, and telephone all use electric current for communications. This chapter discusses electric current and the energy sources that drive it. Such current-causing energy sources are called *electromotive forces*, or *emfs*.

True forces act locally and have units of newtons (N). The definition of emf, however, is not local. It is the work per unit charge for some specific path, and thus involves all the points along the path, not a single point. Its algebraic symbol is $\mathcal{E}$, and it has units of J/C = V (volts).

7.1.1 *A Brief History of Electrical Conduction*

Until the development of the voltaic cell (1791), the only type of emf was due to electrostatic energy, produced by static electricity devices and used directly, or stored by capacitors. Since capacitors and static electricity devices typically discharge very quickly, it was difficult to perform controlled experiments. In the 1770s, Cavendish compared the "conducting power" of different metals by the intensity of the shocks he received on discharging an electrical device through a circuit that included a length of metal and his tongue. He thereby gained valuable comparative information, but it was not yet quantitative science. Scientists had neither a long-lasting source of emf nor a quantitative means to measure electric current.

The absence of a long-lasting source of emf ended with the discovery of voltaic cells (using chemical energy), beginning in the 1790s with the work of Galvani, and then of Volta. In 1822, Seebeck discovered thermoelectric devices (using thermal energy). A voltage electrometer, as in Figure 5.8, could then characterize the strength of a long-lasting source of emf by putting the emf in an open circuit (so there is no current flow). One could next measure current by using a long-lasting source of emf in a circuit of very low "conducting power" to slowly charge a bank of capacitors in parallel. From the capacitor charge Q (monitored with a charge electrometer, as in Figure 2.4) and the charging time t, the average electric current $\bar{I} = Q/t$ could be determined. Such a procedure is tedious. Without a simple and reliable method to measure the instantaneous value $I = dQ/dt$ of the electric current, it was not practical to determine the relationship between the electric current and the emf.

In 1820, Oersted discovered that an electric current (driven by a battery of voltaic cells) could cause a magnetic needle to deflect. It was found that the deflection was proportional to the electric current. (This effect will be discussed in Chapter 10.) Here, at last, was the instantaneous current-measuring device that scientists had been seeking; what Ampère named the *galvanometer*.

With this device to measure current, and a thermoelectric power source (more stable than a voltaic power source), Ohm studied how the current through a wire depends upon the emf, the material, and the length and cross-sectional area of the wire. In 1827, he published what we now call *Ohm's law*. Had Nobel prizes in physics been awarded at that time, Ohm surely would have been a recipient.

7.1.2 *Ohm's Law in a Nutshell*

Consider a material with two electrodes, across which there is a voltage difference ΔV, and through which passes a current I. Then $I = \Delta V/R$ defines the *electrical resistance R*. When Ohm's law applies, two results hold: (1) the electric current I through an object passes from high to low voltage; (2) I and ΔV are proportional, so that R is independent of ΔV. Nearly all materials satisfy Ohm's law for a small enough range of I and ΔV. Moreover, for many materials, Ohm's law is an exceedingly good approximation for a wide range of I and ΔV. Nevertheless, no material satisfies Ohm's law for *all* values of I and ΔV.

Consider Figure 7.1. The material of Figure 7.1(a) satisfies Ohm's law for a wide range of I and ΔV, and is thus called *ohmic*; the material of Figure 7.1(b) does not satisfy Ohm's law for a wide range of I and ΔV, and is thus called *non-ohmic*.

When Ohm's law applies, R depends on the geometry of the object and the type of material, but not on the current or emf. For wires, where the electrodes are attached at the ends, R is proportional to the length and inversely proportional to the cross-sectional area. Thus, overall, R is inversely proportional to the linear dimension (compare with capacitance C, which in Chapter 6 was found to be proportional to the linear dimension).

Metals, such as copper and aluminum, are ohmic over a wide range of currents and voltage differences, as in Figure 7.1(a). However, silicon and other semiconductors (so called because they conduct electricity better than an insulator like

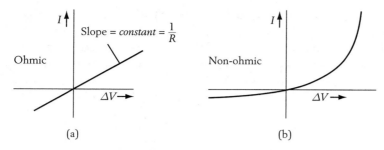

Figure 7.1 Current response I of a circuit element to a voltage difference ΔV placed across two points. Resistance $R \equiv \Delta V/I$ is the inverse slope of the tangent to the curve. (a) Ohmic response, where I is proportional to ΔV. (b) Non-ohmic response, where I is not proportional to ΔV.

window glass, but worse than a conductor like copper), are non-ohmic, as in Figure 7.1(b). (There are an infinite number of I versus ΔV curves that can be non-ohmic, but only straight lines with positive slope and passing through the origin are ohmic.) Many materials and devices (e.g., vacuum tubes) are non-ohmic. Modern electronics uses ohmic materials to transport current, and non-ohmic materials for amplifiers and other nonlinear devices. We will concentrate upon ohmic materials; the study of non-ohmic materials and their uses is an advanced topic.

7.1.3 A Historical Sidelight
Optional

Ohm's contemporary Ampère understood the concepts of voltage and of current. Further, Ampère understood the idea of an electric fluid—actually, he thought in terms of a positive and a negative fluid—driven by emf and limited by some sort of resistance. However, do not feel sorry for Ampère for not discovering Ohm's law. He was thoroughly occupied studying the magnetism of electric currents, which forms the basis of Chapters 10 and 11. The 15 years from 1819 to 1834 provided enough discoveries about electricity and magnetism that Chapters 7 to 12 are devoted to them.

If anything, feel sorry for Ohm. First, he published his careful and detailed experimental work. Then he published a very mathematical work, based on his

Medical Applications

Electrocardiography is a well-established method to study the heart by measuring its electrical output. A method called electrical impedance imaging, still under development, provides electrical inputs, measures electrical outputs, and interprets the results using Ohm's law. Healthy lungs, filled with air, show up in an impedance image as regions of low conductivity, but lungs suffering from pulmonary edema, and thus partially filled with fluid, show up in an impedance image as regions of much higher conductivity. This technique may eventually enable noninvasive monitoring of the blood circulatory system.

experiments, which developed the analogy between current flow and heat flow. Next, because one group thought his mathematical work was not based on experiment, he was refused a university appointment. Finally, because another group noticed what they considered to be his philosophically unacceptable reliance on experiment, he lost his position teaching secondary school! For six years, he scraped out a meager existence, until gradually his work became known, and he began to receive honors from abroad. Ohm at last received a major professorial appointment in 1849, 22 years after the publication of his book.

7.2 **Electric Current**

Electric current I is the rate at which charge crosses a given cross-sectional area per unit time. That is,

$$I = \frac{dQ}{dt}. \quad \text{(definition of electric current)} \quad (7.1)$$

It has units of C/s, which is the ampere (A). As discussed in Section 7.1, I can be determined from the deflection of the magnetic needle in a galvanometer. When there is only one type of charge carrier, the current is the product of their charge q and the rate dN/dt at which they cross a given cross-section of the circuit.

Consider the general case, where a wire might have a bend, or a nonuniform cross-section. Then both the direction and magnitude of the current per unit area need not be uniform. (The vector specified both by the local direction and by the magnitude of the current per unit area is written as $\vec{J}$ and is called the *current density*.) If the local normal to the cross-section is specified by $\hat{n}$, then the current per unit area along the normal $\hat{n}$ is $dI/dA = \vec{J} \cdot \hat{n}$. The total electric current I passing through that cross-section thus is given by

$$I = \int dI = \int \frac{dI}{dA} dA = \int \vec{J} \cdot \hat{n} \, dA, \quad \text{(definition of electric current density)} \quad (7.2)$$

where the integral is over the cross-section of the wire.

Example 7.1   Charge flow and current

Every 50 ms, a charge 1.5 C uniformly crosses a 40 mm^2 area, traveling in the x-direction. Find (a) I; (b) dI/dA; and (c) $\vec{J}$.

Solution: (a) By (7.1), $I = dQ/dt$. Since Q increases in proportion to time,

$$I = \Delta Q/\Delta t = 1.5 \text{ C}/.05 \text{ s} = 30 \text{ C/s} = 30 \text{ A}.$$

(b) Because the charge crosses the area uniformly, by (7.2)

$$\frac{dI}{dA} = \frac{I}{A} = \frac{30 \text{ A}}{40 \times 10^{-6} \text{ m}^2} = 1.5 \times 10^6 \text{ A/m}^2.$$

(c) Since this corresponds to $\hat{n}$ along the x-direction, by (7.2), $J_x = \vec{J} \cdot \hat{n} = dI/dA = 1.5 \times 10^6$ A/m^2, and $J_y = J_z = 0$.

For an ordinary fluid with local velocity $\vec{v}$, $\int \vec{v} \cdot \hat{n} dA$ equals dV/dt, the rate of flow of fluid volume, or volume flux of fluid, across a given cross-section of, for example, a pipe. (Check the units.) Thus $\vec{v} \cdot \hat{n}$ may be interpreted as the volume flux per unit area. Similarly, $\int \vec{J} \cdot \hat{n} dA = I$ equals dQ/dt, the rate of flow of charge, or charge flux across a given cross-section of, for example, a wire. Thus $\vec{J} \cdot \hat{n}$, in addition to its interpretation as current per unit area, also may be interpreted as charge flux per unit area. This strengthens the analogy between flow of fluid and flow of electricity. Moreover, it justifies the use of Φ_E as electric flux and $d\Phi_E/dA = \vec{E} \cdot \hat{n}$ as electric flux per unit area.

Equation (7.2) applies for arbitrary cross-sections. For a cross-section that is parallel to the sides of a wire, so $\vec{J}$ is perpendicular to $\hat{n}$, the current is zero. For a wire of uniform cross-section A and uniform current density $\vec{J}$ along the normal to the cross-section, (7.2) reduces to

$$I = JA, \qquad J = |\vec{J}|. \tag{7.3}$$

See Figure 7.2(a). However, (7.2) applies more generally.

Example 7.2 **Cylinders and radial current flow**

Consider Figure 7.2(b), which represents two concentric cylinders of radius r and length l, with radial current density $J_r \equiv \vec{J} \cdot \hat{r}$ that is independent of position along the z-axis. Moreover, the total radial current is independent of r. (a) For $\hat{n} = \hat{r}$ and a given r, compute the total radial current $I = \int \vec{J} \cdot \hat{n} dA = \int \vec{J} \cdot \hat{r} dA = \int J_r dA$. (b) If I is independent of r and proportional to l, find

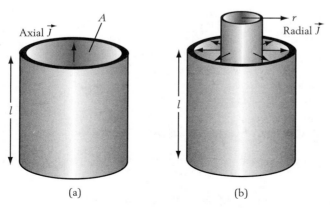

Figure 7.2 Two forms of current flow for cylindrical geometries: (a) axial current flow, (b) radial current flow.

J_r as a function of the radial current per unit length $K = I/l$. (Note: If I depended on r, then there would be charge buildup.)

Solution: (a) For this geometry $I = \int J_r dA = J_r A = J_r(2\pi rl)$. (b) Thus $J_r = I/(2\pi rl) = K/(2\pi r)$. Hence J_r varies inversely with r. The larger the radius r, the greater the area $A = 2\pi rl$ through which the current must pass, and thus the lower the current density J_r.

7.3 Global Form of Ohm's Law

7.3.1 *Ohm's Law*

When an overall voltage difference ΔV is applied between two terminals connected to an ohmic material, leading to a total current I through the material, Ohm's law in its overall, or global, form applies. To repeat what was noted in Section 7.1, Ohm's law states (1) current I flows from higher to lower voltage; (2) in

$$I = \frac{\Delta V}{R}, \qquad \text{Ohm's law (global form)} \qquad (7.4)$$

R is independent of ΔV or I, as in Figure 7.1(a). Note that (7.4) also applies to non-ohmic materials (e.g., silicon), but then R depends on ΔV or I, as in Figure 7.1(b).

Example 7.3 **The resistance across the feet of a duck. (Is the duck ohmic?)**

Two wires are connected across the feet of a solid copper duck. A current of 2.5 A passes into the duck when the voltage drop across its feet is 0.04 V. In addition, when the voltage is halved, so is the current. (a) Find the resistance across the feet of the duck. (b) Is the duck ohmic?

Solution: (a) $R = \Delta V/I = 1.6 \times 10^{-2}$ Ω for currents less than or equal to 2.5 A. (b) Since R is independent of I for this current range, the duck is ohmic in this current range.

For the specific case of a wire of length l and cross-sectional area A, Ohm also found that

$$R = \frac{\rho l}{A}. \qquad \text{(resistance of a wire)} \qquad (7.5)$$

Here ρ is the electrical resistivity; it is material dependent, but independent of ΔV or I or the geometry of the resistor. (Earlier ρ denoted charge per unit volume; context will usually define the meaning of the symbol ρ.) This is consistent with our earlier statement that R^{-1} is proportional to the linear dimension: doubling all dimensions doubles l and quadruples A, so R halves and R^{-1} doubles. Clearly, the resistance R depends on the resistivity ρ and on the geometry of the

Table 7.1 Table of resistivities

Material	Glass	Silicon (pure)	Graphite	Steel	Aluminum	Copper	Silver	Tungsten
ρ (in Ω-m)	10^{12}–10^{13}	640	3.5×10^{-5}	40×10^{-8}	2.8×10^{-8}	1.69×10^{-8}	1.5×10^{-8}	5.6×10^{-8}
$\rho^{-1}d\rho/dT$ (in K^{-1})	-0.7	-0.075	-0.0005	0.0008	0.0039	0.0039	0.0038	0.0045

wire. Take a wire. Mash it, bash it, smash it: its resistance R may change, but the resistivity ρ will not.

The unit of electrical resistance is the *ohm*, or Ω, which by (7.4) has the same units as volts/ampere, so Ω =V/A. The unit of electrical resistivity, by (7.5), is the Ω-m. Table 7.1 shows the resistivities ρ for many common materials, and the *temperature coefficient of resistivity* $\alpha \equiv \rho^{-1}d\rho/dT$ at room temperature (293° Kelvin, or 20° Celsius). α is given in K^{-1}. For temperatures not far from 293 K, the straight-line approximation

$$\rho(T) \approx \rho(293) + d\rho/dT(T - 293) = \rho(293)[1 + \alpha(T - 293)] \quad (7.6)$$

gives $\rho(T)$.

Example 7.4 Resistivity of a wire

Consider a wire of radius $a = 0.05$ inch ($=.00127$ m) and length $l = 8$ m. It passes a current $I = 1.874$ A when $\Delta V = 50.0$ mV is applied across its ends. When the voltage is doubled, so is the current. (a) Is the material ohmic? (b) Find R and ρ. (c) Can you deduce the material of the wire?

Solution: (a) The linear variation of I with ΔV means that the material is ohmic. (b) By (7.4), $R = \Delta V/I = .02668$ Ω. The wire has area $A = \pi a^2 = 5.07 \times 10^{-6}$ m^2, so, by (7.5), $\rho = RA/l = 1.690 \times 10^{-8}$ Ω-m. (c) From the table of resistivities, this corresponds to copper (Cu) at room temperature. (Note: The *gauge* of a wire *approximately* indicates how many wires give 1 inch of thickness. Our wire has thickness $2a = 0.1$ inch; ideally, this would make it 10 gauge, or #10, but in fact wire of radius 0.051 inch is 10 gauge according to the United States Electrical Code.)

Application 7.1 The lightbulb, an apparently non-ohmic material

In some cases, a material will appear to be non-ohmic, with a current and voltage difference that are not proportional. Such is the case with the tungsten filaments of ordinary lightbulbs. The reason for this apparently non-ohmic behavior is that the Joule heating increases the filament temperature, and that significantly changes the resistivity. For sufficiently small currents—usually so small that the bulb won't light—the filament is ohmic.

Optional

Example 7.5 Resistance of a cell membrane

Consider a long nerve axon—a cylindrical shell of thickness d, radius a, and length l, where $d \ll a \ll l$. (a) Find the resistance to current flow along the axis, as in Figure 7.2(a). (b) Find the resistance to radial current flow (across the cell membrane), as in Figure 7.2(b).

Solution: (a) Apply (7.5) with length l and area $A = A_{annulus} = 2\pi ad$. Then

$$R_{axial} = \frac{\rho l}{2\pi ad}.$$

(b) Apply (7.5) with length d and area $A = A_{radial} = 2\pi al$. Then

$$R_{radial} = \frac{\rho d}{2\pi al}.$$

In practice, a and l vary significantly from cell to cell, but ρ and d do not. Hence it is useful to define the specific resistance

$$\tilde{R}_m \equiv R_{radial} A_{radial} = \rho d.$$

For cell membranes, a characteristic value of $\tilde{R}_m$ is 10^3 ohm-cm^2. With $d = 10$ nm, this corresponds to $\rho = 10^9$ ohm-m. Not surprisingly, this is vastly larger than the 1.69×10^{-8} ohm-m of Cu. If $\tilde{R}_m$ is known, and a and l are known, then $R_{radial} = \rho d / 2\pi al = \tilde{R}_m / 2\pi al$ can be determined.

Optional | **Application 7.2** **Resistance across a cell wall**

The resistivity of subcutaneous tissue fluid is about 0.78 ohm-m (not too far from the 4.4 ohm-m of sea water). The specific resistance $\tilde{R}_m$ of the cell interior for a cell of thickness 100 μm thus is $\tilde{R}_m = \rho d = 0.78$ ohm-cm^2. Since the previous example gives $\tilde{R}_m \sim 10^3$ ohm-cm^2 for the cell membrane (i.e., cell wall), the resistance of a cell mostly is due to the cell wall. If we model the human arm by $A = 20$ cm^2 area with resistivity $\rho = 10^9$ ohm-m, to have a resistance of $R = 10^5$ ohm requires, by (7.5), an effective length $l = RA/\rho = 2 \times 10^{-6}$ m. This corresponds only to about 200 cell walls 100 μm thick, much shorter than the length of an arm (about 40 cm). When an electric current passes along an arm, it cannot take a straight-line path, which corresponds to about (40 cm)/(100 μm) = 4000 cell walls. The current finds its way to at least three separate physiological systems: the blood system, the lymph system, and the nervous system. The blood system, with highly conductive blood and relatively wide veins and arteries, has the lowest resistance of these, and thus carries most of the current. Nevertheless, the nervous system carries some of this current, one reason humans are susceptible to electric shocks. The medicinal procedure known as *acupuncture*, where needles placed at one part of the body affect another part of the body (and sometimes are connected to electrodes at different potentials), may use the low resistance of the nervous system.

Resistor Color Codes

Resistors are labeled with a 4 or 5 band code of colors. The last color gives the tolerance, with ±5% for gold and ±10% for silver; the next-to-the-last color gives the exponent; the other colors give the prefactor to the exponent. The numerical values for the colors other than gold and silver are given in Table 7.2. Comparison with this table gives that the 4-band code red-violet-orange-silver means $27 \times 10^3 \pm 10\%$ Ω, and the 5-band code yellow-blue-green-brown-gold means $465 \times 10^1 \pm 5\%$ Ω.

Table 7.2 Code for color bands marked on resistors

0	1	2	3	4	5	6	7	8	9
Black	Brown	Red	Orange	Yellow	Green	Blue	Violet	Grey	White

7.3.2 *Joule Heating*

In 1841, James Joule determined experimentally that electrical energy can be converted into heat, at a rate given by I^2R. As noted in Chapter R, this is called *Joule heating*. It is true whether or not R is independent of I and ΔV. We now derive this result.

By the energy considerations of Chapter 5, when a charge dQ is taken across a voltage ΔV, the change in electrical potential energy $dU = dQ(\Delta V)$ must be converted to another form of energy. See Figure 7.3, where the $+$ and $-$ refer to the high and low voltage sides. If the charge traveled in a vacuum, this other form of energy would be the kinetic energy of the charge carriers. However, since the same current enters and leaves the wire, whatever is carrying the charge does not speed up, and thus there is no increase in kinetic energy. Let us assume that, as quickly as electrical work is done on the charge carriers, energy is lost in collisions with the material of the wire, so all the electrical work goes into heat. Then, using $\Delta V = IR$, the rate of production of heat is given by

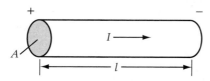

Figure 7.3 Resistor R and charge flow dQ through it in time dt, for calculation of Joule heating rate.

$$P = \frac{dU}{dt} = \frac{dQ}{dt}\Delta V = I(\Delta V) = I(IR) = I^2R. \qquad \text{(rate of Joule heating)}$$

$$(7.7)$$

Example 7.6   **A toaster's resistance**

A toaster is rated at $P = 720$ W for a voltage of 120 V. What is the toaster's resistance?

Solution: By (7.7), 720 W $= I(120$ V$)$, so $I = 6$ A, and $R = \Delta V/I = 20\ \Omega$. Without information about what happens for other currents, we cannot tell if the toaster wires are ohmic or not. They appear to be non-ohmic because on heating their temperature rises significantly, which affects the electrical resistivity ρ. By (7.5), that affects the electrical resistance.

7.4 **Local Form of Ohm's Law**

Energy conservation, applied to a particle moving from point A to point B, is a global statement that does not specify the details of how the particle goes from A to B. Newton's law, however, provides a local statement about the motion of the particle all along its path from A to B. (Thus, Newton's law provides more detailed information than provided by energy conservation.) So far, we have only a global statement of Ohm's law, involving the voltage difference between two terminals connected to a material. We now seek a more local statement of Ohm's law that will apply pointwise throughout the material.

7.4.1 *Ohm's Law Recast*

If Ohm's law holds, then if ΔV doubles, so does I, leaving R unaffected. A doubling of both ΔV and I can be accomplished by doubling both $\vec{E}$ and $\vec{J}$. This leads us to write what we call the local form of Ohm's law:

$$\vec{J} = \sigma \vec{E}. \qquad \text{Ohm's law (local form)} \qquad (7.8)$$

Here the factor σ is called the *electrical conductivity* and is independent of $\vec{E}$ for ohmic materials. (Earlier, σ denoted charge per unit area; context will usually define its meaning.) Equation (7.8) automatically satisfies the first part of Ohm's law, that current flows from high voltage to low voltage. This is because $\vec{J}$ points along the direction of the current, and $\vec{E}$ points from high voltage to low voltage. Let us see how σ of the local form of Ohm's law relates to ρ of the global form of Ohm's law.

Consider a wire (e.g., for a toaster) of conductivity σ, cross-section A, and length l. For example, see Figure 7.3. Let it carry a uniform current density J along its axis. Then, by (7.3), the total current is $I = JA$. Moreover, the electric field is uniform over the length of the wire. Hence, taking ΔV to be positive, and choosing the limits of integration to make $\int \vec{E} \cdot d\vec{s}$ positive, $\Delta V = \int \vec{E} \cdot d\vec{s} = El$ across the wire. Use of $J = \sigma E$, the scalar form of (7.8), then leads to

$$I = JA = (\sigma E)A = \sigma \left(\frac{\Delta V}{l}\right) A = (\Delta V)\frac{\sigma A}{l}. \qquad (7.9)$$

Since (7.4) and (7.5) yield $I = \Delta V(A/\rho l)$, comparison with (7.9) yields

$$\rho = \frac{1}{\sigma}. \qquad \text{(resistivity in terms of conductivity)} \qquad (7.10)$$

Thus conductivity and resistivity are inversely related, one favoring and the other hindering conduction.

Example 7.7   A copper wire

Consider a #10 Cu wire ($A = 5.07 \times 10^{-6}$ m^2) at room temperature, where $\rho = 1.690 \times 10^{-8}$ ohm-m. Let $E = |\vec{E}| = 0.05$ V/m. Find the conductivity of copper, and the current density and the current passing through the wire.

Solution: Equation (7.10) gives $\sigma = 1/\rho = 5.91 \times 10^7$/ohm-m, so (7.8) then gives $J = |\vec{J}| = \sigma|\vec{E}| = 2.96 \times 10^6$ A/m^2. Equation (7.3) then gives $I = JA = 15.0$ A.

7.4.2 *Determining Resistance in General*

If the connections to an object change, the resistance will change. This is clear from the example of the cell membrane, where there is both a parallel and perpendicular resistance. The resistance is different between adjacent fingers on one of your hands, and between the index fingers on both of your hands.

Therefore, consider an ohmic material for a general geometry. Equation (7.4) can still be used to determine R from experiment because we measure both ΔV and I. But how do we determine R from theory? We must use the general forms $\Delta V = \int \vec{E} \cdot d\vec{s}$ (making sure that the limits of integration give ΔV positive) and $I = \int \vec{J} \cdot \hat{n} dA$. Thus (7.4) becomes

$$R = \frac{\Delta V}{I} = \frac{\int \vec{E} \cdot d\vec{s}}{\int \vec{J} \cdot \hat{n} dA}, \qquad \text{(general definition of resistance)} \qquad (7.11)$$

which must be positive. When the local form of Ohm's law, (7.8), also applies, the numerator and denominator both double if the voltage doubles, so the resistance is ohmic. Moreover, if either $\vec{J}$ or $\vec{E}$ can be determined, then by (7.8) the other can be determined. Hence, both the numerator and denominator in (7.11) can be calculated theoretically. For complex geometries, $\vec{E}$ is not easily calculated, so the electrical resistance can only be measured.

Equation (7.11) has the *scaling* property that, if distances are increased a factor of two, then the numerator (proportional to a length) doubles and the denominator (proportional to an area) quadruples, so the resistance halves. Recall that (7.5), for a wire, has this property; if the length scale doubles, then the length l doubles and the area A quadruples. More generally, consider two conducting objects with the same shape, and made of the same material (e.g., solid copper ducks). When measured with respect to corresponding points on those objects (e.g., feet), the ratio of their electrical resistances equals the inverse of the ratio of their characteristic length scales.

Example 7.8 **Comparing the resistance of two similar ducks**

Consider two solid copper ducks of similar geometry. The foot-to-foot resistance of the large one is $R_{large} = 0.04 \ \Omega$. If the large one is 30 cm tall, and the small one is 10 cm tall, determine R_{small}.

Solution: By the scaling property of resistance (inversely with the length), the small one has triple the resistance of the large one, so $R_{small} = 0.12 \ \Omega$.

Example 7.9 **Resistance for radial current flow**

Determine the resistance for a cylindrical geometry with current-conserving radial flow from $r = a$ to $r = b$. The conductivity is σ and the object has length l along its axis. See Figure 7.2(b).

Solution: Equations (7.11) and (7.8) and the example of Section 7.2 give

$$R = \frac{\Delta V}{I} = \frac{1}{I}\int_a^b \vec{E}\cdot d\vec{s} = \frac{1}{I}\int_a^b \frac{\vec{J}}{\sigma}\cdot\hat{r}\,dr = \frac{1}{\sigma I}\int_a^b J_r\,dr$$

$$= \frac{1}{\sigma I}\int_a^b \frac{I}{2\pi rl}\,dr = \frac{1}{2\pi l\sigma}\int_a^b \frac{dr}{r} = \frac{1}{2\pi l\sigma}\ln\frac{b}{a}$$

This satisfies the scaling property that, if all distances double, then the resistance halves.

7.5 Resistors in Series and in Parallel

We now apply Ohm's law to combinations of resistors. The two most common cases are resistors in series and in parallel. The resistors need not be wires; they may be as complex as copper ducks, just as long as they satisfy Ohm's law, (7.4).

7.5.1 *Resistors in Series*

Consider two resistors R_1 and R_2 in series, as in Figure 7.4. This is analogous, in ordinary fluid flow, to two water hoses placed in series. For fixed pressure head at the faucet, the water flow will decrease relative to the case with one hose because the fluid must successively feel a drag force from each hose. Hence the resistance to flow increases. Similarly, the equivalent resistance R of R_1 and R_2 in series should be greater than either R_1 or R_2. Let us see if this is the case.

In order that charge not continually build up anywhere in the circuit, the same current I must flow through each of them. This is a manifestation of the law of charge conservation. Thus

$$I = I_1 = I_2. \tag{7.12}$$

Moreover, because voltage is additive, the sum of the voltages across each resistor is the total voltage across the combination. Thus

$$\Delta V = \Delta V_1 + \Delta V_2. \tag{7.13}$$

(Here, $\Delta V = V_a - V_c$, $\Delta V_1 = V_a - V_b$, and $\Delta V_2 = V_b - V_c$.) The equivalent resistance R is given, from Ohm's law, by $R = \Delta V/I$. As expected, R is greater

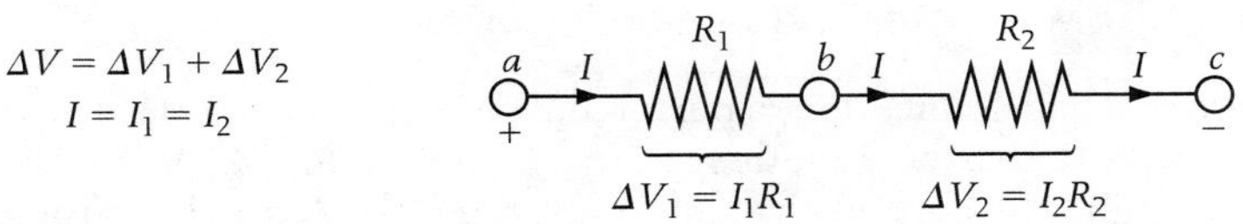

Figure 7.4 Two resistors in series. The same current passes through each and through the equivalent resistance. The voltage across the equivalent resistance is the sum of the voltages across each resistor.

than either R_1 or R_2 because, for fixed current I, the net voltage drop exceeds that across either resistor. We now determine R explicitly.

Applying Ohm's law to the system as a whole, and to each resistor separately, yields

$$I = \frac{\Delta V}{R}, \qquad I_1 = \frac{\Delta V_1}{R_1}, \qquad I_2 = \frac{\Delta V_2}{R_2}. \tag{7.14}$$

Rearranging the equations (7.14), and putting them into (7.13), yields

$$IR = I_1 R_1 + I_2 R_2. \tag{7.15}$$

Dividing (7.15) by I, and using (7.12), the effective resistance R is given by

$$R = R_1 + R_2. \quad \text{(resistors in series)} \tag{7.16}$$

Indeed, R is greater than either R_1 or R_2. Since $\Delta V_1 = I R_1 = \Delta V(R_1/R)$, and $R_1 < R$, by using ΔV as the input and ΔV_1 as the output, this circuit can be used as a *voltage divider*.

Recall the case of fixed current flow along the axis of a wire, with R given by (7.5). It makes sense that the length l appear in the numerator because in this case doubling the length at fixed current also doubles the voltage; it is like adding two wires in series.

Note that, for resistors in series: (1) the current is the same through each resistor and is the same as through the combination; (2) by (7.16) the largest resistance dominates, and the combined resistance is larger than the largest resistance; (3) the voltage across each resistor is proportional to its resistance.

Example 7.10   **Toaster and wire in series**

Consider a wire having $R_1 = 0.1\ \Omega$ and a toaster having $R_2 = 20\ \Omega$. See Figure 7.4. A voltage $\Delta V = 120$ V is available. Compare the current through the toaster alone and when it is placed in series with the wire.

Solution: For the toaster alone, $I = \Delta V / R_2 = 120/20 = 6$ A. For the toaster in series with the wire, (7.16) yields $R = 20.1\ \Omega$, and $I = \Delta V / R = 120/20.1 = 5.97$ A, which is nearly the same as for the toaster alone. For this reason, we usually neglect the electrical resistance of the connecting wires in a circuit. However, if the connecting wire's length were larger by a factor of 100, then its resistance would increase by a factor of 100, to $10\ \Omega$, which is *not* negligible.

Example 7.11   **Toaster and person in series**

Because of a bad connection, a person having $R_1 = 5 \times 10^4\ \Omega$ is in series with the $R_2 = 20\ \Omega$ toaster. Together, they are subject to 120 V. See Figure 7.4. (a) Determine the current through the system. (b) Assess both the effectiveness of the toaster and the likelihood that the person feels a shock.

Solution: (a) Clearly, the person dominates the resistance of this circuit ($R_1 \gg R_2$), so R of (7.16) is very nearly $R_1 = 5 \times 10^4\ \Omega$. Then $I = \Delta V/R \approx 120/(5 \times 10^4) = 2.4 \times 10^{-3}$ A $= 2.4$ mA. (b) The voltage across the toaster is $I R_2 \approx 0.048$ V, so low that it is unlikely to operate effectively. The voltage across the person is approximately $120 - 0.05 = 119.95$ V, certainly large enough to cause a shock. Actually, it is current, rather than voltage, that causes problems for the human body. A current even as small as 1 mA can cause a significant shock and should be avoided.

7.5.2 *Resistors in Parallel*

Consider two resistors R_1 and R_2 in parallel, as in Figure 7.5. We may think of R_2 as a bypass to R_1. An analogous situation in fluid flow comes from medicine, where a clogged coronary artery is bypassed by an artificial artery. From the increased net flow, we expect that the equivalent resistance R of the combination is less than either R_1 or R_2. Let us see if this is the case.

By the path independence of the voltage, R_1 and R_2 are subject to the common voltage difference

$$\Delta V = \Delta V_1 = \Delta V_2. \tag{7.17}$$

Here, $\Delta V = V_a - V_b$.

In order that charge not accumulate anywhere in the circuit, including the vertices a and b, the current I entering and leaving must be the same as the sum of the currents flowing through each. This is a manifestation of the law of conservation of charge. Thus

$$I = I_1 + I_2. \tag{7.18}$$

The equivalent resistance R is given, from Ohm's law, by $R = \Delta V/I$. As expected, this is smaller than either R_1 or R_2 because, for fixed voltage drop ΔV, the currents through each resistor add. We now explicitly determine R.

In (7.18), using (7.4) for each resistor yields

$$\frac{\Delta V}{R} = \frac{\Delta V_1}{R_1} + \frac{\Delta V_2}{R_2}. \tag{7.19}$$

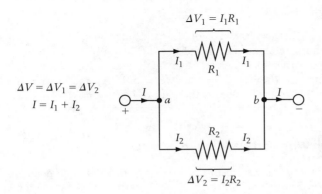

Figure 7.5 Two resistors in parallel. The same voltage difference is across each resistor and across the equivalent resistance. The current through the equivalent resistance is the sum of the currents through each resistor.

Dividing (7.19) by ΔV, and using (7.17), the inverse of the effective resistance is given by

$$\frac{1}{R} = \frac{1}{R_1} + \frac{1}{R_2}. \quad \text{(resistors in parallel)} \quad (7.20)$$

Indeed, R is less than either R_1 or R_2. Since $I_1 = \Delta V/R_1 = I(R/R_1)$, and $R < R_1$, by using I as the input and I_1 as the output, this circuit can be used as a *current divider*.

For some purposes, it is useful to think in terms of what is called the conductance

$$G \equiv \frac{1}{R}. \quad \text{(conductance)} \quad (7.21)$$

Its unit, the *mho*, is the same as an inverse ohm. For resistors in parallel, the conductances add, just as for capacitors in parallel the capacitances add.

Recall Example 7.9, which obtains R for radial flow between two cylinders. It makes sense that the length l appears in the denominator because in this case, doubling the length at fixed voltage difference also doubles the current; it is like adding two wires in parallel.

Note that, for resistors in parallel: (1) the voltage difference is the same across each resistor, and is the same as across the combination; (2) by (7.21), the smallest resistance dominates, and the combined resistance is smaller than the smallest resistance; (3) the current through each resistor is inversely proportional to its resistance.

Example 7.12　Wire and toaster in parallel

Let the 0.1 Ω wire (R_1) and the 20 Ω toaster (R_2) of Example 7.10 be placed in parallel, and let the combination be connected to a source of fixed current $I = 6$ A. See Figure 7.5. Find the voltage across the combination, and the current through each.

Solution: By (7.20), the resistance of the combination is 0.0995 Ω, which is nearly that of the wire alone. Further, $\Delta V = IR = 0.597$ V, much less than the 120 V needed to drive such a current through the toaster alone. Most of the current flows through the wire ($I_1 = \Delta V/R_1 = 5.97$ A), rather than the toaster ($I_2 = \Delta V/R_2 = 0.03$ A). The wire is the path of least resistance.

Example 7.13　Wire and cladding in parallel

If a bare wire has a resistance of $R_1 = 0.1$ Ω, and it is given an insulating cladding of $R_2 = 5 \times 10^8$ Ω, which resistance dominates? See Figure 7.5.

Solution: By (7.20), since $R_2 \gg R_1$, $R \approx R_1$. Hence the resistance of the combination is totally dominated by the wire itself. Clearly it is a good approximation to neglect the effect of the cladding. Again the wire is the path of least resistance.

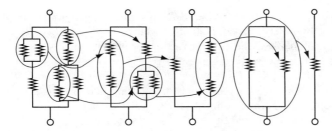

Figure 7.6 Combinations of series and parallel circuits. Parallel and series resistors are replaced by equivalent resistances until there is only a single equivalent resistance. Not all circuits can be analyzed in this way.

7.5.3 *More Complex Circuits*

Many, *but not all*, complex-looking circuits made up only of resistors can be analyzed in terms of resistors in series and in parallel. From Chapter 6, this also is true for circuits made up only of capacitors. Figure 7.6 presents a resistor circuit that can be so analyzed. The voltage across a resistor R is proportional to R ($\Delta V = IR$), whereas the voltage across a capacitor C is inversely proportional to C ($\Delta V = Q/C$). Therefore it should not be surprising that the rules for series and parallel addition are interchanged on going from resistors to capacitors.

Even if a circuit can be analyzed using the method illustrated in Figure 7.6, the equivalent resistance alone does not reveal what is happening in the individual parts of the circuit.

Example 7.14 **A set of three resistors**

Consider a circuit with a resistor $R_1 = 8\ \Omega$ in one arm, and $R_2 = 4\ \Omega$ and $R_3 = 2\ \Omega$ in the other arm. See Figure 7.7(a). The connection b between R_1 and R_2 is at -6 V, and the connection c between R_1 and R_3 is at 18 V. Find the currents through each resistor, and the voltage at connection a between R_2 and R_3.

Solution: Even before finding the equivalent resistance of this circuit, we can answer certain questions. First, the voltage across R_1 is $\Delta V_1 = V_c - V_b = 18 - (-6) = 24$ V, so by Ohm's law the current through R_1 is $I_1 = \Delta V_1/R_1 = 24/8 = 3$ A. Since voltage is path independent, $\Delta V_1 = 24$ V is the same as the total voltage

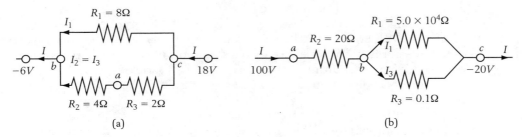

Figure 7.7 Two resistor circuits: (a) two parallel arms, one arm being complex, (b) two series arms, one arm being complex.

across the lower arm, containing R_2 and R_3 in series. The combination of R_2 and R_3 in series is, by (7.16), an equivalent resistance of $R_{23} = R_2 + R_3 = 4 + 2 = 6\ \Omega$. Thus, by Ohm's law, the current through the lower arm is $I_{23} = I_2 = I_3 = 24/6 = 4$ A. Hence the voltage across R_2 is $\Delta V_2 = I_2 R_2 = 4 \cdot 4 = 16$ V. Since point b connecting R_2 and R_1 is at -6 V, then point a connecting R_2 and R_3 is at $V_a = -6 + 16 = 10$ V. Alternatively, the voltage across R_3 is $\Delta V_3 = I_3 R_3 = 2 \cdot 4 = 8$ V, and since point c connecting R_3 and R_1 is at 18 V, then point a is at $V_a = 18 - 8 = 10$ V. Note that the equivalent resistance of R_1 and R_{23} in parallel is, by (7.20), $R_{eq} = (R_1^{-1} + R_{23}^{-1})^{-1} = (24/7)\ \Omega$, so the total current I passing through R_{eq} is given by $I = 24/(24/7) = 7$ A. As should be the case, this equals the sum of the currents through each arm: $I_1 + I_{23} = 3 + 4 = 7$ A.

Example 7.15 Another set of three resistors

To Example 7.11, add a wire of $R_3 = 0.1\ \Omega$ in parallel to the person of $R_1 = 5 \times 10^4\ \Omega$, so together they are in series with the toaster of $R_2 = 20\ \Omega$. A voltage $\Delta V = 120$ V is placed across the combination. See Figure 7.7(b). Find the current through the system. Assess both how effective the toaster would be and the likelihood of the person feeling a shock.

Solution: Here, the wire (R_3) has such a low resistance that it serves as a path of least resistance, "shorting" out the person (R_1). By (7.20), the resistance of this combination in parallel is very nearly the same as the wire itself, so $R_{13} \approx R_3 = 0.1\ \Omega$. By (7.16), the combination thus has a resistance $R = R_2 + R_{13} \approx 20.1\ \Omega$, so $I \approx 120/20.1 = 5.97$ A. This current goes through the toaster, which should operate normally. The voltage across the wire and the person is $\Delta V_{13} = I R_{13} \approx I R_3 = 0.6$ V. This is so low that the person should not feel a shock. The current through the person is $I_2 = 0.6/5 \times 10^4 = 1.2 \times 10^{-5}$ A. The rest of the 5.97 A flows through the wire. This is the basis of how ground wires protect users of electrical devices, even when there are "shorts."

Chapter 8 discusses how to treat circuits that are too complex to be analyzed as combinations of resistors in series and in parallel.

7.6 Meters: Their Use and Design

Optional

Today, currents and voltages from long-lived emfs, as well as capacitance and resistance and other electrical quantities, are often measured with a single device, called a *multimeter*. Changing its settings changes the internal circuitry. Digital multimeters (with liquid crystal displays) are a relatively recent advance involving complex circuitry. They are "black boxes" to most of their users. Whereas the resistance of an analog multimeter (with a needle dial) varies with the dial setting, the resistance of a digital multimeter in voltmeter mode has a fixed resistance (usually about 10 MΩ).

Analog multimeters use the fact that a current can deflect a magnetic needle. They are still commonly available, and they use circuits that the student can easily understand. For these reasons, we describe them in some detail. Their primary measuring circuit is a galvanometer placed in circuits that can be adapted to measure currents and voltages of different magnitudes. The best galvanometers can measure currents as small as a pA (10^{-12} A). The inverse of the full-scale current is called the *current sensitivity* and is given in units of Ω/V.

Example 7.16 **A galvanometer**

Consider a galvanometer with internal resistance $R_g = 25 \ \Omega$ that produces a full-scale deflection of its indicating needle for a current $I_0 = 2$ mA. (a) Find the full-scale voltage difference ΔV_0. (b) Find the sensitivity of this galvanometer.

Solution: By (7.4), $\Delta V_0 = I_0 R = 50$ mV. (b) The sensitivity here is $I_0^{-1} = 500 \ \Omega/V$.

The galvanometer of Example 7.16 will be used in the remainder of this section. Note that two of the three quantities R, I_0, and ΔV_0 fully specify the properties of a galvanometer.

7.6.1 *Current Measurement: The Ammeter*

A device that measures current is called an *ammeter*. To directly measure the current passing through a circuit, all the current must go through the meter. That is, the meter must be put in series with the rest of the circuit. See Figure 7.8(a). Thus we must break the circuit at some point, and take one end of the circuit to one lead of the meter, and the other end to the other end of the meter. To disturb the system's overall resistance as little as possible (here, by minimizing the extra voltage drop due to the meter), the overall ammeter resistance R_A must be very small compared to the resistance R of the circuit whose current is to be measured:

$$R_A \ll R. \quad \text{(current measurement)} \quad (7.22)$$

Internally, an ammeter has a set of parallel (or *shunt*) resistors that change from one ammeter setting to another. See Figure 7.8(b). Thus, with R_{sh} the shunt resistance, (7.20) yields $R_A = (R_g^{-1} + R_{sh}^{-1})^{-1}$. By suitable choice of R_{sh}, the ammeter can be made to read any current that exceeds the full-scale deflection current of the galvanometer. This is an example of a current divider,

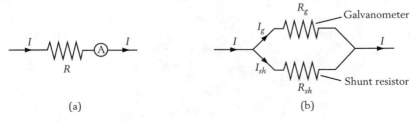

(a)   (b)

Figure 7.8 Measuring current. (a) Circuit with an ammeter, which to measure current must be placed in series with the rest of the circuit. (b) Schematic of an ammeter circuit, with a galvanometer resistance (galvanometer not shown) and a shunt resistance (to permit larger currents to be measured without "blowing" the galvanometer).

mentioned after (7.20). To read currents that are much less than full scale on the galvanometer requires using a more sensitive galvanometer.

Example 7.17 Shunt resistance for ammeter

Find the shunt resistance needed to convert the galvanometer of Example 7.16 into an ammeter with a full-scale current of $I = 100$ mA.

Solution: This requires that the galvanometer current of $I_g = 2$ mA correspond to a total current of 100 mA. Therefore the shunt resistor must carry $I_{sh} = I - I_g = 98$ mA. See Figure 7.8(b). With the equivalent resistance R_A of the ammeter consisting of the galvanometer and the shunt in parallel, we have 50 mV$= I_g R_g = I_{sh} R_{sh} = I R_A$. Hence $R_{sh} = 50/98 = 0.5102$ Ω and $R_A = 50/100 = 0.5$ Ω.

Example 7.18 Accuracy of an ammeter

Let us use the ammeter of Example 7.17 to measure the current through a resistance $R = 20$ Ω. (a) Find the current. (b) Find the percentage of error, if 100 mV is placed across R. (c) Is this a readable current? A safe current?

Solution: (a) The series resistance of R and R_A is 20.5 Ω. (b) If 100 mV is placed across R alone (as in an ideal ammeter, where $R_A^{ideal} \to 0$), a current $I = 100/20 = 5$ mA will flow through it. If 100 mV is placed across R and the actual meter in series, the meter will read $I = 100/20.5 = 4.88$ mA, 2.5% less than for an ideal ammeter. (c) Since $I R_A = (4.88$ mA$) \times (0.5$ $\Omega) = 2.44$ mV, this is readable but well below the 50 mV that produces full deflection. (Since $I R_A = I_g R_g$, we have $I_g = 0.0976$, well below the full-scale deflection of $I_0 = 2$ mA.) If 100 V is placed across the combination, the 2.4 V across the meter might cause the meter to burn up, or the deflecting needle to bend, or a protective fuse to burn up ("blow").

7.6.2 *Voltage Measurement: The Voltmeter*

A device that measures electrical potential differences, or voltages, is called a *voltmeter*. To directly measure the voltage difference between two points of a circuit, the leads of the meter must contact these two points, so the meter is in parallel with the part of the circuit between these two points. See Figure 7.9(a). The voltmeter measures the voltage along the path from one lead, through the voltmeter, and to the other lead. By the path independence of the voltage, this equals the voltage along a path from one lead, through the circuit, and to the other lead. To disturb the system's overall resistance as little as possible (here, by minimizing the extra current flowing to the meter), the overall voltmeter resistance R_V must be very large compared to the resistance R of the part of the circuit to be measured:

$$R_V \gg R. \quad \text{(voltage measurement)} \quad (7.23)$$

Internally, an analog voltmeter has a set of series resistances R_{ser} that change from one voltmeter setting to another. See Figure 7.9(b). Thus, if R_{ser} is the series

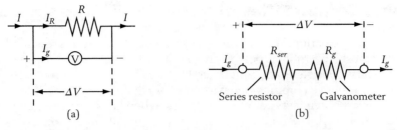

Figure 7.9 Measuring voltage difference. (a) Circuit with a voltmeter, which to measure voltage difference must be placed in parallel with the rest of the circuit. (b) Schematic of a voltmeter circuit, with a galvanometer resistance (galvanometer not shown) and a series resistance (to permit larger voltages to be measured without "blowing" the galvanometer).

resistance, (7.16) yields $R_V = R_g + R_{ser}$. By suitable choice of R_{ser}, the voltmeter can be made to read any voltage that exceeds the full-scale deflection voltage of the galvanometer. This is an example of a voltage divider, mentioned after (7.16). To read voltages that are much less than full scale on the galvanometer requires using a more sensitive galvanometer.

Example 7.19 Series resistance for voltmeter

Find the series resistance R_{ser} needed to convert the galvanometer of Example 7.16 into a voltmeter with a full-scale deflection of $\Delta V = 500$ mV.

Solution: This requires that a current of $I_g = 2$ mA passing through R_g must produce a total voltage drop of 500 mV across R and R_{ser} in series. See Figure 7.9(b). Then the voltmeter resistance is $R_V = \Delta V / I_g = 500/2 = 250$ Ω. (This equals the product of ΔV and the sensitivity.) Hence $R_{ser} = R_V - R_g = 250 - 25 = 225$ Ω.

Example 7.20 Accuracy of a voltmeter

In Example 7.19, let the voltmeter be used to measure the voltage across a resistance $R = 20$ Ω. (a) How accurate will the reading be if the power source provides a fixed 100 mV across R, even when the voltmeter is in place? (b) If the power source provides a fixed 5 mA current?

Solution: (a) For a fixed 100 mV, both R and the voltmeter are subject to 100 mV, so the voltage reading will be completely accurate. (b) For a 5 mA constant current source, the parallel resistance R_{par} of R and $R_V = 250$ Ω is 18.52 Ω. Then $I R_{par} = 5 \times 18.52 = 92.6$ mV, a 7.5% decrease from the value $I R = 100$ mV without the voltmeter in the circuit.

Take A Break!

You've just completed the first part of this chapter. Next we'll discuss sources of emf.

7.7 Some Complexities of Voltaic Cells: The Car Battery

As noted in Section 7.1, current-causing energy sources are called emfs. One of the most common types of emf is chemical in nature: the voltaic cell. A car battery consists of six voltaic cells in series. As we know from experience, car batteries have complex behavior. It is worth discussing and interpreting some commonly known facts about them so that when later we present a simple model for a voltaic cell, the reader will have a sense for its limitations.

Such a discussion must include some chemistry. Each voltaic cell contains two electrodes (for the lead-acid cell used in car batteries, Pb and PbO_2) separated by electrolyte (for the lead-acid cell at full charge, the electrolyte is a 35% solution by weight of sulfuric acid, H_2SO_4, in water). Chemical reactions at each electrode provide the emf for the cell. These reactions involve ions from the electrolyte and from the electrodes. Ions cross the voltaic cell to transfer charge to the electrodes, where electrons carry the charge. *If we must make an analogy to a water pump, then the electrical pump of a voltaic cell should be thought of as localized at its two electrode-electrolyte surfaces.*

7.7.1 *Starting a Car: Fast Discharge*

When you use a car battery to start the electrical motor of an automobile (which then starts the gasoline engine), it provides about 600 A. If the points on the spark plugs are wet, the car may not start, even if you continue to turn the starter key. After about two minutes, the battery appears to go dead. However, if you wait a few more minutes, the battery (if it was well charged initially) will recover. If during the wait, you wipe the moisture off the spark plugs, the car then will start.

What is happening here? Initially, the ion density in each voltaic cell of the car battery was nearly uniform. After about two minutes of this fast discharge, the ions near the electrodes have been used up by chemical reactions at the electrodes. (See Figure 7.10(a), which depicts the ion density in a single voltaic cell. The Pb electrode is negative and the PbO_2 electrode is positive.) At this point, there are no ions adjacent to the electrodes, so there is no chemical reaction, and negligible emf. However, after waiting a few minutes, ions from the bulk diffuse to the electrodes, giving a new density that is not quite as large as initially, but certainly large enough to start the car. Figure 7.10(b) shows the profile after a long wait.

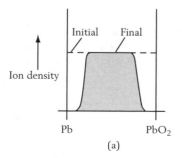
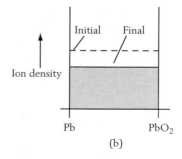

Figure 7.10 Ion density profile for a car battery initially at full charge, subject to a partial discharge: (a) fast discharge, (b) fast discharge followed by a long wait.

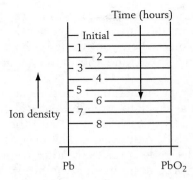

Figure 7.11 Ion density profile for a car battery, for a slow and complete discharge.

7.7.2   *Leaving the Car Lights on Overnight: Slow Discharge*

Now consider what happens if you've left the car lights on overnight. As will be shown in Chapter 8, the car lights draw about 6 A. This rate of discharge is sufficiently slow that ions can be provided from the bulk of each voltaic cell at a rate fast enough to keep up with the chemical reactions at the electrode, so the ionic density profile falls nearly uniformly. See Figure 7.11. Hence, after slow discharge, when the battery appears to be dead, it really is dead.

7.7.3   *Fast Recharge*

Now assume your car battery has gone dead, and you now start up your car with jumper cables and a good battery. (The circuit for this is discussed in Chapter 8.) You drive the car for a few minutes and then park it, but just to make sure that the battery still has some "juice," you try to start the car again. It works. You go into the house. Hours later you return to your car, and it won't start. What went wrong?

During that five-minute drive you charged up the battery by increasing the density of ions near the electrodes, but not in the bulk of the electrolyte. See Figure 7.12(a). When you started the car to test it for "juice," you depleted many of these ions near the electrode. And the hours-long wait gave the ions

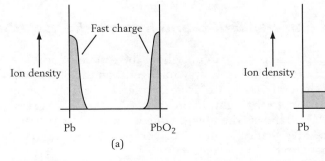

Figure 7.12 Ion density profile for a car battery initially discharged: (a) after a fast charge, (b) after a fast charge and a waiting period so that the ion density has "relaxed" to a more uniform distribution.

that remained near the electrode enough time to spread out to the bulk of the electrolyte, giving a rather uniform and rather low ion density, even near the electrodes. See Figure 7.12(b). Thus, when you needed a high density of ions near the electrodes, they weren't there, and the battery wouldn't start the car.

In practice, there are always plenty of ions available in the electrolyte. The true limitation on the ability of a lead-acid cell to provide electrical power is the number of Pb and PbO_2 ions on the electrodes, in contact with the electrolyte. Chemical reactions convert them to $PbSO_4$ on the electrode surfaces, so new Pb and PbO_2 ions must diffuse to the electrode surfaces in order to provide new reactive material. This relatively slow process limits the amount of power a car battery can provide.

Battery rechargers provide perhaps 2 to 6, or even 12 amps; this is not enough to start a car. You have to use the recharger to build up a "charge" on the battery itself before you can start the car. Note that a good car battery loses nearly 1% per day of its "charge" to non–current-producing chemical reactions at the electrodes; this is why batteries eventually go dead even when they are disconnected.

Recently manufactured, and more compact, car batteries have a greater rate of self-discharge than less compact batteries manufactured previously. If not charged up before a car owner leaves for a few weeks, such batteries are often found to be dead when the car owner returns. The same thing occurs if they are not used regularly.

7.8 Emf and Ohm's Law

Current-producing chemical reactions drive electric charge across the electrolyte of a voltaic cell, putting an excess of charge on one electrode and a deficit of charge on the other electrode. This buildup of charge on the electrodes tends to oppose the current flow with a "back voltage" ΔV. (ΔV is also called the *terminal voltage*.) The value of ΔV depends on the "load" attached to the electrodes of the cell. On an open circuit, $I = 0$; the value of ΔV for which $I = 0$ is defined as the emf $\mathcal{E}$ of the cell. That is, $\mathcal{E} \equiv \Delta V_{I=0}$. The chemistry provides $\mathcal{E}$, and the physics (electric charge driven by chemical reactions to the cell's electrodes) provides the ΔV in opposition to $\mathcal{E}$.

When $\Delta V \neq \mathcal{E}$, a cell can also drive an electric current, and the cell must be described by more than its emf $\mathcal{E}$.

7.8.1 *A Source of Emf Has Emf $\mathcal{E}$ and Internal Resistance r*

Figure 7.13(a) shows a specific circuit where the source of emf is a voltaic cell. One electrode of the voltaic cell is drawn longer than the other, to indicate that, within the cell, the emf $\mathcal{E}$ tends to drive current $I > 0$ from the smaller to the larger electrode; here, to the right. In order that $\Delta V|_{I=0} = \mathcal{E}$ correspond to zero current flow, $\Delta V > 0$ means the higher voltage must be on the right, to oppose the current flow. This sign convention for voltage is opposite that for a resistor, where the voltage drives the current.

Let us apply some theoretical reasoning to the energetics of the source of emf. Consider small deviations of ΔV from $\mathcal{E}$, for which there is a small nonzero

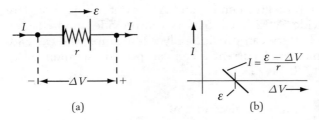

Figure 7.13 Determining the emf of a battery or other source of emf. (a) The large electrode is taken to be positive, and the small electrode is taken to be negative. The electrolyte, which contacts both electrodes, is represented by the internal resistance r. Both the current and the voltage difference are measured. (b) Current versus voltage difference for a source of emf. The current goes to zero when ΔV equals the emf $\mathcal{E}$, and the internal resistance r is determined from the slope.

current I. In Figure 7.13(a), during a time dt, charge $dQ = I\,dt$ enters the source of emf at the left, and (as long as charge doesn't build up within the source of emf) an equal charge exits at the right.

Here is what the source of emf does: (1) it raises the voltage across the cell by ΔV so that the electrical potential energy of the charge dQ increases by $(\Delta V)dQ$; (2) because it has its own electrical resistance r, called *internal resistance*, it also provides the Joule heating energy $I^2 r\,dt = Ir\,dQ$. These two types of energy must come at the expense of the source of emf and must be proportional to dQ. Calling the proportionality constant D, the source of emf provides energy $D\,dQ$. This goes into energies $(\Delta V)dQ$ and $Ir\,dQ$. Hence

$$D\,dQ = (\Delta V)dQ + Ir\,dQ, \quad \text{so } D = \Delta V + Ir.$$

Since D must be consistent with $\Delta V = \mathcal{E}$ for $I \to 0$, we deduce that $D = \mathcal{E}$. Thus we expect that

$$\mathcal{E} = \Delta V + Ir. \tag{7.24}$$

Now compare with experiment. The typical experimental form of the I versus ΔV curve is given in Figure 7.13(b). Consistent with (7.24), for ΔV near $\mathcal{E}$, Figure 7.13(b) satisfies

$$I = \frac{\mathcal{E} - \Delta V}{r}. \qquad \text{(current–voltage relationship for voltaic cell)} \tag{7.25}$$

We will assume that (7.25) applies for *all* ΔV, not merely for ΔV near $\mathcal{E}$. (For complex sources of emf, r can vary with I and ΔV.) Equation (7.25) is equivalent to (7.24).

We also will assume (7.25) to hold for sources of emf other than voltaic cells—such as batteries of voltaic cells, thermoelectric devices or photovoltaic cells. From (7.25), for a given emf $\mathcal{E}$ and ΔV, r determines how large a current the emf will provide. A 12 V car battery can provide hundreds of amps, but a 12 V electronics battery can provide only a few amps; the car battery has the smaller r. The maximum current a cell can provide spontaneously occurs for $\Delta V = 0$, and by (7.25) is $\mathcal{E}/R$.

Example 7.21 **Characterizing an emf**

Let $I = 0$ for $\Delta V = 1.45$ V, and let $I = .25$ A for $\Delta V = 1.35$ V. Find $\mathcal{E}$ and r.

Solution: By (7.25), since $I = 0$ for $\Delta V = 1.45$ V, we have $\mathcal{E} = 1.45$ V. Next, since $I = .25$ A for $\Delta V = 1.35$ V, (7.25) yields $r = 0.4\ \Omega$.

7.8.2 *Potentiometers Measure Emf*
Optional

Just as R characterizes a resistor, so $\mathcal{E}$ and r characterize a source of emf. From (7.25), measurement of I versus ΔV yields $\mathcal{E}$. This leads to the principle of the emf-measuring device called the *potentiometer*.

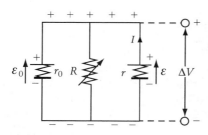

Figure 7.14 Potentiometer circuit. The known, stable emf $\mathcal{E}_0$, and the unknown emf $\mathcal{E}$ are opposed to each other. The variable resistance R is adjusted until the current I through the unknown emf is zero. The voltage difference across the unknown emf then equals $\mathcal{E}$.

Consider the right arm of the circuit in Figure 7.14. It contains a voltaic cell of unknown emf $\mathcal{E}$ and internal resistance r, and there is a voltage difference ΔV in opposition to $\mathcal{E}$. Usually the arrow associated with the direction of the chemical emf is not specified because that information is contained in the relative size of the two electrodes. The upper arms of the circuit have positive charge, and the lower arms have negative charge, since both voltaic cells tend to drive current from bottom to top. The upper arm acts as a reservoir at a positive potential, and the lower arm acts as a reservoir at a negative potential. We do not address the difficult question of how to obtain the charge distribution around the circuit that produces the electric field and the electrical potential.

By (7.25), when the current I is zero, the total emf acting on the right arm in Figure 7.14 is zero, and $\Delta V = \mathcal{E}$. In practice, for potentiometers a very stable chemical emf $\mathcal{E}_0$, whose specific value isn't important, drives current through a variable resistance R (represented in Figure 7.14 as a dial—the arrow), and R is adjusted until $I = 0$. The measured value of ΔV then corresponds to $\mathcal{E}$.

Sometimes a variable resistor, as in Figure 7.14, is referred to as a potentiometer, or "pot," for short. It is specified by its maximum resistance.

7.8.3 *On How Emf's Drive Current*
Optional

Consider a situation, as might occur for a wire of copper, where there is only one type of charge carrier (electrons), but the average force $\vec{F}$ acting on the electrons is partly electrical ($q\vec{E}$) and partly nonelectrical ($\vec{F}'$). For example, the wire might not be at a uniform temperature, so there might be a tendency for the electrons to drift to higher (or lower) temperature. Write the total average force per unit charge as $\vec{F}/q \equiv (\vec{E} + \vec{E}')$, where $\vec{E}' \equiv \vec{F}'/q$ is due to the nonelectrical force. Now (7.8) must be generalized to

$$\vec{J} = \sigma(\vec{E} + \vec{E}'). \qquad \text{Ohm's law (local)} \qquad (7.26)$$

When the effect of all sources of emf are included, Ohm's law becomes

$$I = \frac{\sum_i \mathcal{E}_i}{R}, \qquad \text{Ohm's law (global)} \qquad (7.27)$$

where the sum is over each type of emf $\mathcal{E}_i$ along the path of the current, and the resistance R might be due to many resistances. See Figure 7.15. When properly interpreted, (7.27) includes both (7.4) for a wire and (7.25) for an emf. When applied to a circuit as a whole, (7.27) contains no contribution from the electrostatic force (due to $\vec{E}$) since $\Delta V = 0$ for a circuit as a whole. Hence, for a circuit as a whole (but *not* for *part* of a circuit), voltage change does not contribute to the net emf. Only the nonelectrostatic forces drive current around a circuit.

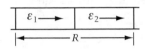

Figure 7.15 Two (localized) sources of emf, within a region of total resistance R.

7.8.4 *Some Nonelectrostatic Types of Emf*

Besides nonelectrostatic emf from voltaic cells, the many possible types of emf include

1. Electromagnetically induced electric fields (Faraday's law of electromagnetic induction), to be studied in Chapter 12. This is a *true* electric field because $\vec{F}/q$ is independent of the type of charge carrier. This can drive current around a circuit, which is why the power company makes electrical generators based on it.

2. Chemical-specific diffusion force due to chemical density gradients within an ionic solution (this force is purely statistical in nature; the diffusion force acting on air molecules defeats gravity's attempts to pull the atmosphere to the surface of the earth). Because it depends on each ion seeing its own density gradient, this is charge carrier specific and is not a true electric field. Diffusion causes the density to even out in Figures 7.10(a) and (b) and 7.12(a) and (b).

3. Thermoelectric force due to thermal gradients, usually within a wire. Because this is charge carrier specific, it is not a true electric field. In our previous considerations, this effect was neglected because it was assumed that the wire temperature was uniform.

Thermoelectric Effect Consider a circuit of two metals, with the two junctions at different temperatures. See Figure 7.16(a). If one metal is split in two pieces (so there is no current flow), there will be a voltage across the split. Switching

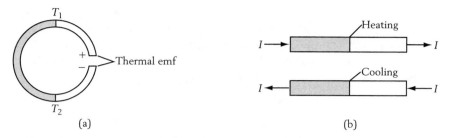

Figure 7.16 Thermoelectric effect, where a nonuniform temperature across a material produces a voltage difference between its ends. (a) A ring geometry, with two materials, one of them split. Their two contacts are maintained at different temperatures. If the split is closed, a steady-state current will flow, either clockwise or counterclockwise (according to the materials and the temperatures at which they are maintained). (b) The Peltier effect, where current flow causes a temperature gradient to develop at the junction between two materials. According to the direction of current flow, either heating or cooling can occur.

the junction temperatures switches the voltage. If there is no split, then a current flows. The thermoelectric effect, although small, is stable. Because of that stability, Ohm used it as a source of emf, rather than the chemical emf from a battery. In 1834, Peltier discovered the inverse effect, wherein a current flowing through a circuit of two metals will heat or cool the junction, according to its flow direction. See Figure 7.16(b). The Peltier effect is used to cool small computers, and refrigerator-coolers driven by the 12 V output of automobile batteries are now commercially available.

Example 7.22 **Gravity as an emf**

For a particle of charge q and mass M, the effective field is $\vec{E}' = m\vec{g}/q$. In the atmosphere, consider both a small ion of mass m and a large water droplet of mass M, both with the charge $q = 2e$. What is the relative magnitude of the effective fields acting on the ion and the droplet?

Solution: They are in proportion to the mass, so the effective field on the ion is much smaller than for the droplet.

Example 7.23 **Muons, electrons, and emfs**

The *muon*, an unstable particle with the same charge but approximately 207 times the mass of the electron, does not have the same chemical binding as the electron. Do the muon and the electron have the same emf?

Solution: Because the release of chemical energy at the electrodes of a voltaic cell would be different for muons than for electrons, despite their identical charges the muon and electron emfs would differ. This example shows that the emf of a voltaic cell cannot be due to a true electric field.

7.9 Energy Storage by Voltaic Cells

We now discuss voltaic cells in more detail.

7.9.1 *Electrolytes and Electrodes Make Up Voltaic Cells*

Circuits with wires and *electrolytes* have two wire–electrolyte interfaces, the *electrodes*, where chemical reactions transfer electric charge from electrons to ions,

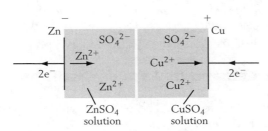

Figure 7.17 A simple voltaic cell, with Zn and Cu electrodes. The electrolyte is actually two separated solutions of $ZnSO_4$ and $CuSO_4$. In spontaneous operation, Zn atoms from the Zn electrode tend to go into solution as Zn^{2+} ions, driving other Zn^{2+} ions into the $CuSO_4$ region. This drives Cu^{2+} ions onto the Cu electrode. For these processes to occur, the Zn electrode must gain electrons and the Cu electrode must lose electrons. Hence the Zn electrode is more electronegative than the Cu electrode.

or vice versa. See Figure 7.17, where the electrodes are Zn and Cu. The Zn should not be directly in contact with the $CuSO_4$ solution, or else the Zn surface will become covered with a blackish substance, so it is impractical to have a single electrolyte solution in this case. One way to separate the Zn from the $CuSO_4$ solution (P. Heller, private communication) is to use two blotters, one soaked in $ZnSO_4$ and placed directly against the Zn, and the other soaked in $CuSO_4$ solution and placed directly against the Cu. (Thus, near the Cu electrode the electrolyte is $CuSO_4$ solution, and near the Zn electrode the electrolyte is $ZnSO_4$ solution.) The $ZnSO_4$ blotter should be relatively thick so that the Cu^{2+} ions cannot easily diffuse across it. Zn^{2+} ions from Zn go into solution preferentially over Cu^{2+} ions from Cu. Thus, at the Zn electrode, Zn^{2+} ions go into solution, and at the Cu electrode, the $CuSO_4$ provides a source of Cu^{2+} ions that come out of solution and plate onto the Cu electrode. (When $CuSO_4$ solution is in contact with a Zn electrode, Zn goes into solution as Zn^+ ions and Cu^+ ions plate onto the Zn electrode as Cu. Dissolved oxygen then reacts with the Cu to yield Cu_2O, presumably the blackish substance.)

The electrode–electrolyte–electrode combination is called a *voltaic cell*. It produces what is called a *chemical emf*. (If the cell provides energy, or "discharges," as in Figure 7.17, it is called a *galvanic cell*; if it absorbs energy, or "charges," it is called an *electrolytic cell*. We will use *voltaic cell* for both.) Batteries use voltaic cells, both in series (for higher voltage) and in parallel (for higher current). Voltaic cells have the following fundamental properties:

1. They store a finite amount of energy E_{cell}, proportional to the number of ions that can react at the electrodes, often called their "charge" Q_{cell}. Q_{cell} is limited by the least abundant active component. Of two cells made of the same materials (e.g., a AAA cell and a D cell), the larger D cell has the larger charge.

2. Their emf, which drives electric current, is due to chemical reactions at the two electrode–electrolyte interfaces. These interfaces serve as "surface pumps," and constitute the "seats of emf" of the voltaic cell. AAA and D cells, having the same chemistry, have the same emf.

3. Their electrical resistance is dominantly determined by the material and geometry of the electrolyte. The D cell, with a much larger electrode area than the AAA cell, provides a larger current than the AAA cell, which means it has less internal resistance.

For a voltaic cell nearly all the voltage change occurs at the surface pumps, located at each of the two electrode–electrolyte interfaces. Only for large electric currents is there a significant voltage drop across the electrolyte separating the electrodes.

7.9.2 *Energy of a Voltaic Cell and Its Rate of Discharge*

As indicated, E_{cell} is proportional to Q_{cell}. The proportionality constant, as will be shown in Section 7.9.3, is the emf $\mathcal{E}$. Hence we write

$$E_{cell} = \mathcal{E} Q_{cell}. \quad \text{(energy stored by voltaic cell)} \tag{7.28}$$

$\mathcal{E}$ depends on the chemical reactions at the two electrode–electrolyte interfaces. When the electrodes are the same, they pump with equal strengths in opposite directions, so there is no net emf. On the schematic of the voltaic cell of Figure 7.13(a), $\mathcal{E}_2$ from electrode 2 tends to push current out of the larger plate into the wire, and $\mathcal{E}_1$ from electrode 1 tends to push current out the smaller plate into the wire. Since $\mathcal{E}_2$ and $\mathcal{E}_1$ oppose, the net emf is

$$\mathcal{E} = \mathcal{E}_2 - \mathcal{E}_1. \quad \text{(net emf of voltaic cell)} \tag{7.29}$$

It immediately follows from (7.28) that the rate of change of the energy of a cell is given by

$$\frac{d E_{cell}}{dt} = \mathcal{E}\frac{d Q_{cell}}{dt}. \tag{7.30}$$

When the cell discharges, so $d Q_{cell}/dt < 0$, it causes a current I to flow in the external circuit. Taking $I > 0$ when the cell is discharging, we have $I = -d Q_{cell}/dt$. From (7.30) and the principle of energy conservation, the power $\mathcal{P}$ provided to the rest of the circuit is given by

$$\mathcal{P} = \mathcal{E} I. \quad \text{(rate of discharge of voltaic cell)} \tag{7.31}$$

This can go into charging up a capacitor, or heating up a wire, or both.

Example 7.24 Properties of an AA cell

Consider a AA alkaline voltaic cell with $\mathcal{E} = 1.2$ V, charge $Q_{cell} = 2.45$ A-hr $= 8820$ C, and negligible internal resistance relative to a resistor $R = 4$ Ω. Take it to have volume $V_{cell} = .0432$ cm^3. Find (a) the cell's energy,

(b) its energy density u_{cell}, (c) its rate of discharge through R, and (d) how long it will discharge, assuming this constant rate, before it is "dead."

Solution: (a) By (7.28), $E_{cell} = 10,584$ J. (b) $u_{cell} = E_{cell}/V_{cell} = 2.45 \times 10^8$ J/m³. (c) Applying Ohm's law to the circuit as a whole, where the net emf is due only to the voltaic cell, the current is $I = \mathcal{E}/R = 1.2/4 = 0.3$ A, so (7.31) gives $\mathcal{P} = 0.36$ W. (d) The time T to totally discharge is given by $T = E_{cell}/\mathcal{P} = 29,400$ s, or 8 hours and 10 minutes.

7.9.3 *Charging a Capacitor with a Voltaic Cell*

Now consider how a voltaic cell charges a capacitor. See Figure 7.18. By (6.20), the capacitor energy is given by

$$E_{cap} = \frac{1}{2}\frac{Q^2}{C}. \tag{7.32}$$

When the cell and the capacitor are connected, the cell will discharge, and the capacitor will charge. If the cell has an initial charge $Q_{cell}^{(0)}$, and the capacitor $Q^{(0)} = 0$, then by charge conservation $Q_{cell} = Q_{cell}^{(0)} - Q$ after the capacitor had received a charge Q. In equilibrium, as a function of Q the total energy $E_{cell} + E_{cap}$ is a minimum. Thus, using (7.28) and (7.32),

$$0 = \frac{d}{dQ}(E_{cell} + E_{cap}) = \frac{d}{dQ}\left[\mathcal{E}\left(Q_{cell}^{(0)} - Q\right) + \frac{Q^2}{2C}\right] = \left(-\mathcal{E} + \frac{Q}{C}\right). \tag{7.33}$$

Equation (7.33) implies that in equilibrium the voltage drop across the capacitor is

$$\Delta V = \frac{Q}{C} = \mathcal{E}. \tag{7.34}$$

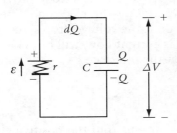

Figure 7.18 Charging a capacitor with a voltaic cell.

This agrees with (7.25) for $I = 0$, thus justifying our use of $\mathcal{E}$ in (7.28) through (7.31). Within experimental error, measurements of the $\mathcal{E}$ of (7.33) with an electrometer, and the $\mathcal{E}$ of (7.25) with a potentiometer, should yield the same value. Note that, on connecting the capacitor to the emf, the capacitor gains energy $Q^2/2C = C\mathcal{E}^2/2$, but the voltaic cell loses energy $\mathcal{E}Q = Q^2/C = C\mathcal{E}^2$. The difference goes into heating the connecting wires, as happens when two capacitors are connected.

Example 7.25 Charging a capacitor

Let $\mathcal{E} = 1.2$ V, $r = 0.4\ \Omega$, and $C = 12\ \mu$F. Find ΔV, Q, and the capacitor energy.

Solution: The capacitor voltage is $\Delta V = \mathcal{E} = 1.2$ V, its charge is $Q = C\Delta V = 14.4$ μC, and its energy stored is $Q^2/2C = 8.64$ μJ. Note that the specific values of both C and r are irrelevant to ΔV, and that the specific value of r is irrelevant to any of the system's final properties. The next chapter, however, shows that the value of the resistance R in an RC circuit is relevant to the rate at which the capacitor charges up. Not surprisingly, the larger the C (e.g., the greater the area of the capacitor plates) and the larger the R (e.g., the thinner the wire), the longer it takes to charge up.

7.10 Voltaic Cells in Simple Circuits

Now consider voltaic cells in circuits where there is current flow.

7.10.1 *Ideal (Resistanceless) Voltaic Cell*

By (7.25), a voltaic cell with emf $\mathcal{E}$ and internal resistance r, with a voltage difference across its terminals of ΔV, has a current $I = (\mathcal{E} - \Delta V)/r$. Thus $\Delta V = \mathcal{E} - Ir$. For any finite current I, as $r \to 0$ this implies

$$\Delta V = \mathcal{E}. \qquad \text{(ideal source of emf)} \qquad (7.35)$$

Hence, *when the internal resistance of a voltaic cell can be neglected, the terminal voltage is the same as the emf,* no matter what the current. Another way to say this is that, for an ideal voltaic cell, the voltage gain ΔV on crossing from the negative electrode (the small plate) to the positive electrode (the large plate) equals $\mathcal{E}$. An ideal voltaic cell is drawn without the resistance r.

Real voltaic cells always have an internal resistance, but when the other resistances in a circuit are much larger than r, the latter may be neglected. We will assume that the internal resistance associated with each electrode of a voltaic cell is negligible compared to the resistance of the electrolyte. Thus the voltage drop across each electrode–electrolyte interface will equal the emf. Charge distributes itself in such a way that the voltage jump across each electrode provides an emf that cancels the emf of the chemical reaction at the electrode. Specifically, if there is an emf $\mathcal{E}_1$ associated with electrode 1, the voltage drop across electrode–electrolyte interface 1, even when there is current flow, will be $\Delta V_1 = \mathcal{E}_1$. (This neglects resistance in the electrode and at the interface.)

7.10.2 *"Shorted" Voltaic Cell*

In Figure 7.19(a), the voltaic cell drives current from the positive electrode to the exterior circuit to the negative electrode. In this figure the opposite terminals of the voltaic cell are directly connected, or "shorted," so $\Delta V = 0$ in (7.25). If (7.25) is valid even for ΔV not close to $\mathcal{E}$, then $I = \mathcal{E}/r$. This is the largest current that a voltaic cell can provide spontaneously. It cannot provide this current for very long because it rapidly depletes the ions in the vicinity of the electrodes; the present discussion, with fixed R, is valid only for slow discharge.

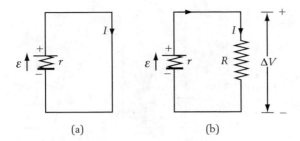

Figure 7.19 Two circuits for discharging a voltaic cell: (a) "shorting"—directly connecting the two electrodes of the cell, (b) connecting the electrodes with a resistor in series.

7.10.3 *Voltaic Cell in Series with a Resistor*

Now consider a circuit consisting of a resistor R and a voltaic cell ($\mathcal{E}, r$). See Figure 7.19(b). Let the voltage across R be ΔV, and let the current I flow through R from the positive to the negative side. For R, the only emf is ΔV, so (7.4) yields

$$I = \frac{\Delta V}{R}. \qquad (7.36)$$

From (7.36) for the arm of the resistor, and (7.25) for the arm of the voltaic cell, eliminate I, solve for ΔV, and then solve for I. This gives

$$\Delta V = \frac{\mathcal{E}R}{r + R}, \qquad I = \frac{\mathcal{E}}{r + R}. \qquad (7.37)$$

When a cell discharges, as is the case here, the terminal voltage ΔV is less than the chemical emf $\mathcal{E}$. However, when a cell charges, so $I < 0$, by (7.25) $\Delta V > \mathcal{E}$.

Another way to obtain the second part of (7.37) is to apply Ohm's law to the circuit as a whole. This can be done because the same current goes through each resistance. We treat the circuit as having the resistance $R_{eff} = r + R$ of two resistors in series. For the circuit as a whole there is no term in ΔV because $\Delta V = 0$ for the circuit as a whole. Then, by (7.27) the chemical emf $\mathcal{E}$ alone drives the current, given by $I = \mathcal{E}/R_{eff}$.

This circuit corresponds to what Ohm actually studied. He measured the deflection of a galvanometer for a circuit of fixed emf $\mathcal{E}$ and fixed resistance r in series with a resistor R whose length and cross-sectional area he varied.

Example 7.26 **Voltage profile for a closed circuit**

Figure 7.20(a) shows a circuit of a voltaic cell, including the effects of each electrode. Take $\mathcal{E}_2 = 1.4$ V and $\mathcal{E}_1 = -0.6$ V, with $R = 0.4\ \Omega$ and $r = 0.1\ \Omega$, and set $V = 0$ at point a. (a) Find the emf of the voltaic cell. (b) Find the current through the circuit. (c) Plot the voltage profile on going around the circuit.

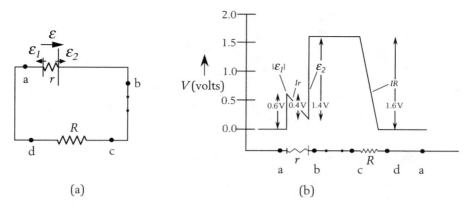

Figure 7.20 A battery may be thought of as two surface pumps $\mathcal{E}_2$ and $\mathcal{E}_1$ (one at each electrode–electrolyte interface) and an internal resistance r. (a) A circuit with a battery discharging through a resistor. (b) The voltage profile across the circuit. The net voltage increase at the electrodes compensates for the net voltage decrease across the resistances, both internal and external.

Solution: (a) By (7.29), $\mathcal{E} = \mathcal{E}_2 - \mathcal{E}_1 = 2$ V. (b) By (7.37), $I = 2/(0.1 + 0.4) = 4$ A. (c) To plot the voltage profile requires the voltage changes across the resistors, given by $Ir = 0.4$ V and $IR = \Delta V = 1.6$ V. Because the connecting wire has negligible resistance, the adjacent circuit points a and d have the same voltage, so $V_d = V_a = 0$ V. Lets start at a and go counter clockwise. Since the current flows clockwise, and current flow through wires goes from high to low voltage, c has a higher voltage than d by IR, so $V_c = V_d + IR = IR = 1.6$ V. Because the connecting wire has negligible resistance, $V_b = V_c = 1.6$ V. At b there is a downward voltage jump of 1.4 V (to exactly oppose the $\mathcal{E}_2$), to 0.2 V just within the electrolyte at electrode 2. Then the voltage rises by $Ir = 0.4$ V on crossing the internal resistance r, to 0.6 V. Finally, there is a downward voltage jump of 0.6 V on crossing electrode 1 just within the electrolyte at electrode 1, giving the starting value of 0 V at a. See Figure 7.20(b). (You might find it helpful to trace the voltage changes going clockwise from a.)

Note the jumps in voltage that occur at each of the electrode–electrolyte interfaces. Also note the linear variations across the resistances. Since in this example both chemical emfs tend to drive current from a to b through the interior of the cell, both of the corresponding electrode voltage jumps tend to drive current from b to a through the interior of the cell. (Diffusion effects, discussed earlier, make the voltage profile across the electrolyte within a real voltaic cell more complex than a straight line.)

7.10.4 *Voltaic Cell on Open Circuit*

Figure 7.21(a) depicts a voltaic cell $(\mathcal{E}, r)$ in a circuit whose switch is open. It may be thought of as being connected to a very small capacitance C, where the capacitance is associated with the ends of the switch. By (7.34), this corresponds to $\Delta V = \mathcal{E}$, so that *for a voltaic cell on open circuit, the terminal voltage is the same as the emf.*

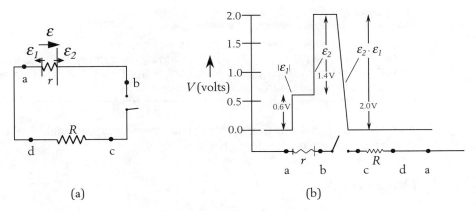

Figure 7.21 A battery may be thought of as two surface pumps $\mathcal{E}_2$ and $\mathcal{E}_1$ (one at each electrode–electrolyte interface) and an internal resistance r. (a) A circuit with a battery and a resistor on open circuit. (b) The voltage profile across the circuit. The net voltage increase at the electrodes compensates for the net voltage decrease across the open switch.

Example 7.27 **Voltage profile for an open circuit**

In Figure 7.21(a), take $\mathcal{E}_2 = 1.4$ V and $\mathcal{E}_1 = -0.6$ V, with $R = 0.4\ \Omega$ and $r = 0.1\ \Omega$, and set $V = 0$ at point a. (a) Find the emf of the voltaic cell. (b) Find the current through the circuit. (c) Plot the voltage profile on going around the circuit.

Solution: (a) By (7.29), $\mathcal{E} = \mathcal{E}_2 - \mathcal{E}_1 = 2$ V. (b) Since the circuit is open, no current flows, so $I = 0$. (c) The voltage drops across r and R are zero. As discussed earlier, across the switch $\Delta V = \mathcal{E}$, so $\Delta V = 2$ V. See Figure 7.21(b). Note the jumps in voltage that occur at each of the electrodes, and the change in voltage across the switch. When the switch is open, it acts like a capacitor of very low capacitance; small amounts of charge (equal and opposite) go to the two sides of the switch, producing a voltage difference.

Take Another Break!

You've just completed the second part of this chapter. When you're ready, we'll move on to charge carriers within wires and voltaic cells.

7.11 **Drag Force**

One of the last topics in Chapter 6 was the decrease of electrical potential energy when two conductors, initially at different voltages, are connected by a conducting wire, and charge is transferred until the voltages equalize. Just as parachutists in the air are subject to the force of gravity and a drag force, so electrons in a wire are subject to the electrical force and a drag force. In both cases, this leads to a limiting velocity. For electrons in most materials, the relationship between drag force and this limiting velocity is a simple proportionality. We will show that such proportionality leads to Ohm's law.

7.11.1 *Drag Force*

Consider an object of mass m acted on by a constant downward force F and a drag force F_{drag}. Let y be the downward direction, and let the velocity in this direction be v, with $v = 0$ at time $t = 0$. See Figure 7.22. By Newton's law of motion,

$$m\frac{dv}{dt} = F + F_{drag}. \tag{7.38}$$

The drag force F_{drag} opposes the velocity (i.e., resists the motion). For low velocity motion through a fluid, drag is proportional to velocity. We assume such proportionality in the present case. The coefficient of proportionality has units of mass divided by time, which we write as m/τ, where τ is called the *relaxation time*. (If our object is a massive parachutist, τ is due to many collisions with the tiny molecules in the air. If our object is an electron in a metal, or an ion in an electrolyte, τ can be due to a single collision; then τ may be called the *collision time*.) Thus

$$F_{drag} = -m\frac{v}{\tau}. \tag{7.39}$$

Because F_{drag} is the rate of decrease of momentum due to collisions, and mv is the momentum, τ^{-1} is the rate of decrease of momentum, divided by the momentum. For an object in air (a parachutist or an oil drop), if the air becomes denser, collisions become more frequent, and τ becomes shorter. For an electron in a metal, if the temperature of the metal is raised, collisions also become more frequent, and again τ becomes shorter.

Consider a single electron in a metal, without an external force to give it a net acceleration. For a time much less than τ, the electron would move rapidly in a random direction. After a time on the order of τ, the electron would collide

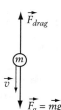

Figure 7.22 Forces acting on a spherical particle of mass m, moving downward under gravity with velocity $\vec{v}$, and subject to air drag.

with a massive ion in the metal, going off at about the same high speed, but in another random direction. The velocity then averages to zero.

Using (7.39), (7.38) can be written as

$$m\frac{dv}{dt} = F - m\frac{v}{\tau}.$$ (7.40)

7.11.2 *Terminal Velocity*

To find the velocity at large times, called the terminal velocity v_∞, we do not need the full time dependence of the solution to (7.40). After enough time elapses, the velocity v increases to the point where the drag force $-mv/\tau$ is large enough to balance the constant force F so that $dv/dt = 0$. Said another way, (7.40) becomes $0 = F - mv_\infty/\tau$, or

$$v_\infty = \frac{F\tau}{m}.$$ (7.41)

For a parachutist at terminal velocity, none of the power provided by gravitational potential energy goes into increasing kinetic energy. Where does this energy go? Into heating the atmosphere. This power is given by the force of gravity mg times the terminal velocity $v_\infty = g\tau$. Thus

$$P = mgv_\infty = mg^2\tau = \frac{mv_\infty^2}{\tau}, \qquad v_\infty = g\tau. \qquad \text{(parachutist)} \quad (7.42)$$

Similarly, as shown in the next section, electrons heat up the wire in which they move. Detailed study yields an equation analogous to (7.42) from which τ can be deduced.

Example 7.28 **Parachutist**

Consider a parachutist of mass 80 kg and terminal velocity 5 m/s. Estimate (a) the relaxation time τ; (b) the rate of heating of the atmosphere.

Solution: (a) For a parachutist, in (7.41) $F = mg$, so $F/m = g$. Taking the value $v_\infty = 5$ m/s, and $g = 9.8$ m/s$^2 \approx 10$ m/s^2, (7.41) yields that $v_\infty = g\tau$, so $\tau = v_\infty/g \approx 0.5$ s. This is an appropriate value (perhaps you have seen movies in which parachutists quickly reach terminal velocity on opening their parachutes). Thus the relaxation time τ is also the characteristic time for the terminal velocity to be reached. (b) Equation (7.42) gives $P \approx (80)(10)(5) = 4000$ W. This example does not constitute an endorsement of skydiving.

7.11.3 *Measuring the Electron Charge: Millikan's*
Optional ## *Oil-Drop Experiment*

This discussion of terminal velocities leads to an important story in the history of science—Millikan's 1909 determination of the *quantum of charge e*. Millikan studied the falling of charged oil drops—the mass in Figure 7.22 is more like an

> **Friction vs. Frictionless, Aristotle vs. Galileo**
>
> In your mechanics course it may have been noted that the ancients (in particular, Aristotle) believed that an object moves only when an external force is applied to it. When the frictional forces finally were minimized or eliminated, scientists (in particular, Galileo) saw that this is not correct. The Aristotelian view, although not fundamental, nevertheless gives a very good description of what happens to charge carriers inside an ohmic conductor, where there are unavoidable frictional forces, and the average velocity of the charge carriers is proportional to the electrical force.

oil drop than a parachutist. He made two measurements on each oil drop, whose charge q and mass m were unknown beforehand: (1) its terminal velocity under gravity; (2) the special field $\vec{E}_s$ needed to keep the drop stationary, so $q\vec{E}_s + m\vec{g} = 0$. Here is the complex train of reasoning: (1) For a droplet of unknown radius R and known mass density ρ_{oil}, the mass is $m = (4\pi R^3/3)\rho_{oil}$. From the theory of fluids, with η the known viscosity of air, $F_{drag} = mv/\tau = 6\pi\eta Rv$, so $6\pi\eta R = m/\tau = (4\pi R^3/3)(\rho_{oil}/\tau)$. Hence $\tau = (2/9)(\rho_{oil}R^2/\eta)$. Measurement of $v_\infty = g\tau$ gave $\tau = v_\infty/g$, and thus R. R and ρ_{oil} gave $m = (4\pi R^3/3)\rho_{oil}$. (2) Measuring the field needed to keep the droplet stationary yielded $\vec{E}_s$, and $|\vec{g}|/|\vec{E}_s|$ gave q/m. Finally, m and q/m gave q. The charge was found to take on many values, both positive and negative, but it was always an integer multiple of 1.6×10^{-19} C. This was taken to be the magnitude of the charge e of the electron. From J. J. Thomson's 1897 electron deflection experiments in cathode ray tubes, the ratio e/m was known to be 1.75×10^{11} C/kg, for all cathode materials. Hence the electron mass was found to be 9.1×10^{-31} kg.

Historical Note   From Faraday's work in the 1830s, the q/m ratio was known for ions in electrolysis (measure the charge q passing through the circuit and divide by the mass m of the electroplated material). Since atomic weights were known, and estimates of Avogadro's number were becoming available in the 1900s, at that time it was possible to estimate ionic masses m, and thus to estimate the charge q in the ionic q/m ratio. By atomic neutrality, the electron charge also would be on the order of this value. However, there was no direct measurement of the unit of charge until Millikan's work. Recall that the nuclear model of the atom was not established until two years later, in 1911.

7.12 Conductivity of Materials—I

This section uses a drag model to derive the local form of Ohm's law, with an explicit form for the conductivity σ that explains the difference between conductors and insulators. Current in a wire is carried by many electrons, moving at an average velocity. Let us first study the more general problem of the number dN of objects that cross a given cross-section per unit time dt. This defines the *number current* dN/dt. From the number current and the amount of charge (or mass, etc.) that each object carries we can determine the charge current (or mass current, etc.). A change in sign for a current means a change in its direction (e.g., if rightward is positive, then leftward is negative).

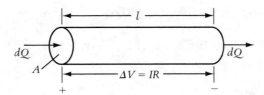

Figure 7.23 Schematic of current flow in a wire, due to charge carriers (not shown) of density n, moving with drift velocity v_d. The wire has length l and cross-sectional area A.

Consider a wire of length l and cross-sectional area A, with n objects per unit volume, as in Figure 7.23.

Example 7.29 *n* **for conduction electrons in copper**

Copper (Cu) has a mass density of 8.95 g/cm^3 and an atomic weight of 63.5 g/mole; each mole contains 6.02×10^{23} atoms. Take Cu to have one conduction electron per atom. Estimate n, the number of conduction electrons per unit volume for copper.

Solution: We have

$$n = \left(6.02 \times 10^{23} \frac{\text{atoms}}{\text{mole}}\right)\left(\frac{\text{electron}}{\text{atom}}\right)\left(\frac{\text{mole}}{63.5 \text{ g}}\right)\left(\frac{8.95 \text{ g}}{\text{cm}^3}\right)\left(\frac{\text{cm}}{.01 \text{ m}}\right)^3$$

$$= 8.48 \times 10^{28} \frac{\text{electrons}}{\text{m}^3}.$$

Our choice of one conduction electron arises from chemistry; a free copper atom with 29 electrons has a ground state orbital configuration of $1s^2 2s^2 2p^6 3s^2 3p^6 3d^{10} 4s$. (Thus, for example, six electrons are in the $3p$ orbital.) Only the last—the $4s$—shell, with one electron per atom, is not completely filled. When the copper atoms come together to form solid copper, the individual atomic $4s$ orbitals become delocalized orbitals that extend over the entire crystal of copper. Alternatively, a Hall effect measurement (Chapter 10) will yield n for any conducting or semiconducting material that has only a single type of charge carrier.

7.12.1 *Number Current through Uniform Tube*

To determine dN/dt requires the number of objects N within the wire of Figure 7.23, and the time T they take to cross it.

The number of objects in the wire is $N = nAl$, where Al is the wire volume. If they move with an average velocity v_d (where the d stands for *drift*), then it takes a time $T = l/v_d$ for all of them to cross the length of the wire. Hence

$$\frac{dN}{dt} = \frac{N}{T} = \frac{nAl}{(l/v_d)} = nAv_d. \qquad \text{(flow through uniform wire)} \qquad (7.43)$$

This result can be applied to a salmon run. For a sluice of cross-sectional area 15 m^2, if the average salmon velocity is 0.8 m/s and $dN/dt = 3$ salmon/s, (7.43) gives a salmon density of $n = 0.25$ salmon/m^3.

7.12.2 *Electric Current through Uniform Wire*

Since each of the wire's N electrons has a charge $-e$, by (7.43) the electric current is

$$I = \frac{dQ}{dt} = -e\frac{dN}{dt} = -enAv_d. \tag{7.44}$$

By (7.44), from a measurement of I and a knowledge of (n, A, e), we can determine v_d. Also by (7.44), if the negatively charged electrons go to the right, the electric current goes to the left. For no drift $(v_d = 0)$, there is no current $(I = 0)$.

Example 7.30 **Electron drift velocity in a copper wire**

#10 copper wire (typical building wire) has a radius a of 0.05 inch $(= .00127$ m). (a) Taking $n = 8.48 \times 10^{28}/m^3$, find the drift velocity for a 10 A current. (b) Find how long it will take an electron, on average, to cross 1 m of this wire.

Solution: (a) The wire has $A = \pi a^2 = 5.07 \times 10^{-6}$ m^2. From (7.44) a 10 A current then has $|v_d| = I/neA = 1.45 \times 10^{-4}$ m/s. (Don't confuse area A with amperes A!) (b) $t = d/v_d = 6900$ s, or nearly 2 hours!

7.12.3 *Current Density for Electrons*

Related to the electric current I is the electric current per unit area $\vec{J}$, or current density. By (7.3), for uniform $\vec{J}$, its component along I is given by

$$J = \frac{I}{A}. \tag{7.45}$$

Placing (7.44) in (7.45) gives $J = -nev_d$, which yields, on including its direction,

$$\vec{J} = -ne\vec{v}_d. \tag{7.46}$$

Observe that, for electrons, $\vec{J}$ is opposite to $\vec{v}_d$. Negatively charged electrons moving rightward means both that the right side becomes more negative and that the left side becomes more positive.

7.12.4 *For the Drag Model, $\vec{J}$ Is Proportional to $\vec{E}$*

Applying (7.41) to electrons, and including direction, so $v_\infty \to \vec{v}_d$ and $\vec{F} \to -e\vec{E}$, yields

$$\vec{v}_d = -\frac{e\tau}{m}\vec{E}. \tag{7.47}$$

Placing (7.47) into (7.46) then yields

$$\vec{J} = (-ne)\left(-\frac{e\tau}{m}\right)\vec{E} = \frac{ne^2\tau}{m}\vec{E}. \tag{7.48}$$

By (7.48), the drag model with independent charge carriers predicts that the current density $\vec{J}$ is proportional to the electric field $\vec{E}$. Hence, for a conductor in equilibrium, where $\vec{J} = \vec{0}$, we have $\vec{E} = \vec{0}$, as assumed in our discussion of electrical conductors in equilibrium (Section 4.6). Even if the charge carriers are not independent, the proportionality between $\vec{J}$ and $\vec{E}$ remains valid.

Comparison of (7.48) and (7.8) yields a specific, drag model–dependent form for the conductivity (due to Debye), written as

$$\sigma_D = \frac{ne^2\tau}{m}. \qquad \text{(drag model for conductivity)} \tag{7.49}$$

The factors in (7.49) have the following origin: n is the charge carrier density, e is their charge, e/m gives their acceleration per electric field, and τ gives the time they can accelerate in the field before colliding with a massive ion and losing their memory of how they had been moving.

Example 7.31 **The relaxation time for copper**

For copper at room temperature, a previous example gave $\rho = \sigma^{-1} = 1.7 \times 10^{-8}$ Ω-m, and from yet another example, $n = 8.48 \times 10^{28}$ electrons/m^3. Find the relaxation time, τ, for the conduction of electrons in copper.

Solution: From ρ, n, and $e = 1.6 \times 10^{-19}$ C, $m = 9.1 \times 10^{-31}$ kg, (7.49) yields $\tau = 2.5 \times 10^{-14}$ s. Thus we have finally obtained the relaxation time τ for the conduction electrons in copper, at room temperature, even though we could not determine it by a measurement on any individual electron.

Equation (7.49) explains that the difference in conductivities between a conductor and an insulator is due to the vastly different values of their charge carrier density n: *good conductors have a high density of charge carriers, and poor conductors have a low density of charge carriers.*

7.12.5 *Joule Heating: Microscopic Viewpoint*

The conductivity has already been obtained for the drag model, but it is also instructive to obtain the conductivity by considering the rate at which an individual electron contributes to the heating of the wire. This is given by the scalar product of the electric force $-e\vec{E}$ and the drift velocity $\vec{v}_d$. Application of (7.47) yields

$$\mathcal{P}_{el} = (-e\vec{E}) \cdot \vec{v}_d = (-e\vec{E}) \cdot \left(-\frac{e\vec{E}\tau}{m}\right) = \frac{e^2E^2\tau}{m} = \frac{mv_d^2}{\tau}. \qquad \text{(one electron)} \tag{7.50}$$

[Compare this with (7.42) for the rate of heating of the atmosphere by a parachutist.] For $N = nAl$ electrons in the wire, (7.50) yields, with $E = \Delta V/l$,

$$P = (nAl)\frac{e^2 E^2 \tau}{m} = nAle^2 \left(\frac{\Delta V}{l}\right)^2 \frac{\tau}{m} = (\Delta V)^2 \left(\frac{ne^2\tau}{m}\right)\left(\frac{A}{l}\right) \qquad \text{(wire)} \tag{7.51}$$

Equations (7.7) and (7.9) can be combined to give $P = I(\Delta V) = (\Delta V)^2 \sigma A/l$. Comparison with (7.51) gives the same σ as (7.49).

7.12.6 *Limitations of the Model*

A number of assumptions are made by the drag model, most of them not providing significant limitations. First, it assumes that the charge carriers would all have zero velocity in the absence of an electric field. That is not really true: they can have nonzero velocity either from thermal fluctuations (i.e., random collisions of the electrons with one another or with the nuclei of the solid) or because the current-carrying electrons (i.e., the conduction electrons) are in delocalized orbitals, characterized by relatively high energies and velocities. The characteristic velocity of the highest energy-occupied state in some metals can be as high as 3×10^6 m/s—on the order of 1/100th of the speed of light! Thus, even without an electric field, electrons are zipping around incredibly quickly. However, just as many are moving one way as the other, so they carry zero net electric current. An applied electric field shifts their average velocity from zero to the drift velocity v_d, which is typically negligible ($\sim 10^{-4}$ m/s from a previous example) compared to the large characteristic (but random) electron velocity ($\sim 10^6$ m/s). The drag model also assumes that the state, or orbital, of each charge carrier is just like every other. *Despite these apparent limitations*, by interpreting τ as an average over the charge carriers, we can use (7.49).

On the other hand, the drag model also assumes that the electric field $\vec{E}$ is uniform. If $\vec{E}$ changes significantly over a distance that is shorter than the average distance an electron travels between collisions (the *mean-free path* λ), then the local relations (7.47) and (7.48) will *not* be valid. The breakdown of the local form of Ohm's law can occur for very pure metals at low temperatures.

7.13 **Conductivity of Materials—II**

Optional

Figure 7.24 illustrates the full temperature dependence of the conductivity σ for a few characteristic types of materials, including glass, ordinary salt (NaCl), the intrinsic (pure) semiconductor Si, the semimetal graphite, the metal Cu, the metal Pb (which goes superconducting), and the high-temperature superconductor $YBa_2Cu_3O_7$.

One of the goals of the theory of condensed matter (which includes all liquids and solids) is to explain, in detail, the temperature dependence of the resistivity of all matter. This is not a trivial problem. John Bardeen, Leon Cooper, and J. Robert Schrieffer received the Nobel Prize in 1972 for their theory of the phenomenon of superconductivity. Nevertheless, Sir Neville Mott and Phillip Anderson were awarded a Nobel Prize in 1977 for their work on the much older

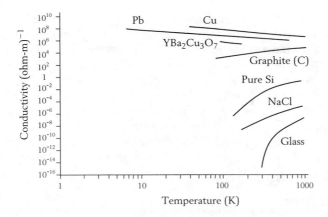

Figure 7.24 Temperature dependence of the conductivity σ for some representative types of material: metals (Cu and Pb), insulators (NaCl and glass), a semiconductor (Si), a semimetal (graphite C), and the high-temperature superconductor $YBa_2Cu_3O_7$. For $T < 7$ K, Pb is a superconductor, and for $T < 92$ K, $YBa_2Cu_3O_7$ is a superconductor. In the insulators, conduction takes place by motion of the ions or by site-to-site hopping of electrons in localized orbitals.

phenomenon (hearkening back to Stephen Gray's 1729 discovery) that most materials are either conductors or insulators. Moreover, the 1956 Nobel Prize went to John Bardeen, William Shockley, and Walter Brattain, for their discovery of the transistor effect in the semiconductor germanium. Thus, researchers who have studied all types of electrical conductors have received Nobel Prizes.

What follows is a discussion, largely based on (7.49), of the electrical properties of materials.

7.13.1 *Two Ordinary Means of Electrical Conduction*

One conduction process involves *delocalized orbitals*, as in metals. The other conduction process involves electrons *hopping* between *localized orbitals*, when there is enough thermal energy. Two mechanisms can produce localized orbitals, or *localization*. Mott localization occurs when electrons interact with one another, via the Coulomb interaction, in such a way that they prevent one another from moving. Anderson localization occurs when structural disorder in the material causes deep attractive wells from which the electrons have difficulty escaping. When localization occurs, a material has a low density of charge carriers n; thus, by (7.49), it is a poor conductor. In Franklin's experiments on the location of electric charge on the Leyden jar and on glass panes, discussed in Chapter 1, the orbitals associated with the glass were localized.

7.13.2 *Metals*

For ordinary metals, the density n of charge carriers is nearly independent of temperature, but the characteristic relaxation time τ decreases as the temperature increases because collisions become more frequent. From (7.49), such a temperature-dependent τ explains qualitatively why, in Cu and Pb, the conductivity σ decreases with increasing temperature. The decrease of τ with temperature is striking. It is understandable only when we take into account that the electrons carrying the electric current have orbitals that extend over the entire crystal. At very low temperatures, the low-energy orbitals are nearly all occupied, so there are not many orbitals in which the electrons can scatter. Thus τ becomes very large at low temperatures, decreasing as the temperature is raised. Below some temperature that depends on the purity of the sample, τ saturates because all the scattering is due to collisions with impurities.

7.13.3 *Semiconductors*

For intrinsic (pure) semiconductors at zero degrees Kelvin (absolute zero), the delocalized orbitals with the lowest energy are all filled, forming the equivalent of a filled atomic shell. The full set of such orbitals is called an *energy band*. When the band orbitals are all occupied, the electrons in them cannot carry a net electric current. Moreover, the *band gap*—the energy to excite an electron in the highest energy occupied band (the valence band) to an orbital in the next higher energy band (the conduction band)—is on the order of 0.5 eV or more (depending on the material). Hence the carrier density n is zero, so by (7.49) the conductivity is zero. However, as the temperature is increased, the next higher band becomes more occupied, thereby increasing n. This effect dominates any decrease in τ with increasing temperature and explains why, for pure semiconductors, the conductivity σ increases with increasing temperature.

For nonintrinsic (*doped*) semiconductors, such as silicon (Si) doped with gallium (Ga), the energy to excite electrons from orbitals around the Ga-dopant is much less than in pure Si, so the carrier density n increases with increasing temperature much more rapidly than in pure Si. Thus the conductivity at first increases with increasing temperature. Once all the orbitals localized around the dopant have been thermally excited, the remaining temperature dependence of the conductivity is due to the decrease of τ with increasing temperature.

A major reason semiconductors are useful is that, even at a fixed composition and temperature, their conductivity can be changed significantly. For example, shining light on them increases their conductivity. The light energy excites an electron from the valence band to the conduction band, thus making more electrons in the conduction band, and more "holes" in the otherwise-filled valence band; both electrons and holes can conduct electricity. The electrical conductivity of a semiconductor can also be changed by applying a large enough electric field to attract or repel a significant number of electrons from the sample.

7.13.4 *Superconductors Have a Resistanceless Means to Carry Electric Current*

One of the more puzzling topics in the electrical resistivity of metals has been the problem of superconductors, which lose their electrical resistance below a characteristic critical temperature T_c. Kamerlingh Onnes discovered this phenomenon in 1911, when he cooled mercury (Hg) below 4.2 K. See Figure 7.24, which shows the resistivity versus temperature for lead (Pb) and for the high-temperature superconductor $YBa_2Cu_3O_7$. The 1986 Nobel Prize in Physics went to Alex Müller and Georg Bednorz for finding a new class of materials with a relatively high T_c near 35 K. Materials with T_c's as high as 135 K have been found; they can be cooled on immersion in readily available liquid nitrogen, which is fluid from 21 K to 77 K.

When a system goes superconducting, besides the usual mode of conduction via delocalized orbitals, which leads to electrical resistance, it develops a special superconducting mode of conduction. For steady current flow (dc, or direct current), this superconducting mode truly has no electrical resistivity. Because the system carries current using the "mode of least resistance," for direct current it uses the superconducting mode and does not use the normal means of carrying current. However, at finite frequencies (i.e., for an emf that oscillates in time, such as that provided by the power company) the current-carrying effectiveness of the superconducting mode is limited, in proportion to the frequency, by the inertia of the electrons. In that case, the normal means of carrying current—with its associated electrical resistance—is partially used, its current being added in parallel to the current provided by the superconducting mode of carrying current. Later we will discuss the magnetic properties of superconductors and show they imply that superconductors do indeed have such a new "mode" of zero dc resistivity. We do not yet have a good theoretical understanding of the temperature dependence of the resistivity for the high T_c materials.

Conduction by Superconductors

The special mode of conduction in superconductivity involves (1) pairs of electron orbitals (known as *Cooper pairs*) with opposite momentum, become slightly correlated, something like having distant dancing partners, and (2) the average motion of each electron pair is the same, something like having all pairs of distant dancing partners simultaneously circling around a large dance hall at the same rate. Individual behavior, which is responsible for electrical resistance, is not present in this correlated motion.

7.13.5 *Non-Ohmic Behavior*

At low current density J, all nonsuperconducting materials are ohmic. However, for large J, Ohm's law is no longer satisfied; the J at which this occurs depends on the material. Generally, the lower the carrier density n, the lower the critical current J_c at which the system becomes non-ohmic. Thus, J_c is high for metals but low for semiconductors. Non-ohmic behavior has many applications. With a so-called *reverse bias* voltage, we can nearly turn off the current; with a so-called *forward bias* voltage, we can produce a very large current. This is illustrated in Figure 7.1(b). Such an equivalent of a valve is called a *diode*. The first diode was based on the *cathode ray tube*. This is a vacuum tube having a negatively charged cathode that emits electrons, and a positively charged anode that absorbs them. The cathode is heated to increase the rate of emission.

If an intermediate metallic grid screen is added to the diode, small changes ΔV_g in the grid voltage (the input) can produce large changes ΔI in the current to the anode (the output). This is the basis of the electronic amplification device called the *triode*. See Figure 7.25.

The first amplifiers and nonlinear devices used vacuum tubes. They were bulky, slow to respond, expensive, fragile, and unreliable. In 1948, Bardeen, Brattain, and Shockley produced the first semiconductor amplifier, called a *transistor*. At low voltages, the resistance was high, and at high forward bias voltages, the resistance was low: much smaller than for a vacuum tube. Thus there was a *trans*fer of re*sistance*; this is the origin of the word *transistor*. Semiconductor triodes employ the functional equivalents of the cathode (emitter), the anode (collector), and the grid (base). In 1958 and 1959, Jack Kilby and Robert Noyce independently showed that it was possible to fabricate and interconnect tiny transistors, resistors, and capacitors on a single semiconductor chip. This was the *integrated circuit*. The compactness, speed, complexity, and low price that thus became possible have changed the world of electronics and made possible microcomputers (e.g., in automobiles, microwave ovens, and copying machines), powerful personal computers, and rapid and reliable worldwide communications. Kilby (but not the deceased Noyce) shared of the 2000 Nobel Prize in physics.

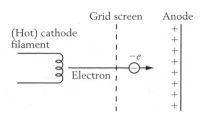

Figure 7.25 Fundamental parts of a vacuum tube triode, which has a non-ohmic response. Although modern electronics does not use vacuum tubes, they are more easily visualized and understood than the corresponding modern devices.

7.13.6 *Quantum of Conductance*

Electron orbitals actually are *waves*. On atoms, they are localized because of the electrical attraction to the nucleus, but conduction electrons are in orbitals that extend over the entire crystal. In 1957, Landauer showed that, if the length of a narrow wire connecting two bulk conductors is smaller than the size of the orbitals that carry electric current across the wire, then the electrical conductance is the same thing as wave transmission. Whenever the wave properties of electrons become important, the system must be described by quantum mechanics, and Planck's quantum constant $h = 6.624 \times 10^{-34}$ J-s can appear. In the present case, because C^2/J-s = C^2/(C-V-s) = (C/s)/V = A/V = $1/\Omega$, the quantity

$$\frac{e^2}{h} = 0.36 \times 10^{-4} (\Omega)^{-1} \tag{7.52}$$

has the same units as an inverse ohm. Thus it might come as no surprise that, for very small systems, the electrical conductance $G = R^{-1}$ is on the order of e^2/h. What is more interesting, and somewhat surprising, is that for very small systems and very low temperatures, the electrical conductance has plateaus at integer and some rational multiples of e^2/h. This is discussed in Section 10.7.

Problems

7-1.1 A 300-ohm resistor is placed side by side with a vertical D cell (1.5 V), the top side of the D cell being positive. (a) How much current flows through the resistor? (b) The top end of the resistor is connected to the top end of the D cell. See Figure 7.26. How much current flows through the resistor? (c) The bottom end of the resistor is now connected to the bottom end of the D cell, while keeping the previous connection in place. How much current flows through the resistor?

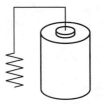

Figure 7.26 Problem 7-1.1.

7-1.2 A 300-ohm resistor is placed side by side with two vertical D cells, the top sides of the D cells being positive. (a) How much current flows through the resistor? (b) The top end of the resistor is connected to the top end of one D cell. How much current flows through the resistor? (c) The bottom end of the resistor is now connected to the bottom end of the other D cell, while keeping the previous connection in place. See Figure 7.27. How much current flows through the resistor?

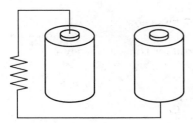

Figure 7.27 Problem 7-1.2.

7-2.1 A 12 V car battery provides 6 A to the headlights for half an hour while the tire is replaced. Find the amount of charge transferred and the rate at which electrons are transferred.

7-2.2 The electric current to the screen of a computer monitor is 300 μA. To how many electrons per second does this correspond?

7-2.3 You are to electroplate 20 g of Ag using a cell of AgCl. (a) How much charge must flow through the circuit to do this? (Both the Ag^+ and the Cl^- must be included.) (b) If the current is 150 mA, how long does this take? The atomic weights of Ag and Cl are 107.9 and 35.5, respectively.

7-2.4 On an assembly line, 45 cardboard boxes per minute pass a checkpoint. Each cardboard box contains 24 boxes of Flakies breakfast cereal. Each box of Flakies contains about 9600 individual Flakies. Find the "Flakies current:" the rate at which Flakies pass the checkpoint.

7-2.5 In a cyclotron, a proton moves in an orbit of radius 4 cm at velocity 8×10^7 m/s. Find the average electric current due to the proton.

7-2.6 For a situation with spherical symmetry, a radial current flux and constant total current I cross any concentric sphere. Find J_r in terms of I and the radius r.

7-2.7 Every 60 ms, a charge 0.8 C uniformly crosses a 28 mm^2 area, traveling in the xy-plane at clockwise 20° to the x-axis. Find (a) I; (b) dI/dA; (c) $\bar{J}$; (d) the current crossing a 2 mm^2 area whose normal is at a clockwise 70° angle to the x-axis.

7-2.8 There are strong analogies between electric current flow through wires, fluid flow through pipes, and heat flow through solids. For each of these, to the best of your knowledge indicate what is flowing and what is causing the flow.

7-2.9 Based on a two-fluid model (one positive and one negative), in Oersted's time the term "electric conflict" was used to describe electric current flow. Would the fluids flow in the same direction? Discuss the complications associated with calculating the electric current due to two oppositely charged carriers. (This can occur in semiconductors and in electrolytes.)

7-2.10 A hose is aimed at a box with a narrow hole of area 2.4 mm^2, and water collects at the bottom of the box (of cross-section 48 cm^2) at the rate of 6.7 mm/s. If the angle that the hose makes to the narrow hole is varied, the collection rate can get as high as 9.2 mm/sec. Find (a) the mass current of water into the box; (b) the mass current per unit area initially entering the box; (c) the mass current per unit area in the stream of the hose; (d) the angle that the hose made initially to the small hole.

7-2.11 12 A flows radially inward toward the axis of a 24 cm long cylinder. If the current is constant crossing each concentric cylinder, find the current per unit area for a radius of 8 mm.

7-3.1 Consider the following I versus V data for one particular circuit element: (2 A, 8 V), (4 A, 20 V). (a) What is the resistance at each voltage? (b) Is the system ohmic?

7-3.2 A 2.58 A current flows through a uniform resistor, across which a voltmeter reads 438 mV. (a) Find its resistance. (b) If the resistor is ohmic, and the voltage across it is doubled, find the new current and the new resistance. (Assume a fixed temperature.) (c) A resistor made of the same material is larger by a factor of 4.6 in all directions. Find its resistance.

7-3.3 Consider Equation (7.4). Let a wire of resistance $R = 5\ \Omega$ be along the x-axis, its right end at a higher voltage than its left end by $\Delta V = 10$ V. (a) If leftward current is considered to be positive, find the numerical value of I, including its sign. (b) If rightward current is considered to be positive, find the numerical value of I, including its sign.

7-3.4 (a) A gold wire of length 8 cm has radius 0.24 mm. Find its resistance. (b) Repeat for a copper interconnect (for integrated circuits) that is 120 nm thick, 0.5 cm wide, and 0.8 cm long.

7-3.5 Consider a Cu cable of cross-sectional area 5.4 cm². (a) Find its electrical resistance per meter. (b) Find the resistance of 4 km of this cable.

7-3.6 A 0.5 nF parallel-plate air capacitor is filled with a fluid of resistivity 12 Ω-m. Determine the resistance between the plates.

7-3.7 A cylindrical sample of soil is of length 18 cm and radius 1.4 cm. Its end-to-end electrical resistance is 2470 Ω. Determine its electrical resistivity.

7-3.8 A piece of copper at room temperature has length 8 cm and radius 4 mm. Find its resistance. At 20 K above room temperature, find its length, radius, and its resistance. $L = L_0(1 + \alpha_L \Delta T)$ relates room temperature lengths L_0 to lengths at nearby temperatures. The coefficient of thermal expansion α_L of Cu is 1.7×10^{-5} K^{-1}.

7-3.9 Alternate slices of iron and carbon (of the same radius) are to be used to make resistors that are nearly temperature independent. If the iron slices are to be 2.4 mm long, find the desired length of the carbon slices. Use $\alpha = 6.5 \times 10^{-3}$ K^{-1} for iron. Consider the carbon to be graphite.

7-3.10 A wire is drawn out through a die so that its new length is twice as great as its original length. If its original resistance was 0.4 Ω, find its new resistance.

7-3.11 A 420 mA current flows through an electric cable with 60 strands of wire, each of resistance 1.2 $\mu\Omega$. (a) Find the current in each strand. (b) Find the applied voltage difference. (c) Find the cable resistance.

7-3.12 A microwave oven is rated at 1400 W for 120 V power. What current does it draw, and what is its resistance?

7-3.13 (a) Find the resistances of a 60 W bulb for use in the home and a 60 W bulb to be used with a 12 V car battery. (b) Which is higher?

7-3.14 (a) Find the resistances of a 60 W bulb and a 100 W bulb intended for use in the home. (b) Which is higher?

7-3.15 A ventricular defibrillator, attached by low-resistance paddles placed above and below the heart, provides 20 A and 10^5 W. (a) Find the voltage drop across the paddles. (b) Find the electrical resistance. (Normally, the body's electrical resistance is much larger, but here the contact area of the electrodes is very large, and the electrical contact is very good.)

7-3.16 Assume that a 60 W lightbulb (at 120 V) has a room temperature resistance of 120 Ω. (a) Find the current and power usage when the bulb is first turned on. (b) Find the resistance and current when the bulb is at operating temperature.

7-3.17 A 5 W portable radio using four resistance-less 1.5 V batteries in series was left on from 9 am to 11 am. (a) Find the total energy consumption. (b) Find the average current. (c) Find the amount of charge that passed through the circuit.

7-3.18 A space heater produces 1400 W at 120 V. Find (a) its resistance; (b) the current; and (c) the cost to run it for two hours, at 8.9 cents/kW-hr.

7-3.19 Nichrome wire is used for a heating element to produce 500 W at 120 V. (a) Find its resistance. (b) Find the ratio of its length to its cross-section. Take $\rho = 1.00 \times 10^{-6}$ Ω-m.

7-3.20 Two cows are 100 ft away from a tree, the body of one facing the tree and the body of the other sideways to the tree. A lightning bolt hits the tree. Assume that the current spreads out radially along the surface. Which cow is likely to get the larger shock, and why?

7-3.21 A 120 V motor uses 15 A while lifting a 10-ton elephant vertically at a rate of 20 cm/s. (a) Find the rate at which it provides power lifting the elephant, in horsepower (1 hp = 745.7 W). (b) Find its efficiency of power use (efficiency is useful energy or power divided by the total energy or power). (c) Does this efficiency make sense?

7-3.22 Under normal circumstances, the resistance of the body is on the order of 10^5 Ω. (a) Find the current flow for $\Delta V = 120$ V. (b) With sweaty hands, the resistance might be as low as 2000 Ω. (b) Find the current flow for $\Delta V = 120$ V. (c) Assume that the resistance between a pair of forceps, on the left and right sides of the heart during surgery, is 8 Ω. Find the stray voltage difference to cause a potentially fatal current of 200 mA.

7-4.1 A 24 A current flows rightward through an Al wire with cross-sectional area 2.4 mm^2 and length 1.9 m. (a) Find the electric field in the wire, in magnitude and direction. (b) Find the voltage drop across the wire, and indicate which side has the higher voltage.

7-4.2 Two solid brass Frankenstein monsters differ in weight by a factor of 6.85. A 25 A current passes through the less massive one when 6 V are placed across its electrodes. Find the electrical resistances of both.

7-4.3 A solid copper sphere has radius 1 cm. A second solid sphere of uniform but unknown composition has radius 2 cm. Electrodes are placed at opposite poles of each. The first sphere has a resistance 1.21 times that of the second sphere. (a) Estimate the resistivity of the material of the second sphere. (b) Of what might the second sphere be made?

7-4.4 (a) Show that the resistance of a uniform (*homogeneous* is the technical term) spherical shell of radii $a < b$ and conductivity σ, for radial current flow, is $R = (a^{-1} - b^{-1})/4\pi\sigma$. (b) Show that, as $b - a = d \to 0$, and taking $A = 4\pi a^2 \approx 4\pi b^2$, R reduces to (7.5).

7-4.5 An otherwise solid spherical steel shell of inner radius 1.4 mm and outer radius 2.2 cm has a small hole to accommodate a 46 cm long Cu wire of diameter 0.04 cm and a thin layer of insulation.

The copper brings a 2 A current uniformly to all parts of the inner surface. See Figure 7.28. (a) Assuming that the current density is uniform in both the copper and the steel, calculate the resistance of the copper and of the steel. (b) Calculate the voltage drops across each. You may use the result of the previous problem for a spherical electrode.

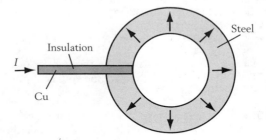

Figure 7.28 Problem 7-4.5.

7-4.6 For Cu wire, here are some *gauge* numbers, their diameters (in inches), and their safe current (in amperes): (4, 0.20431, 95), (6, 0.16202, 75), (8, 0.12849, 55), (10, 0.10189, 30), (12, 0.080808, 20), (14, 0.064064, 15), (16, 0.050820, 10), (18, 0.040303, 5). Find the product of the gauge and the diameter. (In an early definition of gauge, this product was a constant. Presently, gauge also is determined by the desire to have a multiple of five value for the maximum safe current.)

7-4.7 Ampere actually considered that the electromotive force acted within a wire, whereas the chemical emf is localized at the electrodes of the voltaic cell. If there is no chemical emf in a wire, what causes electric current to flow in a wire?

7-4.8 In previous chapters, we took $\vec{E} = \vec{0}$ inside a conductor. In this chapter, we take $\vec{E} \neq \vec{0}$ inside a conductor. Explain the difference.

7-4.9 Explain how the local form of Ohm's law implies that current flows along the direction from higher voltage to lower voltage.

7-4.10 Consider a 10-foot length of copper tubing with an inner diameter (ID) of $\frac{1}{2}$ inch and an outer diameter (OD) of $\frac{5}{8}$ inch. Find the electrical resistance for (a) axial current flow and (b) radial current flow.

7-5.1 Resistors $R_1 = 4$ Ω and $R_2 = 9$ Ω are in series. A current of 3 A passes through R_2. Find (a) the equivalent resistance R of the combination, (b) the voltage drops across R_1 and R_2, (c) the total voltage drop, (d) the current through R_1, and (e) the current through the combination.

7-5.2 Resistors $R_1 = 4\ \Omega$ and R_2 are in series. A current of 3 A passes through R_2 and the voltage drop across the combination is 33 V. Find (a) R_2, (b) the voltage drops across R_1 and R_2, (c) the current through R_1, (d) the current through the combination, and (e) the equivalent resistance.

7-5.3 Resistors $R_1 = 4\ \Omega$ and R_2 are in series. A voltmeter reads 14 V across R_1 and 21 V across R_2. Find (a) R_2 and (b) the current through R_1 and R_2. For the combination, find (c) the voltage drop, (d) the current, and (e) the equivalent resistance.

7-5.4 Resistors $R_1 = 4\ \Omega$ and $R_2 = 8\ \Omega$ are in parallel. A current of 3 A passes through R_1. Find (a) the current through R_2 and (b) the voltage drops across R_1 and R_2. For the combination, find (c) the voltage drop, (d) the current, and (e) the equivalent resistance.

7-5.5 Resistors $R_1 = 4\ \Omega$ and R_2 are in parallel. A current of 3 A passes through R_1 and a current of 8 A passes through R_2. Find (a) R_2 and (b) the voltage drops across R_1 and R_2. For the combination, find (c) the voltage drop, (d) the current, and (e) the equivalent resistance.

7-5.6 Resistors $R_1 = 4\ \Omega$ and R_2 are in parallel. A voltmeter reads 2 V across R_1. An ammeter reads 4 A through R_2. Find (a) R_2 and (b) the current through R_1. For the combination, find (c) the voltage drop, (d) the current, and (e) the equivalent resistance.

7-5.7 Resistors R_1 and R_2 are in parallel, and together they are in series with R_3. A 10 A current passes through the combination, a 4 A current passes through R_1, 12 V is across R_3, and 16 V is across R_2. Find (a) each resistance, (b) the currents through R_2 and R_3, and (c) the voltage across R_1. For the combination, find (d) the voltage drop and (e) the equivalent resistance.

7-5.8 Resistors $R_1 = 6\ \Omega$ and $R_2 = 8\ \Omega$ are in parallel, and they are in series with resistors $R_3 = 3\ \Omega$ and $R_4 = 9\ \Omega$. (a) Find the equivalent resistance. (b) If 2 A passes through R_1, find the voltage drop across the combination.

7-5.9 Two students were each given four 1 Ω resistors and told to connect them up to yield an equivalent resistance of 1 Ω. Their circuits were different. Draw each circuit.

7-5.10 You are told to find sets with the minimum number of integer-valued resistors that in various combinations will provide integer-valued resistances from 1 to 6. (a) How many sets do you find, and what are they? (b) Repeat for 1 to 7.

7-5.11 Design a three-way lightbulb using two filaments that operate singly or in series to produce 60 W at its lowest setting and 150 W at its highest setting. Find the two resistors and the intermediate setting. Assume 120 V power.

7-5.12 Consider a set of identical flashlight bulbs and a D cell of negligible resistance. Answer in terms of *brighter, same,* or *less bright*. Consider the following cases: (1) Bulb A is placed in series with the D cell. See Figure 7.29(a). (2) Bulb B is in parallel with bulb A, and together they are in series with the D cell. See Figure 7.29(b). (a) Compare the visual intensity of bulb A in cases (1) and (2). (b) In (2), how does the intensity of bulb B compare to that of bulb A? (3) Bulbs A and B are placed in series with each other and the D cell. See Figure 7.29(c). (c) How does the intensity of bulb A compare with its intensity in (1)? Now bulb C is placed in parallel with bulb A. See Figure 7.29(d). (d) How does the intensity of bulb A change? (e) How does the intensity of bulb B compare to that of bulb A?

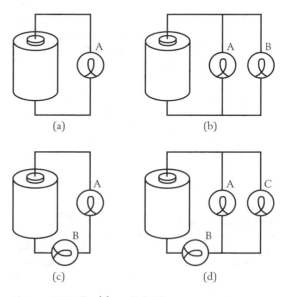

Figure 7.29 Problem 7-5.12.

7-5.13 In the previous problem, let the D cell provide 1.5 V, and let $R = 0.5\ \Omega$ for each bulb.

Find the currents and visual intensities in each case, taking the intensity of each bulb to be 20% of $I^2 R$, and assuming the bulbs to be ohmic.

7-5.14 Consider a circuit with an unknown resistor R_2 leading to resistors $R_1 = 6 \, \Omega$ and $R_3 = 3 \, \Omega$ in parallel with one another. See Figure 7.7(b). In the present problem, let the common point between all three resistors be at $V_b = -4$ V, and let the other end of R_2 be at $V_a = 16$ V. If $I_1 = 4$ A through R_1, find (in any order) (a) R_2, (b) the currents through the other resistors, and (c) the voltage V_c at the other end of the connection between R_1 and R_3.

7-5.15 (a) Compare the rules for resistors in parallel and for capacitors in parallel. (b) Compare the rules for resistors in series and for capacitors in series. (c) What fundamental relationships cause R and $1/C$ to be analogous?

7-5.16 To repair the base of the bulb-holder for his lamp, Steven used a stainless steel dinner knife, but he forgot to turn off the power. The dinner knife accidentally contacted both the side and the base of the lamp-socket. (a) Would you expect a fuse to blow? (b) Would you expect Steven to get a shock?

7-5.17 Researchers on artificial lightning, before setting off a bolt from their 20-foot-high apparatus, are warned 10 seconds early so that they can stand on one foot. Otherwise, they might receive a shock across their legs, more or less in proportion to the separation between their legs. Explain.

7-5.18 To protect yourself against lightning strikes when on a mountain, Flora recommends that you lie down on the ground so that you present no sharp peak that the lightning might strike. Dora, on the other hand, recommends that you crouch with your feet together, because that way you are not too high and if a lightning bolt strikes nearby, the path of the *ground current* that flows along the surface will have to choose between your nearly fixed foot-to-foot resistance and a minimized ground resistance. Discuss.

7-6.1 Consider a galvanometer of resistance 20 Ω and full-scale deflection of 50 μA. (a) To what voltage does this correspond? (b) Find its current sensitivity.

7-6.2 (a) Find the shunt resistance needed to convert a galvanometer of resistance 20 Ω and full-scale deflection of 50 μA into an ammeter with a full-scale current of $I = 10$ mA. (b) This ammeter is used to measure the current through an unknown resistor R, where the power source provides a constant 1.452 V, no matter what the load. If the ammeter reads 8.4 mA, find R and the current without the meter in the circuit.

7-6.3 (a) Find the series resistance needed to convert a galvanometer of resistance 20 Ω and full-scale deflection of 50 μA into a voltmeter with a full-scale voltage of $\Delta V = 100$ mV. This now is used to measure the voltage across an unknown resistor R, where the power source provides a constant 152 mA, no matter what the load. If the meter reads 84 mV, find (b) R and (c) the voltage without the meter in the circuit.

7-6.4 Consider an emf $\mathcal{E}$ with internal resistance r that is in series with a resistor R. You are to determine R using a voltmeter with resistance R_V and an ammeter with resistance R_A. The ammeter is placed in the circuit. (a) Let the voltmeter be connected across R alone. Show that $1/R = I/\Delta V - 1/R_V$. (b) Let the voltmeter be connected across R in series with the ammeter. Show that $R = \Delta V/I - R_A$.

7-6.5 A multimeter contains a galvanometer with sensitivity 25,000 Ω/V, which on its most sensitive scale reads 100 mV. (a) If, on its 10 V scale, it reads 6.5 V across a 200 Ω resistor, find the voltage read by an ideal voltmeter (assuming the current through the circuit is fixed). (b) Give the fractional error.

7-6.6 A multimeter contains a galvanometer with sensitivity 25,000 Ω/V, which on its most sensitive scale reads 20 mA. (a) If, on its 2 A scale, it reads 1.4 A across a 200 Ω resistor, find the current it would read with an ideal ammeter (assuming the voltage across the resistor is fixed). (b) Give the fractional error.

7-7.1 Describe the spacial profile of the ion distribution within one cell of an automobile battery that has been completely discharged, and then is given a quick charge.

7-7.2 (a) Why does a battery go dead when it just sits on a shelf? (b) Consider a 125 A-hr battery that loses chemical charge at a constant rate of 0.05 A. Estimate the amount of time it takes to lose 80% of its "charge."

7-7.3 Lead-acid cells usually go bad because of sulfation of $PbSO_4$ on both electrodes (at nearly 1% of the chemical charge per day) closing up the pores of both electrodes (large pores give a large reaction area, and thus a large current). Compare the chemical charge consumed this way with the chemical charge consumed by starting a car four times a day, each time with a draw of 600 A for 1 s. Assume that the fully charged battery holds a charge of 100 A-hr.

7-7.4 Bertie, who usually bikes to work, often finds that his automobile battery with 80 A-hr capacity is in need of recharging. When that happens, he usually puts it on a trickle charge of 2 A overnight (about 10 hours). Within a few weeks, his battery has gone dead again. What better charging regimen might you recommend to him?

7-7.5 An automobile alternator can provide a current of 130 A when the car is driven at highway speeds. The car battery has a capacity of 120 A-hr, but it has been completely discharged. After a jump-start, how long must the car be driven before it becomes half-charged?

7-7.6 Bertie has been advised by his mechanic that a solar-powered trickle-charger would prevent his infrequently used car battery from going dead. In "bright sun," the charger can provide its maximum current of 0.9 A; at all times the "charge" on the battery decays at a rate of 0.04 A, due to non–current-producing chemical reactions. For a 12-hour day, what fraction of bright sun light must be provided to just compensate for the loss due to these non–current-producing chemical reactions?

7-8.1 On open circuit, a high-resistance voltmeter reads 1.36 V for a voltaic cell. When a low-resistance ammeter reads 0.4 A, the voltmeter reads 1.24 V. Find the emf and internal resistance for the cell.

7-8.2 Consider a black box containing two leads. It is made part of a circuit. When an ammeter reads 0.24 A to the black box, the voltmeter reads 2.47 V across the black box. When the ammeter reads 0.38 A, the voltmeter reads 2.14 V. Find the emf and internal resistance for the black box.

7-8.3 When a person with a metal filling in a tooth places a piece of aluminum foil in her mouth, and the foil contacts the filling, she notices a strong "metallic" taste. (a) What are the elements in the electric circuit that is completed when contact is made? (b) What is the source of energy for this effect?

7-8.4 Here is a set of I versus ΔV data taken for an emf: (0.41 A, −5.6 V), (0.09 A, −3.2 V), (−0.19 A, 0.8 V), (0.51 A, 3.2 V). Estimate $\mathcal{E}$ and r with an eyeball fit to this simulated data. (Note: There is no unique answer, only a range of reasonable values.)

7-9.1 A 12 V car battery is rated at 120 A-hr. (a) To what charge in Coulombs does this correspond? (b) How long can this provide a 2 A current? (c) How long can this provide 60 W?

7-9.2 A Li/SO$_2$ D cell has $Q_{cell} = 8$ A-hr and $\mathcal{E} = 3.3$ V. (a) Find its energy storage. (b) Find its rate of discharge through a 6 Ω resistor. (c) Find how long the discharge will last.

7-9.3 A zinc-carbon AA cell has $E_{cell} = 3456$ J and $\mathcal{E} = 1.2$ V. (a) Find the charge stored, in A-hr. (b) Find its rate of discharge through a 12 Ω resistor. (c) Find how long the discharge will last.

7-9.4 It is conventional to use the term *battery* to describe both the 1.5 V AAA device used for flashlights and the 12 V device used for automobiles. One is a single voltaic cell, and one is a true battery of voltaic cells. Which is which?

7-10.1 Consider a voltaic cell of internal resistance $r = 0.15$ Ω, open circuit voltages across the left and right electrodes of magnitude 0.6 V and 1.2 V, and a net emf of $\mathcal{E} = 1.8$ V. It is in a circuit with a resistor $R = 0.75$ Ω, as in Figure 7.20(a). Let $V_a = 0.5$ V. Assume that the connecting wires have zero resistance. (a) Sketch the voltage around the circuit. If the voltaic cell has a "charge" of 1.2 A-hr, find (b) how long it will take to discharge, and (c) how much energy it has initially.

7-10.2 Repeat the previous problem if everything is the same, except that the net emf is $\mathcal{E} = 0.6$ V.

7-10.3 For a voltaic cell with $\mathcal{E}_2 = 1.3$ V, $\mathcal{E}_1 = 0.3$ V, and $r = 0.2$ Ω, connected to a resistor $R = 0.3$ Ω, the current flows clockwise. (a) Find the current flow for this circuit. (b) Plot the voltage profile around the circuit, taking $V = 0$ at electrode 1.

7-10.4 A lemon cell (two different metal electrodes placed in a lemon) is tested with two voltmeters. One voltmeter, with resistance 20,000 Ω, reads 0.65 V. The other voltmeter, with resistance 1000 Ω, reads 0.43 V. Find the internal resistance of the cell and its emf.

7-11.1 Consider two well-separated water droplets of $M = 10^{-20}$ kg, one with an excess electron and the other with a deficit of an electron. They are near the surface of the earth in a downward electric field of magnitude $E = 100$ V/m. (1) Find the total force on each, including both electricity and gravity. (2) Let the droplets have the same velocity for the same net force. Assuming that $\tau = 2.0 \times 10^{-9}$ s, find their terminal velocities.

7-11.2 A marble of radius 0.8 cm and density 2.5 g/cm³ falls through a container of shampoo, with density 1.0 g/cm³. If the marble reaches a terminal velocity of 0.9 cm/s, find the viscosity of the shampoo.

7-11.3 In the Millikan oil-drop experiment, $F_{drag} = -6\pi \eta R v$. Including the effect of the buoyancy of air (to be more accurate), the effective mass of the droplet is $m_{eff} = (\rho_{oil} - \rho_{air})(4\pi/3)R^3$. (a) Show that $\tau \sim R^2$. (b) Let $\rho_{oil} = 0.92$ g/cm³, $\rho_{air} = 1.293 \times 10^{-3}$ g/cm³, and $\eta_{air} = 0.0182 \times 10^{-3}$ N-s/m². Find τ/R^2. (c) Find the value of R that will give $v_\infty = 0.1$ mm/s.

7-11.4 A nighttime video accidentally records a flaming meteorite crashing into the ocean with a terminal velocity of 64 m/s. (a) Estimate τ for the meteorite. (b) If the meteorite is approximated as a sphere made of iron, estimate its radius. (c) Estimate its mass.

7-12.1 For a current density of 5.6 A/cm², how long would it take an electron to cross a 25-foot-long extension cord? *Hint:* What is the cord likely to be made of?

7-12.2 A 20 cm long rod with 4-mm-by-4-mm cross-section carries 4 A when a voltage difference of 0.5 V is placed across its ends. (a) Find the resistivity. (b) Find the electric field within the rod. (c) Estimate the drift velocity of the charge carriers, taken to be of density $n = 4.5 \times 10^{28}/m^3$.

7-12.3 A plasma globe has an inner radius of 0.6 cm and an outer radius of 9 cm. The total radially outward current is 25 μA and is due to a uniform density n of charge carriers of charge e, with drift velocity 40 m/s at the outer radius. Find n and the drift velocity at the inner radius.

7-12.4 At a proton storage ring, a beam of protons is uniformly distributed around a ring of circumference 1050 m and moves at nearly the speed of light. It produces a current of 40 A. The beam is then directed to a 75 kg block of copper of specific heat $C_v = 0.093$ cal/g-K, whose temperature rises by 0.215 K on absorbing the beam. (a) Find the time it takes for the Cu to absorb the beam. (b) Find the rate of temperature increase of the copper block while it absorbs the kinetic energy of the proton beam. (c) Find the number of protons in the beam. (d) Find the kinetic energy of the proton beam.

7-12.5 A disk with a uniform distribution of surface charge Σ, and of radius a, turns at an angular velocity ω. (a) Find the effective electric current dI of the charge between r and $r + dr$. (b) Find the total effective electric current I.

7-12.6 Consider a 0.3 mm diameter wire made of semiconductor with two charge carriers, electrons and holes (having charge $+e$). (a) If 5.4×10^{15} electrons/sec pass one way and 2.4×10^{15} holes/sec pass the other way, what is the total electric current? (b) If the electron density is twice the density of the holes, and the electron drift velocity is 4×10^{-3} m/s, what is the hole drift velocity?

7-12.7 An electrostatic generator is provided 850 nA by a belt that moves with velocity 8 m/s and is 4.4 cm wide. Find the charge per unit area on the belt.

7-12.8 (a) Find the maximum safe current densities for gauge #4, #10, and #16 wire. (b) To what drift velocities do these correspond? See Problem 7-4.6 for information about gauges.

7-12.9 Our analysis of current flow through a wire can be applied to traffic flow along a highway. A tunnel of length L and width w (proportional to the number of lanes) is like a wire of length L and area A. The areal density of cars n_c is like the volume density of charge carriers n. The average velocity of cars v_c is like the drift velocity v_d. Show that the number of cars per unit time crossing the width is given by $n_c v_c w$.

7-13.1 Arrange in order of increasing density of conduction electrons: conductors, insulators, semiconductors.

7-13.2 Consider a typical metal. (a) Explain why its conductivity decreases with increasing temperature. (b) Explain why the greater the number of impurities, the lower the conductivity at low temperatures.

7-13.3 Suggest why semiconductors show nonlinear behavior at smaller current densities than do metals.

7-13.4 Explain how, by having an additional low resistivity mode of conduction, superconductors can bypass the finite resistivity of its conduction electrons.

7-G.1 Consider a slab-shaped wire of cross-sectional area A along the horizontal that is much less than the cross-sectional area A' along the vertical. See Figure 7.30. (a) Sketch the current flow pattern for current·flow along the horizontal, from a to a'. (b) Repeat for current flow along the vertical, from b to b'. (c) Which case should have the larger resistance?

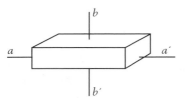

Figure 7.30 Problem 7-G.1.

7-G.2 (a) The voltage across a wire is quadrupled. If the drift velocity had been 2×10^{-4} m/s, find the new drift velocity. (b) The voltage across the accelerating grid screen in a TV tube is quadrupled. If the velocity of the electrons hitting the screen had been 4×10^6 m/sec, find their new velocity when they hit the screen. *Hint:* Recall that for a TV tube, the electrons have relatively low velocity on emission by the cathode and before acceleration.

7-G.3 When the electric field lines for a resistor and for a capacitor are the same, the resistance R of one and the capacitance C of the other are related. (a) For the capacitor problem, with the voltage in the dielectric decreased by the factor κ (the dielectric constant) appropriate to that material, show that $\Delta V = Q/C = \oint \vec{E} \cdot \hat{n} dA/(4\pi k\kappa C)$. See Figure 7.31(a). (b) For the resistor problem, show that $\Delta V = IR = (\int \vec{J} \cdot \hat{n} dA)R = \sigma(\int \vec{E} \cdot \hat{n} dA)R$. See Figure 7.31(b). (c) Show that $RC =$

$(4\pi k\kappa\sigma)^{-1}$. (Since current normally is confined to a wire, the analogous capacitance problem must have a material whose large dielectric constant confines the electric field lines to the same region as the wire, or else the geometry must be very simple, such as concentric spheres, concentric cylinders, or parallel plates.)

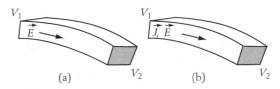

Figure 7.31 Problem 7-G.3.

7-G.4 Consider two nonconcentric spherical electrodes of radii a and b, at a large separation $r \gg a, b$. (a) Show that $C = Q/(V_a - V_b) = [k(a^{-1} + b^{-1} - 2r^{-1})]^{-1}$. (b) Show that, for the resistor case, $R = (1/a + 1/b - 2/r)/(4\pi\sigma)$.

7-G.5 Consider a reduced portion of two concentric spheres, given by the difference between a spherical cone of radius b and one of radius a, of common half-angle θ. See Figure 7.32. (a) Show that, for $\kappa \gg 1$, $C = Q/(V_a - V_b) = (1 - \cos\theta)[\kappa k(a^{-1} - b^{-1})]^{-1}$. (b) Show that $R = (a^{-1} - b^{-1})/[2\pi\sigma(1 - \cos\theta)]$. (c) Why must $\kappa \gg 1$?

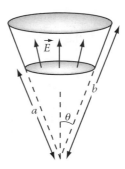

Figure 7.32 Problem 7-G.5.

7-G.6 (a) Discuss the changes in the field lines if the spherical cone resistor from a to b is cut off to yield a flat base. (b) For small cone angles θ, calculate the resistance. Assume a uniform resistivity ρ. (c) Is your approximation valid for all cone angles?

7-G.7 Compute the product of resistance and length for radial flow between two infinitely long cylinders with uniform resistivity ρ, radii $a < b$, and angle α rather than 2π. See Figure 7.33.

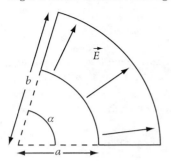

Figure 7.33 Problem 7-G.7.

7-G.8 For a wire, when the electric field is due only to electric charge, we can write $\vec{E} = -\vec{\nabla}\phi$. In that case, the local form of Ohm's law becomes $\vec{J} = \sigma\vec{E} = -\sigma\vec{\nabla}\phi$. For heat flow, Fourier showed that the heat current $\vec{Q}$ (an energy per unit time per unit area) satisfies $\vec{Q} = -\kappa\vec{\nabla}T$, where κ is the thermal conductivity and T is the temperature. Hence there are similarities between heat flow and current flow. By analogy with (7.5), find an expression for the thermal resistance $R_H \equiv \Delta T/I_Q$ for a wire of length l and cross-section A, where $I_Q = |\vec{Q}|A$ is an energy per unit time.

7-G.9 Ohm actually measured a magnetic needle deflection $X = a/(b + x)$, where a and b were fitting constants, and x was the length of his wire. This he took to be proportional to the current through the wire wrapped about the needle in this primitive form of galvanometer. During a given run, where he would change x but not anything else, the values of a and b did not change. To what extent does such a form establish what we now call Ohm's law? *Hint:* What does this result say about linearity in the emf? In the current? On the dependence on properties of the wire?

7-G.10 In 1820, Ampere wrote that "The currents of which I am speaking are accelerated until the inertia of the electric fluids and the resistance which they encounter... make equilibrium with the electromotive force, after which they continue indefinitely with constant velocity so long as this force has the same intensity." Thus Ampere commits himself to these ideas: (a) a two-fluid model of electricity; (b) that inertia matters in determining the drift velocities of the "electric fluids;" and (c) that resistance (in the sense of drag) matters in determining the drift velocities of the "electric fluids." Which, if any, of these views are correct for ordinary metals?

7-G.11 Assume that the solar wind is due to a uniform spherical flux of protons, which undergo no collisions. (a) If their speed is 550 km/s just outside the sun, what is their speed at the earth? (Neglect electric fields and the earth's gravity, but include the sun's gravity.) (b) If their net current is I just outside the sun, and there is no charge buildup between the earth and the sun, what is I at the earth? (c) How does their density vary with distance from the sun? (d) If their density is $8.5/\text{cm}^3$ at the earth, what is their density just outside the sun? (e) Compute I. (f) Compute the current density at the earth and just outside the sun. (g) If the sun starts out uncharged on January 1 of this year, and then becomes charged negatively because of the solar wind, how large an electric field would there be at the earth on January 1 of next year? (h) Does this model appear to be realistic?

7-G.12 Heating rust (Fe_2O_3) in the presence of CO yields $Fe_2O_3 + 3CO \Rightarrow 2Fe + 3CO_2$. Thinking only in terms of weight, why is this process called *reduction*? Why is the inverse process, or the process $4Fe + 3O_2 \Rightarrow 2Fe_2O_3$, called *oxidation*? Given the ionic states Fe^{+3} and O^{-2}, discuss electron transfer to the iron under oxidation and reduction. (Although heating rust in air is ineffective at reducing the iron, some oxides can be reduced simply by heating them in air.)

7-G.13 A resistor with the color code violet-red-orange-silver is connected to a 12 V battery. (a) Find the radial resistance. (b) Find the expected current. (c) Find the largest anticipated current.

Chapter 8

Batteries, Kirchhoff's Rules, and Complex Circuits

Chapter Overview

Section 8.1 provides a brief introduction. Section 8.2 reviews some of the history of Galvani's discovery of the *galvanic cell* (which is now called the voltaic cell), and of Volta's invention of the *voltaic pile* (which is now called the battery). (Neither of these workers understood the science of their very real discoveries.) Section 8.3 discusses a battery of identical voltaic cells, and Section 8.4 discusses the relative cost of electrical power obtained from batteries and from the electric company. Section 8.5 considers the distinct issues of maximizing the actual amount of power transfer (relevant to high-power applications) and maximizing the efficiency of that power transfer (relevant to low-power applications). Section 8.6 presents Kirchhoff's rules. Section 8.7 presents some nontrivial applications of Kirchhoff's rules, including the jumper-cable problem. Section 8.8 considers the short-time and long-time behavior of circuits with both capacitors and resistors. This is a prequel for the discussion, in Section 8.9, of the charge and discharge of *RC* circuits. Section 8.10 considers the charge on the surface of a circuit, which produces the electric field within the volume of the circuit. It also shows how the buildup of surface charge on a wire can be analyzed in terms of *parasitic capacitance* in parallel with the resistance of the wire; the sharper the turns on the electronic superhighway, the slower the circuit can respond. Section 8.11 contains an optional discussion of the *bridge circuit*. Section 8.12 presents an optional discussion of *plasma oscillations,* a collective of motion of the electrons in metallic conductors. Section 8.13 provides a brief Interlude, summarizing the material already studied, and indicating what yet has to be treated. It discusses the limitations on the validity of a circuit analysis, which assumes instantaneous action at a distance rather than including the finite speed of light, and discusses what happens when we first throw a switch in a circuit, or when we send an electrical signal down a cable. A full treatment requires that we pursue the relationship of electricity to magnetism, and that we study electromagnetic radiation. This provides a lead-in for the remainder of the book. ▪

8.1 Introduction

The concept of an *electric circuit* did not exist when the only type of emf was electrostatic. Moreover, not until Volta's invention of the voltaic pile, or battery, could large electric currents be sustained. In this chapter, as we progress to the study of complex circuits, we first study the properties of the battery as a combination of individual voltaic cells.

Most circuits cannot be analyzed solely in terms of series and parallel resistors with a battery here or there. To deal with bridge circuits, or systems with multiple arms containing sources of chemical emf, or systems with both resistors and capacitors, we must apply the ideas of the previous chapter in a more systematic fashion, using *Kirchhoff's rules*, first stated around 1850. This systematic approach to circuits uses solely the macroscopic properties of circuits: resistances, capacitances, and emfs (both electrostatic and nonelectrostatic).

8.2 Discovery Must Include Reproducibility: It Need Not Include Understanding
Optional

The history of the voltaic cell and of the battery illustrates that a great discovery or invention can be made even when the discoverer or inventor does not understand how the phenomenon or device works.

The anatomist Galvani, in 1780, noticed that frogs' legs went into spasms when stimulated by an external electrostatic source. However, it took him much effort to make the work reproducible. By 1786, he had learned that frogs' legs would also go into spasms when made part of a circuit with two dissimilar metals. He concluded that the frogs' legs themselves were a source of "animal" electricity, of the same nature as ordinary electricity. He thought in terms of the discharge of something like a Leyden jar internal to the frog, which through an unknown biological process could recharge. Galvani's explanation was wrong, but he had made a great discovery, which he published in 1791. It had taken him many years to make the effect reproducible. Galvani is the father of the field of *electrophysiology*.

The physicist Volta—already renowned for the discovery of methane, for the invention of the electrophorus (see Chapter 1), and for his studies of capacitance—began to study Galvani's effect. He realized that the frogs' legs were serving as sensitive detectors, rather than as sources of electricity. By 1792, he had established that the two metals ("dry conductors") were necessary to cause an electric current to flow, but that the frog's leg was not. It simply served as what he called a "moist conductor" (i.e., an electrolyte, like salt water). He went on to discover the electrochemical series (e.g., silver electrodes are higher in voltage by 1.55 V than zinc electrodes). He also made the observation that placing intermediate metals in the circuit (e.g., between the silver and the zinc) had no effect on the current produced.

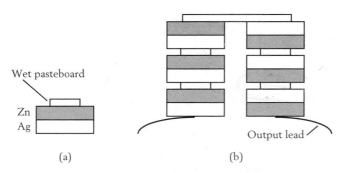

Figure 8.1 A simple battery. (a) A single voltaic cell consisting of a piece of Zn and a piece of Ag separated by wet pasteboard (which provides ions). (b) A battery consisting of many such voltaic cells in series. Note that the Zn electrode on the bottom right, and the Ag electrode on the bottom left, have no effect on the net emf.

In 1800, Volta invented what came to be known as the voltaic pile, a battery of voltaic cells connected in series, which made the effects that he had been studying much more intense. See Figure 8.1. One of his early piles had 32 Zn–Ag voltaic cells in series, with about 50 volts across the terminals of the pile, a powerful battery indeed. Almost immediately, other scientists used the voltaic pile to decompose water, collecting hydrogen at one electrode and oxygen at the other. Thus was born the subject of electrochemistry.

Using his considerable experimental skill, in 1797, Volta discovered another effect related to voltaic cells—the *contact potential* between two dissimilar metals. Unlike the chemical emf, which is associated with transfer of both electrons and ions, the contact potential is associated with transfer only of electrons. The contact potential is a measure of the work function of a metal. (Recall that the work function is the energy to remove an electron from a metal. It is related to the energy to remove an electron from an isolated atom of that metal.)

Although Volta was aware that oxidation occurred at the silver electrode, he focused only on the electric current aspect of his cells, considering the associated flow of chemicals to be a mere side effect of no fundamental significance. His view was that the contact potential at the metal–metal interface was the power source that drove the electric current in his cells, and that the electrolyte served only to bring the other ends of the metals to the same electrical potential. In other words, to Volta the voltaic cell was a perpetual motion machine. (This was some 50 years before conservation of energy was an established principle.) He was wrong, but his reputation, justly deserved on the basis of his many contributions to the science of electricity, kept the scientific community from seriously challenging his viewpoint for many years. Even after Faraday's work on electrolysis established that the electrical and chemical effects are inextricable, many prominent scientists continued to accept the contact potential as the energy source for the voltaic cell.

A simple experiment would have demonstrated that Volta was wrong about the source of energy in the voltaic cell. As 1915 Nobel Laureate in physics

W. L. Bragg (in *Electricity*, Macmillan, New York, 1936, p. 50) wrote:

> *The energy for driving the current [in Galvani's frog experiments] comes from the slight chemical action between the metals and the muscles or nerves they touch. In Volta's Pile it comes from an action of the liquid upon the metals themselves, and if you look at illustrations of his pile you will see that there is an extra silver plate at one end and zinc plate at the other which are really unnecessary and do not help to increase the strength of the pile.*

See Figure 8.1. As Bragg indicates, removal of these extra plates would have yielded no change in the chemical emf, in contradiction to Volta's view.

Ultimately, it was accepted that the energy source for the voltaic cell was chemical in origin. However, there was no full-fledged theory until chemistry itself had become more developed, toward the end of the 19th century. A full discussion of voltaic cells, including what happens within the electrolyte and the electrodes, must consider both their chemical and their physical aspects. Physicists and chemists are still trying to develop a quantitatively accurate microscopic theory of the voltaic cell. The hardest part is the description of what happens at the electrode–electrolyte interface.

8.3 Batteries Are Combinations of Voltaic Cells

8.3.1 *Identical Voltaic Cells in Series*

Consider n identical voltaic cells each with chemical emf $\mathcal{E}_1$ and internal resistance r_1, in series with a resistor R. We wish to find the current and terminal voltage. Figure 8.2(a) depicts six of them, as would be found within a common 9 V battery used for electronics (containing six 1.5 V cells inside; open one up if you don't believe this). As in the previous chapter's discussion of a resistor R in series with a single cell, this will be analyzed in two ways. First, Ohm's law

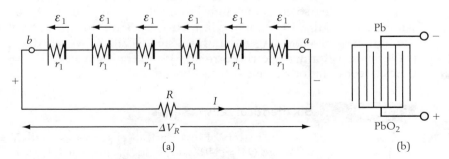

(a)   (b)

Figure 8.2 Identical voltaic cells. (a) A schematic of identical voltaic cells connected in series across a load resistor R. The net emf is proportional to the number of cells in series. (b) A schematic of the connections for a lead-acid cell, where many nominally identical voltaic cells are placed in parallel. The maximum current is proportional to the number of cells in parallel.

will be applied to each circuit element. Next, Ohm's law will be applied to the circuit as a whole.

Because electric charge is conserved, for steady current flow, all the cells and the resistor R have the same current I passing through them. By symmetry, each cell has the same voltage ΔV_1 across its terminals. Moreover, by uniqueness of the voltage difference, $V_b - V_a$, the voltage drop across the upper arm $(n\Delta V_1)$ is the same as that across the lower arm (ΔV_R), so $n\Delta V_1 = \Delta V_R$. Also, by Ohm's law, with $\mathcal{E}_1$ and ΔV_1 driving current in opposite directions, each cell satisfies

$$I = \frac{\mathcal{E}_1 - \Delta V_1}{r_1}, \qquad \Delta V_1 = \frac{1}{n}(V_b - V_a), \tag{8.1}$$

and the resistor satisfies

$$I = \frac{\Delta V_R}{R}, \qquad \Delta V_R = (V_b - V_a) = n\Delta V_1. \tag{8.2}$$

With this information, we can obtain the currents and voltages associated with the circuit. Solving (8.1) and (8.2) for ΔV_R yields

$$\Delta V_R = IR = n\Delta V_1 = n(\mathcal{E}_1 - Ir_1). \tag{8.3}$$

Solving for I and then ΔV_R gives

$$I = \frac{n\mathcal{E}_1}{nr_1 + R}, \qquad \Delta V_R = \frac{n\mathcal{E}_1 R}{nr_1 + R}. \tag{8.4}$$

Hence, the effective chemical emf $\mathcal{E}_{eff}$ and effective internal resistance r_{eff} are

$$\mathcal{E}_{eff} = n\mathcal{E}_1, \qquad r_{eff} = nr_1. \tag{8.5}$$

Thus, putting the cells in series increases the net internal resistance and the net chemical emf. It also increases the net energy that the system can provide.

Alternatively, consider the system as a whole. The net chemical emf is due to n cells in series, so $\mathcal{E}_{eff} = n\mathcal{E}_1$. The net resistance is due to n internal resistances r and R in series, so $R_{eff} = nr + R$. From Ohm's law applied to the circuit as a whole, $I = \mathcal{E}_{eff}/R_{eff}$, (8.4) follows immediately. Thus, for the battery as a whole, the emf $\mathcal{E}_{eff}$ is n times larger than for a single cell. However, the "charge" Q_{eff} is the same as for a single cell. By (7.28), the energy storage $E_{cell} = \mathcal{E}_{eff} Q_{eff}$ is n times larger than for a single cell, as expected.

Example 8.1   **Six identical cells in series**

Consider an electronics battery of six identical cells in series, with $\mathcal{E}_1 = 1.5$ V, $r_1 = 0.12$ Ω, and charge $Q_1 = 240$ C. (a) Find $\mathcal{E}_{eff}$ and r_{eff}. (b) Find the current I through and the voltage ΔV_R across a 2.28 Ω resistor connected to this battery. (c) Find the energy stored by this battery.

Solution: (a) By (8.5), $\mathcal{E}_{eff} = 6(1.5) = 9$ V and $r_{eff} = 6(0.12) = 0.72$ Ω. (b) By (8.4), $I = 3$ A and $\Delta V_R = 6.84$ V. (c) By (7.28), $E = 9(240) = 2,160$ J. This is six times the energy storage $1.5(240) = 360$ J for a single cell.

8.3.2 *Voltaic Cells in Parallel*

Open up a 12 V car battery and you will find six 2 V lead storage cells in series. Moreover, each cell really consists of many (negative) Pb electrodes connected in parallel, interlocking with (positive) PbO_2 electrodes connected in parallel. See Figure 8.2(b), with five Pb and four PbO_2, which corresponds to a common motorcycle battery. This is equivalent to eight 2 V cells in parallel; both sides of the four PbO_2 electrodes contribute.

Example 8.2 **Eight identical cells in parallel**

Consider eight identical cells ($\mathcal{E}_1$, r_1) in parallel, with a terminal voltage ΔV and charge Q_1. (a) Find the current I and effective resistance r_{eff} of the system in terms of $\mathcal{E}_1$, ΔV, and r_1. (b) For the battery as a whole, find its effective charge Q_{eff}, its effective emf $\mathcal{E}_{eff}$, and its energy storage E relative to the values for a single cell.

Solution: (a) For eight identical cells ($\mathcal{E}_1$, r_1) in parallel, as in Figure 8.2(b), the current adds, so $I = 8(\mathcal{E}_1 - \Delta V)/r_1 = (\mathcal{E}_1 - \Delta V)/(r_1/8)$. Thus $r_{eff} = r_1/8$. (b) This use of eight cells in parallel gives an effective emf that is the same as for a single cell: $\mathcal{E}_{eff} = \mathcal{E}_1$. However, the effective charge and the energy stored increase by a factor of eight: $Q_{eff} = 8Q_1$, $E_{eff} = \mathcal{E}_{eff}Q_{eff} = 8\mathcal{E}_1 Q_1 = 8E_1$.

The electric eel uses a different design than a car battery: instead of cells in series, each cell having many subcells in parallel, the eel has many cells in parallel, each cell having many subcells in series. Moreover, the source of energy in the eel's specialized cells, called *electrocytes*, is not due to chemical reactions. Rather, they employ *ion pumps* that selectively and actively transfer certain ions across the cell wall. *Ion channels* also selectively but passively permit certain ions to cross the cell wall.

8.3.3 *Terminal Voltage of a Battery in Use*

As for an individual voltaic cell, the *terminal voltage* ΔV_T of a battery is—not surprisingly—defined as the voltage across the terminals of the battery. The sign is chosen so that ΔV_T opposes the chemical emf $\mathcal{E}$. See Figure 8.3. By Ohm's law, if positive current flow is taken to be caused by the emf $\mathcal{E}$, then

$$I = \frac{\mathcal{E} - \Delta V_T}{r}, \tag{8.6}$$

so

$$\Delta V_T = \mathcal{E} - Ir. \tag{8.7}$$

Equation (8.6) is equivalent to (7.25) of the previous chapter, with ΔV relabeled ΔV_T. It has also been used as (8.1).

On open circuit, so there is no current flow, $\Delta V_T = \mathcal{E}$. On discharging, where $I > 0$, by (8.7) ΔV_T is less than $\mathcal{E}$. On charging, where $I < 0$, by (8.7) ΔV_T exceeds $\mathcal{E}$. However, the terminal voltage is not always described by (8.7). This

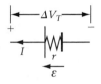

Figure 8.3
Terminal voltage
ΔV_T of a battery.

is because the voltaic cells within batteries really are more complex internally, as discussed in Chapter 7.

Batteries that can be recharged (such as car batteries) are called *secondary* batteries; those that cannot be recharged (such as Zn–Ag cells) are called *primary* batteries. Chemical reactions are not always, in practice, reversible (e.g., a gas might form on an electrode), or the byproduct at an electrode might get consumed by another reactant within the electrolyte, so not all batteries are rechargeable. In recent years, the shelf life of alkaline batteries has dramatically increased, in part because improvement in separator materials within the electrolyte (MnO_2) has decreased internal chemical reactions.

8.4 The "Charge" on a Battery, and Its Cost

We now estimate the amount of "charge" that an automobile battery can provide. (Remember, this is not the transfer of charge from one material to another, as occurs with rubbing, but rather the transfer of charge around an electric circuit.) When the two automobile headlights are on (using about 36 W each), by (7.31) the current that they draw is about $2 \cdot 36W/12V = 6$ A. If our automobile battery can be discharged after about four hours, we deduce that its charge is about $6 \text{ A} \cdot 4 \text{ hr} = 6 \text{ A} \cdot 14,400 \text{ s} = 86,400$ C. Since the charge should not depend on how the cells are connected, it should be the same when they are in parallel, in which case it is clear that each of the 48 cells has a charge of 1800 C. No wonder that, when voltaic cells were first invented, they were such a marvel; contrast that with the 10^{-9} C associated with static electricity from a comb pulled through clean, dry hair. (The original batteries did not have a charge as large as 1800 C, but even a charge of 10 C is enormous compared to what could be obtained by static electricity.) It would take a considerable time to produce 1800 C of charge by electrostatic methods, even by some of the sophisticated electrostatic induction-based "charge-doubling" devices of the late 18th century. Moreover, such charge would be stored at high voltages, and thus be subject to electrical breakdown. Figure 8.4 presents terminal voltage versus time for slow and rapid discharge, for a lead-acid cell. Not only is the discharge time shorter on a fast discharge (by definition), but the terminal voltage is less, following (8.6). In comparing voltaic cells of the same type, a good relative measure of how long

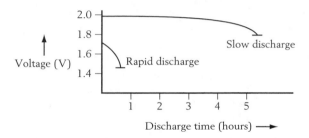

Figure 8.4 Characteristic discharge properties of a voltaic cell.

they will last (their charge) is their relative mass. Thus there are 1.5 V AAA (11 g), AA (23 g), C (66 g), and D (138 g) alkaline cells with charges of about 1.2, 2.5, 6.8, and 14.0 A-hr of low-current (10 mA) usage.

Example 8.3   **Electricity cost: batteries versus the power company**

Find the cost per hour to run a 100 W lightbulb both for 80 D cells in series (to provide 120 V), at $0.50 per D cell, and for 120 V ac power, at $.16 per kilowatt-hour. Neglect internal resistance. (In practice, the internal resistance of the batteries would dominate the lightbulb resistance, so the lightbulb probably would not even light.)

Solution: For 100 W at 120 V, the current is $I = 100$ W/120 V $= 0.83$ A. 80 D cells in series are equivalent to a 120 V dc cell. They will last a time 5.2A-hr/0.83A $= 6.24$ hr, at a cost of $(80)(\$0.50) = \$40$. On the other hand, at $.16 per kilowatt-hour, the electric company will charge for $(100$ W$)(6.24$ hr$) = 0.624$ kW-hr, or $1.00. Thus, for these prices battery power is 40 times as expensive as power from the electric company; about $12.80 per kW-hr.

Finally, note that heavy-duty alkaline cells last nearly twice as long for high current usage as do general-purpose alkaline cells although the general-purpose cells last nearly as long for low current usage. Typical alkaline cells use a Zn cathode, an MnO_2 anode, and a KOH electrolyte.

8.5 Maximizing Power Transfer versus Maximizing Efficiency of Power Transfer

Now consider a battery of emf $\mathcal{E}$ and internal resistance r in series with a resistor R, and consider the power transfer to R, called the *load*, or *load resistor*. (The usage comes from mechanics, where we speak of machines driving a mechanical load.) See Figure 8.5. Two different questions often considered are (1) how to get the maximum power transfer to R (independent of how much power is lost in the internal resistance r); (2) how to get the most efficient power transfer to R (thus minimizing the power loss in the internal resistance r).

8.5.1 *Impedance Match (R = r) for Maximum Power Transfer*

Is it possible to play tennis with a ping-pong paddle, or ping-pong with a tennis racket? Yes. However, no championship tennis player uses a ping-pong paddle

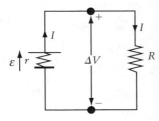

Figure 8.5 A battery in series with a load resistor, to determine how to maximize power transfer and how to maximize efficiency.

(even one made of steel, so it won't break). For a given ball, we must ask the question "What racket gives the maximum power transfer?" (This neglects the issue of *control*.) The answer is that we should *match the impedance* of the racket to that of the ball. The principle of *impedance matching* is one of the most useful design principles that a scientist or engineer can employ. More generally, impedance matching tells us to match a property of the power source—the impedance (whatever that is in a given situation)—to the corresponding property of the object receiving that power. We now establish this principle for an electric circuit with resistors, where *impedance* means *electrical resistance*, and we then apply this principle to estimate the electrical resistance of both the starting motor and the battery of a car.

Consider a chemical emf $\mathcal{E}$ with internal resistance r in series with a resistor R, as in Figure 8.5. From (8.4) with $n = 1$, $\mathcal{E} = \mathcal{E}_1$, and $r = r_1$, the current is

$$I = \frac{\mathcal{E}}{r + R}. \qquad (8.8)$$

Then the rate of heating $\mathcal{P}$ of R is

$$\mathcal{P} = I^2 R = \frac{\mathcal{E}^2 R}{(r + R)^2} = \frac{\mathcal{E}^2}{r} \frac{R/r}{(1 + R/r)^2}. \qquad (8.9)$$

Figure 8.6 plots two dimensionless quantities, $\mathcal{P}/(\mathcal{E}^2/r)$ versus R/r, showing that a maximum occurs near $R/r = 1$. Such a maximum makes sense: for small R/r the load (R in $I^2 R$) is so small that it does not use very much power, and for large R/r the load is so large that the current (I in $I^2 R$) is not very large. For some intermediate value of R/r we thus expect a maximum, which occurs when its derivative with respect to the "load" R is zero. Using the second equality in (8.9),

$$\frac{d\mathcal{P}}{dR} = \frac{\mathcal{E}^2}{(r + R)^2} - 2\frac{\mathcal{E}^2 R}{(r + R)^3} = \frac{\mathcal{E}^2(r - R)}{(r + R)^3}. \qquad (8.10)$$

Clearly the maximum occurs for $r = R$: when this condition is satisfied, we say that there is impedance matching. In this case $I^2 R = I^2 r$ so that the power provided by the battery goes equally to the load and the internal resistance.

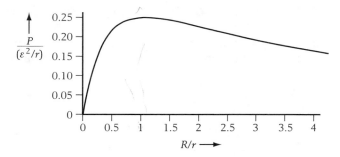

Figure 8.6 Normalized power transfer to the load resistor, as a function of the ratio of load resistance to internal resistance.

Example 8.4 **Power consumption by a starting motor**

Consider a 12 V car battery that provides $I = 600$ "cold-cranking amps" (CCA) on startup. Find the internal resistance r that gives maximum power to the starting motor, and find that maximum power.

Solution: Starting the car is the greatest power use of the battery, so for maximum power the starting motor resistance R should be impedance matched to the battery resistance r: $R = r$. With $\mathcal{E} = 12$ V and $I = 600$ cold-cranking amps, (8.8) gives $\mathcal{E} = I(R+r) = I(2R)$, or $12 = 600(2R)$, so $R = r = 0.01$ ohm for both the battery and the starting motor. The power consumed by the starting motor, taken to be impedance matched, is $\mathcal{P} = I^2 R = 3600$ W. A rule of thumb in the car industry is that a battery should have about 1.5 CCA per cubic inch of engine. Note that an electrical motor in operation produces a so-called *back emf* that opposes the driving emf, and thus causes the current to decrease as the motor comes into operation. This back emf is a consequence of Faraday's law, to be discussed in Chapter 12. It has nothing to do with chemical emfs.

8.5.2 *Low Internal Resistance ($r \ll R$) for Efficient Power Transfer*

There is another, distinct, principle involved when we want *maximum efficiency* of power transfer, so that nearly all the energy provided by the chemical emf goes to the load R rather than to the internal resistance r. Here, we want to maximize the ratio of the power to the load ($I^2 R$) relative to the total power ($I^2(r + R)$). That is, we want to maximize

$$\text{Efficiency} = \frac{I^2 R}{I^2(r + R)} = \frac{R}{r + R} \tag{8.11}$$

with respect to r, for fixed R. It does not take a rocket scientist (i.e., calculus) to determine that this occurs for $r \to 0$. This principle is used in low-power applications.

Example 8.5 **Current draw by a headlight**

Consider a headlamp, consuming $\mathcal{P} = I^2 R = 36$ W. (a) What current must the battery produce to run the headlamp, assuming maximum efficiency of power transfer? (b) Find the rate of heating of the internal resistance.

Solution: Assume that $R \gg r$. Then (8.8) gives $I \approx \mathcal{E}/R$, so $\mathcal{P} \approx \mathcal{E}^2/R$. Thus $R \approx \mathcal{E}^2/\mathcal{P} = 144/36 = 4$ Ω. As assumed, this is indeed much more than $r = 0.01$ Ω. The battery can easily provide 36 W, and the rate of heating of the internal resistance, $I^2 r = I^2 R(r/R) \approx 36(.01/4) = 0.09$ W, is negligible. Note that $I = \mathcal{E}/R = 12/4 = 3$ A. Since batteries are less effective at low temperatures, in cold climates some people turn on the headlights before trying to start their car. As this example shows, that can hardly warm the battery. The most effective way to heat the battery is to short its terminals so that all its power goes into the battery itself. That will provide $I = \mathcal{E}/R = 12/0.01 = 1200$ A, a huge current that can cause sparking and can be provided for only a few seconds at a time.

8.6   Kirchhoff's Rules Tell Us How to Analyze Complex Circuits

Properly speaking, Kirchhoff's rules are a set of two rules that describe any sort of circuit with connecting wires of zero resistance, resistors R, capacitors C, and nonelectrostatic sources of emf $\mathcal{E}$. However, before they can be applied to any circuit, the currents and voltages for that circuit must be defined precisely. Therefore, we precede Kirchhoff's rules with rule 0.

> **Rule 0**   *Draw a schematic of the circuit, including a sign convention for what you mean by positive current flow I through each arm.*

In addition, if specific voltages will later appear in your analysis, define them explicitly. We have already applied the zeroth rule in Figure 8.2(a), Figure 8.3, and Figure 8.5, but it may have slipped past you that *someone* had to decide what to call positive and negative. This is like setting up a coordinate system, where you have to choose which direction is positive. (Precision is valuable even in ordinary conversation: proper names lead to less confusion than "he," "she," or "it"; specific directions, like "up" or "down," are better than "this way" and "that way.")

We now apply Rule 0 to the basic circuit in Figure 8.7(a), where only the nonelectrostatic emf and the resistances are given. Because a battery of positive emf drives current up through the battery and down through the resistors, we take the conventions for positive current flow as in Figure 8.7(b). For a given arm of the circuit the direction drawn for the current doesn't mean that the current actually flows in that direction, only that a positive current would flow in that direction. Similarly, because we expect that the battery will pump positive charge to the top, the positive voltage difference is defined as in Figure 8.7(b). The positive side doesn't mean that the voltage is higher there, only that a positive voltage would be higher there. If the battery were reversed, we could still employ the same sign conventions, but the currents and voltage difference would all have negative sign.

For the circuit of Figure 8.7(a), there are two ways to define positive for each of the three currents as well as for the one voltage difference. Hence there are a total of $2^4 = 16$ possibilities for defining the sign conventions; Figure 8.7(b) is only one of them. All 16 possibilities will yield the same (correct) answers once the different sign conventions are accounted for. (We could even

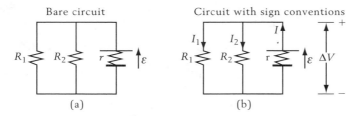

Figure 8.7 Circuit to illustrate Rule 0 of Kirchhoff's rules: define your sign conventions. (a) Basic circuit, with no sign conventions given. (b) Same circuit, but with a set of sign conventions chosen.

choose sign conventions by flipping a coin. What counts is sticking to the sign convention.)

In a complex problem, with 20 resistors and 10 batteries, there is no *a priori* way to tell how any of the currents flow or the sign of any of the relative voltages. It is satisfactory to arbitrarily choose a set of sign conventions and then stick to them. If I_9 turns out to be negative, then I_9 actually flows opposite to the direction of positive I_9. If ΔV_7 turns out to be negative, then the actual sign of ΔV_7 is opposite to positive ΔV_7.

For a related problem (traffic flow), defining the direction of positive current flow is like deciding, for a north–south street, whether to call northward or southward the positive direction; if northward is positive, then southward is negative. *Because your lab partner might not choose the same conventions, it is important to show in your figure your conventions for positive directions.*

Positive or Negative

Recall that, for a ball starting at the origin, at rest, and falling, if positive y is upward, then $y = -\frac{1}{2}gt^2$; and if positive y is downward, then $y = \frac{1}{2}gt^2$. In each case, the ball falls down, even though "down" is negative in one case and positive in the other.

Rule 1 *Apply charge conservation to the circuit.*

(This can be done only after defining the sign conventions for current, in Rule 0.) For each *node* (or *junction*) of the circuit, this amounts to ensuring that *the current into each node equals the current out of that node.* This is known as Kirchhoff's first rule, and is often called the *nodal rule*, or the *junction rule.* That is, at each node,

$$\Sigma I_{in} = \Sigma I_{out}. \tag{8.12}$$

For Figure 8.7(b), (8.12) applied to the top node implies that $I = I_1 + I_2$; applied to the bottom node, it implies the equivalent result $I_1 + I_2 = I$. [More generally, for a closed circuit with n nodes, there are only $n - 1$ independent applications of (8.12).] For Figure 8.8(a), (8.12) implies that $I_1 + I_2 = I_3$.

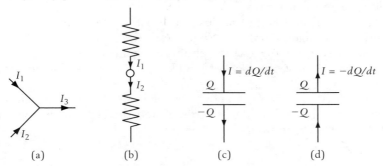

(a) (b) (c) (d)

Figure 8.8 Circuit to illustrate Rule 1 of Kirchhoff's rules: apply charge conservation. (a) Current in = current out ($I_1 = I_2$) at the node where three arms meet. (b) Current in = current out ($I_1 = I_2$) for a node that separates two parts of a single wire. (c) Relationship between current and charge when positive current enters positive plate of a capacitor. (d) Relationship between current and charge when positive current enters negative plate of a capacitor.

Example 8.6 **Current leaving one resistor and entering another**

Between any two circuit elements, we can take an arbitrary point to be a node. See Figure 8.8(b), which shows the two resistors R_1 and R_2. Relate I_1 and I_2.

Solution: Here, (8.12) implies that $I_1 = I_2$.

Example 8.7 **Current and charge for a capacitor**

Relate I in a wire leading to a capacitor having charge Q, to dQ/dt for the capacitor, in Figure 8.8(c) and Figure 8.8(d).

Solution: For Figure 8.8(c),

$$I = \frac{dQ}{dt} \tag{8.13}$$

gives the relationship between positive I and the charge on the *positive* plate of the capacitor. For Figure 8.8(d) $I = -dQ/dt$ gives this relationship because a positive current *decreases* the charge on the positive plate.

Rule 2 *Apply the path independence of the voltage: since you return to the same point, the voltage change on circulating around each loop in the circuit must be zero.*

(This is often called the *loop rule.*) That is, for each circuit,

$$\sum_{loop} \Delta V = 0. \qquad \text{(path independence of voltage)} \tag{8.14}$$

Equivalently, the voltage difference between two points, via any given path, is independent of that path.

Figure 8.2(a) and Figure 8.7 used the path independence of the voltage in defining the voltages. Applied to Figure 8.9, the path independence of the voltage means that beginning at d and ending at b will give the same answer whether we circulate clockwise—$V_b - V_d = (V_b - V_c) - (V_d - V_c)$—or counterclockwise—$V_b - V_d = (V_b - V_a) - (V_d - V_a)$.

To use Rule 2, we must know, for each circuit element, the relationship between the voltage differences across it and the current through it. For completeness, here is a summary.

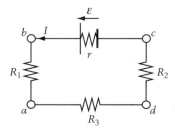

Figure 8.9 Circuit to illustrate Rule 2 of Kirchhoff's rules: apply uniqueness of the voltage so that the net voltage change on going around a circuit is zero.

(a) For a resistor R (or a circuit arm with resistance R), apply Ohm's law, (7.38),

$$I = \frac{\sum \mathcal{E}}{R},\qquad(8.15a)$$

where $\sum \mathcal{E}$ is the sum of all the emfs associated with that resistor: the electrostatic voltage across the arm (which may be unknown), chemical emfs, or other emfs. This common treatment of both batteries and resistors implies that a resistor can be thought of as a battery with no chemical emf.

If an emf tends to cause positive I, the emf is given a positive sign; if it tends to cause negative I, the emf is given a negative sign. Therefore, if in one circuit a battery emf is reckoned as positive in (8.15a), on reversing the battery terminals the emf must now be reckoned as negative in (8.15a). For a simple resistor, the only emf is its voltage difference ΔV, so if ΔV tends to drive current in the positive direction, $I = \Delta V/R$. For the battery of Figure 8.7(b), (8.15a) gives

$$I_1 = \frac{\Delta V}{R_1}, \qquad I_2 = \frac{\Delta V}{R_2}, \qquad \text{and} \qquad I = \frac{(\mathcal{E} - \Delta V)}{r}.$$

For Figure 8.9, (8.15a) gives $I = \mathcal{E}/(r + R_1 + R_2 + R_3)$.

Example 8.8 **Current for, and voltage around, a circuit**

In Figure 8.9, let $\mathcal{E} = 6$ V, $r = 4$ Ω, $R_1 = 1$ Ω, $R_2 = 2$ Ω, $R_3 = 3$ Ω, and $V_a = 3.4$ V. Find the current I, and V_b, V_c, V_d.

Solution: $r + R_1 + R_2 + R_3 = 10$ Ω. Hence $I = (6V)/10\Omega = 0.6$ A. Since I flows from b to a, by Ohm's law V_b is higher than V_a. Specifically, $V_b = V_a + IR_1 = 3.4 + 0.6 = 4.0$ V. Similarly, $V_d = V_a - IR_3 = 3.4 - 1.8 = 1.6$ V, and $V_c = V_d - IR_2 = 1.6 - 1.2 = 0.4$ V. Alternatively, for the emf, $I = [\mathcal{E} - \Delta V]/r = [6 - (V_b - V_c)]/4$, so $V_c = V_b - 6 + 4I = 4.0 - 6.0 + 2.4 = 0.4$ V.

As shown in Section 7.10, for an ideal battery (internal resistance $r \to 0$), the terminal voltage is the same as its chemical emf, the high-voltage side associated with the larger plate. The current through an ideal battery has no effect on its voltage since its Ir value is zero.

Caveat (Warning)

It is conventional to assume that the connecting wires in a circuit have negligible resistance. This is certainly not literally true; it is very wrong if we use many extension cords in series (i.e., to bring power to a place far from one's house). In such cases, wire resistance must be included.

(b) For a capacitor, apply Volta's law, which can be written as

$$\Delta V = \frac{Q}{C}.\qquad(8.15b)$$

Circuits with capacitors are discussed in Section 8.8.

Treat Ideal Batteries with Care

We often consider ideal batteries, which have negligible internal resistance r. As shown in the previous chapter, the voltage difference ΔV associated with such an idealized battery equals its chemical emf $\mathcal{E}$. However, the internal resistance of a battery cannot always be neglected. Consider two ideal batteries with different emfs, in parallel with one another, as in Figure 8.10. The emfs $\mathcal{E}_1 = 12$ V and $\mathcal{E}_2 = 10$ V would give 12 V across the left arm and 10 V across the middle arm. This would contradict Rule 2—the uniqueness of the voltage. Including either or both of the internal resistances would resolve this problem. Typically, batteries of very different emfs are not connected in parallel with one another: that will cause the higher emf battery to discharge rapidly, due to the low internal resistances.

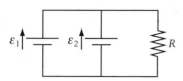

Figure 8.10 Circuit with ideal batteries (having no internal resistance). If the emfs of these batteries aren't the same, analysis of this circuit will lead to contradictions.

Without having given an explicit statement of them, we used Kirchhoff's rules in our analysis of resistors in series and parallel, a voltaic cell in series with a resistor, and a battery of voltaic cells in series. The next two sections will analyze successively more complex circuits. The chapter concludes with a study of a circuit consisting of a battery, a resistor, and a capacitor. A later chapter will consider what happens when there is an electromagnetically induced emf (Faraday's law), associated with which is a circuit element called an *inductor*. *Note:* The number of unknown (variables) must equal the number of equations (constraints), or else the problem is unsolvable.

8.7 Applications of Kirchhoff's Rules

A circuit in the laboratory certainly does not look like the clean schematics presented here. Instead of perfectly straight wires, real circuits have sloppily placed wires that run over and under one another. It can be a difficult task to disentangle and trace their connections. But these connections are precisely what are needed to apply Kirchhoff's rules. The discussion that follows assumes we have already analyzed the connections of our circuit.

The examples of the present section consider some geometries that can, in some sense, be thought of as series or parallel circuits, even though there are resistors *and* batteries in the arms. Just as series resistors have a common current, so a series of batteries has a common current; *for series circuit problems, focusing on the common current helps us to solve them.* Just as parallel resistors have a common voltage difference, so batteries in parallel have a common voltage difference; *for parallel circuit problems, focusing on the common voltage difference helps us to solve them.* This section will consider both types of circuits.

8.7.1 *Batteries in Series (Common Current I)*

Earlier in the chapter, symmetry considerations were used to study the case of many *identical* voltaic cells in series. What happens when the batteries are not

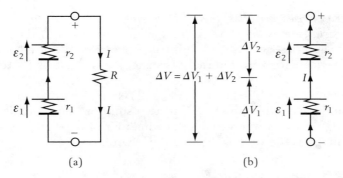

Figure 8.11 Two batteries in series: (a) connected to a load resistor, (b) batteries and their terminal voltages.

identical? Consider an arm of a circuit with two nonidentical batteries, each with its own internal resistance, in series. See Figure 8.11(a), which applies Rule 0 and Rule 1 for the current, by indicating the direction of current flow and that there is a common current I passing through each circuit element. That is, defining I_1, I_2, and I as the currents through the batteries and the resistor R, Rule 1 gives

$$I = I_1 = I_2. \tag{8.16}$$

Figure 8.11(a) also indicates the two nodes $+$ and $-$ that can be used to define the voltage difference $\Delta V = V_+ - V_-$. Although I and ΔV cannot be known without specifying the rest of the circuit (which here is the resistor R), a relationship between the two can be found just by analyzing the arm with the batteries. To do this, consider Figure 8.11(b). By Ohm's law, with the battery emfs and the voltages driving current in opposite directions,

$$I_1 = \frac{\mathcal{E}_1 - \Delta V_1}{r_1}, \qquad I_2 = \frac{\mathcal{E}_2 - \Delta V_2}{r_2}, \tag{8.17}$$

which leads to

$$\Delta V_1 = \mathcal{E}_1 - Ir_1, \qquad \Delta V_2 = \mathcal{E}_2 - Ir_2. \tag{8.18}$$

By Rule 2—the uniqueness of the voltage—the net voltage across the two batteries is the sum of the individual voltages, so

$$\Delta V = \Delta V_1 + \Delta V_2. \tag{8.19}$$

Now, use of (8.18) in (8.19) leads to

$$\Delta V = (\mathcal{E}_1 + \mathcal{E}_2) - I(r_1 + r_2), \tag{8.20}$$

so solving for I

$$I = \frac{(\mathcal{E}_1 + \mathcal{E}_2) - \Delta V}{r_1 + r_2}. \tag{8.21}$$

Thus, *for real batteries in series, the effective emf $\mathcal{E}$ is the sum of the emfs, and the effective resistance r is the same as the sum of the resistances.* This is true for many batteries in series, not just two; it is true even when the arm contains ideal batteries (batteries with zero internal resistance) and resistors (batteries with zero emf).

Elimination of I_1 and I_2 in favor of ΔV_1 and ΔV_2 by equating the expressions in (8.17) would have led, with (8.19), to two simultaneous equations in ΔV_1 and ΔV_2. Using current as the variable, treating four or five batteries in series is easy; using voltage as the variable, it would be awful.

Example 8.9   **Batteries in series I**

In Figure 8.11(b), let $\mathcal{E}_1 = 12$ V, $r_1 = 0.01$ Ω, $\mathcal{E}_2 = 8$ V, $r_2 = 0.01$ Ω, and let 10 V be across an unknown resistor R. Find the current I and the resistance R.

Solution: Rule 0 tells us to establish a sign convention for the current and voltage difference for each circuit element. Positive voltage differences are taken as shown in Figure 8.11(b), and positive current is taken as circulating clockwise around the circuit, as in Figure 8.11(a). Charge conservation, expressed by Rule 1, is automatically satisfied for Figure 8.11(a), because the current into each circuit element (two batteries and one resistor) equals the current out. Rule 2, that the voltage is path independent, means that the ΔV for the left arm, to which (8.21) is appropriate, is the same as ΔV for the right arm. To obtain the current, (8.21) yields $I = [(\mathcal{E}_1 + \mathcal{E}_2) - \Delta V]/(r_1 + r_2) = [(12 + 8) - 10]/(0.01 + 0.01) = 500$ A. Now, for the resistor R, ΔV drives I so, according to Ohm's law,

$$I = \frac{\Delta V}{R}. \qquad (8.22)$$

This gives $R = \Delta V/I = 10 \text{ V}/500\text{A} = 0.02$ Ω.

Example 8.10   **Batteries in series II**

Now let ΔV be unknown, and let the batteries be connected to a known resistor $R = 0.06$ Ω, as in Figure 8.11(a). Find I and ΔV.

Solution: Rules 0, 1, and 2 were already satisfied in the earlier discussion. All that is left is application. Substituting numerical values into (8.20) and (8.22) gives

$$\Delta V = (12 + 8) - I(0.01 + 0.01) \text{ and } \Delta V = I(0.06).$$

Eliminating ΔV then gives $20 - I(0.02) = I(0.06)$, so $I = 20/0.08 = 250$ A. Then $\Delta V = IR = 250(0.06) = 15$ V.

Example 8.11   **Batteries in series III—reversing a battery**

In Example 8.9, let $\mathcal{E}_2$ be reversed, and replace R by a 10 V battery of negligible internal resistance (an ideal battery). Find the current I.

Solution: Rules 0, 1, and 2 were already satisfied in the earlier discussion. We can use (8.21) if we change the sign of $\mathcal{E}_2$. This gives $I = (12 - 8 - 10)/0.02 = -300$ A. The sign change means that the current actually flows counterclockwise in Figure 8.11(b). As long as we include the sign of the current, we don't have to redraw the figure.

These examples yield rather large currents, characteristic of those produced on starting a car.

8.7.2   *Batteries in Parallel (Common Voltage ΔV)*

Section 8.3.2 considered the case of many *identical* voltaic cells in parallel. What happens when the cells in parallel are not identical? Consider an arm of a circuit with two nonidentical batteries, of known emfs and internal resistances, in parallel with each other. See Figure 8.12(a). As usual, our goal is to relate I and ΔV.

By the path independence of the voltage—Rule 2—each arm has the same voltage drop ΔV across it, so

$$\Delta V = \Delta V_1 = \Delta V_2. \tag{8.23}$$

Next, by Ohm's law applied to each battery,

$$I_1 = \frac{\mathcal{E}_1 - \Delta V}{r_1}, \qquad I_2 = \frac{\mathcal{E}_2 - \Delta V}{r_2}. \tag{8.24}$$

The nodal rule—Rule 1—applied to either of the two common nodes yields

$$I = I_1 + I_2. \tag{8.25}$$

Using (8.24) in (8.25) yields

$$I = I_1 + I_2 = \frac{(\mathcal{E}_1 - \Delta V)}{r_1} + \frac{(\mathcal{E}_2 - \Delta V)}{r_2} = \left(\frac{\mathcal{E}_1}{r_1} + \frac{\mathcal{E}_2}{r_2} \right) - \Delta V \left(\frac{1}{r_1} + \frac{1}{r_2} \right). \tag{8.26}$$

Elimination of ΔV_1 and ΔV_2 in favor of I_1 and I_2 by equating the expressions in (8.24) would have led, with (8.25), to two simultaneous equations in I_1 and

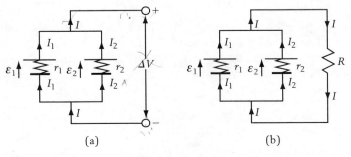

(a)   (b)

Figure 8.12 Two batteries in parallel: (a) with terminal voltage ΔV; (b) connected to a load resistor.

I_2. Using voltage as the variable, treating four or five batteries in parallel is easy; using current as the variable would be awful.

Example 8.12 Batteries in parallel I

Consider the same batteries and resistors as in Example 8.9: $\mathcal{E}_1 = 12$ V, $r_1 = 0.01$ Ω, $\mathcal{E}_2 = 8$ V, $r_2 = 0.01$ Ω. Now take $\Delta V = 10$ V. (This 10 V can be produced by a 10 V emf, or by an emf of more than 10 V in series with a resistor, but not by any resistor alone.) Find the current I that flows through the circuit. Find the current through each battery, and indicate if they are charging or discharging.

Solution: The analysis of (8.23) through (8.26), which satisfies Rules 0, 1, and 2, applies here. Equation (8.26) requires the quantities

$$\frac{\mathcal{E}_1}{r_1} + \frac{\mathcal{E}_2}{r_2} = \frac{12}{0.01} + \frac{8}{0.01} = 2000 \text{ A} \quad \text{and} \quad \frac{1}{r_1} + \frac{1}{r_2} = \frac{1}{0.01} + \frac{1}{0.01} = 200 \text{ Ω}^{-1}.$$

Using these in (8.26) gives $I = 2000 - 10(200) = 0$, so there is zero net current flow. Note that $I_1 = 1200 - (10/0.01) = 200$ A and $I_2 = 800 - (10/0.01) = -200$ A; battery 1 is discharging and battery 2 is charging.

Example 8.13 Batteries in parallel II

In Example 8.12, let ΔV be unknown, but due to a known resistor $R = 0.005$ Ω, as in Figure 8.12(b). Find the current through the resistor, the voltage across the resistor, and the current through each battery. Verify current conservation.

Solution: The analysis of (8.23) through (8.26), which satisfies Rules 0, 1, and 2, applies here. Substituting numerical values into (8.26) and (8.22) gives

$$I = 2000 - 200\Delta V \quad \text{and} \quad I = \frac{\Delta V}{0.005} = 200\Delta V.$$

Eliminating I gives $2000 - 200\Delta V = 200\Delta V$, so $\Delta V = 2000/400 = 5$ V. Then $I = 5/0.005 = 1000$ A, $I_1 = 1200 - (5/0.01) = 700$ A, and $I_2 = 800 - (5/0.01) = 300$ A. Note that $I_1 + I_2 = I = 1000$ A; current is indeed conserved.

For batteries in parallel, it is useful to think of each battery as a *current source*, where battery 1 in (8.24) provides a source current $\mathcal{J}_1 = \mathcal{E}_1/r_1$, and so on. With this interpretation, the two batteries in parallel have an effective source current $\mathcal{J}$, and effective resistance $\mathcal{R}$, where

$$\mathcal{J} = \mathcal{J}_1 + \mathcal{J}_2 = \frac{\mathcal{E}_1}{r_1} + \frac{\mathcal{E}_2}{r_2}, \qquad \frac{1}{\mathcal{R}} = \left(\frac{1}{r_1} + \frac{1}{r_2} \right). \tag{8.27}$$

Note that a large emf with a large resistance can be less effective at producing current than a moderate emf with a small resistance.

Using (8.27), rewrite (8.26) as

$$I = \mathcal{J} - \frac{\Delta V}{\mathcal{R}}. \tag{8.28}$$

Thus, *for real batteries in parallel, the strength $\mathcal{J}$ of the effective current source is the sum of the effective currents produced by each emf, and its effective resistance $\mathcal{R}$ is the same as the emf resistances in parallel.* This is true for many batteries in parallel, not just two.

Application 8.1   The jumper-cable problem

Consider two batteries in parallel, with the net current I going through the resistance R, which represents the starting motor of the car with the "bad" battery. Figures 8.13(a) and 8.13(b) give equivalent circuits that can both represent this problem. One of the batteries is "good" (1) and one is "bad" (2). The car with the bad battery is started by using jumper cables to put the good battery in parallel with the bad battery. This configuration is similar to that of Example 8.13, but for computational purposes we will use values more appropriate to the jumper-cable problem.

Note that the last jumper-cable connection should be to the car-body ground, which is connected to the ground post of the battery. If there is a spark on connection, it will be far from the battery, where outgassing of (explosive) H_2 can occur when the battery becomes overcharged or undercharged.

An earlier estimate, for a good car battery of emf $\mathcal{E}_1 = 12$ V, gave 0.01 Ω for the internal resistance of the battery (and of the starting motor of the car), so let $r_1 = R = 0.01$ Ω. To model the bad battery, note that when a battery goes bad it loses some emf, but more important is that its internal resistance goes up significantly. (Of course, it loses charge, but that doesn't appear in our circuit equations.) To be specific, take $r_2 = 1.0$ Ω. Also take battery 2 to have only five good 2 V cells, so $\mathcal{E}_2 = 10$ V.

Applying (8.27) to this problem, $\mathcal{J} = 1210$ A and $\mathcal{R}^{-1} = 101$ mhos. [A *mho* is an (ohm)$^{-1}$.] Then (8.28) yields $I = 1210 - 101\Delta V$, and (8.22) yields $I = (\Delta V/0.01) = 100$ ΔV. Elimination of I gives $1210 - 101\Delta V = 100$ ΔV, so $\Delta V = 1210/201 = 6.02$ V (accurate to three decimal places). Then (8.22) yields $I = 6.02/0.01 = 602$ A, and (8.24) yields $I_1 = 1200 - 6.02/0.01 = 598$ A and $I_2 = 10 - 6.02/1 = 4$ A. Note that $I_1 + I_2 = I = 602$ A, so that current is conserved.

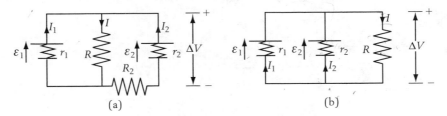

(a)   (b)

Figure 8.13 Two circuits with three arms. They look different, but for $R_2 = 0$ they are identical from the point of view of circuit analysis. They are equivalent to the circuit used when jumper cables are used to start a car with a "dead" battery. In that case, the load resistor R represents the resistance of the starting motor. (a) R in the middle arm. (b) R in the right arm.

By the previous chapter, the batteries provide power at a rate $P = \mathcal{E}I$ when $I > 0$ corresponds to discharge. Thus $P_1 = \mathcal{E}_1 I_1 = 7176$ W and $P_2 = \mathcal{E}_2 I_2 = 40$ W, giving a net discharge rate of 7216 W. The rates of heating of the resistors are $I_1^2 r_1 = 3576$ W, $I_2^2 r_2 = 16$ W, and $I^2 R = 3624$ W, so the net rate of heating is 7216 W. In one second, the batteries use 7216 J of chemical energy, and 7216 J of heat is produced. Energy, too, is conserved.

Alternative analysis: We can also solve for the circuit of Figure 8.13(a) by using current conservation at the upper node $(I_1 + I_2 = I)$ and requiring zero voltage change on going around each loop. Going clockwise around the left loop beginning in the lower-left corner yields $0 = \mathcal{E}_1 - I_1 r_1 - IR$, or $0 = 12 - I_1(0.01) - I(0.01)$. [Here a voltage $\mathcal{E}_1 = 12$ V is gained on crossing the electrodes of the battery, $I_1 r_1 = I_1(0.01)$ is lost on crossing the internal resistance r, and $IR = I(0.01)$ is lost on crossing the resistor R.] Similarly, going counterclockwise around the right loop gives $0 = -I_2(R + r_2) + \mathcal{E}_2 - IR$, or $0 = -I_2(0.01 + 1) + 10 - I(0.01)$. This yields a total of three equations for the three unknowns, and thus is a well-defined problem.

Application 8.2　A mixed circuit

Now consider the circuit given in Figure 8.14. Here, the resistors are given and one of the emfs is given. Our goal is to find the unknown emf and the current through each resistor.

By Rule 1—charge conservation—the current I_r leaving point a, passing counterclockwise through the right circuit, and then reentering a from above, must be the same throughout the circuit. Applying Rule 1 again, no current can enter or leave a from the left. Again by Rule 1, no current enters or leaves point b from the right. Hence the same current must enter b from the left and leave it going upward. This means that the circuit on the right (which includes point a) and the circuit on the left (involving point b) are independent.

Circulating clockwise around the circuit on the right gives a voltage change $\Delta V = 0$, by Rule 2, since we return to our starting point. But with current I_r in this circuit, this voltage drop must be $8I_r$. Hence $I_r = 0$; there is no current flow through the circuit on the right.

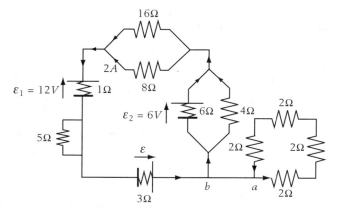

Figure 8.14　Complex circuit to illustrate the use of all of Kirchhoff's rules.

To analyze the circuit on the left (which includes point b), first note that the voltage drop across the 8 Ω resistor is $\Delta V_8 = 2(8) = 16$ V, and that this must be the same as across the 16 Ω resistor, so 16 V $= 16 I_{16}$. Hence the current through the 16 Ω resistor is 1 A, and the total current circulating counterclockwise through the left circuit is $I_l = 2 + 1 = 3$ A.

Because the wire in parallel with the 5 Ω resistor is taken to have zero resistance, the full 3 A current flows through the wire (path of least resistance). Thus $I_5 = 0$, and the 5 Ω resistor has no effect on the circuit. For the 6 V, 6 Ω battery in parallel with the 4 Ω resistor, the common voltage is

$$\Delta V_4 = 4 I_4 = 6 I_6 - 6 = 6(3 - I_4) - 6.$$

(Here Rule 1 says that $I_l = I_4 + I_6$.) Solving for I_4, we obtain $I_4 = 1.2$ A, so $\Delta V_4 = 4 I_4 = 4.8$ V across the 4 Ω resistor. Then $I_6 = I_l - I_4 = 3 - 1.2 = 1.8$ A.

Circulating clockwise around the left circuit beginning at point b, the successive voltage increases across the 3, 1, 8, and 4 Ω resistances sum to $3(3) + 3(1) + 2(8) + 4.8 = 32.8$ V. This must be compensated by a net 32.8 V voltage increase on circulating counterclockwise across the electrode–electrolyte interfaces of $\mathcal{E}_1$ and $\mathcal{E}$. Because, from Chapter 7, the voltage difference across an electrode–electrolyte interface equals the emf for that interface, there must be a net counterclockwise emf of 32.8 V from $\mathcal{E}_1$ and $\mathcal{E}$. Since $\mathcal{E}_1$ provides a 12 V clockwise emf, $\mathcal{E}$ must provide a $32.8 + 12 = 44.8$ V counterclockwise emf.

8.8 Short- and Long-Time Behavior of Capacitors

Chapter 6 considered capacitors that were charged by unspecified means. Chapter 7 showed that a capacitor, when connected to a battery of emf $\mathcal{E}$, would develop a voltage difference $\Delta V = \mathcal{E}$. We now consider how a battery charges a capacitor in real circuits that contain resistance, and how a capacitor discharges through a resistor. *A circuit with a resistor and a capacitor is called an RC circuit.* The present section considers the charge and current for an RC circuit at short times (where we start) and at long times (where we end). The next section considers the details of the time development from short to long times.

Although a capacitor initially may be uncharged ($Q = 0$), for short times after a battery is switched into a circuit the capacitor immediately can start the charging process. Charge can enter one plate and leave the other, yielding a nonzero current ($I \neq 0$) associated with the capacitor.

At long times, if the power source is steady and there are resistors to absorb energy, the system will come to an equilibrium, and any point of the circuit will settle down to a constant voltage. Thus, the voltage ΔV across any circuit element will be a constant. For a resistor R, this means that the current $I = \Delta V / R$ becomes a constant. For a capacitor, this means that the charge $Q = C \Delta V$ becomes a constant, so $d_Q / dt \to 0$. Hence, since $I = dQ/dt$, at long times there is no current to or from a capacitor. That is,

$$I^{cap}(t \to \infty) = 0. \qquad \text{(long-time behavior of capacitor current)} \qquad (8.29)$$

Let us see how this works for a few examples.

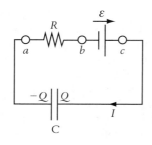

Figure 8.15 An RC circuit with an ideal emf $\mathcal{E}$.

Consider an RC circuit with a resistor R, a capacitor C, and an ideal battery $\mathcal{E}$ with zero internal resistance. See Figure 8.15.

The only emf acting on the resistor is the voltage $\Delta V_R = V_a - V_b$ across its ends, which by the path independence of the voltage is the sum of the voltages across the battery and across the capacitor. Because the battery is taken to have zero resistance, its terminal voltage ΔV_T is $\mathcal{E}$, tending to drive the current one way, and the capacitor voltage

$$\Delta V_C = V_c - V_a = \frac{Q}{C} \tag{8.30}$$

tends to drive the current the other way. Thus (8.15a) and (8.30) yield

$$I = \frac{dQ}{dt} = \frac{\Delta V_R}{R} = \frac{V_a - V_b}{R} = \frac{(V_a - V_c) + (V_c - V_b)}{R}$$
$$= \frac{-\Delta V_C + \Delta V_T}{R} = \frac{-Q/C + \mathcal{E}}{R}. \tag{8.31}$$

Initially, the capacitor is uncharged, so $Q_0 = 0$, but it has an initial current, given by $I_0 = \mathcal{E}/R$. After a long time, the capacitor has charged up, so $I_\infty = 0$. By (8.31), it then has the charge

$$Q_\infty = C\mathcal{E}, \tag{8.32}$$

in agreement with the previous chapter's discussion of charge transfer from a battery to a capacitor. One way of describing an uncharged capacitor is to say that it behaves like a "short" because there is no voltage drop across it. Once it develops a charge, its voltage drop is no longer zero.

Example 8.14 **Short- and long-time behavior I**

Consider a circuit with a chemical emf $\mathcal{E}$ with resistance r in one arm, a resistor R in a second arm, and a capacitor C in a third arm. See Figure 8.16, where the currents in each arm are labeled. Find the initial and final currents, and the final charge on the capacitor.

Solution: By Kirchhoff's first rule, $I = I_1 + I_2$ and $I_2 = dQ_2/dt$. At all times, the voltage must be the same across C and R (Kirchhoff's second rule). Thus $I_1 R = Q_2/C$, so $I_1 = (Q_2/C)/R$. Since $Q_2 = 0$ initially, this says that $I_1 = 0$ initially; all the current flows to the capacitor. Hence, initially, this circuit behaves like a battery and a capacitor in series. The initial current is $I = I_2 = I_0 = \mathcal{E}/r$. On the other hand, after a long time, the capacitor is fully charged, and no current flows through it. Hence after a long time, this circuit behaves like a battery and a resistor

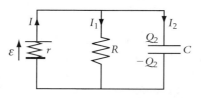

Figure 8.16 A battery driving a resistor in parallel with a capacitor.

in series. The final current is $I = I_1 = I_\infty = \mathcal{E}/(r + R)$. The capacitor develops a charge determined by $I_\infty R = Q_2/C$.

Example 8.15 **Short- and long-time behavior II**

Consider a *bridge circuit*, with a capacitor C as the bridge connecting four resistors, as in Figure 8.17. Give a qualitative analysis of its behavior at short times and at long times after the switch S is connected.

Solution: (a) For short times, the capacitor remains uncharged, so its ends are at the same voltage ($V_b = V_c$). This means that the midpoints are at the same potential (i.e., the capacitor behaves like a "short"). Thus we may consider that R_1 and R_2 are in parallel, yielding R_a; and R_3 and R_4 are in parallel, yielding R_b. Then R_a and R_b are added in series. From this we can also find the current through each resistor, and by considering either of the nodes to the capacitor, we can find the current to the capacitor.

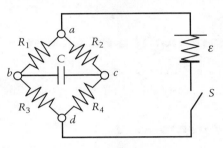

Figure 8.17 A bridge circuit of resistors, with a capacitance across the bridge, and a switch to connect the emf $\mathcal{E}$.

(b) For long times, the capacitor is charged up, and no current flows through it. In that case, R_1 and R_3 are in series with each other, as are R_2 and R_4, and the combinations are in parallel with one another. A knowledge of the current through each arm can give the voltage difference across the capacitor at long times, and by (8.15b) the equilibrium charge on the capacitor.

Example 8.16 **Short- and long-time behavior III**

Consider a circuit with $R_1 = 2\ \Omega$ and $R_2 = 3\ \Omega$ in parallel, leading to $C = 3\ \mu F$, leading to $R_3 = 5\ \Omega$ and $R_4 = 1\ \Omega$ in parallel. See Figure 8.18. Give a quantitative analysis of its behavior (a) at short times and (b) at long times if at $t = 0$, $V_a - V_b$ suddenly goes from 0 to 5 V. Take all charges and current to be zero until $t = 0$.

Solution: (a) For short times, since the capacitor is uncharged, the capacitor has no effect on the circuit, so the initial behavior of this circuit is the same as the initial behavior of the circuit in Figure 8.17. R_1 and R_2 are in parallel and have

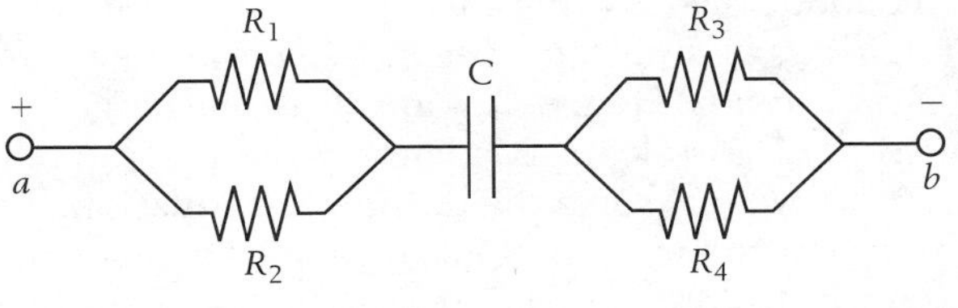

Figure 8.18 A circuit with two sets of resistors in parallel, separated by a capacitor.

an equivalent resistance $R_l = (1/2 + 1/3)^{-1} = 1.2\ \Omega$; R_3 and R_4 are in parallel and have an equivalent resistance $R_r = (1/5 + 1)^{-1} = 0.833\ \Omega$. R_l and R_r are in series, so $R_{equiv} = R_l + R_r = 1.2 + 0.833 = 2.033\ \Omega$. Since $\Delta V = V_a - V_b = 5$ V, by Ohm's law $I = \Delta V / R_{equiv} = 5/2.033 = 2.459$ A. This goes into charging up C. Note that the voltage drop across R_1 and R_2 is $\Delta V_l = I R_l = 2.049$ V, and the voltage drop across R_3 and R_4 is $\Delta V_r = I R_r = 2.951$ V. These sum to 5 V, as expected. From ΔV_l and ΔV_r we can obtain the individual currents. For example, $I_1 = \Delta V_l / R_1 = 1.025$ A. (b) For long times, the capacitor is fully charged, and its voltage stops all current flow. It develops a charge $Q = C \Delta V$, where $\Delta V = V_a - V_b$ is the voltage across the system as a whole. Thus $Q = (3 \times 10^{-6})(5) = 1.5 \times 10^{-5}$ C.

8.9 Charging and Discharging: The RC Circuit

Having discussed the short-time and long-time behavior of capacitors and resistors, we are now in a position to study their behavior at any time.

8.9.1 *Charging: Numerical Integration I*

Let us rewrite (8.31) as

$$\frac{dQ}{dt} = \frac{\mathcal{E}}{R} - \frac{Q}{RC}. \tag{8.33}$$

This is called a *first-order differential equation* because the highest derivative is the first derivative. Equation (8.33) gives the slope at any instant of time, from a knowledge of the charge at that time. Therefore, assume that the charge is known at some time (call this $t = 0$). To find Q at a short time Δt later, the straight-line approximation gives

$$Q(\Delta t) = Q(0) + \frac{dQ}{dt}\Delta t, \tag{8.34}$$

where dQ/dt is obtained from (8.33) using $Q(0)$. Repeating this procedure, now with $Q(2\Delta t) = Q(\Delta t) + dQ/dt|_{t+\Delta t}\Delta t$, yields the solution to (8.33) at all times. By making the time step Δt shorter and shorter, we can test for convergence. In principle, such a numerical procedure will solve any first-order differential equation, even when it is more complicated than (8.33). See Figure 8.19.

Take time to think about this. Solving this differential equation predicts the future for this circuit. *That, in a nutshell, is what differential equations involving the time do. They predict the future.* Hence they are an astonishingly powerful tool. We don't need to consult an astrologer, Madame X the fortune teller, or a network of "psychic friends." We just solve the differential equation. This powerful idea began with physics, but is now applied in many other areas. Economists (meteorologists) attempt to predict the economic (meteorological) future using economic (meteorological) models expressed in the language of differential equations. The better the model, the better the prediction.

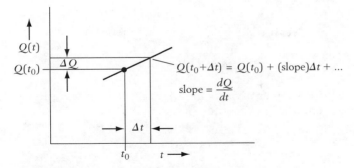

Figure 8.19 Charge as a function of time. If the charge and the current (which is the slope of the charge versus time curve) are known at time t_0, then the charge can be found at time $t_0 + \Delta t$.

8.9.2 *Charging: Numerical Integration II*

Another approach that can be taken to (8.33) is to think of it in terms of the ultimate goal of finding $Q(t)$. Then rewrite (8.33) as

$$dQ = \frac{dQ}{dt}dt = f(Q)dt, \tag{8.35}$$

where for (8.33) $f(Q) = \mathcal{E}/R - Q/RC$. In general $f(Q)$ can be thought of as any function of Q, or for our case, as the right-hand side of (8.33). (More generally, the right-hand side could also depend on t, but we will not consider that possibility.) Now rewrite (8.35) as

$$\frac{dQ}{f(Q)} = dt. \tag{8.36}$$

The right-hand side of (8.36) trivially integrates to t. If we can also integrate the left-hand side, either numerically or analytically, giving some function $g(Q)$, then inverting $g(Q) = t$ will yield $Q(t)$.

8.9.3 *Charging: Analytical Integration*

In the present case, with $f(Q) = \mathcal{E}/R - Q/RC$, the integral in (8.36) can be done analytically. First, introduce the notation

$$I_0 = \frac{\mathcal{E}}{R}, \tag{8.37}$$

$$\tau_{RC} = RC, \qquad (RC \text{ time constant}) \tag{8.38}$$

where RC has the dimensions of a time. [dQ/dt and Q/RC of (8.32) have the same dimensions.] Then, writing

$$f(Q) = I_0 - \frac{Q}{\tau_{RC}},$$

(8.36) becomes

$$\frac{dQ}{I_0 - Q/\tau_{RC}} = dt. \tag{8.39}$$

Now take the positive quantity $u = I_0 - Q/\tau_{RC}$ as the integration variable. Then $dQ = -\tau_{RC}du$, and (8.39) becomes

$$-\frac{\tau_{RC}du}{u} = dt, \tag{8.40}$$

which can be integrated easily. Writing (8.40) as a definite integral, with $Q = 0$ at $t = 0$, and replacing u by $I_0 - Q/\tau_{RC}$, yields

$$-\tau_{RC}\ln(I_0 - Q/\tau_{RC})|_0^Q = t|_0^t. \tag{8.41}$$

On switching right- and left-hand sides, (8.41) becomes

$$t = -\tau_{RC}[\ln(I_0 - Q/\tau_{RC}) - \ln(I_0)] = -\tau_{RC}\ln(1 - Q/I_0\tau_{RC}). \tag{8.42}$$

With $I_0\tau_{RC} = (\mathcal{E}/R)(RC) = C\mathcal{E} = Q_\infty$, the solution of (8.42) for Q is

$$Q = I_0\tau_{RC}[1 - \exp(-t/\tau_{RC})] = Q_\infty[1 - \exp(-t/\tau_{RC})]. \quad Q_\infty = C\mathcal{E} \tag{8.43}$$

This starts at zero charge and rises to full charge at time infinity. $\tau_{RC} = RC$ is the time it takes for a capacitor to charge to 63% of its long-time value. Differentiation of the charge, (8.43), gives

$$I = \frac{dQ}{dt} = I_0\exp(-t/\tau_{RC}). \tag{8.44}$$

See Figure 8.20 for both the charge of (8.43) and the current of (8.44).

Example 8.17 **Measurement of the capacitor charging time gives the capacitance**

A 2 V emf is applied to a circuit with a 450 Ω resistor and an unknown capacitor. After 3.8 ms, the current is 2.2 mA. Find the time constant τ_{RC} and the capacitance C.

Solution: First, at $t = 3.8$ ms, the voltage across the resistor is $IR = 0.99$ V. Then (8.31), with $\mathcal{E} = 2$ V, gives $Q/C = 2 - 0.99 = 1.01$ V. From (8.43), $Q/C = \mathcal{E}[1 - \exp(-t/\tau_{RC})]$, so

$$\exp(-t/\tau_{RC}) = 1 - Q/C\mathcal{E}, \text{ giving } t/\tau_{RC} = -\ln(1 - Q/C\mathcal{E}).$$

For our example, at $t = 3.8$ ms, $(1 - Q/C\mathcal{E}) = 0.495$, so $\tau_{RC} = -t/\ln(1 - Q/C\mathcal{E}) = 5.40 \times 10^{-3}$ s. Then, by (8.38), $C = \tau_{RC}/R = 1.200 \times 10^{-5}$ F. Multimeters measure unknown capacitances by such a method.

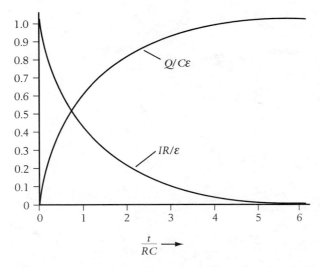

Figure 8.20 Charge and current as a function of time for a capacitor that is initially uncharged and is then connected in series with a battery and a resistor.

8.9.4 *Discharging: Trial Solution*

On removal of the emf [let $\mathcal{E} \to 0$ in (8.31)], the discharge is described by the solution to

$$\frac{dQ}{dt} = -\frac{Q}{RC}, \tag{8.45}$$

subject to some charge $Q = Q_0$ at $t = 0$. We can solve this equation by the same method as for the previous case, but let us try yet another method, now that we have some experience with this type of equation.

Since exponential decay seems to be characteristic of the charging problem, let us assume an exponential time dependence for the decay of the charge. Moreover, since $\tau_{RC} = RC$ was the time constant for charging, let's try it as the time constant for discharging. Thus, try the form

$$Q = Q_0 \exp(-t/\tau_{RC}), \tag{8.46}$$

which satisfies the initial condition that $Q = Q_0$, and has the expected uncharged final state. However, we must verify that this is a solution of (8.45). Differentiating (8.46) yields

$$I = \frac{dQ}{dt} = -\frac{Q_0}{\tau_{RC}} \exp(-t/\tau_{RC}). \tag{8.47}$$

Putting (8.46) and (8.47) into (8.45) yields

$$-\frac{Q_0}{\tau_{RC}} \exp(-t/\tau_{RC}) = -\frac{Q_0}{RC} \exp(-t/\tau_{RC}), \tag{8.48}$$

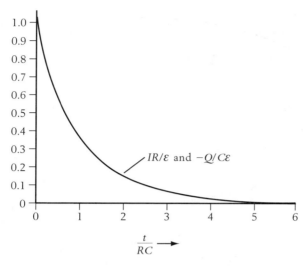

Figure 8.21 Charge and current as a function of time for a capacitor that is initially charged and is then discharged through a resistor.

from which we deduce that

$$\frac{1}{\tau_{RC}} = \frac{1}{RC}.$$

(8.49)

This is consistent with (8.38). Thus, τ_{RC} is characteristic of the exponential for both charge and discharge. In discharge, $\tau_{RC} = RC$ is the time for the charge on a capacitor to reduce to 37% of its initial value. The reason for the sign change in the current, relative to the charging process, is that the current now flows in the opposite direction. See Figure 8.21 for both the charge and the current.

These results can be applied directly to a number of interesting cases, including a charged person who discharges to ground upon jumping from an insulated platform; a person who charges when, standing on an insulated platform, she touches a Van de Graaf generator; or the electrostatic discharge of an electronic chip.

Example 8.18 **Discharge of a capacitor**

For the circuit of the previous example, 2.9 ms after discharge begins, find the current, the charge, and the rate of heating.

Solution: Since $\mathcal{E} = 2$ V and $C = 1.200 \times 10^{-5}$ F, the initial charge is $Q_0 = C\mathcal{E} = 2.40 \times 10^{-5}$ C. Since $\tau_{RC} = 5.40 \times 10^{-3}$ s, substitution into (8.47) for $t = 2.9$ ms yields $I = -2.60$ mA, the negative sign indicating discharge. Either (8.45) or (8.47) then yields $Q = 1.402 \times 10^{-5}$ C. The rate of heating is then $I^2 R = 3.04 \times 10^{-3}$ W.

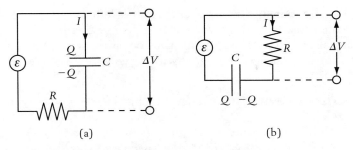

Figure 8.22 *RC* circuit with emf $\mathcal{E}$: (a) Voltage across capacitor read by high resistance voltmeter. (b) Voltage across resistor read by high resistance voltmeter.

8.9.5 Applications
Optional

Another application of the *RC* circuit is as a *filter.* Since the voltage $\Delta V_C = Q/C$ across the capacitor takes time (on the order of *RC*) to build up, it is sensitive to emfs that vary slowly relative to *RC* (low frequencies), but it is insensitive to emfs that vary quickly relative to *RC* (high frequencies). On the other hand, the voltage $\Delta V_R = IR$ across the resistor builds up immediately, but it dies out after a time on the order of *RC*. Hence it is sensitive to emfs at high frequencies, but it is insensitive to emfs at low frequencies. As a consequence, in an *RC* circuit the capacitor can be used to pass low frequencies but filter out high frequencies, and the resistor can be used to filter out low frequencies but pass high frequencies. See Figure 8.22, which shows the output leads in each case, where it is assumed that the output resistance (also called the impedance) is so high that it draws no current (as for an ideal voltmeter). The emf is circled to indicate that it might be something other than a battery—a power supply or signal generator, for example. Filters are discussed in more detail in Chapter 14.

Related to this are two more applications of an *RC* circuit. If the voltage $\Delta V_C = Q/C$ across the capacitor is thought of as the input, then the voltage ΔV_R across the resistor can be thought of as a *differentiator* because $\Delta V_R = Rd Q/dt = RCd(\Delta V_C)/dt$. Similarly, if the voltage across the resistor is thought of as the input, then the voltage across the capacitor can be thought of as an *integrator* because $\Delta V_C = Q/C = \int Idt/C = \int \Delta V_Rdt/RC$.

Note that when a capacitor discharges, its electrical energy goes into heat. We can see how this occurs at each instant. Consider a resistor and capacitor in series. In a time dt the charge $dQ = Idt$ passes through the circuit. The capacitor loses energy $-(Q/C)d Q$ (note that $dQ < 0$). The resistor heats up by $I^2 Rdt$. Equating the loss in electrical energy to the heat loss gives $-(Q/C)d Q = I^2 Rdt = IRdQ$, which leads to $I = (-Q/C)/R$, as in (8.45).

8.10 Surface Charge Makes the $\vec{E}$ Field
Optional ## That Drives the Current

Kirchhoff's rules do *not* specify the surface charge that produces the electric field within a wire or resistor. However, from a knowledge of the circuit elements

(R's and C's), we can find the current, the current density, and the electric field within the resistors and the wires. Then, from a detailed knowledge of the configuration of the wires, it is difficult (but not impossible) to find the surface charge distribution producing the field within the wires.

8.10.1 *Surface Charge and the Field in the Wire*

Here's the argument showing that, for steady current flow through a wire, even of variable cross-section, $\vec{E}$ is due to charge distributed over the surface: (1) Pick out a volume within the conductor. (2) For steady current, the net flux $\oint \vec{J} \cdot d\vec{A}$ of electric current density $\vec{J}$ through the surface surrounding our volume is zero. (If it weren't zero, then charge would pile up inside the conductor, thus making the electric field and the electric current non-steady, in violation of our assumption of steady current.) (3) Since $\vec{E} = \vec{J}/\sigma$ for a uniform material, by $\oint \vec{J} \cdot d\vec{A} = 0$ we have $\oint \vec{E} \cdot d\vec{A} = 0$. (4) By Gauss's law, since $\oint \vec{E} \cdot d\vec{A} = 0$, the charge enclosed is zero for steady current. (5) Hence there is no charge in our volume, which can be anywhere in the bulk of the wire. (6) Therefore the charge must be distributed over the surface of the wire. (The details of this surface charge distribution depend on the shape of the circuit; as Kirchhoff knew, typically there is surface charge all over the surface of the circuit.)

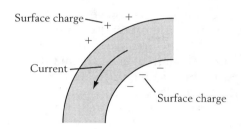

Figure 8.23 Surface charge produces the electric field that drives current through a wire. Near bends in the wire, the surface charge can be relatively large. The electric field depends on the charge everywhere in the circuit (and, indeed, everywhere in the universe); do not think that the surface charge depicted here is solely responsible for the electric field that drives the current near the bend in the wire.

We are aware of no nontrivial example for which the surface charge density Σ_S can be solved exactly. However, we can estimate the order of magnitude of the Σ_S needed to move the current around a bend in a wire. See Figure 8.23. First, note that $J = \sigma E = E/\rho$, where σ is the conductivity and ρ is the resistivity. Next, note that, if there is a surface charge density Σ_S, near the surface it looks like a sheet of charge, so it produces an electric field with $E = |\vec{E}|$ on the order of $4\pi k \Sigma_S$. Thus $J \sim 4\pi k\sigma \Sigma_S$; solving for Σ_S gives

$$\Sigma_S \sim \frac{J}{4\pi k\sigma} = \frac{J\rho}{4\pi k}. \qquad \text{(surface charge density } \Sigma_S \text{ at bend)} \qquad (8.50)$$

Example 8.19 Surface charge on a wire

For a #10 copper wire (radius $a = 0.05$ inch), carrying 10 A, find J, and estimate Σ_S and the surface charge Q_S. Compare to the case of graphite carbon wire.

Solution: $J = I/(\pi a^2) = 1.972 \times 10^6$ A/m², so with resistivity $\rho = 1.690 \times 10^{-8}$ Ω-m, (8.50) gives $\Sigma_S \sim 2.95 \times 10^{-13}$ C/m². For an area of $\pi a^2 = 5.07 \times$

10^{-6} m^2, this corresponds to a surface charge $Q_S = 1.49 \times 10^{-18}$ C, or about 9 electrons. Had the wire been made out of a much poorer conductor, such as graphite carbon, with $\rho = 3.5 \times 10^{-5}$ Ω-m, the surface charge density would be larger by the ratio of the resistivities, or nearly a factor of 2000, to 18,000 electrons.

Just as for electrical screening, surface charge is *not* caused by individual electrons moving to or from the surface, but rather by *all* the electrons in the vicinity collectively distorting their orbitals. For an electron to go truly to or from the surface, it would have to go into or out of an orbital localized at the surface. That does *not* happen for ordinary conductors.

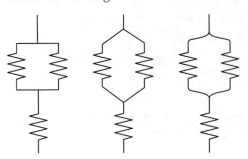

Figure 8.24 Three microscopically different circuits having the same equivalent resistance.

The electric field that causes current to flow within a wire is due to charge distributed over the surface and depends upon details of the circuit. Consider different circuits with the same resistors, as in Figure 8.24. If their wiring is not the same, the surface charge distribution won't be the same. This also means that jiggling the wires of any of the circuits in Figure 8.24 will cause some charge to rearrange.

8.10.2 *Surface Charge and Parasitic Capacitance*

To initiate current flow through the wire, the power source must not only drive current through the wire, it must also drive current to charge up the surface of the wire. As shown earlier, this is because, within the wire, the surface charge sets up the field that causes current flow. These processes of current flow along the wire, and of charge from one part of the surface of the wire to another, are in parallel. Thus, *the wire has a parasitic capacitance C_p that is in parallel with the resistance R of the wire itself.* See Figure 8.25.

Let's consider this idea within the context of the circuits of Figure 8.24. They all have the same resistance, but they have different parasitic capacitances C_p. The circuit on the far left, with its sharp corners, has the largest C_p; the circuit on the far right, with its less sharp corners, has the smallest C_p. Hence, the RC time constant for the circuit on the right is smaller than that on the left. Therefore the circuit on the right will "start up" and "turn off" in less time than the circuit on the left.

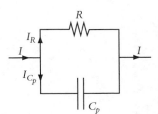

Figure 8.25 Schematic of how a wire "loads up" with surface charge. When there is no current, the initial current transfers surface charge from one part of the surface to another; once the current has reached steady state, so has the surface charge.

Parasitic Capacitance

Parasitic capacitance is not merely a subtle theoretical point. Modern computer chips use integrated circuits that must carry electrical signals very quickly. If the interconnecting "wires" (interconnects) have sharp corners, that increases the parasitic capacitance, and slows the response of the system, because the corners must charge up before the interconnect can carry a current.

Properly, the way to determine the amount of parasitic capacitance C_p is to determine the additional electrical energy associated with it, via (6.21), written as

$$U = \frac{1}{2} C_p (\Delta V)^2. \tag{8.51}$$

Here, U represents the difference between the electrical energy when the current is flowing and when it is not flowing. U can be calculated using (6.25), or

$$U = \int u_E \, d\mathcal{V} = \frac{1}{8\pi k} \int E^2 \, d\mathcal{V}, \tag{8.52}$$

where $d\mathcal{V}$ is a volume element. Since the electric field is due to some complicated surface charge distribution, this is not easily calculated; we know of no simple examples. A good rule of thumb is that C_p is proportional to the length l of the wire: by (6.6), $C_p \sim l/k$ if we neglect the logarithmic factors. For a 10 cm long wire, this predicts $C_p \sim 0.1$ pF. However, this estimate need not always be correct. For a wire that connects two plates of a parallel-plate capacitor, if the wire is between the two plates, and goes normally outward from one plate, straight across to the other, no surface charge is needed to drive the current. Thus in this case, the extra electrical energy, and the parasitic capacitance, are both zero. Nevertheless, in almost any other situation, the parasitic capacitance will be nonzero.

8.11 The Bridge Circuit

Optional

Bridge circuits present a situation where an analysis in terms of series and parallel is not applicable since such an analysis does not account for some internal nodes. Bridge circuits require the full formalism of Kirchhoff's rules. In Figure 8.26(a), the resistances on the left side are R_1 and R_3, and on the right side they are R_2 and R_4, with R_5 being the bridge resistance. A current I enters the 1–2 end and leaves the 3–4 end. The voltage difference between these two ends is ΔV. Our object is to determine the equivalent bridge resistance

$$R_{bridge} \equiv \frac{\Delta V}{I}. \tag{8.53}$$

First consider a special case. If the voltage at a is the same as the voltage at b, then

$$I_1 R_1 = I_2 R_2, \quad \text{and} \quad I_3 R_3 = I_4 R_4. \tag{8.54}$$

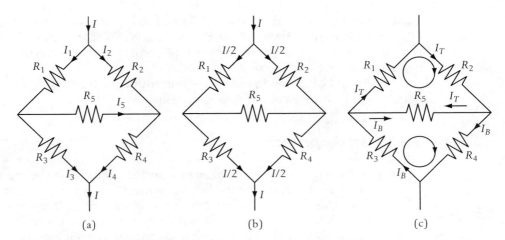

Figure 8.26 A bridge circuit of resistors. (a) Analysis in terms of current I in and out, with individual currents for each arm. (b) Partial analysis where I splits up equally for each arm, with no current across the bridge. (c) Partial analysis where current I_T circulates clockwise through top loop and I_B circulates clockwise through bottom arm.

Moreover, in that case no current flows through the bridge resistor, so by Rule 1 (charge conservation) $I_1 = I_3$ and $I_2 = I_4$. The ratio of the second to the first of these current equations yields

$$\frac{I_3}{I_1} = \frac{I_4}{I_2}.$$ (8.55)

Combining (8.53) and (8.54) yields

$$\frac{R_1}{R_2} = \frac{R_3}{R_4}.$$ (8.56)

When (8.56) holds, the current through the bridge resistor (R_5) is zero, so its numerical value is immaterial, as in Figure 8.27 where it has been take to be 0. Bridge circuits are often used to measure unknown resistances (e.g., R_4) by using

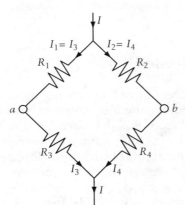

Figure 8.27 Circuit of Figure 8.26, with bridge resistor removed.

fixed values for two of the resistors (R_1 and R_3), and varying the other resistor (R_2) until no current flows across the bridge. When that occurs, the unknown resistance can be obtained from (8.56). (Compare with the capacitance bridge discussed in Chapter 6.)

We now indicate how to obtain the current through each arm in the general case, when the bridge condition is *not* satisfied.

8.11.1   *Standard Approach*

Solving for each of the currents through a bridge circuit with the input current I specified involves solving for the five unknown values of the currents through the resistors. This requires finding five linear equations in the five unknowns, driven by the known input current. These five equations come from current conservation at three nodes (the fourth node condition is a linear combination of the others) and from zero voltage change on going around any two independent closed loops (such as the two internal loops).

The current equations at three nodes yield

$$I = I_1 + I_2, \qquad I_1 = I_3 + I_5, \qquad I_2 + I_5 = I_4, \qquad (8.57)$$

and the requirements of zero voltage change on going around each of the two internal loops are

$$I_1 R_1 + I_5 R_5 = I_2 R_2, \qquad I_5 R_5 + I_4 R_4 = I_3 R_3. \qquad (8.58)$$

Once these currents are known in terms of the input current I, the equivalent bridge resistance R_{bridge} can be determined by the requirement that the voltage drop for the equivalent resistance be the same as the total voltage drop across one arm, or

$$\Delta V = I R_{bridge} = I_1 R_1 + I_3 R_3. \qquad (8.59)$$

Solving five simultaneous equations can be time consuming, especially without a computer. Maxwell invented the idea of *loop currents* to reduce the number of unknowns.

8.11.2   *Maxwell's Loop Current Approach*

Instead of introducing currents for each arm, and then using these currents at each vertex, where current is conserved, we use loop currents. These automatically satisfy current conservation. In Figure 8.26(b), consider that the known current I enters on top and leaves at the bottom, arbitrarily taking the current to split equally along each arm, with no current through the bridge resistor. This choice cannot be correct in general, but we can make the correct current split occur at the upper junction by introducing a clockwise circulating current I_T for the top loop, and a clockwise circulating current I_B for the bottom loop.

See Figure 8.26(c). When I_T and I_B are chosen properly, the correct currents

$$I_1 = \frac{I}{2} - I_T, \quad I_2 = \frac{I}{2} + I_T, \quad I_3 = \frac{I}{2} - I_B, \quad I_4 = \frac{I}{2} + I_B, \quad I_5 = I_B - I_T$$

$$\text{(8.60)}$$

are obtained. If another split for the current I had been taken, the relationships for the I_n's would have been different, but the final results will be independent of the choice of variables.

Instead of (8.58), we now find the loop equations (i.e., voltages):

$$\left(\frac{I}{2} - I_T\right)R_1 + (I_B - I_T)R_5 = \left(\frac{I}{2} + I_T\right)R_2,$$

$$(I_B - I_T)R_5 + \left(\frac{I}{2} + I_B\right)R_4 = \left(\frac{I}{2} - I_B\right)R_3. \qquad \text{(8.61)}$$

These are two equations in the two unknowns I_T and I_B, which can be solved straightforwardly. In expressing the answers, it is useful to introduce the definition

$$D = R_5(R_1 + R_2 + R_3 + R_4) + (R_1 + R_2)(R_3 + R_4). \qquad \text{(8.62)}$$

Solving (8.61) then gives

$$I_B = I\frac{(R_3 - R_4)(R_1 + R_2 + R_5) + R_5(R_1 - R_2)}{2D},$$

$$I_T = I\frac{(R_1 - R_2)(R_3 + R_4 + R_5) + R_5(R_3 - R_4)}{2D}. \qquad \text{(8.63)}$$

Placing (8.63) into the last of (8.60) yields the bridge current

$$I_5 = I\frac{R_3 R_2 - R_4 R_1}{D}. \qquad \text{(8.64)}$$

Note that $I_5 = 0$ when the bridge condition, (8.56), is satisfied.

From (8.63) and (8.64) placed in (8.60), the currents appearing in (8.59) for ΔV are

$$I_3 = I\frac{R_5(R_2 + R_4) + R_4(R_1 + R_2)}{D}, \qquad I_1 = I\frac{R_5(R_2 + R_4) + R_2(R_3 + R_4)}{D}.$$

$$\text{(8.65)}$$

Finally, from (8.65) and (8.59), the equivalent resistance becomes

$$R_{bridge} = R_5\frac{(R_1 + R_3)(R_2 + R_4)}{D} + \frac{R_1 R_2(R_3 + R_4) + R_3 R_4(R_1 + R_2)}{D}.$$

$$\text{(8.66)}$$

How can we be sure of this complicated expression? If Laserbrain Software tried to sell you a program that claimed to solve all possible circuit problems, and gave (8.66) as an example of what the program could do, how would you test it? One way, of course, is to trace through all the algebra yourself. But it's easy to get the algebra wrong, right? Your grandmother would know what to do: perform some simple, commonsense tests to verify that this general expression reduces to some simpler cases where you think you know the answer. Here are some of these tests:

1. For $R_5 = 0$, (8.66) should reduce to the expected result of R_1 and R_2 in parallel, and R_3 and R_4 in parallel, with their combinations in series.

2. For $R_5 \rightarrow \infty$, (8.66) should reduce to the expected result of R_1 and R_3 in series, and R_2 and R_4 in series, with their combinations in parallel.

3. For $R_1 \rightarrow \infty$, all the current should flow through R_2, and none should flow through R_1.

Note on recognizing old friends. One of the most important reasoning tools is the method of analogy. This can be used best when you can easily recognize similarities and differences. An analogy can be made between circuits with resistors and circuits with capacitors. From our results for resistors, we can obtain the corresponding results for capacitors by replacing the resistances R by the inverses of the capacitances C. This is because $1/C = \Delta V/Q$ holds for capacitors, whereas $R = \Delta V/I$ holds for resistors. Clearly, ΔV plays the same role in each case, Q and I play similar roles, and $1/C$ and R play similar roles.

8.12   Plasma Oscillations

Optional

When electrons within a good electrical conductor are disturbed from equilibrium, the electric field that they produce tends to restore them to equilibrium. Just as a mass on a spring oscillates when disturbed from equilibrium, so too does the electron charge density oscillate when disturbed from equilibrium. This is independent of the drag effect and occurs on a shorter time scale than the momentum relaxation time. This oscillation also occurs in a gas of atoms (such as on the sun) that is heated so much that it is ionized. Such a gas is called a *plasma*, and for that reason, this sort of oscillation is called a *plasma oscillation*. Just as the mass on the spring eventually stops moving because of friction, so too does the electron charge density on a good electrical conductor stop changing because of friction.

We now obtain the so-called *plasma frequency* for a slab of thickness d and area A, and containing equal and opposite amounts of charge at density $\pm ne$. (The final result, however, is independent of geometry.) The positive charge (due to the ion cores) is assumed to be immobile. The negative charge (due to the conduction electrons in delocalized orbitals) is assumed to move as a unit by a displacement x, where $x \ll d$. At each end of the slab, a volume Ax gets an excess charge $\pm Q$, where $Q = ne Ax$. See Figure 8.28. Thus, the field inside is like that of a capacitor, for which the electric field is given by $E = 4\pi k\sigma = 4\pi k(Q/A) = 4\pi knex$.

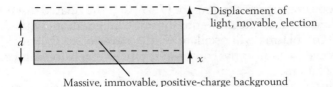

Figure 8.28 Slab of conductor, and uniform displacement (exaggerated, for emphasis) of its conduction electrons. In a real plasma oscillation, the displacement at the surface is more complex, but the plasma oscillation frequency is unaffected by the electron behavior in this relatively small region.

The equation of motion for the displacement x of an electron (of mass m) within the slab is thus

$$m\frac{d^2x}{dt^2} = -eE = -4\pi kne^2x. \tag{8.67}$$

(The relatively small amount of electric charge in the surface region feels a force that drags it along with the rest of the charge. It is like the tail following the dog.) In mechanics, you have already studied the motion of a mass m attached to a spring of spring constant K; its position x satisfies

$$m\frac{d^2x}{dt^2} = -Kx \tag{8.68}$$

and oscillates like $x = A \cos \omega t + B \sin \omega t$, with the resonance frequency

$$\omega = \sqrt{\frac{K}{m}}. \tag{8.69}$$

Comparison of (8.67) and (8.68) yields $K \to 4\pi kne^2$. By (8.69), what is called the *plasma frequency* is given by

$$\omega_p = \sqrt{\frac{4\pi kne^2}{m}}. \tag{8.70}$$

Example 8.20 **Plasma frequency of copper**

For copper, find ω_p. Use Example 7.29 for n.

Solution: With $n = 8.48 \times 10^{28}$ m^{-3}, as appropriate to Cu, (8.70) yields $\omega_p \approx 1.6 \times 10^{16}$/sec. Note that $\omega_p \gg \tau^{-1}$, where Example 7.31 estimated that $\tau = 2.5 \times 10^{-14}$ s for Cu at room temperature.

A good conductor supports plasma oscillations, with the plasma period

$$T_p = \frac{2\pi}{\omega_p}. \tag{8.71}$$

After a relaxation time τ passes [compare (7.47) and the related discussion], the plasma oscillations relax to the new equilibrium configuration, with charge only on the surface. The plasma will oscillate on the order of $\tau / T_p = \omega_p \tau / 2\pi$ times before it has decayed significantly. For Cu at room temperature, this corresponds to about 60 oscillations.

A poor conductor (i.e., a carbon resistor), which does not support plasma oscillations, within a time τ will directly relax to its new equilibrium configuration. As for a good conductor, for steady currents a poor conductor also has charge only on the surface.

8.13 Interlude: Beyond Lumped Circuits (R's and C's)

Optional

"It may be more interesting and instructive not to go by the shortest logical path from one point to another. It may be better to wander about, and be guided by circumstances in the choice of paths, and keep eyes open to the side prospects, and vary the route later to obtain different views of the same country."

—Oliver Heaviside, Vol. 2 of *Electrical Theory* (1893)

"I roamed the countryside searching for answers to things I did not understand."

—Leonardo da Vinci

8.13.1 *Where We Stand*

So far we have studied electrostatics (Chapters 1 to 6) and current-carrying circuits (Chapters 7 and 8). We have introduced what are called *lumped* circuit elements: capacitors and resistors. Nevertheless, there are limitations to this method of analysis. This becomes apparent on studying the behavior of circuits at very short times and very high frequencies. One of these limitations is due to parasitic capacitance, discussed in Section 8.10. Another is that we have not yet included the effects of the magnetic fields, either those produced by magnets (Chapter 9) or those produced by the electric currents through the circuit (Chapters 10 and 11). Chapter 12 introduces yet another lumped circuit element: the inductor, which stores magnetic energy. It also discusses another type of electric field, produced by a time-varying magnetic field, for which the field lines close on themselves. This electromagnetically induced electric field has a nonzero circulation, and thus can drive current around a circuit. It is the basis of all electrical power generators. Chapter 13 is devoted to motors and generators.

8.13.2 *Limitations on the Validity of Circuit Theory*

It would seem that we have studied enough physics to understand the behavior of ordinary circuits. And indeed we have, as long as we understand the limitations to what we have studied. *Our most critical simplification was to assume action at a distance: that the electric field is transmitted instantaneously.* As we will show in Chapter 15, the electric field does not propagate instantaneously. Associated with it is a magnetic field, and together they progagate at the speed of light

$c \approx 3 \times 10^8$ m/s. A description of what happens when we switch on a circuit that neglects the finite speed of light will not be completely valid.

There are additional complications. On making and unmaking circuit connections by hand, the changes are determined to a large extent by muscle movement, and this is much slower than the electronic response. When a switch is first "thrown," if the voltage is large enough there also can be sparking. Let us neglect these effects, and simply consider what must happen to get the system to behave in a manner that is described by simple circuit theory.

Four Time Scales When a switch is thrown for an RC circuit, there are four time scales of relevance:

1. The material-dependent time T_p corresponding to the period of a plasma oscillation, given by (8.71). This time is determined by the inertial mass m of the free electrons, their charge e, and their density n. For copper, it is on the order of 10^{-16} s.

2. The material-dependent relaxation time τ for electrons, due to collisions with the ionic background of the material (on the order of 10^{-14} s for copper at room temperature).

3. The circuit geometry–dependent time t_d for an electrical signal to cross the circuit. For a circuit of characteristic dimension d,

$$t_d \sim \frac{d}{c}, \tag{8.72}$$

where c is the velocity of light ($c \approx 3 \times 10^8$ m/s). For $d \sim 0.1$ m, t_d is on the order of 3×10^{-10} s.

4. The circuit element-dependent time constant τ_{RC}. From (8.38), $\tau_{RC} = RC$. For $R = 0.1 \ \Omega$ and $C = 10$ nF, we have $\tau_{RC} = 10^{-9}$ s.

Let us now consider how these times are relevant when the power is turned on in a circuit. First, the information that the circuit has been closed must be transmitted. This is propagated at the speed of light. Hence, after a time t_d, the circuit has had a chance to adjust to the closing of the switch. During this time, the plasma oscillations have undergone many oscillations of period $T_p \ll t_d$, and they have died down after the relaxation time $\tau \ll t_d$. Thus, once the information that the switch has been closed has propagated across the circuit, Kirchhoff's rules and ordinary circuit theory become applicable. Hence, for circuit theory to be valid, we must have $t_d \ll \tau_{RC}$, or

$$\frac{d}{c} \ll RC. \qquad \text{(condition for validity of simple circuit theory)} \tag{8.73}$$

It is not difficult to design circuits that violate the preceding condition. For $R = 0.1 \ \Omega$ and $C = 10$ pF, we have $RC = 10^{-11}$ s, so (8.73) requires that d be smaller than $cRC = 0.03$ m, a somewhat restrictive condition.

8.13.3 *What Happens for Times Too Short for Circuit Theory to Be Valid*

At times too short for circuit theory to be valid, the finite speed of light must be considered. This occurs, for example, when signals are sent using cable (which has both an inner wire that carries the signal, and an outer ground wire). The area of electrical engineering dealing with such time delays is both active and important.

Consider what happens when a switch is thrown and we wish to follow the signal as it travels around the circuit. See Figure 8.29. Let the circuit contain three resistors in series with a battery and a switch. When the switch is thrown, two different electromagnetic signals are emitted, both traveling at the speed of light. First, there is a (usually weak) signal that travels more or less radially outward from the switch, at the speed of light. In addition there is a signal that moves along the wire of the circuit. (The wire can be thought of as guiding the wave.) As it travels along from both terminals of the power source, the distribution of surface charge density rapidly adjusts to the signal. If the signal reaches the end resistors R_1 and R_3 at the same time, then it begins to dissipate energy in them. Only after the signal has reached the middle resistor R_2 does the middle resistor begin to dissipate energy. However, once the signal has traversed the system, and circuit theory has become valid, the current will be the same through each resistor. Because the time to cross the circuit (10^{-10} s) is so small compared to our visual reflexes (not faster than 10^{-2} s), for all intents and purposes we never notice this transient regime. Note that signals along human nerve axons can be analyzed in a way that is very similar to how one analyzes signals along a wire; however, the propagation velocity for a nerve axon is on the order of 10 m/s, much less than the velocity of light.

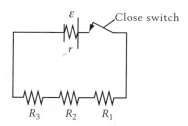

Figure 8.29 Circuit to illustrate circuit response on closing the switch.

8.13.4 *A Misconception about Circuits*

From the preceding discussion, it follows that, when a power supply is switched on, for times much less than the *very* short light-crossing time t_d, it provides power only to the part of the circuit nearest the power supply. For short enough times, current may only flow in the nominally resistanceless wires near the power supply. For somewhat longer times, the current may flow in a resistor nearest the power supply, but not in farther resistors. Finally, for times that exceed t_d, the current circulates around the circuit, heating resistors and charging up capacitors, as described by the circuit theory of Chapters 7 and 8. Some people have the misconception that, in a series circuit the nearest resistor gets more current than the others, and thus that it "uses up" the current. Only for very short times does the nearest resistor get more current than the others; this is not becauses it uses up the current, but because the current hasn't had enough time to reach the other resistors.

8.13.5 *Bringing It All Together*

Now that you have a better understanding of the individual details of the electric fluid model, you may find it helpful to reread Sections R.1 through R.8.

Problems

8-3.1 A boat battery consists of twelve 2 V "cells" in series. Each cell consists of ten 2 V voltaic cells in parallel. Each "cell" has an internal resistance of 0.02 Ω, and a charge of 10.5 A-hr. Find (a) the overall emf, (b) the internal resistance, and (c) the charge of the battery.

8-3.2 An electric eel shocks via many simultaneously excited cells, called *electroplaques* (or *electrocytes*). During discharge, these cells have $\mathcal{E} = 150$ mV and $r = 0.25$ Ω. (The emf may be thought of as electrostatic in origin, from the difference in potential between inside and outside the cell. Beause of the temporary opening of ion channels in the cell wall, this resistivity during cell discharge is much lower than for a cell at rest.) About 5000 electroplaques are in series (to give high voltage), and about 140 of these sets of 5000 are in parallel (to give high current). (a) Find the equivalent emf and the equivalent resistance of the electric eel's "power pack." If the eel discharges through a 500 Ω resistance (e.g., salt water), find the current that it provides. Note that it takes some electric eels (really, they are fish) only a few ms to recharge. Some have been known to kill a horse.

8-3.3 A single fully charged car battery (120 A-hr) is to be used to start 150 cars every day. Each start draws an average of 80 A for 1.2 s. (a) Neglecting self-discharge due to non–current-producing chemical reactions, how many days will the battery last without recharging? (b) If self-discharge occurs at a rate that will discharge a fully charged battery in 85 days, how long will the battery last when used to start all 150 cars every day?

8-3.4 The state of a discharged battery is to be determined. To be acceptable, it must be able to provide 200 A to a starting motor with $R = 0.01$ Ω. On open circuit, its terminal voltage is 12.2 V. On being charged by 8 A, its terminal voltage is 14.4 V. (a) Find its internal resistance. (b) Find the current that will flow through the starting motor. (c) Is the battery acceptable?

8-3.5 For load resistors of 0.4 Ω and 0.8 Ω, the terminal voltage of a battery reads 10.1 V and 10.4 V. Find its emf and internal resistance.

8-3.6 You have two load resistors for use with a battery. With one, the current is 40 A for a terminal voltage of 10.4 V; with the other, the current is 25 A for a terminal voltage of 10.8 V. (a) Find the battery's emf and internal resistance. (b) Find the values of the load resistors.

8-3.7 A 2.1 V cell of internal resistance 0.05 Ω is charged by an ideal dc power source set at 2.36 V. (a) Find the rate of charging of the battery. (b) Find the rate of Joule heating of the battery. (c) Find the rate at which the battery gains energy. (d) Find the efficiency of the use of energy provided by the power source (ratio of useful energy or power to energy or power provided).

8-3.8 A 3 V battery provides 6 W to a load for a 2.5 A current. Find the internal resistance of the battery.

8-3.9 A flashlight bulb is powered by three 1.5 V batteries in series, each with internal resistance 0.25 Ω. The batteries each discharge their chemical energy at the rate 4.5 W. (a) Determine the resistance of the lightbulb. (b) Determine the current.

8-4.1 An automobile battery recharger has two settings: fast (6 A) and slow (2 A). (a) At the slow rate, how long will it take to recharge completely a dead battery with a full capacity of 80 A-hr? (b) At 10 cents/kW-hr and 80% efficiency of power utilization, how much will this cost? *Note:* Part (a) is independent of the efficiency in part (b).

8-4.2 Assume that an 80 A-hr battery is fully discharged and can take the full 40 A current provided by the alternator when the car is moving at 60 miles/hr. Compute the cost to recharge the battery when the car goes at this speed, where it uses 25 miles/gallon, at $1.40/gallon. *Note:* This is not the least expensive method to recharge a battery.

8-5.1 Light of intensity 0.14 mW/cm² shines on a solar cell of area 20 cm². When connected to a 200 Ω load resistor, its terminal voltage is 0.15 V; when connected to a 400 Ω load resistor, its terminal voltage is 0.24 V. (a) Find its emf and internal resistance. (b) For the 200 Ω load resistor, find the efficiency of conversion from light energy to heat. (c) Repeat for the 400 Ω load.

8-5.2 A 6 V battery with internal resistance $r = 1\ \Omega$ drives a 70% efficient motor that lifts a 8 N box at velocity 0.5 m/s. (a) Find the current. (b) Is the answer unique?

8-5.3 A 6 V battery with internal resistance $r = 1\ \Omega$ drives an 80% efficient motor. If the motor provides 3 W of useful energy per amp, find the current.

8-5.4 Two 1.5 V cells, each of internal resistance 0.4 Ω, are to be used to power a small resistance heater. What is the maximum power they can provide, and what is the corresponding heater resistance?

8-5.5 A high-voltage transmission line cable is to have resistance per unit length 0.204 Ω/km and carry an 80 A current. (a) Find the resistance of a 16 km length that carries power 8 km to and 8 km from the load. (b) Find the voltage drop across this length. (c) For Cu wire (mass density 8.98 g/cm³), what is the mass of the cable? (d) If the power is provided to a 12 Ω load resistor, find the efficiency of energy transmission.

8-5.6 A 500 W immersion heater is placed in a quart of water at 20°C. Assume a 95% efficiency for the heater. (a) How long will it take to raise the temperature to boiling? (b) How much longer will it take to boil away half of the water? (c) At 16 cents/kW-h, how much does it cost to raise the temperature to boiling? (d) At 16 cents/kW-h, how much does it cost to boil away half of the water?

8-5.7 Consider a 60 W lightbulb for 12 V usage. (a) Neglecting the internal resistance of the 12 V source, find the resistance of the lightbulb. (b) For an internal resistance of 0.01 Ω, find the efficiency of usage of the lightbulb (to three decimal places). (c) Repeat for an internal resistance of 0.1 Ω.

8-5.8 Consider a power source of emf $\mathcal{E}_{in}$ that provides a fixed power of 0.2 MW, and a power user of emf $\mathcal{E}_{out}$. To transfer this power, they must be connected by wires that go from the power source to the power user and back. Let each of these input and output wires have resistance $R/2$. Which would be more efficient at transferring power to the user, a voltage difference across the wire of 1000 V or of 10,000 V?

8-6.1 In Figure 8.7(a), use the conventions that $I_1 > 0$ is upward through R_1, $I_2 > 0$ is downward through R_2, $I > 0$ is downward through r, and $\Delta V > 0$ to the upper side of the battery is positive. (a) Write down current conservation. (b) Write down (8.15a) for each arm.

8-6.2 (a) In Figure 8.8(a), if $I_2 = 2$ A and $I_3 = 6$ A, find I_1. (b) In Figure 8.8(b), if $I_2 = 4$ A, find I_1. (c) In Figure 8.8(c), if $I = 6$ A, is the upper plate charging or discharging, and at what rate? (d) Repeat (c) for Figure 8.8(d).

8-6.3 A current source is one for which the current is nearly independent of load. You are given an emf of 1.4 V. What value of resistance R in series with the emf and the load would make the current constant to within 1% for loads of 4 Ω and 12 Ω? See Figure 8.30.

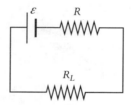

Figure 8.30 Problem 8-6.3.

8-6.4 Design a constant current source of 2 mA (±1%), intended to be used with series loads of up to 100 Ω. It uses an ideal source of emf $\mathcal{E}$ and a resistance R, whose values are at your disposal.

8-6.5 A voltage divider consists of a fixed ideal source of emf $\mathcal{E}$ and a fixed resistance R, where the output voltage is taken with a high-resistance device placed across a variable resistance r that is a part of R. See Figure 8.31. It is desired to produce a voltage source whose output varies from 0.2 V to 2.2 V. (a) For $\mathcal{E} = 4$ V, what range of values must

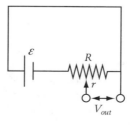

Figure 8.31 Problem 8-6.5.

r/R take on? (b) For $\mathcal{E} = 6$ V, what range of values must r/R take on?

8-6.6 A voltage divider is contructed from a 2.0 V emf with internal resistance of 0.4 Ω in series with a variable resistance R from zero to 15 Ω. The output voltage is measured across R. Find the range of output voltages.

8-6.7 The current driven by an ideal battery $\mathcal{E}$ through a resistor R is 8 A. A 5 Ω resistor is added in series, and the current drops to 6 A. Find R and $\mathcal{E}$.

8-7.1 Here a "good" battery charges a "bad" battery. In Figure 8.13(b), let the battery $\mathcal{E}_1 = 12$ V, $r_1 = 0.01$ Ω be in series with, but opposed to $\mathcal{E}_2 = 10$ V, $r_2 = 0.08$ Ω. To neglect the third arm, consider that $R \to \infty$. For each battery, find (a) the current, (b) the terminal voltage, (c) the rate at which energy goes into chemical charge or discharge of each battery, and (d) the rate of Joule heating in the internal resistances.

8-7.2 For the circuit of Figure 8.13(b), take $\mathcal{E}_1 = 6$ V, $\mathcal{E}_2 = 10$ V, $r_1 = 0.01$ Ω, $r_2 = 0.02$ Ω, $R = 0.01$ Ω. (a) From scratch, and with numbers, analyze the circuit using Kirchhoff's rules. (b) Solve for the voltage across R. (c) Find the current through R and the currents provided by each of the batteries. (d) Find the rate of heating of each resistor and the rate at which energy goes into chemical charge or discharge of each battery.

8-7.3 For the circuit of Figure 8.13(b), take $\mathcal{E}_1 = 6$ V, $\mathcal{E}_2 = 10$ V, $r_1 = 0.01$ Ω, $r_2 = 0.02$ Ω, $R = 0.01$ Ω. However, reverse the terminals of battery 2. (a) From scratch, and with numbers, analyze the circuit using Kirchhoff's rules. (b) Solve for the voltage across R. (c) Find the current through R and the currents provided by each of the batteries. (d) Find the rate of heating of each resistor and the rate at which energy goes into chemical charge or discharge of each battery.

8-7.4 For the circuit of Figure 8.13(b), take $\mathcal{E}_1 = 6$ V, $\mathcal{E}_2 = 10$ V, $r_1 = 0.01$ Ω, $r_2 = 1.2$ Ω, $R = 0.01$ Ω. (a) From scratch, and with numbers, analyze the circuit using Kirchhoff's rules. (b) Solve for the voltage across R. (c) Find the current through R and the currents provided by each of the batteries. (d) Find the rate of heating of each resistor and the rate at which energy goes into chemical charge or discharge of each battery.

8-7.5 See Figure 8.32, where a bulb B_2 and switch S_2 are placed in parallel with switch S_1. Using

responses of bright, dim, and off, categorize the bulb responses for the following switch configurations: (a) S_1 open, S_2 open; (b) S_1 open, S_2 closed; (c) S_1 closed, S_2 open; (d) S_1 closed, S_2 closed.

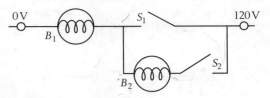

Figure 8.32 Problem 8-7.5.

8-7.6 In Figure 8.33, let all the resistors be 30 Ω. Let 8 A enter at D and 8 A leave at C. (a) Find the voltage difference between C and D. (b) Find the equivalent resistance of the circuit. (c) If the resistors have resistance R, find the equivalent resistance between C and D.

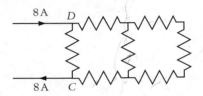

Figure 8.33 Problem 8-7.6.

8-7.7 In Figure 8.33, let all the resistors be 30 Ω. If the squares of resistors were to repeat indefinitely to the right, find the resistance between C and D. *Hint*: The resistance is unchanged if you add one more "unit" to this infinite ladder of resistors.

8-7.8 Consider the five-resistor bridge circuit of Figure 8.34. Each resistor has $R = 6$ Ω. Find the equivalent resistance between (a) A and B; (b) B and C; (c) A and D. *Hint*: Let current I enter at one node and leave at another, and find the voltage difference between these nodes in terms of I.

Figure 8.34 Problem 8-7.8.

8-7.9 For an infinite square network of identical resistors R, find the resistance between two adjacent nodes. *Hint*: Use superposition on adjacent nodes A

and B. Let current I leave A and find $V_A - V_B$; then let current I enter B and find $V_A - V_B$.

8-7.10 For the circuit of Figure 8.35, find the equivalent resistance between A and B. *Hint*: Let current I enter at A and leave at B, and find the voltage difference between these nodes in terms of I.

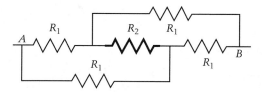

Figure 8.35 Problem 8-7.10.

8-7.11 For the circuit of Figure 8.36, find the equivalent resistance between A and B. *Hint*: Let current I enter at A and leave at B.

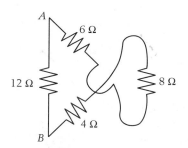

Figure 8.36 Problem 8-7.11.

8-7.12 For the circuit of Figure 8.37, find the equivalent resistance across the electrodes of the battery, and find the current provided by the battery. All resistors (including the internal resistance) are 2 Ω.

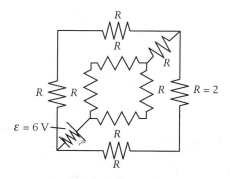

Figure 8.37 Problem 8-7.12.

8-7.13 Consider a set of batteries in series, as in Figure 8.38.

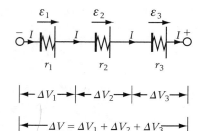

Figure 8.38 Problem 8-7.13.

Show that

$$I = \frac{\sum_i \mathcal{E}_i - \Delta V}{\sum_i r_i} = \frac{\mathcal{E}_{eff} - \Delta V}{R_{eff}},$$

$$\mathcal{E}_{eff} \equiv \sum_i \mathcal{E}_i, \quad R_{eff} \equiv \sum_i r_i.$$

8-7.14 Consider a set of batteries in parallel, as in Figure 8.39.

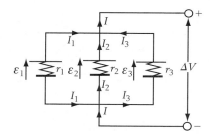

Figure 8.39 Problem 8-7.14.

Show that

$$I = \mathcal{J} - \frac{\Delta V}{\mathcal{R}}, \quad \mathcal{J} = \sum_i \left(\frac{\mathcal{E}_i}{r_i}\right), \quad \frac{1}{\mathcal{R}} = \sum_i \left(\frac{1}{r_i}\right).$$

8-7.15 If a resistor R is responsible for ΔV in Figure 8.39, then $I = \Delta V/R$. Show that $\Delta V = \mathcal{J}/(R^{-1} + \mathcal{R}^{-1})$.

8-7.16 For three batteries in parallel with a resistor R, let $r_1 = 0.01$ Ω, $r_2 = 0.02$ Ω, $r_3 = 0.1$ Ω, $R = 0.025$ Ω, $\mathcal{E}_1 = 6$ V, $\mathcal{E}_2 = 8$ V, $\mathcal{E}_3 = 12$ V. (a) Find the voltage ΔV across R. (b) Find the currents through each arm. (c) Verify current conservation.

8-7.17 For three batteries in parallel with a resistor R, let $\mathcal{E}_1 = 6$ V, $\mathcal{E}_2 = 10$ V, $\mathcal{E}_3 = 12$ V, $r_1 = 0.01$ Ω, $r_2 = 0.02$ Ω, $r_3 = 0.05$ Ω, $R = 0.01$ Ω. (a) Analyze the circuit using Kirchhoff's rules. (b) Solve for the voltage across R. (c) Find the current through R and the currents provided by each of the batteries. (d) Verify current conservation.

8-8.1 In Figure 8.16, let $\mathcal{E} = 12$ V, $r = 2\ \Omega$, $R = 6\ \Omega$, and $C = 4.5\ \mu$F. The capacitor is uncharged initially. The battery is connected to the circuit at $t = 0$. (a) At $t = 0^+$ find the initial current through each resistor and the charge on the capacitor. (b) Find the current through each resistor and the charge on the capacitor a long time after the battery is connected to the circuit.

8-8.2 In Figure 8.16, let a resistor R_2 be placed in the same arm as C. Let $\mathcal{E} = 6$ V, $r = 1\ \Omega$, $R = 6\ \Omega$, $R_2 = 3\ \Omega$, $C = 6\ \mu$F. The capacitor is uncharged initially. The battery is connected to the circuit at $t = 0$. (a) Find I, Q_2, I_1, and I_2 just after the battery is connected to the circuit. (b) Find I, Q_2, I_1, and I_2 a long time after the battery is connected to the circuit. (c) Sketch I as a function of time.

8-8.3 In Figure 8.17, let $R_1 = 3\ \Omega$, $R_2 = 6\ \Omega$, $R_3 = 12\ \Omega$, and $R_4 = 6\ \Omega$, and $C = 4\ \mu$F. An ideal 12 V battery suddenly is switched on across the top and bottom connections. (a) Just after switch-on, give the charge on the capacitor, the currents through each resistor, and the current to the capacitor. (b) After a long time, give the final currents through each resistor, the charge on the capacitor, and the current to the capacitor.

8-8.4 Reconsider Example 8.16 of Section 8.8 with respective resistances of 4, 1, 3, 2 (in ohms). (a) Find all the currents through the resistors and to the capacitor. (b) Verify current conservation for both plates of the capacitor.

8-8.5 Consider a resistor R and a capacitor C_p connected in parallel, as in our model circuit for parasitic capacitance (see Figure 8.25). Let a constant current source I_0 bring current in at one end and out at the other. Initially, $Q_{C_p} = 0$. (a) Find the initial value of I_R. (b) Find the final value of I_R.

8-9.1 An RC series circuit, as in Figure 8.15, is driven by an emf $\mathcal{E} = 4.2$ V, with $R = 5$ MΩ and $C = 2.4\ \mu$F. (a) Find the time constant. (b) Find the initial current and the current at long times. (c) Find the initial charge and the charge at long times. (d) Find the total amount of heating of the resistor during the time it takes to reach half charge. (e) Repeat for full charge.

8-9.2 For a series RC circuit, the initial current on connecting the emf is 2.5 A. After 3.4 ms, the current is 0.8 A. After a long time, the voltage across the capacitor is 6 V. Find R, C, and the emf $\mathcal{E}$.

8-9.3 A 12 μF capacitor in a series RC circuit is charged to 4 V. A time 12 ms after the switch is thrown, the capacitor voltage is 0.05 V. (a) Show by integration that the total amount of heating in the resistor up to that time equals $(1/2)C(V_0^2 - V^2)$. (b) Evaluate this numerically. (c) Obtain (a) by energy conservation.

8-9.4 An automobile blinker circuit contains a 3 V battery with $r = 0.24\ \Omega$ that drives a circuit with two arms. One of the arms contains a neon bulb, and the other contains a resistance R and a capacitance C. See Figure 8.40. The neon bulb goes on when the voltage exceeds 1.8 V, at which time its resistance effectively goes to zero. The neon bulb goes off when the voltage is less than 0.6 V, at which time its resistance effectively goes to infinity. Determine how long the blinker is on and how long it is off.

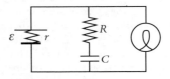

Figure 8.40 Problem 8-9.4.

8-9.5 (a) Find the resistance of a 1.6 cm long glass rod of 2 mm diameter. (b) Find τ_{RC} if this is placed in series with a 4 nF capacitor. (c) Find the series capacitance C needed to make $\tau_{RC} = 10^3$ s. (It is not trivial to measure large resistances.)

8-10.1 Discuss how best to make a 90° turn of a lithographed-on wire for an integrated circuit board: one 90° turn, two 45° turns, and so on. Consider the cost of making a complex turn.

8-10.2 Consider a resistor R and a capacitor C_p connected in parallel, as in our model circuit for parasitic capacitance (see Figure 8.25). Let a constant current source I_0 bring current in at one end and out at the other. Find Q_{C_p} and I_R as a function of time.

8-10.3 Consider a resistor R and a capacitor C_p connected in parallel, as in Figure 8.25. Let a constant emf $\mathcal{E}$ with internal resistance r bring current into this system. (a) In what limit of r/R does this problem reduce to the previous problem? (b) For arbitrary r/R, find Q_{C_p} and I_R as a function of time.

8-10.4 Imagine that the three circuit elements in Figure 8.24 are attached to identical external circuits with a battery to drive current. (a) Will the surface charge densities be different even for the three identical external circuits? (b) How does proximity of the three external circuits to their associated circuit elements affect the surface charge density of the external circuits? (That is, are the surface charge densities on the external circuits more or less alike the farther we get from the circuit elements?)

8-10.5 When the current through a wire is constant in time, the wire has no bulk charge. (a) If the surface charge distribution is Σ_S for current I, give the surface charge distribution for current $2I$. (b) Discuss how the surface charge can change by current flow through the bulk and current flow along the surface. (c) Which provides the least resistance: pure surface currents, pure bulk currents, or a combination of the two?

8-10.6 Consider a resistor of cross-sectional area 2 mm², carrying 1 A. (a) Find the current density. (b) If the resistor is of copper, find the electric field. (c) Repeat for the resistor made of aluminum. (d) If two such wires, one of copper and the other of aluminum, are in series, what is the discontinuity in electric field where they come in contact? This discontinuity is caused by electric charge on the interface. (e) If the current goes left to right, from copper to aluminum, find the sign of the charge, and the charge per unit area. *Hint:* Use Gauss's law. *Note:* This approach doesn't tell how much charge is on the copper and how much is on the aluminum; a more advanced theory is needed to tell that.

8-10.7 Consider a wire of uniform cross-sectional area A, made of two different materials. Let current flow from the material with conductivity σ_1 to the material with conductivity σ_2. See Figure 8.41. The current density J doesn't change on crossing the surface. (a) Explain why the charge per unit area Σ_S at the interface satisfies $\Sigma_S > 0$ if $\sigma_1 < \sigma_2$. (b) Using Gauss's law, find Σ_S explicitly.

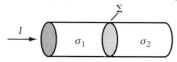

Figure 8.41 Problem 8-10.7.

8-10.8 Let current I pass from one wire of cross-section A to another cross-section A. Let them have conductivities σ_1 and σ_2, with $\sigma_1 > \sigma_2$. Let there be an intermediate region between the two, of thickness d, where σ^{-1} varies linearly between σ_1^{-1} and σ_2^{-1}. Take the relative permittivity $\kappa = 1$. (a) Determine the electric field in each wire and in the intermediate region. (b) Using Gauss's law, determine the charge density in the intermediate region, and explain its sign.

8-11.1 Someone says that, instead of (8.65), the solutions are

$$I_3 = I\frac{R_3(R_1 + R_5) + R_1(R_4 + R_5)}{D},$$

$$I_1 = I\frac{R_1(R_3 + R_5) + R_3(R_2 + R_5)}{D}.$$

$$(8.65')$$

Find a check, based upon a specific set of values for the resistors, that would distinguish between (8.65) and (8.65').

8-11.2 Consider a circuit like that in Figure 8.33, but with arbitrary values for each resistance. (a) How many loop currents must be solved for in this case? (b) If all the resistors are 2 Ω, the problem simplifies. Explain why. (c) If all the resistors are 2 Ω, find the current through each resistor, and the equivalent resistance.

8-12.1 Consider a material with $\tau = 1.2 \times 10^{-13}$ s. For what electron density would $\omega_p\tau = 2\pi$?

8-12.2 A plasma oscillation is observed for which $\omega_p\tau = 42$. If $\tau = 5.8 \times 10^{-13}$ s, find the electron number density.

8-G.1 You have a blender and a computer, each of which takes a 12 A fuse. One is a fast-blow fuse and one is a slow-blow fuse. Which device gets which fuse, and why?

8-G.2 Consider a large square network (e.g., 100 by 100) of identical resistors R. Using the nodal rule, show that the voltage at any interior node is the average of the voltages at the surrounding four nodes.

8-G.3 An ideal 400 V emf drives a current through a load resistor R and a 30 km long twisted-pair copper cable (15 km each way), each strand of resistance R_w and radius 1.6 mm. At some point along the cable, an unknown fraction α from the emf, there is a short, with unknown finite resistance R_S, across the two strands of the cable. See Figure 8.42. With the load resistance disconnected (e.g., consider that $R \to \infty$), the source provides 2.60 A. (Call the circuit resistance R_c in this case.) With the load resistance shorted out (e.g., consider that $R \rightarrow 0$), the source provides 5.15 A. (Call the circuit resistance R_b in this case.) (a) Find the resistance R_w. (b) Find α and R_S. Probably this is solved most easily by numeric means. (c) Show that $2\alpha R_w = R_c - \sqrt{(2R_w - R_b)(R_c - R_b)}$. On solving for αR_w, we find R_S by $R_S = R_b - 2\alpha R_w$. This elegant solution is due, independently, to the 19th-century telegraph engineers Blavier and Heaviside (who had no calculators).

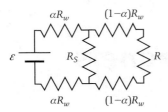

Figure 8.42 Problem 8-G.3.

8-G.4 A bridge circuit of capacitors has capacitances 1 and 2 in the upper arms, 3 and 4 in the lower arms, and 5 in the bridge (capacitances in units of μF). Find the overall capacitance of this circuit.

8-G.5 Verify that the equation for R_{bridge} gives the expected results when $R_5 = 0$; $R_5 = \infty$; $R_1 = 0$; $R_1 = \infty$.

8-G.6 Consider a point P of a circuit with an emf $\mathcal{E}$ and a resistor R in series. Initially, it is at the same voltage as an external ground point. Let point P now be connected to the positive terminal of a 12 V battery whose other terminal is at ground. See Figure 8.43. (a) What effect can there be on the

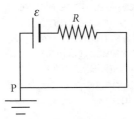

Figure 8.43 Problem 8-G.6.

charges on the wire in the circuit? (b) What is the long time effect on the current in the circuit itself?

8-G.7 A 12 V power supply with unknown internal resistance r recharges a battery with unknown emf $\mathcal{E}$ and internal resistance 1 Ω with 5 A in parallel with an indicator bulb of resistance 3 Ω that draws 1 A. (Figure 8.44 draws the internal resistances externally, as is commonly done.) Find: (a) $V_b - V_a$; (b) the current through the power supply (and its direction); (c) r; (d) $\mathcal{E}$.

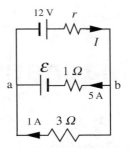

Figure 8.44 Problem 8-G.7.

8-G.8 Consider that the exterior of a cellular membrane of area A has an electrical potential that is ΔV higher than the interior potential. Take the membrane to have specific capacitance C_m (units of F/m^2). (a) Relate the charges on the interior and exterior surfaces to ΔV and the properties of the membrane. (b) Considering that this ΔV causes current to flow across all parts of the membrane surface, of specific resistance R_m (units of Ω-m^2), relate the current flow through the membrane to this voltage. (c) Show that the voltage difference goes to zero expoentially, with relaxation time $\tau = R_m C_m$. *Hint:* To be specific, consider a membrane shaped like a spherical shell.

8-G.9 Find the unknown currents in Figure 8.45.

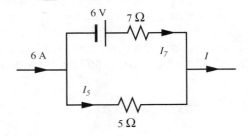

Figure 8.45 Problem 8-G.9.

*"We regard the end of the needle which points to the north as having a charge of positive magnetism, the end of the needle which points to the south as having a charge of negative magnetism. . . . It must be distinctly understood that this method of regarding the magnets and the magnetic field is only introduced as affording a convenient method of **describing** briefly the phenomena in that field and not as having any significance with respect to the constitution of magnets or the mechanism by which the forces are produced."*

—J. J. Thomson,
Elements of the Mathematical Theory of Electricity and Magnetism, fifth edition,
Cambridge University Press (1921)

Chapter 9

The Magnetism of Magnets

Chapter Overview

Section 9.1 provides a brief introduction to the magnetism of magnets. Section 9.2 summarizes the analogy between magnetic poles and electric charge, and considers the interaction of two identical magnets and the magnetic properties of magnetic dipoles. Section 9.3 studies the relationship between magnetic dipole moment $\vec{\mu}$ and magnetization $\vec{M}$, and considers the practical problems of finding the force required to pull a magnet off a refrigerator door and the disturbance of a compass reading due to a distant magnet. Section 9.4 discusses the two types of magnetic sources and shows how to obtain the magnetic field within a magnet. Section 9.5 distinguishes the types of magnetic materials, according to their differing responses in an external field. Section 9.6 discusses ferromagnets, in particular, and the magnetization process, as described by *hysteresis loops*. Section 9.7 considers the *demagnetization field,* which is of practical importance for ferromagnets. Section 9.8 applies the results for the demagnetization field to analyze particle-deflection experiments. These show that magnets behave as if they contained microscopic current sources—as if the electrons themselves contain tiny electric currents—rather than magnetic poles. Section 9.9 discusses magnetic oscillations, both for large magnets (e.g., compass needles) and small magnets (e.g., nuclei); the latter applies to nuclear magnetic resonance (NMR), the basis of the powerful diagnostic called magnetic resonance imaging (MRI). Section 9.10 considers why only some materials are magnets, why some magnets are "hard" and some are "soft," and how the world's best hard magnets are designed. ■

9.1 Introduction

Consider a magnet that attaches a note to a refrigerator door. The magnet, because it retains its magnetic properties both in isolation and in the presence of other magnets, is known as *permanent*, or hard. The refrigerator door, because it responds strongly to the magnetism of the bar magnet, but does not retain its magnetic properties in isolation, is known as a soft magnet. Ordinary iron nails and paper clips are other examples of soft magnets.

Earth's Magnetic Field

In general, orientation in a plane is only one of the three attributes of a magnetic field (which is a vector): in addition, a magnetic field also has an orientation relative to the vertical, and a *magnitude*. For the earth's magnetic field, all three vary over the earth, and tables of these, as a function of position, specified by latitude (we'll use the symbol θ^*) and longitude (ϕ), have been used in navigation. See Figure 9.1(a). True north is defined to be along the earth's rotation axis, which passes through the pole, or North, star. In practice, the easterly deviation from true north, called the *declination* angle (see Figure 9.1b), the *dip*, or *inclination* angle (positive dip is downward relative to the local horizontal plane; see Figure 9.1c), and the magnitude of the field in the local horizontal plane are most conveniently measured. Magnetic north presently is about 1300 miles south (about 11°) of true north. Magnetic north changes slowly with time, reversing, on average, every 300,000 years or so. Navigation has also been done by observing the stars and knowing the time of day. It is now done by signals emitted by earth-orbiting man-made satellites. Both because the rotation axis of the earth changes with time and because the earth moves relative to the stars, even the pole star changes (slowly) with time.

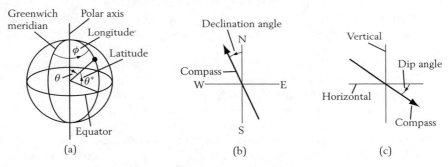

Figure 9.1 The earth's magnetic field. (a) Angles defining longitude and latitude. (b) Declination angle in local plane of the earth. (c) Dip angle relative to the local vertical.

Lodestone (the mineral magnetite, Fe_3O_4) was the first material found to display what we would call magnetic properties. Taking advantage of its permanent magnetism, it was used in China over 2000 years ago as a navigational aid. It has been used for over 1000 years in the form of compass needles that orient along the earth's magnetic field.

A characteristic value for the magnitude of the earth's magnetic field (which includes both the horizontal and vertical components) is 0.5×10^{-4} tesla (T). The SI unit for the magnetic field, the tesla $= T = N/A$-m, is named after Nikola Tesla, who during the 1880s developed the first ac (alternating current) motor. In cgs-emu units, the unit of magnetic field is the gauss (G), where $1\ G = 10^{-4}\ T$.

Some History

The earliest known work on magnetism in the western literature is by Pierre de Maricourt, also known as Peter Peregrinus, in 1269. He describes how to locate a good natural magnet, or lodestone, how to shape it into a sphere, how to locate its *poles*, and how to locate its north by placing it on a small wooden vessel floating within a large vat of water. He also gave a rule for the interaction of the poles of two different magnets: *opposite poles attract*. Surprising to moderns, there was no corresponding statement that like poles repel: rather, according to Peregrinus they only *seem* to repel:

> ...*the Northern part in a stone attracts the Southern part in another stone, and the Southern the Northern. But if you do the opposite, namely, bring the Northern part toward the Northern, the stone which you are carrying in your hand will seem to repel the floating stone, and if you apply the Southern part to the Southern, the same will happen. The reason is that the Northern part feels the Southern, which makes it seem to repel the Northern; of this there is a token in the fact that the Northern part will in the end join itself to the Southern.*

Peregrinus's hypothesis certainly satisfies the spirit of Occam's razor—simplicity—but it does not satisfy the fact that like poles really do repel, as can be seen more clearly with long bar magnets than with spherical magnets.

In 1600, Gilbert published his opus *De Magnete*, which discussed many properties of permanent magnets, noting that they have no net *pole strength*, and including the statement that "opposite poles attract and like poles repel." Thus, by 1600, it was known that the force between two poles acted along their lines of centers, and could be either attractive or repulsive. The inverse square law between poles of two long bar magnets was established first by the Englishman John Michell around 1750, and independently by Coulomb, around 1785, both of them using torsion balances. At this point, it became possible to give a quantitative description of magnetism in terms of the interactions of magnetic poles.

On the other hand, in 1820, Oersted discovered that electric currents can deflect a compass needle. Almost immediately, Ampère realized how permanent magnetism could be described in terms of electric currents. There is a fundamental distinction between these two descriptions of magnetic sources.

9.1.2   *Two Ways to Treat Magnets*

A magnetic pole produces magnetic flux but no magnetic circulation, and thus may be called a *flux source*. Its field line drawing rules are like those for electric fields due to electric charges at rest: field lines originate (terminate) on positive (negative) poles, and field lines do not close on themselves. Magnets behave as if they contain magnetic poles that sum to zero, so that zero net magnetic flux leaves any magnet. Figure 4.2(a) illustrates an electric field and a volume from which a net electric flux emerges.

Chapter 11 shows that the magnetic field due to an electric current produces magnetic circulation but no magnetic flux, and thus may be called a *circulation source*. Its field-line drawing rules are that there are no poles, and that field lines

Magnetic Poles

Although true magnetic poles have not been observed in any laboratory, they are presented for two reasons. First, the formalism of the magnetism of magnetic poles, which is a reprise of the electricity of electric charge, is much simpler than the formalism of the magnetism of electric currents. (For that reason, in practice the magnetic pole formalism is employed to design both large magnets used in magnetic resonance imaging and in particle accelerators, and the microscopic magnetic particles used in magnetic recording tape and in computer hard drives.) Second, the next two chapters, which obtain all the laws of the magnetism of electric currents, begin with the magnetic pole formalism and the equivalence between a magnet and a current loop (when both are viewed from a distance).

can close on themselves. Figure 5.18 illustrates an electric field and a circuit for which there is a net electric circulation.

Only by performing experiments that actually probe the interior of a magnet can we distinguish between the two types of source; the experiments favor the electric current viewpoint. Nevertheless, as long as we are outside a magnet, we cannot tell whether it is a flux source with zero net *magnetic charge* or a current source. Therefore it is legitimate to treat the exterior of the magnet using the formalism of magnetic poles, which has a similar structure to the already studied formalism of electric charges.

The present chapter considers the magnetism of magnets as if it is due to magnetic poles, with an inverse square law like that of electricity. Chapters 10 and 11 discuss the magnetism of magnets as if it is due to electric currents, which requires a more complex formalism.

9.2 Outside a Magnet We Can Use Magnetic Charge (Poles)

It is an experimental fact that outside a magnet its magnetic properties can be obtained by treating it as if it contained a distribution of magnetic poles, or charges, whose numerical value summed over the magnet is zero. From Michell's and Coulomb's work on long, narrow bar magnets, the interaction between two poles varies as the inverse square of their separation. To be more precise, we must specify a unit of magnetic charge q_m, which is analogous to electric charge q, and a magnetic force constant k_m, which is analogous to the electric force constant k. We will choose units of magnetic charge, or pole strength, such that $k_m \equiv \frac{\mu_0}{4\pi} = 1.0 \times 10^{-7}$ N/A^2, exactly. Here μ_0 is called the *permeability* of free space. With this set of units, Table 9.1 summarizes the basic correspondences between electric and magnetic poles.

A literal analogy between electricity and magnetism would, by analogy to Chapter 1, consider conservation of magnetic charge and provide examples where magnetic charge is transferred between two objects but overall is conserved. However, because each magnet has zero net magnetic charge, there can be no such transfer of magnetic charge between magnets. The remainder of this section and the next section applies the analogy between electricity and magnetism to magnetic charge, magnetic force, and magnetic field.

Table 9.1 Electric charge and magnetic pole equivalences

Quantity	Electricity	Magnetism
Charge	q (C)	q_m (A-m)
Field	$\vec{E}$ (N/C = V/m)	$\vec{B}$ (T = N/A-m)
Force	$q\vec{E}$ (N)	$q_m\vec{B}$ (N)
Coupling constant	$k \equiv \frac{1}{4\pi\epsilon_0} \approx 9.0 \times 10^9$ N-m^2/C^2	$k_m \equiv \frac{\mu_0}{4\pi} = 1.0 \times 10^{-7}$ N/A^2
Point source	$\vec{E} = kq\frac{\hat{r}}{r^2}$	$\vec{B} = k_m q_m \frac{\hat{r}}{r^2}$
Charge/area	σ (C/m^2)	σ_m (A/m)
Sheet source	$|\vec{E}| = 2\pi k|\sigma|$	$|\vec{B}| = 2\pi k_m|\sigma_m|$
Dipole moment	$p = ql$	$\mu = q_m l$
Dipole moment/volume	polarization $P = p/V$	magnetization $M = \mu/V$
Torque on dipole in field	$\vec{p} \times \vec{E}$	$\vec{\mu} \times \vec{B}$
Energy of dipole in field	$-\vec{p} \cdot \vec{E}$	$-\vec{\mu} \cdot \vec{B}$

9.2.1 *Force between Magnetic Poles*

Consider the force $\vec{F}$ on a pole of pole strength q_m at position $\vec{r}$, due to another pole q_{m1} at $\vec{r}_1$, as in Figure 9.2.

By Table 9.1, $\vec{F}$ satisfies a rule analogous to Coulomb's law, (2.4), for electric charges. That is,

$$\vec{F} = \frac{k_m q_m q_{m1}}{R_1^2}\hat{R}_1. \qquad \text{(force on pole } q_m\text{, toward } \hat{R}_1,\ \vec{R}_1 = \vec{r} - \vec{r}_1) \quad (9.1)$$

Example 9.1 shows that the unit of magnetic pole strength is A-m.

Note: For a magnet, the magnetic poles typically are not concentrated only at the ends; we can verify this by sprinkling iron filings onto a piece of paper that covers the magnet. Nevertheless, for simplicity, we will make this idealization.

Example 9.1 **Interaction of two identical long bar magnets: estimating q_m**

Consider two bar magnets with length l and cross-sectional area A, and with the same pole strength q_m (determined by their ability to pick up the same number of nails, etc.). Let $l \gg \sqrt{A}$, so the magnets are very long. (a) Give

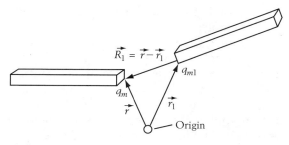

Figure 9.2 Two idealized magnets and their nearest poles.

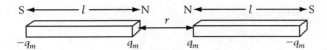

Figure 9.3 Two collinear bar magnets, at a separation where their interaction is dominated by the nearest poles, and those poles can be treated approximately as points.

an equation for estimating the force between the magnets when they are collinear, with two like poles so close that they provide the dominant force, but farther away than $\sqrt{A}$. See Figure 9.3. (b) For $l = 20$ cm $= 0.2$ m, and a square cross-section of side $a = \sqrt{A} = 0.5$ cm $= 0.005$ m, let $|\vec{F}| = 0.34 \times 10^{-1}$ N when $r = 0.04$ m. Estimate q_m.

Solution: (a) If the separation r between the two nearest poles greatly exceeds the width of the poles, then it is a good approximation to consider only the interaction between these poles, treated as points. Hence, from (9.1), with $q_{m1} = q_m$,

$$|\vec{F}| \approx \frac{k_m (q_m)^2}{r^2}. \quad \text{(equal strength poles interacting)} \quad (9.2)$$

(b) Solution of (9.2) yields $q_m = r(|\vec{F}|/k_m)^{1/2} = 23.3$ A-m. Including the forces between the other poles (two attractive interactions at separation $r + l \approx l$, and a repulsive interaction at separation $r + 2l \approx 2l$) would give a more accurate estimate for q_m.

9.2.2 *Measuring the Magnetic Field*

Just as for electricity the force on a charge q in an electric field $\vec{E}$ is $\vec{F} = q\vec{E}$, so for magnetism the force on a magnetic pole q_m in a magnetic B field is

$$\vec{F} = q_m \vec{B}. \quad (9.3)$$

Using this equation, with a known pole strength q_m, a measurement of $\vec{F}$ gives $\vec{B}$, with units of tesla (T). Example 9.2 shows that T = N/A-m. Hence the unit of magnetic charge has units of A-m = N/T.

Example 9.2 **Measuring the field due to a magnetic pole**

Find the field due to one pole at the other in Example 9.1.

Solution: By (9.3), $|\vec{B}| = |\vec{F}/q_m| = 0.34 \times 10^{-1}$ N/(23.3 A-m) $= 1.46 \times 10^{-3}$ T.

9.2.3 *Field of a Monopole*

Comparing (9.1) and (9.3), the field set up at $\vec{r}$ by q_{m1} at $\vec{r}_1$ is given by

$$\vec{B} = \frac{k_m q_{m1}}{R_1^2} \hat{R}_1. \quad (\hat{R}_1 \text{ to observation point, } \vec{R}_1 = \vec{r} - \vec{r}_1) \quad (9.4)$$

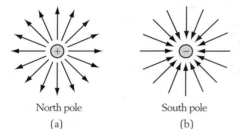

North pole   South pole

(a)   (b)

Figure 9.4 Magnetic field direction and sign of poles: (a) Positive poles make outward magnetic field. (b) Negative poles make inward magnetic field.

See Figure 9.2. As for the electric field $\vec{E}$, the magnetic field $\vec{B}$ points away from positive (north) poles, and toward negative (south) poles. See Figure 9.4.

9.2.4 *Magnetic Dipoles*

Consider a magnet of length l and pole strengths $\pm q_m$. Let us find the field on its axis, a distance r from its center, as in Figure 9.5(a). By (9.4), at the center of the compass needle the magnetic field due to the magnet points along its axis $\hat{x}$, with magnitude due to both q_m and $-q_m$, given by

$$B_{mag} = \frac{k_m q_m}{(r - l/2)^2} + \frac{k_m(-q_m)}{(r + l/2)^2} = k_m q_m \frac{(r + l/2)^2 - (r - l/2)^2}{(r^2 - l^2/4)^2} = \frac{2k_m q_m r l}{(r^2 - l^2/4)^2}.$$

(9.5)

Example 9.3   **A magnet can disturb a compass reading**

Consider a magnet with the same pole strength as in the previous example, so $q_m = 23.3$ A-m, but let its length be only $l = 1$ cm. Let the earth's field point along true north and, as shown in Figure 9.5(a), let the the magnet be 20 cm west of, and point toward, the compass needle. Determine by how much the magnet disturbs the orientation of a compass needle.

Solution: Consider each pole separately, so that (9.5) applies. Let the in-plane (horizontal) component of the earth's magnetic field be B_{Eh}, and let it point along

(a)   (b)

Figure 9.5 (a) Compass needle along axis of bar magnet. (b) Decomposition of total magnetic field acting on the compass needle into its components due to the earth and due to the bar magnet. Effect of a bar magnet on the local magnetic field. Without the bar magnet, the needle would point toward the top of the page.

$\hat{y}$. Since $r = 20.5$ cm, (9.5) gives $B_{mag} = 0.541 \times 10^{-5}$ T. The needle aligns along the direction of the net magnetic field (see Figure 9.5b). Hence this field will cause the needle to rotate from $\hat{y}$ toward $\hat{x}$ by an angle θ satisfying

$$\tan \theta = \frac{B_{mag}}{B_{Eh}}. \tag{9.6}$$

Taking $B_{Eh} = 5.0 \times 10^{-5}$ T, this gives $\tan \theta = 0.116$, corresponding to an angular deflection of 6.7°! In centuries past, no wonder sailors were warned to keep magnets and iron from the vicinity of the navigator's compass. Today, global navigation systems employing satellites have made the navigator's compass obsolete; however, magnets don't stop responding if the satellites stop functioning.

An electric dipole with charges $\pm q$ separated by l has an electric dipole moment $p = ql$, and the vector $\vec{p}$ points from the negative to the positive charge. Similarly, a magnetic dipole with poles $\pm q_m$ (in units of A-m) separated by l, has a magnetic dipole moment

$$\mu = q_m l. \tag{9.7}$$

The vector $\vec{\mu}$ points from the negative to the positive pole. It has units of A-m^2. See Figure 9.6(a). Some field lines for it are drawn in Figure 9.6(b).

Consider (9.5) when $r \gg l$, so $r/(r^2 - l^2/4)^2 \approx r/r^4 = 1/r^3$. Then, using (9.7), Equation (9.5) becomes

$$B_{mag} \approx \frac{2k_m \mu}{r^3}. \qquad \text{(on axis)} \tag{9.8}$$

The field of a magnetic dipole falls off with distance as r^{-3}, just as for an electric dipole [compare (3.12)]. Applied to Example 9.3, where $q_m = 23.3$ A-m and $l = 1$ cm, (9.7) yields $\mu = 0.233$ A-m^2. With $r = 20.5$ cm, (9.8) then yields 5.41×10^{-6} T, in three-decimal place agreement with the exact calculation of Example 9.3.

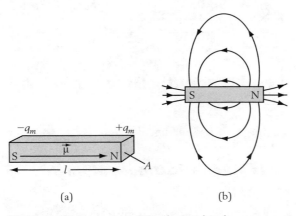

(a)   (b)

Figure 9.6 Bar magnet: (a) relationship between poles and direction of magnetic moment, (b) field-line pattern.

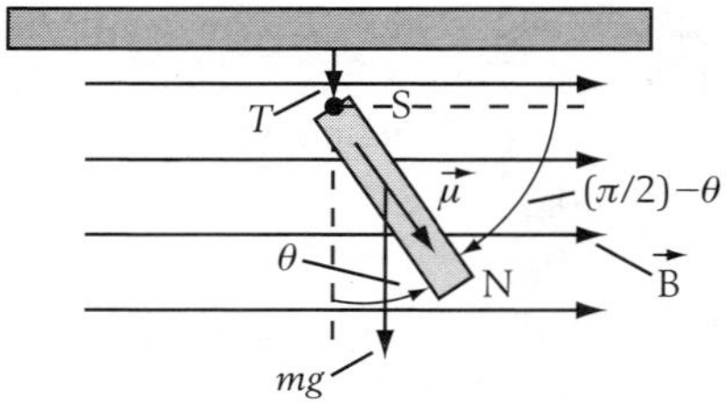

Figure 9.7 A bar magnet suspended by a string in the earth's gravity and a uniform horizontal magnetic field.

Example 9.4 **Suspended magnet in a horizontal $\vec{B}$ field**

Consider a magnet of mass m, length l, and magnetic moment $\vec{\mu}$, where $\mu = |\vec{\mu}|$. It is suspended from the ceiling by a string, its N pole toward ground, due to gravity. A horizontal magnetic field $\vec{B}$ is now applied, where $B = |\vec{B}|$. See Figure 9.7. (a) Find the condition that determines the equilibrium angle. (b) Find the change in magnetic energy on going from $\theta = 0°$ (aligned with the field) to the equilibrium θ.

Solution: (a) The field acts to make $\vec{\mu}$ point rightward, along $\vec{B}$, at angle θ with respect to the vertical. Since the angle between $\vec{\mu}$ and $\vec{B}$ is $\pi/2 - \theta$, the torque from $\vec{\mu} \times \vec{B}$ is counterclockwise, and of magnitude $\mu B \sin(\pi/2 - \theta) = \mu B \cos\theta$. By taking the torque with respect to the point of contact with the string, the torque from the string tension T can be neglected. However, the torque from gravity is clockwise, and of magnitude $mg(l/2) \sin\theta$. When the magnet is in equilibrium, these two torques are equal, or

$$\mu B \cos\theta = mg(l/2) \sin\theta,$$

leading to the condition that $\tan\theta = (2\mu B)/(mgl)$. (b) The change in magnetic energy on going from $\theta = 0°$ to this θ is given by the final energy minus the initial energy, or (from Table 9.1)

$$-[-\mu B \cos(\pi/2 - \theta) - (-\mu B \cos(0))] = -\mu B(1 - \sin\theta).$$

This is negative. On the other hand, the change in gravitational energy, $\frac{1}{2}mgl(1 - \cos\theta)$, is positive. For the magnet to go spontaneously from $0°$ to θ, the overall change in energy must be negative.

9.3 Magnetization and Magnetic Dipole Moment

A magnet is characterized by its *magnetization* $\vec{M}$, which we now define and relate to a number of important properties of magnets.

9.3.1 *Magnetization $\vec{M}$ Is Magnetic Dipole Moment per Unit Volume*

Although we did not use this terminology earlier, the electric dipole moment per unit volume is called the *polarization* $\vec{P}$, where $\vec{P} = \vec{p}/V$. Similarly, the magnetic dipole moment per unit volume is called the *magnetization* $\vec{M}$, where

$$\vec{M} = \frac{\vec{\mu}}{V}.$$ (9.9)

$\vec{M}$ has units of magnetic pole strength per unit area, or $N/(T\text{-}m^2) = A/m$. Because $\vec{M}$ is independent of the volume of the system, for many purposes it is a more fundamental quantity than $\vec{\mu}$.

9.3.2 *"Magnetic Charge" per Unit Area σ Equals Magnetization M*

Consider a bar magnet of uniform magnetization $M = |\vec{M}|$, length l, and cross-sectional area A (so its volume $V = Al$), as in Figure 9.6(a). By (9.9), it has

$$\mu = MV = MAl.$$ (9.10)

Comparison of (9.7) and (9.10) yields the length-independent "magnetic charge," or pole strength,

$$q_m = \frac{\mu}{l} = \frac{MV}{l} = MA.$$ (9.11)

Hence the charge densities on the surfaces are $\pm\sigma_m = \pm q_m/A$. Use of (9.11) then yields

$$\sigma_m = \frac{q_m}{A} = M. \quad \text{("magnetic charge" per unit area)}$$ (9.12)

This result can be stated more generally: at a surface with outward normal $\hat{n}$, the magnetic surface charge density σ_m is related to the magnetization $\vec{M}$ at that surface by

$$\sigma_m = \vec{M} \cdot \hat{n}.$$ (9.12′)

For a bar magnet, $\vec{M}$ usually is along its axis. Since the normal $\hat{n}$ to a side of a magnet is perpendicular to the axis, there usually is no magnetic charge along the sides of a bar magnet.

Example 9.5   **From magnetic moment to magnetization**

Consider the bar magnets of Example 9.1 ($q_m = 23.3$ A-m, $A = 0.25 \times 10^{-5}$ A/m^2, $l = 0.2$ m). Find their magnetic moment and magnetization.

Solution: By (9.7), $\mu = q_m l = (23.3$ A-m$)(0.2$ m$) = 4.66$ A-m^2. By (9.11), $M = q_m/A = 9.32 \times 10^5$ A/m. This is slightly less than for the alloy alnico V (sometimes used for loudspeakers).

9.3.3 *Magnetic Field Due to a Sheet of "Magnetic Charge" q_m*

In the analogy between electricity charges and magnetic poles, we replace k by k_m, $\vec{E}$ by $\vec{B}$, and q by q_m. Hence the electric field due to a sheet of uniform charge per unit area σ, $|\vec{E}| = 2\pi k\sigma$, has as its magnetic analog,

$$|\vec{B}| = 2\pi k_m \sigma_m. \qquad \text{(field due to sheet of "magnetic charge")} \qquad (9.13)$$

By (9.12), this may be rewritten as

$$|\vec{B}| = 2\pi k_m M. \qquad (9.14)$$

Example 9.6 $\vec{B}$ **just outside the end of a long magnet of alnico V**

Consider a long bar magnet made of the alloy alnico V, of length $l = 20$ cm and square cross-section of area $A = 1$ cm^2. Let its M be along its axis. See Figure 9.6(a). At room temperature in zero applied field, alnico V has $M = 9.95 \times 10^5$ A/m. Estimate the field on the axis of the magnet, 1 mm outside the north pole, and find the magnetic field of the distant pole.

Solution: At 1 mm outside the north pole, which is nearly 20 cm from the south pole, the total magnetic field is dominated by the contribution from the north pole. Moreover, since the pole faces each have width $a = \sqrt{A} = 1$ cm, which is much larger than 1 mm, the north pole appears to be a sheet of magnetic charge density $\sigma_m = M_r = 9.95 \times 10^5$ A/m. From (9.13), $B = 2\pi k_m \sigma_m = 0.625$ T, which is more than 10,000 times the earth's magnetic field. Note that, by (9.4), the magnetic field of the distant pole of this bar magnet is $k_m q_m/r^2 = k_m \sigma_m A/l^2 = 2.49 \times 10^{-4}$ T, which adds a contribution much smaller than that due to the near pole, but about five times that of the earth.

9.3.4 *Lifting Strength of a Magnet: Parallel-Plate Capacitor Analogy*

When a permanent magnet of magnetization M along its axis is placed against the surface of a refrigerator (a soft magnet), the permanent magnet is attracted by a force we will call its *lifting strength*. Only if we pull harder than the lifting strength will the magnet come off. It can be estimated as follows. When a permanent magnet is brought up to a refrigerator door, the refrigerator responds as if it had an image magnet of the opposite polarity within. See Figure 9.8. Hence there is

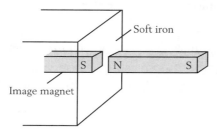

Figure 9.8 Permanent magnet in shape of a bar, placed very near the face of a large soft iron magnet. This is related to the geometry of a magnet on a refrigerator. The soft iron responds as if there were an image bar magnet of opposite polarity and (nearly) equal strength.

an attraction between the permanent magnet and the refrigerator. (The dominant effect arises from the near end of the magnet and its image.) However, if you look inside the refrigerator there will be, perhaps, a tuna sandwich, but certainly no image magnet. This attraction is similar to what happens with the amber effect, discussed in Section 1.2; a more precise analogy is to a small electric charge that is brought up to a large sheet of electrical conductor.

Let the magnet have its N pole, a flat surface of area A, against the refrigerator. Then there is an image magnet (S pole) just against the magnet; for Figure 9.8, consider that the gap between the actual N pole and the image S pole is very small. This geometry is just like that for two nearby capacitor plates. (Magnetic poles can't move, so contact can't cause discharge.) Neglecting edge effects, the magnetic field acting on the image pole, due to the actual magnet, is given by (9.14). Use of (9.3), (9.12), and (9.14) yields

$$|\vec{F}| = |-q_m||\vec{B}| = (\sigma_m A)(2\pi k_m M) = (2\pi k_m)M^2 A. \qquad \text{(lifting strength)} \qquad (9.15)$$

For the alnico V magnet of the previous example, $q_m = \sigma_m A = (9.95 \times 10^5 \text{ A/m})$ $(10^{-4} \text{ m}^2) = 99.5 \text{ A-m}$, giving a force $|\vec{F}| = (99.5 \text{ A} - \text{m})(0.625 \text{ T}) = 62.2 \text{ N}$. In terms of the force per unit area, this amounts to $F/A = 62.2 \text{ N}/10^{-4} \text{ m}^2 = 6.22 \times 10^5 \text{ N/m}^2$, or about six times atmospheric pressure.

Example 9.7 **Estimating the magnetization from the lifting strength**

A 28 cm long permanent magnet has a 0.2-cm-by-0.2-cm cross-section. Its north pole is held 0.1 mm from a refrigerator door (made of soft iron). A force of 0.018 N is required to pull the magnet off the door. Estimate its magnetization (magnetic moment per unit volume). Briefly explain your reasoning.

Solution: Since the refrigerator door is made of soft iron, it responds as if it had an image magnet inside it. Since the door and the magnet are much closer (0.1 mm) than the magnet width (0.2 cm), the near pole of the magnet produces a nearly uniform field within the image magnet region of the door. Moreover, since the near pole is much closer to the refrigerator (0.1 mm) than the distant pole (28 cm), the near pole dominates. The near pole and its image may be treated as sheets, so (9.15) applies. Solving for M yields $M = \sqrt{|\vec{F}|/2\pi k_m A}$. With $A = 0.04 \text{ cm}^2$ and $|\vec{F}| = 0.018 \text{ N}$, this yields $M = 8.46 \times 10^4 \text{ A/m}$.

9.4 Inside a Magnet There Really Are No Poles

Although, outside a magnet, its properties can be described in terms of the magnetism of *magnetic charges*, or *poles*, no one has yet observed any *isolated* magnetic pole: *the sum of the pole strengths distributed on any magnet sums to zero.* Break a magnet into two or more pieces, and each piece will have zero net pole strength, poles seeming to appear at the point of the break. See Figure 9.9. Place the pieces near one another, and they will spontaneously attract, tending to resume the original shape, with the poles at the break canceling one another.

Figure 9.9 Effect of breaking a bar magnet: (a) Before breaking it, it appears to have two poles. (b) After breaking it, there appear to be two smaller magnets, each with the same pole strength as the original magnet.

9.4.1 *The Name of the Field If Magnets Really Contained Magnetic Poles: $\vec{H}$*

The only way to tell if there are poles inside a real magnet is to do an experiment—such as discussed in Section 9.8—that actually probes the interior of the magnet. For comparison with such experiments, it is convenient to have a name for the quantity computed as if magnetic poles really did exist. Although we have been calling the symbol $\vec{B}$ the magnetic field, in the early 19th century, physicists called a symbol $\vec{H}$ the magnetic field. In the SI system, $\vec{B}$ and $\vec{H}$ have different units, which helps distinguish between them: in free space,

$$\vec{H} = \frac{\vec{B}}{\mu_0}, \quad \mu_0 = 4\pi k_m = 4\pi \times 10^{-7} \frac{\text{N}}{\text{A}^2}. \quad \text{(free space)} \quad (9.16)$$

For $\vec{B}$, the rules given in Table 9.1 hold only outside a magnet. However, on replacing $\vec{B}$ by $\mu_0 \vec{H}$, these rules hold for $\mu_0 \vec{H}$ both inside *and* outside a magnet. As we will see shortly, $\vec{B}$ and $\vec{H}$, within a magnet, have a more complex relationship than (9.16).

When physicists say "magnetic field," they usually mean $\vec{B}$, which technically is called the *magnetic induction*. To avoid this possible ambiguity, we sometimes employ the usage magnetic $\vec{B}$ field or simply $\vec{B}$ field, and magnetic $\vec{H}$ field or simply $\vec{H}$ field.

By analogy to Gauss's law relating the flux of $\vec{E}$ through a closed surface to the electric charge enclosed by that surface, or

$$\Phi_E = \oint \vec{E} \cdot \hat{n} dA = 4\pi k Q_{enc}, \quad (9.17)$$

with our choice of $\vec{H}$ due to magnetic charge Q_m, we have

$$\oint \mu_0 \vec{H} \cdot \hat{n} dA = 4\pi k_m Q_{m,enc}, \quad \text{or} \quad \Phi_H \equiv \oint \vec{H} \cdot \hat{n} dA = Q_{m,enc}. \quad (9.18)$$

Going from $\vec{B}$ to $\vec{H}$

With $B_{earth} \approx 0.5 \times 10^{-4}$ T, by (9.16), we have $H_{earth} \approx 39.8$ A/m. Sometimes the unit for H is given as A-turns/m. The dimensionless unit "turns" is appended to distinguish the SI unit for the magnetic $\vec{H}$ field from that of the SI unit for the magnetization, which is A/m. "Turns" arises because, as shown in Chapter 11, current-carrying coils wound with many turns of wire also can produce a magnetic $\vec{H}$ field (and a magnetic $\vec{B}$ field).

For electrostatics and for permanent magnets, there is zero circulation in the sense that, on integrating around a closed path with path element $d\vec{s} = \hat{s}\,ds$,

$$\Gamma_E = \oint \vec{E} \cdot \hat{s}\,ds = 0, \qquad \Gamma_H = \oint \vec{H} \cdot \hat{s}\,ds = 0. \qquad \text{(zero circulation)} \quad (9.19)$$

Thus $\vec{E}$ due to static charges, and $\vec{H}$ for permanent magnets, have *flux sources* but no *circulation sources*. Chapter 11 shows that electric currents serve as circulation sources for $\vec{B}$ and $\vec{H}$. Chapter 12 shows that time-varying $\vec{B}$-fields serve as circulation sources for $\vec{E}$.

9.4.2 Since $\vec{B}$ Has Zero Flux, Its Normal Component Is Continuous

Since there are no isolated magnetic poles, $\vec{B}$ cannot have flux as its source, so $\Phi_B = 0$ for all closed surfaces. Thus, whatever flux enters one part of a closed surface, the same amount must leave another part of it. Consider a pancake-shaped Gaussian surface (a "pillbox") enclosing a magnet face, as in Figure 9.10, with outward normal $\hat{n}$. By this argument, the normal component of $\vec{B}$, or $B_n = \vec{B} \cdot \hat{n}$, must be continuous across the face. We now use this continuity of B_n to determine $\vec{B}$ within the magnet.

Consider a bar magnet so long that near one end we can obtain the field $\vec{H}$ by considering only the magnetic charge at that end. From (9.13) and (9.16), $\mu_0 H = 2\pi k_m \sigma_m$ just outside this "sheet" of magnetic charge, with the direction of $\vec{H}$ changing as we go from inside to outside the magnet. See Figure 9.11(a). Thus, at its poles there is a discontinuity in $\mu_0 \vec{H} \cdot \hat{n}$ (from $2\pi k_m \sigma_m$ to $-2\pi k_m \sigma_m$). Hence $\mu_0(\vec{H}_{out} - \vec{H}_{in})$ has magnitude $4\pi k_m \sigma_m = 4\pi k_m M = \mu_0 M$. Since we want $\vec{B}_{in} = \vec{B}_{out}$ at the interfaces, this suggests the following definition for $\vec{B}$. With $\vec{H}$ obtained, both inside and outside the magnet, by the rules of Table 9.1 in Section 9.3 (with $\vec{B} \to \mu_0 \vec{H}$), and with $\vec{M}$ known, we set

$$\vec{B} = \mu_0(\vec{H} + \vec{M}). \qquad \text{(general definition of } \vec{B}) \qquad (9.20)$$

In free space, where $\vec{M} = \vec{0}$, (9.20) reduces to (9.16). In general, (9.20) also

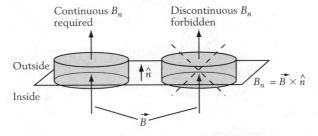

Figure 9.10 The normal component of the magnetic field $\vec{B}$ is continuous on going from within a magnet to outside a magnet. An allowed $\vec{B}$ and a disallowed $\vec{B}$ are depicted.

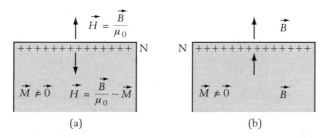

(a)   (b)

Figure 9.11 Comparison of the behavior of $\vec{B}$ and $\vec{H}$ on crossing from within a magnet to outside the magnet, for a surface with positive magnetic poles. (a) $\vec{H}$, whose normal component is *not* continuous. (b) $\vec{B}$, whose normal component *is* continuous.

makes the normal component of $\vec{B}$ continuous at *all* interfaces because it adds the missing $\mu_0\vec{M}$ within the magnet. See Figure 9.11(b). Because the normal component of $\vec{B}$ is continuous, $\vec{B}$ does not have magnetic charge as a flux source. Hence

$$\oint \vec{B} \cdot \hat{n}\, dA = 0. \qquad \text{(Gauss's law for magnetism)} \qquad (9.21)$$

If true, or "free," magnetic charges are discovered, (9.21) can be modified to incorporate them. Free magnetic charge would be a flux source for both $\vec{B}$ and $\vec{H}$. Following the analogy to electricity, we can compute $\vec{H}$ for a given set of q_m's, and then we can compute $\vec{B}$.

9.5 Types of Magnetic Materials

It is well known that, unlike the hard and soft magnets we have discussed so far, most materials have no *obvious* magnetic properties. However, when put in a (large) nonuniform magnetic field, most materials are either attracted (*paramagnets*) or repelled (*diamagnets*). See Figure 9.12. Dielectrics, when placed in a nonuniform electric field, are attracted (*paraelectrics*), but there are no electrical analogs of diamagnets. [The attraction is a consequence of the amber effect, for which there is a magnetic analog. One may think of diamagnetism as the magnetic analog of an (imaginary) antiamber effect.]

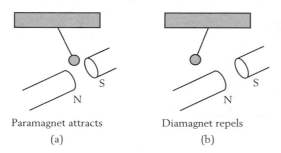

Paramagnet attracts   Diamagnet repels
(a)   (b)

Figure 9.12 Two weak types of magnetism. (a) Paramagnetism (material attracted to strong field regions; magnetic analog of the amber effect). (b) Diamagnetism (material repelled from strong field regions; no electrical analog).

Paramagnets and diamagnets develop small magnetizations that are proportional to the applied field $\vec{H}$, so that

$$\vec{M} = \chi \vec{H}. \tag{9.22}$$

Here χ ("chi") is called the *magnetic susceptibility*. Since $\vec{M}$ and $\vec{H}$ have the same units, χ is a dimensionless quantity. It is material dependent and temperature dependent. For paramagnets $\chi > 0$, and for diamagnets $\chi < 0$. In small fields, (9.22) also holds for soft ferromagnets ($\chi > 0$) and for perfect diamagnets ($\chi = -1$), provided that account is taken of the demagnetization field produced by the material itself, as discussed in Section 9.7.

When (9.22) holds, Equation (9.20) can be rewritten as

$$\vec{B} = \mu_0(\vec{M} + \vec{H}) = \mu_0(1 + \chi)\vec{H} = \mu_0\mu_r\vec{H} = \mu\vec{H}. \tag{9.23}$$

Here we introduce the notation

$$\mu \equiv \mu_0\mu_r, \qquad \mu_r \equiv (1 + \chi) \tag{9.24}$$

for the *permeability* μ and the *relative permeability* μ_r. Permeability has the same symbol as the magnetic moment, so beware of the possibilities of notational confusion.

For low to moderate fields, (9.22) also holds for soft ferromagnets (such as used in transformers). However, hard ferromagnets, and soft ferromagnets in large fields, are not described by (9.22).

9.5.1 *Paramagnets Are Attracted by Magnetic Fields*

As indicated, paramagnets, when placed in a nonuniform external magnetic field, are *attracted to the large field region*. See Figure 9.12(a). Moreover, because they have $\chi > 0$, their magnetization $\vec{M}$ points along the applied field $\vec{H}$. Paramagnets have a weak tendency to concentrate the magnetic field.

Examples of paramagnetic materials are Al, Ba, U, CuO, with $\chi \sim 10^{-5}$. Some materials, such as $CuSO_4$ and the rare earth atoms (and their compounds), are "strongly" paramagnetic, with χ as large as 10^{-3}. At the microscopic level, the properties of paramagnets are due to the permanent magnetism of the tiny electronic magnetic moments of electrons, and their local environment within the material. However, unless a magnetic field is applied, these electronic magnetic moments typically point randomly relative to one another, and therefore yield no net magnetic moment. Applying a small magnetic field tends to make them have a small component of their magnetic moment along the field, thus explaining why χ in paramagnets is usually small but positive.

9.5.2 *Diamagnets Are Repelled by Magnetic Fields*

As indicated, diamagnets, when placed in a nonuniform external magnetic field, are *repelled from the large field region*. See Figure 9.12(b). Moreover, because they have $\chi < 0$, their magnetization $\vec{M}$ points opposite to the applied field $\vec{H}$. Diamagnets have a weak tendency to "expel" the applied magnetic field. Diamagnetic materials include Cu, Cu_2O, water, the noble gases, and ionic crystals. At

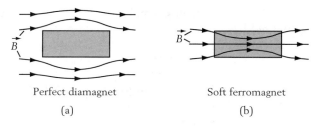

Perfect diamagnet

(a)

Soft ferromagnet

(b)

Figure 9.13 Effect on the magnetic field of two strong types of magnetism. (a) A perfect diamagnet repels magnetic field. (b) A soft ferromagnet attracts a magnetic field. Here the magnetic field is $\vec{B}$, whose normal component is continuous at surfaces.

room temperature, most materials are diamagnetic, with $\chi \sim -10^{-5}$. Graphite has a χ that is 20 times this large. At the microscopic level, diamagnetism is due to the magnetic field of the electric currents from electron orbitals. We will discuss the magnetism of electric currents in Chapters 10 and 11.

Perfect diamagnetism is an extreme form of diamagnetism. A perfect diamagnet can nearly completely expel an applied $\vec{B}$ field. See Figure 9.13(a). (By "completely expel," we mean that, within its interior, a perfect diamagnet produces a magnetic field that cancels the applied magnetic field. This is similar to how conductors "screen" electric fields from their interior.) To yield $\vec{B} \approx \vec{0}$ in (9.23), perfect diamagnets must have $\chi \approx -1$. They are invariably superconducting. At the very low temperatures where helium can be liquified (4 K), the elements Pb and Hg are superconductors—and perfect diamagnets.

9.6 Ferromagnetic Materials

As already indicated, there are two classes of ferromagnetic materials.

1. Permanent, or magnetically hard materials, which *retain most of their magnetization* when the external magnetic field is removed. Examples are the alnico alloys and the recently developed rare earth magnets—such as $Nd_2Fe_{14}B$ (NEO, for neodymium).

2. The magnetically soft materials, like iron, which *become strongly magnetized and concentrate an applied magnetic field*. See Figure 9.13(b). However, they *lose most of their magnetization* when the external magnetic field is removed. Soft magnets are like paramagnets that have a huge χ (often exceeding 1000, and sometimes as large as 10^6). They are useful in electromagnets, in transformers, in magnetic recording heads, as magnetic "screens," and as "keeper" magnets that circulate the magnetic field from one pole to the other on a permanent magnet, to prevent the permanent magnets from demagnetizing. See Figure 9.14. As already noted, refrigerator doors are made of a soft magnetic material.

Real ferromagnetic materials have properties associated with both of these extreme categories. For example, ordinary iron nails, which are often magnetized by wrapping wire around them and passing an electric current through the wire (thus producing an electromagnet), retain a small amount of their magnetism

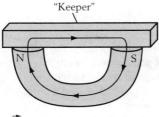

"Keeper"

N   S

$\vec{B}$ circulates through "keeper"

Figure 9.14 A "keeper" magnet is made of soft magnetic material and is used to retain the magnetic field of a permanent magnet, thereby helping preserve the magnetism of the permanent magnet. (Despite their name, permanent magnets can eventually demagnetize.)

after the electric current is turned off, but they are not very good permanent magnets. The names *soft* and *hard* arose because it is easier to magnetize iron (which is relatively malleable, or soft) than the less malleable, or hard, iron alloys.

9.6.1 *Hysteresis Loops Characterize Magnetic Materials in Detail*

A complete characterization of a magnetic material is given in terms of what is called a *hysteresis loop*. This can give either B versus H or M versus H, where H is the field *within* the material. By (9.20), from B versus H, we can determine M versus H, and vice versa. (For ordinary paramagnets or diamagnets, the hysteresis loops are simply straight lines through the origin; paramagnets have a very small positive slope, and diamagnets have a very small negative slope. Such hysteresis loops are relatively uninteresting.) For materials with a significant magnetization (including soft and hard magnets), it is essential to know the hysteresis loops, which have complex behavior not described by (9.22). Examples of M versus H for various types of magnetic material are given in Figure 9.15.

Consider the hysteresis loop in Figure 9.15(a). It starts with path 1, known as the *initial curve*, which begins at $M = 0$ and $H = 0$, corresponding to an unmagnetized material in zero applied field. Now H is increased to a large value, M increasing until it attains the saturation magnetization M_s. For path 2, H is decreased, and the magnetization follows along path 2 until H reaches a large

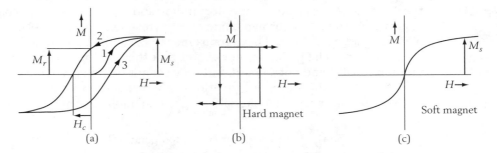

Figure 9.15 Hysteresis loops for various types of magnetic materials. (a) A magnet that can be easily magnetized and tends to retain its magnetism, but not perfectly. (b) A hard magnet, which retains its magnetism until the demagnetization field is reached. (c) A soft magnet, which magnetizes and demagnetizes easily. (Even the last kind of magnet can show a certain amount of hysteresis, wherein the magnetization is not a single-valued function of the applied field. However, there is no hysteresis on the scale of this figure.)

Magnetism's Fundamental Unit

Wilhelm Weber, in the mid-19th century, considered that the fundamental unit of magnetism is something like an indivisible atomic magnet; normally, the positive pole of one atomic magnet cancels the negative pole of an adjacent magnet, leaving only poles at the ends of the magnet. A significant advantage of this picture is that it explains why the magnetization of magnets saturates. The magnetic poles on a magnet can appear to redistribute when subjected to external magnetic fields (or when heated nonuniformly). We interpret this in terms of a rearrangement of the atomic magnets rather than a motion of true magnetic poles within the magnet. The characteristic magnetic moment of an atomic magnet is the *Bohr magneton* $\mu_B = eh/4\pi m_e = 9.27 \times 10^{-24}$ A-m^2. Here $h = 6.63 \times 10^{-34}$ J-s is *Planck's constant*, and $m_e = 9.11 \times 10^{-31}$ kg is the electron mass. One μ_B per cube 2×10^{-10} m on a side gives a magnetization $M = 9.27 \times 10^{-24}$ A-m^2/8×10^{-30} m$^3 = 1.16 \times 10^6$ A/m, which is very close to the saturation magnetization of common magnets.

reversed value. (Notice that when $\vec{H} = \vec{0}$, the remanent magnetization M_r is significant.) For path 3, H is increased from a large reversed field to a large positive value. Paths 2 and 3 are known as *major loops*.

Although only the complete hysteresis loop can fully characterize a magnet, three aspects of the loop are of special interest (see Figure 9.15a): (1) the *saturation magnetization* M_s, the magnetization in a large applied field; (2) the *remanent magnetization* M_r, or *remanence*, the magnetization after a large magnetic field has been removed; (3) the *coercive force* $\vec{H}_c$, applied opposite to the magnetization, needed to cause $\vec{M} = -\vec{H}$, or $\vec{B} = 0$. A slightly larger reversed field (not indicated on path 3) gives a magnetization that returns to the origin when the field is removed. It is called the *remanent coercive force* and lies along a backwards extension of the initial curve, 1.

Good permanent magnets have large values for M_r and H_c. Ideal permanent magnets have square hysteresis loops characterized by $M_r = M_s$ and H_c. See Figure 9.15(b). Good soft magnets have large values for M_s and χ, and small values for M_r and H_c. Ideal soft magnets have $M_r = 0$ and linear M versus H with slope χ until $M \approx M_s$, where the slope approaches zero. See Figure 9.15(c). Magnets of intermediate magnetic hardness are used in computer hard drives ("hard" in "hard drive" refers to the physical rigidity of the magnetic disk within the hard drive). Magnetic tape, for audio- and videocassettes, is coated with elongated grains of the brown oxide of iron (maghemite), γ-Fe$_2$O$_3$, of intermediate magnetic hardness. These grains magnetize along their long dimension, of about 0.5 μm. The grains are not so soft that they can be readily demagnetized unintentionally, and not so hard that they cannot be remagnetized intentionally (i.e., to change the stored information). The properties of some common soft and permanent magnetic materials are given in Tables 9.2 and 9.3.

Table 9.2 Properties of soft magnets (room temperature)

	M_S(A/m)	H_c(A/m)	χ
Iron	17.0×10^5	80	5000
Mu metal	8.5×10^5	4	10^5
Supermalloy	5.1×10^5	0.16	8×10^5

Table 9.3 Properties of permanent (hard) magnets (room temperature)

	$M_r(A/m)$	$H_c(A/m)$
Anisotropic barium ferrite	37.0×10^5	20.0×10^5
NEO ($Nd_2Fe_{14}B$)	10.3×10^5	16.0×10^5
Alnico V	9.95×10^5	4.4×10^4
Carbon steel	8.0×10^5	4.0×10^3
γ-Fe_2O_3	3.7×10^5	2.4×10^4

All magnetic materials lose their magnetization on being heated above the material-dependent *Curie* temperature T_c (after Pierre Curie). For iron, $T_c = 1043$ K, which is well above room temperature, but many compounds are ferromagnetic only at much lower temperatures. See Figure 9.16.

9.7 Demagnetization Field $\vec{H}_{demag}$

The field within the material, H, plotted along the abscissa of Figure 9.15, must include both the external field $\vec{H}_{ext}$ (due to other magnets and to the magnetic field produced by electric currents) and the *demagnetization field* $\vec{H}_{demag}$ (caused by the magnet being studied). Thus

$$\vec{H} = \vec{H}_{ext} + \vec{H}_{demag}. \tag{9.25}$$

The demagnetization field is important only for materials exhibiting strong magnetic properties, because it is proportional to the magnetization. It is negligible for paramagnets and diamagnets.

For a magnet with remanence, $\vec{H}_{demag}$ acts even without an externally applied magnetic field. One reason magnets tend to lose their magnetization with time is this demagnetization field. To avoid such demagnetization fields, and thus to retain (i.e., "keep") the magnetization, *keeper magnets* are employed, as in Figure 9.14.

For uniformly magnetized, ellipsoidally shaped magnets, which include spheres, pancakelike objects, and cigar-shaped objects, $\vec{H}_{demag}$ is uniform within the magnet. For other shapes, $\vec{H}_{demag}$ is nonuniform.

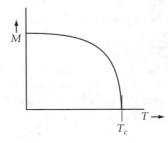

Figure 9.16 Magnetization versus temperature for a typical magnet. The transition temperature T_c is known as the Curie temperature. At high temperatures, magnets tend to lose their magnetism.

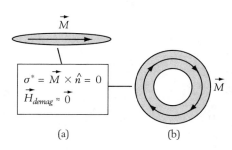

Figure 9.17 Zero demagnetization field for needle or toroid magnetized normal to its surface. (a) The needle geometry, where the magnetization is along the direction of the needle. Here the magnetic poles occupy only the ends of the magnet, and the corresponding demagnetization field is very small. (b) The toroidal geometry, where the magnetization is along the tangent to the toroid. Here there are essentially no magnetic poles, and the corresponding demagnetization field is negligible.

9.7.1 $\vec{H}_{demag} = \vec{0}$ for Needle or Toroid Magnetized Parallel to Surface

For a needle-shaped magnet magnetized along its axis, the magnetization at the surface is nearly perpendicular to the surface normal (except perhaps in the tiny region at the end of the needle), so $\sigma_m = \vec{M} \cdot \hat{n} \approx 0$. See Figure 9.17(a). For a toroidal magnet magnetized tangentially, $\sigma_m = \vec{M} \cdot \hat{n} = 0$ exactly. See Figure 9.17(b). Since $\sigma_m = 0$, there is no flux source for $\vec{H}$, so $\vec{H} = \vec{0}$, both inside and outside. Since $\vec{B} = \mu_0 \vec{H}$ outside the magnet, $\vec{B} = \vec{H} = \vec{0}$ outside. Clearly,

$$\vec{H}_{demag} = \vec{0} \qquad (\vec{M} \text{ along needle and toroid}) \qquad (9.26)$$

in this case. However, inside the magnet $\vec{B} = \mu_0 \vec{M} \neq \vec{0}$. Although the normal component of $\vec{B}$ is continuous $(= \vec{0})$, its tangential component is not continuous, on crossing from the interior (where $\vec{B} = \mu_0 \vec{M}$) to the exterior (where $\vec{B} = \vec{0}$).

From the hysteresis loop for a toroidal magnet, where $\vec{H}_{demag} = \vec{0}$, we can deduce the hysteresis loop for magnets of other shapes by including the appropriate demagnetization field. Keeper magnets of soft iron, as in Figure 9.14, are often placed across the poles of horseshoe magnets to direct the magnetic flux of one pole directly to the other, thus making the combination behave like a toroidal magnet, with small demagnetization field.

For shapes other than those discussed, the demagnetization field is nonzero. We now consider $\vec{H}_{demag}$ for the shape that gives the *largest* demagnetization field.

9.7.2 $\vec{H}_{demag} = -\vec{M}$ within a Thin Slab Magnetized along Its Normal

By (9.12′), for a slab magnetized uniformly along its normal, the surface charge densities are $\sigma_m = \pm M$. See Figure 9.18. Now recall that, when edge effects are neglected, the electric field $\vec{E}$ within a capacitor of charge density $\pm \sigma$ on its plates has magnitude $4\pi k \sigma$. Likewise, $\mu_0 \vec{H}$ within a magnetic slab has magnitude $4\pi k_m \sigma_m = 4\pi k_m M = \mu_0 M$. This gives $|\vec{H}| = |\vec{M}|$. Since $\vec{H}$ points from the positive to the negative side—that is, opposite to $\vec{M}$—here we have $\vec{H} = -\vec{M}$.

This field is precisely $\vec{H}_{demag}$, so for a slab

$$\vec{H}_{demag} = -\vec{M}. \qquad (\vec{M} \text{ normal to slab}) \qquad (9.27)$$

From (9.27), it follows that $\vec{B} = \mu_0(\vec{H} + \vec{M}) = \mu_0(\vec{H}_{demag} + \vec{M}) = \vec{0}$ inside the slab. Furthermore, since $\vec{H} = \vec{0}$ outside the slab (just as $\vec{E} = \vec{0}$ outside a capacitor, when edge effects are neglected), and since $\vec{M} = \vec{0}$ outside the slab, we have $\vec{B} = \mu_0(\vec{H} + \vec{M}) = \vec{0}$ outside the slab, just as it was zero inside. Again, the normal component of $\vec{B}$ is continuous across an interface. (Edge effects cause the field outside a slab magnetized along its normal to be small but nonzero. We will discuss this in Chapter 11.)

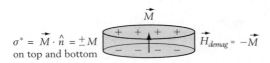

Figure 9.18 A thin magnetic slab magnetized along its normal has a large demagnetization field because it has the maximum pole strength on its slab faces. The magnetic field $\vec{B}$ is nearly zero both within and outside the magnet.

The slab and the needle are the two extreme geometries: no geometry will produce less demagnetization than a needle, and no geometry will produce more demagnetization than a slab. For a sphere, $\vec{H}_{demag} = -\frac{1}{3}\vec{M}$.

Example 9.8 **Demagnetization fields, coercive forces, and permanent magnets**

Consider the hard magnetic materials materials NEO ($Nd_2Fe_{14}B$) and carbon steel. Determine the demagnetization field of these good permanent magnets in the thin slab geometry of Figure 9.18.

Solution: We use material constants from Table 9.3. By (9.27), for NEO $|\vec{H}_{demag}| = M_r = 10.3 \times 10^5$ A/m, so an additional $H_{ext} = H_c - H_{demag} = (16.0 - 10.3) \times 10^5 = 5.7 \times 10^5$ A/m, or $B = 0.72$ T, must be applied opposite to the magnetization to demagnetize (in this case, to reverse) the magnetization. This is very large; it is difficult to accidentally demagnetize a thin slab of NEO with magnetization along its normal. Again by (9.27), for carbon steel $|\vec{H}_{demag}| = M_r = 8.0 \times 10^5$ A/m, which exceeds H_c for carbon steel. Thus the sample will not "take" a magnetization normal to the slab; although it is not obvious, the magnetization will rotate into the plane of the slab, for which there is a negligible demagnetization field.

Example 9.9 **Demagnetization fields for soft magnets—needles and slabs—in an applied field**

Consider the soft magnetic material known as mu metal, which is used for magnetic shielding. Determine the effect of the demagnetization field on the magnetization of a mu metal in the earth's magnetic field $B_{earth} \approx 0.5 \times 10^{-4}$ T. Consider both the thin needle and the thin slab geometries.

Solution: We use material constants from Table 9.2. With the field and magnetization along the axis of a needle-shaped sample, for which $H_{demag} = 0$, we have $H = H_{ext} = H_{earth} = 39.8$ A/m. Hence, if (9.22) holds, the magnetization is $M = \chi H_{ext} = 3.98 \times 10^6$ A/m. This exceeds $M_s = 8.5 \times 10^5$ A/m, which is not consistent with (9.22). Hence the magnet is in the large field regime of the hysteresis loop, where $M = M_s$. With the field and magnetization along the axis of a thin slab, by (9.27), $|\vec{H}_{demag}| = M$, so (9.22) gives $M = \chi(H_{ext} - M)$. Solving for M in terms of H_{ext} yields $M = [\chi/(\chi + 1)]H_{ext}$. Since χ is so large, this gives $M \approx H_{ext} \approx 39.8$ A/m. This value is much smaller than M_s, so that use of (9.22) is valid. Observe that this M is smaller by about a factor of $\chi \approx 10^5$ than for the needle geometry. Clearly, demagnetization effects can be very large for soft ferromagnets.

9.8 How We Know $\vec{B}$ Is Truly Fundamental

In the next chapter, we will learn that the force $\vec{F}$ on a particle of charge q as it moves with velocity $\vec{v}$ through a region of magnetic field $\vec{B}$ is

$$\vec{F} = q\vec{v} \times \vec{B}. \tag{9.28}$$

It is not clear that $\vec{B}$, rather than $\mu_0\vec{H}$, should be used in this equation. The toroid and slab geometries suggest two ways to distinguish between how $\vec{B}$ and $\vec{H}$ cause the deflection of charged particles:

1. Let high-energy charged particles pass through a tangentially magnetized toroidal magnet, where $\vec{B} \neq \vec{0}$ but $\vec{H} = \vec{0}$ inside. If $\vec{B}$ is the fundamental field, they will be deflected, whereas if $\vec{H}$ is the fundamental field, they will be undeflected. Experimentally, they are deflected in an amount quantitatively described by (9.28). This supports the fundamental nature of $\vec{B}$. See Figure 9.19(a). Such experiments are used as a tool in particle physics, where

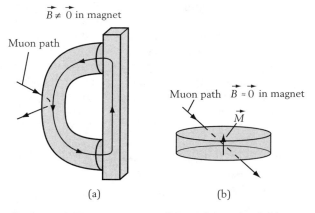

Figure 9.19 Two geometries for studying the field within a magnet using muons, which can enter matter without interacting strongly: (a) toroidlike geometry, with a large field, (b) thin magnetic slab, with a small field.

deflection by the $\vec{B}$ field of the magnet helps determine the momentum of charged particles.

2. Let high-energy charged particles pass through a slab magnetized along its normal, where $\vec{B} = \vec{0}$ but $\vec{H} = -\vec{M} \neq \vec{0}$ (and large) inside. If $\vec{B}$ is the fundamental field, they will be undeflected, whereas if $\vec{H}$ is the fundamental field, they will be deflected. See Figure 9.19(b). Therefore, such a geometry is not expected to assist in the analysis of particle motion. It is not clear that any such experiments have been performed.

Chapter 12 will show that $\vec{B}$ (and hence, not $\vec{H}$) also determines the electromotive force associated with Faraday's law. The magnetic induction $\vec{B}$ is the quantity of true fundamental significance.

Particle Tracks and Muons

To have clean particle tracks in the deflection experiments described earlier, the electrically charged muons (particles similar to electrons, but about 200 times more massive) are used. This is because their strongest interactions with other particles occur via electromagnetism, and because when they do interact, the momentum transfer is small since they are mass-mismatched with both the much lighter electrons and the much heavier nuclei.

9.9 Magnetic Oscillations

Optional

Just as the torque on an electric dipole moment $\vec{p}$ in an electric field $\vec{E}$ is given by $\vec{\tau} = \vec{p} \times \vec{E}$, so the torque on a magnetic dipole moment $\vec{\mu}$ in a magnetic field $\vec{B}$ is given by

$$\vec{\tau} = \vec{\mu} \times \vec{B}. \qquad \text{(torque on a magnetic dipole)} \qquad (9.29)$$

Thus

$$|\vec{\tau}| = |\vec{\mu}||\vec{B}|| \sin \theta|, \qquad (9.30)$$

where θ is the angle between $\vec{\mu}$ and $\vec{B}$.

From classical mechanics, we know that the torque drives the angular momentum $\vec{L}$:

$$\frac{d\vec{L}}{dt} = \vec{\tau}. \qquad (9.31)$$

Equations (9.29) and (9.31) can be used to explain the operation of a compass needle, leading to a torque that drives the magnet to orient $\vec{\mu}$ along $\vec{B}$, and to have oscillations about that equilibrium. They also can be used to explain the oscillations around equilibrium of tiny electronic and nuclear magnets; these can be observed using the techniques of *electron spin resonance* (ESR) and *nuclear magnetic resonance* (NMR). The word *spin* refers to angular momentum that appears to be intrinsic to the object. It is to be distinguished from any *orbital* angular motion that arises because the object is in an orbital (either an electron about the atom or a nucleon about the nucleus). Associated with spin (a vector) is a magnetic moment (also a vector). For simplicity, we will use $\vec{L}$ for all these types of angular momentum.

9.9.1 *Oscillation of a Macroscopic Magnet: Compass Needle*

A compass needle, of magnetic moment $\vec{\mu}$ and moment of inertia I (not to be confused with electric current), oscillates about its equilibrium position in a magnetic field $\vec{B}$. We wish to find the dependence of the oscillation frequency ω on $\mu = |\vec{\mu}|$, I, and $B = |\vec{B}|$.

Let $\vec{\mu}$ and $\vec{B}$ lie in the xy-plane, and let θ be the angle of the magnet with respect to the external field. See Figure 9.20(a). By (9.30), the z-component of the torque on the magnet is given by $\tau = -\mu B \sin \theta$; the minus sign indicates that the torque tends to restore equilibrium. For small angles θ, where $\sin \theta \approx \theta$, the torque becomes

$$\tau = -\mu B \theta. \tag{9.32}$$

For a macroscopic object of moment of inertia I about its z-axis, the angular momentum component L_z is given by

$$L_z = I \omega_z = I \frac{d\theta}{dt}. \qquad \text{(macroscopic object)} \tag{9.33}$$

Then (9.31) becomes

$$\frac{dL_z}{dt} = I \frac{d^2\theta}{dt^2} = \tau. \tag{9.34}$$

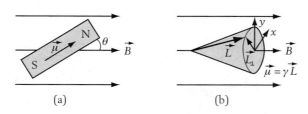

(a)   (b)

Figure 9.20 Magnetic oscillations: (a) a macroscopic magnet, (b) a microscopic magnet. Both satisfy the same rule for magnetic torque, and both can have angular momentum. The macroscopic magnet has a relatively small amount of angular momentum along its axis, causing its rotational dynamics to be dominated by its moment of inertia. The microscopic magnet has a relatively large amount of angular momentum along its axis, causing its rotational dynamics to be dominated by the angular momentum along its axis.

For small θ, use of (9.32) in (9.34) yields

$$I\frac{d^2\theta}{dt^2} = -\mu B\theta. \quad \text{(macroscopic magnet)} \tag{9.35}$$

This is just like the harmonic oscillator equation, with $(x, K, m) \to (\theta, \mu B, I)$. Hence the angle θ oscillates. Instead of the harmonic oscillator frequency $\omega = \sqrt{K/m}$, here we have

$$\omega = \sqrt{\frac{\mu B}{I}}. \tag{9.36}$$

If μ and I are known, a measurement of ω from the oscillation of a compass can yield the magnetic field in the plane of oscillation of the compass. Thus an ordinary compass resting on a table can be used to obtain the horizontal component of the earth's magnetic field B_{Eh}. If neither μ nor B_{Eh} is known, then measurement of the torque (at a known angle) and the resonance frequency, by (9.30) and (9.36), would enable us to determine both μ and B_{Eh}.

9.9.2 *Oscillation of a Microscopic Magnet: Magnetic Resonance*

Most atoms and their nuclei have a magnetic moment, whose value serves as a fingerprint. In this section, we show that, for a given magnetic field, the frequency at which an atom or nucleus resonates is proportional to the magnetic moment. This forms the basis of electron spin resonance (ESR) and of nuclear magnetic resonance (NMR). The 1952 Nobel Prize in physics was awarded to Felix Bloch and Edward Purcell, for their independent discovery of NMR.

For a nuclear or atomic magnet, the magnetic moment $\vec{\mu}$ is related to the angular momentum $\vec{L}$ by

Equation (9.37) implies that $\vec{L} = \vec{\mu}/\gamma$ for nuclear and atomic magnets, rather than $\vec{L} = I\vec{\omega}$ of (9.33), which holds for the angular momentum of rotating macroscopic magnets. Molecules and atomic clusters have a total angular momentum that is a combination of both $\vec{\mu}/\gamma$ and $I\vec{\omega}$.

$$\vec{\mu} = \gamma\vec{L}, \quad \text{(microscopic magnet)} \tag{9.37}$$

Resonant Absorption of Electromagnetic Energy

With both NMR and ESR, the intensity of the absorption of electromagnetic energy at the resonance frequency is proportional to the number of absorbing nuclei. Moreover, the frequency of absorption can depend on magnetic fields produced by the local environment. Thus, these methods may be used as diagnostics for the presence of, and the study of the local environments of, various atoms, ions, and nuclei. NMR is the basis for the powerful medical diagnostic technique of magnetic resonance imaging (MRI). The intensity of absorption in a known magnetic field gradient provides a local measure of the density of different types of nuclei. Modern computers (and, in particular, powerful desktop workstations and personal computers) make it possible to analyze a wealth of data taken in many orientations, and then to deduce the position in the body where the energy was absorbed.

where γ, the gyromagnetic ratio, depends upon the atom or nucleus. Specifically, $\gamma_{electron} = -1.759 \times 10^{11}$ $s^{-1}T^{-1}$ and $\gamma_{proton} = 2.675 \times 10^{8}$ $s^{-1}T^{-1}$. As a consequence of their lower $|\gamma|$, protons oscillate (resonate) at much lower frequencies than do electrons.

We will solve (9.29) and (9.31) to obtain the resonance frequency of a given magnetic moment. Unlike the case for the compass needle, for an atom or nucleus the angular momentum $\vec{L}$ has a fixed magnitude. Combining (9.31), (9.29), and (9.37) gives

$$\frac{d\vec{L}}{dt} = \gamma \vec{L} \times \vec{B}. \qquad \text{(microscopic magnet)} \qquad (9.38)$$

This is very different from (9.35), and it has a very different solution. To be specific, let $\vec{B}$ point along $\hat{z}$. Then, rather than an oscillation (in which only L_x might participate), there is a *precession* (in which L_x and L_y participate equally). See Figure 9.20(b).

Since the vector cross-product is perpendicular to both vectors, the right-hand side of (9.38) is perpendicular to both $\vec{B}$ and $\vec{L}$, so

$$\vec{L} \cdot \frac{d\vec{L}}{dt} = 0, \qquad \vec{B} \cdot \frac{d\vec{L}}{dt} = 0. \qquad (9.39)$$

Since $\vec{B}$ points along $\hat{z}$, $\vec{L} \times \vec{B}$ is normal to $\hat{z}$. Hence (9.38) implies that $dL_z/dt = 0$, so L_z does not change. Also, using (9.38) and (9.39) yields

$$\frac{d|\vec{L}|^2}{dt} = \frac{d(\vec{L} \cdot \vec{L})}{dt} = 2\vec{L} \cdot \frac{d\vec{L}}{dt} = 0 \qquad (9.40)$$

so that $|\vec{L}|$ is fixed. With both L_z and $|\vec{L}|$ fixed, so is

$$\sqrt{L_x^2 + L_y^2} = \sqrt{L_x^2 + L_y^2 + L_z^2 - L_z^2} = \sqrt{|\vec{L}|^2 - L_z^2}. \qquad (9.41)$$

Hence the vector

$$\vec{L}_\perp = L_x \hat{x} + L_y \hat{y} \qquad (9.42)$$

has fixed magnitude $|\vec{L}_\perp| = \sqrt{L_x^2 + L_y^2}$. Thus, as stated previously, we may write

$$L_x = L_\perp \cos\phi, \qquad L_y = L_\perp \sin\phi, \qquad (9.43)$$

where ϕ can depend on time. With $B_z = B$ and $B_y = 0$, the x-component of (9.38) yields

$$\frac{dL_x}{dt} = \gamma(L_y B_z - L_z B_y) = \gamma L_y B. \qquad (9.44)$$

Differentiating the first part of (9.43) and placing it in (9.44) yields

$$-\frac{d\phi}{dt}L_\perp \sin\phi = (\gamma B)L_\perp \sin\phi. \tag{9.45}$$

Thus

$$\frac{d\phi}{dt} = -\gamma B, \tag{9.46}$$

which can also be obtained from the y-component of (9.38). The solution of (9.46) is

$$\phi = -\omega t + \text{constant}, \tag{9.47}$$

where

$$\omega = \gamma B \tag{9.48}$$

is the *magnetic resonance frequency* in radians/s. In terms of cycles/sec, or Hz, we use

$$f = \frac{\omega}{2\pi} = \frac{\gamma B}{2\pi}. \tag{9.49}$$

Such rotation of $\vec{L}$ about $\vec{B}$ is called *precession*.

A magnetic moment will absorb energy when driven at its magnetic resonance frequency. For a known magnetic field, a measurement of ω will yield γ. Correspondingly, if γ is known, a measurement of ω will determine B. Equation (9.48) is the basis of MRI.

Example 9.10   NMR for MRI

Consider hydrogen in water. Find the NMR resonance frequency for $B = 1$ T.

Solution: By (9.48), the protons absorb energy at the frequency $f = \gamma B/2\pi = (2.675 \times 10^8)(1)/2\pi = 42.6$ MHz. This is somewhat below the FM band, which begins around 88 MHz.

Here are some technical aspects of NMR and ESR. For NMR: (1) it is easy to build tunable measuring devices in this frequency range; (2) there are many types of nuclei with distinct magnetic moments; and (3) the resonances are very sharply defined in frequency. On the other hand, for ESR: (1) the resonant frequencies can be inconveniently high for measurement devices; (2) there is only one type of electron and not many radically different environments for it; and (3) different ESR resonances are not as sharply defined in frequency. All these factors lead to the choice of NMR over ESR for use in MRI.

Magnetic resonance experiments on electron and nuclear spins within materials provide evidence that, in the torque law of (9.29), it is $\vec{B} = \mu_0(\vec{H} + \vec{M})$, rather than the $\mu_0 \vec{H}$ due to magnetic poles, that applies.

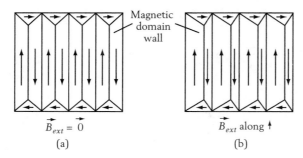

Figure 9.21 Magnetic domains: (a) zero external field, (b) moderate external field. By causing the domain walls to move, the field causes domains aligned with (against) the field to grow (shrink).

9.10 How Permanent Magnets Get Their Permanent

Optional

Real magnets are not as simple as we have described them. Typically, they are very nonuniform unless they are in such a large magnetic field that they have become saturated. Locally, they take on the saturation magnetization appropriate to their temperature, $M = M_s$, but the magnetization in different parts of the magnet does not point in the same direction. Regions of a given direction of magnetization are called *magnetic domains*. See Figure 9.21(a). Domains are separated by thin regions called *domain walls*, across which the local magnetization rotates from one domain orientation to the other. The magnetic moment gets larger in an applied field either because the domain walls move (mostly at low fields) or because the magnetization within the domains rotates (mostly at higher fields). See Figure 9.21(b). Only at very large fields do all the domains become oriented in the same direction, in which case the magnetization is truly uniform. For smaller fields, the domain pattern develops, and the magnetization is not truly uniform.

We now address a number of important issues.

9.10.1 Why Do Some Materials Have a Magnetization Whereas Others Do Not?

The answer is that the electrostatic energy of electron orbitals can sometimes be lowered if the electron magnetic moments develop a preferential ordering relative to one another. This is associated with the fact that, by what is called the Pauli principle, no two electrons can occupy the same orbital, unless they have opposite intrinsic—or *spin*—angular momentum. In some cases, there is a preference for the electron magnetic moments to counteralign, leading to what are called *antiferromagnets*. Antiferromagnets are poor magnets, but they are interesting just the same, with their own magnetic resonances. The tendency of many materials to develop magnetic order is so weak that it is easily overwhelmed by thermal agitation of the atomic magnets; magnetic order occurs only at low temperatures. Many complex types of magnetic order exist, one of which is glasslike: so-called spin-glasses, with local magnetic order that is random to the eye.

> **Crystal Field Energy**
>
> In technical terms, the orientation of the orbitals relative to the crystal is attributed to the *crystal field energy;* the orientation of the magnetic moments relative to the orbitals is attributed to the *spin-orbit energy;* and the overall orientation of the magnetization relative to the crystal is attributed to the *magnetic anisotropy energy.*

9.10.2 *What Determines the Direction of the Magnetization?*

The answer involves two steps. First, to minimize their electrical energy the orbitals of the magnetic electrons orient with respect to the crystal (e.g., the plane of the orbital may orient along or perpendicular to one of the crystal axes). Second, to minimize their magnetic energy, the magnetic moments of the magnetic electrons orient with respect to the electron orbital. The net effect is that the magnetic electrons preferentially orient their magnetic moments with respect to the crystal. For high anisotropy, where there is a strong preference for certain orientations, the domain wall thickness is very small in order for the magnetization to avoid these unfavorable orientations.

9.10.3 *Why Does the Magnetization Depend on Temperature?*

The answer to this question requires comparing the energy of ordering of the magnetic state, and the energy of thermal fluctuations (which have a disordering influence that is proportional to the temperature). For iron, at low temperatures the energy of ordering completely wins, and all the spins locally point in the same direction (although there can be a domain structure over a larger spatial scale). However, as the temperature increases, the spins start to point in a less ordered fashion, and eventually, at the Curie temperature T_c, they become disordered. The magnetization $M_s(T)$ decreases smoothly to zero at T_c, as in Figure 9.16.

9.10.4 *Designing a Strong Permanent Magnet: NEO*

Iron has a rather low magnetic anisotropy, but a rather large magnetic moment. It is easily demagnetized, and therefore it is not very good as a permanent magnet, although it is a very good soft magnet. NEO ($Nd_2Fe_{14}B$) is a man-made material that has been designed to take advantage of the magnetic anisotropy of the Nd (neodymium), and the magnetic moment of the Fe (iron), to produce a large permanent magnetic moment. The B (boron) is needed mainly to form the crystal. The Fe and the Nd order magnetically. Because of a strong antiferromagnetic interaction between the Fe and the Nd, they align with opposing magnetic moments. The axis of alignment is not random: because of the strong magnetic anisotropy acting on the Nd, there is a preferred axis for the Nd and Fe. The Fe dominates the magnetic moment of this system because there are seven Fe for each Nd.

In practice, a well-magnetized sample of NEO begins as a powder of many small crystallites. In a strong magnetic field, it is then heated (technically, *sintered*), which partially melts together the magnetized crystallites. For a sample prepared in this way, if there is a defect site at which the local magnetic moment reorients, this reorientation only propagates to the boundary of the crystallite because the coupling between crystallites is relatively weak. Thus the use of crystallites helps limit the amount of demagnetization that can be seeded by a given crystalline defect.

Problems

9-2.1 Two 5 cm long, 2-mm-by-2-mm magnets are uniformly magnetized. One will lift twice as much iron as the other. When placed on the same axis at 3 cm nearest distance, as in Figure 9.3, their force of attraction is 0.04 N. Estimate their pole strengths. *Note:* As shown in Example 9.6, the amount of iron a magnet can lift is proportional to the square of its pole strength.

9-2.2 For $|\vec{B}| = 0.5$ T in a region of space, the force is 0.75 N on a single pole q_m of a long magnet ($l = 10$ cm). (a) Find q_m. (b) Find the field due to the magnet 5 cm along its perpendicular bisector.

9-2.3 A small magnet, collinear with the orientation of a compass needle, is moved toward the compass needle. At 20 cm, the compass needle switches direction. If the horizontal component of the earth's field is 0.41 G, find the moment of the magnet.

9-2.4 Find the field strength 50 cm along the axis of a small bar magnet of moment 6 A-m. Repeat for 50 cm along the perpendicular bisector.

9-2.5 The torque is 0.04 N-m on a magnet with $\mu = 2.4$ A-m at a 40° angle to $\vec{B}$. Find $|\vec{B}|$.

9-2.6 A dipole of moment $\mu_1 = 0.96$ A-m, at the origin and pointing along $\hat{x}$, is subject to a torque 0.0075 N-m along $\hat{z}$. This torque is due to magnet 2 pointing along $\hat{y}$, and located in the xy-plane a distance 2.4 cm along a line making a 30° counterclockwise angle to the x-axis. See Figure 9.22. Find μ_2.

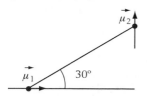

Figure 9.22 Problem 9-2.6.

9-2.7 From some unspecified "rude observations," Newton indicated that the magnetic force varied as

the inverse cube of the distance. Show how such an inverse cube law might be obtained by measuring the force on the pole of one long magnet due to another short magnet.

9-2.8 In the *astatic balance*, two thin magnetic needles of equal moment are mounted on a fiber with their moments normal to the fiber axis. See Figure 9.23. The upper needle (in the $z = a$ plane) has its magnetic moment along $\hat{x}$, and the lower needle (in the $z = 0$ plane) has its magnetic moment along $-\hat{x}$. (a) Will there be a net torque in a uniform field (such as that of the earth)? (b) Let a nonuniform field $\vec{B}$ be applied along the y-direction, with the applied field larger for the needle in the $z = a$ plane. How will the astatic balance twist?

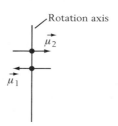

Figure 9.23 Problem 9-2.8.

9-2.9 (a) Describe the response of a permanent magnet when an unmagnetized rod of soft iron is brought up to its center, its north, and its south poles. (b) Describe the response of an unmagnetized rod of soft iron when the north pole of a permanent magnet is brought up to the center and the ends of the soft iron rod.

9-2.10 The earth's field at a measuring site is specified as follows: horizontal field 0.145 G, with declination (relative to north) 11.4° west, and inclination (relative to the vertical out of the surface of the earth) 67°. Find the field strength and its component in the meridional plane (i.e., one that passes through the polar axis and the observation point).

9-2.11 The magnetic line poles λ_m and $-\lambda_m$ are parallel to z, intersecting the $z = 0$ plane at $(x_0, 0)$ and (x_0, l), respectively. Thus the magnetic dipole moment is parallel to the y-axis. With finite dipole moment per unit length $\vec{v}$ (where $|\vec{v}| = \lambda_m l$), but $l \to 0$, show that $l\vec{B} = 2k_m[-\vec{v} + 2(\vec{v} \cdot \hat{r})\hat{r}]/r^2$, where r is the nearest distance to the dipole lines.

9-2.12 Consider a dipole sheet, with dipole moment per unit area $d\mu/dA = Ml_0$. To obtain such a sheet, in the previous problem let $\vec{v} \to d\vec{v} = (d\mu/dA)\hat{y}dx$, and integrate on x from $-\infty$ to 0. For an observer at (x_0, y_0), show that $\vec{B} = -[2k_m Ml_0/r_0](\hat{z} \times \hat{r}_0)$, where $r_0 = \sqrt{(x_0^2 + y_0^2)}$ and $\hat{r}_0 = \hat{y}\sin\theta_0 + \hat{x}\cos\theta_0$ points from the edge of the dipole sheet to the observer.

9-3.1 Consider a uniformly magnetized magnet with $M = 3.2 \times 10^5$ A/m, $l = 5$ cm, and $A = 16$ mm^2. It is magnetized along its axis. (a) Find q_m, σ_m, and μ. (b) Estimate the field on the axis at 0.1 cm, 2 cm, and 10 cm from the north pole.

9-3.2 Two identical magnets of length l, cross-section A (where $\sqrt{A} \ll l$), magnetization M, and mass m are placed in a tube, one above the other, with like poles near each other. (a) If the magnets are very strong, find the value of the separation s to suspend the upper one against gravity. (b) If the magnets are weak, estimate the minimum magnetization needed to suspend the upper one against gravity.

9-3.3 The lifting strength of a long magnet with 3 cm^2 cross-section is 30 N. Find its magnetization.

9-3.4 The magnetization of a long magnet with 8.4 cm^2 cross-section is $M = 2.2 \times 10^5$ A/m. Find its lifting strength.

9-3.5 Consider a long bar magnet of magnetization M, length l, and square cross-section (with side $a \ll l$). It is oriented normal to a large sheet of soft magnetic material (e.g., a refrigerator), with its near pole a distance r from the sheet. Find the leading dependence on r of the force of attraction for (a) $r \ll a$, (b) $a \ll r \ll l$, (c) $l \ll r$.

9-4.1 For arbitrary α, determine the discontinuity of the normal component of the artificially defined quantity $\vec{C} = \mu_0(\vec{H} + \alpha\vec{M})$ on crossing the pole of a very long bar magnet. Treat $\vec{H}$ as being due to magnetic poles.

9-4.2 Consider a nonuniformly heated bar magnet of length l and cross-sectional area A, magnetized along its axis, with $M = M_1$ at its north pole, and $M = M_2 < M_1$ at its south pole. Since the magnetic charges on the ends do not sum to zero, there must be some bulk magnetic charge. (a) Assuming that the magnetic charge density is uniform, find its magnitude. (b) Find the field along the axis at a distance $4l$ from its north pole, under the assumption that the magnet is very narrow, so that the bulk magnetic charge can be thought of as a line charge. (c) Compare that field with the field at the same position if the magnet had a uniform magnetization M_1.

9-5.1 How would you use a soft magnet to focus the magnetic field of a permanent magnet?

9-5.2 How would you use a soft magnet to create a region of weakened magnetic field?

9-5.3 Relate field expulsion and magnetic repulsion; field concentration and magnetic attraction. (See Figures 9.12 and 9.13.)

9-5.4 In scalar form, for SI units we have $B = \mu_0(H + M)$, whereas for cgs-emu units we have $B = H + 4\pi M$. The SI units are tesla (T) for B, ampere-turn/meter (A-turn/m) for H, and ampere/m (A/m) for M. The cgs-emu units are gauss (G) for B, where 1 T $= 10^4$ G, oersted (Oe) for H, and "magnetic moment unit"/cm^3, or mmu/cm^3 for M. (a) By taking $M = 0$, show that 1 A-turn/m $= 4\pi \times 10^{-3}$ Oe. (b) By taking $H = 0$, show that 1 A/m $= 10^{-3}$ mmu/cm^3. (c) Show that $\chi_{emu} = M/H$ gives values $1/4\pi$ as large as in SI units.

9-5.5 A soft magnetic material has $M = 2500$ A/m in a field $B = 0.005$ T. (a) Find the SI values for H and χ. (b) Find the cgs-emu values for M, B, H, and χ. *Hint:* See Problem 9-5.4.

9-5.6 For a small applied magnetic field, sketch M versus H for a paramagnet, a diamagnet, a soft ferromagnet, and a perfect diamagnet.

9-6.1 A magnet has $M_s = 0.84 \times 10^6$ A/m and $M_r = 0.53 \times 10^3$ A/m. Would this make a good permanent magnet? Discuss.

9-6.2 (a) Of the materials listed in Tables 9.2 and 9.3, which would make the best hard magnet? (b) Which would make the best material for magnetic focusing?

9-7.1 (a) For a magnet of length $l = 16$ cm, cross-section 0.25 cm^2, and magnetization $M = 7.8 \times 10^5$ A/m, estimate $\vec{H}$ and $\vec{B}$ at the midpoint. (b) Is $\vec{H}_{demag}$ large or small within the volume of a long magnet?

9-7.2 Consider a needle-shaped sample of permalloy in the earth's magnetic field of 0.5×10^{-4} T. Find its magnetization. *Hint:* Is it due to susceptibility or to saturation?

9-7.3 In Example 9.8, how might a "keeper" magnet (of what shape?) maintain the magnetization of a thin slab of carbon steel along the normal?

9-7.4 Determine the magnetization for an iron needle that is oriented (a) along the earth's magnetic field, (b) normal to the earth's magnetic field.

9-8.1 Explain why $\vec{H} \approx \vec{0}$ in Figure 9.19(a) and $\vec{B} \approx \vec{0}$ in Figure 9.19(b).

9-9.1 The period of oscillation of a magnet of moment of inertia 4.2 g-cm^2 is 2.5 s. It has a magnetic moment of 0.4 A-m^2. Find the magnetic field it is in.

9-9.2 A compass needle oscillates 8 times per minute outdoors, but only 7.4 times per minute indoors. For the field component B_h in the plane of the surface of the earth, find the ratio of $B_{h,out}$ outdoors to $B_{h,in}$ indoors.

9-9.3 A small magnet, when mounted as a compass needle, makes five oscillations per minute. When mounted as a dipping needle (so it can measure the declination from the vertical of the magnetic field), it makes a maximum of nine vibrations per minute. Find the dip angle of the field.

9-9.4 For a molecule, $\vec{L} = I\vec{\omega}$, but the value of the moment of inertia I is dependent upon the axis about which the molecule is rotating. (a) Discuss the relative values of I for a cigar-shaped molecule. (b) Repeat for a pancake-shaped molecule.

9-9.5 Show that, for magnetic resonance, study of dL_y/dt, instead of dL_x/dt, also leads to (9.48).

9-10.1 Discuss the effect on the magnetization of NEO if the interaction between Fe and Nd were ferromagnetic rather than antiferromagnetic.

9-10.2 In your own words, explain why the magnetization in NEO is oriented relative to the crystal axes.

9-G.1 Devise an experiment to establish that the interaction between two magnets is not due to electrostatic forces.

9-G.2 Magnetic attraction and the amber effect appear to differ. In the amber effect, the comb has a net charge, and attracts small objects. A magnet has no net pole strength, but it attracts soft iron, like nails. Show how the effects are similar (a) for the amber effect, by including the charge left behind on your hair after rubbing the comb through it; (b) for magnets, by neglecting the effect of the distant magnetic pole.

9-G.3 Of two geometrically identical bar magnets, it is determined that one of them is magnetized and one is not, by their effect on iron filings. Without iron filings, how can you determine which magnet is magnetized and which is not? You can move the magnets.

9-G.4 In the early 19th century, Poisson proposed, by analogy to electricity, that two types of fluid (north pointing or austral, and south pointing or boreal) could explain the magnetism of magnets. However, such a theory needed the additional and arbitrary hypothesis that neither type of fluid can leave the magnet (except in equal quantities). Discuss the plausibility of this model.

9-G.5 Poisson's magnetic fluid model predicts that a magnet in a strong magnetic field can develop a very large magnetization (just as an electrical conductor can develop a very large electric polarization when placed in a large electric field). Experimentally, all magnets have a limit to their magnetization. How can we explain the limit to magnetization using the fluid theory? Using the atomic magnet theory? Is there a limit to the polarization of a conductor? (Consider the depletion layer discussed at the end of Chapter 5.)

9-G.6 Here is how to measure χ for paramagnets and diamagnets, which both have such a small magnetic susceptibility χ that their demagnetization field is negligible relative to the external field. A measurement of magnetization and field then yields χ. (Similar considerations hold for the electrical analog, to determine a small electric permeability—as for gases, but not for wood—when the depolarization field is negligible relative to the external field.) This method is due to Faraday, who discovered diamagnetism. The sample, of length l, is suspended vertically in the midplane $y = 0$ between the N and S poles of a source magnet, and the force on the sample is measured with a balance. At the sample, the field due to the magnet is nearly vertical, so $M = \chi H \approx \chi B_y(0)/\mu_0$, where $B_y(0)$ is shorthand for $B_y(x, 0, z)$. This induces charges $\pm q_m = \sigma_m A = MA$ on the top and bottom of the sample. See Figure 9.24. (a) Is the sample, depicted in Figure 9.24, paramagnetic or diamagnetic? (b) Explain why there is no net force along y. (c) Let $B_x(\pm l/2)$ be shorthand for $B_x(x, \pm l/2, z)$, and note that $B_x(x, 0, z) = 0$ by symmetry. Show that $F_x = q_m[B_x(l/2) - B_x(-l/2)] \approx q_m l (dB_x/dy)$, using the straight-line approximation for differences. (d) With $q_m = MA = \chi HA \approx \chi(B_y/\mu_0)A$, show that $F_x = Al(\chi/\mu_0)B_y(dB_x/dy)$. (e) Evaluate F_x for $\chi = 0.8 \times 10^{-5}$, $V = 3.8$ mm^3, $B_y = 0.05$ T, and $dB_x/dy = 4.2$ T/m.

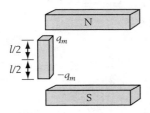

Figure 9.24 Problem 9-G.6.

9-G.7 Consider a small square loop of dimension $a \times a$, in the xy-plane. See Figure 9.27. (a) Show that, in the absence of any sources of $\vec{B}$, $0 = \oint \vec{B} \cdot d\vec{s} = a^2(dB_y/dx - dB_x/dy)$, where $d\vec{s}$ is taken counterclockwise. Use $f(x + a) \approx f(x) + (df/dx)a$. Hence $dB_y/dx = dB_x/dy$. (b) Applying this to the previous problem, for volume $V = Al$, show that $F_x = V(\chi/\mu_0) \cdot B_y(dB_y/dx) = (1/2\mu_0)d(\vec{M} \cdot \vec{B})/dx$. Hence, a knowledge of the properties of the field component B_y and of the volume of the sample permit χ to be determined from a measurement of F_x.

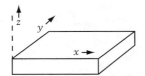

Figure 9.25 Problem 9-G.7.

9-G.8 Consider a thin disk normal to the z-axis, of thickness dz and radius ρ centered on the axis of a dipole. See Figure 9.26. (a) Show that $0 = \oint \vec{B} \cdot d\vec{A} \approx (2\pi\rho dz)B_r + (\pi\rho^2)a(dB_z/dz)$,

so $B_\rho \approx (-\rho/2)(dB_z/dz)$. (b) With B_z for a dipole, from (9.8), find B_ρ near the axis (small ρ).

Figure 9.26 Problem 9-G.8.

9-G.9 Hard disks, or hard drives, used for magnetic storage of information in the form of magnetized regions, have been made from aluminum platters in the following way. The surface is given a hard coating of nickel-phosphorus, and then is given circumferential grooving. A chromium underlayer, a cobalt alloy magnetic layer, and a hydrogenated carbon overlayer are then sequentially sputtered on, and the disk finally is lubricated. In 1996, track widths were of the order of 5 μm, and bit lengths (the region defining a region of well-defined magnetization) were of order 0.2 μm. By definition, 8 bits equals 1 byte. Estimate the storage density. [Answer: Of order 0.08 Gb/in^2, where Gb is a gigabyte.]

9-G.10 M for aligned magnetite (Fe$_3$O$_4$) is 5×10^5 J/T-m^3. This is about the same as for the $4\mu_B$ of the Fe^{+2} ion in the unit cell volume. The Fe^{+3} ions in the unit cell have opposing moments, which cancel. (a) Estimate the volume of the unit cell. (b) Estimate the critical volume V_c of magnetite at which the energy of magnetic alignment in the earth's magnetic field equals the thermal energy $k_B T$, where $k_B = 1.38 \times 10^{-23}$ J/K, and room temperature is $T = 293$ K. (c) Certain field-sensitive bacteria contain about 20 magnetite spheres of $d \approx$ 50 nm. Is this enough magnetic moment to align the bacteria despite the randomizing influence of the thermal energy?

9-G.11 Earnshaw's theorem, related to fixed electric charges (Section 3.11), also applies to permanent magnets, which alone cannot be used to levitate an object stably. However, a flat magnet resting on a tabletop can support stably one end of a similar magnet whose other end is in contact with the table. Discuss why this situation does or does not violate Earnshaw's theorem.

9-G.12 A magnet is in a vertical tube on the earth's surface. (a) If another, attractive magnet, is fixed above the first magnet, will the vertical equilibrium position of the first magnet be stable? (b) If another, repulsive magnet, is fixed below the first magnet, will the vertical equilibrium position of the first magnet be stable?

9-G.13 A small magnet can be levitated below a large magnet with two pieces of (diamagnetic) graphite, one above and one below the smaller magnet. (a) Can one permanent magnet below another permanent magnet be in equilibrium? (b) Can one permanent magnet below another permanent magnet be in *stable* equilibrium? (c) Why doesn't Earnshaw's theorem apply to diamagnets and paramagnets? (d) How can the diamagnets make the equilibrium stable?

9-G.14 A theorem about storage of information states, loosely, that the ratio of energy-to-read to energy-to-write can be made very small. Discuss this in the context of reading and writing printed matter, compact disks, and hard disks.

9-G.15 Two magnets, of moments $\vec{\mu}_1$ and $\vec{\mu}_2$, lie in the horizontal plane and are free to rotate about the vertical axis, their bases attached to a frictionless turntable. One magnet is centered at $(-a, 0, 0)$ and has its moment pointing along $\hat{y}$. The other is centered at $(a, 0, 0)$ and has its moment pointing along $\hat{x}$. (a) Indicate how each magnet will twist. (b) Using conservation of angular momentum, indicate how the turntable will twist. (c) Compute the torque on each magnet. (d) With I the moment of inertia of the turntable, compute its angular acceleration.

9-G.16 Figure 9.27 shows two refrigerator-type permanent magnets placed against a refrigerator ("soft iron"), including the domain structure of the magnets. Indicate the magnetization (if any) of the part of the refrigerator near each magnet. Determine which of the magnets is held on more strongly.

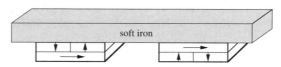

Figure 9.27 Problem 9-G.16.

"It has been shown by numerous experiments, of which the earliest are those of Ampère, and the most accurate those of Weber, that the magnetic action of a small plane circuit at distances which are great compared with the dimensions of the circuit is the same as that of a magnet whose axis is normal to the plane of the circuit, and whose magnetic moment is equal to the area of the circuit multiplied by the strength of the current."

—James Clerk Maxwell,
A Treatise on Electricity and Magnetism

Chapter 10

How Electric Currents Interact with Magnetic Fields

Chapter Overview

Section 10.1 gives a brief introduction to Oersted's 1820 discovery of the magnetism of electric currents, and of discoveries that quickly followed, especially by Ampère. Section 10.2 discusses Maxwell's concise statement of Ampère's results, and Section 10.3 works out some of its consequences. Section 10.4 applies it to obtain the magnetic force on a current-carrying wire in a magnetic field, and then considers a few important applications. Section 10.5 applies it to obtain the magnetic force on a single electric charge moving in a magnetic field. Section 10.6 considers a few important applications, including its use to "discover" the electron. Section 10.7 discusses another application of these magnetic force laws, the Hall effect. This effect has revealed much about the behavior of electrons in solids. It was recently discovered that the Hall effect can provide both a universal resistance standard and a method to determine fundamental constants with a very high degree of accuracy. Finally, Section 10.8 shows that the magnetic work done on the center of mass of a circuit comes at the expense of magnetic work done on the charge carriers within that circuit. ∎

10.1 Introduction

In 1820, H. C. Oersted, in Copenhagen, tried a classroom demonstration that had previously yielded only negative results. Intending to see if heating of a wire by current flow might cause deflection of a nearby compass needle, he used current driven by a battery of 20 Cu-Zn voltaic cells. See Figure 10.1. The compass needle did not deflect toward or away from the wire, the radial directions he expected from a heating effect. It did not deflect parallel or antiparallel to the axis of the wire. It deflected in the tangential direction, a most unexpected and counterintuitive result. When connected so that the current flowed along the upper path (A-A′-A″), the magnet deflected one way; when

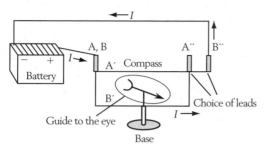

Figure 10.1 Oersted's experiment. The compass is fixed in place, but the connections can be changed so that the current flows along either the upper path (A-A'-A") or the lower path (B-B'-B").

connected so that the current flowed along the lower path (B-B'-B"), the magnet deflected the other way. Oersted's magnet would only work when placed in a horizontal plane, so he could not trace the magnetic field all around the wire, but he concluded that the magnetic field must circulate around the wire. That is, the field lines traverse circles that close on themselves.

We will call the rule giving the direction of circulation of the magnetic field due to the wire.

Oersted's right-hand rule: *Point the thumb of your right hand along the direction of current flow. Curl your fingers, which will now be part of a circle. (Do this! It is an order! Your fingers have to learn this unusual rule as much as does your conscious brain.) The direction in which your fingers curl gives the direction of circulation of the magnetic field around a long current-carrying wire.* See Figure 10.2(a).

Everywhere along the axis of the wire, the field circulates about the axis; Figure 10.2(a) gives only three of the infinite number of circles associated with the field lines. Figure 10.2(b) gives the field lines for a long wire carrying current into the page. Check that they have the correct direction of circulation by pointing the thumb of your right hand into or out of the page, and curling your fingers.

Example 10.1 **Quantitative application of Oersted's right-hand rule**

Oersted's right-hand rule is of more than qualitative significance. From it we can also arrive at some quantitative conclusions. Consider two long wires, carrying the same current I, #1 out of and #2 into the page, and separated by a distance $2a$, as in Figure 10.2(c). At a point a distance y along their perpendicular bisector, by symmetry they produce magnetic fields of the same magnitude, say, 0.005 T. For $a = 2$ cm and $y = 1$ cm, find the net magnetic field at this point.

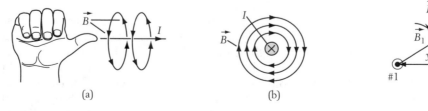

Figure 10.2 (a) Oersted's right-hand rule. The thumb points along the direction of the current I and the fingers circulate along the direction of the magnetic field $\vec{B}$.
(b) Schematic of clockwise-circulating field lines around a long wire carrying current into the page. (c) Superposition of the fields of two wires, one into the page and one out of the page.

Solution: The arcs in Figure 10.2(c) are centered around wires #1 and #2. Using Oersted's right-hand rule, wire #1 carrying current out of the page (#2 carrying current into the page) produces a field $\vec{B}_1$ ($\vec{B}_2$) that circulates counterclockwise (clockwise). This determines the directions of $\vec{B}_1$ and $\vec{B}_2$ at the point in question. We then perform vector addition to find the *total* field. From Figure 10.2(c), the total field is upward and of magnitude $2|\vec{B}_1|\cos\theta = 2(0.005)(a/\sqrt{y^2+a^2}) = 0.00894$ T.

Within months after Oersted's discovery, experimenters in Paris were hard at work studying the phenomenon. Biot and Savart used two techniques discussed in the previous chapter (measuring the disturbance of orientation by, and the oscillation frequency of, magnetized steel needles suspended by silk fibers). They determined that the field due to the wire falls off inversely with distance r, and that it varies linearly with the strength of the electric current. Ampère found that wires carrying current in the same direction attract, and that wires carrying current in the opposite direction repel. From this he deduced that each wire, as part of a larger current loop, was behaving like a magnet. Arago observed that iron filings were attracted to a current-carrying wire. He showed his results to Ampère, who proposed winding a wire in a helical shape and placing a steel needle within. In this way the two succeeded in magnetizing the needle. Soft iron did not retain a magnetization.

In his 1826 opus on the magnetism of electric currents, Ampère notes that the terminology *electromagnetic* had been used to describe the phenomena observed by Oersted. This involved an electric current (*electro*) and a magnet (*magnetic*). To describe the interaction of two electric currents, Ampère proposed the terminology *electrodynamic*. To contrast with the electrical attractions and repulsions due to static electricity, he proposed the terminology *electrostatic*. Scientists have used this terminology ever since.

10.2 The Magnetism of Electric Currents—Ampère's Equivalence

As the quotation from Maxwell shows, Ampère established that, when viewed from a distance, a small current loop is equivalent to a magnet of appropriate strength and orientation. That is, in an external magnetic field, it moves just as its equivalent magnet would move; further, it and its equivalent magnet cause the same force and torque on a distant magnet. This equivalence has extraordinary implications: *it will enable us to derive all the important results describing how electric currents interact with and produce magnetic fields*. As a consequence, we call this Ampère's equivalence. It leads to another right-hand rule, which we will apply and then show to be equivalent to Oersted's.

10.2.1 *Ampère's Current Loop Decomposition*

Before proceeding further, we discuss a theoretical idea of Ampère, based upon the principle of superposition. It enables us to apply this equivalence to large current loops and even when we are close to small current loops. The idea is that we can decompose any large current loop I into a large number of (very small)

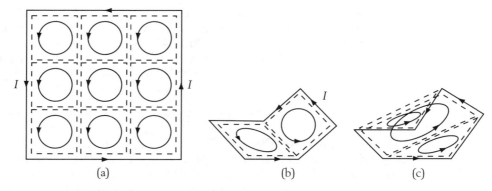

Figure 10.3 (a) Ampère's current loop decomposition into subloops.
(b) Decomposition of a nonplanar current loop into planar loops.
(c) Alternative decomposition of a nonplanar current loop into planar loops.

Ampèrian current loops, each of which has the same current I. See Figure 10.3(a). This can be done because the nearest sides of adjacent subloops carry currents in opposite directions, and thus cancel. This decomposition is not unique, as is especially obvious if the large current loop does not lie in any single plane. See Figures 10.3(b) and 10.3(c).

10.2.2 *Ampère's Equivalence*

Consider a small current loop, within a single plane, of current I and area A. See Figure 10.4(a). Ampère established that the magnetic dipole moment of this small current loop has magnitude

$$\mu = I A. \qquad \text{(moment of equivalent current loop)} \qquad (10.1)$$

The direction of the equivalent magnet is determined by what we call Ampère's right-hand rule.

> **Ampère's right-hand rule:** *Curl the fingers of your right hand in the direction of the current flow; your thumb then points in the direction of the equivalent magnetic dipole moment $\vec{\mu}$.* See Figure 10.4(b). The magnetic field $\vec{B}$ at the center of the loop also points in this direction.

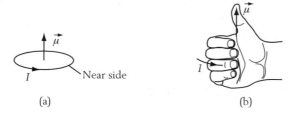

Figure 10.4 (a) Ampère's right-hand rule for replacing a current loop I by an equivalent magnet, of moment $\vec{\mu}$. (b) The fingers circulate along the current I and the thumb points along $\vec{\mu}$.

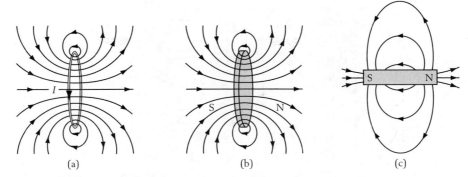

Figure 10.5 (a) Field lines of a current loop with $\vec{\mu}$ pointing to the right. (b) Field lines of the equivalent magnet with $\vec{\mu}$ pointing to the right. At large distances from both, the field-line patterns are the same. (c) Field-lines of a bar magnet with $\vec{\mu}$ pointing to the right.

Thus, the magnetic dipole moment is normal to the plane of the loop and is proportional to the current and to the area of the loop. If there are ten turns, the moment will be ten times larger. If there are seven turns one way and three turns the other way, the net effect will be a moment that is $7 - 3 = 4$ times larger than the moment of a single turn, and the moment will point in the direction of the moment of the majority (seven) of turns. Sometimes we will employ the terms *magnetic moment*, or even *dipole moment*, in place of *magnetic dipole moment*.

See Figure 10.5(a) for the field lines due to a current loop; see Figure 10.5(b) for the exterior field lines of the equivalent magnet. (The interior field lines, due to the demagnetization field, point opposite to the magnetic moment, as in Figure 9.18.) At a distance large compared with the loop itself, the field lines of a current loop are indistinguishable from those for a bar magnet, as in Figure 10.5(c).

Example 10.2 **Some current loops and their equivalent magnets**

Just as a current loop is equivalent to a thin disk-shaped magnet magnetized along its normal (called a *magnetic sheet*), so a disk-shaped magnet magnetized along its normal is equivalent to a current loop.

The equivalence of a current loop to a magnet enables us to predict qualitatively what the magnetic field of a current loop is like, and how two current loops will interact. We simply replace each loop by its equivalent magnet, and then use opposites attract, likes repel. Consider Figure 10.6(a). Loop B, in the center, is fixed in place, loop A is free to translate, and loop C is free to rotate. How do loops A and C respond?

Solution: By Ampère's right-hand rule, Figure 10.6(a) becomes equivalent to Figure 10.6(b). Then A is repelled by B, and C rotates clockwise, so that its S faces the N of B.

Figure 10.6 (a) Current loops. (b) Equivalent magnets and their magnetic moments.

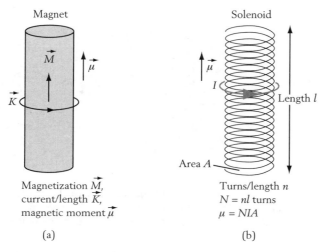

Magnet

Solenoid

Magnetization $\vec{M}$,
current/length $\vec{K}$,
magnetic moment $\vec{\mu}$

Turns/length n
$N = nl$ turns
$\mu = NIA$

(a)

(b)

Figure 10.7 (a) Magnet with moment $\vec{\mu}$ pointing upward. (b) Equivalent solenoid.

Application 10.1 **The bar magnet and its equivalent surface current**

Figure 10.7(a) depicts a bar magnet of magnetic moment μ, length l, and uniform cross-sectional area A. The equivalence of a magnet to a current loop enables us to replace the magnet by an equivalent cylindrical sheet of surface current circulating around that cross-section. Just as the thin magnetic disks of Figure 10.2(b) are equivalent to the current loops of Figure 10.2(a), here the long magnet of Figure 10.7(a) is equivalent to a sheet of surface current with current per unit length K circulating as indicated in Figure 10.7(a). For length l the current is $I = Kl$, so by (10.1) the magnet has magnetic moment $\mu = I A = (Kl) A = KV$, where $V = Al$ is the volume of the magnet. Because the magnetization is the dipole moment per unit volume, or $M = \mu/V$, for the current sheet

$$M = \frac{\mu}{V} = \frac{KV}{V} = K. \tag{10.2}$$

Realizing a current sheet experimentally can be difficult. This geometry can also be produced by using a long, tightly wound coil, or *solenoid*, of cross-section A and length l, with n turns per unit length, carrying current I. See Figure 10.7(b). This has $N = nl$ turns, each contributing to the magnetic moment, so $\mu = (nl) I A$. The magnetization of this solenoid satisfies

$$M = \frac{\mu}{V} = \frac{(nl) I A}{Al} = nI. \tag{10.3}$$

Thus, when viewed from the magnetic pole viewpoint, by the previous chapter the magnet has charge density $\sigma_m = \pm M$ and magnetic pole strength $q_m = \pm\sigma_m A = \pm MA$ on its poles, and when viewed from this chapter's equivalent current viewpoint, the magnet has current density $K = M$ circulating about its axis. The direction of the current is determined by Ampère's right-hand rule. From Chapter 9, we know that a characteristic magnetization for a permanent magnet is $M \approx 10^6$ A/m. By (10.2), the corresponding current per unit length is thus $K \approx 10^6$ A/m, corresponding to a wire of diameter

1 mm (so $n = 10^3/\text{m}$) carrying $I = 1000$ A, wrapped around a cylinder. No wonder magnets are such powerful sources of magnetic field!

10.2.3 *Magnetic Moment of a Current-Carrying Parallelogram*

Let us apply Ampère's equivalence to a circuit shaped like a parallelogram, where the current I successively goes along side $\vec{a}$ and then along side $\vec{b}$. See Figure 10.8(a). The area of the parallelogram is given by

$$A = |\vec{a}||\vec{b}|| \sin \theta_{ab}| = |\vec{a} \times \vec{b}|, \qquad (10.4)$$

where θ_{ab} is the angle between $\vec{a}$ and $\vec{b}$ in their common plane, and $\vec{a} \times \vec{b}$ is the vector cross-product of $\vec{a}$ with $\vec{b}$.

By (10.1) and (10.4),

$$\mu = IA = I|\vec{a} \times \vec{b}|. \qquad (10.5)$$

Moreover, applying Ampère's right-hand rule to get the direction $\hat{\mu}$ of the moment $\vec{\mu}$, we find that $\hat{\mu}$ coincides with the direction of $\vec{a} \times \vec{b}$. Thus $\vec{\mu}$ can be written as

$$\vec{\mu} = I\vec{a} \times \vec{b}. \qquad (10.6)$$

For N turns circulating in the same direction, $\vec{\mu}$ is larger by the factor N.

Optional   **Vector Product Right-Hand Rule Review**

This was discussed previously, in Sections R.9 and R.10. First, recall that $\vec{a}$ and $\vec{b}$ define a plane, and that the vector cross-product is perpendicular to that plane; this provides a check on your application of the rule for the vector cross-product. (You can think of this as the *zeroth rule* for the vector cross-product. If you get this part correct, the worst that can happen is that you get the wrong sense for the normal to the plane.) Thus, sweeping your right hand in this plane, from $\vec{a}$ to $\vec{b}$ through an angle of less than 180°, your thumb will point in the direction of the vector cross-product. See Figure 10.8(b), where a different $\vec{a}$ and $\vec{b}$ are given, compared to Figure 10.8(a). We call this the vector product right-hand rule to distinguish it from the other right-hand rules that we have studied.

At large distances from the loop, the details of how the magnetic moment is produced do not matter. However, to make the Ampère approach valid at short distances, we should use the Ampère current loop decomposition into an infinite number of tiny Ampèrian loops, and replace each loop by a tiny magnet. The

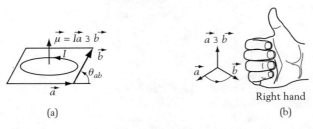

(a)   (b)

Figure 10.8 (a) Current loop I and its magnetic moment $\vec{\mu}$. (b) Vector product right-hand rule.

Since, due to electrical resistance, currents ordinarily decay with time, the surface current *K* flowing around the outside of a magnet must be subject to zero resistance. That would explain why no current passes through you when you touch a magnet—the magnet is the path of least resistance. But if you were to shave off the surface of the magnet (as in Figure 10.3a), you wouldn't get a shock either! There really is no true current flowing around the outside of the magnet. Associated with every magnetic atom is a microscopic Ampèrian current loop, subject to no electrical resistance. Together (as in Figure 10.3a) the Ampèrian current loops add up to one net surface current flowing around the outside. When you shave off the surface of the magnet, you only shave off a thin layer of atomic magnets, leaving behind a new surface with its own effective current flowing around the outside. If you accept the picture of electrons in orbits around an atom, nearly oblivious to the external world, then you should be able to accept the picture of microscopic Ampèrian currents associated with atoms. Ultimately, the Ampèrian currents can be traced to the magnetism of electrons themselves.

net effect is that the current loop is replaced by a uniformly magnetized, thin slab-shaped magnet whose perimeter is the same as that of the loop. Thus, a current loop the size of a penny can be replaced by a thin magnetic sheet shaped like a penny, as in Figure 10.2. By (10.1), the magnetic moment per unit area is given by

$$\frac{\mu}{A} = I. \tag{10.7}$$

Hence, by considering the magnetic field to be due to a sheet of magnetic dipoles, if we know the magnetic field of a single dipole, we can then sum over the fields $d\vec{B}$ due to the sheet dipoles $d\vec{\mu}$ to obtain the magnetic field, even at points close to the loop. In the next chapter, we will show another way to compute the $\vec{B}$ field at any distance from a current loop.

10.2.4 *Equivalence of Oersted's and Ampère's Right-Hand Rules (RHRs)*

Showing that Ampère's RHR implies Oersted's RHR. Consider a long wire, to which Oersted's right-hand rule applies. Figures 10.9(a) and 10.9(b) give

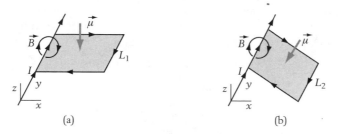

(a)   (b)

Figure 10.9 (a) Long wire and current loop that has part of the wire along one arm. (b) Long wire and an alternative current loop that has part of the wire along one arm. The field-line patterns are the same in each case.

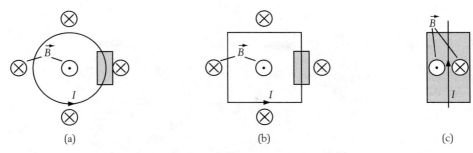

Figure 10.10 (a) Circular loop and its field-line patterns, with a closeup of part of the loop. (b) Square loop and its field-line patterns, with a closeup of part of the loop. (c) Closeup appropriate to a small part of either the circular loop or the square loop, showing the field lines due to that part of the loop.

the long wire and the associated field $\vec{B}$. Now consider that the wire is actually finite, and a part of a closed current loop. There are an infinite number of ways to choose such a closed current loop, so we show two possibilities (L_1 and L_2) in Figures 10.9(a) and 10.9(b), and their associated magnetic moments. The $\vec{B}$ field within either loop can be obtained by using Ampère's right-hand rule. It is in the same direction as obtained from Oersted's right-hand rule for the long wire. To summarize, we have shown that Ampère's RHR for a current loop yields equivalent results to Oersted's RHR for a long wire.

Showing that Oersted's RHR implies Ampère's RHR Consider a current loop, to which Ampère's right-hand rule applies, and get so close to it that locally it looks straight. Then the field direction is the same as we would obtain using Oersted's right-hand rule. See Figure 10.10, where a circular loop and a square loop are shown in parts (a) and (b), and a closeup of part of them, to which Oersted's right-hand rule applies, is shown in part (c). The $\otimes$ and $\odot$ denote the direction of $\vec{B}$. To summarize, we have shown that Oersted's RHR for a long wire yields equivalent results to Ampère's RHR for a current loop.

10.3 Some Consequences of Ampère's Equivalence

Quantitatively, Ampère's equivalence states that, in an external field, the current loop behaves like a permanent magnet of moment $\vec{\mu}$. This has a number of important consequences.

10.3.1 *Torque on a Current Loop*

As for a permanent magnet, the torque $\vec{\tau}$ on a current loop is given by (9.29), or

$$\vec{\tau} = \vec{\mu} \times \vec{B}, \qquad |\vec{\tau}| = |\vec{\mu}||\vec{B}||\sin\theta_{\vec{\mu},\vec{B}}|. \tag{10.8}$$

One application of this equation is to the torque that turns the needle of a galvanometer. Here $\vec{B}$ is due to a permanent magnet, and $\mu = |\vec{\mu}| = NIA$ is due to N turns of wire carrying current I and having cross-sectional area A.

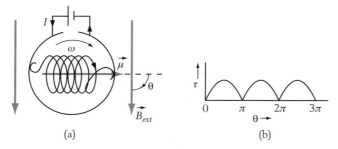

(a)   (b)

Figure 10.11 (a) Schematic of a slip-ring. A battery drives current I through the circuit, which includes multiple loops of wire attached to a rotor in sliding contact with the fixed stator. The magnetic moment $\vec{\mu}$ instantaneously points to the right. A fixed external field $\vec{B}_{ext}$ points downward. (b) The torque $\vec{\tau}$ on the rotor as a function of orientation of the stator. The reversal of the connections at the gaps causes the torque always to cause rotation in the same direction.

Application 10.2   The slip ring and motors

Equation (10.8) also applies to all motors involving rotational motion. One of the necessary developments was the invention of the slip-ring in 1832, by Ampère. See Figure 10.11(a) for a schematic. Here a permanent magnet produces $\vec{B}_{ext}$, and a battery produces a current that passes through two half-rings separated by two gaps. Semicircular sliding contacts attached to a movable axle close the circuit. The electric current passing through the wire wound around the axle produces a magnetic moment $\vec{\mu}$ that is subject to a torque $\vec{\mu} \times \vec{B}_{ext}$ causing the axle to rotate toward the direction of $\vec{B}_{ext}$. If the connection of the windings to the battery were fixed, then the axle would oscillate about $\vec{B}_{ext}$, just like a permanent magnet. However, when $\vec{\mu}$ is nearly aligned with $\vec{B}_{ext}$ so the torque is nearly zero, the angular momentum of the axle takes the sliding contacts past the gaps and the current through the axle changes direction, thus causing $\vec{\mu}$ to reverse. Hence the torque continues to rotate the motor in the same direction. See Figure 10.11(b). Because of its ubiquitous presence in motors and generators, the slip-ring has had a significant impact on our lives; it perhaps is one of the most important but unappreciated electrical inventions.

Example 10.3   Torque on a current loop

Consider a square current loop, of side $a = 0.04$ m, with one arm on the z-axis, about which it can rotate. A current $I = 1.5$ A goes up this arm. Its lower adjacent side is at an angle $\theta = 30°$ away from the x-axis, toward the y-axis. A field of magnitude $B = 4 \times 10^{-2}$ T points along the y-axis. See Figure 10.12(a). Find the torque on the magnet.

Solution: First we obtain $\mu = IA = 2.4 \times 10^{-3}$ A-m^2. Second, by Ampère's right-hand rule, we have that $\vec{\mu}$ lies in the xy-plane, at a $120°$ angle to x. Third, by (10.8), $\vec{\tau}$ points along $-\hat{z}$, with magnitude $\tau = \mu B \sin 30° = 4.8 \times 10^{-5}$ N-m. The torque is such as to align $\vec{\mu}$ with $\vec{B}$, thus minimizing the energy.

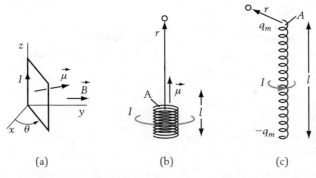

Figure 10.12 (a) A current loop in an external field $\vec{B}$. (b) A squat solenoid. At a distance it can be approximated as a dipole. (c) A long thin solenoid. Near (but not too near) each end it can be treated as a sum of monopoles.

10.3.2 *Orientation Energy of a Current Loop*

In an electric field $\vec{E}$, the orientation energy of an electric dipole of moment $\vec{p}$ is given by (3.43), or $U = -\vec{p} \cdot \vec{E}$. Similarly, in a magnetic field $\vec{E}$, the orientation energy of a magnetic dipole of moment $\vec{\mu}$ (due either to a magnet or to a current loop) is given by

$$U = -\vec{\mu} \cdot \vec{B}. \qquad (10.9)$$

Example 10.4 **Energy to align a current loop**

Consider the previous example. How much work is done in rotating the circuit from $\theta = 0°$ to $\theta = 30°$?

Solution: For $\vec{\mu}$ aligned with $\vec{B}$ as our initial configuration ($\theta = 0$, so $\cos\theta = 1$), by (10.9) $U = -\mu B = -9.6 \times 10^{-5}$ J. For $\theta = 30°$ in our final configuration, (10.9) gives $U = -\mu B \cos\theta = -8.31 \times 10^{-5}$ J. Thus it takes work $W = 1.29 \times 10^{-5}$ J to bring the magnet to its final configuration. This is about the same as the work it takes to lift a mass $m = 10^{-3}$ kg (about a penny) by 1.3 mm in the earth's gravity.

10.3.3 *Equivalent Magnets*

Here we use the fact that the magnetic field produced by the current loop is the same as the magnetic field of the equivalent magnetic sheet, with magnetic moment per unit area given by (10.7).

Application 10.3 **A coil is like a dipole, and the tip of a long thin solenoid is like a monopole**

(a) Consider a solenoid with n turns per unit length, length l, and uniform cross-section A, each turn carrying current I, so $K = nI$. By either (10.2) or (10.3), it has magnetic moment

$$\mu = MV = KV = (nI)(Al).$$

At a distance $r \gg l$ along its axis, by (9.8) the magnetic field has the dipole form

$$B = |\vec{B}| = \frac{2k_m\mu}{r^3},$$

no matter what the shape of the magnet or its equivalent circuit. See Figure 10.12(b).

(b) Now consider a magnet and equivalent circuit that is long and thin, as in Figure 10.12(c), so that its radius $a \ll l$. For large distances r from either tip of the solenoid, such that $a \ll r \ll l$, by Ampère's equivalence the tips can be treated as monopoles $q_m = \pm\mu/l = \pm nIA$ with field strength $B = k_m q_m/r^2$. P. Heller has successfully wound small solenoids with a very high and uniform winding density n and measured their $\vec{B}$ fields. He finds experimentally that the inverse square law is well satisfied, both on and off the axis.

10.3.4 The Bohr Magneton μ_B Relates Atomic Magnetic Moment
Optional and Angular Momentum

We now show how modern views of the atom lead to atomic magnetic moments of limited magnitude, thus providing a way to understand why the magnetization of a magnet is of limited magnitude (saturation magnetization). Atomic magnetic moments are limited because they are proportional to atomic angular momenta, which are themselves limited.

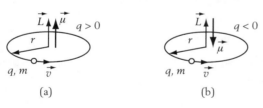

Figure 10.13 (a) Orbital motion and magnetic moment of a positive charge $q > 0$. (b) Orbital motion and magnetic moment of a negative charge $q < 0$.

A particle with mass m and charge q, moving in a circle of radius r with velocity $\vec{v}$, has both angular momentum $\vec{L} = m\vec{r} \times \vec{v}$ normal to the orbit and magnetic moment $|\vec{\mu}| = IA$ normal to the orbit. See Figure 10.13, where both $q > 0$ and $q < 0$ are depicted. Here, $|\vec{L}| = mrv\sin 90° = mrv$.

For $q > 0$, $\vec{\mu}$ is parallel to $\vec{L}$. Let us relate $|\vec{\mu}|$ and $|\vec{L}|$. First note that $v = 2\pi r/T$, where T is the period of the motion. Thus $|\vec{L}| = mrv = 2m(\pi r^2)/T$. Now note that $I = q/T$ and $A = \pi r^2$. Thus $|\vec{\mu}| = IA = q(\pi r^2)/T$. Hence, $|\vec{\mu}| = |(q/2m)\vec{L}|$. Including direction, this is

$$\vec{\mu} = \frac{q}{2m}\vec{L}. \qquad (10.10)$$

We now relate this to the magnetic properties of matter.

In 1910, Bohr gave a *quantization rule* for electrons moving in atoms, relating the magnitude of $\vec{L}$ to Planck's constant h. With l an integer, Bohr's

> **Motion**
>
> Motion at the atomic level has different properties than motion on a larger scale; for one thing, there is no friction. This explains why the magnetism of magnets doesn't decay with time. Weber's conception of indivisible magnets, and Ampère's conception of microscopic atomic currents not subject to friction (suggested to him by Fresnel), are both supported by modern work. Although the Bohr model of the atom cannot be taken literally, it provides a good way to make order-of-magnitude estimates of the properties of atoms.

relationship is

$$|\vec{L}| = \frac{h}{2\pi}l. \qquad (l \text{ an integer}) \qquad (10.11)$$

From (10.11), with $q = e$ in (10.10) we deduce that, for electron orbits in atoms,

$$|\vec{\mu}| = \frac{eh}{4\pi m}l. \qquad (l \text{ an integer}) \qquad (10.12)$$

We call

$$\mu_B = \frac{eh}{4\pi m} \qquad \text{(Bohr magneton)} \qquad (10.13)$$

the *Bohr magneton*. Its value is 9.27×10^{-24} A-m^2.

In practice, only diamagnetic atoms (compare Figure 9.12b) derive their magnetism from orbital motion. For paramagnetic (compare Figure 9.12a) and ferromagnetic atoms, there is a source of magnetism associated with the intrinsic angular momentum $\vec{S}$ of each electron, as if each electron were a little gyroscope. For that reason $\vec{S}$ is called the electron *spin*, and it, too, comes in quantized units. The intrinsic magnetic moment of the electron is twice that given by μ_B. Nevertheless, the Bohr magneton gives the characteristic value of atomic magnetic moments.

10.4 Magnetic Force on a Current-Carrying Wire

Consider (10.6) for the magnetic moment $\vec{\mu} = I\vec{a} \times \vec{b}$ of a current-carrying parallelogram, and (10.9) for the energy $U = -\vec{\mu} \cdot \vec{B}$ of a current-carrying loop in a $\vec{B}$ field. With them we will determine the magnetic force $d\vec{F}$ due to an applied field $\vec{B}$ on a piece of wire of length $d\vec{s} = \hat{s}ds$ pointing along the direction of the current I. See Figure 10.14, where a bar magnet serves as the $\vec{B}$ source. Our derivation follows Maxwell.

To obtain $d\vec{F}$ on $d\vec{s}$, we use (10.9) to find the energy variation δU when there is a change in the magnetic moment $\delta\vec{\mu}$:

$$\delta U = -\delta\vec{\mu} \cdot \vec{B}. \qquad (10.14)$$

Now consider the element $d\vec{s}$ of the circuit in Figure 10.14. Displacement of $d\vec{s}$ by $\delta\vec{r}$ adds to the circuit a parallelogram-shaped loop of sides $\delta\vec{r}$ and $d\vec{s}$, where the current I successively goes through sides $\delta\vec{r}$ and $d\vec{s}$. (Ampère's experimental

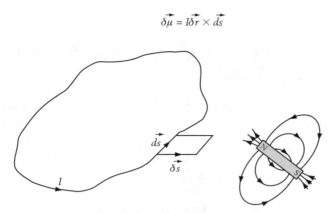

$$\vec{\delta\mu} = I\vec{\delta r} \times \vec{ds}$$

Figure 10.14 Current loop I in an arbitrary external field (with source represented by the magnet). The element $\vec{ds}$ of the loop, in an imaginary deformation by $\delta\vec{r}$, leads to an additional magnetic moment $\delta\vec{\mu} = I\delta\vec{r} \times \vec{ds}$.

setup, similar to Figure 10.14, had a slide-wire arm $\vec{ds}$ that could be displaced by $\delta\vec{r}$. Hence no elastic energy was expended to distort the circuit.) By (10.6), the magnetic moment of the circuit changes by

$$\delta\vec{\mu} = I\delta\vec{r} \times \vec{ds}, \tag{10.15}$$

and thus (10.14) becomes

$$\delta U = -\delta\vec{\mu} \cdot \vec{B} = -I(\delta\vec{r} \times \vec{ds}) \cdot \vec{B}. \tag{10.16}$$

We will now need a simple mathematical manipulation. Just as

$$(\hat{i} \times \hat{j}) \cdot \hat{k} = 1 = \hat{i} \cdot (\hat{j} \times \hat{k}) = (\hat{j} \times \hat{k}) \cdot \hat{i},$$

so too can the vectors in (10.16) be rearranged. Thus

$$\delta U = -I(\vec{ds} \times \vec{B}) \cdot \delta\vec{r}. \tag{10.17}$$

Neglecting elastic energy, consider how we might actually make such a displacement $\delta\vec{r}$. If there is really a force $d\vec{F}$ acting on $\vec{ds}$, then the wire would start moving spontaneously, thereby gaining kinetic energy. To prevent that from occurring, we act with a force from our hand, $\vec{F}_{hand}$, that nearly cancels $d\vec{F}$. Thus we do work

$$\delta W = \vec{F}_{hand} \cdot \delta\vec{r} \approx -d\vec{F} \cdot \delta\vec{r}. \tag{10.18}$$

This work does not go into elastic energy or kinetic energy; it goes into a change δU in the magnetic energy. Thus, equating δW in (10.18) to δU yields

$$\delta U = -d\vec{F} \cdot \delta\vec{r}. \tag{10.19}$$

Equating the terms multiplying $\delta\vec{r}$ in (10.19) and (10.17) gives

$$d\vec{F} = I\vec{ds} \times \vec{B}. \quad \text{(force on a current-carrying element)} \tag{10.20}$$

The qualitative statement that the force is normal to both $d\vec{s}$ and to $\vec{B}$ was made by Ampère, so (10.20) has been called *Ampère's force law*. This important result is well verified experimentally. It tells us how to find the force on a current-carrying wire in a magnetic field.

Integrating the force law (10.20) over any circuit yields

$$\vec{F} = I \oint (d\vec{s} \times \vec{B}). \tag{10.21}$$

10.4.1 The Force on a Straight Wire in a Uniform $\vec{B}$

If $\vec{B}$ is uniform over an arm of length l, then $\vec{B}$ can be brought outside the integral, so (10.20) applied *to a single arm* integrates to

$$\vec{F}_{arm} = I \int_{arm} d\vec{s} \times \vec{B} = I \left(\int_{arm} d\vec{s} \right) \times \vec{B} = I\vec{l} \times \vec{B}. \tag{10.22}$$

> **Note**
>
> If $\vec{B}$ is uniform, then so are each of its components (B_x, B_y, B_z). However, although a pole q_m (the tip of a long narrow magnet or solenoid, as in Figure 10.12c) can produce, over a sphere of radius r, a constant magnitude $|\vec{B}| = kq_m/r^2$, its components will not be constant over the sphere because its *direction* varies over the sphere.

Here $\vec{l}$ has length l and points along the direction of current flow.

One application of (10.22) is to the "magnetic blow-out" of a welder's arc. If the arc between the two welding electrodes corresponds to a current along x, and a magnetic field suddenly is applied along y, then by (10.22) there will be a force on the arc that is along z. If the force is large enough, the arc will be deflected enough from its path along x (between the electrodes) that it will break the electrical contact and the arc will be extinguished.

Example 10.5 **Electromagnetic balance**

Consider an electric circuit carrying a current I, with an arm of length l, in an external magnetic field $\vec{B}$. Let the current be rightward, and the magnetic field be into the paper, as in Figure 10.15. (a) Find an algebraic expression for the force acting on the arm l. (b) Evaluate the field for $l = 0.25$ m, $I = 2$ A, and $F = 0.004$ N.

Solution: (a) Equation (10.22) can be applied to each arm. The field acting on the upper arm (not shown) is zero, so the force on the upper arm is zero also. The forces on the side arms cancel because the currents in these arms are in opposite directions to each other. For the lower arm, the angle θ between $\vec{l}$ and $\vec{B}$ is $90°$.

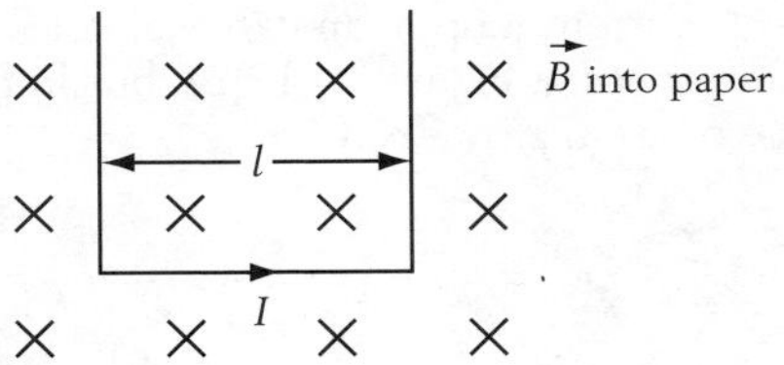

Figure 10.15 Part of a current loop I in a uniform field $\vec{B}$. There is an upward force on the lower arm.

Thus (10.22) gives

$$F = IlB. \tag{10.23}$$

Moreover, $\vec{F}$ points upward. By measuring the change in the "weight" of the object, we can determine F, and thus deduce either I or B. (b) For the given values of l, I, and F, (10.24) yields $B = F/(Il) = 0.008$ T.

If $\vec{B}$ in Figure 10.15 were rotated to partly point along $\hat{x}$ (the direction of the current), the force on the lower arm would still point upward. However, its magnitude would change. For a $25°$ angle of rotation, the force would be $IlB \sin 65°$.

10.4.2 *The Force on a Closed Circuit in a Uniform $\vec{B}$ Is Zero*

Equation (10.21) simplifies considerably if the vector $\vec{B}$ is uniform over the *entire* circuit. Then $\vec{B}$ can be brought outside the integral, so

$$\vec{F} = I \left(\oint d\vec{s} \right) \times \vec{B} = \vec{0}. \qquad (\vec{B} \text{ uniform}) \tag{10.24}$$

This is zero because, for any closed loop (such as a quarter-mile track), the integral over the vector elements is zero, or $\oint d\vec{s} = \vec{0}$. (The runner has returned to her starting point.) Hence the force on the loop must be zero. Note that the integral $\oint ds$ over the *length* ds of the vector elements is nonzero. (The runner has run a quarter-mile.)

Thus, to obtain a force on an electric circuit, $\vec{B}$ must be nonuniform. Of course, even for a uniform $\vec{B}$, there can be a torque on the circuit, from (10.8).

We now turn to some examples related to this section as a whole.

Example 10.6 Force and torque on a current loop

Determine the force and torque on the square current loop of Example 10.3, with current I and side a. See Figure 10.16.

Solution: Because the field is uniform, by (10.24) there is no net force. As a consequence, by a theorem of mechanics, in calculating the torque it doesn't matter which point is used to determine the moment arms. Let's use the center of the loop. Applying (10.22) to the top arm gives a force $\vec{F}_{top}$ that is along the upward vertical, as is the moment arm $\vec{r}_{top}$, so the torque $\vec{\tau}_{top} = \vec{r}_{top} \times \vec{F}_{top}$ is zero. For the bottom arm $\vec{F}_{bot} = -\vec{F}_{top}$ and $\vec{r}_{bot} = -\vec{r}_{top}$, so $\vec{\tau}_{bot} = \vec{\tau}_{top}$ is also zero. For the near arm, with current along $-\hat{z}$, the force is along $+\hat{x}$, has magnitude IaB, and

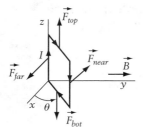

Figure 10.16 A current loop in an external field $\vec{B}$. This is the same geometry as in Figure 10.12(a), but in more detail, with the force on each arm given.

has moment arm of length $a/2$ in the xy-plane at θ to the x-axis. This gives a torque $\tau_{near} = (a/2)(IaB)\sin\theta$ along $-\hat{z}$. For the far arm, $\vec{\tau}_{far} = \vec{\tau}_{near} = (a/2)(IaB)\sin\theta$, again along $-\hat{z}$. Hence the total torque has magnitude

$$\mu B \sin\theta, \qquad \text{where } \mu = IA = Ia^2, \qquad (10.25)$$

just as for (10.8). For $\theta = 30°$ this agrees in magnitude and direction with the analysis of Example 10.3.

Example 10.7   Current loop in flaring magnet field: the audio speaker

In an audio speaker, the speaker coils are attached to a movable speaker cone and are fed current I by an amplifier. Consider that the magnetic field is produced by a bar magnet that is fixed in place (although in practice the magnet geometry is more complex). See Figure 10.17(a). Assume that for each loop of speaker wire the magnetic field is uniform in magnitude, but flares outward by an angle θ, as in Figure 10.17(b). [Thus $\vec{B}$ is a nonuniform vector, so by the note following (10.22), the force on the loop can be nonzero.] Take a loop to be centered at the origin, in the yz-plane, and have radius a. It carries current I that is clockwise as viewed from the $+x$-axis. (a) Find the force on the loop of speaker wire. (b) Find the optimal value of θ. (Note: Both for its physics and its mathematics, this is a very instructive example.)

Solution: (a) Before calculating the force on the loop, let's use a qualitative argument to find its direction. Since the current in the speaker coil is clockwise to an observer on the right, by Ampère's right-hand rule, the speaker coil behaves like an equivalent magnet whose magnetic moment points to the left. Thus the actual magnet and the equivalent magnets have their N poles nearest, so the equivalent magnet (and the speaker coil) should feel a force of repulsion—to the right.

Now consider a more quantitative analysis, based on (10.20). Break up the loop into many parts. For the i element, $d\vec{s}_i$ is normal to the field $\vec{B}_i$, so the angle $\theta_{d\vec{s}_i, \vec{B}_i}$ between $d\vec{s}_i$ and $\vec{B}_i$ is $\phi_i = 90°$, yielding $\sin\phi_i = 1$. (The symbol θ has already been used for the flare angle.) Equation (10.20) then yields

$$|d\vec{F}_i| = I|d\vec{s}_i \times \vec{B}_i| = I|d\vec{s}_i||\vec{B}_i||\sin\phi_i| = I(ds_i)(B_i)(1) = I(ds_i)(B). \quad (10.26)$$

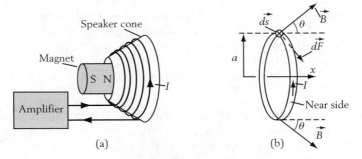

Figure 10.17 (a) A speaker cone, including its permanent magnet and the current loops glued to it, driven by current I from the amplifier. (b) Analysis of a loop of speaker wire, focusing on the force $d\vec{F}$ on one element $d\vec{s}$ in the flaring field $\vec{B}$.

However, only a fraction $\sin\theta$ of this, where θ is the flaring angle of the field, points along the x-axis. See Figure 10.17(b). There also are force components toward the axis, which tend to compress the loop (a result not predicted by our Ampère's force law analysis). However, they lie in the yz-plane; by symmetry they integrate to zero. Thus the contribution along the x-axis is all that matters, so we need compute only the component of $d\vec{F}_i$ along the x-direction. From (10.26), that entry is

$$(d\vec{F}_i)_x = I(ds_i)B\sin\theta. \qquad (10.27)$$

On performing the sum, we can take out the constant factors I, B, and $\sin\theta$. The sum is then given by

$$F_x = \Sigma_i(d\vec{F}_i)_x = IB\sin\theta\,\Sigma_i ds_i = IB\sin\theta\,(2\pi a), \qquad (10.28)$$

since the total length of the loop is $2\pi a$. Thus, we performed this complicated-looking integral as a sum, hardly using any more than the concept of calculus.
    We may rewrite (10.28) to read

$$F_{loop} = I(2\pi a)B\sin\theta. \qquad \text{(force on loop in flaring field)} \qquad (10.29)$$

In practice, an audio speaker uses many turns N of wire, so the total force is N times (10.29). If the flare angle $\theta = 0$, so that $\vec{B}$ is uniform, (10.29) gives no net force at all, consistent with (10.25). (b) Clearly, F_{loop} of (10.29) is maximized if θ is 90°. Indeed, this is how speaker magnets are designed. Efficient speakers have almost no stray magnetic field.

Example 10.8   Force on a speaker cone

As an explicit example of a flaring field in the context of a speaker coil, consider that the field $\vec{B}$ is due to a long magnet or long solenoid, where only one pole dominates. For the loop in Figure 10.17(b), centered at the origin, imagine a pole q_m on the x-axis at $(-d,0,0)$ producing the field $\vec{B}$. For $a = 0.005$ m, $d = 0.00866$ m, $I = 0.02$ A, and $N = 50$ turns, find the net force on the speaker cone.

Solution: From the geometry of the problem, $r = \sqrt{d^2 + a^2} = 0.01$ m, $\sin\theta = a/r = 0.5$, and $2\pi a = 0.0314$ m. For a pole q_m, $B = k_m q_m/r^2$ gives $B = 2 \times 10^{-2}$ T. Finally, with $I = 0.02$ A and $N = 50$, 50 times (10.29) yields a force of 3.14×10^{-4} N. This acts on the speaker cone, causing it and the air surrounding it to move. Because I is time dependent (coming from the amplifier, which in turn is fed by a compact disk or a radio tuner), the force, and thus the motion of the speaker cone, is time dependent. This motion, as we all know, produces sound.

The force described in (10.20), $d\vec{F} = I d\vec{s} \times \vec{B}$, acts on the charge carriers in a direction normal to the wire. When the charge carriers start to move in that direction, they collide with the atoms of the wire, as described in Chapter 7. In this way, the momentum transferred by $d\vec{F}$ to the charge carriers is rapidly transferred to the more massive (or *ponderous*) circuit. Hence the circuit itself feels this force. For that reason, in the 19th century, it was called a *ponderomotive force*. For a rigid object, the ponderomotive force effectively acts on the center of mass as a whole. We will find it useful to distinguish between ponderomotive forces that act on macroscopic objects and *local electromotive forces* that act on electric currents within a macroscopic object. (As indicated in Chapter 7, the terminology *electromotive force* is reserved for the work on taking a unit charge around a circuit.) In the next section, we will derive a local electromotive force whose origin lies in the same physics as (10.20).

10.5 The Force on a Charge Moving in a Magnetic Field

10.5.1 *Derivation*

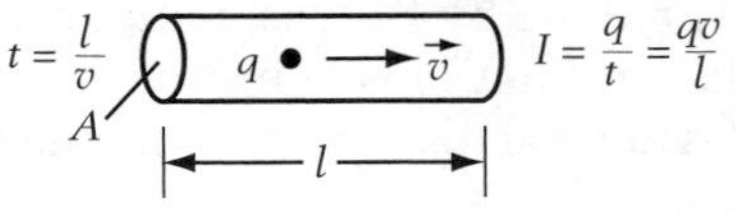

Figure 10.18 Force $d\vec{F}$ on a charge q moving in a magnetic field. The imaginary line segment S is of length $l = vt$. During the time t, q passes completely across S.

We can use (10.20) to derive the force on a single charge q moving with velocity $\vec{v}$ in a field $\vec{B}$. See Figure 10.18.

The idea is, in (10.20), or $d\vec{F} = I\,d\vec{s} \times \vec{B}$, to consider a line of identical charges q at imaginary spacing l, moving with velocity $\vec{v}$. They cross l in time $t = l/v$, where $v = |\vec{v}|$. The current thus is $I = q/t = qv/l$, and it is associated with length l. Thus $I\,d\vec{s}$ has magnitude $(qv/l)l = qv$. Moreover, $I\,d\vec{s}$ points along $q\vec{v}$, so $I\,d\vec{s} = q\vec{v}$. Hence, in $d\vec{F} = I\,d\vec{s} \times \vec{B}$, use $q\vec{v}$ for $I\,d\vec{s}$ and $\vec{F}$ for $d\vec{F}$ to obtain

$$\vec{F} = q\vec{v} \times \vec{B}. \qquad \text{(force on a moving charge in a } \vec{B} \text{ field)} \qquad (10.30)$$

Note that the imaginary spacing l has disappeared. This result was discovered theoretically, both by Heaviside (1889) and by Lorentz (1895), many years after Ampère's discovery of (10.20). Note that the microscopic viewpoint of this derivation was not available to Ampère or Maxwell.

Since (10.30) involves the vector cross-product of $\vec{v}$ and $\vec{B}$, the magnetic force $\vec{F}$ on the particle is normal to the plane defined by both $\vec{v}$ and $\vec{B}$. For any cross-product, the magnitude of the resultant vector equals the product of the magnitudes of the two vectors (here $|\vec{v}|$ and $|\vec{B}|$) and the sine of the angle between the two vectors.

Example 10.9 Direction of magnetic force on moving electron

Consider an electron ($q = -e$) with $\vec{v}$ along $\hat{\imath}$ (left to right on the page), in a $\vec{B}$ along $\hat{\jmath}$ (bottom to top of the page). Find the direction of the magnetic force acting on it.

Solution: $\vec{v} \times \vec{B}$ is along $\hat{\imath} \times \hat{\jmath} = \hat{k}$ (out of the page). By (10.30), since $q < 0$, $\vec{F}$ is along $-\hat{k}$ (into the page).

Example 10.10 Magnetic force on moving electron in the earth's magnetic field

Consider an electron with energy 10 eV, moving in the earth's magnetic field, taken to have the local magnitude $|\vec{B}_{earth}| = 0.5 \times 10^{-4}$ T. (a) Find the maximum force acting on the electron. (b) Find the corresponding acceleration, and compare with that of gravity.

Solution: The maximum force occurs when the electron is moving perpendicular to the direction of $\vec{B}_{earth}$. The velocity is given by $\frac{1}{2}mv^2 = eV$, where $V = 10$ V, so $v = \sqrt{2eV/m}$, which yields $v = 1.875 \times 10^6$ m/s. Then $|\vec{F}|_{max} = ev|\vec{B}_{earth}| = 1.5 \times 10^{-17}$ N. This corresponds to an acceleration of $a = |\vec{F}_{max}|/m = 1.60 \times 10^{13}$ m/s^2, which completely overwhelms the 9.8 m/s^2 of gravity.

10.5.2 *The Magnetic Force Does Zero Work*

The magnetic force on a moving charge has a curious property; it does no work. That is because the work $dW_{mag} = \vec{F} \cdot d\vec{r}$ done by the magnetic force in time dt when q moves by $d\vec{r}$ involves the scalar product of two mutually perpendicular vectors. On the one hand, $\vec{F} = q\vec{v} \times \vec{B}$ is normal to $\vec{v}$; on the other hand, $d\vec{r} = \vec{v}dt$ is along $\vec{v}$. Hence $dW_{mag} = \vec{F} \cdot d\vec{r} = 0$.

The force described by (10.30), when it is applied to individual charge carriers in a wire, can cause current flow even when there is no battery in the circuit, and thus it can be thought of as an electromotive force. We will call it the *local electromotive force*. Chapter 12 will develop this idea extensively.

Example 10.7 involves a current-carrying speaker coil in a magnetic field. Using a term introduced in Example 10.8, we can say that there is a net ponderomotive magnetic force (integrated over the loop) acting on the coil. Let the velocity of a charge carrier in the coil be written as $\vec{v} = \vec{v}_{CM} + \vec{v}_d$, where $\vec{v}_{CM}$ is the velocity of the center of mass, and $\vec{v}_d$ is the drift velocity of the charge carriers relative to the center of mass (thus $\vec{v}_d$ is along or against the current density $\vec{J}$). Example 10.7 implicitly took $\vec{v}_{CM} = \vec{0}$. If $\vec{v}_{CM} \neq \vec{0}$, then the ponderomotive force does work ΔW_{pmf} on the center of mass of the coil. Since $dW_{mag} = 0$ on each charge, there must be another sort of work that is equal and opposite to ΔW_{pmf}. This is the work ΔW_{emf} done by the emf associated with the component of magnetic force $q\vec{v}_{CM} \times \vec{B}$ along the wire. If the coil speeds up ($\Delta W_{pmf} > 0$), then the current slows down ($\Delta W_{emf} < 0$). (A similar argument applies to the electromagnetic balance of Example 10.5.) This is discussed in detail in Section 10.8.

10.6 Applications of the Magnetic Force Law

Three important applications of (10.30) are (1) circular motion of electric charges, as employed in particle accelerators (e.g., cyclotrons) and mass spectrometers (sensitive detectors of the charge-to-mass ratio of atoms and molecules); (2) magnetic deflection of electric charges, as in vacuum tubes (e.g., TV tubes and computer monitors); and (3) magnetic mirrors, which confine charged particles with magnetic fields, as used in fusion devices (such devices, if they become practical, would use deuterium from water to provide abundant nuclear power that could be converted to electricity).

10.6.1 *Motion Perpendicular to Uniform $\vec{B}$ Is Circular*

Consider a uniform $\vec{B}$, and an initial $\vec{v}$ that is perpendicular to $\vec{B}$. Figure 10.19(a) depicts the motion of positive charges, and Figure 10.19(b) depicts the motion

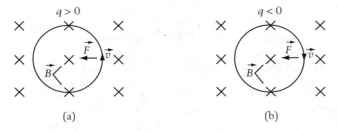

Figure 10.19 (a) Motion of a positive charge $q > 0$ in a field $\vec{B}$ pointing into the page. (b) Motion of a negative charge $q < 0$ in a field $\vec{B}$ pointing into the page.

of negative charges, with deflections perpendicular to their motion, according to (10.30). Since the magnetic force is perpendicular to the velocity, this is very similar to two familiar situations where there is circular motion. The first case is that of planetary motion in a circle. Here the force of gravity due to the sun causes the planet to move in a circle, provided that (1) there is no radial component to the velocity, and (2) the attractive gravitational force F_g precisely equals the product of the mass m and the inward radial acceleration v^2/R. The second case is a bucket swung by a rope. Here the bucket moves in a circular orbit when the tension T in the rope matches the product of the mass m and the inward radial acceleration v^2/R. In both of these cases, the force is perpendicular to the velocity and there is an obvious center to the orbit, determined by the nature of the force (gravity in one case and the rope in the other).

For a charged particle moving in a magnetic field, the force is perpendicular to the velocity, but there is no obvious center to the orbit. Indeed, we can have many particles having orbits about different centers. However, that does not affect the fact that there is a circular orbit.

Let us now equate the magnitude of the magnetic force $\vec{F}$ to the product of the mass m and the radial acceleration v^2/R. Then $|\vec{F}| = |q||\vec{v}||\vec{B}||\sin\theta| = |q|vB$, since $\vec{v}$ and $\vec{B}$ are at an angle of $90°$. Thus

$$|q|vB = \frac{mv^2}{R},\qquad(10.31)$$

which implies

$$R = \frac{mv}{|q|B}. \quad \text{(radius of cyclotron orbit)}\qquad(10.32)$$

This result for R has all the qualitative dependences that might be expected: it increases as the momentum mv increases, and it decreases as the charge q and field B (which produce the deflecting force) increase. However, without the derivation we would not know the powers to which each variable appears, nor that the constant of proportionality is unity. For a proton ($q = e$ and $m = 1.67 \times 10^{-27}$ kg) with $v = 4.5 \times 10^5$ m/s, in a field $B = 3 \times 10^{-5}$ T (comparable to that found just outside the earth's upper atmosphere), (10.32) gives $R = 156.6$ m. This is sufficiently small that a proton from the solar wind can circle around the earth's magnetic field lines.

> ### Mass Spectrometer and Momentum Analyzer
>
> The bending of charged particles moving in a magnetic field is the basis of the *mass spectrometer*. Unknown atoms, of mass M, are ionized, accelerated, passed through a velocity selector (see Section 10.6.4), and then enter a region of known $\vec{B}$. They are bent in a semicircle by $\vec{B}$, and then leave the field region with reversed velocity. From the radius of their circular motion, by (10.32) their q/M value can be determined. Since q is always an integer times e, the radius of the largest circle should correspond to $q = e$. Hence M can be determined. An alternative use of the same device is to employ particles of known mass and charge, and by their bending in a known magnetic field, to determine their velocity. This is called a *momentum analyzer.*

We can rewrite (10.32) to obtain v/R, which is the same as the angular frequency ω. That is,

$$\omega = \frac{v}{R} = \frac{|q|B}{m} \qquad \text{(cyclotron frequency)} \qquad (10.33)$$

is independent of the radius of the orbit. Equation (10.33) is called the cyclotron frequency. The corresponding period is given by

$$T = \frac{2\pi}{\omega} = \frac{2\pi m}{|q|B}. \qquad (10.34)$$

The period is the "year" for the particle. Hence, for a charged particle moving in a circle under the influence of a uniform magnetic field, the year is independent of the radius of the orbit. This is a surprising result, given our experience with gravity: planets within the solar system have nearly circular orbits with a year that increases with the radius R (as $R^{3/2}$). For a proton in a field $B = 3 \times 10^{-5}$ T [as in the paragraph after (10.32)], (10.33) gives $\omega = 2.87 \times 10^3$ rad/s, and (10.34) gives $T = 2.18 \times 10^{-3}$ s. Thus protons above the earth are constantly spiraling around the local magnetic field, constrained by it from developing sustained motion in any direction but that of the field line.

10.6.2 *For a Charge in a Uniform $\vec{B}$, Motion Is Circular or Spiral*

Again consider the case of uniform $\vec{B}$. Since the magnetic force on a charge q and velocity $\vec{v}$ is normal to the plane defined by $\vec{v}$ and $\vec{B}$, there is no force along $\vec{B}$. Hence a nonzero component of $\vec{v}$ along $\vec{B}$ (given by $\vec{v} \cdot \hat{B}$) will remain unchanged. Thus, the motion along $\vec{B}$ will proceed with a uniform value given by $\vec{v} \cdot \hat{B}$, independent of the motion perpendicular to $\vec{B}$. Since the motion in the plane perpendicular to $\vec{B}$ is a circle, the net motion is a spiral. (This also may be seen as follows. Start in the reference frame where $\vec{v}$ is normal to $\vec{B}$, so q moves in a circle about $\vec{B}$. Then go to the reference frame where q has a component along $\vec{B}$.) See Figure 10.20, where $\vec{v}$ has components in all three spacial directions. The

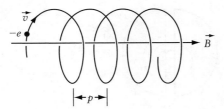

Figure 10.20 Electron with velocity $\vec{v}$ subject to the uniform magnetic field $\vec{B}$, causing motion in a spiral of pitch p.

distance that q moves in the direction of $\vec{B}$, per revolution about $\vec{B}$, is called the *pitch p.*

10.6.3 *The Cyclotron*

The radius independence of the period in (10.34) is the key to devices called *cyclotrons*, which inject charged particles into a magnetic field, and then "hit" them with an electric field at just the right place and time in their orbit (determined by the cyclotron frequency) to accelerate them. Without this radius independence of the period, it would be much harder to control the acceleration timing. On acceleration, the velocity and [by (10.32)] the radius grow. The limiting values are determined by the magnetic field that can be produced. The cyclotron was invented by E. O. Lawrence and M. S. Livingston in 1934. It was the first man-made "atom smasher," accelerating charged particles to high enough energies that, when they collided, they split apart or otherwise produced results that revealed the inner workings of matter.

When a particle moves, its mass increases, as shown by Einstein in his special theory of relativity. (Here, "special" means "restricted.") He found that $m = m_0/\sqrt{1 - v^2/c^2}$, where m_0 is the *rest mass*. This increase in mass decreases the cyclotron frequency of (10.33), an effect that is unimportant unless the velocity approaches the speed of light. In that case, the timing of an acceleration device must adjust to the particle velocity, which makes for more complex cyclotron design. For that reason, such accelerators are called *synchrocyclotrons*.

Cyclotron

The cyclotron consists of two half-cans (tunafish cans in shape), technically called "dees" because they look like the letter D from above. See Figure 10.21(a) for an x-ray view. The two dees A and B are given opposite voltages such that when a particle is in A, and crosses to B, it gains energy. Then, while in B, the voltages are switched, so on crossing from B to A the particle again gains energy. See Figure 10.21(b). This use of timing so that linear acceleration occurs every half-cycle bears some resemblance to the way slip-rings (compare Section 10.3) cause rotational acceleration of motors every half-cycle. The dees can be thought of as 180° pieces of pie; more recent designs, to increase the rate of acceleration, use six 60° pielike regions (still called dees), the odd-numbered dees being grounded, and the even-numbered dees having a voltage that is adjusted.

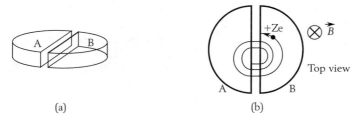

(a)   (b)

Figure 10.21 Cyclotron. (a) Perspective of the two dees, with uniform field $\vec{B}$ pointing downward. (b) Top view of the two dees, including the motion of a charged particle in $\vec{B}$, which points into the page. The particle is injected and then accelerated as it passes from dee to dee.

10.6.4 *Magnetic Deflection with Crossed $\vec{E}$ and $\vec{B}$ Fields: e/m for the Electron*

In the 1890s, many physicists were studying sparking in evacuated tubes (*vacuum tubes*) because such studies promised to provide information about the microscopic nature of matter. It was learned, both by direct measurement and by their deflection in a magnetic field, that the charged particles emitted by all types of cathodes, known as *cathode rays*, were negatively charged. In 1897, J. J. Thomson performed a crucial experiment on cathode rays. Making a tube with a better vacuum than others had made, he found that cathode rays in his vacuum tubes did not collide with any residual gas, and he could now study how the path of the cathode rays was affected by both electric and magnetic fields. From this, he determined their charge-to-mass ratio, e/m.

> ### Modern Particle Accelerators
>
> The cyclotron is the grandfather of modern particle accelerators. In the first cyclotrons, the particle energies achieved were comparable to the energies of naturally occurring atomic nuclei and revealed much about nuclear physics. [Nuclear physics is daily applied in smoke alarms (which detect smoke particles ionized by a small amount of radioactive isotope in the alarm), medical diagnostics (where small amounts of radioactive isotopes, once swallowed, can then be imaged in the body), archeological dating (with radioactive isotopes), and nuclear power (using energy released from nuclear fission).] As cyclotron technology improved, the achievable particle energies increased to the point that the constituents of the nuclei—the protons and neutrons—could themselves be blown apart, and the products studied. The study of such subnuclear particles, as with astronomy (the study of ultralarge and ultradistant objects), requires the construction of large and sensitive detectors. Unlike astronomy, however, where the universe is already present and waiting to reveal its secrets, in the area of subnuclear physics the universe (of subatomic particles) also must be created. Subnuclear physics (also called high-energy physics) has been at the forefront of data collection and data storage techniques, and many important software developments—from computational methods to electronic mail, electronic publishing, and the World Wide Web—have been made by workers associated with this area.

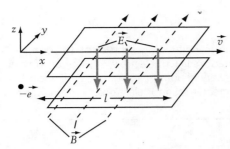

Figure 10.22 Motion of a charged particle q in crossed electric and magnetic fields $\vec{E}$ and $\vec{B}$, subject to the no-deflection condition $|\vec{E}| = |\vec{v}||\vec{B}|$.

The total force on a particle of charge q and velocity $\vec{v}$ is the sum of the electric force $q\vec{E}$ and the magnetic force $q\vec{v} \times \vec{B}$, giving what is known as the Lorentz force

$$\vec{F} = q(\vec{E} + \vec{v} \times \vec{B}). \qquad \text{(Lorentz force law)} \qquad (10.35)$$

Thomson first chose $\vec{v}$ and $\vec{B}$ to be mutually perpendicular. Then he took $\vec{E}$ to be opposite $\vec{v} \times \vec{B}$, so the electric and magnetic forces were made to oppose one another. Figure 10.22 takes $\vec{v}$ along $\hat{x}$, $\vec{B}$ along $\hat{y}$, and $\vec{E}$ along $-\hat{z}$. Thus $|\vec{v} \times \vec{B}| = vB$ and $\vec{v} \times \vec{B}$ is along $\hat{z}$, which is opposite to $\vec{E}$.

Thomson then made the electric and magnetic forces cancel out completely, giving $\vec{F} = \vec{0}$ in (10.35). Explicitly, this gives $q(\vec{E} + \vec{v} \times \vec{B}) = q[(-\hat{z})E + \hat{z}vB] = q(-\hat{z})(E - vB) = 0$, so

$$v = \frac{E}{B}. \qquad \text{(no-deflection condition)} \qquad (10.36)$$

This is the principle of a *velocity selector*. It enabled Thomson to determine the particle velocity from measurements of E and B.

Thomson then turned off the $\vec{B}$ field and studied the deflection of the cathode rays due to E alone. Chapter 3 considers the deflection of a particle of charge q and mass m moving along the x-axis with velocity v_x and subject to a y-directed electric field E due to plates of width L. Referring to Figure 3.26, the deflection angle θ was given by (3.51) as $\tan\theta = a_y T/v_x$, where the acceleration $ay = qE/m$ from (3.50), and the crossing time $T = L/v_x$. Combining these results, and using (10.36) with v_x for v, yields

$$\tan\theta = \frac{qE}{m}\frac{(L/v)}{v} = \frac{q}{m}\frac{EL}{v^2} = \frac{q}{m}\frac{B^2L}{E}. \qquad (10.37)$$

A measurement of θ, L, E, and B then yielded q/m. Thomson found that cathode rays—from *all* cathode materials—yielded only one value for q/m, which he called $-e/m$, where e/m was about 1.75×10^{11} C/Kg. This suggested that only a single type of object came from all cathodes.

Shortly after Thomson's discovery of the unique value of q/m for cathode rays, Fitzgerald remarked that "we are dealing with free electrons in these cathode rays." It took a few more years to convince skeptics that the word *electron* (invented by Stoney in 1894 to describe the charge associated with ions) actually corresponds to a real object. This is known as Thomson's e/m experiment.

This q/m ratio was much larger in magnitude (by a factor of 10^4 or more) than the range of values obtained from Faraday's law of electrolysis. (A range of values of q/m occurs in electrolysis because ions vary in both m and q, and q/m is much smaller for ions because m for ions is much greater than for electrons.) For example, Cu^{2+} yields $q/m = 3.02 \times 10^6$ C/Kg.

10.6.5 *Motion in a Flaring $\vec{B}$ Field: The Magnetic Mirror*

Optional

The complete theory of this effect, whereby a moving charged particle is contained by a flaring $\vec{B}$ field, is complex. However, some qualitative considerations can be used that make the point quite clear. See Figure 10.23. If the field is gently flaring over the orbit of the particle, and if the particle's velocity is mostly normal to the field, its orbit is nearly circular. The force on the particle, integrated over an orbit, then is similar to the force on the current loop in the audio speaker problem of Example 10.7, where there is a net force on the current loop. This force can either speed up the particle or slow it down; if it slows down the particle enough, the particle can reverse direction, yielding the magnetic mirror effect.

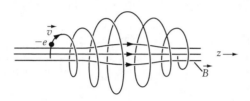

Figure 10.23 Electron with velocity $\vec{v}$ subject to the nonuniform magnetic field $\vec{B}$, causing motion in a spiral of changing radius and pitch p. If the field is strong enough, and the velocity is low enough, the electron can be confined within a "magnetic mirror."

Consider a charged particle moving in a magnetic field $\vec{B}$ with $B_z = B\cos\theta > 0$ and a component $B\sin\theta$ that flares radially out of the xy-plane, as in Figure 10.17(b). Let the particle, of charge $q > 0$, circulate clockwise as viewed from the $+z$-axis, and let $v_\perp$ be the magnitude of the velocity in the xy-plane, so that the motion is nearly circular, with approximately constant radius r. The z-component of the force on the particle is then

$$F_z = -qv_\perp B \sin\theta. \tag{10.38}$$

[We neglect F_x and F_y because, integrated over an orbit, they cancel out, as in the case of the audio speaker, described by (10.29).] If $v_z > 0$ and $\theta > 0$, the force opposes v_z; this is the principle of the magnetic mirror. (A more complete theory shows that, with r the local radius of the orbit, $B_z r^2$ is a constant, so that larger B_z causes a tighter orbit, a not unreasonable result.) Note that, in Figure 10.23, in the left (right) part of the magnetic mirror θ is positive (negative), so the magnetic mirror pushes the particle to the right (left). If the velocity along the axis is not too high, the magnetic mirror can confine the particle.

Magnetic Mirror Effect

The magnetic mirror effect happens to both electrons and protons in the upper atmosphere of the earth, where they spiral back and forth from the south pole to the north and back again. There are two distinct beltlike regions (*Van Allen belts*), with mostly electrons in the outer belt and mostly protons in the inner belt. Near the poles, where the field is largest, the orbits are tightest, by (10.32). Moreover, that is where the electrons and protons get closest to gas in the lower atmosphere. Collisions with the gas "excites" the gas atoms; when they recombine, they emit light. In the northern hemisphere the *aurora borealis,* and in the southern hemisphere the *aurora australis,* are due to electron–atom recombination in the ionosphere. The electrons originate with the sun (part of the *solar wind*, which contains an equal number of protons).

10.7 **The Hall Effect**

The Hall effect, discovered in 1880, established the sign of the charge carriers for ordinary materials. The experiment is straightforward. Consider a magnetic field pointing into the paper and a current I going to the right through a thin strip of width w and thickness t into the page. By $d\vec{F} = I d\vec{s} \times \vec{B}$, the current is subject to a force that points upward. The charge carriers thus move to the top of the wire. If charge carriers are positive, the top develops a positive voltage relative to the bottom. See Figure 10.24(a). If they are negative, the top develops a negative voltage relative to the bottom. See Figure 10.24(b).

These considerations about the Hall effect apply only when there is a single type of charge carrier. The interpretation is more complex for materials with more than one type of charge carrier (e.g., a semiconductor, which has both conduction-band electrons and valence-band "holes" as its carriers—see Section 7.13.3).

10.7.1 *The Hall Voltage*

The magnetic force $q\vec{v} \times \vec{B}$ acting on a charge q moving with velocity $\vec{v}$ in a magnetic field $\vec{B}$ can be interpreted as a force $q\vec{E}_{mot}$ due to what is called the

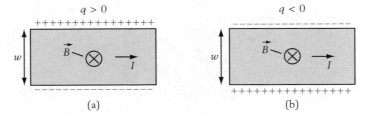

Figure 10.24 The Hall effect for rightward current I in a uniform field $\vec{B}$ that points into the page, so the force on the current is toward the upper surface. (a) Positive charge carriers, which go to the upper surface, leading to a downward Hall field $\vec{E}$. (b) Negative charge carriers, which go to the upper surface, leading to an upward Hall field $\vec{E}$.

motional electric field

$$\vec{E}_{mot} = \vec{v} \times \vec{B}, \qquad (10.39)$$

which tends to deflect the charge carriers q normal to the direction of the current flow. This drives charge to the top or bottom of the wire, and this charge produces a so-called *Hall field* $\vec{E}_H$. The current stops flowing upward when $\vec{E}_{mot}$ and $\vec{E}_H$ cancel. This leads to the same no-deflection condition (10.36) as in the Thomson e/m experiment. We write this as

$$E_H = v_d B. \qquad (10.40)$$

For positive (negative) charge carriers $\vec{E}_H$ is down (up), as in Figure 10.24. The voltage difference between the top and bottom of the wire, separated by w, is the *Hall voltage*

$$V_H = E_H w = v_d B w. \qquad (10.41)$$

For positive (negative) charge carriers V_H is higher on the upper (lower) side of the wire. See Figure 10.24.

   The voltmeter circuit for the Hall probe is not in the magnetic field. Hence the only emf driving current through the Hall probe voltmeter is the voltage difference $\int_{vm} \vec{E} \cdot d\vec{s}$ across the two sides of the wire through the voltmeter. This $\vec{E}$ is truly electrostatic; there is no $\vec{v} \times \vec{B}$ term, so we write $\vec{E}_{es}$ for $\vec{E}$. By the path independence of integrals over electrostatic fields, $\int_{vm} \vec{E}_{es} \cdot d\vec{s}$ equals the voltage difference $\int_{wire} \vec{E}_{es} \cdot d\vec{s}$ across the two sides of the wire through the wire itself, which is what we want to know. If the Hall probe has resistance R_{probe}, then a current $I_{probe} = V_H/R_{probe}$ passes through the Hall probe. Recall that a voltmeter must draw very little current to avoid disturbing the circuit under study, so that R_{probe} must greatly exceed the resistance across the strip.

   When applied to soft ferromagnets, Hall effect experiments support $\vec{B} = \mu_0(\vec{H} + \vec{M})$, rather than the $\mu_0 \vec{H}$ of magnetic poles, as the quantity that applies in the force law (10.30).

10.7.2 *Hall Resistance, Hall Voltage, and Hall Coefficent*

Optional

Recall from Chapter 7 that the current density $J = nqv_d$, where v_d is the drift velocity of the charge carriers. Thus

$$v_d = \frac{J}{nq}. \qquad (10.42)$$

Using (10.41) and (10.42), $V_H = E_H w$ takes the form

$$V_H = \frac{J B w}{nq}. \quad = V_d B w \qquad (10.43)$$

We define the *transverse*, or *Hall*, resistance as

$$R_t = \frac{V_H}{I}. \qquad (10.44)$$

Note that R_t is nonzero even for a superconductor. Hence, unlike the usual *longitudinal* resistance R, it is not dissipative in nature. (Longitudinal because there the voltage is *along* the direction of the current. We write R for what could be called R_l.) With $I = JA$, $A = wt$, and (10.43), Equation (10.44) becomes

$$R_t = \frac{JBw}{nq(JA)} = \frac{B}{nqt}. \qquad (10.45)$$

From measurements of B and t, nq can be determined, and from $q = \pm e$, the charge carrier density n can be determined. The sign of the charge carriers is determined from the sign of the voltage, as in Figure 10.24. One use of the Hall effect is to measure B. In that case, by (10.45) the factor qnt is determined from a measurement of R_t in a known B. With this calibration of qnt, a future measurement of R_t can yield B, by (10.45).

For the alkali metals (Na, K, etc.) and the noble metals Cu, Ag, and Au, the Hall voltage is negative, as in Figure 10.24(b). However, for many materials, including Co, Zn, and Fe, the Hall voltage is positive, as in Figure 10.24(a). Moreover, for semiconductors, the sign can be either positive or negative, depending on how the sample is *doped* with impurities. The fact that both signs are observed may be understood in terms of nearly—but not completely – occupied *bands* of electron orbitals in solids.

An energy band is a set of orbitals for electrons in a solid, somewhat like a shell of orbitals in an atom. When an energy band is totally occupied, it has zero net momentum. When it is missing only one electron, of momentum $\vec{p}$, the momentum of the remaining electrons is $-\vec{p}$. We can describe the system as having a *hole* of momentum $-\vec{p}$. Similarly, we can describe the system as having a hole with charge $-(-e) = e$. In some materials, electrical conduction takes place via electrons in nearly filled energy bands. These are describable in terms of holes, thus explaining the positive sign of the Hall resistance.

> **Example 10.11** **Finding the carrier density and dopant type for a semiconductor**
>
> A 120 μm thick sample of Si that has been doped either with the donor As (which gives electrons) or the acceptor Ga (which gives holes) has $V_H = -0.651$ V in a 2 T field, for $I = 0.25$ A. (a) Determine the dopant. (b) Find the Hall resistance. (c) Deduce the value of nq.
>
> **Solution:** (a) Since the Hall voltage is negative, the carriers are negatively charged, so the dopant is As. (b) From (10.44), the Hall resistance is $R_t = -2.604$ Ω. (c) From (10.45), $nq = B/tR_t = -0.64 \times 10^4$ C/m^3. With $q = -e$, this gives $n = 4.0 \times 10^{22}$ m^{-3}.

Note that the so-called *Hall coefficient*

$$R_H = \frac{E_y}{J_x B_z} \qquad (10.46)$$

takes on the value $-1/nq$ for our single charge carrier model. For Figure 10.24, $B_z < 0$; in Figure 10.24(a), $E_y < 0$; and in Figure 10.24(b), $E_y > 0$. For Example 10.11, $R_H = 1.563 \times 10^{-4}$ m^3/C.

Scientists using this technique must distinguish between three similar-sounding quantities: Hall resistance, Hall voltage, and Hall coefficient. Only the last depends on the intrinsic properties of the sample.

10.7.3
Optional
The Quantum Hall Effect: A Quantized Resistance

In two-dimensional geometries, when analyzing the Hall resistance of (10.45), we should use n_s, the number per unit area of charge carriers, rather than nt. Then we expect, with $q \to e$,

$$R_t = \frac{B}{en_s}. \qquad \text{(Hall resistance in two dimensions)} \qquad (10.47)$$

In 1980, a study of the Hall resistance R_t for charge carriers in a two-dimensional geometry found that R_t, at a fixed magnetic field B, stayed at a constant value even when the apparent density n_s was varied over a wide range of values. See Figure 10.25(a). It did not change when the temperature was changed, or when the sample was changed. It was found, quantitatively, that the Hall resistance was quantized in fractions of the quantity

$$R_0 = \frac{h}{e^2} = 25812.806\ \Omega, \qquad \text{(quantum of resistance)} \qquad (10.48)$$

where Planck's constant $h = 6.626 \times 10^{-34}$ J-s. Specifically, R_t remained at one of the plateau values

$$R_t = \frac{R_0}{i}, \qquad (i = \text{integer}) \qquad (10.49)$$

where i is an integer. The presence of Planck's constant in R_o is an indication of quantum effects, wherein the electrons must be described in terms of orbitals. In Chapter 7, we mentioned other quantum effects in the ordinary electrical resistance.

The phenomenon described by (10.49), and known as the integral *quantum Hall effect* (QHE), earned Klaus von Klitzing the 1985 Nobel Prize in physics.

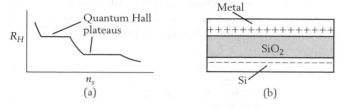

Figure 10.25 Quantum Hall effect. (a) For the quantum Hall effect regime, Hall coefficient as a function of the charge carrier density n_s per unit area. (b) Side view of an experimental sample that displays the Hall effect. The Hall effect occurs within the Si, which has an excess of electrons, taken from the metal on the other side of the insulating SiO_2 layer.

The charge carriers in his experiments were electrons at the surface of a sample of very pure Si, attracted to (but separated from) a positive metallic plate by a layer of insulating material. See Figure 10.25(b). By changing the voltage of this tiny capacitor, the amount of charge associated with it could be varied. The fact that the apparent value of n_s does not seem to change at the plateau values of R_t is an indication that under certain conditions the additional charge does not go into conducting states. Essential to the success of these experiments were (1) very large magnetic fields (to make the cyclotron frequency high), (2) high purity of the Si (to decrease the chances of a collision with an impurity during a cyclotron orbit), and (3) very low temperatures (to decrease the chances of a collision with a thermal excitation during a cyclotron orbit). These collision-suppressing conditions are similar to the collision-suppressing high-vacuum condition that Thomson needed to observe the e/m ratio for the electron.

R_0 can be measured so accurately that it provides a resistance standard. Moreover, it also provides a method to determine the combination $R_0 = h/e^2$ of the fundamental constants e and h. Since the discovery of (10.49), quantization of R_t has been found with simple fractions in place of the integer i. Thus, both the integral and the *fractional* quantum Hall effects (FQHE) have been discovered. The 1998 Nobel Prize in physics was awarded to Horst Störmer and Daniel Tsui for their discovery of the FQHE, and to Robert Laughlin for his detailed explanations of both the QHE and the FQHE. Systems that exhibit the FQHE are quantum liquids with the property that they have a preferred uniform electron density. When an extra electron is added, rather than the density going to a larger (nonpreferred) uniform density, all the electrons adjust to give the total density three (or five, or seven, etc.) localized but movable bumps, each of which has a third (or a fifth, or a seventh, etc.) of an electron charge. The rest of the system remains at the preferred uniform density. Isn't nature mysterious and unpredictable?

10.8 On Magnetic Work

Optional

In Section 10.5.2, we showed that the work dW done by the magnetic force on an individual charge q is zero. On the other hand, when we considered the audio speaker and the electromagnetic balance in Section 10.4, the integrated magnetic force $d\vec{F} = I\,d\vec{s} \times \vec{B}$ acting on the center of mass was nonzero. Because this force can change the state of *motion* of the massive, or *ponderous*, center of mass, we called it a *ponderomotive force*, or *pmf*. The work it does is called dW_{pmf}.

10.8.1 *Work by the Ponderomotive Force and Work by the Electromotive Force*

The audio speaker and the electromagnetic balance appears to contradict the general argument of Section 10.5.2 that $dW_{mag} = 0$. However, a careful analysis leads to the conclusion that, even for these cases, the *net* magnetic work is zero. What happens is this: in addition to the work dW_{pmf} done by the ponderomotive magnetic force, there is another magnetic force that does an equal and opposite amount of work on the charge carriers. Because this force can change the state of *motion* of the *electric* charge carriers, it can be thought of as an electromotive force; in Section 10.5.2, we called it the *local electromotive force*. Let the associated work done on the charge carriers be written as dW_{emf}. Thus, we claim that

$$dW_{mag} = dW_{pmf} + dW_{emf} = 0.$$

According to this equation, if the ponderomotive force does work $dW_{pmf} = 2 \times 10^{-3}$ J (i.e., moves the wire of the electromagnetic balance), then the work done on the charge carriers is $dW_{emf} = -2 \times 10^{-3}$ J (i.e., the current through the wire must decrease).

10.8.2 *Proof That Net Magnetic Work Is Zero*

For simplicity, consider only positive charge carriers, so the charge carriers and the current are in the same direction. Let the wire have center-of-mass velocity $\vec{v}_{CM}$. (Recall that most of the wire's mass is concentrated in the background ions, not in the very light charge carriers, so $\vec{v}_{CM}$ is basically independent of how the charge carriers are moving.) Let the wire contain N charge carriers with drift velocity $\vec{v}_d$, defined relative to the material of the wire.

The *net* velocity of the charge carriers is thus

$$\vec{v} = \vec{v}_{CM} + \vec{v}_d. \tag{10.50}$$

Recall that, in a time dt, a charge q moving at velocity $\vec{v}$ moves a distance $d\vec{r} = \vec{v}dt$. Thus, as in Section 10.5, the magnetic work done on this charge is

$$dW = q\vec{v} \times \vec{B} \cdot d\vec{r} = q(\vec{v} \times \vec{B}) \cdot \vec{v}dt = 0. \tag{10.51}$$

In words, this is zero because the cross-product of $\vec{v}$ with $\vec{B}$, $\vec{v} \times \vec{B}$, is perpendicular to $\vec{v}$.

Use of (10.50) in (10.51) then gives four terms:

$$0 = q(\vec{v}_{CM} + \vec{v}_d) \times \vec{B} \cdot (\vec{v}_{CM} + \vec{v}_d)dt = q(\vec{v}_{CM} \times \vec{B}) \cdot \vec{v}_{CM}dt + q(\vec{v}_d \times \vec{B}) \cdot \vec{v}_d dt$$
$$+ q(\vec{v}_d \times \vec{B}) \cdot \vec{v}_{CM}dt + q(\vec{v}_{CM} \times \vec{B}) \cdot \vec{v}_d dt. \tag{10.52}$$

The first two terms in the second equality of (10.52) are zero, for the same reason that (10.51) is zero. Since the total is zero, the third and fourth terms must cancel. We now develop their physical interpretation.

From the third term in (10.52), when all N charge carriers in the wire are included, the work dW_{pmf} done by the ponderomotive force on the center of mass, discussed in Section 10.4, is

$$dW_{pmf} = \vec{F}_{pmf} \cdot \vec{v}_{CM}dt, \qquad \vec{F}_{pmf} \equiv Nq\vec{v}_d \times \vec{B}. \tag{10.53}$$

Here $\vec{F}_{pmf}$ is the ponderomotive magnetic force acting on the center of mass, due to the motion of the charge carriers relative to the wire. Applied to a wire of length l, $\vec{F}_{pmf}$ yields (10.22), $I\vec{l} \times \vec{B}$. To see this, in $\vec{F}_{pmf}$ replace $Nq\vec{v}_d$ by $I\vec{l}$, as in the derivation of (10.30).

From the fourth term in (10.52), when all N charge carriers in the wire are included, the work dW_{emf} done by the local electromotive force on the charge carriers is

$$dW_{emf} = \vec{F}_{emf} \cdot \vec{v}_d dt, \qquad \vec{F}_{emf} \equiv Nq\vec{v}_{CM} \times \vec{B}. \qquad (10.54)$$

Here $\vec{F}_{emf}$ is the total force on the charge carriers due to the motion of the wire. We will have more to say about this local electromotive force in Chapter 12, where we will also discuss the more general case when the magnetic field varies with time.

With the definitions of (10.53) and (10.54), the last two terms of (10.52) sum to give

$$0 = dW_{mag} = dW_{emf} + dW_{emf}. \qquad (10.55)$$

Thus, *when a current-carrying wire moves in a magnetic field that does not vary with time, the work done by the ponderomotive force on the center of mass is equal and opposite to the work done by the local electromotive force on the charge carriers.* For the electromagnetic balance, the center of mass is lifted up, but to compensate energetically, the charge carriers are slowed down. Similarly, for the audio speaker, if the center of mass is speeded up, then the charge carriers are slowed down.

Problems

10-1.1 Two long wires carry equal currents in opposite directions (as for power companies). Let them intersect normal to the page at $(0, 0.5$ m$)$ and $(0, 1.4$ m$)$, the latter carrying current into the page. See Figure 10.26. At their midpoint, the magnetic field of each has magnitude 0.028 T. (a) Find the direction and magnitude of the total field. (b) Repeat if the currents reverse direction.

Figure 10.26 Problem 10-1.1.

10-1.2 Two long wires carry current, 1 into and 2 out of the paper. Wires 1 and 2 intersect the page at $(0, 0)$ and $(0, 2$ cm$)$. At the point P $(2$ cm, 2 cm$)$, the field of each has magnitude 0.06 T. See Figure 10.27. (a) Show the directions of $\vec{B}_1$ and $\vec{B}_2$. (b) Find the direction and magnitude of the total field $\vec{B}$ at P.

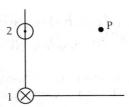

Figure 10.27 Problem 10-1.2.

10-1.3 Wire 1, carrying current I into the page, intersects the page at (a, b) and produces a field of magnitude $.04$ T at the origin. (Take $a = 2$ cm and $b = 5$ cm.) Wire 2, carrying current $2I$ out of the page, intersects the page at $(a, -b)$. Find: (a) $\vec{B}_1$, (b) $\vec{B}_2$, and (c) the total magnetic field $\vec{B}$ at the origin.

10-1.4 Wire 1 carries 7 A perpendicularly into the page, intersecting it at (3 cm, 4 cm), and producing a field of magnitude 0.035 T at the origin. Wire 2 is added, also carrying 7 A perpendicularly into the page, and also a distance 5 cm from the origin. If the total field has magnitude 0.052 T, find where the second wire can be located.

10-2.1 A current loop lies on the page, with clockwise current as seen from above. A magnet is below the page, oriented vertically, its north pole closest to the current loop. (a) Give the orientation of the magnetic moment of the magnet that is equivalent to the current loop. (b) Give the direction of the force on the current loop, due to the magnet. (c) Give the direction of the force on the magnet, due to the current loop. (d) Give the direction of the torque on the current loop.

10-2.2 A current loop lies on a table, with clockwise current as seen from above. A magnet lies to its right, its magnetic moment pointing radially outward from the current loop. See Figure 10.28. (a) Give the orientation of the magnet that is equivalent to the current loop. (b) Give the direction of the torque on the current loop, due to the magnet. (c) Give the direction of the torque on the magnet, due to the current loop. (d) Give the direction of the force on the current loop.

Figure 10.28 Problem 10-2.2.

10-2.3 Two current loops lie on a table. See Figure 10.29. The one on the left (right) has clockwise (counterclockwise) current as seen from above. (a) Replace the loops by equivalent magnets. (b) Give the direction of the force on the current loop on the right.

Figure 10.29 Problem 10-2.3.

10-2.4 Give the direction of the force on the current loop in Figure 10.30, due to the two magnets.

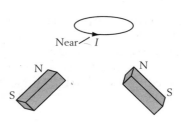

Figure 10.30 Problem 10-2.4.

10-2.5 A long wire carries current into the page, intersecting it at the origin. A horseshoe magnet (as in Figure 9.14, with the keeper removed) has its N and S poles at $(0, \pm a)$. Determine the direction of the force on the wire. *Hint:* Complete the circuit of the wire, and then think of the circuit as a magnet.

10-2.6 A wire is wound ten turns around a cylinder in one direction and eight turns in the other direction. (a) Compare its magnetic moment to that of a single turn of wire. (b) Compare if all 18 turns are in the same direction.

10-2.7 (a) Find the magnetic moment of a 24-turn rectangular loop carrying 0.76 A, of sides $\vec{a} = 0.45\hat{i} - 0.65\hat{j} + 0.18\hat{k}$ and $\vec{b} = 0.68\hat{i} - 0.23\hat{j} + 0.33\hat{k}$, with distances in m. (b) Find the torque on the loop due to a 0.034 T field along $\hat{i}$.

10-2.8 Find how much current a 35-turn square loop, 12 cm on a side, must carry in order to cancel, at a distance, the field of a magnet with moment 24.5 A-m². (Wrap the loop around the magnet.)

10-2.9 Consider two separated co-axial single-turn coils of the same radius. See Figure 10.31. (a) If they carry current in the same direction, determine the type of force (attractive or repulsive) between them. (b) Repeat for the case when the currents flow in opposite directions. (c) If the coils are identical, and are very close to each other, what does this imply about the direction of the force between two parallel current-carrying wires?

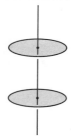

Figure 10.31 Problem 10-2.9.

10-3.1 A current loop of radius 2 mm has 12 turns. It sits on a table and carries 3 A clockwise looking down on it. It is in a 0.004 T magnetic field pointing to the right. See Figure 10.32. Find (a) its magnetic moment, (b) the torque on the loop, (c) the torque on the loop if the magnetic field tips out of the page by 23°.

Figure 10.32 Problem 10-3.1.

10-3.2 A 10-turn trapezoidal loop of sides 2 cm and 3 cm at a 48° angle to one another sits on a table and carries 400 mA counterclockwise looking down on it. It is in a 0.04 T field pointing to the left. See Figure 10.33. Find (a) its magnetic moment, (b) the magnitude of the torque on it, (c) the change in orientation energy if it could align its $\vec{\mu}$ with $\vec{B}$.

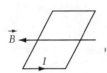

Figure 10.33 Problem 10-3.2.

10-3.3 A 200-turn solenoid 12 cm long with a 0.4-mm-by-0.4-mm square cross-section carries 3.2 A. (a) Find its magnetic moment and its pole strength. (b) Estimate the field along its axis at 3 cm beyond one end. (c) Estimate the field along its axis at 34 cm from its center. (d) Estimate the field 8 cm along its perpendicular bisector.

10-3.4 A compass needle of length 9 cm and mass 25 g is in an external magnetic field of 0.05 T. The torque on it is 8 N-m when its axis is at 70° to the magnetic field. Determine its pole strength.

10-3.5 (a) Estimate the magnetization for a hydrogen atom with angular momentum quantum $l = 1$. Use volume $V = (2a)^2$ with $a = 0.05$ nm. (b) Compare this with the magnetization of a good magnet.

10-4.1 The Ampère force law has been applied to the pumping of blood in heart–lung machines and artificial kidney machines. Treat blood as if it has a single carrier (Na$^+$) with $n = 7.5 \times 10^{25}$ m^{-3}, in addition to lots of water molecules. (a) If a 0.024 A

current flows along z transverse to the direction of arterial blood flow along y, and a 2 T field is directed along x, find the ponderomotive force driving the flow. See Figure 10.34. For calculational purposes, approximate the artery as a square of side 1.70 mm. (b) What happens if the transverse current or the field is reversed? (c) What happens if both the current *and* the field in part (a) are reversed?

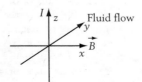

Figure 10.34 Problem 10-4.1.

10-4.2 The Ampère force law has been applied to the pumping of liquid Na coolant for nuclear reactors. (Liquid Na is a good conductor, with both electrons and Na$^+$ ions; the electrons do most of the conducting.) Consider a 20-cm-by-20-cm conduit of liquid Na forced along $+x$, with a 0.8 T field along $+z$. At some point in the conduit, two electrodes will cause an electric current to flow, normal to the direction of liquid Na flow. See Figure 10.35. (a) In what specific direction must the current flow? (b) To make a force of 260 N, how much current must flow?

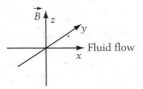

Figure 10.35 Problem 10-4.2.

10-4.3 In Figure 10.15, does the total magnetic force tend to compress or expand?

10-4.4 Find the force per unit length on a power line carrying 3.8 kA eastward in a magnetic field of magnitude 300 μT downward and magnitude 220 μT in the horizontal plane at 25° south of east.

10-4.5 A wire carrying current I is in a uniform magnetic field $\vec{B}$ that points out of the page. The current comes in from negative infinity along the x-axis, forms a semicircle of radius a in the first and second quadrants, and then continues on to positive infinity along the x-axis. See Figure 10.36. Find the force (a vector) on the semicircular part of the wire.

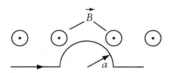

Figure 10.36 Problem 10-4.5.

10-4.6 A cylinder of radius a and length l is on a plane inclined at an angle θ to the horizontal, and subject to a $\vec{B}$ field that points vertically upward. As shown in Figure 10.37, it has N turns of wire wrapped around it, mass M, and moment of inertia $\mathcal{I}$. If the surface is rough (so there is no slipping), what current I (magnitude and direction) will keep the cylinder in static equilibrium with its turns of wire parallel to the incline?

Figure 10.37 Problem 10-4.6.

10-4.7 A coil of radius a has N turns. It lies in the xy-plane, centered at the origin, and carries a current I that is clockwise as seen from above the page. It is in a field $\vec{B} = A(2z\hat{z} - x\hat{x} - y\hat{y})$. (a) Find an algebraic expression for the total force on the coil. (b) Evaluate this numerically for $I = 3.4$ A, $a = 2.6$ cm, $N = 500$, and $A = 0.035$ T/m.

10-4.8 A 2 cm long rod carries 1.6 A along $0.36\hat{i} - 0.48\hat{j} + 0.64\hat{k}$. It is in a magnetic field $\vec{B} = -.18\hat{i} + 0.25\hat{j} + 0.32\hat{k}$. (a) Find the force on the rod. (b) Repeat if the magnetic field is rotated about the rod axis by 180°. *Hint:* It is not necessary to perform the rotation.

10-4.9 In a uniform field $\vec{B}$, a flexible conducting wire carries a current from A to C. (a) Show that the total force on the wire is independent of its shape and total length. (b) If $\vec{B} = 0.4\hat{x}$ T, $I = 3.5$ A, and the vector from A to C is along $\hat{z}$ and has length 35 mm, find the force on the flexible conducting wire.

10-4.10 A 4-cm-by-6-cm rectangular loop carrying counterclockwise current I is in a field $\vec{B}$. Although $\vec{B}$ always points into the page, $|\vec{B}|$ varies quadratically with y, being 0.04 T at the bottom ($y = 0$) and 0.38 T at the top ($y = 6$ cm) of the loop. Find (a) the force on each arm, and (b) the net force on the loop.

10-4.11 A 3-cm-by-5-cm rectangular loop carrying clockwise current I is in a field $\vec{B}$. Although $\vec{B}$ always points out of the page, $|\vec{B}|$ varies linearly with y, being 0.1 T (0.2 T) at the bottom (top) of the loop. See Figure 10.38. (a) Find the forces on the top and bottom of the loop. (b) Find the forces on the sides. (c) Find the net force and indicate whether the loop tends to compress or expand.

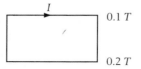

Figure 10.38 Problem 10-4.11.

10-4.12 In the crucial part of a D'Arsonval galvanometer, shown in Figure 10.39, the magnetic field does not depend on position along the symmetry axis z (perpendicular to the page). However, $\vec{B}$ does vary in the perpendicular xy-plane: from the left pole face the field points radially inward; in the soft magnetic core of the central region, the field points to the right; to the right of the central core, the field lines point radially outward to the right pole face. Thus, outside the core the field is nonuniform. Nevertheless, for a coil of magnetic moment $\vec{\mu}$ wrapped around the core (see Figure 10.39, where the current goes into the page for the lower part of the coil), $\vec{\tau} = \vec{\mu} \times \vec{B}$ applies, with $\vec{B}$ the uniform rightward field within the core. Applying the Ampère force law to each arm of the coil, taken to be a square of side a, derive this result.

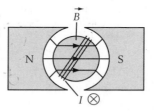

Figure 10.39 Problem 10-4.12.

10-4.13 A rail gun lies in the plane of the page, in a field $\vec{B}$ that is out of the page, and with $|\vec{B}| = 20$ T. It is designed to accelerate a mass $m = 14$ g from rest to 10^4 m/s in only 10^{-3} s. The mass is a conducting rod that can slide horizontally on rails of separation $l = 2.5$ mm, which provide the current. See Figure 10.40. Find (a) the acceleration of the mass, (b) the magnitude of the force on the mass, (c) the minimum current needed to provide this force.

Figure 10.40 Problem 10-4.13.

10-4.14 A loop of radius a, carrying current I counterclockwise as seen from $+x$, is in a magnetic field $\vec{B}$ whose local magnitude B is fixed, but whose direction always makes a flaring angle θ away from the axis of the loop. See Figure 10.41. Draw and evaluate the force $d\vec{F}$ on an element $d\vec{s}$ that is at the top of the loop. Find the component of $d\vec{F}$ along the x-axis. Integrate to obtain the total force on the loop.

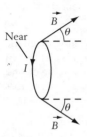

Figure 10.41 Problem 10-4.14.

10-4.15 A long magnet lies on the z-axis, its N pole (of pole strength q_m) at the origin and its S pole along $-z$. Parallel to the xy-plane, a distance r from the origin, is a semicircular loop of radius a, carrying current I clockwise as seen from $+z$. See Figure 10.42. Determine the force (including direction) on the loop, due to the N pole of the magnet.

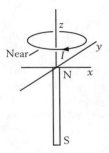

Figure 10.42 Problem 10-4.15.

10-5.1 A wire carries current into the page, intersecting the page at the origin. An electron at (2 cm, 0) has velocity making an angle of 40° clockwise from the x-axis. See Figure 10.43. (a) Indicate the direction of the force $\vec{F}$ on the electron. (b) If the field of the wire has magnitude 0.003 T and $|\vec{v}| = 6 \times 10^4$ m/s, find $|\vec{F}|$. (c) Find the acceleration in terms of $g = 9.8$ m/s^2.

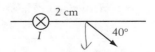

Figure 10.43 Problem 10-5.1.

10-5.2 An electron beam moving from the bottom of the page toward the top of the page deflects upward out of the page. (a) If this deflection is due to a magnetic field, in what direction is that field? (b) Repeat for an electric field.

10-5.3 For an electron of energy 2 eV, how large a magnetic field will produce a maximum magnetic force equal in magnitude to the electron's weight?

10-5.4 The screen of a cathode ray tube is dark except for a single bright dot at the center. If a wire is placed just above the top of the screen, carrying current to the right, in what direction will the dot deflect? *Hint:* Consider the electron during its path to the screen.

10-5.5 Find the force on a positron (the antiparticle of the electron, with charge $+e$ and mass m_e) with velocity $\vec{v} = (0.76\hat{i} - 0.26\hat{k}) \times 10^5$ m/s in a magnetic field $\vec{B} = 0.34\hat{i} - 0.96\hat{j} + 0.63\hat{k}$ T.

10-5.6 A proton with velocity $\vec{v} = (0.35\hat{i} - 0.86\hat{k}) \times 10^5$ m/s is subject to a net force $\vec{F} = (2.5\hat{i} + 4.7\hat{j} - 1.4\hat{k}) \times 10^{-15}$ N. Can this be due solely to a magnetic field? Explain your reasoning.

10-5.7 A magnetic field lies in the yz-plane. If the instantaneous magnetic force on an electron moving with velocity $\vec{v} = (0.24\hat{i} + 0.96\hat{k}) \times 10^6$ m/s is given by $\vec{F} = (0.44\hat{i} - 0.29\hat{j} - 0.11\hat{k}) \times 10^{-14}$ N, find the field $\vec{B}$.

10-5.8 An electron at (0,3 cm) moves at 25° counterclockwise to the y-axis. A wire carries current into the paper, intersecting the paper normally at the origin. See Figure 10.44. (a) Indicate the direction of the force $\vec{F}$ on the electron. (b) If the field of the wire has magnitude 0.004 T and $|\vec{v}| = 5 \times 10^4$ m/s, find $|\vec{F}|$.

Figure 10.44 Problem 10-5.8.

10-5.9 Derive the sign of the magnetic force law, (10.30), for $q < 0$.

10-6.1 (a) Describe the orbit of an electron that moves into the page $(-\hat{z})$, where it encounters a magnetic field that points along $\hat{y}$. (b) Repeat for a proton.

10-6.2 An electron moving in a magnetic field has a cyclotron period of 3 μs. (a) Find the magnetic field. (b) Find the magnetic field needed to make a proton have the same cyclotron period.

10-6.3 A singly ionized ^{7}Li atom, of mass 1.16×10^{-26} kg, moving along the x-axis, enters a region of uniform magnetic field $\vec{B}$ pointing along z. It then orbits in a semicircle of radius $R = 2$ cm. (a) Indicate the direction in which it deflects. (b) If $|\vec{B}| = 0.06$ T, find the velocity of the atom.

10-6.4 An electron beam with energy 8 keV moves in a circle of radius 2.4 cm. Find (a) its velocity, (b) the period of the motion, (c) the magnitude of the magnetic field it is in, (d) the magnetic force, (e) the radial acceleration.

10-6.5 Two particles both make clockwise circles in a magnetic field, one having twice the period of the other, and a larger orbit than the other. A mechanics student says the larger period must be due to the larger orbit. How do you respond?

10-6.6 Find the current associated with a proton in a cyclotron orbit when $B = 1$ T.

10-6.7 An alpha particle makes 150 revolutions before emerging from a cyclotron of radius 9.5 cm with energy 25 MeV. Find (a) the cyclotron period, (b) the magnetic field, (c) the voltage across the dees. Assume two accelerations per revolution.

10-6.8 In a mass spectrometer, an ion of charge $+Ze$ and mass M enters a region of uniform field B, and travels in a semicircle until it hits the wall a distance d from where it entered. See Figure 10.45. If the ion achieved its energy by being acceler-

ated through a voltage difference V, show that $M = (B^2 d^2 Ze/8V)$.

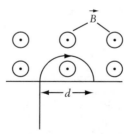

Figure 10.45 Problem 10-6.8.

10-6.9 Consider a mass spectrometer. Find the deflections of a deuteron (a ^{2}H, or deuterium, nucleus), a triton (a ^{3}H, or tritium nucleus), a ^{3}He nucleus, and a ^{4}He nucleus, if the deflection for a proton of the same kinetic energy is 1.8 cm.

10-6.10 Using crossed electric and magnetic fields that produce canceling forces, we can design a velocity selector. For a 0.4 T magnetic field, find the electric field needed to produce a 2×10^5 m/s beam.

10-6.11 A deuteron (ionized ^{2}H) moves at $v = 0.5c$ in a 12 cm radius. (a) Find the field B using the deuteron's rest mass. (b) Find the field B, using the fact that, at high speeds, the mass increases by the factor $(1 - v^2/c^2)^{-1/2}$.

10-6.12 A heavy ion has charge Ze and mass M. It is to be accelerated in a cyclotron to $v = 0.1c$, where c is the speed of light. The accelerating voltage V is 50 kV per crossing of the dees, in $B = 5$ T. (As a function of time, the accelerating voltage, ideally, is a square wave flat-top, but in practice it is a sine wave whose peak matches the particle's crossing of the dee.) (a) If the initial radius is small compared to the final radius R, find the number of trips N before the ion is ejected. (b) Evaluate N with $Z = 31$ for ^{129}Xe at 25 MeV/nucleon, with $R = 0.7$ cm. Note that a characteristic separation between the dees is 3 cm.

10-6.13 Figure 10.19 depicts a charge q and mass m in spiral motion about a field B. (a) Explain why v_B, its velocity component parallel to B, does not change during the motion. (b) Explain why $v_\perp$, the magnitude of its velocity component perpendicular to B, does not change during the motion. (c) Show that the period T to make one turn around the field direction satisfies $T = 2\pi m/qB$, just as when $v_B = 0$. (d) Show that, projected normal to B, q makes a circle of radius $R = v_\perp/2\pi$. (e) Show that the pitch p satisfies $p = v_B T$.

10-6.14 The isotope ^{12}C is quadruply ionized to become ^{12}C^{4+}, with energy 12 MeV per nucleon. It moves at a $37°$ angle to the earth's magnetic field, of local magnitude 420 μT. (a) Find the period. (b) Find the radius. (c) Find the pitch of the spiral.

10-6.15 A proton has velocity $\vec{v} = (5\hat{x} - 4\hat{y}) \times 10^5$ cm/s and is in a magnetic field $\vec{B} = (-3\hat{x} - 2\hat{y}) \times 10^{-3}$ T. (a) Find the angle between $\vec{v}$ and $\vec{B}$. (b) Determine the radius of the spiral on which the proton moves. (c) Determine the pitch of the spiral on which the proton moves. See Problem 10-6.13.

10-6.16 A magnetic field B points along $\hat{y}$ in the half-space $z < 0$. There is no magnetic field for $z > 0$. A particle with charge $q > 0$ and mass m is injected at the origin with initial velocity $\vec{v} = -v_0\hat{z}$. See Figure 10.46. (a) Describe its trajectory in the field; (b) determine how far it penetrates into the field region; (c) determine how long it remains in the field region; (d) determine where it exits the field region.

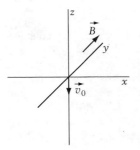

Figure 10.46 Problem 10-6.16.

10-6.17 A magnetic field B points along $\hat{y}$ in the half-space $z < 0$. There is no magnetic field for $z > 0$. A particle with charge $q > 0$ and mass m is injected at the origin with initial velocity $\vec{v} = v_0(-\hat{z} + \hat{x})$. (This case is similar to that of Figure 10.46.) (a) Describe its trajectory in the field; (b) determine how far it penetrates into the field region; (c) determine how long it remains in the field region; (d) determine where it exits the field region.

10-6.18 A magnetic field B points along $\hat{y}$ in the half-space $z < 0$. There is no magnetic field for $z > 0$. A particle with charge $q > 0$ and mass m is injected at the origin with initial velocity $\vec{v} = v_0(-\hat{z} + \hat{y})$. (This case is similar to that of

Figure 10.46.) (a) Describe its trajectory in the field; (b) determine how far it penetrates into the field region; (c) determine how long it remains in the field region; (d) determine where it exits the field region.

10-6.19 For $\vec{B}$ making an angle θ to z, a charge q moves with $v_z \ll v_\perp$ as in Figure 10.23. Let T be the period of the cyclotron orbit in (10.34). (a) Show that the effective current $I_{eff} = (q/T)$ and that $v_\perp = (2\pi r/T)$. (b) Hence show that, in (10.38), $qv_\perp = (I_{eff}T)(2\pi r/T) = I_{eff}2\pi r$. (c) Compare (10.38) with (10.29) for the audio speaker.

10-6.20 Consider a point mass m moving in a circle of radius r due to the gravitational force from a much more massive point mass M. (a) Show that $\omega = v/r = \sqrt{GM/r^3}$. (b) Show that the "year" $T = 2\pi/\omega = 2\pi r^{3/2}/(GM)^{1/2}$. Hence in this case the year depends on the radius, as determined in detail by Kepler.

10-6.21 Consider an object of mass m moving in a circle of radius r due to the force from a rope of (variable) tension F. (a) From $F = mv^2/r$, show that $\omega = v/r = \sqrt{F/mr}$. (b) Show that the "year" $T = 2\pi/\omega = 2\pi(mr/F)^{1/2}$.

10-7.1 Maxwell wrote "the mechanical force which urges a conductor carrying a current across the lines of magnetic force, acts, not on the electric current, but on the conductor which carries it. . . . The only force which acts on electric currents is electromotive force." Discuss how the Hall effect contradicts this assertion.

10-7.2 One of Maxwell's objections to the electric fluid model was that it was not possible, given the state of knowledge of his time, to tell whether there was one fluid or two (or more!), and if one fluid, its sign. He also objected that it was not possible to determine the velocity of flow of the fluid. How did the Hall effect change this situation? Relate the drift velocity (to be inferred) to measureable quantities. Can the Hall effect give the magnitude of the charge on the charge carrier?

10-7.3 A conducting wire of width 0.68 cm carries a current I to the right. A voltmeter reads that the top side is 0.3 μV higher than the bottom side. There is a uniform 0.05 T magnetic field pointing into the paper. (a) Find the sign of the charge carriers. (b) Find the drift velocity of the charge carriers.

10-7.4 A conducting rod of length 0.48 m, oriented parallel to y, is in the xy plane and is moving along the x-axis with velocity 0.2 m/s. There is a uniform 0.4 T magnetic field along the z-axis. (a) Find the motional $\vec{E}$ field acting on the rod. (b) After any transients have died down, which end of the rod is at the higher voltage, and by how much? (c) Find the electrostatic field $\vec{E}_s$ in the rod.

10-7.5 (a) Taking $n = 8.5 \times 10^{28}$ m^{-3} for Cu, find the Hall coefficient. (b) If a Cu wire has width 5.6 mm and thickness 0.35 mm, in a 0.05 T field, find the Hall resistance. (c) Find the amount of probe misalignment (along the direction of current flow) to make the ordinary resistance between the probes equal the Hall resistance.

10-7.6 Consider a conducting rod in the xy-plane at a counterclockwise angle of 20° to the x-axis. It moves along $\hat{y}$ with velocity 20 cm/s in a 0.005 T magnetic field that points along $\hat{z}$. See Figure 10.47. (a) Which end of the rod is positively charged? (b) Find the motional electric field and the electrostatic electric field. (c) Find the voltage difference between the two ends.

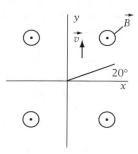

Figure 10.47 Problem 10-7.6.

10-7.7 Consider a conducting rod of length l parallel to the y-axis, moving along the x-axis with velocity v. It is in a magnetic field $\vec{B}$ along the z-axis. (a) Which side has the higher voltage? (b) Find the motional electric field and the electrostatic electric field. (c) Find the voltage difference between the two ends. (d) How do the answers change if the rod tips by θ toward the x-axis?

10-7.8 Consider a wire made of soft iron, with relative permeability of 1200. If the Hall voltage is measured to be 0.058 V, what would the Hall voltage be if the relative permeability were one? *Note:* The Hall voltage of soft magnets shows experimentally that in the magnetic force $q\vec{v} \times \vec{B}$ it is indeed $\vec{B}$ due to electric currents (both macroscopic and microscopic) that appears. An alternative and reasonable possibility, but one contradicted by experiment, would be that the magnetic force is $q\vec{v} \times \mu_0\vec{H}$, where $\mu_0\vec{H} = \vec{B} - \mu_0\vec{M}$ is due to magnetic poles and $\vec{M}$ is the magnetic moment per unit volume.

10-8.1 If $dW_{pmf} = 0.084$ J on a loop that moves in a constant field $\vec{B}$, find dW_{emf}.

10-8.2 For an ordinary metal, the ponderomotive force acts directly on the charge carriers, not the ions. Nevertheless, the wire in Figure 10.15 can move, which means that the ions move. (a) Explain how this occurs. (b) If the ratio of the ion mass to the charge carrier mass is 5.6×10^4, and the ponderomotive force is 2.40 N, find the drag force acting on the electrons.

10-G.1 Devise an experiment to determine the magnetic moment of a magnet that has been embedded in a potato.

10-G.2 Following (6.52), take $l = 1/na^2$ for the mean-free path l of a particle in a gas of number density n, where a is of the order of an atomic dimension. Taking $a = 0.2$ nm and $T = 300$ K, find the density and pressure $P = nk_BT$ needed to make $l = 10$ cm. This indicates how good a vacuum Thomson needed for the e/m measurement described in Section 10.6.4.

10-G.3 Find the force between two co-axial current loops at constant I_1 and I_2 and radii R_1 and R_2, that are far from each other. Use $U = -\vec{\mu} \cdot \vec{B}$, with $\vec{\mu}$ due to one magnet and $\vec{B}$ due to the other magnet (in the dipole limit), and $\vec{F} = -\vec{\nabla}U$ to find the force on either magnet. Make sure that in $\vec{F}$ you vary the position of the magnet for which you want the force. Verify that the result gives the correct sign.

10-G.4 In a current-carrying wire subject to a magnetic field, the local magnetic forces on the charge carriers would cause them to accelerate. (a) What force causes the charge carriers to reach terminal velocity? (b) Discuss the reaction force associated with the previous part, and argue that it must be accompanied by a force acting on the ions, which have the vast majority of the mass of the wire.

10-G.5 For the current loop in a flaring magnetic field, as in Figure 10.17(b), Elmo says that $|\vec{F}| = \Sigma_i |d\vec{F}_i| = \int |d\vec{F}_i|$. What is wrong with this? Think in terms of a sum on a spreadsheet. Note that (10.29) gives zero force for $\theta = 0$, as we expect for a uniform field, but Elmo's approach gives a nonzero value.

10-G.6 Derive $d\vec{F}/dA = \vec{K} \times \vec{B}$ for the force per unit area on a current sheet with current per unit length $\vec{K}$ in a magnetic field $\vec{B}$.

10-G.7 Derive $d\vec{F}/dV = \vec{J} \times \vec{B}$ for the force per unit volume on a volume distribution of current with current per unit area $\vec{J}$ in a magnetic field $\vec{B}$.

10-G.8 Place a small bar magnet between the poles of a large horseshoe magnet, N to N and S to S. Let the magnetic field of the large horseshoe magnet be uniform. (a) Is there a net force on the small magnet? (b) Does the small magnet tend to compress or expand? (c) If the small magnet is replaced by a small current loop, is there a net force on the current loop? (d) Does the loop tend to compress or expand?

10-G.9 Consider a circuit carrying a current I. If the wires are jiggled without disturbing the contacts, give at least two reasons (one discussed in Chapter 7 and one discussed in this chapter) why this might change the current or voltage reading for the circuit.

10-G.10 Let a 40 eV electron move perpendicularly to a field $B = 2 \times 10^{-10}T$, appropriate to interstellar space. Take the hydrogen atom number density to be $n \approx 1/m^3$. (a) Find the electron's cyclotron frequency and cyclotron period. (b) Find its velocity and orbit radius about the field. (c) Estimate the mean-free path for collisions of the electron with hydrogen atoms, and the characteristic number of cyclotron orbits the electron will make before it collides with a hydrogen atom. *Hint:* See Problem 10-G.2.

10-G.11 The earth has a magnetic moment of magnitude. Given that the earth's geographic north is a magnetic south, if this magnetic moment is due to a current around the equator, find the magnitude and direction of this current.

"The word "electromagnetic" [has been] used to characterize the phenomena produced by the conducting wires of the voltaic pile. I have determined to use the word "electrodynamic" in order to unite under a common name all of these phenomena [of the sort that Oersted discovered], and particularly to designate those which I have observed between voltaic conductors."

—André Marie Ampère (1820)

"These attractions and repulsions between electric currents differ fundamentally from the effects produced by electricity in repose."

—André Marie Ampère (1822)

Chapter 11

How Electric Currents Make Magnetic Fields: The Biot–Savart Law and Ampère's Law

Chapter Overview

There are three major parts to this chapter: the Biot–Savart Law, Ampère's Law, and applications of these laws to superconductors and electromagnets. Section 11.1 gives a brief introduction to this chapter and a brief history of the discovery of how electric currents make magnetic fields. The first part begins with Section 11.2, which states the Biot–Savart law, giving the magnetic field due to any current-carrying circuit (somewhat like Coulomb's law for the electric field due to electric charge). Section 11.3 derives the Biot–Savart law using Ampère's equivalence. Section 11.4 shows how to use the Biot–Savart law. Section 11.5 applies it, using the principle of superposition. Section 11.6 finds forces due to these magnetic fields derived from the Biot–Savart law. This leads to the definition of the *ampère* in terms of the force between two parallel, current-carrying wires.

The second part begins with Section 11.7, which states Ampère's law. Section 11.8 derives Ampère's law using Ampère's equivalence, which relates the circulation around an arbitrary closed circuit and whatever current may pass perpendicularly through it. Such an arbitrary closed circuit is called an *Ampèrian circuit*. This relationship is somewhat like Gauss's law relating electric flux to the charge enclosed by an arbitrary closed surface, called a *Gaussian surface*. Section 11.9 shows how Ampère's law can be used to make a noninvasive measurement of the current passing through an Ampèrian circuit. Section 11.10 obtains the magnetic fields of very symmetrical current sources by application of Ampère's law.

The third part is optional. Section 11.11 discusses *field expulsion* by perfect diamagnets (which are superconductors). Here, surface currents make a magnetic field that, within the diamagnet, cancels any externally applied magnetic field. Section 11.12

discusses soft magnets, and how they intensify applied magnetic fields with their Ampèrian currents. Section 11.13 discusses two experiments establishing that the electric currents associated with perfect diamagnetism flow over macroscopic paths, whereas those associated with magnets flow over microscopic current paths. In both cases, the currents do not decay with time, so they must flow without resistance. Section 11.14 discusses electromagnets, which are used to produce very large magnetic fields. Superconductors are also used to produce very large magnetic fields. ∎

11.1 Introduction

This is the third and final chapter on static magnetic fields. Chapter 9 studied magnetic materials and used an analogy to electrostatics to study the properties of magnets. Chapter 10 showed how Ampère's equivalence between magnets and current loops yields the force and the torque on current-carrying wires in magnetic fields. The present chapter discusses how electric currents make magnetic fields: the Biot–Savart law (pronounced *bee-oh-suh-var*) and Ampère's law. Applications run from the microscopic (the magnetic field produced at the atomic nucleus, due to electrons within the atom) to the macroscopic (the earth's magnetic field, due to electric currents deep within the earth). More important for daily life, these laws also have been used on the human scale to design motors, electromagnets, write-heads for magnetic memory, and a vast array of other devices. In the late 1800s, electrical engineering departments were founded to teach students how to develop and exploit such applications. Because there is so much material, we organize it into three parts.

11.1.1 *Some History*

Arago, while traveling, learned of Oersted's work. On returning to France, he described Oersted's experiments on the interaction between a magnet and a current-carrying wire to the Académie des Sciences on September 11, 1820. From this, Ampère deduced that two current-carrying wires would also interact. Within a week, he had confirmed this result, showing that two parallel wires carrying current in the same (opposite) direction attract (repel). On October 30, Biot and Savart presented the results of their experiments on the interaction of a magnet and a long straight wire. They showed that the direction of the magnetic field of a long straight wire is as given by what we have called Oersted's law, and they showed that the field falls off with radial distance r from the wire as r^{-1}. They went on to study the magnetic field produced when the wire was bent into two semiinfinite straight sections making an angle of 2α ($2\alpha = \pi$ being a straight wire). Ampère pointed out two errors in their initial theoretical analysis of their experimental results. However, the mathematical physicist Laplace, using their data, deduced the correct form of the law for the field produced by any section of the wire; this is what we know as the Biot–Savart law. Ampère's work, both experimental and theoretical, consisted of studies of the interaction between two current-carrying wires, and led to the Ampère force law of the previous chapter. What is called Ampère's law actually may be due to Maxwell (in the 1860s), rather than Ampère (in the 1820s).

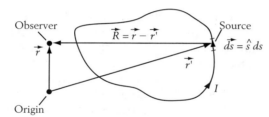

Figure 11.1 Geometry for the Biot–Savart law giving the magnetic field $\vec{B}$ at $\vec{r}$ due to a current-carrying wire, with vector length elements $d\vec{s}$ at $\vec{r}'$. Here $\vec{R} = \vec{r} - \vec{r}'$ is the vector pointing from the source at $\vec{r}'$ to the observer at $\vec{r}$.

11.2   Magnetic Field of a Current-Carrying Wire

11.2.1   *Statement of the Biot–Savart Law*

The Biot–Savart law tells us how to compute the magnetic field $\vec{B}$ at the observation point $\vec{r}$ due to an electric circuit–carrying electric current I (i.e., a current loop). First some definitions. As usual, $\vec{R} = \vec{r} - \vec{r}'$ is the vector pointing from the source at $\vec{r}'$ to the observer at $\vec{r}$. Define $R = |\vec{R}|$ so that $\hat{R} = \vec{R}/R$. The current loop consists of segments of length ds that point along $\hat{s}$ defined by the local direction of the electric current I. Thus the vector length element is $d\vec{s} = \hat{s}\,ds$, where $ds = |d\vec{s}| > 0$. See Figure 11.1.

The Biot–Savart law states that

$$\vec{B} = \int d\vec{B}, \qquad d\vec{B} = k_m I \frac{d\vec{s} \times \hat{R}}{R^2} = k_m I \frac{d\vec{s} \times \vec{R}}{R^3},$$

$$k_m \equiv 10^{-7} \frac{\text{T-m}}{\text{A}}. \quad \text{(Biot–Savart law)} \tag{11.1}$$

(The second form for $d\vec{B}$ follows from $\hat{R}/R^2 = \vec{R}/R^3$.) Except for the vector cross-product $d\vec{s} \times \hat{R}$, (11.1) is not much more complicated than the expression for the electric field due to electric charges distributed along a line. Nevertheless, because of the cross-product, it is worth examining in more detail.

11.2.2   *Numerical Analysis*

In part because of the vector cross-product, students often find the Biot–Savart law to be complicated and difficult to apply. In the spirit of "know your enemy," it is useful to analyze the Biot–Savart law from the viewpoint of a spreadsheet analysis, imagining how you would calculate $\vec{B}$ if you had some source-measuring elves who would approximate it by many tiny elements of length. The elves would then determine the midpoint $\vec{r}'$ of each source element, and the vector $d\vec{s}$ along which the current flows for that element. In addition, they would measure the observation position $\vec{r}$. Alternatively, they could measure the four quantities ds and $\hat{s}$, and then $d\vec{s} = \hat{s}\,ds$ could be computed within the spreadsheet. However, we have chosen to make it easier on them, since $d\vec{s}$ involves only three quantities. To be specific, let the index i on the elements go from 1 to 28. See Figure 11.2.

In what follows, refer to Table 11.1, which shows the first two rows of a spreadsheet calculation. The first column (unlabeled) would contain the entries 1 to 28, one for each source element. Columns A to C contain the three components of the source position vectors $\vec{r}'$. Columns D to F contain the three components of the source element vector $d\vec{s}$. A separate part of the spreadsheet stores the three components of the observer vector $\vec{r}$, as well as k_m and I. Once the input entries have been made, the computations begin.

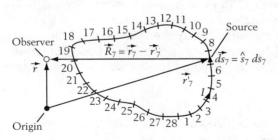

Figure 11.2 Discretized version of Figure 11.1, with circuit broken up into 28 elements.

Columns G to I contain the components of the vector $\vec{R} = \vec{r} - \vec{r}'$ from the source point to the observation point. For example, $R_x = r_x - r'_x$. Column J contains the length $R = |\vec{R}|$, obtained by the Pythagorean theorem with the entries in columns G to I. Columns K to M contain the components of $d\vec{s} \times \vec{R}$. Finally, columns N to P contain the components of $d\vec{B} = k_m I (d\vec{s} \times \vec{R})/R^3$. Thus, summing column N yields B_x; similarly, column O yields B_y, and column P yields B_z.

Computational versus Analytical

From a computational point of view, it is preferable to work with $d\vec{B} = k_m I (d\vec{s} \times \vec{R})/R^3$, rather than $d\vec{B} = k_m I (d\vec{s} \times \hat{R})/R^2$, since then we do not need to have separate columns for $\hat{R}$ and $d\vec{s} \times \hat{R}$. However, for analytic work the inverse square form is often preferable. Note that $\vec{r}'_{i+1} = \vec{r}'_i + d\vec{s}_i$ connects consecutive entries of $\vec{r}'$, so the relationship $d\vec{r}'_{i+1} = \vec{r}'_{i+1} - \vec{r}'_i = d\vec{s}_i$ holds.

Example 11.1 Calculation of $d\vec{B}$

In the preceding circuit, let $I = 4$ A, $\vec{r} \equiv (-0.5, 2.2, 0)$ m, $\vec{r}'_7 \equiv (7, 2, 0)$ m, and $d\vec{s}_7 \equiv (-0.2, 0.4, 0)$ m. Determine $d\vec{B}_7$.

Solution: $\vec{R}_7 = \vec{r} - \vec{r}'_7 \equiv (-7.5, 0.2, 0)$ m, from which $|\vec{R}_7| = 7.503$ m and $d\vec{s}_7 \times \vec{R}_7 \equiv (0, 0, 2.96)$ m^2. Hence

$$d\vec{B}_7 = \frac{k_m I d\vec{s}_7 \times \vec{R}_7}{R_7^3} = 2.80 \times 10^{-9} \text{ T}\hat{k}. \tag{11.2}$$

Columns N to P in the seventh row successively receive the three entries corresponding to $(dB_x, dB_y, dB_z) = (0, 0, 2.80 \times 10^{-9})$ T.

Table 11.1 Spreadsheet calculation of $d\vec{B}$

	A	B	C	D	E	F	G	H	I	J	K	L	M	N	O	P
1	r'_x	r'_y	r'_z	ds_x	ds_y	ds_z	R_x	R_y	R_z	$|\vec{R}|$	$(d\vec{s} \times \vec{R})_x$	$(d\vec{s} \times \vec{R})_y$	$(d\vec{s} \times \vec{R})_z$	dB_x	dB_y	dB_z

This procedure is complicated but completely well defined, and it works even when the circuit is so complex that analytic methods fail. Once the source entries are made in columns A to F, and the observer entries are made elsewhere, all the other results can be obtained completely automatically by the spreadsheet. If the observation point is changed, only the observer position need be changed.

Sections 11.4 and 11.5 present some situations that do not require numerical methods. First, however, we derive the Biot–Savart law using Ampère's equivalence.

11.3 Derivation of Biot–Savart Law: Field of Current-Carrying Wire

To find the field $\vec{B}$ at the position $\vec{r}$, due to a current loop I, we first use the definition of $\vec{B}$ in terms of the force $\vec{F}_{q_m}$ on a magnetic pole q_m placed at $\vec{r}$. See Figure 11.3.

Thus

$$\vec{F}_{q_m} = q_m \vec{B}. \tag{11.3}$$

Now note that, by action and reaction, the force $\vec{F}_I$ on a circuit due to the monopole q_m is equal and opposite $\vec{F}_{q_m}$:

$$\vec{F}_{q_m} = -\vec{F}_I. \tag{11.4}$$

Combining (11.3) and (11.4) yields

$$\vec{B} = \frac{\vec{F}_{q_m}}{q_m} = -\frac{\vec{F}_I}{q_m}. \tag{11.5}$$

Hence, from the force $\vec{F}_I$ on the current loop, we can deduce $\vec{B}$. For example, if $q_m = 2$ A-m and $|\vec{F}_I| = 0.08$ N, then $|\vec{B}| = 0.04$ T. Now consider the general case.

From Ampère's force law, (10.20), the force $\vec{F}_I$ on the current-carrying loop in the field $\vec{B}_{q_m}$ due to the monopole is

$$\vec{F}_I = \oint d\vec{F} = I \oint d\vec{s} \times \vec{B}_{q_m}. \tag{11.6}$$

Note that the vector $-\vec{R} = -(\vec{r} - \vec{r}\,')$ points from $\vec{r}$ to $\vec{r}\,'$. Then (9.4) gives the

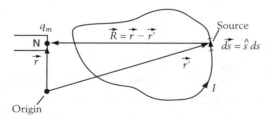

Figure 11.3 Geometry for proof of the Biot–Savart law. Here a long magnet, with nearby pole q_m, is acted on by the current loop, and vice versa.

field $\vec{B}_{q_m}$ at position $\vec{r}$, due to q_m at $\vec{r}\,'$, as

$$\vec{B}_{q_m} = \frac{k_m\, q_m(-\hat{R})}{R^2}. \qquad (\vec{R} = \vec{r} - \vec{r}\,') \qquad (11.7)$$

Placing (11.7) into (11.6) yields

$$\vec{F}_I = I \oint d\vec{s} \times \frac{k_m\, q_m(-\hat{R})}{R^2}, \qquad (11.8)$$

Finally, using (11.8) in (11.5) gives the Biot–Savart law for the field $\vec{B}$ at $\vec{r}$ due to the current loop I:

$$\vec{B} = k_m I \oint \frac{d\vec{s} \times \hat{R}}{R^2} = k_m I \oint \frac{d\vec{s} \times \vec{R}}{R^3}. \qquad (\vec{R} = \vec{r} - \vec{r}\,') \qquad (11.9)$$

For surface currents, we use $\vec{K}dA$ rather than $Id\vec{s}$ (where $\vec{K}$ includes the direction of the surface current K), and we integrate over the element of area dA. Similarly, for volume currents, we must use $\vec{J}dV$ rather than $Id\vec{s}$, and we must integrate over the element of volume dV.

11.4 Applications of the Biot–Savart Law

11.4.1 *Circuit in Plane, Observer in Plane*

Figures 11.1 through 11.3 depict a circuit in a plane and an observer in that same plane. At the observer, the field is perpendicular to the plane. This follows immediately because, when $d\vec{s}$ and $\hat{R}$ are in the same plane, their cross-product $d\vec{s} \times \hat{R}$ is perpendicular to the plane. In our earlier numerical calculation of (11.2), the field was along $\hat{z}$. This simplifies the problem of finding $\vec{B}$ because we only have to add up the dB_z column (P) with nonzero entries. The B_x and B_y columns sum to zero because each dB_x and dB_y is zero.

11.4.2 *Field at Center of Circular Current Loop*

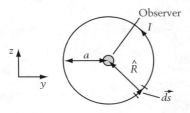

Figure 11.4 Geometry for magnetic field at the center of a circular current loop of radius a.

Circular current loops are an important and commonly encountered geometry. Consider a loop of radius a in the yz-plane, centered at the origin, with its normal along $\hat{x}$, which we take to be out of the page. Let the observer be at the center of the loop; this is a specific example of the previous general case of the observer in the plane of the circuit. See Figure 11.4. With the current counterclockwise as viewed from the positive x-axis, $d\vec{s} \times \hat{R}$ points along $\hat{x}$, and $|d\vec{s} \times \hat{R}| = |d\vec{s}||\hat{R}|| \sin 90°| = (ds)(1)(1) = ds$.

By (11.1) with $R = a$, all source $d\vec{s}$'s produce $dB_x = k_m I \, ds/a^2$. With $\oint ds = 2\pi a$, this integrates to

$$B_x = \frac{2\pi k_m I}{a}. \qquad \text{(field at center of circular loop)} \qquad (11.10)$$

If $I = 2$ A and $a = 2$ cm $= .02$ m, then (11.10) gives $B_x = 6.28 \times 10^{-5}$ T, approximately the strength of the earth's field. Ten turns of wire (all wound in the same direction) would give ten times this field.

Example 11.2 **Field on axis of circular current loop**

Consider the same loop as in Figure 11.4, but let the observer be anywhere on the axis (x) of the loop. See Figure 11.5. The current is counterclockwise as viewed from the positive x-axis. At point P on the current loop, $d\vec{s}$ points along $\hat{y}$, $\hat{R}$ lies in the xz-plane, and the angle between $d\vec{s}$ and $\hat{R}$ is $90°$. (a) For the observer, indicate the direction of the total field $\vec{B}$. (b) For the $d\vec{B}$ produced by $I d\vec{s}$ at P, indicate which components are nonzero. (c) Compute the total field $\vec{B}$ expressed in terms of the current. (d) Check your result at $x = 0$. (e) Compute the total field $\vec{B}$ expressed in terms of the magnetic moment μ. (f) Check your result for large x/a. (g) Evaluate B_x for $I = 1$ A and $a = 4$ cm $= 0.04$ m, at $x = 3$ cm $= 0.03$ m.

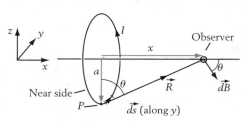

Figure 11.5 Geometry for magnetic field on the axis of a circular current loop of radius a.

Solution: (a) By rotational symmetry about z, B_y and B_z are zero. (Thinking of this as a magnet, by Ampère's equivalence the field points along the axis of the loop.) (b) As indicated in Figure 11.5, $d\vec{B}$ has nonzero components along x and z. In general, all three components are nonzero. (c) Because only B_x is nonzero, we need only consider $dB_x = |d\vec{B}| \cos\theta$, where $\cos\theta = a/R = a/(x^2 + a^2)^{\frac{1}{2}}$. By (11.1), $|d\vec{B}| = k_m I |d\vec{s} \times \hat{R}|/R^2$. Also,

$$|d\vec{s} \times \hat{R}| = |d\vec{s}||\hat{R}||\sin 90°| = ds(1)(1) = ds,$$

where $ds = |d\vec{s}|$. Thus

$$dB_x = |d\vec{B}| \cos\theta = \frac{k_m I}{R^2} ds \frac{a}{R} = \frac{k_m I (ds) a}{R^3}.$$

Since R is a constant, the integral yields

$$B_x = \frac{k_m I a}{R^3} \oint ds = \frac{k_m I a}{R^3} 2\pi a = \frac{2\pi k_m I a^2}{(x^2 + a^2)^{\frac{3}{2}}}. \qquad \text{(field on axis of loop)}$$

$$(11.11)$$

(d) For $x = 0$, so $R = a$, (11.11) gives $B_x = 2\pi k_m I/a$, in agreement with (11.10).

(e) Using $\mu = IA = I\pi a^2$, (11.11) takes the form

$$B_x = \frac{2k_m\mu}{(x^2 + a^2)^{\frac{3}{2}}}. \qquad \text{(field on axis of a circular loop)} \qquad (11.12)$$

(f) For large x, so $R \to x$, (11.11) and (11.12) go to $B_x = 2k_m\mu/x^3$. This is as expected for a dipole, by Ampère's equivalence; see (9.8). For comparison, far along the axis of an electric dipole of moment p, $E_x = 2kp/x^3$; see (3.12).
(g) For our current loop, $\mu = IA = 0.503$ A-m^2. At a distance $x = 3$ cm $= 0.03$ m, so $\sqrt{x^2 + a^2} = 0.05$ m, (11.12) gives $B_x = 8 \times 10^{-6}$ T, about a fifth of the earth's magnetic field. Ten turns of wire wound in the same direction will increase the μ and the $\bar{B}$ by a factor of ten.

This example has some important generalizations. (1) By taking two identical co-axial coils, with equal currents, and by spacing them just right, it is possible to make a field that is very uniform in the vicinity of their midpoint. This simple method to produce a nearly uniform field along the x-axis is due to Helmholtz. (2) By taking two identical co-axial coils, with equal and opposite currents, and by spacing them just right, it is possible to make a field that is zero at the midpoint and has a very uniform slope along the x-axis in the vicinity of their midpoint. MRI often employs magnetic fields with very uniform slopes. (3) By winding a wire around a long cylinder, we can simulate a long set of equally spaced co-axial coils, with equal currents. As indicated in Chapter 10, this is known as a solenoid. It is useful because it produces a uniform field everywhere within the solenoid. (In fact, to produce a uniform field, the cross-section need not be circular so long as it is the same all along the axis of the tube.)

Example 11.3　Field on axis of disk-shaped magnet

Consider a magnetic disk of thickness $l = 2$ mm, radius $a = 2$ cm, and with $M \approx 1.0 \times 10^6$ A/m (the remanent magnetization M_r of the permanent magnetic material NEO). See Figure 11.6. Find the field a distance $x = 3$ cm along its axis.

Solution: By Ampère's current loop decomposition (Section 10.2), a finite current loop can be decomposed into many tiny current loops. By Ampère's equivalence, at a distance we may consider each tiny current loop to be a tiny magnet. These magnets add up to a thin magnetic disk with the same perimeter and thickness as the current loop of Figure 11.5. Since the disk really is thin ($l \ll a$), (11.12) for the current loop applies. For the parameters given, $\mu = MV = M(\pi a^2 l) = 2.51$ A-m^2. Then at $x = 3$ cm, (11.12) gives $B_x = 0.0107$ T.

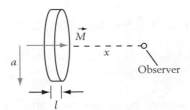

Figure 11.6 Geometry for magnetic field on the axis of a thin, circular magnetic disk of radius a, magnetized normal to the slab. With $\mu = IA = I(\pi a^2)$, this gives the same field as for the current loop of Figure 11.5.

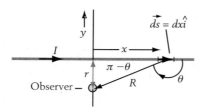

Figure 11.7 Geometry for magnetic field due to an infinitely long current-carrying wire.

Example 11.4   Field due to long current-carrying wire

Long wires are an important and commonly encountered geometry. Let a wire be along the x-axis, carrying current along $+\hat{x}$, so $d\vec{s} = \hat{x}\,dx$. Let the observer be a distance r below the origin. See Figure 11.7. (a) Find the direction and magnitude of the field due to the wire. (b) Evaluate this for $I = 10$ A and $r = 1$ cm $= 0.01$ m.

Solution: Figure 11.7 shows that the circuit and observer are in the same plane. By Section 11.4.1, the field is perpendicular to this plane. Let θ be the angle between $\hat{x}$ and $\hat{R}$ (to the observer). From Figure 11.7, if $x \to -\infty$, then $\theta \to 0$, and if $x \to +\infty$, then $\theta \to \pi$. Also from Figure 11.7, $d\vec{s} \times \hat{R}$ points along $-\hat{z}$, and $d\vec{s} \times \hat{R} = -\hat{z}dx\sin\theta$, where we have $ds = dx$. To evaluate the integral in (11.1), use θ as the variable so that x and R must be expressed in terms of θ and r. Since $\cot(\pi - \theta) = -\cot\theta$,

$$x = r\cot(\pi - \theta) = -r\cot\theta, \qquad dx = -r\frac{d}{d\theta}(\cot\theta)d\theta = r\csc^2\theta\,d\theta,$$

and $R^2 = x^2 + r^2 = r^2\csc^2\theta$. Then (11.1) becomes

$$B_z = k_m I \int \frac{(d\vec{s} \times \hat{R})_z}{R^2} = -k_m I \int_{-\infty}^{\infty} \frac{dx\sin\theta}{R^2}$$

$$= -k_m I \int_0^{\pi} \frac{r\csc^2\theta\,d\theta\sin\theta}{r^2\csc^2\theta} = -k_m I \int_0^{\pi} \frac{d\theta\sin\theta}{r} = -\frac{2k_m I}{r}. \quad (11.13)$$

Since $\hat{z}$ points out of the paper, the field points into the paper, in agreement with Oersted's right-hand rule. This procedure also applies to finite lengths of wire, with appropriate changes in the limits of integration.

More generally, at a distance r from a long current-carrying wire, the field has magnitude

$$B = \frac{2k_m I}{r}. \qquad \text{(field for long, thin wire)} \qquad (11.14)$$

Its direction is given by Oersted's right-hand rule.

11.5   Applications of the Principle of Superposition

The magnetic fields for many configurations can be determined by superimposing some of the preceding results.

Example 11.5   Field along plane midway between two long wires

This example, a reprise of the configuration of Figure 10.2(c), is related to the magnetic field produced by power lines or ordinary line cord, along their

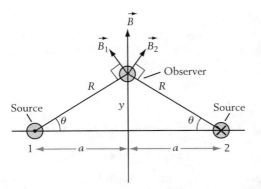

Figure 11.8 Use of superposition to find the field on the perpendicular bisector between two long wires carrying equal currents in opposite directions.

perpendicular bisector. Let each wire carry current I, one into and one out of the paper. See Figure 11.8, where we use Oersted's right-hand rule for their directions. (a) Find the total field due to these wires. (b) Find the equivalent magnet. (c) Evaluate the field due to power lines carrying $I = 2000$ A, with wire separation $a = 0.8$ m, at a distance $|y| = 9$ m. (d) Evaluate the field due to an electric blanket carrying $I = 2$ A with wire separation $a = 2$ mm at a distance $|y| = 2$ cm.

Solution: (a) By symmetry, $|\vec{B}_1| = |\vec{B}_2|$. Now add the fields vectorially. If the wires are at $\pm a$ along the x-axis, then each is a distance $R = \sqrt{y^2 + a^2}$ away. $(\vec{B}_1)_y$ is smaller than $|\vec{B}_1|$ by a factor of $\cos\theta = a/R$, and similarly for wire 2. The x-components cancel. Thus, using (11.14) for their magnitudes, with $r \to R$,

$$\vec{B} = \vec{B}_1 + \vec{B}_2 = 2\hat{y}|\vec{B}_1|\cos\theta = 2\hat{y}\left(\frac{2k_m I}{R}\right)\frac{a}{R} = \hat{y}\frac{4k_m I a}{y^2 + a^2}. \qquad (11.15)$$

(b) The circuit of Figure 11.8 can be converted to an infinite magnetic slab geometry on using (11.15), with $I \to Ml$. See Figure 11.9, where Ampèrian surface currents $K = I/l = M$ must circulate around the outside of the magnet. Magnets of this shape (but not magnetized normal to their plane) are used to seal refrigerator doors.

(c) For the power lines, (11.15) gives $|\vec{B}| = 7.84 \times 10^{-6}$ T. (d) For the electric blanket, (11.15) gives $|\vec{B}| = 3.96 \times 10^{-6}$ T. These fields are about a tenth of the earth's magnetic field. By twisting a pair of wires carrying equal and opposite current, the net magnetic field can be decreased even more. This is the origin of the term *twisted pair*, often used to describe wires employed for communications purposes.

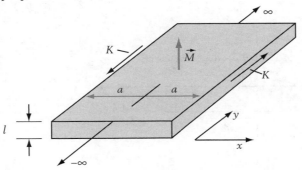

Figure 11.9 Infinite magnetic slab magnetized normal to the slab. It is equivalent to the two wires of Figure 11.8.

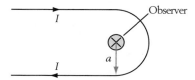

Figure 11.10 Current-carrying hairpin. It can be decomposed into two semi-infinite current-carrying wires and a current-carrying semicircle.

Example 11.6   Field at center of current-carrying hairpin

In Figure 11.10, find the field at the observer's position.

Solution: We decompose the hairpin into two semi-infinite wires and a half-loop of radius a. See Figure 11.10. The field due to the half-loop is half that due to the full loop, given by (11.10). Its direction is into the paper. In addition, the two semi-infinite wires each produce half the effect of a full wire, together yielding the same effect, (11.14), as a full wire. Their field also points into the paper. Thus the net field is into the page, with magnitude

$$B = \frac{1}{2}\frac{2\pi k_m I}{a} + \frac{2k_m I}{a} = \frac{k_m I(\pi + 2)}{a}. \qquad (11.16)$$

Example 11.7   Field due to current sheet

Let a current sheet lie in the xz-plane, with current into the paper $(-\hat{z})$, and current per unit length K measured along x. See Figure 11.11. Find the magnetic field above and below the current sheet.

Solution: A thickness dx of the sheet is like a subwire with current $I \to dI = K\,dx$ directed into the paper. We will add up the effects of each of these subwires. Oersted's right-hand rule gives the direction of the field $d\vec{B}$ of the subwire, as shown in Figure 11.11. For an observer along the y-axis, the distance to the subwire, of thickness dx, is $R = \sqrt{y^2 + x^2}$. By (11.14), $|d\vec{B}|$ due to the subwire is

$$|d\vec{B}| = \frac{2k_m K\,dx}{R}. \qquad (11.17)$$

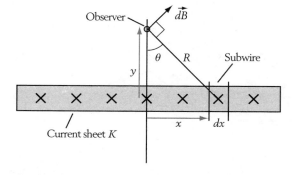

Figure 11.11 Cross-section of infinite current sheet carrying current per unit length K, where the length is measured along the x-direction.

Two variables appear: x and θ. Let us work with θ, using the fact that y is a constant. Then $x = y \tan\theta$, so $dx = y \sec^2\theta \, d\theta$ and $R^2 = y^2 + x^2 = y^2 + y^2 \tan^2\theta = y^2 \sec^2\theta$. Then

$$dB_x = |d\vec{B}|\frac{y}{R} = \frac{2k_m K y dx}{R^2} = \frac{2k_m K y dx}{y^2 + x^2} = \frac{2k_m K y^2 \sec^2\theta \, d\theta}{y^2 \sec^2\theta} = 2k_m K d\theta.$$

This integrates on θ from $-\pi/2$ to $+\pi/2$ to yield (with $B_x \to B$)

$$B = 2\pi k_m K. \qquad \text{(field of current sheet)} \qquad (11.18)$$

By symmetry, this is the only nonzero component of the field. By Oersted's right-hand rule, $\vec{B}$ is along $\hat{x}$ for $y > 0$, and along $-\hat{x}$ for $y < 0$. In each of these regions, $\vec{B}$ is uniform—that is, it does not weaken as one moves away from the current sheet.

Uniformity is often desireable. By connecting a single ribbon of conductor as in Figure 11.12(a), a region of uniform field can be produced. Let us neglect edge effects and superimpose the fields of the top and bottom of the ribbon. Then, using (11.18), the total field within the ribbon has magnitude $B = 4\pi k_m K$. To produce $B = 2 \times 10^{-4}$ T then requires that $K = 1.592 \times 10^2$ A/m= 1.592 A/cm. The direction of the field within the conducting ribbon is in agreement with what we would expect using Ampère's right-hand rule.

We will shortly show that a uniform field can also be produced by the cylindrical current sheet of Figure 11.12(b), which also gives $B = 4\pi k_m K$. As indicated earlier, this solenoidal geometry often is realized by a wire wound as in Figure 11.12(c). This can be thought of as a superposition of a current sheet (as in Figure 11.12b) that produces a field along the axis, and of a long wire that produces a field that circulates about the axis. As a consequence, the field lines do not close on themselves, but spiral out to infinity. (In Figure 11.12c we do not draw the complex spiral for $\vec{B}$.)

11.6 Forces on Magnets and Current-Carrying Wires

Up to this point, the discussion has been only of the magnetic fields produced by electric currents. We now consider the forces that such fields produce on magnetic poles and current-carrying-wires.

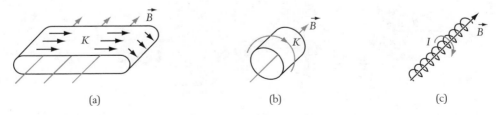

(a)   (b)   (c)

Figure 11.12 Sequence of related geometries with the same current per unit length K that produce the same uniform field within them: (a) bent-around ribbon, (b) cylindrical sheet, (c) solenoid with n turns per unit length and current I, where $K = nI$.

Biot–Savart Law

The Biot–Savart law originally was developed to find the magnetic field of a long wire, in order to explain the torque on a permanent magnet (a compass needle). Ampère's work was more ambitious: to determine the force between two current-carrying circuits. By theoretical reasoning, he found a form for $\vec{B}$, which we do not present, that enforced action and reaction between every pair of current elements in the two circuits. Although this form correctly gives the *net* force between any current element and a closed current-carrying circuit, it is not accepted today. Although the Ampère force law and the Biot–Savart law do not enforce action and reaction between every pair of current elements in the two circuits, overall momentum is conserved when the momentum of the electromagnetic field (Section 15.10) is included.

To obtain the force on circuit #1 due to circuit #2, we use the Biot–Savart law to find $\vec{B}_2$ everywhere on circuit #1 due to circuit #2, and then use the Ampère force law in the form, $d\vec{F}_1 = I_1 d\vec{s}_1 \times \vec{B}_2$, to find the net force on circuit #1. We now apply the Biot–Savart law to find both the force on a magnet due to a current loop and the force between two long wires.

11.6.1 *Force on a Monopole along Axis of Current Loop*

Consider the force on a pole q_m of a long narrow magnet, at the observer's position in Figure 11.5. Then (11.3) for F_{q_m} and (11.12) for B_x give that the force is along the x-axis, with magnitude

$$F_{q_m} = q_m B_x = \frac{2k_m q_m \mu}{(x^2 + a^2)^{3/2}}. \tag{11.19}$$

This has the same magnitude as, but is opposite to, the force on the current loop due to the magnetic pole, discussed in the previous chapter as an example of a flaring magnetic field.

11.6.2 *Force between Two Parallel Wires*

Consider a pair of overhead wires that carry current to and from a power station. See Figure 11.13(a). What is the magnetic force between these wires, if

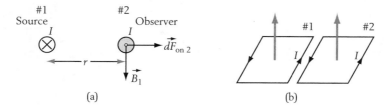

(a)   (b)

Figure 11.13 (a) Two parallel wires carrying current in opposite directions. Their interaction is repulsive. (b) Two closed-circuit versions of the parallel wires, showing their magnetic moments, which lead to repulsion.

they carry equal currents $I_1 = I_2 = I$ in opposite directions? Let us first obtain qualitative information about this interaction. Figure 11.13(b) illustrates two adjacent rectangular current loops carrying I circulating in the *same* direction. Their nearest arms, like the wires in Figure 11.13(a), carry current in opposite directions, and dominate the interaction between the two current loops. Using Ampère's equivalence, these circuits are equivalent to two magnets oriented in the same direction so that the force will be repulsive. Thus antiparallel currents repel; correspondingly, parallel currents add.

We now determine the force on the wire to the right due to the field of the wire to the left. Both wires are normal to the paper, one passing through the origin with current into the paper, and the other coming out of the paper at $x = r$. By (11.14), the field due to wire #1 on the left has magnitude $|\vec{B}_1| = 2k_m I_1/r$ and points down the page (along $-\hat{y}$). Now apply the Ampère force law, in the form $d\vec{F}_2 = I_2 d\vec{s}_2 \times \vec{B}_1$, to wire #2 on the right, with $d\vec{s}$ pointing along the current (out of the paper, along $\hat{z}$). Since $d\vec{s}_2$ and $\vec{B}_1$ are perpendicular, their cross-product $d\vec{s}_2 \times \vec{B}_1$ has magnitude $|d\vec{s}_2||\vec{B}_1| \sin 90° = (ds_2)(B_1)$. Thus $dF \equiv |d\vec{F}_2| = I_2(ds_2)B_1$. Moreover, $d\vec{F}_2 = I_2 d\vec{s}_2 \times \vec{B}_1$ points to the right, by the vector cross-product right-hand rule or by $\hat{z} \times (-\hat{y}) = \hat{x}$. The force is repulsive. Finally, the force per unit length on circuit #2 is given by

$$\frac{dF}{ds} = I_2 B_1 = I_2 \frac{2k_m I_1}{r} = \frac{2k_m I_1 I_2}{r}. \tag{11.20}$$

For $I_1 = I_2 = 100$ A and $r = 1$ cm, (11.20) gives $dF/ds = 0.2$ N/m.

11.6.3 *Defining the Ampère: SI Units*

If the two wires carry current in the same direction, the force per unit length has the same magnitude as (11.20), but instead of being repulsive, it is attractive. The international standard defining the ampere arises from this expression for the force. Placing two long wires a meter apart, and adjusting the current I in each until the force per unit length is 2×10^{-7} N/m, gives by definition a current of 1 ampere. This electromagnetic definition is the basis of the SI unit of current, and thus the basis of the SI unit of charge, for which k must be measured. Numerous other approaches have been taken to define the unit of charge. In electrostatic-cgs units, $k = 1$ serves to define the unit of electric charge, and k_m must be measured. In magnetostatic-cgs units, $k_m = 1$ serves to define the unit of magnetic pole strength, and k must be measured. We will work only with SI units, but be aware of the existence of these other, equally valid, sets of units.

11.6.4 *Magnetic Pressure*

Optional

A current-carrying circuit is subject to self-stresses that tend to expand it. Consider, for example, the ribbon of conductor in Figure 11.12(a). Neglecting edge effects, the field on the lower surface produced by the upper surface is given by (11.18). In $d\vec{F} = I d\vec{s} \times \vec{B}$, with $I d\vec{s}$ replaced by $\vec{K} dA$, the force on an area dA of the lower surface becomes $d\vec{F} = \vec{K} dA \times \vec{B}_l$. As for two wires carrying current in opposite directions, this force is repulsive. Since $\vec{K}$ and $\vec{B}$ are perpendicular, the force per unit area is given by $dF/dA = KB_l$. This field B_l is half of the total field $B = 4\pi k_m K$. Hence $dF/dA = \frac{1}{2}KB = B^2/8\pi k_m$. This can be interpreted as a magnetic pressure

$$P_{mag} = \frac{B^2}{8\pi k_m}. \qquad \text{(magnetic pressure)} \qquad (11.21)$$

For $B = 1$ T, (11.21) gives $P_{mag} \approx 4 \times 10^5$ N/m^2, or about four atmospheres.

Although it was derived only for the case of a ribbon conductor, (11.21) is true more generally. Note that for the ribbon conductor, the field lines are parallel to the ribbon. They can be thought of as exerting pressure on one another. In (11.21), letting $k_m \to k$ and $B \to E$ gives the corresponding pressure between electric field lines, in agreement with Section 6.10 on electric flux tubes.

External magnetic fields tend to orient a circuit so that the field of the circuit points along the applied field (as for a current loop, thought of as a magnetic sheet). In addition, self-fields tend to make a circuit expand (i.e., pressure), to enclose the maximum amount of magnetic flux. (Think of a square circuit, where opposite sides repel.) Hence, a flexible circuit of fixed length in a uniform field $\vec{B}$ will form a circle whose enclosed area has its normal aligned with $\vec{B}$.

You've now finished the discussion of the Biot–Savart law, which is part one of the chapter. This is a good place to take a break. Then we will continue with Ampère's law (of magnetic circulation), which is part two of the chapter.

11.7 Statement of Ampère's Law

Ampère's law relates the magnetic circulation around a closed path, called an Ampèrian circuit, to whatever current passes through the area defined by that closed path. *Any closed path can be used with Ampère's law, just as any closed surface can be used with Gauss's law.* Ampèrian circuits are purely imaginary. They can be made to correspond to real electric circuits, but such Ampèrian circuits are rarely useful. In Section 11.7.1, we discuss how Ampèrian circuits can be applied.

11.7.1 *Magnetic Circulation Γ_B*

Magnetic circulation Γ_B, for a given closed path defined by a set of directed line elements $d\vec{s} = \hat{s}\,ds$, is a measure of the "swirliness" of the magnetic field around that path. Here $ds = |d\vec{s}| > 0$. Specifically, Γ_B is defined as

$$\Gamma_B = \oint \frac{d\Gamma_B}{ds}\,ds \equiv \oint \vec{B}\cdot\hat{s}\,ds = \oint \vec{B}\cdot d\vec{s}, \qquad \frac{d\Gamma_B}{ds} \equiv \vec{B}\cdot\hat{s}.$$

$$\text{(magnetic circulation)} \quad (11.22)$$

Here $d\Gamma_B/ds \equiv \vec{B}\cdot\hat{s}$, the component of $\vec{B}$ along the path direction $\hat{s}$, is the *magnetic circulation per unit length*. It has units of T, whereas the magnetic circulation has units of T-m.

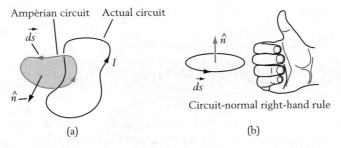

Figure 11.14 (a) An Ampèrian circuit that encloses an actual, current-carrying circuit. (b) The *circuit-normal right-hand rule* relating the direction of circulation $d\vec{s}$ around, and the normal $\hat{n}$ to, an Ampèrian circuit.

For a given Ampèrian circuit, the magnetic field due to a magnetic pole q_m has zero net magnetic circulation. This follows by analogy to electrostatics, where by (5.35) the electric field due to a charge q has zero circulation for any circuit. However, if electric current in an actual physical circuit passes through that same Ampèrian circuit, the magnetic field $\vec{B}$ due to an electric current can have nonzero magnetic circulation. See Figure 11.14(a), where the closed path we refer to is the Ampèrian circuit. Ampère's law implies that a magnetic field $\vec{B}$ due to electric currents differs from a magnetic field $\vec{B}_{q_m}$ due to magnetic charges.

For the current in Figure 11.14(a), by Oersted's right-hand rule the magnetic field circulates counterclockwise. Let the magnetic circulation be calculated for the Ampèrian circuit of Figure 11.14(a), which encloses this current. With $d\vec{s}$ counterclockwise (clockwise), by (11.22) the circulation is positive (negative).

11.7.2 *Current Enclosed I_{enc}*

From Chapter 7, the current I_{enc} passing through the Ampèrian circuit defined by $d\vec{s}$ is given by an integral over the associated cross-section:

$$I_{enc} = \int dI = \int \frac{dI}{dA} dA = \int \vec{J} \cdot \hat{n} \, dA. \tag{11.23}$$

The relationship between the normal $\hat{n}$ in (11.23) and the Ampèrian circuit element $d\vec{s}$ in (11.22) is given by the circuit-normal right-hand rule.

Circuit-normal right-hand rule: *Curl the fingers on your right hand in the direction of $d\vec{s}$; your thumb will then point along $\hat{n}$.* See Figure 11.14(b). For the Ampèrian circuit of Figure 11.14(a), $\hat{n}$ would tend to point toward you, and the cross-section is the hatched region. The current is positive when it is in the same direction as your thumb.

11.7.3 *Ampère's Law*

With all this defined, we now state Ampère's law:

$$\Gamma_B = 4\pi k_m I_{enc}, \quad \text{or} \quad \oint \vec{B} \cdot d\vec{s} = \int \vec{J} \cdot \hat{n} dA, \quad \text{(Ampère's Law)} \quad (11.24)$$

with $d\vec{s}$ and $\hat{n}$ related by the circuit-normal right-hand-rule of Figure 11.14(b). Typically, in using Ampère's law, we select an area through which current flows, and the perimeter defines the Ampèrian circuit. Thus, the actual physical circuit, which provides I_{enc}, typically passes through the Ampèrian circuit.

11.7.4 *Uses of Ampère's Law*

Recall that Gauss's law, which employs Gaussian surfaces, has three primary uses: (1) noninvasive measurement of the charge Q_{enc} within a closed surface; (2) relationship between surface charge density σ_S and the normal component $\vec{E}_{out} \cdot \hat{n}$ of the electric field just outside a conductor in equilibrium (for which $\vec{E}_{in} = \vec{0}$ inside); (3) determination of the electric field $\vec{E}$ when the charge distribution is so symmetrical that it produces a uniform electric flux density.

Similarly, Ampère's law, which employs Ampèrian circuits, has three primary uses: (1) noninvasive measurement of the current I_{enc} through a closed circuit (Section 11.9); (2) relationship between surface current density $\vec{K}$ and the transverse component $\hat{n} \times \vec{B}_{out}$ of the magnetic field just outside a perfect diamagnet, for which $\vec{B}_{in} = \vec{0}$ inside (Section 11.11); (3) determination of the magnetic field $\vec{B}$ when the current distribution is so symmetrical that it produces a uniform magnetic circulation per unit length (Section 11.10).

But before using Ampère's law, first we derive it using Ampère's equivalence.

11.8 Derivation of Ampère's Law for Magnetic Circulation: The Magnetic Shell

The derivation is performed in steps. First we consider a trivial case. Next we introduce the concept of the magnetic shell. Then we outline the strategy of the proof, using a numerical example. Finally, we give the proof itself, followed by some additional applications of the magnetic shell concept.

11.8.1 *Circuit Not Enclosing Current*

Consider an Ampèrian circuit for which $I_{enc} = 0$, as in Figure 11.15(a). If Ampère's law holds, then by (11.24), $\Gamma_B = 0$. Let us derive this result.

By Ampère's equivalence, at a distance from the current loop, the field $\vec{B}$ produced by the current loop is equivalent to the field $\vec{B}_m$ produced by the magnetic poles q_m of a distribution of equivalent magnets. Because the corresponding electric field produced by electric dipoles has zero circulation for this and any other circuit, the magnetic circulation for this circuit is also zero. That is,

$$\Gamma_B = \oint \vec{B} \cdot d\vec{s} = \oint \vec{B}_m \cdot d\vec{s} = 0. \quad \text{(distant sources)} \quad (11.25)$$

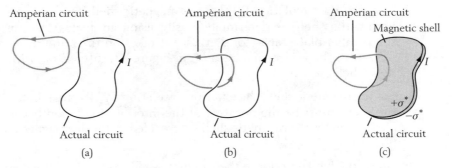

Figure 11.15 (a) An Ampèrian circuit that does not enclose an actual, current-carrying circuit. (b) An Ampèrian circuit that encloses an actual, current-carrying circuit. (c) An Ampèrian circuit that pierces the magnetic shell that is equivalent to an actual, current-carrying circuit.

11.8.2 *Ampère's Magnetic Shell*

Now consider the magnetic circulation for an Ampèrian circuit with $I_{enc} \neq 0$, as in Figure 11.15(b). This can be obtained using another idea of Ampère, called a *magnetic shell*. It has positive poles on one side and negative poles on the other, and is a generalization of the magnetic disk of Section 11.4.4. The argument bears repeating.

First, decompose the actual current loop, which can be quite irregular, into a set of tiny equivalent circuits, as discussed in Section 10.2. (This decomposition is not unique, so we are free to employ a useful one.)

Second, use the equivalence, at a distance, between each tiny current loop and a tiny magnet. Then add up the tiny magnets, which, placed side by side, simulate the actual circuit of Figure 11.15(a) or Figure 11.15(b). This yields the *magnetic shell* of Figure 11.15(c), of small and unspecified thickness l; shortly, we will take the limit $l \to 0$. Like the current loop, the magnetic shell can be irregular. See Figure 11.15(c), where the top of the magnetic shell is positive and the bottom is negative, by Ampère's right-hand rule.

We now find the magnetic moment that the magnetic shell must have to produce the same field as the actual current loop. From $\mu = IA$ of Chapter 10 and $\mu = q_m l = \sigma_m Al$ of Chapter 9, the magnetic moment per unit area is

$$\frac{\mu}{A} = \sigma_m l = I, \qquad (11.26)$$

just as in (10.7). Thus $\sigma_m l = I$, for all l, no matter how small. Equivalently, from Chapter 10 $K = M = \sigma_m$, so $Kl = I = \sigma_m l$. Thus the current loop has been replaced by the magnetic shell.

11.8.3 *Strategy for Proving Ampère's Law*

We now show that, in Figure 11.15(c), integration of $d\Gamma_{B_m}$ only through the shell from the bottom to the top equals, in Figure 11.15(b), the integral of $d\Gamma_B$ for *all* the Ampèrian circuit.

1. The interior circulation $\Gamma_{B_m}^{int}$ for the magnetic shell (taken from $-\sigma_m$ to $+\sigma_m$ within the shell) is determined easily using an analogy to our old friend the parallel-plate capacitor. Let's say it takes on the numerical value $\Gamma_{B_m}^{int} = -50$ T-m, the minus sign because in Figure 11.15(c) $\vec{B}_{m,int}$ within the magnetic shell opposes $d\vec{s}$.

2. For the magnetic shell, the exterior circulation $\Gamma_{B_m}^{ext}$ (from $+\sigma_m$ to $-\sigma_m$ outside the shell) must be the negative of the interior circulation because the total circulation for the magnetic shell is zero. Hence, for the magnetic shell, $\Gamma_{B_m}^{ext} = 50$ T-m.

3. $\vec{B}_{m,ext}$, the field exterior to the magnetic shell, equals $\vec{B}_{ext}$, the field exterior to the current loop. Thus the exterior circulation $\Gamma_{B_m}^{ext}$ for the magnetic shell (see Figure 11.15c) and the exterior circulation Γ_{B}^{ext} for the current loop (see Figure 11.15b) must be the same. Hence, for our example, $\Gamma_{B}^{ext} = 50$ T-m.

4. For the current loop, the exterior circulation Γ_{B}^{ext} is essentially all the magnetic circulation Γ_B; it neglects only terms proportional to l, which go to zero as $l \to 0$. Hence, for our example, $\Gamma_B = 50$ T-m.

Working backward from 4 to 1, this argument is equivalent to the sequence of equalities

$$\Gamma_B = \Gamma_B^{ext} = \Gamma_{B_m}^{ext} = -\Gamma_{B_m}^{int}. \tag{11.27}$$

11.8.4 *Proof of Ampère's Law*

Let's now obtain $\Gamma_{B_m}^{int}$ algebraically. We must integrate over $\vec{B}_{m,int}$ for the magnetic shell, taking the integral from $-\sigma_m$ to $+\sigma_m$. This is like taking the integral of the electric field $\vec{E}$ within a capacitor from $-\sigma$ to $+\sigma$. There $E_{int} \equiv |\vec{E}_{int}| = 4\pi k\sigma$, and the voltage difference (electric circulation) for plate separation d is $E_{int}d = 4\pi k\sigma d$. Here $B_{m,int} = 4\pi k_m \sigma_m$ within the magnetic shell. Moreover, as noted above, since for our path $d\vec{s}$ opposes $\vec{B}_{m,int}$,

$$\Gamma_{B_m}^{int} = -B_{m,int}l = -4\pi k_m \sigma_m l = -4\pi k_m I. \tag{11.28}$$

Hence, applying (11.28) to (11.27) yields

$$\Gamma_B = 4\pi k_m I. \tag{11.29}$$

This is Ampère's law, which, written out in even more detail than (11.24), is

$$\oint \vec{B} \cdot d\vec{s} = 4\pi k_m I_{enc} = 4\pi k_m \int \vec{J} \cdot \hat{n} dA. \tag{11.30}$$

As usual, the circuit-normal right-hand rule, given in Figure 11.14, relates $d\vec{s}$ and $\hat{n}$.

We have only proved Ampère's law for a single current-carrying wire. However, by the principle of superposition, if we add other wires, their circulation adds to the left-hand side of (11.30), and their I_{enc} adds to the right-hand side.

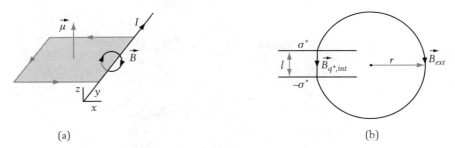

(a)   (b)

Figure 11.16 (a) Long wire and finite circuit that has a common arm with that wire, showing the field and the equivalent magnetic moment $\vec{\mu}$ of the circuit. (b) Field outside the magnetic sheet that is equivalent to the finite circuit of part (a). Imagine that the sheet thickness l is infinitesimal.

11.8.5 *Field of a Long, Thin Current-Carrying Wire*

The magnetic shell can be used to find the already known $\vec{B}$ field of an infinitely long current-carrying wire I. Consider this wire to be part of a large rectangular circuit with one side fixed, whose other sides can be chosen at an arbitrary angle. For simplicity, let the shell lie in the xy-plane for positive y, so the equivalent magnetic moment points along the $+z$-axis. See Figure 11.16(a), which shows a finite rectangular circuit. By Oersted's right-hand rule, the magnetic field must circulate about the axis of the wire, taken to be y. We will now find the magnetic field due to the wire by finding the exterior magnetic field of the magnetic shell.

We determine $\int \vec{B} \cdot d\vec{s}$ for two paths starting on the positive side of the magnetic shell, and ending on the negative side. See Figure 11.16(b), where the shell thickness is finite for clarity. One is an exterior path of radius r, chosen symmetrically, so that the circulation per unit length, $d\Gamma_B/ds = \vec{B}_{ext} \cdot \hat{s}$, is uniform along the path, if the infinitesimal thickness l of the shell is neglected. Then $\int \vec{B} \cdot d\vec{s}$ gives $B_{ext}(2\pi r)$. The other path goes through the shell from positive to negative, and is the negative of (11.28), or $4\pi k_m I$. We thus conclude that $B_{ext}(2\pi r) = 4\pi k_m I$, or $B_{ext} = 2k_m I/r$. With $B_{ext} \to B$, this is the same as (11.14).

11.8.6 *Fringing Field of a Capacitor*

The equivalence of the magnetic shell and a long wire can be turned into the electrical problem of the fringing field of a parallel-plate capacitor. For distances much larger than the plate separation, a parallel-plate capacitor looks very much like the electrical equivalent of a magnetic shell. See Figure 11.17(a).

From Ampère's equivalence, the current I of Figure 11.16(a) goes to $Kl = Ml = \sigma_m l$ of Figure 11.16(b). Thus $(B, k_m, I) \to (B, k_m, \sigma_m l)$. Making the analogy to electricity, we let $\sigma \to \sigma_m$, so $(B, k_m, \sigma_m l) \to (E, k, \sigma l)$. Then from $B = 2k_m I/r$ for the long wire, the analogy yields

$$E_{ext} = \frac{2k\sigma l}{r} \qquad (11.31)$$

for the fringing field of a closely spaced capacitor. See Figure 11.17(b). Note that

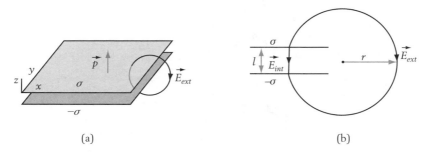

Figure 11.17 (a) Parallel-plate capacitor, showing fringing field. (b) Side view of parallel-plate capacitor, showing fringing field that is analogous to that of Figure 11.16(b).

within such a capacitor $E_{in} = 4\pi k\sigma$, so (11.31) becomes

$$E_{ext} = \frac{E_{in}l}{2\pi r} = \frac{\Delta V}{2\pi r}. \tag{11.32}$$

If $\Delta V = 100$ V and $l = 0.001$ m, for $r = 0.02$ m (11.32) gives $E_{ext} = 796$ V/m, compared with $E_{int} = |\Delta V|/l = 10^5$ V/m. P. Heller has verified (11.32) experimentally.

11.9 Ampère's Law Implies That Circulation Yields Current

If we can compute or measure the circulation [the left-hand side of Ampère's law, (11.24)] around some Ampèrian circuit, then we can deduce [by the right-hand side of (11.24)] the current I_{enc} passing through that circuit, including its direction of flow. This is the basis of the Rogowski coil and the clip-on ammeter.

Example 11.8   **Using the circulation**

Consider an irregularly shaped Ampèrian circuit of length $L = 12$ cm (e.g., a circuit shaped like the Ampèrian circuit in Figure 11.15b). For $d\vec{s}$ circulating counterclockwise, let $\Gamma_B = 0.042$ T-cm. Find (a) how $\vec{B}$ circulates (on average); (b) the average component of $\vec{B}$ along the circuit; (c) the direction of I_{enc}; and (d) the magnitude of I_{enc}.

Solution: (a) Since the circulation is positive, $\vec{B}$ and $d\vec{s}$ must circulate in the same direction, on average. Thus both $d\vec{s}$ and $\vec{B}$ circulate counterclockwise. (b) The component of $\vec{B}$ along the circuit is $\vec{B} \cdot \hat{s} = d\Gamma_B/ds$. Hence, the average value of $\vec{B} \cdot \hat{s}$ is the average value of $d\Gamma_B/ds$, which equals Γ_B divided by its length L, or $(0.042$ T-cm$)/(12$ cm$) = 0.0035$ T. (c) Since $\vec{B}$ circulates counterclockwise, Oersted's right-hand rule tells us that I_{enc} points out of the page, as in Figure 11.15b. (d) Finally, use of Ampère's law as $\Gamma_B = 4\pi k_m I_{enc}$ gives $I_{enc} = \Gamma_B/4\pi k_m = 3.34 \times 10^2$ A. Thus we have determined I in Figure 11.15. If the sign of the circulation were negative, the current would reverse in direction.

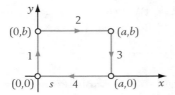

Figure 11.18 Ampèrian circuit for the computation of the magnetic circulation, and the subsequent calculation of the current I_{enc} that passes through it.

Example 11.9 **Computing the circulation**

Consider a magnetic field

$$\vec{B} = (2 - x^2)\hat{j}, \qquad (11.33)$$

where x in cm gives $\vec{B}$ in 10^{-4} T. Take as the Ampèrian circuit a rectangle in the $z = 0$ plane, with corners $(0,0)$, $(a,0)$, (a,b), $(0,b)$. Let $a = 7$ cm and $b = 4$ cm, and consider clockwise circulation. See Figure 11.18. Determine the circulation for this Ampèrian circuit, and determine the amount and direction of current flow through it.

Solution: To find the total circulation, consider each arm separately.

1. The first arm goes from $(0,0)$ to $(0,b) = (0,4$ cm$)$, so $d\vec{s} = \hat{j}dy$, with $ds = dy > 0$. Then

$$\vec{B} \cdot d\vec{s} = (2 - x^2) \times 10^{-4}dy = 2 \times 10^{-4}dy,$$

since $x = 0$ for that arm. It gives a contribution of

$$2 \times 10^{-4} \int_0^b dy = (2 \times 10^{-4} \text{ T})(4 \text{ cm}) = 0.08 \times 10^{-4} \text{ T-m}$$

to the total circulation.

2. The second arm goes from $(0,b) = (0,4$ cm$)$ to $(a,b) = (7$ cm, 4 cm$)$ so that $d\vec{s} = \hat{i}dx$, with $ds = dx > 0$. This $d\vec{s}$ is perpendicular to $\vec{B}$, so it gives zero contribution to the total circulation.

3. The third arm goes from $(a,b) = (7$ cm,4 cm$)$ to $(a,0) = (7$ cm,0$)$ so that $d\vec{s} = -\hat{j}ds$, where $ds = -dy > 0$, from the limits of integration. Thus $\vec{B} \cdot d\vec{s} = (2 - x^2)dy = -47 \times 10^{-4}dy$, since $x = a = 7$ cm for that arm. It gives a contribution of

$$(-47 \times 10^{-4}) \int_b^0 dy = (-47 \times 10^{-4} \text{ T})(-4 \text{ cm}) = 1.88 \times 10^{-4} \text{ T-m}$$

to the total circulation.

4. The fourth arm goes from $(a,0) = (7$ cm,0$)$ to $(0,0)$ so that $d\vec{s} = -\hat{i}ds$, where $ds = -dx > 0$, from the limits of integration. This $d\vec{s}$ is perpendicular to $\vec{B}$, so it gives zero contribution to the total circulation.

Summing all four terms gives a total circulation $\Gamma_B = 1.96 \times 10^{-4}$ T-m. By (11.24), it must equal $4\pi k_m I_{enc}$, where I_{enc} is positive into the paper for a positive clockwise circulation. Thus $I_{enc} = \Gamma_B/4\pi k_m = 156$ A; we have non-invasively determined I_{enc}, both in magnitude and in its sense relative to the page.

Application 11.1   **Measuring the circulation: Rogowski coils and clip-on ammeters**

Devices known as *Rogowski coils* surround a current-carrying wire and permit the current to be read *noninvasively* (i.e., without breaking the circuit to include an ammeter). Although they have a small gap, they may be thought of as closed Ampèrian circuits. (Electric current can only circulate around a real circuit that is closed, but a magnetic field can circulate around an Ampèrian circuit that is nearly closed.) Rogowski coils do the equivalent of measuring the circulation Γ_B. They are calibrated, by Ampère's law, (11.30), to display the current I_{enc}. For example, a Rogowski coil could surround the current loop in Figure 11.15(b), like the Ampèrian circuit in that figure. A more common but less precise device, which measures a characteristic magnetic field due to the wire, is the *clip-on ammeter:* it clips on (properly, around) a current-carrying wire. The Hall voltage yields I_{enc} for *dc* clip-on ammeters, and the emf from Faraday's law—to be discussed in the next chapter—yields I_{enc} for *ac* clip-on ammeters. P. Murgatroyd has developed Rogowski coil devices that measure local current densities.

11.10   Applications of Ampère's Law and Symmetry

Gauss's law can be used to determine $\vec{E}$ where the electric charge distribution has a high degree of symmetry (spherical, cylindrical, or planar) and is known. Similarly, Ampère's law can be used to determine $\vec{B}$ where the electric current distribution has a high degree of symmetry and is known.

To see this, note that for a circuit of length s, the average circulation per unit length is given by

$$\overline{\frac{d\Gamma_B}{ds}} = \overline{\vec{B} \cdot \hat{s}} = \frac{\oint \vec{B} \cdot d\vec{s}}{s} = \frac{4\pi k_m I_{enc}}{s}. \qquad (11.34)$$

When the circulation per unit length is uniform along the circuit, or part of a circuit, the average circulation per unit length is the same as the local circulation per unit length and both equal $|\vec{B}|$. Both then are given by (11.34).

Example 11.10   **Field inside a wire with uniform current density**

Consider a long, straight wire of radius a carrying a uniform current density $\vec{J}$ into the paper, with $J = |\vec{J}|$. See Figure 11.19. It carries total current $I = J(\pi a^2)$, so $J = I/(\pi a^2)$. Use Ampère's law and symmetry to find the magnetic field inside the wire $(r < a)$.

Solution: By Oersted's right-hand rule, for $r > a$ the magnetic field $\vec{B}$ will circulate clockwise around the center of the current distribution. Moreover, by

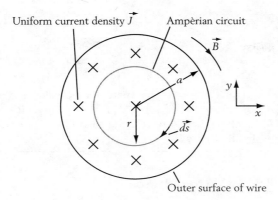

Uniform current density $\vec{J}$

Ampèrian circuit

$\vec{B}$

Outer surface of wire

Figure 11.19 Cross-section and field $\vec{B}$ of a long wire of radius a with uniform current per unit area. The current is into the page and area is measured normal to the page.

the circular symmetry of the current distribution, $|\vec{B}|$ should depend only upon r. This also should be true for $r < a$. Let $\hat{\theta}$ represent the clockwise tangential direction. Then $\hat{s} = \hat{\theta}$ and $\vec{B} = B\hat{\theta}$, where $B = |\vec{B}|$. Because the problem has so much symmetry, we can choose an Ampèrian circuit in Figure 11.19 for which the circulation per unit length $d\Gamma_B/ds = \vec{B} \cdot \hat{s} = B\hat{\theta} \cdot \hat{\theta} = B$ has the same value all along the Ampèrian circuit. Such Ampèrian circuits are circles concentric with the current distribution. (A nonconcentric circle wouldn't give a uniform circulation per unit length, nor would a noncircular circuit.) The total circulation, with path length $2\pi r$, is thus $\Gamma_B = (d\Gamma_B/ds)(2\pi r) = B(2\pi r)$.

The corresponding enclosed current is $I_{enc} = J(\pi r^2)$. Hence Ampère's law, (11.24), yields

$$B(2\pi r) = 4\pi k_m I_{enc} = 4\pi k_m J(\pi r^2), \qquad (11.35)$$

so

$$B = 2\pi k_m J r = \frac{2k_m I r}{a^2}. \qquad \text{(field within wire of finite radius)} \quad (11.36)$$

This result could have been obtained from the Biot–Savart law only with great effort. Alternatively (11.34), with $I_{enc} = J(\pi r^2)$ and $s = 2\pi r$, reproduces (11.36). Note that, for $r = a$, where $I_{enc} = J(\pi a^2)$, (11.36) gives $B = (2\pi k_m)(I/\pi a^2)a = 2k_m I/a$, which matches (11.14).

Example 11.11 **Field outside long current-carrying wire of finite radius**

Consider a clockwise-circulating Ampèrian circuit with radius $r > a$. See Figure 11.20. Use Ampère's law and symmetry to find the magnetic field outside the wire ($r > a$).

Solution: The circulation per unit length $d\Gamma_B/ds = \vec{B} \cdot \hat{s}$ for this symmetric circuit takes on the uniform value B, and the total circulation is $B(2\pi r)$. By (11.24),

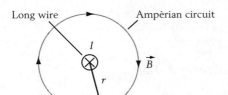

Long wire

Ampèrian circuit

I

$\vec{B}$

r

Figure 11.20 Field $\vec{B}$ of a long, thin wire carrying current into the page.

this must equal $4\pi k_m I_{enc}$, and here $I_{enc} = I$. Hence $B(2\pi r) = 4\pi k_m I$, or

$$B = \frac{2k_m I}{r}, \qquad \text{(field outside wire of finite radius)} \qquad (11.37)$$

in agreement with (11.14). For $r = a$, this matches the value obtained in the previous problem. Alternatively, (11.37) follows from (11.34) with $I_{enc} = I$ and $s = 2\pi r$. Equation (11.37) is much more general than (11.14), which was only derived for an infinitesimally thin wire.

Example 11.12 Field inside toroidal coil

Consider a finite solenoid (compare Figure 11.12.c) with N turns of wire bent into a circle. See Figure 11.21. This is called a toroidal coil, or toroid. (a) Use Ampère's law and symmetry to find the magnetic field within the coil, of inner radius a and outer radius b, and height w normal to the page. (b) For $a = 3$ cm and $b = 4$ cm, and 250 turns, evaluate the field at the mean radius.

Solution: (a) By Ampère's right-hand rule, for the current in Figure 11.21, $\vec{B}$ circulates counterclockwise. (Neglect the nonuniformities due to the turns not being infinitesimally close to one another, and consider only the average field.) Take an Ampèrian circuit that is a circle of radius r, concentric with the toroid, with $d\vec{s}$ circulating counterclockwise, so $\hat{s}$ points along $\vec{B}$. Here $B = |\vec{B}|$ is uniform because the Ampèrian circuit is concentric with the toroid. Then $d\Gamma_B/ds = \vec{B} \cdot \hat{s} = B$ is uniform. Hence $\Gamma_B = \int (d\Gamma_B/ds)ds = B \oint ds = B(2\pi r)$. Since the toroid has N turns, each carrying current I, we have $I_{enc} = NI$. Thus, by (11.24), $B(2\pi r) = 4\pi k_m I_{enc} = 4\pi k_m (NI)$, so

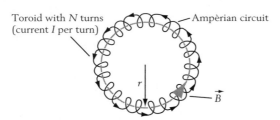

Figure 11.21 Toroid with inner radius a and outer radius b, with N turns, each carrying current I. The field $\vec{B}$ is confined to the interior of the toroid.

$$B = \frac{2k_m NI}{r}. \qquad \text{(field within toroidal coil)} \qquad (11.38)$$

Alternatively, (11.34) with $I_{enc} = NI$ and $s = 2\pi r$ reproduces (11.38). This result could not have been obtained from the Biot–Savart law without a great deal of effort. Although the toroid of Figure 11.21 has a small circular cross-section, (11.38) holds for more general cross-sections. (b) For our parameters, at the mean radius of 3.5 cm, (11.38) yields $B = 2.86 \times 10^{-3}$ T.

Example 11.13 Field outside current sheet

Consider a current sheet with current per unit length K that points into the paper, with the sheet centered about the x-axis. See Figure 11.22. Use Ampère's law and symmetry to find the magnetic field.

Solution: By Oersted's right-hand rule, the field above the sheet will be to the right, and below the sheet it will be to the left. Consider Ampèrian circuit A_1, where $d\vec{s}$ circulates clockwise. It has $I_{enc} = 0$. For arm 1, $\int \vec{B} \cdot d\vec{s} = B_1 l$, since $\vec{B}$ and $\hat{s}$ are parallel. For arm 2, $\int \vec{B} \cdot d\vec{s} = 0$, since $\vec{B}$ and $\hat{s}$ are perpendicular. For arm 3, $\int \vec{B} \cdot d\vec{s} = -B_3 l$, since $\vec{B}$ and $\hat{s}$ are antiparallel. For arm 4, $\int \vec{B} \cdot d\vec{s} = 0$, since $\vec{B}$ and $\hat{s}$ are perpendicular. Thus the total circulation is $\Gamma_B = (B_1 - B_3)l$. By Ampère's law, (11.24), this must be zero because $I_{enc} = 0$. Hence $B_1 = B_3$. Hence the field above the current sheet is uniform, an echo of the uniform field of a sheet of charge (but the direction is different for the two cases!). Call the field magnitude B. By symmetry, B is the same above and below the sheet.

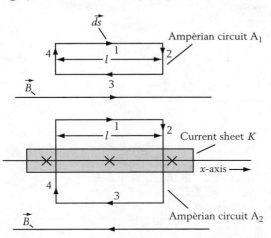

Figure 11.22 Finite current sheet carrying current per unit length K into the page, with length measured along the x-direction. Ampèrian circuit A_1 contains no current, and thus has zero circulation, which implies that the field is the same on its top and bottom arms. Ampèrian circuit A_2 contains a finite current, and thus has nonzero circulation.

Now consider Ampèrian circuit A_2, where $d\vec{s}$ circulates clockwise. It has $I_{enc} Kl$. For arm 1, $\int \vec{B} \cdot d\vec{s} = Bl$, since $\vec{B}$ and $\hat{s}$ are parallel. For arm 2, $\int \vec{B} \cdot d\vec{s} = 0$, since $\vec{B}$ and $\hat{s}$ are perpendicular. For arm 3, $\int \vec{B} \cdot d\vec{s} = Bl$, since $\vec{B}$ and $\hat{s}$ are again parallel. For arm 4, $\int \vec{B} \cdot d\vec{s} = 0$, since $\vec{B}$ and $\hat{s}$ are perpendicular. Thus the total circulation $\Gamma = 2Bl$. By Ampère's law, (11.24), $\Gamma_B = 4\pi k_m I_{enc}$, so $2Bl = 4\pi k_m (Kl)$. Thus

$$B = 2\pi k_m K, \quad \text{(field due to current sheet)} \quad (11.39)$$

as in (11.18), derived by direct integration. Alternatively, (11.34) with $I_{enc} = Kl$ and $s = 2l$ reproduces (11.39).

Example 11.14   Field inside and outside infinitely long solenoid

Consider a solenoid with n turns per unit length, each carrying current I, as in Figure 11.23. Use Ampère's law and symmetry to find the magnetic field.

Solution: By Ampère's right-hand rule, within the solenoid $\vec{B}$ points to the right. Outside the solenoid, if $\vec{B}$ is nonzero, it should point along the axis. Consider Ampèrian circuit A_1, which is outside and above the solenoid. $d\vec{s}$ for arm 1 is along the field $\vec{B}_{top}$, and $d\vec{s}$ for arm 3 is opposed to the field $\vec{B}_{bot}$. Since arms 2 and 4 do not contribute (because $\vec{B}$ is perpendicular to $d\vec{s}$), from (11.22) the total circulation then is $\Gamma_B = B_{top}l - B_{bot}l$. Since the total current enclosed by the Ampèrian circuit is zero, by Ampère's law $\Gamma_B = 0$. This implies $B_{top} = B_{bot}$. Since the separation of the arms could have been anything—including

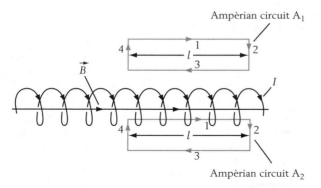

Figure 11.23 Solenoid with n turns per unit length, each carrying current I. Ampèrian circuit A_1 contains no current, and thus has zero circulation, which implies that the field is the same on its top and bottom arms. Ampèrian circuit A_2 contains a finite current, and thus has nonzero circulation.

infinity—B_{top} could have corresponded to the field at infinity, which we assume to be zero. Hence, the field outside the solenoid is both uniform *and* zero.

Now consider Ampèrian circuit A_2, which is partially inside the solenoid. Since the field outside is zero, and since sides 2 and 4 don't contribute, the only contribution to the circulation comes from side 1, within the solenoid, which gives Bl. In this case, I_{enc} is due to $N = nl$ turns of wire, each carrying current I. By the circuit-normal right-hand rule, the sign of I_{enc} is positive so that $I_{enc} = (nl)I$. By (11.24), $Bl = 4\pi k_m(nl)I$, so

$$B = 4\pi k_m nI = \mu_0 nI. \qquad \text{(field anywhere within solenoid)} \qquad (11.40)$$

Alternatively, (11.34) with $I_{enc} = (nl)I$ and $s = l$ reproduces (11.40). Besides this result being independent of position within the solenoid, it is also independent of the solenoid cross-section, which could be rectangular or even irregular.

If the solenoid is enclosed by a concentric solenoid at infinity, the field is not zero at infinity, so the field everywhere within the region we have considered will be larger by the field of the solenoid at infinity.

This concludes of the second part of this chapter. For those of you who will go on to study surface currents for perfect diamagnets and magnets, and the electromagnet, this is good a place to take a break.

11.11 Surface Currents and Perfect Diamagnetism

Optional

Chapter 9 showed that there is a class of materials, called *perfect diamagnets*, that can completely expel applied $\vec{B}$ fields, and that these materials also have zero electrical resistance, so that they are also called *superconductors*. This section shows, for a simple example, how within a perfect diamagnet an applied magnetic field can be cancelled by a magnetic field due to the perfect diamagnet itself. It does this by setting up an electric current that is localized at its surface over a distance on the order of a few atoms to perhaps a few hundred atoms, depending on the material. We will not concern ourselves with the distribution of the current near the surface, simply assuming that, for our purposes, it is literally localized at the surface. Close to the surface, we treat the surface as locally flat. See Figure 11.24.

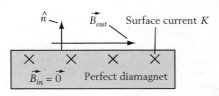

Figure 11.24 Closeup of the surface of a perfect diamagnet, which is characterized by having zero field inside: $\vec{B}_{in} = \vec{0}$. This is accomplished by the material developing just the right surface current density K.

Consider a point on the surface of an arbitrary material, with $\hat{n}$ the local outward normal. Let $\vec{B}_{far}$ be due to distant sources, and thus it does not change on moving from just outside to just inside the surface. Let $\vec{B}_s$ be due to the local current density K on the surface, so by (11.39) $|\vec{B}_s| = 2\pi k_m K$. $\vec{B}_s$ changes sign on crossing the surface, so $\vec{B}_s^{out} = -\vec{B}_s^{in}$. Hence, just outside the material $\vec{B}_{out} = \vec{B}_{far} + \vec{B}_s^{out}$. Likewise, just inside the material $\vec{B}_{in} = \vec{B}_{far} + \vec{B}_s^{in}$. Subtracting these gives $\vec{B}_{out} - \vec{B}_{in} = \vec{B}_s^{out} - \vec{B}_s^{in} = 2\vec{B}_s^{out}$. Again using (11.39), $2|\vec{B}_s^{out}| = 2(2\pi k_m) K = 4\pi k_m K$, so $|\vec{B}_{out} - \vec{B}_{in}| = 4\pi k_m K$.

For a perfect diamagnet $\vec{B}_{in} = \vec{0}$, so $|\vec{B}_{out}| = 4\pi k_m K$. With $\vec{K}$ the (vector) surface current per unit length, in vector form this becomes

$$\hat{n} \times \vec{B}_{out} = 4\pi k_m \vec{K}. \qquad \text{(outside perfect diamagnet)} \qquad (11.41)$$

The directions associated with (11.41) may be verified by considering $\vec{K}$ into the paper, for a semi-infinite perfect diamagnet with $\hat{n} = \hat{j}$, as in Figure 11.24. Then, by Oersted's right-hand rule, $\vec{B}_{out}$ must be along $\hat{i}$.

Equation (11.41), relating the magnetic field $\vec{B}_{out}$ outside a perfect diamagnet to the surface current $\vec{K}$, describes what we will call *electromagnetic shielding*. It is a consequence of the properties of perfect diamagnets and Ampère's law. It is similar to

$$\vec{E}_{out} \cdot \hat{n} = 4\pi k\sigma_S \qquad \text{(outside conductor)} \qquad (11.42)$$

for *electrostatic screening* of the electric field. This relates the electric field $\vec{E}_{out}$ outside a conductor to the surface charge density σ_S. Equation (11.42) is a consequence of the properties of conductors and Gauss's law.

Example 11.15   Monopole above a perfect diamagnet

Consider a monopole q_m at a distance h above an infinite sheet of perfect diamagnet, as in Figure 11.25(a). (A monopole can be simulated by either one end of a long, narrow magnet, or by one end of a long, narrow solenoid.) What do the magnetic field lines look like? Since the field is expelled by the perfect diamagnet, field lines that approach the material will be bent away. See Figure 11.25(a). The simplicity of the geometry suggests using the method

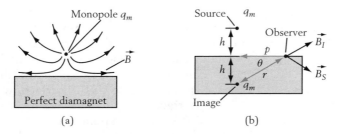

(a)   (b)

Figure 11.25 (a) A monopole q_m above a perfect diamagnet. The field lines are due to both the monopole and the perfect diamagnet. (b) The monopole and its "image" within the perfect diamagnet, as viewed from outside.

of images: from outside the material, it appears that the perfect diamagnet responds with an image monopole q_m a distance h beneath the surface of the sheet. See Figure 11.25(b), where $\vec{B}_S$ is due to the source and $\vec{B}_I$ is due to the image. The field due to both the actual and to the image monopoles never enter the sheet: the field is expelled. (a) Find the field just outside the surface. (b) Find the local current density K.

Solution: (a) At a distance ρ from the center of the sheet, and just outside the sheet, in Figure 11.25(b) the total magnetic field is given by

$$B_{out} = 2|\vec{B}_{q_m}|\cos\theta = 2\frac{k_m\,q_m}{(\rho^2 + h^2)}\frac{\rho}{(\rho^2 + h^2)^{1/2}} = 2k_m\,q_m\frac{\rho}{(\rho^2 + h^2)^{3/2}}.$$

$$(11.43)$$

The factor of 2 arises because of the equal contributions of both the actual and the image monopole. The factor $|\vec{B}_{q_m}| = k_m\,q_m/(\rho^2 + h^2)$ is the magnitude of the field due to either pole at that position just above the sheet. The factor $\cos\theta = \rho/(\rho^2 + h^2)^{1/2}$ gives the component of the field along the surface.

(b) Since $|\hat{n} \times \vec{B}_{out}| = B_{out}$, (11.41) gives $K = B_{out}/4\pi k_m$; substitution of B_{out} from (11.43) then yields

$$K = \frac{q_m}{2\pi}\frac{\rho}{(\rho^2 + h^2)^{3/2}}.$$

$$(11.44)$$

This gives the surface current K. It is zero just below the actual monopole q_m

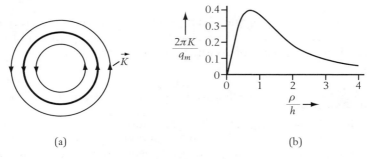

(a)   (b)

Figure 11.26 (a) Pattern of surface currents K for a monopole q_m above a perfect diamagnet. (b) Surface current K as a function of ρ/h.

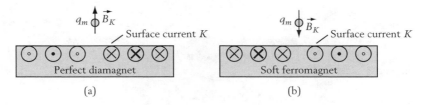

Figure 11.27 (a) Surface current for a monopole q_m above a perfect diamagnet. (b) Surface current for a monopole q_m above a soft ferromagnet.

($\rho = 0$). For small ρ, K varies as ρ, but for large ρ, K varies as ρ^{-2}. For a positive monopole q_m above the sheet to be repelled by the magnetic field due to K, the magnetic field due to K must point upward at q_m. Thus, by Ampère's right-hand rule, the surface current K must circulate counterclockwise as viewed from above. This is illustrated in Figure 11.26(a). The thicker the line, the greater the surface current. For K as a function of ρ/h, see Figure 11.26(b).

Figure 11.27(a) gives the surface currents from the same viewpoint as in Figure 11.25. The larger dots and x's represent larger surface currents. Also drawn is the field $\vec{B}_K$ due to the surface currents, at the site of the monopole q_m.

11.12 Ampèrian Surface Currents and Magnets

Optional

The surface current of (11.44) can be used to reveal how a soft magnetic material develops its magnetic field. Infinite sheets of perfect diamagnet and perfectly soft ferromagnet with monopoles above them both behave as if there were image monopoles within them, but the image monopoles are of opposite sign. Hence the surface current K for the soft ferromagnet will be of the same magnitude but in opposite direction to that for the perfect diamagnet. See Figure 11.27(b). Moreover, the field $\vec{B}_K$ due to these surface currents will be in opposite directions, repelling the monopole for the perfect diamagnet (Figure 11.27a), and attracting the monopole for the soft ferromagnet (Figure 11.27b). In short, for a perfectly soft ferromagnet, (11.44) also describes the magnitude of the surface current.

The next time you see a permanent magnet against a refrigerator, think about the Ampèrian surface currents set up in the refrigerator to hold the magnet in place. Why don't these Ampèrian surface currents die down because of electrical resistance? Chapter 10 argued that such currents involved resistanceless microscopic current loops throughout the magnet, rather than a single ordinary current loop circulating around the outside of the magnet. We now present additional evidence for the absence of electrical resistance.

11.13 How We Know

Optional

The following experiments establish that (1) the surface current for perfect diamagnets is due to macroscopic current loops, and (2) the surface current for ferromagnets is due to microscopic current loops. Consider two long, cylindrical

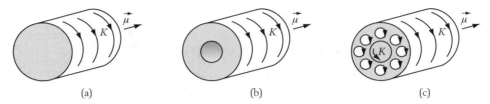

Figure 11.28 Current for the following cases: (a) solenoid, (b) perfect diamagnet with a hole, (c) ferromagnet with a hole.

samples, one a perfect diamagnet with an external field applied along its axis, and the other a hard ferromagnet. (From our conclusions for a hard ferromagnet we will extrapolate to all types of magnets, including soft ferromagnets.)

Experiment 11.1 Drilling a hole in a cylindrical magnet: effect on magnetic moment

Consider a solid, cylindrical perfect diamagnet with a surface current K that produces a magnetic moment $\vec{\mu}$, as for the solenoid in Figure 11.28(a). The current arises in response to a uniform external field $\vec{B}_{ext}$ that points opposite to $\vec{\mu}$. Within the diamagnet $\vec{B}_{tot} = \vec{0}$, so the field due to the diamagnet $\vec{B}_{dia} = -\vec{B}_{ext}$. Treating the diamagnet as a solenoid, with $K = nI$, (11.40) gives $K = B_{ext}/4\pi k_m$. If a hole then is drilled through the center of the perfect diamagnet, as in Figure 11.28(b), the magnetic moment $\vec{\mu}$ does not change. Moreover, the current on the outside is unaffected, and no current flows around the inner surface. This is what we expect for electric currents with a macroscopic circuit that is localized around the outer surface. These are true macroscopic electric currents, and they must flow with zero electrical resistance. Thus perfect diamagnetism implies superconductivity.

For the hard ferromagnet, there is also a surface current flowing around its outside, as in Figure 11.28(a). As for the perfect diamagnet, $K = B/4\pi k_m$. If a hole is then drilled through its center, the magnetic moment $\vec{\mu}$ decreases in proportion to the volume removed. Moreover, the current on the outside is unaffected, but an equal and opposite current flows around the inner surface. See Figure 11.28(c). The net current is due to microscopic Ampèrian current loops that fill the entire sample, and the current in these loops must flow with zero electrical resistance.

Experiment 11.2 Drilling a hole in a cylindrical magnet: effect on magnetic circulation

Another way to establish that the current due to a perfect diamagnet is macroscopic is to use a Rogowski coil or a clip-on ammeter to measure the circulation around the hole in the solenoid. See Figure 11.29(a). By Ampère's law, the circulation is proportional to the current; these are nonzero for the perfect diamagnet. For the permanent magnet with the hole in it, the Rogowski coil or clip-on ammeter will read zero circulation around the hole, since the Ampèrian current on the inner surface cancels the Ampèrian current on the outer surface. See Figure 11.29(b).

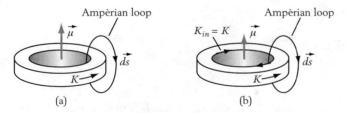

Figure 11.29 (a) Perfect diamagnet. (b) Ferromagnet.

11.13.1 *Perfect Diamagnets Carry Macroscopic Currents and Are Superconductors*

We have just established that perfect diamagnets carry currents that circulate over macroscopic circuits. These currents do not decay in time. Therefore, at a macroscopic scale, a perfect diamagnet has zero electrical resistance, and hence is what is called a *superconductor*. *Its primary property is perfect diamagnetism, not zero electrical resistance.* All materials that are called superconductors are perfect diamagnets, at least for weak enough applied fields. When the applied magnetic field gets too strong, the materials cannot produce large enough surface currents to completely expel the field, and there is some field penetration. For even larger magnetic fields, it is too energetically costly for the material to produce any surface currents at all, and therefore it can no longer "afford" to be superconducting.

11.13.2 *Ordinary Magnets Carry Microscopic Ampèrian Currents*

This still leaves us with the peculiar conclusion that the *microscopic* Ampèrian current loops in a magnet have zero electrical resistance. Does this imply that *macroscopic* electric currents passing through a magnet have zero electrical resistance? Although recently some materials have been discovered that display both superconductivity and magnetism, it appears that the electrons responsible for the superconductivity are in different orbitals than those that cause the magnetism. How, then, do we deal with zero electrical resistance at the microscopic level? This should not be difficult to accept, since we have already accepted the Bohr picture of the atom, with electrons perpetually circling around the nucleus, without any resistance.

Further study of the magnetism of magnets shows that it is associated with electrons in certain orbitals (e.g., electrons in the d orbitals are responsible for the powerful magnetic properties of Fe). The magnetism due to moving charge (the electrons in the orbitals) does not provide any further conceptual difficulties: by the Biot–Savart law, we know how to calculate the magnetic field for a given electric current, such as that associated with a specific orbital.

There is a difficulty, however, with understanding the magnetism of the electron itself. What might be the microscopic current circulating within the electron? We have no real physical picture for what such a microscopic current is like: on the scale of 10^{-15} m, within which they have been probed so far, electrons seem to be point particles. This is still one of the great mysteries of nature.

11.14 The Electromagnet

Our society uses electromagnets in a variety of ways, most of them hidden from plain sight. They are inside electrical transformers; they are used to produce magnetic forces to lift heavy objects (automobiles in junkyards); and they are used to move light objects (doorbell ringers).

There are two ways to compute the total $\vec{B}$ when there are both free electric currents (from ordinary electric circuits) and Ampèrian electric currents (from magnets). The first way is to find all the currents (free and Ampèrian) and to use the Biot–Savart law. The second way is to find the $\vec{B}_{free}$ of the free currents using the Biot–Savart law, and the $\vec{B}_m$ due to magnetic poles q_m, as discussed in Chapter 9. In discussing the electromagnet, it is useful to employ this latter viewpoint.

Recall that, in Chapter 9, the quantity $\vec{H}$ was introduced, via

$$\vec{B} = \mu_0(\vec{H} + \vec{M}). \qquad (\mu_0 = 4\pi k_m) \tag{11.45}$$

We will need $\vec{H}$ because, for soft magnets, the magnetization is determined via $\vec{M} = \chi \vec{H}$.

$\vec{H}$ has two types of sources. Magnetic poles q_m, as discussed in Chapter 9, produce what we will call $\vec{H}_m$, and free currents produce what we will call $\vec{H}_I$. The free currents contribute via the Biot–Savart law:

$$\mu_0 \vec{H}_I = k_m I^{free} \oint \frac{d\vec{s} \times \hat{r}}{r^2},$$

or

$$\vec{H}_I = \frac{I^{free}}{4\pi} \oint \frac{d\vec{s} \times \hat{r}}{r^2} = \frac{I^{free}}{4\pi} \oint \frac{d\vec{s} \times \vec{r}}{r^3}. \tag{11.46}$$

Comparison with (11.24) yields an Ampère's law for the contributions of free currents to $\vec{H}$ as

$$\Gamma_H \equiv \oint \vec{H}_I \cdot d\vec{s} = I_{enc}^{free}. \tag{11.47}$$

11.14.1 *Field within Soft Transformer Core*

The next chapter discusses Faraday's law. It is the basis of electrical transformers that "transform" ac voltages (e.g., from 4000 V on power lines to 120 V or 240 V in houses). An essential part of a transformer is its magnetically soft core, a solid toroidal piece of magnetic material. See Figure 11.30(a). The directions of the current I and the magnetic field $\vec{B}$ are related by Ampère's right-hand rule.

Consider a wire toroid, thought of as a bent-around solenoid, filled with a soft magnetic core. From (11.40) for the solenoid, the toroid alone produces $B_I = 4\pi k_m nI = \mu_0 nI$, to which corresponds $H_I = B_I/\mu_0 = nI$. As shown in Chapter 9, within a solid toroid magnetized along its axis, $H_m = 0$ because there

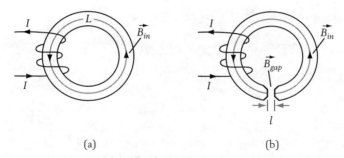

Figure 11.30 (a) Soft magnet used for transformer core.
(b) Soft magnet with a gap.

are no pole charges q_m. Hence, within the toroid, $H = H_I + H_m = H_I$. Then, from the definition of the magnetic susceptibility χ,

$$M = \chi H = \chi H_I = \chi nI. \quad \text{(transformer core)} \quad (11.48)$$

Thus

$$B_{in} = \mu_0(H + M) = \mu_0(1 + \chi)H_I = \mu_0(1 + \chi)nI. \quad \text{(transformer core)}$$
$$(11.49)$$

For $\chi \gg 1$, this gives an enormous enhancement of the B field over an air-core solenoid (for which $\chi \approx 0$).

This result was derived only for a uniformly wound core, not the locally wound core of Figure 11.30(a). However, because soft iron retains nearly all magnetic flux produced by the windings, each winding produces a uniformly circulating flux in proportion to its current, no matter where the winding is placed. Hence the windings can be localized around one part of the electromagnet.

11.14.2 *Field in Gap of Electromagnet*

Consider a horseshoe-shaped piece of soft iron, wrapped with N turns of wire, each carrying current I. See Figure 11.30(b).

Let the gap be l and let the magnet have length L, with $l \ll L$. In addition, let the magnet taper from a cross-sectional area A far from the poles to a smaller area A_{gap} in the gap. See Figure 11.30(b). We take both B_{in} within the magnet and B_{gap} within the gap to be essentially uniform. Because the $\vec{B}$ field lines do not originate or terminate on any magnetic poles, the magnetic flux within a flux tube is conserved. Hence the flux $B_{in}A$ leaving the bulk of the electromagnet equals the magnetic flux $B_{gap}A_{gap}$ entering the gap. Thus

$$B_{in}A = B_{gap}A_{gap}. \quad (11.50)$$

Since $A_{gap} \ll A$, by (11.50) $B_{gap} \gg B_{in}$, so the field in the gap is enhanced. The adjustment from B_{in} to B_{gap} takes place within the tapered region. We need an additional relationship to obtain both B_{in} and B_{gap}. First, however, we introduce

H_{in} and H_{gap}. From (11.49), with H_l rewritten as H_{in},

$$B_{in} = \mu_0(\chi + 1)H_{in}. \tag{11.51}$$

From (11.45), with $M = 0$ in the gap,

$$B_{gap} = \mu_0 H_{gap}. \tag{11.52}$$

The additional relationship we need is Ampère's law for $\vec{H}$, (11.47). With $I_{enc}^{free} = NI$, this yields

$$LH_{in} + lH_{gap} = NI. \tag{11.53}$$

Employing (11.51) and (11.52), Equation (11.53) becomes

$$\frac{LB_{in}}{\chi + 1} + lB_{gap} = \mu_0 NI. \tag{11.54}$$

Equations (11.50) and (11.54) enable us to solve for B_{gap}. Explicitly, with $\mu_0 = 4\pi k_m$,

$$B_{gap} = \frac{4\pi k_m NI}{l + L(A_{gap}/A)/(1 + \chi)}. \qquad \text{(gap of electromagnet)} \tag{11.55}$$

Often we may neglect the second term in the denominator of (11.55), yielding

$$B_{gap} \approx \frac{4\pi k_m NI}{l}. \qquad \left(\frac{L}{l}\frac{A_{gap}}{A} \ll 1 + \chi\right) \tag{11.56}$$

This result may also be obtained by neglecting the first term on the left-hand side of (11.54) and then solving for B_{gap}. Equation (11.56) is similar to (11.38) for the field within a toroid, where for comparison we should replace r by $L/2\pi$. However, instead of the magnet length L, (11.56) has the much smaller gap length l in the denominator, leading to a very large enhancement. This is a very effective way to produce a large magnetic field without actually being inside the magnet!

Equation (11.56) is not valid if l is so small that $l \ll L(A_{gap}/A)/(1 + \chi)$. Then the field approaches the value

$$B_{gap}^{max} = \frac{4\pi k_m NI(1 + \chi)}{L(A_{gap}/A)}. \qquad \left(\frac{L}{l}\frac{A_{gap}}{A} \gg 1 + \chi\right) \tag{11.57}$$

This result may also be obtained by neglecting the second term on the left-hand side of (11.54), solving for B_{in}, and then using (11.50) to obtain B_{gap}.

Recall that H has two types of source: free currents I and magnetic poles q_m (at the poles of the magnet). The magnetic poles make H change discontinuously on crossing from the magnet to the gap: within the magnet, it is like being outside a parallel-plate capacitor; within the gap, it is like being inside a parallel-plate capacitor.

Problems

11-2.1 Let $d\vec{s}$ point along $\hat{j}$, have length 0.05 mm, and be centered at $4\hat{i} + 6\hat{j} - 8\hat{k}$ (in cm). If $I = 1.7$ A, find $d\vec{B}$ at the origin.

11-2.2 Let $d\vec{s}$ point along $\hat{j}$, have length 0.05 mm, and be centered at $4\hat{i} + 6\hat{j} - 8\hat{k}$ (in cm). If $I = 1.7$ A, find $d\vec{B}$ at $2\hat{i} - 5\hat{j} + 6\hat{k}$ (in cm).

11-3.1 Consider a surface current $\vec{K}$, with $|\vec{K}|$ in units of A/m, and let dA be the surface element. Show that $B = k_m \int \vec{K} \times \hat{R} dA / R^2$.

11-3.2 Consider a volume current $\vec{J}$, with $|\vec{J}|$ in units of A/m^2, and let dV be the volume element. Show that $B = k_m \int \vec{J} \times \hat{R} dV / R^2$.

11-4.1 A wire that can safely carry a steady 0.68 A current, and can be formed into a coil of radius no smaller than 0.3 cm, is to be used to form a multi-turn coil to produce a maximum magnetic field of 0.04 T at its center. (a) Find the minimum length of wire that will do this. (b) Find the number of turns needed.

11-4.2 The field is 0.00046 T at 6.5 cm along the axis of a 12-turn coil with radius 1.4 cm. Find the current through the coil.

11-4.3 A uniformly magnetized disk of thickness 1.4 mm and magnetization 0.8×10^5 A/m pointing along its axis produces a 0.025 T field just above its center. Find its radius.

11-4.4 A long thin wire is hidden within and parallel to a pipe of 3.4 cm inner diameter and 4.0 cm outer diameter. The wire makes a 0.004 T field at a position just outside the pipe, and a field 0.0025 T at a position 2 cm outside the pipe. See Figure 11.31. Determine the distance between the pipe's axis and the wire, and the current in the wire.

Figure 11.31 Problem 11-4.4.

11-4.5 A wire carries current I along a straight line from $(-\infty, 0)$ to $(-a, 0)$, then along a semicircle

of radius a from $(-a, 0)$ to $(a, 0)$, and finally from $(a, 0)$ to $(\infty, 0)$. (a) Find the direction and magnitude of the magnetic field $d\vec{B}$ at the origin, due to an element $d\vec{s}$ on the semicircle. (b) Determine the total magnetic field $\vec{B}$ at the origin.

11-4.6 A long, thin solenoid of circular cross-section has 200 turns/cm and 3 A through each turn. (a) If it is 15 cm long and has a radius of 2 mm, determine its magnetic moment. (b) If the solenoid is approximated by a long magnet, determine the magnet's pole strength. (c) Estimate the magnitude of the magnetic field 2 cm away from either pole.

11-4.7 Find the current needed to cancel the earth's field of 0.5×10^{-5} T at the center of a coil of 1.2 cm radius and 26 turns.

11-4.8 (a) Find the magnetic field at (x, y) due to a lightning discharge along the y-axis carrying current I from $y = 0$ to $y = l$. (b) For $l = 2$ km and $I = 1600$ A due to a cloud-to-ground lightning bolt, find the field at $(10$ km$, 1$ km$)$. See Figure 11.32.

Figure 11.32 Problem 11-4.8.

11-4.9 Two identical wires of length l are used to make circular current loops. One is bent into a single circle, and the other is bent into a two-loop circle. If both loops carry the same current I and the field at the center of the one-loop circle is 1.5×10^{-5} T, find the field at the center of the two-loop circle.

11-4.10 Two small, charged conducting balls are separated by a distance l. A straight wire connects them. See Figure 11.33. If the current is I, find the magnetic field at a distance a along the perpendicular bisector of the wire.

Figure 11.33 Problem 11-4.10.

11-5.1 Three sheets are stacked normal to the y-axis at $y = 0$, $y = d$, and $y = 2d$. They carry parallel currents per unit length K, $2K$, and $-3K$ along $-\hat{z}$. See Figure 11.34. (a) Find the field on the planes $y = -0.5d$, $y = 0.5d$, and $y = 1.5d$. (b) Evaluate the field numerically on the plane for $y = 0.5d$, $d = 2$ cm, and $K = 25$ A/cm.

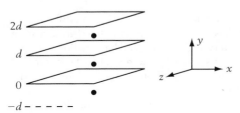

Figure 11.34 Problem 11-5.1.

11-5.2 A current sheet carries 5 A/m into the page; it is normal to the y-axis and passes through the origin. A long wire carries 2 A into the page; its center is 2 cm above the origin. See Figure 11.35. (a) Find the magnitude and direction of the magnetic field in the plane of the page at point $P_1 = (0, 1$ cm $)$; (b) repeat, but at point $P_2 = (1$ cm, 1 cm$)$.

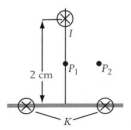

Figure 11.35 Problem 11-5.2.

11-5.3 Two concentric coils (Helmholtz coils) carrying the same current I, in the same direction, can give a very uniform field at their midpoint. Let the coils be of radius R, with normal along the y-axis, one centered at $(0, -s/2)$ and the other at $(0, s/2)$. See Figure 11.36. Consider only the field

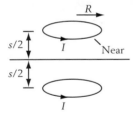

Figure 11.36 Problem 11-5.3.

on the y-axis. (a) Give a qualitative argument that, for $s \ll R$, the field at the origin is a local maximum, as a function of y. (b) Give a qualitative argument that, for $s \gg R$, the field is a local minimum, as a function of position x. (c) Compute, as a function of y, the field along the axis. (d) Find the optimum separation s, where the field is nearly uniform at $y = 0$. (This corresponds to $d^2 B_y/dy^2 = 0$.)

11-5.4 Consider a wire shaped like a partial annulus with radial coordinates $(a, 0)$, (a, θ), (b, θ), $(b, 0)$, the current I flowing clockwise. Take $b > a$. See Figure 11.37. (a) Find the field at the origin. (b) For small θ and small $(b - a)/a$, compute the magnetic moment μ. (c) Verify that the dipole field normal to the axis of a magnet of moment μ, as in part (b), agrees with the result of part (a), at large distances from the magnet.

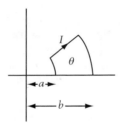

Figure 11.37 Problem 11-5.4.

11-5.5 Consider a square-shaped current loop of side l, centered at the origin and carrying current I in the xy-plane. See Figure 11.38. Show that, on its normal axis at a distance z from the center, $B_z = 2k_m I l^2 / [(z^2 + l^2/4)\sqrt{z^2 + l^2/2}]$.

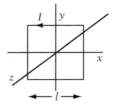

Figure 11.38 Problem 11-5.5.

11-5.6 Consider a rectangle-shaped current loop with corners at $(-a, -b)$, $(-a, b)$, (a, b), $(a, -b)$. It has N turns, each carrying counterclockwise current I. (a) Find the field a distance z along the z-axis. (b) Find the magnetic moment μ of the loop. (c) Verify that part (a) gives the appropriate dipole field for $z \gg \sqrt{a^2 + b^2}$.

11-5.7 Consider two long wires, both with I into the page, one crossing it at $(a,0)$, and the other crossing it at $(-a,0)$. Show that $\vec{B} = \hat{x}(4k_m Iy)/(y^2 + a^2)$ at $(0, y)$.

11-5.8 Consider a wire carrying current I rightward from negative infinity along the x-axis, at $r = a$ bending from $180°$ to $120°$, then going radially out to $r = b$, then bending from $120°$ to $60°$, then going radially in to $r = a$, then going from $60°$ to $0°$, and finally going to positive infinity on the x-axis. See Figure 11.39. (a) Find $\vec{B}$ at the origin. (b) Evaluate this for $I = 0.6$ A, $a = 3.2$ cm, $b = 4.8$ cm.

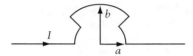

Figure 11.39 Problem 11-5.8.

11-5.9 Two long parallel wires, separated by $s = 3.8$ cm, carry 8 A in opposite directions. See Figure 11.40. Find the magnetic field, including direction, (a) at the midpoint, (b) at 4 cm above the midpoint, and (c) at 30 cm below the midpoint.

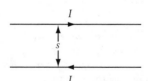

Figure 11.40 Problem 11-5.9.

11-5.10 A long strip of thickness 1.2 mm and width 2.4 cm has magnetization 0.4×10^6 A/m normal to its plane. Find the magnetic field 8 cm above the midpoint.

11-5.11 Consider a long strip of width w and current per unit length K into the page, intersecting the page between $(-w/2,0)$ and $(w/2,0)$. See Figure 11.41. Find $\vec{B}$ at $(x,0)$ for $x > w/2$.

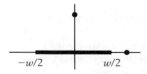

Figure 11.41 Problem 11-5.11.

11-5.12 Consider a long strip of width w and current per unit length K out of the page, intersecting the page between $(-w/2,0)$ and $(w/2,0)$. See Figure 11.41. Find $\vec{B}$ at $(0,y)$.

11-5.13 Consider a thin, rectangular slab centered at the origin, of dimensions (a,b,d) along (x,y,z), respectively. Take the magnetization $\vec{M}$ along z, and take $d \ll a, b$. See Figure 11.42. (a) Find the equivalent surface current. (b) Using the Biot–Savart law, find the field along the z-axis.

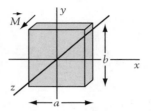

Figure 11.42 Problem 11-5.13.

11-5.14 Consider an annular magnetic slab of inner radius a and outer radius b, with thickness $d \ll a$ and magnetization $\vec{M}$ along its axis. (a) Find the equivalent surface currents. (b) Find the field along its axis.

11-5.15 A finite solenoid of radius a has N turns and is of length L. It is centered at the origin with its axis along $\hat{x}$. See Figure 11.43. (a) Find the field a distance x from its midpoint. (b) Verify that this reduces to the expected results as $L/a \to 0$ and as $L/a \to \infty$.

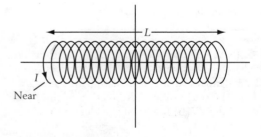

Figure 11.43 Problem 11-5.15.

11-5.16 An annulus of inner radius a and outer radius $a + d$, where $d \ll a$, rotates at an angular velocity ω. It has a uniform charge density σ (half on each surface). This can be thought of as a coil carrying current I. (a) Find I. (b) Find the field a distance x along its axis. (c) Repeat (a) for arbitrary d. (d) Repeat (b) for arbitrary d.

11-5.17 A disk of radius a rotates at an angular velocity ω. It has a uniform charge density σ (half on each surface). This can be thought of as a set of coils. See Figure 11.44. (a) Find the current per unit radial length, dI/dr. (b) Find I. (c) Find the field a distance x along its axis. This so-called Rowland disk (1876), suggested by Helmholtz, was significant because it showed that electric currents from moving charges can produce a $\vec{B}$ field.

Figure 11.44 Problem 11-5.17.

11-5.18 A long, thin wire carrying current I is wrapped around itself in a circle from $r = 0$ to $r = a$, with N turns. See Figure 11.45. (a) Find the current per unit radial length. (b) Find the field a distance x along its axis. (c) Compare with the rotating disk in Problem 11-5.17.

Figure 11.45 Problem 11-5.18.

11-6.1 An infinite wire carrying 2 A from left to right coincides with the x-axis. An electron at $(0,2$ cm$)$ moves at a speed of 950 m/s along $\hat{x}$. (a) Find the magnitude and direction of the force on the electron. (b) Repeat if the electron velocity is along $\hat{y}$; (c) along $\hat{z}$.

11-6.2 An infinite sheet whose normal is along $\hat{y}$ and that intersects the page along the x-axis carries current density K into the page. A wire carries current I normally out of the page and intersects the page at $(0,a)$. (a) Indicate the direction of the force $\vec{F}$, of magnitude $F = 0.5$ N, on the current sheet. (b) If K triples, I quadruples, and a doubles, find the new force.

11-6.3 Find the force on an electron a distance r from a long wire and moving with velocity $\vec{v}$ parallel to the current I of the wire. See Figure 11.46.

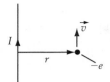

Figure 11.46 Problem 11-6.3.

11-6.4 A long wire carries current I_1 upward along the y-axis. A rectangular loop with sides of length a parallel to the long wire, and sides of length b perpendicular to the long wire, carries clockwise current I_2. Its near arm is s from the long wire. See Figure 11.47. (a) Find the magnetic force on the rectangular loop. (b) Find the magnetic force on the long wire.

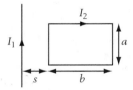

Figure 11.47 Problem 11-6.4.

11-6.5 Consider a rectangular loop of width a and length $b \gg a$. It is made of wire with radius $r \ll a, b$, carrying clockwise current I. See Figure 11.48. (a) Estimate the magnetic force $\vec{F}$ on one of the short sides due to the two long sides, and indicate whether it is compressive or expansive. (If one of the short sides can slide, this corresponds to the geometry of a rail gun.) (b) For $I = 12$ A, $r = 0.35$ mm, $a = 0.84$ cm, and $b = 18$ cm, evaluate $|\vec{F}|$.

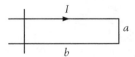

Figure 11.48 Problem 11-6.5.

11-6.6 Consider two parallel wires carrying equal currents in opposite directions. One, with mass per unit length λ, is above the other. (a) Find the height h at which the upper wire would be supported. (b) For $I = 10$ A and copper wire of radius $a = 1$ mm, determine h. The mass density of Cu is 8.93 g/cm³.

11-6.7 A long, thin solenoid of radius a with n turns per unit length, and current I_1 through each turn, sits on a table. It is co-axial with, and one end is at the center of, a large loop of radius $b \gg a$, carrying current I_2 opposite to I_1. (a) Find the magnetic field due to the current loop, at the end of the solenoid. (b) Find the force on the solenoid. (c) Evaluate the magnetic field and the force for $n = 300/\text{cm}$, $I_1 = 2$ A, $I_2 = 4$ A, $a = 0.5$ mm, and $b = 1$ cm.

11-6.8 (a) Show that the magnetic force between opposite sides of a current-carrying square loop tends to expand the loop. (b) Give a qualitative argument that the magnetic force on part of a circular current loop, due to the rest of the loop, tends to make it expand. In general, the magnetic force tends to cause loops to expand.

11-6.9 Two long, parallel wires, separated by $2a$ along the x-axis, both carry I into the page. Use the line tension $B^2/8\pi k_m$ to find the force per unit length between the wires.

11-6.10 Two long, parallel wires are separated by $2a$ along the x-axis, the right wire carrying I into the page and the left wire carrying I out of the page. Use the line pressure $B^2/8\pi k_m$ to find the force per unit length between the wires.

11-7.1 An Ampèrian circuit of fixed shape encloses a wire carrying current $I = 13.7$ A. (a) Find the magnetic circulation. (b) The Ampèrian circuit is moved far from the wire. Find the magnetic circulation.

11-7.2 For a square loop of side 6 cm, the magnetic field points along the y-axis, with $B_y = 0.06$ T on the right arm, and $B_y = -0.02$ T on the left arm. (a) Determine the magnetic circulation $\Gamma_B = \oint \vec{B} \cdot d\vec{s}$ for $d\vec{s}$ circulating clockwise. (b) Find I_{enc}.

11-7.3 The magnetic circulation is 0.025 T-m for an Ampèrian circuit that wraps once around a current-carrying wire. (a) Find the current enclosed by the Ampèrian circuit. (b) Find the magnetic circulation if the Ampèrian circuit turns a second time, in the same sense, around the wire. (c) Find the magnetic circulation if the second turn is in the opposite sense to the first turn.

11-7.4 A wire carries current I into the page at the origin. (a) For a square-shaped Ampèrian circuit in the plane of the paper, centered around the origin, explicitly compute Γ_B. (b) Compare with what is expected from Ampère's law.

11-7.5 For an Ampèrian circuit sitting on the page, of total length 2 cm, $\Gamma_B = 0.0008$ T-m when

$d\vec{s}$ circulates counterclockwise. (a) Indicate how the magnetic field circulates. (b) Find the average magnetic circulation per unit length. (c) Find the amount of electric current associated with the Ampèrian circuit. (d) Indicate the direction of the current flow.

11-7.6 Let $B_x = B_z = 0$, and let B_y depend only on x. Consider a rectangular Ampèrian circuit of dimension 3 cm by 5 cm, oriented parallel to the x- and y-axes. A current of 5.8 A flows into the page. If $B_y = 0.000048$ T for the right-hand side of the circuit, find B_y for the left-hand side of the circuit.

11-8.1 Prove Ampère's law by starting with an infinitesimally thin wire surrounded by an infinitesimally thin concentric Ampèrian circuit, and then deforming the Ampèrian circuit to give it an arbitrary shape.

11-8.2 In (11.28), we take $B_{m,\,int} = 4\pi k_m \sigma_m$. Using $\sigma_m = I/l$, with I finite but $l \to 0$, show that this field dominates any finite correction terms due to the more distant parts of the circuit.

11-8.3 Consider $d\vec{s}$ for an arbitrary Ampèrian circuit and $d\vec{s}$ for a physical, current-carrying circuit. (a) Distinguish between them. (b) For which is the current usually in the direction of $d\vec{s}$? Usually not in the direction of $d\vec{s}$?

11-8.4 A capacitor consists of two circular plates of radius 5 cm and separation 1.2 mm. Let the field along the midplane, 8 mm outside the edge, be 25 V/m. Estimate the charge on the capacitor plates. See Figure 11.17(b).

11-9.1 Let $\vec{B} = (3 + 5y^2)\hat{i}$ for y in m and $\vec{B}$ in T. (a) Find the circulation for a loop defined by the sides of the rectangle (1,2), (3,2), (3,−4), (1,−4). (b) Find the current enclosed, including its direction. (c) Find the average current per unit area.

11-9.2 Let $\vec{B} = (3 + 5y^2)\hat{i}$ for y in m and $\vec{B}$ in T. (a) Find the circulation for a loop defined by the sides of the rectangle (1,2), (1.1,2), (1.1,1.9), (1,1.9). (b) Find the current enclosed, including its direction. (c) Find the average current per unit area.

11-9.3 Let $\vec{B} = (3 + 5y^2)\hat{i}$ for y in m and $\vec{B}$ in T. (a) Find the circulation for a loop defined by the sides of the rectangle (x,y), $(x+dx,y)$, $(x+dx,y+dy)$, $(x,y+dy)$. (b) Find the current enclosed, including its direction. (c) Find the current per unit area.

11-9.4 A magnetic field points along the x-axis, with $B_x = (-0.02 + 0.06y)$ T for y in cm. (a) For a square loop of side 5 cm, its top-left corner at (2 cm, 9 cm), determine the magnetic circulation $\Gamma_B = \oint \vec{B} \cdot d\vec{s}$ for $d\vec{s}$ circulating counterclockwise. (b) Find the current enclosed, including its direction.

11-9.5 A field circulates clockwise, with symmetry about the origin, and $B_\theta = 0.057\rho^2$, where $\rho = \sqrt{x^2 + y^2}$ in m gives B_θ in T. (a) Find the magnetic circulation for an Ampèrian circuit that is a circle of radius 2.5 cm, centered about the origin. (b) Find the average circulation per unit length for this circuit. (c) Find the current I and current per unit area dI/dA passing through this loop. (d) Find the direction of the current passing through this loop.

11-9.6 Consider a 3-cm-by-5-cm rectangular Ampèrian circuit oriented parallel to the x- and y-axes, its lower-left corner at (1 cm, 2 cm). Let $\vec{B}$ point only along the y-axis, with $B_y = (0.04 - 0.06y^2)$ T when y is expressed in cm. (a) Find the magnetic circulation taken clockwise. (b) Determine the direction of the current enclosed by the loop and its magnitude. (c) Find the average circulation per unit length, and interpret that in terms of an average tangential component of $\vec{B}$.

11-10.1 Consider a long wire of radius 0.05 mm, carrying a uniformly distributed 6 A out of the page. (a) Find the magnetic field a distance 0.02 mm from its center. (b) Find the field at 0.08 mm from the center of the wire. (c) Indicate how $\vec{B}$ circulates.

11-10.2 A toroid with inner radius of 6 cm has a square cross-section with side 2.3 cm. Every one of its 180 turns is wound at $2°$ relative to the one before it. (a) If the current is 4.2 A, estimate the field at the center of the toroid hole ($r = 7.15$ cm); at 6.8 cm on the axis of the toroid hole. (b) If the toroid is filled with soft iron of relative permeability 2250, estimate the field at 6.8 cm on the axis of the toroid hole.

11-10.3 A co-axial cable has an inner wire of radius a, plastic separator out to radius b, and then outer casing to radius c, where $b \approx c$. A current I comes out of the page for the inner wire, and a current $4I$ goes into the page for the outer casing. See Figure 11.49 for a situation where b and c are very different. (a) Indicate the direction of the magnetic field for $r < a$, $a < r < b$, and $r > c$. (b) Sketch an Ampèrian circuit that would be useful for finding

the field for $r > c$. Take $d\vec{s}$ to circulate clockwise. (c) For this Ampèrian circuit, find the circulation by direct computation, in terms of the symbol $|\vec{B}|$. (d) Find the circulation using Ampère's law. (e) Determine $|\vec{B}|$ for $r > c$.

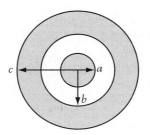

Figure 11.49 Problem 11-10.3.

11-10.4 A toroid with N turns of wire, each turn carrying current I, has a cross-section of area A, and inner radius a, outer radius b. See Figure 11.50. Indicate the direction of the magnetic field $\vec{B}$ at point P, a distance $a < \rho < b$ from the axis of the toroid. (a) If the magnetic field has magnitude $|\vec{B}|$ at P, determine the magnetic circulation for an Ampèrian circuit that is a circle of radius ρ that is concentric with the toroid and passes through P. (b) Directly determine how much current passes through this Ampèrian circuit. (c) Deduce $|\vec{B}|$ at P.

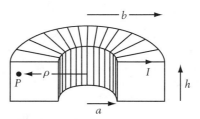

Figure 11.50 Problem 11-10.4.

11-10.5 Consider a co-axial conductor with inner radius a carrying current I out of the page and a sheath for $b < r < c$ carrying the return current I. See Figure 11.49. The current densities are uniform for $r < a$ and for $b < r < c$. (a) Find the current density in the core and in the sheath. (b) Find the field at all radii.

11-10.6 Consider a co-axial cable with inner current $2I$ into the page and uniformly distributed over $r < a$, and outer current I out of the page and uniformly distributed over $b < r < c$. See Figure 11.49. Find the field for (a) $r < a$, (b) $a < r < b$, (c) $b < r < c$, (d) $c < r$.

11-10.7 Show that, within a cylindrical solenoid of arbitrary cross-section (e.g., triangular), the field points only along the axis z and is uniform. Take n turns per length, each carrying current I. (a) Consider two turns, 1 and 2, symmetrically spaced above and below the observation point. Show that for two corresponding elements (so they have the same $d\vec{s}$, and $|\vec{R}_1| = |\vec{R}_2|$), the average field $d\vec{B}' = \frac{1}{2}(d\vec{B}_1 + d\vec{B}_2)$ points along the axis of the solenoid. (b) Fix the circumferential angle of the loops and then integrate over $dI = nI\,dz$. (c) Finally, integrate over the circumferential angle to obtain $\vec{B} = 4\pi n k_m I \hat{z}$. (J. D. Jackson.)

· **11-10.8** Consider two thin, vertical concentric cylinders of length l and radii $a < b$. The inner one carries current upward, and the outer one carries current downward. On the top and bottom the current is radial. There is no buildup of charge anywhere in the circuit. See Figure 11.51. (a) Show that on the top and bottom the radial current per unit length is $K = I/2\pi r$. (b) Show that on the inner and outer surfaces the axial currents per unit length are $K = I/2\pi a$ and $K = I/2\pi b$. (c) Argue that, by symmetry, $\vec{B}$ can only circulate about the axis. (d) Using Ampère's law and symmetry, argue that $\vec{B} = 0$ outside the cylinders ($z < 0, z > l, r < a, r > b$). (e) Using Ampère's law and symmetry, argue that $|\vec{B}| = 2k_m I/r$ within the cylinders. (f) Show that $\vec{B}$ has the expected discontinuity relative to the surface current $\vec{K}$ over the surface of the conductor.

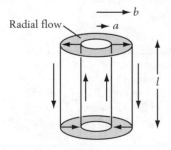

Figure 11.51 Problem 11-10.8.

11-11.1 A long magnet with square cross-section a^2, length $l \gg a$, and magnetic moment μ is placed normal to a superconductor, its lower pole (N) a distance h from the superconductor. (a) For $l \gg h$ (so the near pole dominates), find the force on the magnet. (b) If the magnet, of mass M, levitates, find the equilibrium value of h.

11-11.2 A levitation device using two permanent magnet disks (magnetized along their normals) and two diamagnetic disks (made of graphite) works as follows. Frame A holds the larger magnetic disk A on top. Below frame A, frames B and D hold diamagnetic disks B and D. Magnetic disk C floats in the space between frames B and D. Explain.

11-11.3 Consider a monopole q_m a distance h above a superconductor whose top surface is the $y = 0$ plane, as in Figure 11.25(a). By (11.44), the surface current is $K = (q_m/2\pi)[\rho/(\rho^2 + h^2)^{3/2}]$, with direction indicated in Figure 11.27(a). (a) Show that, on the monopole axis at any $y > 0$ (above the surface), K leads to $B_z = k_m q_m/(y + h)^2$, as you would expect the image charge to produce. (b) Find the field due to K for $y < 0$. (c) Find the force on the source charge q_m.

11-11.4 A magnetic slab with square cross-section a^2, length $l \ll a$, and magnetic moment μ is placed normal to a superconductor, its lower pole (N) a distance h from the superconductor. (a) For $a \ll h$ (so the slab may be treated like a dipole), find the force on the magnet. (b) If the magnet, of mass M, levitates, find the equilibrium value of h. (c) Find the current per unit length K that causes the force.

11-12.1 Consider a monopole q_m above a soft magnet whose top surface is the $y = 0$ plane. (a) Describe the response of the magnet, as seen from the outside. (b) Determine the current per unit length K on the surface, due to Ampèrian currents. (c) Find the force on q_m.

11-12.2 Consider a dipole μ above a soft magnet whose top surface is the $y = 0$ plane. Take its moment to be along the outward normal. (a) Describe the response of the magnet, as seen from the outside. (b) Determine the current per unit length K on the surface, due to Ampèrian currents. (c) Find the force on q_m.

11-13.1 Describe two experiments indicating that, if the magnetism of magnets is due to electric currents, then these currents are not macroscopic currents subject to a finite electrical resistance.

11-13.2 A long, hollow cylinder has uniform magnetization $\vec{M}$ along its axis (z) for $a < \rho < b$. (a) Find the surface currents. (b) Find $\vec{B}$ for all ρ.

11-14.1 Consider an electromagnet with a gap. Let $l = 4$ mm, $L = 22$ cm, $I = 4$ A, $N = 14$ turns, and $\chi = 2500$. (a) For $A_{gap} = A = 1.2$ cm^2, find B within the magnetic material and within the gap. (b) For what value of A_{gap}/A are the two terms in the denominator of (11.55) equal?

11-14.2 N turns of wire carrying current I are wrapped around a toroidal-shaped soft magnetic material of mean radius R and known χ, as in Figure 11.30(a). (a) Estimate B and H inside. (b) Show that, if $\chi = 1000$, then $B = 1001\mu_0 H$, so $B \gg \mu_0 H$.

11-14.3 Consider a permanent magnet with magnetization M, in the shape of a toroid with a gap. Let the magnet have cross-section A_m and length L, and let it have a gap of width $d \ll \sqrt{A_m}$. Use NEO, with a remanence of $M_r = 1.03 \times 10^6$ A/m and a coercive field of $H_c = 1.60 \times 10^6$ A-turn/m. Set $A_m = 2$ cm^2, $L = 12$ cm, and $d = 0.2$ cm. (a) Find the B field within the bulk of the magnet. (b) Find the field within the gap.

11-14.4 Denote by I^{Amp} the Ampèrian current (due to currents that are not free, but rather that circulate about atoms and molecules). (a) Show that $I^{Amp} = KL = ML$. (b) Show that Ampère's law for $\vec{B}$ yields $B_{in}L + B_{gap}l = \mu_0(NI + ML)$. (c) Show that this reduces to (11.54).

11-G.1 A magnetic field has the form $\vec{B} = 6x\hat{i}$, where x in m gives $\vec{B}$ in T. For a cube of side $a = 0.5$ cm that is in the first octant with one corner at the origin, find the flux through each side, and the total flux. Is this an allowable $\vec{B}$ field?

11-G.2 Consider a long wire of finite radius b carrying uniform current density J into the page, its center intersecting the page at the origin. The wire contains a long nonconcentric hole of radius $a < b$, its center intersecting the page at $(c,0)$, with $c + a < b$. Find the magnetic field (a vector) within the hole.

11-G.3 A wire carrying current I is wrapped N times, tightly and uniformly, around a toroid of rectangular cross-section, with inner radius a and outer radius b, and height h. Estimate the field outside the toroid, at a distance s normal to its axis. *Hint:* It is nonzero, and independent of N.

11-G.4 The axis of a hollow tube of radius a and thickness $d \ll a$ is normal to the page. The tube carries current I into the page, uniformly dis-

tributed around its perimeter. (a) Verify that the field at the tube center is zero. (b) Find the field magnitude and direction if only the right-half of the tube (i.e., the first and fourth quadrants) carries current I. (c) Find the field magnitude and direction if only an arc of angle α, symmetrically placed around the y-axis, carries current I.

11-G.5 A current density with axial symmetry flows along z (out of the page), so $J_z = J_z(\rho)$. By symmetry, it makes a tangential magnetic field component B_θ that depends only on ρ ($B_\theta > 0$ is counterclockwise). (a) For a concentric annulus of radius ρ and thickness $d\rho$, show that $dI/d\rho = (2\pi\rho)J_z$. (b) For that annulus, show that Ampère's law gives $d(B_\theta(\rho))/d\rho = 2k_m dI/d\rho = 4\pi k_m \rho J_z$. (c) If $J_z = \alpha\rho^2$, and $B_\theta = 0$ for $\rho = 0$, find $B_\theta(\rho)$.

11-G.6 A monopole of strength q_m is at the center of a loop of radius b in the plane of the page, centered at the origin, carrying current I clockwise. (a) Find the magnetic field due to the current loop at the monopole. (b) If $I = 2$ A, $b = 1$ cm, and the force on q_m points into the page and is of magnitude 0.0005 N, find q_m.

11-G.7 A long, straight conducting wire is bent so that its two parts make an angle $2\theta < \pi$ relative to each other. The wire is placed in the xy-plane with its bend at $(0,s)$, and each half making an angle θ to the y-axis. A current I flows in from the half-wire in the first quadrant and out of the half-wire in the second quadrant. Show that, at the origin, $\vec{B}$ points along $\hat{z}$, and that $B_z = 2k_m I(1 - \cos\theta)/s \sin\theta$. This experiment was performed by Biot and Savart, and analyzed by Laplace.

11-G.8 Consider the field B on the axis of a finite disk of radius R, magnetized normal to its surface. It may be thought of as due to the difference of the fields of two sheets of magnetic charge with charge densities $\pm\sigma_m = M$ and thickness D. (a) Show that $B = D(dB_{sheet}/dx)$, where B_{sheet} is the field on the axis of a single sheet of charge density σ_m. (b) Use (11.12) to obtain B, and then integrate to obtain B_{sheet}. (c) Compare this to the result for the electric field on the axis of a circular sheet of charge density σ.

11-G.9 Problem 9-2.11 considered two line poles $\pm\lambda_m$ parallel to y, intersecting the $z = 0$ plane at $(x_0,0)$ and (x_0,l), respectively. The magnetic dipole moment per unit length is $\nu = \lambda_m l$. Letting $l \to 0$ and $\lambda_m \to \infty$ at constant ν then led to a line of infinitesimal dipoles with field $\vec{B}_m = 2k_m[-\vec{\nu} + 2$

$(\vec{v} \cdot \hat{r})\hat{r}]/r^2$, where r is the nearest distance to the dipole lines. (a) From Ampère's equivalence, show that the circuit corresponding to these dipole lines is a pair of wires separated by a small distance w along x, carrying current I along and against $\hat{z}$, with $v = Iw$ held constant as $w \to 0$ and $I \to \infty$. (b) Using superposition with $w \to 0$ and $I \to \infty$ at constant v, find the field $\vec{B}_I$ due to this pair of wires, and compare with $\vec{B}_m$.

11-G.10 Consider a semi-infinite dipole sheet (e.g., a magnetic shell) that is normal to the plane of the page and intersects the page along the negative x-axis. Take its dipole moment to point along $\hat{y}$, and let it have a finite dipole moment per unit area $d\mu/dA$, which is the same as the magnetization M. Such a sheet may be obtained by building up a set of dipole lines, as in the previous problem, and letting $\vec{v} \to d\vec{v} = (d\mu/dA)\hat{y}dx = M\hat{y}dx$. From Problem 9-2.12, for an observer at (x_0, y_0), $\vec{B} = -[2k_m M l_0/r_0](\hat{z} \times \hat{r}_0) \to -[2k_m I/r_0](\hat{z} \times \hat{r}_0)$, where $r_0 = \sqrt{x_0^2 + y_0^2}$ and $\hat{r}_0 = \hat{y}\sin\theta_0 + \hat{x}\cos\theta_0$. (a) Using Ampère's equivalence, find the equivalent circuit. (b) Find the magnetic field of this circuit and compare with $\vec{B}$ for a semi-infinite dipole sheet.

11-G.11 The Rogowski coil (1912) was originally conceived by Chattock (1889), a student of J. J. Thomson. Here a long rod of length l and cross-section A is made of soft magnetic material. Wrapped around it are many turns of nonmagnetic wire. The i turn of wire intercepts a magnetic flux $B_i A$, where B_i is the field component along the axis of the rod. Thus the rod is like a toroid that corresponds to the Ampèrian circuit, and the nonmagnetic wire wraps around the rod. Show that for a tightly wound coil the sum over turns gives a total flux Φ_B that is A/l times the total circulation Γ_B.

11-G.12 Consider an infinite solenoid along the x-axis, of n turns per unit length and current I. This problem finds the force of attraction between its two halves $x > 0$ (1) and $x < 0$ (2). If $\vec{B}^{(2)}$ is the field due to 2, then the force $F_z^{(1)}$ on 1 ($x > 0$) is given by a sum over the force on each turn i of 1, which can be converted to an integral. Let $\int ds$ represent an integral over the arc length ds of a single turn. (a) Show that

$$F_z^{(1)} = \Sigma_i \int I\,ds\,B_\rho^{(2)} = \int_0^\infty n\,dx \int I\,ds\,B_\rho^{(2)}$$

$$= nI \int_0^\infty dx \int ds\,B_\rho^{(2)} = nI \int_1 \vec{B}^{(2)} \cdot d\vec{A}_1,$$

where $\int_1 dA_1$ is performed over the cylindrical surface defined by part 1, with normal $\hat{n}_1$, area element $dA_1 = |d\vec{A}_1|$, and $d\vec{A}_1 = \hat{n}_1 dA_1$. (b) Using $\oint \vec{B} \cdot d\vec{A} = 0$, show that $\int_1 \vec{B}^{(2)} \cdot d\vec{A}_1$ is the flux through a disk at the mouth of the solenoid, which is half that due to a full solenoid, or $(1/2)B_{sol}A = 2\pi k_m nI A$. (c) Show that $F_z = 2\pi k_m (nI)^2 A$. With $nI = K = M$, this is $F_z = 2\pi k_m M^2 A$, the same as the force of attraction of one infinite magnet on another.

11-G.13 Two thin slab magnets, square shaped of side a and thickness $d \ll a$, have magnetization M normal to their axes. They are co-centered and parallel, at a separation s, where $a \gg s \gg d$. Think of these as nearly parallel wires, with $K = I/d = M$. Neglecting edge effects, show that the force F is given by $F = 8k_m M^2 ad^2/s$. As the magnets are moved closer, the force grows until it saturates at $F_{max} = 2\pi k_m M^2 A = 2\pi k_m M^2 a^2$.

11-G.14 A quadrupole E field in a region can be produced by equal and opposite dipoles placed equally far away from the region. Combining a quadrupole E field, such as $\vec{E} = \alpha(x\hat{x} + y\hat{y} - 2z\hat{z})$, and a uniform B field along z, produces what is called an *ion trap* because it can confine a particle by enforcing harmonic motion. (a) Write the equation of motion in the z-direction. Find the sign of α that causes confinement along z, and give the frequency ω_z of periodic motion along z. (b) Write the equation describing circular motion in the xy-plane with radius r and speed v. Solve for B_z in terms of $\omega_\perp = v/r$. Find the $\omega_\perp$ for which B_z is a minimum, and find the minimum value of B_z in terms of α. (Another form of ion trap uses a quadrupolar potential that oscillates rapidly in time, herding the ions first one way and then the other, so quickly that the average displacement is always opposite to the average force.)

11-G.15 Here is a very sophisticated problem that involves concepts from Chapters 1, 4, and 6, as well as the present chapter. Consider a long wire of radius a, with conductivity σ, carrier density n, and carrier charge $-e$. (a) Show that constant current density $\vec{J} = \sigma(\vec{E} + \vec{v} \times \vec{B})$ implies that at radius $r < a$ there is a radial component of the electric field $E_r = -(2\pi k_m J^2/ne)r$. (Here $\vec{B}$ is due to $\vec{J}$.) (b) Find the corresponding bulk charge density and surface charge density. (E. N. Miranda.)

11-G.16 Not knowing of Ohm's work, Henry found empirically that he could obtain a good electromagnet only by an appropriate (impedance) matching of his battery (with its internal resistance) to his electromagnet. He used the term *quantity battery* to describe a single cell or n cells in parallel, and the term *intensity battery* to describe one with n cells in series. He also used the term *quantity magnet* to describe an electromagnet wound with a short coil, and the term *quality magnet* to describe one with a long coil. Let each cell have resistance r and emf $\mathcal{E}$, let the short coil have resistance R_0, and let the long coil have N turns and resistance NR_0. Further, let the electromagnet have a fixed area A and length l. (a) Explain why Henry could get a large magnetic field with matched coil and battery, but only a low field with an unmatched pair (for example, an intensity battery and a quality magnet). (b) Show that, for quantity matching, $B_{gap} = 4\pi k_m(1/l)[\mathcal{E}/(R_0 + r/n)]$. (c) Show that, for intensity–quality matching, $B_{gap} = 4\pi k_m(N/l)[\mathcal{E}/(NR_0/n + r)]$. (d) Show that adding turns to an intensity magnet does not increase its effectiveness once it has enough turns $(NR_0/n \gg r)$.

11-G.17 Consider a hollow cylinder of radius a with a net uniform current I flowing along its axis. Using Ampère's law and symmetry, show that the field magnitude B is $B = 0$ for $r < a$, and $B = 2k_mI/r$ for $r > a$.

11-G.18 Consider a hollow cylinder of radius a with a net uniform current I flowing along its axis. By direct integration, determine the field inside and outside the cylinder. *Hint:* It may be helpful to use the result that, for $b < 1$, $\int_0^{2\pi} \frac{d\phi}{(1-b\cos\phi)} = 2\pi(1-b^2)^{-1/2}$.
(To establish this integral, substitute $\tan\phi/2 = \gamma\tan\psi$. The special value $\gamma = \sqrt{(1-b)/(1+b)}$ then simplifies the integral over ψ.)

11-G.19 Because of Ampère's equivalence, electric fields and magnetic fields can take on very similar forms in the regions away from the field sources. We now use this idea to learn about magnetic fields from some work on electric fields. Problem 5-G.6 showed, for an electrical potential V that oscillated in one direction and grew or decayed in the other, that by appropriate choice of the relative growth and decay, there would be no charge in the region described by the potential. [Specifically, $V(x,y,z) = A\sin qx \exp(-ky)$, where A is in V, and

q and k are in m^{-1}.] For $q = k$, there was no bulk charge, so the associated electric field could also be a magnetic field. Now consider the problem of the stray field of a real solenoid, with individual wires separated by a distance a. Here, the magnetic field outside and very near the turns of the solenoid oscillates in space as you move along and around the axis. (a) Qualitatively, how does the magnetic field vary radially? (b) What is the spatial scale over which the magnetic field varies radially? Now consider what is called *twisted pair*, a pair of wires carrying current I to and from an electrical device. (c) If the wires were straight, how would the magnetic field fall off with radial distance r? (d) If the wires are wrapped around one another, with characteristic spacial period p, qualitatively how does the magnetic field vary radially, and quantitatively what is the characteristic length over which it varies? (Twisted pair is used to eliminate stray magnetic fields.)

11-G.20 A circuit consists of a straight section from $(-a,0,0)$ to $(a,0,0)$, and then a semicircle of radius a from $(a,0,0)$ to $(-a,0,0)$. If the current circulates clockwise, find the magnetic field at $(0,0,z)$.

11-G.21 Consider a solenoid of arbitrary cross-section normal to the z-axis. Here is another way to show that, at any point within the solenoid, $B = 4\pi k_m nI$. (a) As in Problem 11-10.7, show that the sum of the contributions to $\vec{B}$ from two elements, symmetrically above and below the observation point, points along the axis. (b) Use translational symmetry to argue that the field does not vary along the axis. (c) Use Ampère's law to show that the field does not vary within the solenoid. (d) Use Ampère's law to find $\vec{B}$.

11-G.22 A cylinder of length l and radius a is wrapped with wire of radius $r \ll a$ around its axis, as for a solenoid, beginning at one end and ending at the other with N turns. The wrapping continues back to the original end, and back and forth again and again, until the effective cylinder radius equals $b > a$. The result is, effectively, a superposition of concentric solenoids. Each set has the same N turns, length l, and common ends, but a radius that goes from a to b in steps of $2r \ll b - a$. Find the field on the axis, a distance x from the midpoint.

"When the contact was made, there was a sudden and very slight effect at the galvanometer, and there was also a similar slight effect when the contact with the battery was broken. But when the voltaic current was continuing to pass through the one helix, no galvanometric appearances nor any effect like induction upon the other helix could be perceived, although the active power of the battery was proved to be great, by its heating the whole of its own helix, and by the brilliancy of the discharge when made through charcoal."

*"As the wires approximated [approached], the induced current was in the **contrary** direction to the inducing current. As the wires receded, the induced current was in the **same** direction as the inducing current. When the wires remained stationary, there was no induced current."*

—Michael Faraday
Experimental Researches, Vol. I

Chapter 12

Faraday's Law of Electromagnetic Induction

Chapter Overview

There are two parts to this chapter. The first part discusses Faraday's law. The second part discusses Faraday's law applied to circuits. Section 12.1 provides a brief introduction to electromagnetic induction, an emf due to a time-varying magnetic field.

Section 12.2 places the discovery in its historical context. Section 12.3 discusses some of Faraday's original experiments, introducing the useful terminology of *primary* and *secondary*. The *primary* circuit is the source of the time-varying magnetic flux, usually due to electric current driven by a battery. The *secondary* circuit is the circuit to which Faraday's law is applied, and in which the emf is induced, as determined by the deflection of a galvanometer or some other measuring device. Section 12.4 gives a qualitative rule for the direction of circulation of the induced emf (Lenz's law), and some of the complex possibilities for the forces due to the induced current. Section 12.5 formulates Faraday's law quantitatively, and shows the equivalence of the induced emf and the circulation of the electric field. Section 12.6 shows how the coupling between the primary and the secondary can be described in terms of what is called the *mutual inductance*. Section 12.7 discusses motional emf (motion of the circuit in a magnetic field), which provides another way to calculate the induced emf when there is motion.

The second part discusses Faraday's law applied to circuits. Section 12.8 presents the experience of a young man, known to history only as Mr. William Jenkin, who received an electric shock on disconnecting an electromagnet from a battery. This can be understood in terms of *self-inductance,* which describes how a circuit produces an emf in response to changes in its current. Sections 12.9 and 12.10 discuss mutual inductance and self-inductance. Both forms of inductance are associated with induced emfs and induced currents produced by time-varying magnetic fields.

Circuit Elements

A capacitor has the property called capacitance (given the symbol C), a resistor has the property called resistance (given the symbol R), and an inductor has the property called self-inductance (given the symbol L). These are the three basic types of circuit elements.

Section 12.11 discusses a circuit with both a resistor and an inductor, and Section 12.12 discusses the energy to produce a magnetic field. Section 12.13 discusses induced electric fields and considers an example where both an induced electric field and an electrostatic field are needed to produce uniform current flow through a circuit. Section 12.14 discusses Mr. Jenkin's circuit in some detail. ▪

12.1 Introduction

If a charged rod is moved to one side of a conducting loop, then, by *electrostatic induction*, charge on the conducting loop rearranged. However, there is no current flow circulating around the loop. This is because (cf. Chapter 5) the circulation of the electric field due to fixed electric charges is zero. On the other hand, as discovered by Faraday and Henry, a change in the magnetic field acting on a conductor – such as an electric circuit – leads to a non-zero circulation of the electric field, and thus a current, around the circuit. This process is called *electromagnetic induction*. In the quotation at the chapter head, Faraday spoke in terms of induced currents. For a circuit of resistance R, if there is induced current I then by Ohm's Law there must be an induced emf $\mathcal{E}$, by $I = \mathcal{E}/R$. Therefore we speak both of induced currents and induced emfs. Associated with the induced emf is an induced electric field.

Electromagnetic induction can occur in an electric circuit C in many ways. Here are a few: a magnet moves toward C; C moves toward a magnet; C is within a solenoid and the solenoid current changes with time; C rotates in a steady magnetic field, so that the normal $\hat{n}$ to C changes. In each case, *a time-varying magnetic field causes a circulating electric field*. See Figure 12.1 for an example.

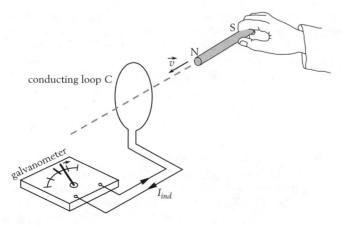

Figure 12.1 Electromagnetic induction caused by a magnet moving toward a conducting circuit C. On the galvanometer, zero current corresponds to the midpoint.

Electromagnetic induction, because it drives currents around circuits, powers modern civilization. Churning river waters (e.g., the Columbia river, from the quote in Chapter 1) and scalding hot steam provide the energy to rotate turbines – laden with copper wires – in external magnetic fields. The resulting time-varying magnetic field induces in the wires electricity that is carried through power lines as alternating current. On a much smaller scale, the signal to the pickup coil of an electric guitar is induced by the time-varying magnetic field of the vibrating steel string (magnetized by a small permanent magnet within the pickup coil.)

12.2 Faraday's Law

12.2.1 *The Discovery of Electromagnetic Induction*

Optional

A. Lost Opportunities After Oersted's 1820 discovery that electric currents produce magnetism, scientists tried to determine if magnetism could produce electricity. Ampère and de la Rive came close in 1822 but, not expecting a *temporary* induced current, misinterpreted their results. Faraday, came close in in 1825, but he connected his secondary circuit only *after* he powered up his primary circuit. Also in 1825, an associate of de la Rive, Colladon, in Geneva, placed the sensitive galvanometer of his secondary in one room and his primary coil in another. After thrusting a powerful bar magnet into the coil (as in Figure 12.1), by the time he would reach the galvanometer, it had returned to its zero point, thus giving the appearance of no response at all. Again, not anticipating a temporary induced current, the effect was missed.

B. Arago's Needle Curiously, in 1824 Arago discovered an effect that indeed was due to electromagnetically induced emfs. He noticed that, on disturbing his compass needle mechanically, it equilibrated more quickly when inside its metal case than when outside the case. He correctly surmised that this was somehow due to relative motion of the magnetic compass needle with the metal of the case. He then built a turntable with a large metal disk. The compass needle, placed above the rotating disk, tended to twist in the sense of the disk's rotation; when turned fast enough, the compass needle would even spin. Others immediately took an interest in this apparent "magnetism of rotation;" finding that a conducting (but nonmagnetic) disk would turn if a powerful magnet was rotated near it. Ampère showed that the effect persisted if a solenoid (i.e. an electromagnet) was substituted for the magnet, thus providing further evidence for the equivalence between conducting loops and magnets. The effect was greatest for those materials with the greatest "conducting power," although there was not yet a quantitative measure — conductivity — of that conducting power. Nevertheless, no explanation was found for Arago's effect, known as "Arago's needle."

C. Discovery by Faraday and Henry Electromagnetic induction finally was discovered independently by two men. One was Michael Faraday, working at the Royal Institution (founded by an expatriate American, Count Rumford) in London. The other was Joseph Henry, working at Albany Academy, in New York. Henry's discovery was made in August of 1830, but he did not publish his results. Faraday's discovery was made in August of 1831, and his results were announced in November of that year. Not until learning of Faraday's work did Henry publish, in 1832. By then he was at the College of New Jersey (since 1896, Princeton University).

Just as the study of *electrostatic* induction of electricity involved the concept of capacitance (both self- and mutual-), so the study of *electromagnetic* induction of electricity involves the concept of inductance (both self- and mutual-). Faraday especially studied the effects of mutual induction between two circuits (the primary and the secondary), whereas Henry especially studied the effects of the self-induction of a circuit. Indeed, Henry remarked upon self-induction in his paper of 1832, whereas Faraday did not discover self-induction until 1834. Faraday's work, however, is more general than Henry's, because Faraday noted that electromagnetic induction occurs when there is relative motion of the primary and the secondary, whereas Henry worked with fixed circuits. Faraday's work is very well-documented because of his habit of taking meticulous notes (now a standard practice in scientific laboratories). On the other hand, many of Henry's notes were burned during a fire in his Washington, DC office of the Smithsonian Institution (in a peculiar symmetry, founded by an Englishman), where Henry had become Director. We present many of Faraday's 1831 experiments in detail, following in his *Experimental Researches, Vol.I*, and his Notes.

12.3 Faraday's Experiments

12.3.1 *Current Induced with Voltaic Cells*

In his published work, Faraday begins by recounting a number of failures to see induced currents in his galvanometer-containing secondary, even on increasing the number of turns of wire in both the primary and secondary. Because insulated wire was not available (except for cotton-covered, soft iron bonnet wire), to use anything but iron wire he had to make his own insulation layer between turns of the wire. As he writes:

> About twenty-six feet of copper wire one-twentieth of an inch in diameter were wound round a cylinder of wood as a helix, the different spires of which were prevented from touching by a thin interposed twine. This helix was then covered with calico [to serve as insulation], and then a second wire applied in the same manner. In this way twelve helices were superposed.

Six of these helices were connected in series and then connected to a galvanometer. The other six helices were connected in series and then connected to a battery consisting, at first, of 10 copper-zinc voltaic cells with 4-inch plates. See Figure 12.2(a). There was no response until Faraday increased the number of copper-zinc voltaic cells in the primary from 10 to 100 (and then to 120). This produced a small effect at the galvanometer, both on connecting and disconnecting the battery; the needle's response on opening was opposite to its response on closing. The first quotation at the chapter head describes this phenomenon.

Faraday wanted to amplify the effect. He argued that the temporary current might be enough to magnetize a steel needle placed within a helix, as he knew occurred in the rapid discharge of Leyden jars (capacitors). See Figure 12.2(b).

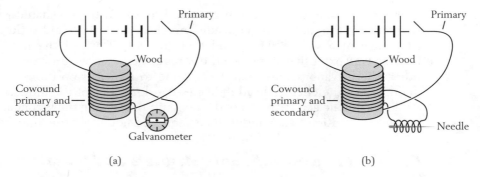

Figure 12.2 Two methods Faraday used to detect induced current in the secondary when the primary circuit was connected or disconnected: (a) a galvanometer, (b) an iron needle, which can magnetize.

He found that the needle magnetized one way on connecting the battery, and the other way on disconnecting it. The magnetization $\vec{M}$ on disconnection was in the same direction as expected if it were produced by a steady current I in the battery circuit, of emf $\mathcal{E}_b$ and resistance R. However, $|\vec{M}|$ on both connection and disconnection was much larger than expected for I given by $\mathcal{E}_b/R$. Hence, during both connection and disconnection, a temporary but large emf was induced, over and above the battery emf $\mathcal{E}_b$.

Faraday next describes a striking variant of the preceding experiment, now employing motion to produce the time variation:

> Several feet of copper wire were stretched in wide zigzag forms, representing the letter W, on one surface of a broad board; a second wire was stretched in precisely similar forms on a second board, so that when brought near the first, the wires should everywhere touch, except that a sheet of thick paper was interposed. [See Figure 12.3.] One wire W was part of a circuit that included a galvanometer, and the other wire W was part of a circuit that included a battery of voltaic cells. The first wire was moved towards the second, and as it approached, the needle was deflected. Being then removed, the needle was deflected in the opposite direction.

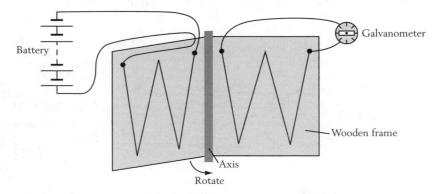

Figure 12.3 Faraday's rotating W circuits. By rotation about its axis, the wooden frame is opened and closed at the resonance frequency of the galvanometer. The current in the left-hand W (the primary) then induces current in the right-hand W (the secondary).

He then made the wires approach and recede at the mechanical resonance frequency of the galvanometer needle (much as we push a child at the resonance frequency of a swing). He found that "the vibrations of the needle...soon became very extensive; but when the wires ceased to move from or towards each other, the galvanometer-needle soon came to its usual position." The second quotation at the chapter head indicates the directions of the induced currents. Adding a voltaic cell to the secondary circuit, thus giving the secondary a steady current in addition to any induced currents, did not affect the induced current.

12.3.2 *Current Induced with Soft Magnets and Voltaic Cells*

In his published work, having described the induction of electric currents by electric currents (what he called "volta-induction"), Faraday then turned to "the evolution of electricity by magnetism" (what he called "magneto-electric induction"). He describes his work with primary and secondary both wrapped around an iron toroid 7/8-inch thick and of outer radius 6 inches. The primary (P) contained three separate helices, each made with 24 feet of copper wire. The secondary (S) contained a helix made with 60 feet of copper wire. With the three helices connected in series, the primary was connected to a battery of only ten copper-zinc cells. See Figure 12.4(a).

> The galvanometer was immediately affected, and to a degree far beyond what has been described when with a battery of tenfold power *without iron* was used; but though the contact was continued, the effect was not permanent, for the needle soon came to rest in its natural position, as if quite indifferent to the attached electro-magnetic arrangement. Upon breaking the contact with the battery, the needle was again powerfully deflected, but in the contrary direction to that induced in the first instance.

When both primary and secondary were wrapped around a hollow pasteboard cylinder, a very weak current could be induced, but when an iron rod of 7/8-inch diameter and 12-inch length was placed within the cylinder, and the primary switched on, "the induced current affected the galvanometer powerfully." See Figure 12.4(b). A copper rod produced no such enhancement of the induced current.

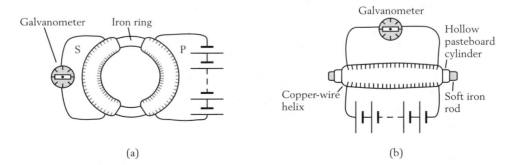

(a)   (b)

Figure 12.4 Current induced with electromagnets. (a) The soft iron of the toroid serves to enhance the common field of the primary and secondary. (b) With the helix filled with soft iron (but not when filled with air or copper), the galvanometer would respond when the primary was connected or disconnected.

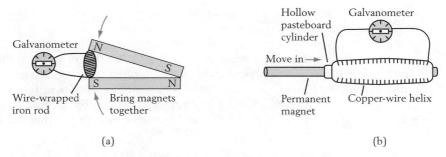

Figure 12.5 (a) When the magnets are brought together or pulled apart, there is an induced current in the galvanometer. (b) When pushed in or out of the core of the helix, the soft iron rod strengthens or weakens the field seen by the secondary.

12.3.3 *Current Induced with Hard Magnets*

Faraday also used a pair of identical 24-inch-long permanent magnets to produce the magnetic field in the iron core. He arranged the magnets with their opposite poles at one end in contact, and then contacted the other poles with the opposite ends of the iron core, "so as to convert it for the time into a magnet." Figure 12.5(a) shows the view looking down on the laboratory bench at the galvanometer. On bringing the magnets together, the induced current was in one direction; on separating them, the induced current was in the opposite direction.

To eliminate the possibility that the preceding results were associated with the magnetization process, Faraday took a permanent magnet 3/4 inches in diameter and 8 1/2 inches long and thrust it into and out of the hollow pasteboard cylinder, successively inducing currents one way and then the other, consistent with turning on and off an electromagnet. See Figure 12.5(b). When the magnet was pushed all the way through, current was induced first in one direction and then the other. Adding a voltaic cell to the secondary circuit, thus giving it a steady current, did not affect the induced current.

Faraday also performed an experiment in which the galvanometer responded to wires from two secondaries made of *different materials*. In this way, he determined that "the currents produced by magneto-electric induction in bodies is proportional to their conducting power." From this and other experiments, he came to the important conclusion that *the amount of induced emf is independent of the material of which the wire is made*. Although he was on the verge of discovering Ohm's law, he did not quite couch his results in terms that would permit a quantitative determination of "conducting power." In modern terms, we would say that if Faraday had used two wires with resistances $R_1 = 1 \ \Omega$ and $R_2 = 100 \ \Omega$, and subjected them to the same induced emf of $\mathcal{E} = 5$ V, then currents $I_1 = 5/1 = 5$ A and $I_2 = 5/100 = 0.05$ A would have been induced.

12.3.4 *The Order of Faraday's Discoveries*

Faraday's notes reveal that he made his first discoveries about electromagnetic induction while using electromagnets and permanent magnets, only later turning to current-carrying electric circuits without soft iron. Since he accepted Ampère's equivalence between magnets and current-carrying circuits, likely this order developed because he was searching for effects associated with magnetic fields, and electromagnets and magnets typically provided more powerful fields than electric circuits. Only after seeing an induced current when thrusting a magnet into a secondary would he have been confident of seeing an induced current on moving one W-shaped circuit toward another. Even then, as indicated, he had to exploit the galvanometer needle's mechanical resonance to cause an appreciable response.

12.3.5 *Summary of Faraday's Results*

Faraday performed a number of other experiments, but we now know enough of them to draw the most important conclusions:

1. A *change of current* in the primary P induces an emf in the secondary.

2. *Relative motion* of the primary P and the secondary S induces an emf in the secondary. (This is true only when there is current in the primary.)

3. *Relative motion* of a magnet and the secondary S induces an emf in the secondary.

In each case, a time variation in a magnetic field causes an electric current to circulate around an electric circuit. More precisely, as Faraday came to see it, *a change in the number of magnetic field lines enclosed by a circuit causes an electric current to circulate around the circuit.* But if more field lines are enclosed by the circuit, in which direction will the current circulate?

Faraday gave a rule for the direction of the circulation of the induced electric current circulating around an electric circuit. He thought in terms of magnetic field lines crossing a circuit. They can point up or down relative to the area enclosed by the circuit. Here is his rule. If field lines due to an external field enter the area enclosed by the circuit, the response is to induce a current in the circuit that makes field lines that leave the circuit; and if external field lines leave the circuit, then the response is to induce a current making field lines that enter. (In doing the field line count, an entering field line oriented one way is equivalent to a leaving field line oriented the other way.)

Figure 12.6(a) presents an example corresponding to relative motion of two circuits, case 2 in the shaded summary box. The primary and secondary are co-axial coils, with the primary carrying current I_P counterclockwise as viewed from the right, and moving rightward toward the secondary. Depicted are the field lines produced by the primary, as well as the time rate of change in the external field "seen" by the secondary, which we call $d\vec{B}_{ext}/dt$. Figure 12.6(b) depicts the response of the secondary, which makes its own field $\vec{B}_{ind}$ by the induced current I_{ind}, where Ampère's right-hand rule relates $\vec{B}_{ind}$ and I_{ind}.

Food for Thought: Convince yourself that if in Figure 12.6(a) the primary moves leftward, but the current I_P remains counter-clockwise, then I_{ind} reverses. Does $\vec{B}_{ext}$ now point → or ←? [My Answer:] Does $d\vec{B}_{ext}/dt$ now point → or ←? [My Answer:] Does $\vec{B}_{ind}$ now point → or ←? [My Answer:]

Food for Thought: Convince yourself that if in Figure 12.6(a) the current I_P reverses, thus reversing $\vec{B}_{ext}$, and if the primary moves rightward,

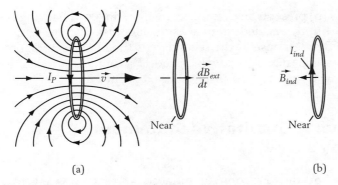

(a)   (b)

Figure 12.6 (a) A primary current loop carrying current
I_P is moved toward the secondary loop, subjecting the
secondary loop to a field $\vec{B}_{ext}$ that is increasing to the
right. (b) The secondary responds with an induced
current I_{ind} that produces a field $\vec{B}_{ind}$ that is to the left
(i.e., opposite to $d\vec{B}_{ext}/dt$).

then I_{ind} reverses. Does $\vec{B}_{ext}$ now point → or ←? [My Answer:] Does
$d\vec{B}_{ext}/dt$ now point → or ←? [My Answer:] Does $\vec{B}_{ind}$ now point → or ←?
[My Answer:]

Food for Thought: If in Figure 12.6(a) the current I_P reverses, thus reversing
$\vec{B}_{ext}$, and if the primary moves leftward, is I_{ind} in the same or the opposite direction relative to I_{ind} of Figure 12.6(a)? [My Answer:] Does $\vec{B}_{ext}$ now point →
or ←? [Answer:] Does $d\vec{B}_{ext}/dt$ now point → or ←? [My Answer:] Does
$\vec{B}_{ind}$ now point → or ←? [My Answer:]

Food for Thought: From these cases, is there any general correlation between
the direction of $\vec{B}_{ext}$ and the direction of $\vec{B}_{ind}$? [My Answer:] Between the
direction of $d\vec{B}_{ext}/dt$ and the direction of $\vec{B}_{ind}$? [My Answer:]

Figure 12.7(a) presents an example corresponding to relative motion of a magnet and the secondary, case 3 in the shaded summary box. An equivalent magnet
moves toward the secondary, whose response is the same as in Figure 12.6(b).
If the equivalent magnet were to move in the opposite direction, then $d\vec{B}_{ext}/dt$,
$\vec{B}_{ind}$, and I_{ind} would all reverse direction.

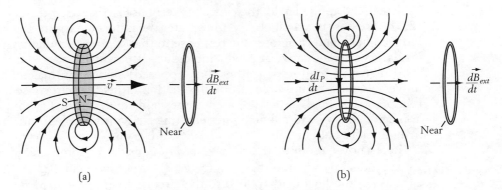

(a)   (b)

Figure 12.7 The same response in the secondary of Figure 12.6 can be produced when
(a) an equivalent magnet moves toward the secondary loop, (b) the current I_P in the
primary increases.

Figure 12.7(b) presents an example corresponding to a changing current in the primary, case 1 in the shaded summary box. Here the current I_P in the primary is increasing. The response of the secondary is the same as in Figure 12.6(b). If I_P were decreasing, then $d\vec{B}_{ext}/dt$, $\vec{B}_{ind}$, and I_{ind} would all reverse. However, $\vec{B}_{ext}$ would continue to point rightward because it is determined by I_P.

12.4 Lenz's Law: A Qualitative Statement of Faraday's Law

Another rule for the direction of induced currents was given by Lenz in 1834. Translating it into modern terminology, Lenz's is our most complete and accurate qualitative statement of Faraday's law. It is formulated in terms of the idea of magnetic flux. For Gauss's law, the *electric* flux through a *closed* surface is what mattered. For Lenz's law, what counts is the *magnetic* flux through an *open* surface, defined by the circuit whose induced emf we seek. One way of visualizing a change in magnetic flux through a circuit is to think, as did Faraday, in terms of field lines crossing the electric circuit.

Lenz's law: *A change in magnetic flux through a circuit induces a circulating electric field; this, in turn, induces a circulating electric current whose own magnetic flux opposes the original change.* (The direction of the magnetic flux caused by the induced current is determined by Ampère's right-hand rule.)

As discussed in Section 9.5 and Figure 9.13(a), a perfect diamagnet in a static applied magnetic field produces its own magnetic field that, within the diamagnet, completely cancels the applied field. Thus we say that it "expels" the magnetic field. It produces this field by setting up (surface) currents (Section 11.11). Lenz's law implies that all materials are like *dynamical* versions of perfect diamagnets, producing induced currents whose field opposes *changes* in the applied magnetic field.

Because Lenz's law only tells us about directions of induced currents and emfs, the *magnitude* of the rate of change of the magnetic flux is irrelevant in Lenz's law. Here is how to maximize the response, that is, to make a material respond most like a perfect diamagnet: (1) for a given material, the *faster* the change in the applied field, the closer the response to that of a perfect diamagnet; (2) the greater the conductivity, the greater the induced current, the greater the induced field, and the closer the response to that of a perfect diamagnet. (Thus, insulators are ineffective at producing induced currents.)

If materials reinforced (rather than opposed) changes in magnetic flux, the universe would be unstable: a tiny fluctuation in field would cause a circuit to reinforce that field, so an infinitely large field (and current) would build up. You can imagine external field lines entering a circuit, and then the circuit itself making more field lines enter the circuit, and so on.

Let us now cut our teeth on Lenz's law by considering some examples. Ideally, we should show separate figures giving (1) the initial field $\vec{B}_{ext}$ applied to the circuit, (2) the final field applied to the circuit, and (3) the field $\vec{B}_{ind}$ produced by the current induced in the circuit. In practice, we will indicate only the directions of $d\vec{B}_{ext}/dt$ and $\vec{B}_{ind}$.

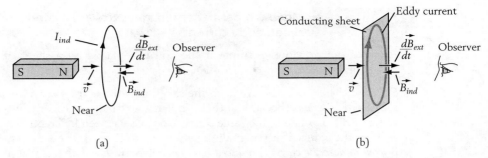

Figure 12.8 Induced currents in two very similar cases, where the N pole of the magnet is moved toward the right. The field acting on the secondaries points rightward and increases. (a) Current loop as secondary. (b) Conducting sheet as secondary (the induced current is called an "eddy current" because it is similar to a swirling eddy in a stream).

Example 12.1   **Moving a permanent magnet toward a coil**

Consider an experiment like that of Figure 12.5(b). Figure 12.8(a) depicts a circular loop in the *yz*-plane at the origin. To the left of the loop, its north pole pointing to the origin, a bar magnet moves toward the loop. Determine how the motion of the magnet affects the field seen by the coil, and how the coil opposes the change in field by its induced emf, the net force on the loop, and whether the loop expands or contracts.

Solution: The motion of the bar magnet increases the applied field $\vec{B}_{ext}$ in the rightward direction. By Lenz's law the loop produces its own induced field $\vec{B}_{ind}$ in the leftward direction. Figure 12.8(a) depicts $d\vec{B}_{ext}/dt$ and $\vec{B}_{ind}$.

By Ampère's right-hand rule, to make such a $\vec{B}_{ind}$ the induced current circulates clockwise as seen by an observer on the right. Further, by Ampère's equivalence, this current loop is like a magnetic disk (not drawn) whose magnetic moment points to the left. Since the N pole of the permanent magnet and the N pole of the equivalent magnetic disk are opposed, the force on the coil is repulsive (i.e., to the right). A similar geometry was discussed in Chapter 10, for the force on a speaker coil due to a permanent magnet (see Figure 10.16). There it was also concluded that a net compressive force acts on the loop. Table 12.1 summarizes our results for the response of the loop in Figure 12.8(a). The directions are given as seen by an observer to the right. (In stating directions, we must always state the position of the observer.)

Properly, this table should be given in terms of magnetic flux, instead of magnetic field. We use magnetic field here because we haven't yet defined magnetic flux in a precise fashion.

Food for Thought: How would table 12.1 change if the magnet were pulled away from the loop? If the loop were pushed toward the magnet? If the loop were pulled away from the magnet? If the magnetic poles were reversed?

Table 12.1 Description of Example 12.1

Observer	$d\vec{B}_{ext}/dt$	$\vec{B}_{ind}$	$\mathcal{E}_{ind}, I_{ind}$	$\vec{F}_{net}$	Compress or expand
To right	⊙	⊗	Clockwise	⊙	Compress

Example 12.2	**Moving a permanent magnet toward a conducting sheet: eddy currents**

Analyze the previous example if the coil is replaced by a sheet of aluminum foil.

Solution: The induced currents in the sheet, now distributed continuously, and called *eddy currents* (because of their similarity to small eddies, or whirlpools, in water), will circulate in the same sense (clockwise as seen by an observer to the right) as the induced current in the coil. See Figure 12.8(b). *The term eddy currents is used to describe currents induced in a solid object rather than a circuit.* Faraday's law applies both for solid objects and for conducting circuits.

The preceding examples lead to a left-hand rule, which we will call Lenz's left-hand-rule.

Lenz's Left-hand rule: *If the thumb of your left hand is pointed along the direction of the rate of change of the magnetic flux, then the fingers of your left hand will curl in the direction of the induced electric field.*

This rule is not often stated, but it is true nevertheless. Some students may find it helpful (different strokes for different folks), but we will not make further use of it.

Example 12.3	**Powering up the primary**

Consider an experiment like that of Figure 12.9(a), which depicts two co-axial coils, their normals along the x-axis. Turning on the switch in the primary (at the origin) causes current to flow counterclockwise as seen by an observer to the right. $\vec{B}_{ext}$ denotes the external field acting on the secondary due to the primary, and $\vec{B}_{ind}$ denotes the induced field acting on the secondary due to the secondary itself. This configuration applies to induction ranges, where the primary is built into the range, and the secondary corresponds to the pot or pan to be heated (so that eddy currents are induced in the secondary). (a) Apply Lenz's law to obtain the directions of the induced emf and the induced current. (b) Find the magnetic forces acting on the secondary coil.

Solution: (a) When the current is switched on in the primary, from Ampère's law this produces a $d\vec{B}_{ext}/dt$ that points *toward* the observer ($\rightarrow$ to us, or $\odot$ to the observer). Since, by Lenz's law, $\vec{B}_{ind}$ must point opposite to $d\vec{B}_{ext}/dt$, then $\vec{B}_{ind}$ points *away from* the observer ($\leftarrow$ to us, or $\otimes$ to the observer). Again by Ampère's law, to produce such a $\vec{B}_{ind}$, the induced emf and the induced current must circulate *clockwise* as seen by the observer. Thus, for an element at the top of the secondary, $d\vec{s}$ points into the page.

(b) To find the force on the secondary due to the induced current, use the force law $d\vec{F} = I d\vec{s} \times \vec{B}$, where $I = I_S$ is the induced current in the secondary and $\vec{B} = \vec{B}_{ext}$ is the applied magnetic field. See Figure 12.9(b). (This is the same geometry as Figure 10.16b.) Consider an element $d\vec{s}$ at the top of the secondary so that it points into the page. A short amount of time has elapsed so that $\vec{B} = \vec{B}_{ext}$ from the primary has had some time to build up; it flares outward and points toward the observer. The cross-product then gives that $d\vec{F}$ points partly toward the observer (repulsion) and partly radially inward (compression). This is consistent with a motion statement of Lenz's law.

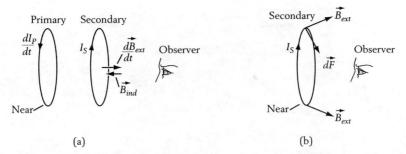

Figure 12.9 (a) Turn-on of a primary causes a rightward field at the secondary, whose induced current tends to make a leftward field. (b) Detailed analysis that gives the force $d\vec{F}$ on an element $d\vec{s}$ due to the induced current in the field $\vec{B}_{ext}$ of the primary. At the top of the loop, $d\vec{s}$ points along $\otimes$ (into the page).

> **Motion statement of Lenz's law**: *The system tends to minimize its flux change, by translation or rotation relative to the primary, and by tending to compress or expand.*

This rule must not be applied too literally. If the system already has a current, there may be preexisting forces on the system; the motion statement of Lenz's law refers only to the forces due to *induced* currents. Thus, in Figure 12.9(a), if there initially had been a current going down the near side, initially there also would be an attractive force between the loop and the magnet, independent of any induced currents and forces.

Another way to obtain the direction of the interaction force is to employ Ampère's equivalence. Replace the primary and its current by its equivalent magnet with its magnetic moment pointing *toward* the observer, and replace the secondary and its current by its equivalent magnet with its magnetic moment pointing *away from* the observer. These two magnets repel; hence the two circuits repel, so the secondary feels a force *toward* the observer.

The force on the primary is equal and opposite to the force on the secondary, and can be obtained from the equivalent magnet viewpoint. The primary also will feel a force of compression, to oppose the change in its own flux.

Table 12.2 summarizes our results for the response of the secondary in Figure 12.9(a). The directions are given as seen by an observer to the right.

Food for Thought: How would table 12.2 change if the primary were powered up in the opposite direction?

Application 12.1   **Magnet above rotating turntable (Arago's needle): eddy current drag, lift, and radial force**

Sometimes, the induced force on an object may require an even more complex description. Consider a magnet held above a rotating aluminum turntable,

Table 12.2 Description of Example 12.3

Observer	$d\vec{B}_{ext}/dt$	$\vec{B}_{ind}$	$\mathcal{E}_{ind}, I_{ind}$	$\vec{F}_{net}$	Compress or expand
To right	$\odot$	$\otimes$	Clockwise	$\odot$	Compress

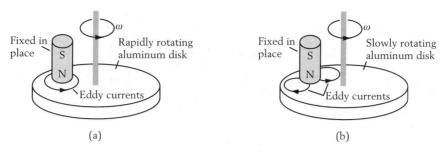

Figure 12.10 Eddy currents induced when a magnet is fixed above a rotating conducting disk. (a) At high rotation rates, the induced currents predominantly expel the field of the magnet (lift, without drag). (b) At low rotation rates, the induced currents predominantly lead to drag (without lift).

where eddy currents are induced, as in Figure 12.10. We now discuss the eddy currents and forces acting on the turntable, as a function of its rate of rotation.

Figure 12.10(a) shows a rapidly rotating conductor. This corresponds to a situation like that of a perfect diamagnet, where the field is expelled and thus the dominant force on the magnet is lift. By pushing the magnet away from the turntable, this decreases the rate of change of magnetic flux seen by the turntable. This lift is called eddy current magnetic levitation (MAGLEV). The eddy currents correspond to an image magnet that repels the actual magnet. Figure 12.10(b) shows a slowly rotating conductor. Here there is relatively little flux expulsion. But there is another way to satisfy Lenz's law. At low velocities, the dominant force on the magnet is drag. This brings the magnet into rotation with the turntable, thereby decreasing the rate of change of magnetic flux seen by the turntable. The eddy currents correspond to image magnets that tend to drag the actual magnet into rotation. Both lift and drag forces occur, to a certain extent, at all rotation rates. In addition, there is a radial force: it is inward when the magnet is near the center (tending to minimize the rate of change of magnetic flux), and outward radial when the magnet is near the edge (tending to minimize the rate of change of magnetic flux). If we were to analyze this system in terms of primary and secondary, the magnet (or its equivalent current) would serve as the primary, and the turntable would serve as the secondary.

These phenomena are closely related to those observed by Arago (see Section 12.2.2) with his compass needle above a rotating metal disk although Arago originally noticed only the equivalent of the drag effect; the two poles of his magnetic needle were dragged in opposite directions, leading to a torque that tended to bring the needle into rotation. They are also closely related to *magnetic braking:* the rotating disk corresponds to the wheel of a train, and the magnet to an electromagnet that is applied when braking is desired.

Food for Thought: How would the eddy currents in Figure 12.10(a) be replaced by a magnet or magnets within the material of the turntable? in Figure 12.10(b)?

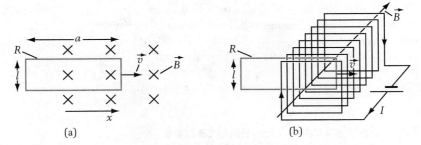

Figure 12.11 (a) A rectangular loop moves into the region of a uniform field $\vec{B}$. (b) A realization of the situation in part (a), where a solenoid of square cross-section produces $\vec{B}$.

Example 12.4 Loop pulled through a $\vec{B}$ field

This resembles the experiment of Figure 12.3. Consider a rectangular loop (the secondary) of resistance R, length a along x, and length l along y. It moves along $+x$ with constant velocity $\vec{v}$ in a region of zero magnetic field, until it reaches a region of uniform field $\vec{B}$ that points into the paper $(-\hat{z})$. Figure 12.11(a) depicts the situation in two dimensions, and Figure 12.11(b) depicts it in three dimensions. From Faraday's law, there will be an emf, and from the motion statement of Lenz's law, there will be an induced magnetic force leftward, opposing the motion. Assume that a rightward external force ensures that the velocity remains constant. Determine the response of the system: (a) on entering the field region; (b) on leaving the field region.

Solution: (a) On entering the field region, the loop gains flux into the paper, so by Lenz's law the loop generates its own flux out of the paper. Thus the induced field is opposite the external field. By Ampère's right-hand rule, the associated current and emf must be counterclockwise. The motion statement of Lenz's law says that the net force due to the induced current opposes the motion, and thus will be to the left. In addition, by the motion statement of Lenz's law, to oppose the increase in flux there must be a compressive force due to the induced current. Once completely within the region of field, the flux is constant, and there is no tendency to generate an emf. Table 12.3 summarizes the response of the loop. The directions are given as seen by the reader.

(b) On leaving the region of field, the flux into the paper due to $\vec{B}_{ext}$ decreases, so by Lenz's law the loop generates its own flux into the paper. By Ampère's law, the associated current and emf must be clockwise. The motion statement of Lenz's law says that the net force due to the induced current will oppose the motion, and thus will be to the left. In addition, by the motion statement of Lenz's law, to oppose the decrease in flux there must be an expansion force due to the induced current. In this case, $\vec{B}_{ind}$ is in the same direction as $\vec{B}_{ext}$.

Table 12.3 Description of Example 12.4

Observer	$d\vec{B}_{ext}/dt$	$\vec{B}_{ind}$	$\mathcal{E}_{ind},\ I_{ind}$	$\vec{F}_{net}$	Compress or expand
Reader	$\otimes$	$\odot$	Counterclockwise	$\leftarrow$	Compress

Another way to look at Lenz's law is to use Faraday's idea that an emf is induced whenever a magnetic field line crosses the circuit. If a line of magnetic field leaves the circuit, then the induced emf drives a current that tends to replace the lost field line. Similarly, if a line of magnetic field enters the circuit, then the induced emf drives a current that tends to cancel the gained field line.

We now turn from a qualitative to a quantitative discussion of Faraday's law.

12.5 Faraday's Law—Quantitative

To determine the forces due to induced currents, we need the currents, and for that we need the induced emf and Ohm's law. Faraday's law relates the induced emf of a circuit to the rate of change of magnetic flux through that circuit.

12.5.1 Defining Electromagnetically Induced EMF

We must first define what we mean by the emf $\mathcal{E}$ around a closed circuit, due to the electric field $\vec{E}$. This is the work per unit charge done by the electric force $\vec{F}_{el} = q\vec{E}$, or

$$\mathcal{E} = \frac{W}{q} = \frac{1}{q}\oint \vec{F}_{el} \cdot d\vec{s} = \frac{1}{q}\oint q\vec{E} \cdot d\vec{s}. \tag{12.1}$$

This may be rewritten as

$$\mathcal{E} = \oint \vec{E} \cdot d\vec{s} = \oint \vec{E} \cdot \hat{s}\, ds \equiv \Gamma_E, \qquad \text{(induced emf equals electric circulation)} \tag{12.2}$$

which is precisely the circulation Γ_E of the electric field. Note that $ds = |d\vec{s}| > 0$, and $\hat{s}$ points along the direction of $d\vec{s}$. *Thus the electromagnetically induced emf and the electric circulation are one and the same.*

Chapter 5, on electric potential, showed that the circulation $\Gamma_E = 0$ for electric fields due to static charges, so that static electric fields produce no emf for a circuit as a whole. However, they can produce an emf across part of a circuit, such as a resistor, as we show explicitly in Example 12.13.

12.5.2 Defining Magnetic Flux Φ_B

The total magnetic flux Φ_B passing through a circuit is defined by

$$\Phi_B = \int \frac{d\Phi_B}{dA}dA = \int \vec{B} \cdot \hat{n}\, dA = \int \vec{B} \cdot d\vec{A}, \qquad d\vec{A} \equiv \hat{n}\, dA. \tag{12.3}$$

Here $dA = |d\vec{A}| > 0$ is defined by the circuit itself, and $\hat{n}$ is along the normal to the circuit. Φ_B is measured in units of $Wb = T\text{-}m^2$, called the weber. The concept of magnetic field lines translates into magnetic flux, just as the concept of electric field lines translates into electric flux.

Faraday's law relates $\mathcal{E}$ around a circuit to the time rate of change of the magnetic flux Φ_B, defined in (12.3), that passes through the open area bounded

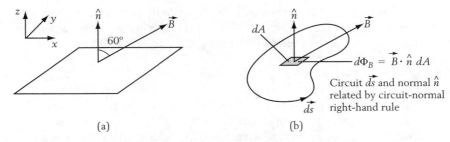

(a) (b)

Figure 12.12 (a) A square loop in a uniform field $\vec{B}$ that is inclined to the loop's normal $\hat{n}$. (b) A more general circuit, with the relation between a circuit element $d\vec{s}$ and the circuit normal $\hat{n}$.

by the circuit. Before stating Faraday's law, let us consider a simple example of how to use (12.3) to compute magnetic flux.

Example 12.5 Square loop in a $\vec{B}$ field

Consider a square loop of area $A = 25 \times 10^{-4}$ m^2 in a uniform $\vec{B}$ field in the xz-plane, with $|\vec{B}| = 0.02$ T. Let $\vec{B}$ make an angle of $60°$ to the normal $\hat{n}$, which is along $\hat{z}$. See Figure 12.12(a). (This situation can be produced by putting the loop within a large solenoid, and tilting the axis of the loop relative to the solenoid by $60°$.) Find the flux Φ_B through the square loop.

Solution: By (12.3),

$$\Phi_B = \int \vec{B} \cdot \hat{n} dA = \int |\vec{B}||\hat{n}| \cos 60° dA = \int |\vec{B}| \frac{1}{2} dA$$

$$= \frac{|\vec{B}|}{2} \int dA = \frac{|\vec{B}|}{2} A = 2.5 \times 10^{-4} \text{ Wb.}$$

On the other hand, if the loop has $N = 200$ turns in the same direction, then the total flux is larger by a factor of 200: $\Phi_B = 0.05$ Wb.

12.5.3 *Quantitative Statement of Faraday's Law*

Faraday's law relates the emf, or electric circulation around a closed path defined by a set of line elements $d\vec{s}$, to the rate of change of the magnetic flux through the area associated with this path, defined by a set of area elements $d\vec{A} = \hat{n} dA$. The same circuit-normal right-hand rule is used as in Ampère's law: when $d\vec{s}$ circulates clockwise (counterclockwise), the direction of $\hat{n}$ is into (out of) the paper. See Figure 12.12(b). With this convention, Faraday's law is

$$\mathcal{E} = \oint \vec{E} \cdot d\vec{s} = -\frac{d}{dt} \int \vec{B} \cdot \hat{n} dA = -\frac{d\Phi_B}{dt}. \quad \text{(Faraday's law)} \quad (12.4)$$

In this chapter, we will often apply (12.4) to circuits that correspond to real physical circuits made of conducting wire, but (12.4) applies to *all* circuits, and thus applies to eddy currents, as in Figure 12.8(b).

From (12.4), there are three independent ways to change a flux: (1) change the field, (2) change the area though which the field passes, and (3) change the angle between the field and the area. Of course, any combination of these three also can change the flux.

Faraday's law gives an induced magnetic field that satisfies Lenz's law. From Chapter 7, for an ohmic material the current density $\vec{J} = \sigma \vec{E}$, where σ is the conductivity. Integration around the current path yields $\oint \vec{J} \cdot d\vec{s} = \sigma \oint \vec{E} \cdot d\vec{s} \neq 0$. Hence, for a uniform system, if the electric field has a circulation, so does the current. Typically, this is true even if the system is not uniform. The electric field may be thought of as having a part that is electrostatic, denoted by $\vec{E}_{es}$, and an electromagnetically induced part, called $\vec{E}_{ind}$. Only $\vec{E}_{ind}$ has nonzero circulation, and only the $\vec{E}_{ind}$ can cause current to be driven around a circuit. However, since $\vec{E}_{ind}$ usually is not uniform around a circuit, usually both $\vec{E}_{ind}$ and $\vec{E}_{es}$ are needed to drive a *uniform* current around a circuit. That means charge must rearrange to produce a nonzero $\vec{E}_{es}$.

Example 12.6 **Square loop in time-varying $\vec{B}$ field of solenoid**

In Example 12.5, the square loop of Figure 12.12(a) is the secondary, and the primary may be taken to be a solenoid (not shown) that encloses the square loop. Let the field of the primary increase at the rate $|d\vec{B}/dt| = 500$ T/s, so the flux increases. (a) Discuss the induced current, forces, and torques. (b) Find the magnitude of the induced emf.

Solution: (a) To oppose the increasing field in Figure 12.12(a), by Lenz's law the induced current is clockwise as seen looking down from the z-axis. By Chapter 10, because the field is uniform, there is no net force on the loop, even when it carries a current. However, because the flux through the square loop is increasing, the motion statement of Lenz's law states that to oppose the change in flux there must be a compressive force on the loop due to the induced current. In addition, the motion statement of Lenz's law says that there will be a torque on the loop, tending to rotate it so that it picks up less flux. From the direction of the induced current, by Ampère's equivalence the magnetic moment $\vec{\mu}$ points opposite to $\hat{n}$, and this leads to a torque that rotates $\hat{n}$ toward $-\hat{x}$. (b) For a single turn, with Wb/s = V,

$$\left| \frac{d\Phi_B}{dt} \right| = \left| \int \frac{d\vec{B}}{dt} \cdot \hat{n} dA \right| = \int \left| \frac{d\vec{B}}{dt} \right| |\hat{n}| \cos 60° dA$$

$$= \left| \frac{d\vec{B}}{dt} \right| \frac{1}{2} \int dA = \left| \frac{d\vec{B}}{dt} \right| \frac{1}{2} A = 0.625 \text{ V}.$$

Together, 200 turns in the same direction would pick up an induced emf of 200 times 0.625 V, or 125 V. Table 12.4 summarizes our qualitative conclusions.

Table 12.4 Description of Example 12.6

Observer	$d\vec{B}_{ext}/dt$	$\vec{B}_{ind}$	$\mathcal{E}_{ind}, I_{ind}$	$\vec{F}_{net}$	$\vec{\tau}$	Compress or expand
Below loop	⊗	⊙	Counterclockwise	None	$\hat{n}$ toward $-\hat{x}$	Compress

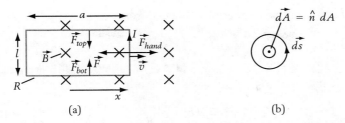

Figure 12.13 (a) The same situation as in Figure 12.11, but in more detail. (b) The circuit-normal relation for part (a).

Example 12.7 Rectangular loop pulled through a $\vec{B}$ field

The previous example considered a situation where the flux change is due to a time variation of the field. Here we consider a situation where the field is constant in time, but the flux changes because the circuit moves into the region of the field so that the area in the field changes. Thus, as in Figure 12.13(a), consider a rectangular loop (our secondary) of resistance R, length a along x, and length l along y. It moves along $+x$ with constant velocity v in a region of zero magnetic field, until it reaches a region of uniform field B that points into the paper $(-\hat{z})$. See Figure 12.13(a). To be specific, take $B = 0.005$ T, $v = 2$ m/s, $l = 0.1$ m, and $R = 0.1$ Ω. (If the loop simply were thrown into the field, by Lenz's law the loop would slow down, so a force is needed to keep the loop moving at a constant velocity.) Find (a) the magnetic flux, (b) the emf, (c) the magnetic force on the loop, and (d) the power needed to pull the loop at a constant velocity.

Solution: To find the magnetic force, we must solve simultaneously for both the current and the velocity. (a) Take $d\vec{A} = \hat{n}dA$ out of the paper (so $d\vec{s}$ circulates counterclockwise, to be consistent with the motional emf analysis shown earlier), as in Figure 12.13(b). Then

$$\Phi_B = \int \vec{B} \cdot \hat{n}dA = \int \vec{B} \cdot \hat{z}dA = -B \int dA = -BA = -Blx \quad (12.5)$$

as the loop enters the field region. (b) Let $v = dx/dt$. Because of the motion, the rate of change of the flux is given by

$$\frac{d\Phi_B}{dt} = -Blv, \quad (12.6)$$

where $v = dx/dt$. This leads to an emf

$$\mathcal{E} = -\frac{d\Phi_B}{dt} = Blv \quad (12.7)$$

that circulates along $d\vec{s}$: counterclockwise, in agreement with the qualitative analysis of Example 12.4. There is a similar analysis as the loop leaves the field. While in the field the flux remains at the constant value Bla, so there is no induced emf.

For our values of B, l, and v, (12.7) gives $\mathcal{E} = (.005$ T$)(0.1$ m$)(2$ m/s$) = 1.0 \times 10^{-3}$ V. Taking (12.7) to be the only emf that acts, this causes a current

$$I = \frac{\mathcal{E}}{R} = \frac{Blv}{R}. \quad (12.8)$$

For $R = 0.1$ Ω, (12.8) gives $I = 1.0 \times 10^{-3}$ V$/0.1$ $\Omega = 0.01$ A. (c) The associated magnetic force $\vec{F}$ on the right arm is obtained from

$$\vec{F} = I \int d\vec{s} \times \vec{B}. \qquad (12.9)$$

(Recall that I is the induced current and $\vec{B}$ is the applied field.) From (12.9), $\vec{F}$ points to the left, and from (12.9) and (12.7), it has magnitude

$$F = IlB = \frac{v(Bl)^2}{R}. \qquad (12.10)$$

For our case, $F = (.01 \text{ A})(.1 \text{ m})(.006 \text{ T}) = 6.0 \times 10^{-6}$ N. (d) To make the loop move at constant velocity v, an external force (e.g., from our hand) of magnitude F must be applied in the opposite direction. Using (12.10), this external force provides power

$$\mathcal{P} = Fv = \frac{(vBl)^2}{R} = \frac{(IR)^2}{R} = I^2 R. \qquad (12.11)$$

This is precisely equal to the rate of Joule heating. For our case, $\mathcal{P} = (.01 \text{ A})^2 (0.1 \text{ }\Omega) = 1.0 \times 10^{-5}$ W. Thus, all the power provided by the hand (Fv) goes into heating ($I^2 R$) of the wire. This generation of electric current by mechanical energy means that the loop is an *electrical generator*. Finally, note that there are also equal and opposite forces on the upper and lower arms, which tend to compress the loop, in agreement with the Lenz's law analysis of Example 12.4. See Figure 12.13(a).

12.6 Mutual Inductance

Since, by the Biot–Savart law, $\vec{B}$ is proportional to the current I_P in the primary, so is Φ_B. Hence it is convenient to write, for the flux through the secondary,

$$\Phi_B = MI_P, \qquad \text{(flux through secondary, current from primary)} \quad (12.12)$$

where the proportionality constant M is called the *mutual inductance* (mutual because one circuit affects the other). (Context will determine whether M refers to mutual inductance or magnetization.) Inductance is given in units of Wb/A $=$ V-s/A $=$ H, called the henry, in honor of Joseph Henry.

Using (12.12), we can calculate M in a number of important cases. The mutual inductance is important because it can be used to calculate the induced emf. Specifically, using (12.12), Equation (12.4) yields for the induced emf

$$\mathcal{E} = -\frac{d\Phi_B}{dt} = -M\frac{dI_P}{dt}. \qquad \text{(induced emf via mutual inductance)} \quad (12.13)$$

Note that the mutual inductance between the primary and the secondary (the flux in the secondary, per current in the primary) is the same as between the secondary and the primary (the flux in the primary, per current in the secondary). M can be of either sign, according to the relative position of the primary and the secondary. If M is positive when two identical coils are one on top of another, then M is negative when they are side by side. This is because when they are one on top of another a clockwise current for the primary makes in the secondary a field that is into the page, but when they are side by side the primary makes in the secondary a field that is out of the page.

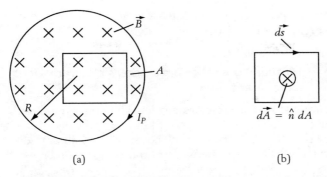

Figure 12.14 (a) A rectangular loop within a circular solenoid. (b) Geometry of the rectangular loop.

Example 12.8 **Loop within a solenoid**

Consider a rectangular loop within a solenoid, the loop axis aligned with the axis of the solenoid. Let the loop have N turns, area A_{loop}, and let the solenoid have n turns per unit length and carry electric current I_P. See Figure 12.14(a). (The final results do not depend either on the shape of the loop or on the cross-section of the solenoid, so long as the solenoid completely encloses the loop.) (a) Find the mutual inductance M in algebraic form; (b) evaluate M for $N = 200$, $n = 6 \times 10^5$/m, and $A_{loop} = 4 \times 10^{-4}$ m^2; (c) evaluate the induced emf for $dI_P/dt = 500$ A/s; (d) discuss the direction of the emf and the induced forces on the loop.

Solution: (a) Figure 12.14(b) gives our $d\vec{s}$ and $d\vec{A} = \hat{n}dA$ conventions, following the circuit-normal right-hand rule. Since the field is uniform within the solenoid, and the normal $\hat{n}$ to the loop is along the field $\vec{B}$, by (12.3) the flux Φ_B is simply BA_{loop}. By (11.40), the solenoid field is $B = 4\pi k_m n I_P$. Each of the N turns picks up this flux, so by (12.12)

$$M = N\frac{\Phi_B}{I_P} = \frac{NBA_{loop}}{I_P} = \frac{N(4\pi k_m)nI_P A_{loop}}{I_P} = 4\pi k_m n N A_{loop}. \quad (12.14)$$

(b) Using $N = 200$, $n = 6 \times 10^5$/m, and $A_{loop} = 4 \times 10^{-4}$ m^2, (12.14) gives $M = 0.060$ H. (c) If $dI_P/dt = 500$ A/s (corresponding to an increasing field), then by (12.4) and (12.12), the emf has magnitude $MdI_P/dt = 30$ V. (d) Since the field is increasing into the page, the induced field is out of the page, necessitating (by Ampère's right-hand rule) a counterclockwise-induced current in the rectangular loop. Because the field is uniform, there is no net force on the loop, but there is a compressive force, by the motion statement of Lenz's law.

Food for Thought: How would the mutual inductance of Example 12.8 change if the axis of the loop were at an angle θ to the axis of the solenoid?

Food for Thought: Qualitatively describe the induced current and the torque, if the loop were rotated about the x-axis, so that the area exposed to the field were to decrease. The rotation can be described using a nonzero value of θ.

If the loop were partly in and partly out of the solenoid, only the part of the coil area in the solenoid would contribute to (12.14). Since n varies inversely with length and A_{loop} varies as length squared, M is proportional to length. Like capacitances, inductances scale linearly with the length. Note that the larger the number of turns in the solenoid (making a larger field) or the coil (making a larger effective area), the larger the M.

Example 12.9 **Mutual inductance in Faraday's electromagnet experiment**

Figure 12.4 presents Faraday's electromagnet. The iron toroid has thickness d and diameter $D \gg d$. The primary coil (P) has n_P turns per unit length, and the secondary coil (S) has N_S turns. Each covers about half of the iron toroid. Find the mutual inductance M between primary and secondary.

Solution: From (9.23) or (11.51), the solenoid produces a field within the core that is larger than $4\pi k_m n_P I_P$ by a factor of $(1 + \chi)$, where the magnetic susceptibility χ is a measure of the Amperian current of the iron core relative to the current of the solenoid. Thus, instead of $B = 4\pi k_m n_P I_P$, we have

$$B_{core} = 4\pi k_m (1 + \chi) n_P I_P, \qquad (12.15)$$

where $\chi \approx 5000$ for soft iron. (This enhancement may be thought of as due to Amperian currents circulating around the iron.) Because the iron nearly completely holds the flux, it doesn't matter how the turns of wire are distributed: only the number of turns and the perimeter πD of the toroid matter. Wrapping the wire of the primary around half the iron toroid hence gives $n_P = N_P/\pi D$. The secondary (S) has area $A_S = \pi d^2/4$ and N_S turns. Thus the mutual inductance is obtained by modifying (12.14), replacing N by N_S, n by $n_P = N_P/\pi D$, A_{loop} by A_S, and multiplying by $(1 + \chi)$. Hence

$$M = 4\pi k_m n_P N_S A_S (1 + \chi) = 4\pi k_m (1 + \chi) N_S N_P \frac{d^2}{4D}. \qquad (12.16)$$

Since M varies as d^2/D, for fixed numbers of turns, M doubles if all length scales double (i.e., d and D double); as noted above, like capacitance, inductance scales linearly with length.

Faraday's description permits an estimate of many of the parameters of his electromagnet. The primary consisted of about 72 feet of copper wire wrapped around an iron bar with $d = 7/8$ inches that had been welded into a ring with $D = 6$ inches, so it had about $N_P \approx [(72)(12)/\pi(0.875)] \approx 314$ turns. The secondary consisted of about 60 feet of copper wire, so it had about $N_S \approx [(60)(12)/\pi(0.875)] \approx 251$ turns. Inserting the parameters of Faraday's

---

On the Source of Magnetic Fields

If the magnetism of soft magnets were due to magnetic poles, then, because for the toroidal geometry there are no magnetic poles, $\vec{B}$ would be due only to $\vec{B}_{sol}$ of the solenoid. Then the factor of $1 + \chi$ in (12.16) would be replaced by 1. However, experiments support the factor of $1 + \chi$. This indicates that the magnetism of soft magnets is due to microscopic Amperian currents rather than magnetic poles. Indeed, the operation of every iron core electrical transformer, throughout the world, is testimony that $\vec{B}$ (due to currents, both macroscopic and microscopic), rather than $\mu_0 \vec{H}$ (due to macroscopic currents and microscopic magnetic poles), applies in Faraday's law.

circuit into (12.16) gives $M = 0.401$ H. If $dI_P/dt = 100$ A/s, the emf would have been, by (12.13), of magnitude 40.1 V.

Note. Even if we cannot compute M, we may be able to measure it. If $\mathcal{E}_S = 12$ V, and $dI_P/dt = 200$ A/s, then (12.16) gives $|M| = 0.06$ H.

Example 12.10   Rectangular loop and a long wire

Consider a long wire and a rectangular loop of sides a and b, with side a parallel to the long wire. See Figure 12.15. They are separated by a distance s. (a) Find their mutual inductance M in algebraic form. (b) Evaluate M for $a = 0.02$ m, $b = 0.04$ m, $s = 0.04$ m. (c) Evaluate M for 200 turns. (d) Evaluate M with a soft iron core and 200 turns.

Solution: (a) M can be obtained either from the flux due to the long wire acting on the loop or from the flux on the long wire (considered to be part of a circuit that connects at infinity) due to the loop. It is much easier to calculate the first. We use (12.12) with $B_{wire} = 2k_m I/r$, taking the circuit normal $\hat{n}$ into the paper, so that $\vec{B}$ and $\hat{n}$ are collinear. With $dA = dr\,dz$, (12.12) yields

$$M = \frac{1}{I}\int_{loop}\vec{B}_{wire}\cdot\hat{n}\,dA = \frac{1}{I}\int_s^{s+b} dr\int_0^a \frac{2k_m I}{r}\,dz$$

$$= 2k_m a\int_s^{s+b}\frac{dr}{r} = 2k_m a\,\ln\frac{s+b}{s}. \tag{12.17}$$

(b) For the stated values of a, b, and s, (12.17) yields $M = 2.89$ nH. (c) A rectangular loop of N turns has a mutual inductance N times larger than given by (12.17). For $N = 200$ turns, the mutual inductance M becomes $(200)(2.89$ nH$) = 577$ nH. This is still rather small. (d) Including a soft iron core causes M to be multiplied by a factor of $(1 + \chi)$, as in (12.16), where for soft iron $\chi \approx 5000$. Thus, with soft iron core *and* $N = 200$ turns of wire, M becomes $(5000)(200)(2.89$ nH$) = 2.89$ mH, a significant value. This enhancement of the mutual inductance explains the presence of soft iron in many electrical devices.

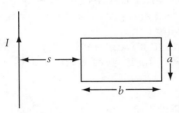

Figure 12.15 A long current-carrying wire and a rectangular loop.

Example 12.11   Loop pulled through a $\vec{B}$ field, via time-varying mutual inductance

Again consider the rectangular loop moving with constant velocity into the region of a uniform magnetic field $\vec{B}$, as in Figure 12.13. Find the emf by thinking of Faraday's law in terms of a changing mutual inductance M.

Solution: As the loop enters the field, with $d\vec{A} = \hat{n}\,dA$ pointing *out* of the paper, as before, we have

$$\Phi_B = -BA = -Blx = -\frac{B}{I}lxI = MI, \qquad M \equiv -\frac{B}{I}lx, \tag{12.18}$$

where B/I is some constant, independent of the current. (For example, if B were due to a solenoid of n turns per unit length, so $B = 4\pi k_m nI$, we would have

$B/I = 4\pi k_m n$.) From (12.4) and (12.18),

$$\mathcal{E} = -\frac{d\Phi_B}{dt} = -I\frac{dM}{dt} = -I\left(-\frac{B}{I}\right)l\frac{dx}{dt} = Blv, \tag{12.19}$$

in agreement with (12.7). This approach, using a time-varying mutual inductance M, is sometimes taken by mechanical engineers. Note that M is negative in this case, due to our opposite sign conventions for positive currents in the loop and in the solenoid. For $d\vec{A} = \hat{n}\,dA$ pointing *into* the page, M would be positive.

12.7 Motional EMF

In Chapter 10, the total force on a charge q moving with velocity $\vec{v}$ in a magnetic field $\vec{B}$ and an electric field $\vec{E}$ was shown to be

$$\vec{F} = q(\vec{E} + \vec{v} \times \vec{B}), \tag{12.20}$$

known as the Lorentz force. It was assumed that $\vec{E}$ was due to static charges, where $\oint \vec{E} \cdot d\vec{s} = 0$, as shown in Chapter 5. In Section 10.7, we interpreted $\vec{v} \times \vec{B}$ as a *motional* electric field:

$$\vec{E}_{mot} = \vec{v} \times \vec{B}. \quad \text{(motional electric field)} \tag{12.21}$$

Equation (12.20), and the interpretation in (12.21), of a motional electric field, were not known until some 60 years after the discovery of Faraday's law. This section shows how $\vec{E}_{mot}$ leads to an emf and provides an alternative approach to some problems solved previously using Faraday's law.

Integrating $q(\vec{E} + \vec{v} \times \vec{B}) \cdot d\vec{s}$ around a loop yields the work W done on q. If $\vec{E}$ is purely electrostatic, then the term in $\vec{E}$ gives zero. Dividing the work by q yields the emf

$$\mathcal{E} = \frac{W}{q} = \oint \vec{E}_{mot} \cdot d\vec{s} = \oint \vec{v} \times \vec{B} \cdot d\vec{s}. \quad \text{(motional emf)} \tag{12.22}$$

Equivalence of Flux Changes due to Motion and to Changes in Current

Maxwell notes that the equivalence of (12.22) and (12.4) was shown experimentally by Felici in 1851, using in the secondary a ballistic galvanometer, which has a slow response to current. Felici started with the primary current off, and the secondary in a position and orientation where the galvanometer did not respond (even after waiting) when the primary current was turned on; this corresponds to $M = 0$. For this configuration, he (1) turned on the current, (2) moved the secondary along an arbitrary path to its final position and orientation, and (3) turned off the primary. Because of its slow response time, the galvanometer could not respond during any relatively fast sequence of operations. However, for this particular sequence, it did not respond at all, even after a wait. This indicated that the time integral of the induced emf from this sequence was zero. Since there was no emf for part (1), and there was no net emf for the entire sequence, the motional emf of part (2) was completely canceled by the turn-off emf of part (3); hence, the motional emf of (12.22) is equivalent to the emf of (12.4) obtained by turning on the primary.

By the above equivalence, when applied to actual situations where there is motion, (12.22) yields precisely the same results as does (12.4).

Example 12.12 Loop pulled through a $\vec{B}$ field, via motional EMF

As in Example 12.7, consider the rectangular loop pulled at a constant velocity v by a force $\vec{F}_{hand}$ into the region of a uniform magnetic field $\vec{B}$ (Figure 12.13). For the part of the loop in the field, there is a motional electric field $\vec{E}_{mot} = \vec{v} \times \vec{B}$. It is of magnitude vB and it points along $+\hat{y}$. Find the emf for this loop using the motional emf approach. Discuss energy conservation.

Solution: We analyze this in pieces.

1. When only the right arm is in the field, this motional electric field drives current counterclockwise, in agreement with our Lenz's law analysis. The emf for the right arm is, by (12.22), with $d\vec{s}$ circulating counterclockwise,

$$\mathcal{E} = \int_0^l vB\hat{y} \cdot \hat{y}dy = vBl. \tag{12.23}$$

Note that for the upper and lower arms, $\vec{v} \times \vec{B}$ is normal to $d\vec{s}$, so these arms do not contribute to (12.22). (For these arms, $\vec{v} \times \vec{B}$ produces a Hall voltage, but that is not relevant to the present considerations.) Of course, $\vec{v} \times \vec{B}$ is zero for the left arm, which is not in the field.

2. When the loop is completely within the field, motional electric fields act on each of the right and left arms—along $+\hat{y}$ in both cases. These will produce canceling emfs, since $d\vec{s}$ changes direction on going from the right to the left arm.

3. When the right arm of the loop leaves the field, a motional electric field acts on the left arm, of magnitude vB and pointing along $+\hat{y}$. It drives current clockwise, in agreement with our Lenz's law analysis. The associated emf is vBl. There is a force on the left arm, of magnitude given by (12.10), pointing to the left. Equal and opposite forces act on the upper and lower arms, tending to expand the loop, again in agreement with the Lenz's law analysis.

4. When only the right arm is in the field, the external force provides power $(IlB)v = I^2R$ to increase the kinetic energy, and the (ponderomotive) force $I\vec{l} \times \vec{B}$ of the magnetic field absorbs equal and opposite power to decrease the kinetic energy. On the other hand, the electromotive force provides power $(vBl)I = I^2R$ to drive current around the circuit, whereas the resistor absorbs all this I^2R energy. Hence, the power absorbed by the ponderomotive force equals the power generated by the electromotive force. This is an example of how, in a magnetic field that does not vary in time, the net work done by the magnetic field is zero.

Example 12.13 Voltage profile around a circuit

See Figure 12.16(a). Let $B = 0.005$ T, $l = 0.1$ m, $v = 2$ m/s, and $R = 0.1\ \Omega$, as in Example 12.7. In addition, let the left and right arms have resistance $R_l = R_r = 0.015\ \Omega$. Take the loop to be of uniform cross-section and conductivity. Denote by $\vec{E}_{es}$ the electrostatic field. (a) Find the voltage profile around the circuit due to $\vec{E}_{es}$. (b) Find $\vec{E}_{es}$ in each arm of the circuit.

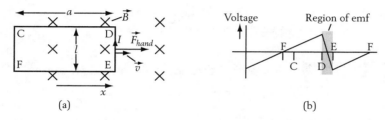

(a)   (b)

Figure 12.16 (a) The circuit of Figure 12.11(a) and Figure 12.13(a). (b) Voltage across this circuit.

Solution: (a) For the left arm of the loop in Figure 12.16(a), $\vec{E}_{mot} = \vec{0}$, and for the top and bottom arms the component of $\vec{E}_{mot}$ along the circuit is zero. However, for the right arm $|\vec{E}_{mot}| = vB = 0.01$ V/m, tending to drive current counterclockwise. By (12.23), this gives a motional emf for the right arm, and an emf for the circuit as a whole, of Blv, driving current counterclockwise. Ohm's law for the circuit as a whole then gives a counterclockwise current $I = Blv/R$, as in Example 12.7, and again $Blv = 1.0 \times 10^{-3}$ V and $I = Blv/R = 0.01$ A. Now consider the voltages. The emf for the left arm, taken directly from C to F, can only be due to a voltage difference, associated with an electrostatic electric field $\vec{E}_{es}$ in that arm. Hence, by Ohm's law, $V_C - V_F = IR_l = 0.15 \times 10^{-3}$ V. Indeed, for the entire counterclockwise path D-C-F-E, of resistance $R - R_r$, the only emf is due to a voltage difference. Hence, by Ohm's law, $V_D - V_E = I(R - R_r) = 0.85 \times 10^{-3}$ V. These values set the scale in Figure 12.16(b), which shows the voltage profile (e.g., due to electric charge) relative to F if there were a set of many leads around the circuit. Compare (and contrast) Figure 12.16(b) to Figure 7.20(b), where the emf is chemical in nature, and the voltage increase across the voltaic cell is not continuous, but (on a scale large compared to atoms) consists of two jumps, one at each electrode. (b) For the left arm of the circuit, $|\vec{E}_{es}| = (V_C - V_F)/l = 0.0015$ V/m. Because the wire is uniform, for the entire path D-C-F-E, including the left arm, $|\vec{E}_{es}| = 0.0015$ V/m. For the direct path D-E, $|\vec{E}_{es}| = (V_D - V_E)/l = 0.0085$ V/m, opposing $\vec{E}_{mot}$, of magnitude 0.01 V/m. The slopes of the voltage profile in Figure 12.16(b) reflect these values of $|\vec{E}_{es}|$. In all parts of the circuit, in the direction of current flow the component of the *total* field, $\vec{E} = \vec{E}_{mot} + \vec{E}_{es}$, has magnitude 0.0015 V/m.

Example 12.14   **Faraday's disk dynamo**

Faraday's disk dynamo (a generator of electricity), shown in Figure 12.17, is particularly easily analyzed from the motional emf viewpoint. It is a conducting disk of radius a that rotates about its axis at a rate ω in a collinear

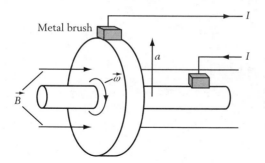

Figure 12.17 Faraday's disk dynamo. By experimentation, Faraday found a configuration of contacts that gave an induced current.

magnetic field $\vec{B}$. Conducting brushes make contact with the circuit at the disk center and at the disk perimeter. Find the emf for Faraday's disk dynamo.

Solution: Let both $\vec{B}$ and the vector angular velocity $\vec{\omega}$ point to the right. Then a point at r from the axis, with velocity $\vec{v}$, has $v = \omega r$, where $v = |\vec{v}|$ and $\omega = |\vec{\omega}|$. $\vec{E}_{mot} = \vec{v} \times \vec{B}$ of (12.21) has magnitude vB and points radially outward. (Hence the current I in Figure 12.17 is radially outward.) With $d\vec{s} = \hat{r}dr$, the motional emf from axis to perimeter is given by

$$\mathcal{E} = \int \vec{E}_{mot} \cdot d\vec{s} = \int (vB\hat{r}) \cdot (\hat{r}dr) = \int_0^a vB\,dr = \int_0^a \omega r\,B\,dr = \frac{1}{2}\omega Ba^2. \quad (12.24)$$

Although this calculation is for a radial path, (12.24) holds for any path from axis to perimeter.

Analyzing Example 12.14 theoretically from the flux change viewpoint is not so straightforward. This is mirrored in the fact that, experimentally, Faraday tried many different sets of contact points (both fixed and moving) before he found one that would give an emf. For $B = 0.005$ T, $\omega = 600$ s^{-1}, and $a = 2$ cm, (12.24) gives $\mathcal{E} = 0.0012$ V. This is small but measureable. (The resistance R_{disk} can be measured between the contact points in the absence of rotation but with a battery to drive the current.) Since $I = \mathcal{E}/(R + R_{disk})$, where R is the resistance of the rest of the circuit, the voltage between the contact points will be $IR_{disk} = \mathcal{E}R_{disk}/(R + R_{disk})$. For rotation rates high enough that the local drift velocity $\vec{v}_d$ of the electrons producing the current is less than the local velocity $\vec{v} = \vec{\omega} \times \vec{r}$ of a point $\vec{r}$ on the rotating disk, the current flow pattern becomes complex, and R_{disk} increases.

In Faraday's original experiment, the magnetic field acted over only a small portion of the disk near the edge. Thus, it was merely a rotated version of the case of a magnet above a rotating turntable, as in Figure 12.16. This example is important because it shows the possibility of using electromagnetic induction to generate constant current.

Optional   **Example 12.15**   **Monopole moving toward conducting loop**

Let us model the situation of Figure 12.8(a) by simplifying the magnet. Consider only its north pole, a monopole q_m moving toward the loop. See Figure 12.18(a). From (9.11), a magnet of magnetization M and area A has $q_m = MA$ at each pole. (a) Using the motional emf approach, analyze the emf of this circuit. (b) For $q_m = 10$ A-m, $a = 4$ cm, $v = 5$ m/s, and $x = 0$, evaluate the emf; repeat for a coil of 100 turns. (c) For $R = 0.1\,\Omega$, find the current in the loop and the force on the loop.

Solution: (a) Rather than consider the monopole to move at velocity $\vec{v}$, and the circuit to be at rest, for the motional emf approach we consider the monopole to be at rest and the circuit to move with velocity $\vec{v}_{loop} = -\vec{v}$. See Figure 12.18(b), where the loop moves in the $-\hat{x}$ direction. The field $\vec{B}$ of the monopole points radially outward from the monopole, at an angle θ to $\vec{v}$, with $|\vec{B}| = k_m q_m/r^2$. Hence $\vec{E}_{mot} = \vec{v}_{loop} \times \vec{B}$ points tangentially in a clockwise direction as seen by an observer to the right (in agreement with our analysis of Figure 12.8(a)), and has magnitude

$$|\vec{E}_{loop}| = |\vec{v}_{loop}|\,|\vec{B}|\,|\sin\theta| = \frac{dx}{dt}\frac{k_m q_m}{r^2}\sin\theta, \qquad \sin\theta = \frac{a}{r}, \qquad r = \sqrt{a^2 + x^2}.$$

$$(12.25)$$

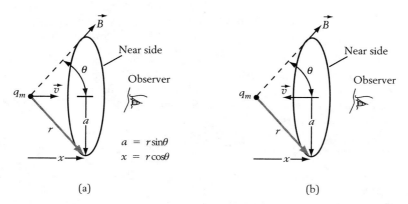

Figure 12.18 Motion of a magnetic pole (representing the N pole of a magnet, as in Figure 12.8) toward a conducting loop. (b) The same situation as seen from the point of view of the magnetic pole, with the conducting loop moving toward the pole.

Integrating around the loop with $d\vec{s}$ clockwise gives

$$|\mathcal{E}| = \left| \oint \vec{E} \cdot d\vec{s} \right| = (2\pi a)|\vec{E}_{loop}| = \frac{2\pi k_m q_m a^2}{(a^2 + x^2)^{3/2}} \frac{dx}{dt}. \tag{12.26}$$

(b) For a given dx/dt, (12.26) is a maximum when $x = 0$. For $dx/dt = -5$ m/s, $x = 0$, $q_m = 10$ A-m, and $a = 4$ cm, (12.26) gives $\mathcal{E} = 7.85 \times 10^{-4}$ V. This is small but measurable. It can be enhanced by multiple turns N of wire; for $N = 100$, $\mathcal{E} = 0.0785$ V. (c) For $R = 0.1$ Ω and $\mathcal{E} = 0.0785$ V, $I = \mathcal{E}/R$ gives $I = 0.785$ A. From (10.29), the force on a loop of N turns in the flaring magnetic field of the pole is

$$F = NI(2\pi a)B \sin\theta, \quad \sin\theta = \frac{a}{\sqrt{a^2 + x^2}}. \tag{12.27}$$

For $N = 100$, $I = 0.785$ A, $q_m = 10$ A-m, $a = 4$ cm, and $x = 0$, $B = k_m q_m/a^2 = 1.125 \times 10^{-4}$ T and $\sin\theta = 1$, so $F = 0.111$ N. This is to the right, repelling the loop from the increasing magnetic field of the magnet. Correspondingly, there must be a repulsive force on the magnet.

Drag Force on a Magnet Falling Down a Copper Tube

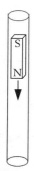

A monopole falling toward a loop is related to a long thin magnet moving down the center of a conducting tube (e.g., of copper). We can think of the magnet as having two point poles, and the copper tube as being a superposition of many thin independent slices of copper, each of which is like a 90° clockwise version of the loop in Figure 12.18. If the magnet is much longer than the inner radius of the tube, the poles act independently. If the force on one pole due to the induced current can be obtained for one slice, as in (12.27), then an integration over the force due to all the thin slices gives the total force on that pole. The total force on the magnet is twice the force on one pole. If the magnet is *not* much longer than the inner radius of the tube, then the calculation becomes more complicated because the effects of the two poles are not independent. Despite that complication, nothing is different in principle.

What follows is a historical incident that motivates the remainder of the chapter.

12.8 Michael Faraday Meets Mr. Jenkin

By 1830, Henry had built powerful electromagnets that could lift up to 650 pounds. By 1832, with a 100-pound electromagnet, he could lift 3500 pounds. That same year, when turning off the power to his magnet, he discovered what we would call self-induction. However, we will pursue the somewhat later discovery of self-induction by Faraday because it is such a good story.

Faraday began his investigation into self-induction in 1834, three years after he had already studied mutual induction. At that time a young man named Mr. Jenkin related to Faraday some electrical experiments in which Jenkin had felt a shock upon disconnecting, with his hands, a battery of voltaic cells connected to an electromagnet. See Figure 12.19. (Faraday wrote that this was the only time he had been led into a fruitful scientific direction by an amateur.) Mr. Jenkin was shocked because the large current that had been passing through the relatively low resistance wire now was passing through him. This can be interpreted in terms of a large induced emf. There was no such induced emf on making the connection.

Suppose that the battery provided $\mathcal{E}_0 = 10$ V, the electromagnet wire had resistance $R_w = 10\ \Omega$, and Mr. Jenkin had resistance $R_J = 10^5\ \Omega$. Then when connected a current $I_w = \mathcal{E}/R_w = 1$ A flowed through the electromagnet, and the much smaller current $I_J = \mathcal{E}/R_J = 10^{-4}$ A flowed through Mr. Jenkin. When he disconnected the battery, the 1 A current through the electromagnet then passed through him. The voltage difference across him was now $I_w R_J = 10^5$ V! An equal voltage had to develop across the electromagnet, by the uniqueness of the voltage. How could such a large voltage develop across the electromagnet without making the current even bigger? In Section 12.9, we will show that the electromagnet provides a self-induced emf that opposes this large voltage.

Faraday began to investigate this phenomenon. He showed that a given length of wire, when coiled up, produces a larger shock on breaking the circuit than when it is not coiled up. He also showed that shorter wires give much less of an

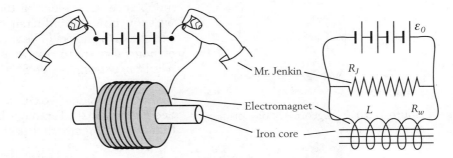

Figure 12.19 The experiment of Mr. Jenkin (represented by R_J), where with his hands he connected and disconnected an electromagnet (represented by R_w and the "electromagnetic inertia," or inductance, L).

> **Iron Core as Parallel Rods**
>
> The iron core in Figure 12.19 is represented in the circuit on the right as a set of parallel lines. This is because a set of parallel rods magnetizes just as effectively as a single large rod, but the eddy currents (which circulate around the rod axes) are much less for a set of parallel rods, and thus less energy is wasted in Joule heating. This principle will be discussed again in Chapter 13.

effect than longer wires. Thus, in some sense the coils and the longer wires have a greater tendency to maintain the current flow. The quantitative measure of this "electromagnetic inertia," as it was later called by Maxwell, is the self-inductance L. We may consider the electromagnet to be a solenoid with a soft iron core. In Section 12.10, we show that adding turns to the coil and filling it with soft iron increases L, just as Example 12.9 shows that these additions increase M.

The value $I R_J = 10^5$ V of the induced emf is independent of the value of L. The emf soon dies down, but the larger the electromagnet—and the larger the L—the slower the decay—and thus the longer-lasting the shock. In Section 12.11, we consider what happens when the battery is turned on and off. The characteristic decay time for a circuit with resistance R and self-inductance L is $\tau = L/R$. We will show how this time relates to Mr. Jenkin's shock on disconnecting the electromagnet. This shock-producing phenomenon is used in automobile ignition systems.

Related to Mr. Jenkin's shock is what happens when a switch is thrown to shut off a large current. For a large self-inductance in the circuit, a very large but temporary current arc may develop in the gap across the switchblades. Even when house lighting is turned off, arcing can occur.

12.9   Mutual Inductance and Self-Inductance

Mutual inductance has been introduced in the context of the electromagnet. We now present a more general discussion of inductance, where a circuit has both mutual inductance and self-inductance. Often a circuit is isolated, in which case it has only self-inductance. At the end of the section, we show how self-inductance explains Mr. Jenkin's shock. (These *coefficients of inductance* are to *current* as the *coefficients of potential*—see Section 6.8—are to *charge*.)

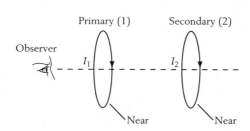

Figure 12.20 The geometry of a primary and a secondary.

Let us consider two circuits 1 and 2. See Figure 12.20. Faraday's law states that the emf in circuit 1 is given by

$$\mathcal{E}_1 = -\frac{d\Phi_B^{(1)}}{dt},$$

(12.28)

where, by (12.3), the total magnetic flux through circuit 1 is given by

$$\Phi_B^{(1)} = \int_1 \vec{B}_1 \cdot d\vec{A}_1 + \int_1 \vec{B}_2 \cdot d\vec{A}_1. \tag{12.29}$$

Here $\vec{B}_1$ is produced by circuit 1, so by the Biot–Savart law it is proportional to I_1. Similarly, $\vec{B}_2$ is produced by circuit 2, and it is proportional to I_2. Thus we may rewrite (12.29) as

$$\Phi_B^{(1)} = L_1 I_1 + M_{12} I_2, \tag{12.30}$$

where L_1 and M_{12} are geometrically determined quantities, independent of current, and are defined by

$$L_1 \equiv \frac{\int_1 \vec{B}_1 \cdot d\vec{A}_1}{I_1}, \qquad M_{12} \equiv \frac{\int_1 \vec{B}_2 \cdot d\vec{A}_1}{I_2}. \tag{12.31}$$

We call L_1 the self-inductance of circuit 1, and M_{12} the mutual inductance between circuits 1 and 2. Similar equations hold for L_2 and M_{21}. Note that

$$M_{12} = M_{21}, \tag{12.32}$$

a result whose proof is assigned as a problem.

Combining (12.28) and (12.30) we obtain, for fixed circuits (so that the inductances don't change),

$$\mathcal{E}_1 = -\frac{d\Phi_B^{(1)}}{dt} = -L_1 \frac{dI_1}{dt} - M_{12} \frac{dI_2}{dt}.$$
(self-induced emf and mutually induced emf) (12.33)

Similarly,

$$\mathcal{E}_2 = -\frac{d\Phi_B^{(2)}}{dt} = -L_2 \frac{dI_2}{dt} - M_{21} \frac{dI_1}{dt}. \tag{12.34}$$

(For a loop pulled through a field, as shown by Example 12.11, the mutual inductance *does* change.) It is usually impractical to compute the inductances for arbitrary circuits (just as it is often difficult to compute the capacitances for arbitrary conductors). However, they can be measured. Thus, by measuring I_1 and I_2 as a function of time, as well as the emf's, it is possible to determine the quantities L_1, L_2, M_{12}, and M_{21}.

For an ideal battery, with negligible resistance, the chemical emf and the terminal voltage are equal in magnitude and tend to drive current in opposite directions. Similarly, for an ideal inductor, the self-induced emf and the terminal voltage are equal in magnitude and tend to drive current in opposite directions.

For Mr. Jenkins, the self-induced emf provided the emf and voltage needed to keep the current going in the electromagnet.

It follows from either (12.33) or (12.34) that, for a circuit in isolation, with self-inductance L but no mutual inductance,

$$\mathcal{E} = -\frac{d\Phi_B}{dt} = -L\frac{dI}{dt}. \qquad \text{(self-induced emf)} \qquad (12.35)$$

If the value of a self-inductance (or a mutual inductance) is needed, it is often preferable to measure it rather than calculate it.

Example 12.16   Measuring self-inductance

A 6 V emf is suddenly switched into a circuit with an inductor and a resistor in series. Just after the switch is thrown, and before there is any time for the current to build up, the current changes at the rate 1200 A/s. (a) Find the self-inductance. (b) Find the emf that would cause the initial current to change at the rate 400 A/s.

Solution: (a) By (12.35), and neglecting signs (since $L > 0$) we have $L = 6/1200 = 5$ mH. (b) Again by (12.35), and neglecting signs, $|\mathcal{E}| = L(dI/dt) = 2$ V.

12.10 Calculating Self-Inductance

Mutual inductance can be calculated for a few of the situations we have already considered. The same is true for self-inductance. However, even for a circuit as simple as a circular coil, calculation of the self-inductance is extremely difficult.

Note that inductance varies as the magnetic flux, which is the product of a field B (which, as we know from the field due to a long wire, varies inversely as the characteristic length) and an area A (which varies quadratically as the characteristic length). Hence, as indicated above, inductance, like capacitance, scales linearly with length.

Example 12.17   Solenoid

Consider a long solenoid of n turns per unit length, length l, and cross-sectional area A. Neglecting edge effects, it produces a uniform field $B = 4\pi k_m nI$. (a) Determine its self-inductance L. (b) Evaluate L for $n = 10^4$ turns/m and $lA = 10^{-4}$ m^3. (c) Discuss the effect of a soft iron core.

Solution: (a) If each of the $N = nl$ turns picks up the same magnetic flux BA, then by (12.31) its self-inductance is

$$L = \frac{N}{I}\int \vec{B}\cdot\hat{n}\,dA = \frac{NBA}{I} = \frac{(nl)(4\pi k_m nI)A}{I} = 4\pi k_m n^2 lA. \qquad (L \text{ of solenoid})$$
$$(12.36)$$

Note that doubling n at fixed l and A quadruples L; the flux per turn doubles and the number of turns doubles. (b) For $n = 10^4$ turns/m and $lA = 10^{-4}$ m^3, (12.36) yields $L = 0.0126$ H, an appreciable value. Geometrically scaling the entire system up by a factor of two, with the number of turns fixed, doubles l (thus halving n) and quadruples A, so L doubles, as expected.

(c) If the solenoid contains a soft iron core with magnetic susceptibility χ, then (as discussed earlier), the field is enhanced by a factor of $(1 + \chi)$, and so is L:

$$L = (1 + \chi)4\pi k_m n^2 l A. \quad (L \text{ of solenoid with magnetically soft core}) \quad (12.37)$$

For soft iron, where $\chi \approx 5000$, this is pertinent to Mr. Jenkin's experiment. The enhancement factor of $(1 + \chi)$ for soft iron causes the inductance of the solenoid described before to become 62.8 H, a large value indeed.

Example 12.18   Co-axial cable

Consider a co-axial cable of length l, with inner current I distributed uniformly within a cylindrical shell of radius a, and return current I distributed uniformly over a thin co-axial cylindrical shell of radius b. Take $l \gg a, b$. See Figure 12.21. (a) Determine its self-inductance L. (b) Determine its characteristic inductance per unit length.

Solution: (a) By Ampère's law and the symmetry of the situation, for $a < r < b$, $B = 2k_m I / r$ and points tangentially if we neglect edge effects. To apply Faraday's law, we take as our circuit a strip of length l that runs from $r = a$ to $r = b$, so that it picks up all the magnetic flux produced by the inner shell. (Figure 12.21 shows a smaller strip, of length w.) For $\hat{n}$ pointing tangentially, $\vec{B} \cdot \hat{n} = B$. Then, by (12.31), with $dA = dr\,dz$,

$$L = \frac{1}{I} \int \vec{B} \cdot \hat{n}\, dA = \frac{1}{I} \int_a^b dr \int_0^l \frac{2k_m I}{r} dz = 2k_m l \ln \frac{b}{a}. \quad (12.38)$$

As expected, this scales linearly with the size of the system. A logarithmic dependence, as in (12.38), can be expected whenever wires are involved, even when they are not straight. (b) Considering the logarithmic factor in (12.38) to be on the order of unity, the prefactor in (12.38) indicates that the inductance per unit length is on the order of 100 nH/m, or 1 nH/cm. As already noted, it is difficult to determine the self-inductance L_{ring} for a wire of radius a that is bent into a ring of radius $R \gg a$, but since R serves both as a measure of length (replace l by $2\pi R$) and outer radius (replace b by R), L_{ring} should be proportional to $k_m R \ln(R/a)$.

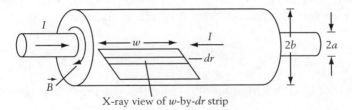

Figure 12.21 A co-axial cable of inner radius a and outer radius b, with only a portion of the outer part, and including an x-ray view of the region $a < r < b$.

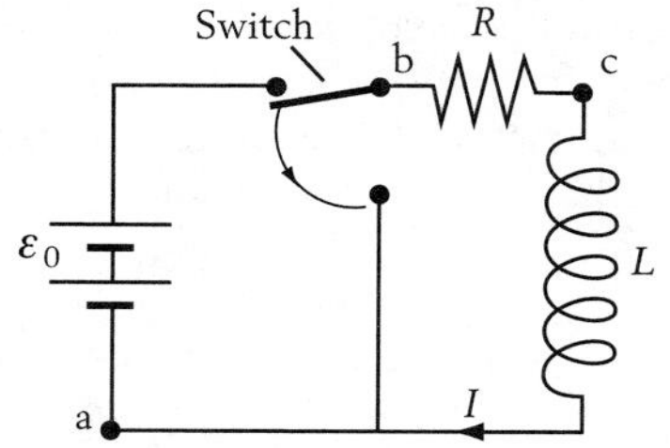

Figure 12.22 Connecting an ideal battery to a circuit with resistance and inductance.

12.11 Self-Inductance and the *LR* Circuit

We now turn to a full analysis of what happens when a chemical emf is switched into and out of an LR circuit.

12.11.1 *Turning On the LR Circuit (Part One)*

Consider a circuit where there is an inductance L (such as a solenoid), a resistor R, a battery of emf $\mathcal{E}_0$, and a switch. See Figure 12.22.

When the battery is part of the circuit as a whole, the net emf is given by

$$\mathcal{E} = -L\frac{dI}{dt} + \mathcal{E}_0. \tag{12.39}$$

From Ohm's law applied to the circuit as a whole, with resistance R, we have

$$I = \frac{\mathcal{E}}{R} = -\frac{L}{R}\frac{dI}{dt} + \frac{\mathcal{E}_0}{R}. \tag{12.40}$$

This may be rewritten as

$$IR + L\frac{dI}{dt} = \mathcal{E}_0. \tag{12.41}$$

12.11.2 *Voltage Drop ΔV_L Across an Inductor*

Before further analyzing (12.41), it is useful to do a voltage analysis for this circuit. From Figure 12.22 and the properties of ideal batteries, $V_b - V_a = \mathcal{E}_0$. Moreover, by Ohm's law, the voltage drop across the resistor in the direction of current flow is $\Delta V_R = V_b - V_c = IR$. From these two results and Figure 12.22, we then deduce that the voltage drop ΔV_L across the inductor in the direction of the current is

$$\Delta V_L = V_c - V_a = (V_c - V_b) + (V_b - V_a) = -IR + \mathcal{E}_0. \tag{12.42}$$

Combined with (12.41), Equation (12.42) yields the important result that

$$\Delta V_L = L\frac{dI}{dt}. \quad \text{(ideal, resistanceless inductor)} \tag{12.43}$$

For Mr. Jenkin, if $L = 1.0 \times 10^{-2}$ H and $\Delta V_L = -\Delta V_R = -1.0 \times 10^5$ V when he disconnected the battery, (12.43) gives $dI/dt = -1.0 \times 10^7$ A/s. The negative sign means that the current is decreasing.

Equation (12.43) is a very general property of ideal (resistanceless) inductors. ΔV_L is equal and opposite to the self-induced emf $\mathcal{E}_{self} = -LdI/dt$; if the self-induced emf tends to drive current one way, then ΔV_L tends to drive current the other way. The voltage profile around the LR circuit of Figure 12.22 would look somewhat like Figure 12.16(b), except that flat regions associated with the resistanceless wires in Figure 12.22 would replace sloped regions associated with wires having resistance in Figure 12.16(a).

In order to produce the voltage ΔV_L, charges must rearrange in the circuit, just as for a resistor, as discussed in Section 8.10. Just as these charges give an extra energy, and thus a parasitic capacitance, to a resistor, so they give a parasitic capacitance to an inductor.

Another way to see that (12.43) applies to an ideal inductor is to use Ohm's law to an inductor of resistance R_L. Then

$$I = \frac{1}{R_L}(\mathcal{E}_{self} + \Delta V_L) = \frac{1}{R_L}\left(-L\frac{dI}{dt} + \Delta V_L\right). \qquad \text{(real inductor)} \quad (12.44)$$

In the limit as $R_L \to 0$, for I to be finite, (12.43) must hold. An equivalent statement is that, within the inductor, for the current density $\vec{J} = \sigma \vec{E}$ to be finite as $\sigma \to 0$, we must have $\vec{E} \to 0$; that is, an infinitesimal $\vec{E}$ will produce a finite $\vec{J}$. Since $\vec{E}$ is the sum of an electrostatic field $\vec{E}_{es}$ (which produces ΔV_L) and an electromagnetically induced field $\vec{E}_{ind}$, this means that $\vec{E}_{es} = -\vec{E}_{ind}$ within the inductor. (If the inductor has a nonzero resistance, and thus a finite conductivity, then $\vec{E}_{es} = -\vec{E}_{ind} + \vec{J}/\sigma$ within the inductor.)

12.11.3 *Turning On the LR Circuit (Part Two)*

Let us now solve (12.41). Those of you who are adept at recognizing old friends—even when in disguise, subject to morning disarray, or with a new "do" or "stache"—will notice that Equation (12.41) is very similar to the equation describing the charging of a capacitor. There we had the form

$$\frac{Q}{C} + R\frac{dQ}{dt} = \mathcal{E}_0, \qquad (12.45)$$

with initial condition $Q = 0$ at $t = 0$. The solution was

$$Q(t) = C\mathcal{E}_0[1 - \exp(-t/\tau_{RC})], \qquad \tau_{RC} = RC, \qquad (12.46)$$

where τ_{RC} is the *capacitive time constant*. When turning on the emf in the RC circuit, after a time τ_{RC} the charge has grown to 63% of its final value.

We wish to solve (12.41) rather than (12.46), with the initial condition that $I = 0$ at $t = 0$, rather than $Q = 0$ at $t = 0$. On replacing the set of symbols

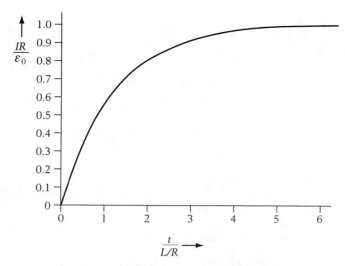

Figure 12.23 The current on turning on the circuit of Figure 12.22, as a function of time. Current is measured in units of the initial current $I_0 = \mathcal{E}_0 R$, and time is measured in units of $\tau_{LR} = L/R$.

$(Q, 1/C, R, \mathcal{E}_0)$ by $(I, R, L, \mathcal{E}_0)$, we transform our old friend (12.45) and its solution (12.46) into our new friend (12.41) and its solution:

$$I(t) = \frac{\mathcal{E}_0}{R}[1 - \exp(-t/\tau_{LR})], \qquad \tau_{LR} = \frac{L}{R}, \qquad \text{(turning on } LR \text{ circuit)}$$

$$(12.47)$$

where τ_{LR} is the *inductive time constant*. When turning on the emf in the LR circuit, after a time τ_{LR} the current has grown to 63% of its final value.

Equation (12.47) gives a current that starts at zero, rises linearly with slope $\mathcal{E}_0/L$, and then saturates at the steady-state value $\mathcal{E}_0/R$. See Figure 12.23. By measuring $\tau_{LR} = L/R$, for a known value of R, we can determine L. If $\tau_{LR} = L/R = 1.0 \times 10^{-5}$ s, and $R = 10\ \Omega$, then $L = 100\ \mu$H. As already noted, it is usually difficult to compute inductances, but they can be measured, as this example makes clear. Multimeters measure inductances by measuring the L/R time constant for a known, built-in R.

The self-inductance dominates the resistance at short times, preventing the current from building up. Only after a time on the order of the inductive time constant does the current build up to a significant value. (The examples of Sections 12.2 to 12.4 correspond to such times.) Initially, the self-inductance completely succeeds in opposing the applied change (the sudden turn-on of the battery emf), but as time goes by, it has less influence.

12.11.4 *Turning Off the LR Circuit*

If we consider that the battery emf $\mathcal{E}_0$ is suddenly set to zero (or if we quickly throw the switch in Figure 12.22), then (12.41) applies if we set $\mathcal{E}_0 = 0$ and start the current at the steady-state value

$$I_0 = \frac{\mathcal{E}_0}{R}. \tag{12.48}$$

Thus we must solve

$$IR + L\frac{dI}{dt} = 0, \tag{12.49}$$

subject to $I = I_0$ at $t = 0$. This is very like the problem of the discharge of a capacitor, with initial condition $Q = Q_0$ at $t = 0$. The solution in that case is

$$Q(t) = Q_0 \exp(-t/\tau_{RC}), \qquad \tau_{RC} = RC. \tag{12.50}$$

When turning off the emf in the RC circuit, after a time τ_{RC} the charge falls to 37% of its initial value.

On replacing the set of symbols $(Q, 1/C, R, Q_0)$ by (I, R, L, I_0), we obtain from (12.50) that

$$I(t) = I_0 \exp\left(-t/\tau_{LR}\right), \qquad \tau_{LR} = \frac{L}{R}. \qquad \text{(turning off } LR \text{ circuit)} \tag{12.51}$$

This equation gives a current that starts at I_0 and decays exponentially, with characteristic decay time $\tau_{LR} = L/R$. See Figure 12.24. When turning off the emf in the LR circuit, after a time τ_{LR} the current falls to 37% of its initial value.

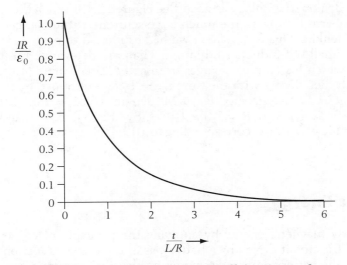

Figure 12.24 The current on turning off the circuit of Figure 12.22, as a function of time. Current is measured in units of the initial current $I_0 = \mathcal{E}_0 R$, and time is measured in units of $\tau_{LR} = L/R$.

Example 12.19   **Current decay for an *LR* circuit**

Let $L = 1$ mH and $R = 1$ Ω. Let $I_0 = 0.24$ A at absolute time 1.86 s. Find the time at which $I = 0.043$ A.

Solution: From (12.51), $\tau_{LR} = 10^{-3}$ s. Now use (12.51) to find t relative to 1.86 s. From (12.51), $\exp(-t/\tau_{LR}) = I(t)/I_0$, so $-t/\tau_{LR} = \ln I(t)/I_0$, or $t/\tau_{LR} = \ln I_0/I(t)$. Plugging in τ_{LR}, I_0, and I yields $t = (0.001 \ s)(1.72) = 0.00172$ s as the time it takes for the current to decay from 0.24 A to 0.043 A. Hence $I = 0.043$ at absolute time $1.86 + 0.00172 = 1.86172$ s.

12.11.5   *Back to the Shocked Mr. Jenkin*

To compare with the result of Mr. Jenkin, it is relevant to determine how much charge has flowed. Integrating (12.51) from $t = 0$ to $t = \infty$ yields

$$Q(t) = I_0 \int_0^\infty \exp(-t/\tau_{LR})dt = -I_0\tau_{LR}\exp(-t/\tau_{LR})|_0^\infty = I_0\tau_{LR} = \frac{I_0 L}{R}.$$

(12.52)

Since the iron core gives a self-inductance that is larger by a factor of $(1 + \chi) \approx 5000$, having an electromagnet in the circuit means that a charge some 5000 times larger will flow, corresponding to a relaxation time τ that is larger by that same factor: the current lasts that much longer. This explains why the shock is greater if the self-inductance is greater. In general, turning off a switch is more likely to lead to a spark than turning it on.

In Mr. Jenkin's experiment, the circuit has three arms. One arm has the battery and its resistance, one arm has the electromagnet and its resistance, and one arm (Mr. Jenkin) has only resistance. See Figure 12.19. When the connection to the battery arm is broken, the much larger current of the coil gets rerouted through Mr. Jenkin. This is I_0, which we had estimated to be 1 A. In fact, Mr. Jenkin has a small self-inductance himself. That will determine the very small time during which his current changes from a very small value (1.0×10^{-4} A) to 1 A, which then decays with the time $\tau_{LR} = L/R$ determined by the L of the electromagnet and the resistance R_J of Mr. Jenkin. For $L = 1.0 \times 10^{-2}$ H and $R_J = 1.0 \times 10^5$ Ω, $\tau_{LR} = L/R = 1.0 \times 10^{-7}$ s. By (12.51), for the current to fall to the safer value of 0.1 mA (corresponding to 10 V) takes a time $t = \tau_{LR}\ln(10^4)$, or $t = 9.2 \times 10^{-6}$ s.

12.12   **Magnetic Energy**

We now study magnetic energy by applying the principle of conservation of energy to an LR circuit driven by a battery, as in the previous section. The rate $\mathcal{E}_0 I$ at which the battery emf $\mathcal{E}_0$ discharges its chemical energy is, by (12.41),

$$\mathcal{E}_0 I = I^2 R + L\frac{dI}{dt}I = I^2 R + LI\frac{dI}{dt}.$$

(12.53)

The first term on the right-hand side is the rate of Joule heating. The second term must be the rate dU_L/dt at which energy is put into the inductor. Thus

$$\frac{dU_L}{dt} = LI\frac{dI}{dt}.$$ (12.54)

Integrating this from zero current gives the energy stored in the inductor:

$$U_L = \frac{1}{2}LI^2. \qquad \text{(magnetic energy stored by inductor)} \qquad (12.55)$$

Let us apply this to a long solenoid of n turns per unit length, length l, and area A, for which L is given by (12.36) and the volume is $V = Al$. Then the magnetic field is uniform, and the magnetic energy per unit volume u_B is given by

$$u_B = \frac{U_L}{V} = \frac{\frac{1}{2}LI^2}{Al} = \frac{1}{2}4\pi k_m n^2 I^2 = \frac{B^2}{2(4\pi k_m)} = \frac{B^2}{8\pi k_m} = \frac{B^2}{2\mu_0},$$

$$\text{(magnetic energy density)} \quad (12.56)$$

where we have used $B = 4\pi k_m nI = \mu_0 I$ for the solenoid. Although (12.56) was obtained only for the case of a solenoid, it is true quite generally. *Energy is stored in the magnetic field.* For $B = 1$ T, a large but achievable value, by (12.56) $u_B = 79.6 \times 10^4$ J/m^3. For soft iron in not too large a field, (12.56) is replaced by

$$u'_B = \frac{u_B}{\mu_r}. \quad \mu_r \equiv 1 + \chi$$

For a field within the iron of $B = 0.2$ T and $\mu_r = 5000$, (12.56) gives $u_B = 1.59 \times 10^4$ J/m^3, but the actual energy density is $u'_B = 3.18$ J/m^3. By (9.23), $B = 0.2$ T corresponds to an applied field of only $0.2/5000 = 0.4 \times 10^{-4}$ T, which is comparable to the earth's magnetic field.

The magnetic energy density is analogous to the electrical energy density derived in Chapter 6:

$$u_E = \frac{\epsilon_0}{2}E^2 = \frac{E^2}{8\pi k}.$$ (12.57)

For $E = 3 \times 10^6$ N/C, the dielectric strength E_d of air, (12.57) gives $u_E = 79.8$ J/m^3. Clearly, it is easier to store energy in the magnetic field (inductors) than in the electric field (capacitors). Note, however, that in many materials the dielectric strength E_d can be much larger than in air. Moreover, when the dielectric properties of a material are included, (12.57) gets multiplied by the dielectric constant κ, which typically exceeds unity by a factor of two or three. For SrTiO$_3$, $\kappa = 330$ and $E_d = 8 \times 10^6$ V/m, by (12.57) giving $u_E = 7.0 \times 10^4$ J/m^3.

For a lead-acid cell, the energy density can be as large as 280×10^6 J/m^3. This explains why lead-acid cells have been used to store excess energy from electric power plants.

Example 12.20 **Inductance of a co-axial cable**

Calculate self-inductance of a co-axial cable with a hollow core, shown in Figure 12.21.

Solution: We first use (12.56) to compute the total magnetic energy

$$U_B = \int u_B dV. \tag{12.58}$$

Then we use (12.55) to determine L from $U_B = U_L$.

By (11.38), $B = 2k_M I/r$ for $r < a$, and by (11.36), $B = 2k_m Ir/a^2$ for $r > a$. Then, with $dV = l(2\pi r)dr$, (12.58) yields

$$U_B = \frac{1}{8\pi k_m} l \int_a^b \left(\frac{2k_m I}{r}\right)^2 (2\pi r)dr = k_m I^2 l \int_a^b \frac{1}{r} dr = k_m l I^2 \ln \frac{b}{a}. \tag{12.59}$$

Comparison of (12.55) with (12.59) gives

$$L = 2k_m l \ln \frac{b}{a}. \tag{12.60}$$

This agrees with (12.37). The energy density approach is particularly helpful in computing the self-inductance when the current is continuously distributed, such as a co-axial cable with a solid core.

12.13 EMF and Electric Field Induced by a Solenoid
Optional

After (12.4) we noted that, when there is an induced emf in a circuit, there usually is not merely an induced electric field $\vec{E}_{ind}$, but also an electric field $\vec{E}_{es}$ due to electric charge. This is needed to produce the same current in all parts of the circuit. Example 12.13 illustrated this point. The first two circuits discussed in the present section have no $\vec{E}_{es}$, but they are symmetric enough that $\vec{E}_{ind}$ can be calculated along these circuits. This information will be used to analyze a third circuit, where there is an $\vec{E}_{es}$.

Consider a solenoid of radius a, n turns per unit length, and current I, whose axis is normal to the paper. We seek the emf for three types of circuits: one within, one outside, and one partially within the solenoid. See Figure 12.25. These circuits need not be real. Let the current increase in a counterclockwise direction so that the rate of change in the field $d\vec{B}/dt$ is out of the page (⊙). By Lenz's law, the induced emf is clockwise.

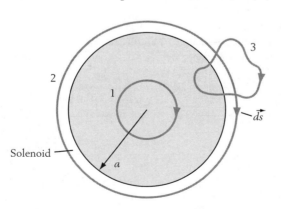

Figure 12.25 Cross-section of a circular solenoid. The three imaginary circuits correspond to one that is concentric and inside (1), one that is concentric and outside (2), and one that is partially inside and partially outside (3).

Let us see this from the mathematics. We neglect the self-inductance of the circuits.

Take $d\vec{A} = \hat{n}\, dA$ into the page $(\otimes)$ so that $d\vec{s}$ and $\hat{s}$ circulate clockwise. To make the emf $\mathcal{E} = \int \vec{E} \cdot d\vec{s}$ clockwise, the induced electric field $\vec{E}$ must circulate with $d\vec{s}$: clockwise. From (12.4), the induced emf is given by

$$\mathcal{E} = \oint \vec{E} \cdot d\vec{s} = -\frac{d\Phi_B}{dt} = -\int \frac{d\vec{B}}{dt} \cdot \hat{n}\, dA = -\frac{d\vec{B}}{dt} \cdot \hat{n}\, A_{flux},$$

$$(\text{uniform } \vec{B}, \hat{n} \text{ is } \otimes) \quad (12.61)$$

where A_{flux} is the area that actually picks up the magnetic flux. As we obtained without the mathematics, the emf corresponds to a clockwise circulation. Let us now apply (12.61).

For two cases of interest, the circuits are circles concentric with the solenoid so that the electric field is tangential and uniform. Then

$$\mathcal{E} = \oint \vec{E} \cdot d\vec{s} = \vec{E} \cdot \hat{s}(2\pi r), \qquad \text{(concentric circle)} \qquad (12.62)$$

so that (12.61) and (12.62) combine to give

$$\vec{E} \cdot \hat{s} = -\frac{1}{2\pi r}\frac{d\vec{B}}{dt} \cdot \hat{n}\, A_{flux}. \qquad \text{(concentric circle, } \hat{s} \text{ is clockwise)} \quad (12.63)$$

1. For circuit 1, a concentric circle of radius $r < a$, the area picking up the flux is given by $A_{flux} = \pi r^2$, so $A_{flux}/2\pi r = r/2$. Then (12.63) becomes

$$\vec{E} \cdot \hat{s} = -\frac{r}{2}\frac{d\vec{B}}{dt} \cdot \hat{n}. \qquad (r < a) \qquad (12.64)$$

If the circuit corresponds to a uniform conducting wire, then this induced electric field is also the total electric field.

2. For circuit 2, a concentric circuit of radius $r > a$, the area picking up the flux is given by $A_{flux} = \pi a^2$, so $A_{flux}/2\pi r = a^2/2r$. Then (12.63) becomes

$$\vec{E} \cdot \hat{s} = -\frac{a^2}{2r}\frac{d\vec{B}}{dt} \cdot \hat{n}. \qquad (r > a) \qquad (12.65)$$

If the circuit corresponds to a uniformly conducting wire, then this induced electric field is also the total electric field.

For $a = r = 0.05$ m and $dB/dt = 10^4$ A/s, (12.64) and (12.65) give $E = 250$ V/m, an appreciable value. Equation (12.65) appears to hold even for $r \to \infty$, thus implying that a localized change in magnetic field can produce a distant change in the electric field. However, implicit in this discussion was the assumption that everything occurs instantaneously. Had we included the fact that the signal takes a finite amount of time to propagate, we would have found that the magnetic field also is nonzero outside the solenoid. Not until Chapter 15 are such effects considered.

3. For circuit 3, consider a planar circuit of arbitrary shape, of length l and of flux-gathering area within the solenoid A_{flux}. By the local form of Ohm's law ($\vec{J} = \sigma \vec{E}$), $\vec{E} \cdot \hat{s}$, the component of the *total* electric field $\vec{E}$ along the physical circuit direction $\hat{s}$, must satisfy $\vec{E} \cdot \hat{s} = J/\sigma$. For a circuit of uniform cross-section and conductivity, J and σ are uniform throughout the circuit, and so is E. Thus (12.61) becomes

$$\mathcal{E} = \vec{E} \cdot \hat{s}\, l = -\frac{d\vec{B}}{dt} \cdot \hat{n}\, A_{flux}, \tag{12.66}$$

so

$$\vec{E} \cdot \hat{s} = -\frac{A_{flux}}{l}\frac{d\vec{B}}{dt} \cdot \hat{n}. \tag{12.67}$$

In general, $\vec{E}$ is the vector sum of both the induced electric field $\vec{E}_{ind}$ and the electrostatic field $\vec{E}_{es}$. To make the current flow along the wire, charge must go to the surface of the wire to produce $\vec{E}_{es}$, in addition to $\vec{E}_{ind}$. In computing (12.67), $\vec{E}_{ind}$ is given by (12.64) for points within the solenoid, where $r < a$, and by (12.65) for points outside the circuit, where $r > a$. Of course, $\oint \vec{E}_{es} \cdot d\vec{s} = 0$, but $\oint \vec{E}_{ind} \cdot d\vec{s} \neq 0$.

12.14 Mr. Jenkin with Self-Inductance

Now consider what would happen if Mr. Jenkin had inductance, or was in series with an inductor, when he connected and disconnected himself from the battery. See Figure 12.26(a).

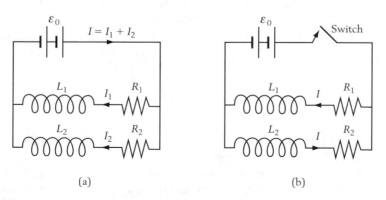

(a)   (b)

Figure 12.26 (a) Circuit where an ideal voltaic cell $\mathcal{E}_0$ drives current through two arms that are in parallel, each arm having both resistance and inductance. (b) Opening the circuit of part (a).

After disconnection, the two arms of the circuit develop a common current I, as in Figure 12.26(b). Take I to be positive when it is in the same direction as the initial current I_1 through the electromagnet, with inductance and resistance L_1 and R_1. Take the initial current through the second arm (containing Mr. Jenkin) to be I_2, with inductance and resistance L_2 and R_2.

How do we determine I? We use Lenz's law in the following form: *If an external magnetic field is changed slowly enough that eddy currents can be set up, then until the eddy currents start to die down a good conductor can nearly completely prevent changes in magnetic flux*. The response of the circuit for very short times, during which $I_1 \to I$ and $I_2 \to -I$, is complicated, involving nonuniform currents and temporary charge buildup in the wires. However, during that time the magnetic flux through the circuit defined by the electromagnet and Mr. Jenkin should be conserved.

This action involving inductors resembles what happens on connecting two charged capacitors. There, the final state had a common but unknown voltage drop across each capacitor, and was subject to charge conservation. We did not worry about the details of *how* it reached that state. Not until Chapter 15 will we touch upon problems that involve details of what is happening along the connecting wires themselves.

In Figure 12.26(b), the common current I circulates counterclockwise. Since the initial current through arm 2, as in Figure 12.26(a), circulates oppositely, the initial magnetic flux is taken to be $\Phi_0 = L_1 I_1 - L_2 I_2$. (Note that the flux $\int \vec{B} \cdot d\vec{A}$ doesn't depend on which way the solenoids have been wound. This is because the directions of both $\vec{B}$ and $d\vec{A}$ change on winding clockwise instead of counterclockwise.) By flux conservation, the final value of the magnetic flux is $\Phi = (L_1 + L_2)I$. Hence flux conservation yields

$$L_1 I_1 - L_2 I_2 = (L_1 + L_2)I. \tag{12.68}$$

Since $I_1 = \mathcal{E}_0/R_1$ and $I_2 = \mathcal{E}_0/R_2$, (12.68) implies that it is not inductance L alone, but rather the inductive time constant L/R that determines which arm of the circuit dominates in determining I. Hence, if $L_1 = 10L_2$, but $R_1 = 100R_2$, arm 2 dominates over arm 1 in determining I.

You can verify (as a problem) that the magnetic energy computed from (12.55) with common current I is less than the initial magnetic energy with I_1 and I_2. The loss of energy is attributed to local currents along the wires that provide no net magnetic flux and that decay quickly. Once the adjustment to I has occurred, decay of I (and of magnetic flux) takes place as for a circuit with inductance $L = L_1 + L_2$ and resistance $R = R_1 + R_2$.

Problems

12-3.1 Consider Figure 12.2(a). (a) Explain why Faraday used calico in this experiment. (b) Explain why Faraday had to employ 100 copper-zinc cells to cause a deflection of the galvanometer. (c) If the galvanometer deflected clockwise on connecting the battery, indicate how the galvanometer deflected on disconnecting the battery.

12-3.2 Consider Figure 12.2(b). (a) Explain what role the magnetic needle played. (b) If the needle magnetized to the left on connecting the battery, indicate how the needle magnetized on disconnecting the battery.

12-3.3 Consider Figure 12.3. (a) If the galvanometer deflected clockwise on moving the wire W's together, indicate how the galvanometer deflected on separating the wire W's. (b) Explain why Faraday moved the wire W's at the resonance frequency of the galvanometer needle.

12-3.4 Consider Figure 12.4(a). (a) Would the galvanometer deflection be larger or smaller if the iron ring were replaced by a copper ring? (b) If the galvanometer deflected counterclockwise on connecting the battery, indicate how the galvanometer deflected on disconnecting the battery.

12-3.5 Consider Figure 12.4(b). (a) Would the galvanometer deflection be larger or smaller if the iron rod were replaced by a copper rod? (b) If the galvanometer deflected counterclockwise on connecting the battery, indicate how the galvanometer deflected on disconnecting the battery.

12-3.6 Consider Figure 12.5(a). (a) Explain what role the permanent magnets play. (b) If the galvanometer deflected clockwise on bringing the magnets together, indicate how the galvanometer deflected on taking the magnets apart.

12-3.7 Consider Figure 12.5(b). (a) Explain what role the permanent magnet plays. (b) If the galvanometer deflected clockwise on bringing the magnet in, indicate how the galvanometer deflected on pulling the magnet out.

12-4.1 For $y < 0$, there is a uniform magnetic field $\vec{B}$ pointing along z (out of the page). For $y > 0$, the field is zero. A rectangular conducting loop, its normal along z, moves with velocity $\vec{v}$ into the field region. See Figure 12.27. Describe, as the loop enters the field region: (a) the change in the magnetic

field due to the motion, (b) the direction of the magnetic field induced in response, (c) the direction of circulation of the induced emf, (d) the direction of circulation of the induced current, (e) the direction of the magnetic force acting on the loop, (f) the tendency of the loop to compress or expand, (g) the tendency of the loop to rotate.

12-4.2 A string hangs from the ceiling, connected to an aluminum ring. Describe a noncontact method to bring the ring into motion, using a cylindrical bar magnet that can fit within the ring.

12-4.3 A two-winding coil produces a clockwise current I when subjected to a time-varying magnetic field. If one of the windings were wound clockwise and the other counterclockwise, how would the current change, and why?

12-4.4 A coil consists of two turns in series: a turn of copper wire wound one way and a turn of iron wire wound the other way. It surrounds an iron core solenoid. (a) When the solenoid is powered up, there is a 2 V emf in the copper wire. Find the induced emf in the iron wire, and overall. (b) Repeat if both turns are made of copper. (c) Repeat if both turns are wound the same way.

12-4.5 Two coils are co-axial along x, so their planes are normal to the page. See Figure 12.28, drawn in perspective. For coil B, a battery drives current clockwise as seen from the left. Coil A is not connected to a battery. Let the variable resistance R increase. Describe (a) the change in the magnetic field through coil A due to the change in R, (b) the direction of the magnetic field induced in coil A, (c) the direction of circulation of the emf induced in coil A, (d) the direction of circulation of the current induced in coil A, and (e) the direction of the magnetic force acting on coil A. (f) How do your answers change if R is decreased?

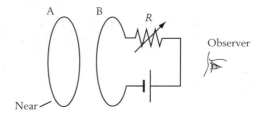

Figure 12.28 Problem 12-4.5.

12-4.6 Two non-co-axial coils rest on the page, their normals along z. See Figure 12.29. For coil B, a battery drives current counterclockwise. Let

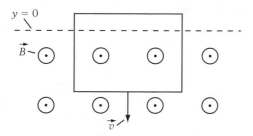

Figure 12.27 Problem 12-4.1.

the variable resistance R increase. Describe (a) the change in the magnetic field through coil A due to the change in R, (b) the direction of the magnetic field induced in coil A, (c) the direction of circulation of the emf induced in coil A, (d) the direction of circulation of the current induced in coil A, and (e) the direction of the magnetic force acting on the loop. (f) How do your answers change if R is decreased?

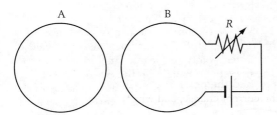

Figure 12.29 Problem 12-4.6.

12-4.7 A uniform magnetic field is directed out of the page. A rectangular circuit lies in the plane of the page. Three sides are rigid, but the right arm is a movable conducting rod. See Figure 12.30. If the right arm slides rightward, for the closed circuit that includes the movable arm describe (a) the change in the magnetic field through the circuit, due to the motion, (b) the direction of the magnetic field induced in response, (c) the direction of circulation of the induced emf, (d) the direction of circulation of the induced currents, and (e) the direction of the magnetic force acting on the loop. (f) How do your answers change qualitatively if this is repeated in a reversed field? (g) How do your answers change qualitatively if this is repeated in an equal field that is tilted relative to the circuit?

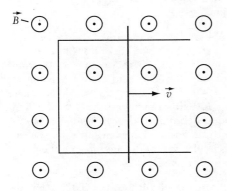

Figure 12.30 Problem 12-4.7.

12-4.8 A copper plate is thrown between the poles of a powerful horseshoe magnet. Neglect gravity. See Figure 12.31. (a) How is the the motion of the plate affected by the presence of the magnet? (b) How would the magnet move if it were not held in place? (c) How could this principle be used for magnetic braking of a train wheel?

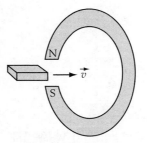

Figure 12.31 Problem 12-4.8.

12-4.9 A sheet of aluminum foil lies on a table. A coil is held in place above, and parallel to, the foil. Suddenly the coil is given a clockwise current, as seen from above the coil. (a) Indicate how the eddy currents in the foil circulate, and explain your reasoning. (b) Indicate the direction of the force on the foil, and explain your reasoning.

For each of the next five problems, indicate your viewpoint as observer, and then give, in a table similar to those in the text: (a) the direction of $d\vec{B}_{ext}/dt$; (b) the direction of $\vec{B}_{ind}$; (c) the direction of circulation of $\mathcal{E}_{ind}$; (d) the direction of circulation of the induced current; (e) the direction of the net magnetic force $\vec{F}_{net}$ (or magnetic torque, as appropriate) acting on each of the electrically conducting circuits; and (f) the tendency of the circuit to compress or expand.

12-4.10 A magnet, its N pole pointing downward, is held above a co-axial conducting loop lying on a table. The magnet is then released.

12-4.11 A square conducting loop lies in the plane of the page. A magnetic field points out of the page. The loop is twisted about y, its right arm coming out of the page and its left arm going into the page.

12-4.12 A magnet, its N pole pointing downward, is held fixed above a conducting loop.

12-4.13 Two co-axial loops have their normal along the vertical. With zero currents, they are held fixed in position. A battery is switched on for the lower loop, giving it a clockwise current as seen from above. See Figure 12.32.

Near

Figure 12.32 Problem 12-4.13.

12-4.14 A current loop in the plane of the page is partially in a uniform magnetic field pointing out of the page. It is pulled out of the field. See Figure 12.33.

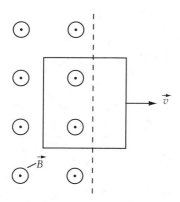

Figure 12.33 Problem 12-4.14.

12-5.1 (a) Characterize $\vec{E}$, $\vec{B}$, emf, magnetic flux, and the time derivative of magnetic flux, as vectors or scalars. (Think of $\vec{B}$ as due to moving charge carriers.) (b) Characterize how they change when $t \to -t$ (don't forget that velocity $\vec{v} = d\vec{r}/dt$ reverses when t reverses). (c) Which are related by Faraday's law?

12-5.2 A square circuit of side 2 cm sits on a table, and at $t = 0$ is subject to a uniform magnetic field that points into the table. Let $|\vec{B}| = 4t + 8t^2$, where $|\vec{B}|$ is in 10^{-3} T for t in ms. (a) At $t = 2$ ms, find the magnetic flux. (b) At $t = 2$ ms, find the rate of change of the magnetic flux. (c) At $t = 2$ ms, find the magnitude and sense of the emf. (d) If the circuit has resistance 5 Ω, at $t = 2$ ms find the force per unit length on the circuit, including direction.

(e) Relate part (d) to the motion statement of Lenz's law.

12-5.3 In the previous problem, let one arm be just outside the field region, and let the circuit be tethered in place by a string that can support 5.7 N. As a function of time, find (a) the magnetic flux, (b) the rate of change of magnetic flux, (c) the magnitude and sense of the emf, (d) the net force on the circuit. (e) Find the time at which the string will break.

12-5.4 A 4-cm-by-4-cm loop of resistance 50 Ω sits on a tabletop, in a uniform magnetic field of 0.2 T that points at 70° from the upward normal. The field starts changing at the rate of -5 T/s. Find (a) the magnetic flux; (b) the rate of change of the magnetic flux; (c) the induced emf, including its direction of circulation; (d) the induced current, including its direction of circulation; (e) the force on each arm; (f) the torque on the loop.

12-5.5 A circular UHF antenna for a portable television has a 15 cm diameter. The magnetic field is oriented at 15° to the antenna's normal and is changing at the rate 0.14 T/s. (a) Find the antenna's emf (which drives current to the amplifier). (b) By what factor would the signal increase if the antenna were at optimal orientation? (c) In part (a), if $R = 3.7$ Ω, find the current I.

12-5.6 Dr. Evil has implanted electrodes into you (take $R_{you} = 10^4$ Ω) and connected them to a rigid 100-turn circuit of dimension 4 m by 2 m, 3 m by 2 m of which is in (and normal to) a 10 T field. [The situation is not unlike that of Figure 12.13(a), except that you are part of the left arm of the circuit.] (a) If a continuous current of 2 mA is the greatest you can stand, how fast can you move out of the field region? (b) How large a force then opposes your motion?

12-5.7 A 4-cm-by-12-cm loop of resistance 25 Ω sits on a tabletop, 4 cm by 5 cm of it sitting in a uniform magnetic field of 4 T that points at 60° from the upward normal. The field starts changing at the rate of -8 T/s. Neglect the self-inductance of the loop. Just after the field starts changing, find (a) the induced emf, including its direction of circulation; (b) the induced current, including its direction of circulation; (c) the net force on the loop; (d) the torque on the loop.

12-5.8 A square loop with $R_{loop} = 0.025$ Ω has 30 turns and is 0.3 cm on a side. It is within a solenoid with 250 turns, 5 cm long, and of radius 0.4 cm. The normal to the loop makes an angle of 25° to the axis of the solenoid. (a) Find the field within the solenoid when it carries a clockwise current $I_{sol} = 6$ A. (b) If $dI_{sol}/dt = 210$ A/s, give the magnitude and direction of the emf induced in the loop. (c) Find the induced current I_{loop} in the loop. (d) Find the force on one arm of the loop, at the instant when $I_{sol} = 6$ A and $dI_{sol}/dt = 210$ A/s, and state whether the force tends to expand or compress. (e) Find the torque on the loop.

12-5.9 Consider a long solenoid of square cross-section, with side $d = 5$ cm, sitting on a table. A rectangular circuit, with $R = 40$ Ω, has sides of length $l = 2$ cm and $b = 6$ cm, parallel to those of the square, with 3 cm of its long side slipped between two turns of the left arm of the solenoid. When $I_{sol} = 4$ A, the field within the solenoid is 0.024 T. (a) If $I_{sol} = 16$ A clockwise, find the magnitude of the field within the solenoid. (b) If I_{sol} starts to decrease at the rate $dI_{sol}/dt = -400$ A/s, find the rate of change of magnetic flux through the rectangular circuit. (c) If the circuit has resistance 40 Ω, find the current induced in the circuit, including direction (as viewed from above the table). (d) Find the direction and magnitude of the net force on the rectangular circuit, for the I_{sol} of part (a).

12-5.10 Consider the monopole moving toward the ring, as in Figure 12.18. Consider an imaginary spherical cap centered at q_m and of radius r, enclosing the physical loop, of radius a, so that $r = \sqrt{a^2 + x^2}$. (a) Show that the magnetic flux is the product of the monopole field $B_{q_m} = k_m q_m / r^2$ and the cap area $A_{cap} = r^2\Omega = 4\pi(1 - \cos\theta) = 2\pi[1 - (x/\sqrt{a^2 + x^2})]$. (b) Show that the emf is given by

$$\mathcal{E} = \oint \vec{E} \cdot d\vec{s} = -d\Phi_B/dt$$

$$= 2\pi k_m q_m a^2 (dx/dt)/(a^2 + x^2)^{3/2}.$$

12-5.11 The flip coil was invented by Faraday in 1851 to show the fundamental relationship between changes in magnetic flux and induced emf. (a) If a coil of area A and N turns is in an unknown field $\vec{B}$, of magnitude B, find the integrated emf on flipping the coil (sitting on an experimental bench) by 180°. (b) If the coil is part of a circuit with resistance R, show that the time-integrated current $\int I dt$ (which can be measured with a ballistic galvanometer) has magnitude $Q = 2NBA/R$.

12-5.12 In a uniform perpendicular $\vec{B}$, consider a rectangular circuit with three arms of iron and a slide-wire of copper, as in Figure 12.30. Deter-

mine the emf at the instant when each iron arm has resistance R_1, and the copper arm has resistance R_2, for the following cases: (a) the copper moves to the right with velocity v; (b) the circuit moves to the right with velocity v; (c) both move to the right with velocity v.

12-5.13 A loop of radius r and resistance R lies in the plane of the page. The loop is compressed at a rate dr/dt in a uniform field $\vec{B}$ that points into the page. See Figure 12.34. (a) Find the induced emf $\mathcal{E}$ and induced current I. (b) Find the force per unit length dF/dl opposing compression. (c) Evaluate this for $B = 2$ T, $R = 0.35$ Ω, $r = 2$ cm, and $dr/dt = 100$ m/s.

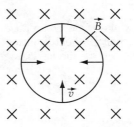

Figure 12.34 Problem 12-5.13.

12-6.1 An emf $\mathcal{E}_1 = 0.36$ V is induced in coil 1 when the current in coil 2 increases at the rate $dI_2/dt = 1.8$ A/s for $dI_1/dt = 0$. (a) Find the mutual inductance M. (b) If the emf in coil 2 is $\mathcal{E}_2 = 5.4$ V, find dI_1/dt if $dI_2/dt = 0$.

12-6.2 A conducting loop of area 4.5 cm² is within and co-axial with a solenoid having $n = 8200$ turns/cm. (a) Find the mutual inductance between the loop and the solenoid. (b) If $dI_{coil}/dt = 435$ A/s, find the induced emf in the solenoid. (c) If the coil is rotated so that its normal makes a 65° angle to the solenoid axis, find the mutual inductance.

12-6.3 Two loops have a mutual inductance of 5.4 mH. If the current in coil 1 is $I = -12 + 6t + 3t^2$, where t in seconds gives I in amps, find the emf in coil 2 at time t.

12-6.4 Two identical coils of 3 cm radius are co-axial, with 2 cm separation. (a) Qualitatively, how does their mutual inductance change if their separation decreases to 1 cm? (b) If one of them is rotated by 90°?

12-6.5 Two torii are concentric, the larger one, of N_1 turns, enclosing the smaller one, of N_2 turns. They both have rectangular cross-sections, but the smaller has a very small cross-section of area

$A_1 = ha$, corresponding to an approximate radial distance ρ and radial thickness a. See Figure 12.35. (a) Compute the mutual inductance of 1 on 2. (b) Compute the mutual inductance of 2 on 1. (The two results should be the same. The result should not depend on details of the larger torus.)

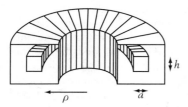

Figure 12.35 Problem 12-6.5.

12-6.6 Two single-turn coils of radii a and b are concentric and lie on the plane of the page. (a) If $a \ll b$, show that the mutual inductance is approximately given by $M = 2\pi^2 k_m a^2/b$. (b) For $a = 0.5$ cm and $b = 6$ cm, evaluate M numerically. (c) If $dI_a/dt = 275$ A/s, find $\mathcal{E}_b$.

12-6.7 Two single-turn coils of radii a and b are co-axial and separated by a large distance R. (a) Find their mutual inductance. (b) If $a = b = 2.4$ cm and $R = 16$ cm, evaluate M.

12-6.8 Two single-turn coils of radii a and b are separated by a large distance R. They are not co-axial, and their normals $\hat{n}_1$ and $\hat{n}_2$ make arbitrary angles to the vector $\vec{R} = \vec{R}_2 - \vec{R}_1$ separating them. Determine their mutual inductance.

12-6.9 A coil of N_c turns and radius b *surrounds* a solenoid of N_s turns, length l, and radius $a < b$. Both are normal to the page. Let clockwise currents I_c and I_s be taken as positive. (a) For $\hat{n}$ pointing into the page, determine an algebraic expression for the magnetic flux $\int \vec{B} \cdot \hat{n}\,dA$ through the coil due to the solenoid. Take $I_s > 0$. (b) If $dI_s/dt > 0$, determine the direction of circulation of the induced emf in the coil. (c) Compute the rate of change of the magnetic flux through the coil, and the magnitude of the emf. (d) Compute the mutual inductance $M = \frac{1}{I_s}\int \vec{B} \cdot \hat{n}\,dA$, first algebraically and then numerically, with $N_c = 20$, $N_s = 800$, $l = 20$ cm, $a = 2$ cm, $b = 4$ cm.

12-6.10 Repeat the previous problem for the coil *within* the solenoid, so $a > b$. Take $N_c = 20$, $N_s = 800$, $l = 20$ cm, $a = 2$ cm, $b = 1$ cm.

12-7.1 A conducting rod of length 0.4 m is in the xy-plane, at a 35° counterclockwise angle to the x-axis. It is moving along the x-axis with velocity 0.2 m/s. There is a uniform 0.4 T magnetic field along z. See Figure 12.36. (a) Find the motional $\vec{E}$ field acting on the rod. (b) After any transients have died down, which end of the rod is at the higher voltage, and by how much?

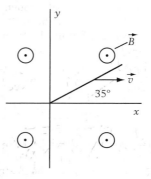

Figure 12.36 Problem 12-7.1.

12-7.2 Consider Example 12.13. (a) In Figure 12.16(a), if all the resistance were in the left arm, what would be the value of $\Delta V \equiv \int \vec{E}_{es} \cdot d\vec{s}$ across the right arm? (b) Across the left arm? (c) Across the top arm?

12-7.3 Consider Example 12.13. (a) In Figure 12.16(a), if all the resistance were in the right arm, what would be the value of $\Delta V \equiv \int \vec{E}_{es} \cdot d\vec{s}$ across the right arm? (b) Across the left arm? (c) Across the top arm?

12-7.4 A conducting rod of length b is normal to a long wire carrying current I, with nearest distance s. The rod moves along the axis of the wire at velocity v. See Figure 12.37. Find the voltage difference between the ends of the rod, and specify which end has the higher voltage.

Figure 12.37 Problem 12-7.4.

12-7.5 At a location where the earth's magnetic field has a vertical component of 2×10^{-5} T and a southward component of 3×10^{-5} T, an airplane flies northward at 900 km/h. If the plane's wingtip-to-wingtip length is 27 m, find the emf induced across the wings, and indicate which wingtip is at the higher voltage.

12-7.6 A ring of radius R lies on the xy-plane, centered at the origin. The ring has conductivity σ and a rectangular cross-section with height h and radial thickness d, where $d, h \ll R$. A monopole q_m falls with velocity v along the axis of the ring, at initial height z. (The geometry is similar to a 90° clockwise version of Figure 12.18.) (a) Find the flux through the wire. (b) Find the emf induced in the wire. (c) Find the current in the wire. (d) Find the field due to the wire, at the position of the monopole. (e) Find the force on the monopole.

12-7.7 A narrow magnet of length 14 cm and magnetic moment 2.8 A-m^2 is co-axial with a loop of radius 1.7 cm. The loop is moved toward the magnet with a velocity 2.5 cm/s. See Figure 12.38. Find the induced emf in the loop when the loop is 3.6 cm from the N pole.

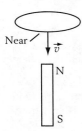

Figure 12.38 Problem 12-7.7.

12-7.8 A long wire carries current I along y. A uniform rectangular loop of sides a and b lies a distance s to the right of the wire in the xy-plane, with a side of length a parallel to the wire. See Figure 12.39. The current in the wire is I. (a) If the loop moves rightward at velocity v, find an algebraic expression for the induced emf in the circuit. (b) Let $a = 6$ cm, $b = 8$ cm, and the circuit resistance $R = 0.56$ Ω. At a given moment of time, let $s = 4$ cm, $v = 14$ cm/s, and $I = 6$ A. Find the emf and the current in the circuit.

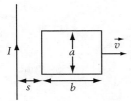

Figure 12.39 Problem 12-7.8.

12-7.9 Consider a uniform disk of conductivity σ, radius a, and thickness d, rotating about its axis at angular velocity ω. There is a field $\vec{B}$ within a

small rectangular region of sides $b, c \ll a$, where b is nearly radial and c is nearly tangential. See Figure 12.40. (a) Determine the motional emf within this region, and indicate how the current flows. (b) Estimate the appropriate electrical resistance (neglect the resistance of the larger area that provides return currents). (c) Determine the force and torque acting on this region. (d) Verify that this force and torque serve to oppose the rotation (i.e., a magnetic brake).

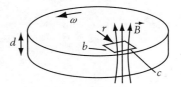

Figure 12.40 Problem 12-7.9.

12-8.1 Consider the circuit in Figure 12.19. (a) Which circuit arm represents Mr. Jenkins's body? Explain why. Now consider that Mr. Jenkins is replaced by a real resistor, and take $\mathcal{E}_0 = 24$ V, $R_J = 500$ Ω, $R_w = 20$ Ω, $L = 0.4$ mH. (b) What is the current through each arm before the emf is disconnected? (c) Just after the disconnection, what is the current through each arm of the circuit? (d) Find the voltage drops across R_w, R_J, and L, just after the disconnection, including the direction they tend to drive current. (e) Sketch, as a function of time, the rate of decay of the current through R_J.

12-8.2 Repeat the preceding problem for $\mathcal{E}_0 = 24$ V, $R_J = 5 \times 10^7$ Ω, $R_w = 20$ Ω, $L = 0.4$ mH.

12-9.1 When a battery of 2 V is suddenly switched on for a circuit containing only a coil, dI/dt has magnitude 740 A/s. Find the self-inductance of the coil.

12-9.2 The self-inductance of a coil is 0.46 mH. (a) Find dI/dt when a 42 V emf is switched on. (b) Find $\mathcal{E}$ when $dI/dt = 142$ A/s.

12-9.3 The self-inductance of coil 1 is 0.46 mH, and its mutual inductance with coil 2 is 240 μH. The coils are co-axial. If the emf in coil 1 is 14.3 V when $dI_1/dt = 890$ A/s, find dI_2/dt. What is the significance of the sign of dI_2/dt?

12-9.4 For a current loop made of a perfect diamagnet, $\Phi_B = $ constant for the flux through its cross-section. Let circuit 1 be a perfect diamagnet, and take $L_1 = 0.64$ mH, $L_2 = 0.45$ mH, and $M = 0.18$ mH. If $I_1 = 2$ A and $I_2 = 0$ at $t = 0$, and then I_2 becomes 0.4 A at $t = 0.034$ s, find I_1 at $t = 0.034$ s.

12-9.5 A metal ring of negligible resistance and self-inductance L has mutual inductance M with a long solenoid connected to 120 V, 60 Hz ac power. See Figure 12.41. The ring, if not held in place, would jump up. (a) What does this fact say about the relative instantantaneous directions of the equivalent magnets of the ring and the solenoid? (b) What does this say about the relative instantantaneous directions of circulation of the currents in the ring and in the solenoid? (c) Explain how your answer to part (b) follows from (12.34).

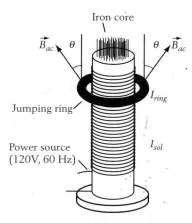

Figure 12.41 Problem 12-9.5.

12-9.6 A current loop of negligible resistance and self-inductance L is in the plane of the page, partially in a uniform but ac magnetic field normal to the page. See Figure 12.42. The loop feels a net force tending to push it out of the field region. (a) What does this fact say about the relative instantantaneous direction of circulation of the current in the loop and the instantaneous direction of the ac magnetic field? (b) If the field is due to a solenoid, as in Figure 12.11(b), explain how your answer to part (a) follows from (12.34).

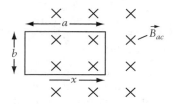

Figure 12.42 Problem 12-9.6.

12-10.1 A solenoid of radius $a = 1.2$ cm and length 24 cm has $L = 12.4$ μH. (a) Find how many turns it has. (b) If $dI/dt = 125$ A/s, find the self-induced emf.

12-10.2 A co-axial cable is 75 m long, with inner and outer radii of 0.35 mm and 0.44 mm. (a) Find its self-inductance. (b) If $dI/dt = 287$ A/s, find the self-induced emf.

12-10.3 The self-inductance of a single coil is L. (a) Find the self-inductance of a coil of the same shape but N turns. (b) If $L = 2.4$ nH for a single turn, evaluate the self-inductance for 240 turns.

12-10.4 A toroid with a soft iron core has a circular cross-sectional area 2.8 cm^2, radius 14 cm, and 780 turns wrapped around it. (a) Estimate its self-inductance. (b) If $dI/dt = 45$ A/s, find the self-induced emf.

12-10.5 Consider a toroid with square cross-section of side a, inner radius b, and outer radius $a + b$. It has N turns. Compute its self-inductance.

12-10.6 Consider a long rectangular circuit with sides a and b, where $b \gg a$. Let the wire radius $R \ll a \ll b$. Neglecting the short ends, find the inductance per unit length. Neglect the field within the wires of the circuit.

12-11.1 A series circuit has $L = 6.5$ μH, $R = 0.23$ Ω, and an ideal battery $\mathcal{E}_0 = 2.4$ V. (a) If $I = 4.2$ A, find dI/dt and the voltages across the resistor and inductor. (b) If the voltage across the inductor is 0.9 V, find I, dI/dt, and the voltage across the resistor.

12-11.2 A series circuit has $L = 3.8$ μH, $R = 0.44$ Ω, and an ideal battery $\mathcal{E}_0 = 4.5$ V. (a) If $dI/dt = 1.1 \times 10^3$ A, find I and the voltages across the resistor and inductor. (b) If the voltage across the resistor is 2.9 V, find dI/dt, I, and the voltage across the inductor.

12-11.3 A real inductor has $L = 8$ μH and $R_L = 5$ Ω (treat them in series). It is connected in series to a resistor $R = 200$ Ω. A switch then connects the circuit to an ideal 12 V battery. (a) If after a time t, $I = 36$ mA, find dI/dt and the voltage across the inductor. (b) Find t.

12-11.4 A circuit has an unknown inductance L and a resistance $R = 6.4$ Ω in series. At $t = 0$, an emf $\mathcal{E}_0 = 3.2$ V is switched on in series with them, and a measurement yields $dI/dt = 580$ A/s. (a) Find L. (b) Find τ_{LR}. (c) Find I after 0.45 ms.

12-11.5 A series circuit has an unknown inductance L, a resistance $R = 16.4\ \Omega$, an emf $\mathcal{E}_0 = 1.4$ V, and a steady current. At $t = 0$, the emf is switched to zero, and a measurement at $t = 0^+$ yields $dI/dt = -208$ A/s. (a) Find L. (b) Find τ_{LR}. (c) Find I after 0.6 ms.

12-11.6 A series circuit has an unknown inductance L, a resistance $R = 22\ \Omega$, an emf $\mathcal{E}_0 = 1.4$ V, and a steady current. At $t = 0$, the emf is switched to 2.4 V, and a measurement at $t = 0^+$ yields $dI/dt = 280$ A/s. (a) Find L. (b) Find τ_{LR}. (c) Find I after 1.2 ms.

12-11.7 An inductor has $L = 8\ \mu\text{H}$ and $R_L = 5\ \Omega$ (treat them in series). It is connected in series to a resistor $R = 200\ \Omega$. At $t = 0$, a switch is thrown connecting the circuit to an ideal 12 V battery. (a) Find the voltage drop across the inductor at $t = 0^+$ and at $t = 25\ \mu\text{s}$. (b) At $t = 0^+$ and at $t = 25\ \mu\text{s}$, sketch the voltage profiles around the circuit, treating the inductor as having a finite length, and neglecting the resistance of the connecting wires. (c) At $t = 0^+$ and at $t = 25\ \mu\text{s}$, what fraction of the electric field in the inductor is electrostatic, and what fraction is electromagnetically induced?

12-11.8 Consider a circuit like that of Mr. Jenkin (Figure 12.19). Let there be a real inductor in one arm, with $L = 2.5\ \mu\text{H}$ and $R_L = 30\ \Omega$, and let there be a resistor $R = 60\ \Omega$ in the other arm. Let the emf $\mathcal{E} = 24$ V, with negligible internal resistance. The current is steady. (a) Find the currents through L and R just before and just after $\mathcal{E}$ is switched out of the circuit at $t = 0$. (b) Find the current in the circuit and the voltage across the inductor at $t = 40\ \mu\text{s}$.

12-12.1 Consider an inductor of 1.6 mH. (a) Find the magnetic energy when $I = 14$ A. (b) If the inductor has a volume 0.68 mm³, find the magnetic energy density. (c) Find the magnetic field.

12-12.2 A solenoid of length 6 cm and area 0.84 cm² produces a field of 0.0043 T for a 12 A current. (a) Find the magnetic energy density. (b) Find the magnetic energy. (c) Find the self-inductance. (d) Find the number of turns.

12-12.3 Evaluate the self-inductance per unit length for a co-axial cable of outer radius 3.6 mm and inner radius 0.45 mm.

12-12.4 (a) Using the energy method, and including the field energy within the core, find the self-inductance per unit length of a co-axial cable of inner radius a and outer radius b. It will differ slightly

from (12.38) because the core is included. (Take a uniform current density in the core.) (b) Evaluate the self-inductance per unit length for outer radius 3.6 mm and inner radius 0.45 mm.

12-12.5 A series circuit has $L = 4$ mH and $R = 0.5\ \Omega$. (a) Find the time constant for the circuit. (b) A switch now connects the circuit to an ideal 12 V battery. Sketch $I(t)$. (c) Find the values of I and dI/dt at $t = 0$, $t = 0.002$ s, and $t = \infty$. For each of these times, find (d) the rate at which the battery discharges its chemical energy, (e) the rate at which heat is dissipated, (f) the rate at which magnetic energy is stored, (g) the rate at which electrical energy is stored, and (h) compare the rate of energy provided by the battery to the sum of the other rates.

12-12.6 Show that $M_{12} = M_{21}$, by using energy considerations (turn on the currents in different orders). Work by analogy to the calculation using the coefficients of potential in Chapter 6.

12-13.1 Consider a solenoid of radius 2 cm. Let $|d\vec{B}/dt| = 440$ T/s within the solenoid. (a) At 0.5 cm from the center, find the magnitude of the induced electric field. (b) At 5 cm from the center, find the magnitude of the induced electric field.

12-13.2 Consider a circuit like 3 in Figure 12.25, made of copper, of length 5.6 cm and radius 0.47 mm, and with 0.6 cm² of it in the field of the solenoid, with $a = 4$ cm. If the induced current density in the copper wire is found to be 520 A/cm², estimate the rate at which the field is changing within the solenoid.

12-13.3 Equation (12.64) gives the tangential field component E_θ for a ring within a large solenoid. It also gives the tangential field of a resistanceless solenoid S′ that extends normal to the page, within the larger solenoid. For S′ there is also an electric field component E_z along the solenoid axis. (a) For a long solenoid of length l and radius a, with N turns, show that $|E_z/E_\theta| = 2\pi Na/l$. (b) Evaluate this for $a = 4$ cm, $l = 20$ cm, and $N = 200$. (c) If the larger solenoid S has 4200 turns/m, and $dI_S/dt = 312$ A/s, find E_θ and E_z. (d) Is E_θ electromagnetically induced or electrostatic? (e) Repeat for E_z.

12-13.4 A circuit with three vertical arms having, from left to right, resistances R_1, R_2, and R_3, encloses two loops. Time-varying solenoid currents give clockwise emfs $\mathcal{E}_l$ and $\mathcal{E}_r$ to the loops.

See Figure 12.43. (a) Find the current through each resistor, in algebraic form. (b) For $R_1 = 2$ Ω, $R_2 = 8$ Ω, $R_3 = 4$ Ω, $\mathcal{E}_l = 3$ V, $\mathcal{E}_r = 5$ V, evaluate these currents numerically. (c) Can the voltage everywhere along the circuit be determined from this information? [Hint: Consider two different configurations (i.e., differently shaped resistors) with the same resistances and emfs.]

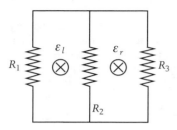

Figure 12.43 Problem 12-13.4.

12-13.5 A circuit has three vertical arms, with resistance R_l in the left arm, R_m in the middle arm, and R_r in the right arm. The right part of this circuit (which includes R_m and R_r) also surrounds a solenoid of area A, length l, and number of turns per unit length n. See Figure 12.44. (a) A voltmeter sits to the right of the circuit. The solenoid current I_s changes with time, giving clockwise emf $\mathcal{E}_r$. Give the voltmeter reading, in algebraic form, if its leads are connected across R_r. (b) The voltmeter now is moved to the left of the circuit. Give the voltmeter reading, in algebraic form, if its leads are connected across R_l. (c) Let $R_l = 2$ Ω, $R_m = 8$ Ω, $R_r = 4$ Ω, $A = 3.6$ cm^2, $l = 42$ cm, and $n = 1200$/m. If $dI_s/dt = 1450$ A/s, numerically evaluate the voltmeter readings of parts (a) and (b).

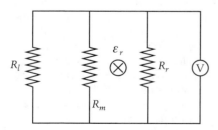

Figure 12.44 Problem 12-13.5.

12-13.6 Consider a square circuit of side $2a$ that surrounds and is concentric with a right-circular solenoid. See Figure 12.45. (a) Show that, along any arm of the square circuit, the induced electric field varies as $\cos^2 \theta / a$, where θ is measured from the distance of closest approach,

which is a. (b) Find how the electrostatic field varies. (c) Show that for a concentric square circuit that is inside a right-circular solenoid, the induced electric field along the circuit is uniform, so the electostatic field is zero.

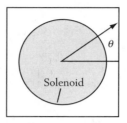

Solenoid

Figure 12.45 Problem 12-13.6.

12-13.7 Consider a square circuit with a long circular solenoid at its center, producing a clockwise emf $\mathcal{E} = 12$ V. Let the top and bottom arms have negligible resistance, and let the right and left arms have resistances $R_r = 1$ Ω and $R_l = 2$ Ω. (a) Find the current. (b) Find the reading of a voltmeter placed across R_l and all the other arms, going clockwise. (c) If the bottom-left corner has voltage $V_D = 0$ (e.g., truly due to electric charge), find the voltages (i.e., due to the electrostatic field alone) at all the other corners A, B, and C, going clockwise. (d) Find the voltage difference across R_r and all the other arms, going clockwise.

12-13.8 Repeat the previous problem where the radius of the solenoid is very small compared to the side of each circuit, and move it to just within the lower-right corner of the circuit (point C of the previous problem).

12-14.1 Two inductors with the same resistance of 2 Ω and $L_1 = 5$ mH, $L_2 = 9$ mH are placed in parallel with a 12 V emf of negligible resistance. (a) Find their steady currents. (b) If the emf is taken out of the circuit, find their initial currents and the current through L_1 after 2 ms.

12-14.2 Two inductors are connected in parallel with a 12 V battery. They have $L_1 = 8$ mH, $R_1 = 0.4$ Ω and $L_2 = 3$ μH, $R_2 = 2$ Ω. (a) Find the currents through each arm before the battery is disconnected, including their directions. (b) Just after the disconnection, find the current through each arm, including their directions. (c) Find how long it takes for the current to decrease by a factor of 8.

12-14.3 Consider a circuit like that of Mr. Jenkin, but now let Mr. Jenkin have both inductance and resistance. Take $L_1 = 5$ mH and $R_1 = 22$ Ω, and $L_2 = 2$ mH and $R_2 = 80$ Ω, and let the two be connected to an ideal 12 V battery. (a) Find the currents through each element when Mr. Jenkin is holding the leads to the electromagnet, but before the battery is connected to the circuit. (b) Find the currents through each element just after the battery is connected to the circuit. (c) Find the currents through each element a few seconds later. (d) Repeat just after the battery is disconnected from the circuit. (e) Find the time constant of the circuit, and find the current 40 μs after the battery is disconnected.

12-14.4 (a) Explain why it was safe for Mr. Jenkin to connect the battery, but hazardous for him to disconnect the battery. (b) Explain why electric arcs occur when large electromagnets (and motors) are turned off, but not when they are turned on.

12-G.1 A rectangular bar of mass M, resistance R, self-inductance L, and width w is pulled at constant velocity v by an external force through a uniform field $\vec{B}$ that makes an angle θ to the normal of the circuit (of negligible resistance). See Figure 12.46. (a) By Lenz's law, there will be a magnetic force on the bar. How will it point, and how will the current circulate? (b) Determine the emf for the circuit, due to the motion of the bar. (c) Determine the current I induced in the circuit. (d) Determine the magnetic force on the circuit. (e) Do you expect the results to be as simple for nonconstant R?

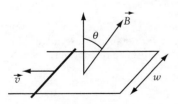

Figure 12.46 Problem 12-G.1.

12-G.2 A wire of mass M, length a, and resistance R can slide without friction on parallel resistanceless rails sitting on a table, closing a circuit containing a constant emf $\mathcal{E}_0$. A uniform magnetic field $\vec{B}$ points into the plane of the circuit. The wire starts from rest. See Figure 12.47. (a) Indicate the direction in which the current will circulate, due to the battery. (b) Compute the magnitude and direction of the magnetic force on the slide wire, and use it in Newton's law of motion. (Force is a vector; don't mix scalars and vectors in your equations.) (c) Compute the motional $\vec{E}$ field on this

arm, and determine the total motional emf $\oint \vec{E} \cdot d\vec{s}$ over the circuit, taking $d\vec{s}$ to circulate counterclockwise. In what sense does *this* emf tend to make current circulate? (d) Use Ohm's law to find the equation for the current through the circuit. Include all important emfs (but neglect the self-inductance). (e) Eliminate the current from Ohm's law to find an equation for the velocity. (f) Solve for the velocity.

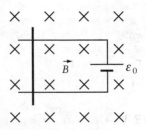

Figure 12.47 Problem 12-G.2.

12-G.3 A superconducting loop of width l, height h, mass M, and self-inductance L is pulled downward by the earth's gravity into a region of uniform magnetic field B that is normal to the loop. See Figure 12.27. The loop starts from rest just outside the field region. (a) Neglecting R (a superconductor), set up the two equations (one for emf, one force) that describe the motion and current of the loop. (b) Show that the position satisfies a harmonic oscillator equation. (c) Taking mass $M = 50$ g, $L = 8$ mH, $h = 5$ cm, $l = 3$ cm, and $B = 1.6$ T, find the characteristic frequency of the motion.

12-G.4 (a) Explain what a motor does. (b) Explain what a generator does.

12-G.5 Explain why eddy currents are to be avoided in motors and generators.

12-G.6 (a) Use the Biot–Savart law to show that the magnetic field varies inversely with the spatial scale of the circuit producing the F. (b) Show that the self-flux of a circuit varies linearly with the spatial scale of the circuit.

12-G.7 A monopole q_m falls vertically with velocity v along the axis of a cylindrical tube of conductivity σ, radius R, and radial thickness d, where $d \ll R$. Consider the tube to be a superposition of rings. Using the result of Problem 12-7.6 for a monopole falling toward a ring, integrate over the rings to obtain the force on the monopole due to the tube.

12-G.8 Repeat the previous problem with the monopole replaced by a dipole of moment μ. This is similar to the situation depicted in Figure 12.18. First find the force due to a ring of radius R, height h, and radial thickness d, then find the force due to an infinite cylinder.

12-G.9 A rectangular superconducting loop of width w along x and length l along y has mass m and self-inductance L. It lies on a frictionless tabletop in the xy-plane. To its right (along the x-axis) is a stripe of width $b > l$ running parallel to y, within which a uniform magnetic field $\vec{B}$ points vertically upward along z. The loop slides with constant velocity v_0 as it enters the field region. [The geometry is similar to that in Figure 12.11(a).] (a) Obtain the equations for the motion and the current. (b) Eliminate the current to show that the motion satisfies an equation with the same structure as a harmonic oscillator. (c) Show that, if the initial velocity is not too large, the loop bounces back out of the magnetic field region. (d) Find the minimum velocity for the loop to pass through the magnetic field region.

12-G.10 It is possible to deduce Faraday's law by using conservation of energy. Consider a magnetic monopole q_m (i.e., the N pole of a very long, narrow magnet) that is constrained to move within a circular tube that is concentric with a wire carrying current I. See Figure 12.48. (a) Show that, on making a single pass around the wire, during time T, the monopole gains energy $q_m \oint \vec{B} \cdot d\vec{s} = q_m \mathcal{M}$, where $\mathcal{M} \equiv \oint \vec{B} \cdot d\vec{s} \equiv \Gamma_B$ is known as the *magnetomotive force*, or mmf. (b) Show that, by energy conservation, the charge $Q = IT$ transported along the wire must have been acted on by an emf $\mathcal{E}$, with $q_m \mathcal{M} + Q\mathcal{E} = 0$. (c) Show that due to the motion of q_m the flux change associated with the long wire, considered to be part of a long, rectangular circuit closed at infinity, is $\Delta\Phi_B = \mu_0 q_m$. (d) Show that $\mathcal{E} = -\Delta\Phi_B / T$. In the limit as $T \to 0$,

Figure 12.48 Problem 12-G.10.

this gives Faraday's law. (By considering a ring with a high density of magnetic charge, and moving dq_m around in a time dt, we can replace T by dt and $\Delta\Phi_B$ by $d\Phi_B$, thus obtaining the time derivative explicitly.)

12-G.11 An electric charge q is fixed at the origin and a monopole q_m moves away from it along the y-axis. Show that the total electric field due to both q and q_m has both circulation and flux.

12-G.12 A moving charge q can be shown to produce both an electric field and a (time-varying) magnetic field. (a) Use the Biot–Savart law, as if it were true for a single charge ($I d\vec{s} \to q\vec{v}$), to find $\vec{B}$. (b) From the time variation of $\vec{B}$ due to the motion show that the magnetic flux gives an electric field whose circulation is nonzero.

12-G.13 (a) Compute the magnetic energy associated with Figure 12.26(a). (b) Compute the magnetic energy associated with Figure 12.26(b). (c) Using (12.68), show that the initial magnetic energy is greater than the final magnetic energy.

12-G.14 Show that the property of perfect diamagnetism (i.e., flux expulsion) is a stronger statement about magnetic flux exclusion than the property of exclusion of *changes* in magnetic flux. (*Hint:* Consider a solid conductor through which a magnetic field passes, and imagine that its conductivity continuously grows to infinity. Does it expel the magnetic field that is already there? What would a perfect diamagnet do?)

12-G.15 Consider a configuration of two coils that has zero mutual inductance. How might this be used to detect the presence of conducting ore?

12-G.16 Analyze the Hall effect (Section 10.7) in terms of motional emf. Compare and contrast with the analysis of that section.

12-G.17 In Figure 12.44, let the voltmeter be moved to the left of the circuit, still making contact across R_r. Recalling that a voltmeter measures $\int \vec{E} \cdot d\vec{s}$ through its own internal circuit, will it read the same value after it is moved? Discuss.

12-G.18 In problem 12-4.4a, find the voltage difference across each coil. Repeat for parts (b) and (c).

"At an industrial exhibition in Vienna, in 1873, a number of Gramme machines [dynamos, or generators] were being placed in position. . . . In making the electrical connections to one of these machines which had not as yet been belted to the engine-shaft [driven by a steam-engine], a careless workman attached to it by mistake a pair of wires which were already connected with another dynamo-machine which was in rapid motion. To the amazement of this worthy artisan the second machine commenced to revolve with great rapidity in a reverse direction . . . ."

"Gramme . . . at once perceived that the second machine was performing the function of a motor, and that what was taking place was an actual transference of mechanical power through the medium of electricity. . . . [Up to that time] almost the only practical use to which the electric motor had been applied was in the operation of dental apparatus . . . ."

—F. L. Pope,
Past President of the American Institute of Electrical Engineers, *Electricity in Daily Life* (1890)

Chapter 13

Mechanical Implications of Faraday's Law: Motors and Generators

Chapter Overview

The previous chapter repeatedly considered the steady-state motion of a loop pulled into a region of magnetic field, finding that such motion produces in the loop an emf and an electric current. Section 13.1 discusses how such emfs and currents became a part of modern civilization's daily life. Section 13.2 discusses the breakthroughs that allowed motor and generator efficiency to become sufficiently high that electromagnetic induction could be taken seriously as a practical source of power, rather than as a mere novelty. Section 13.3 presents some general considerations about *motors* (which convert electrical energy into mechanical energy) and about *generators* (which convert mechanical energy into electrical energy). Both of these are called *electric machines.* To be specific, Section 13.3 introduces a simple model of a *linear machine,* which is the subject of analysis through Section 13.7. The linear machine, which can serve either as a motor or as a generator, is an example of the use and generation of dc—direct current—electricity. (By the use of modern power electronics it is possible to drive ac—alternating current—motors with dc electricity, and to drive dc motors with ac electricity.) Section 13.4 derives the equations that describe the electrical behavior (Ohm's law) and the mechanical behavior (Newton's law of motion) for this system. Section 13.5 solves these equations for the initial and the steady-state response. For the linear generator a *back force* develops (for a rotational generator a *back torque* develops). For a motor a *back emf* develops. Section 13.6 discusses the different types of mechanical and electrical loads to which a motor or generator can be subjected. The discussion of motors and generators closes with Section 13.7, which considers the *transients* that occur when a circuit is first turned on or off. Section 13.8 considers a situation where eddy currents, so unwanted in the case of motors and generators, are useful—magnetic levitation and magnetic drag. ■

13.1 How Electricity Became Part of Daily Life

Optional

The possibilities for using the motive power of electricity (hence *motor*) were realized early on. In 1821, Faraday developed two simple motors using a magnet and a current-carrying wire in a pool of mercury. (Mercury is a good electrical conductor; here it provides a path for electric current.) See Figure 13.1. Electrical energy from a source of emf (not shown) is used to sustain motion (against viscous drag) in the two motors. On the left, a magnet in a bath of liquid mercury turns about the axis of the electric current, which is vertical. In other words, the N magnetic pole turns in the direction of the magnetic field. On the right, the wire turns about the axis of the magnet, which is also vertical. (You are invited to verify that the sense of rotation is correct in each case.) Despite their common current, these motors are independent of each other.

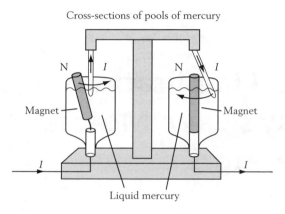

Cross-sections of pools of mercury

Figure 13.1 Faraday's two types of motors. In the motor on the left, the magnet moves about the axis of the current. In the motor on the right, the current moves about the axis of the magnet.

Building on Barlow's 1826 invention of the electromagnet—Arago's earlier magnetizing of a magnetic needle with a helical wire did not include the idea of an electromagnet—Henry improved the lifting strength from 9 to 2300 pounds. Henry then used such an electromagnet to build a motor in 1831. These motors

Beware Mercury

It has been surmised that Faraday developed mercury poisoning from his use of mercury. Even in 1821, at the age of 30, Faraday complained about his fading memory. In 1828, he remarked upon "nervous headaches and weakness." From 1839 to 1844, he was plagued by almost constant and severe headaches. Somehow, he managed to do his studies with what we now call the Faraday ice-pail during 1843. Despite the headaches, he remained in robust physical health, hiking some 30 miles a day even in his mid-50s. By 1845, the giddiness and malaise were gone, but he continued to suffer from a permanent decrease in his powers of memory. By 1855, he was nearly unable to perform research, and in 1861 he had to discontinue his famous Christmas Lectures for children. He died in 1867.

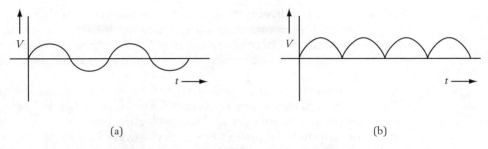

Figure 13.2 (a) Oscillating voltage. (b) Rectified oscillating voltage.

required that the current be produced by a battery, which was very expensive (zinc-copper batteries were used, in which the zinc was consumed): in the 1840s, Joule showed that, for the same amount of energy production, zinc cost 120 times as much as coal.

With Faraday's 1831 discovery of electromagnetic induction, it became possible to think in terms of an alternative source of electrical power for motors. In what may have been the first use of the commutator (i.e., the slip ring), at Ampère's suggestion, Pixii in 1832 rectified current produced by Faraday's law: a handle turned a U-shaped magnet, whose poles passed beneath copper solenoids filled with iron core. The current induced in the solenoids went through a commutator and was rectified. Figure 13.2(a) shows the unrectified signal, and Figure 13.2(b) shows the rectified signal. These early generators of electricity, driven by mechanical power, were called *dynamos*.

Two major forces driving the use of electric power were public lighting and electric trains, both of which used batteries for their electricity. In 1808, Humphrey Davy discovered the carbon arc lamp (driven by a large battery of voltaic cells), which became used in public facilities to provide lighting at night. Around 1851, it was found that electric power could be transmitted effectively through train rails so that a locomotive using an electric motor did not have to carry its own batteries.

The first practical electrical generator of the Pixii type was used in 1858 to power a carbon arc lamp in a lighthouse. (Presumably, the generator was driven by a coal-powered steam engine.) As indicated in the quotation at the chapter head, by 1873 there was considerable interest in electrical generators, primarily for arc lighting, but it seems to have been forgotten that the electricity they generated could also power motors, until the fortuitous rediscovery by a "worthy artisan." (As early as 1842, it was realized that motors and generators were inverse to one another.)

Nevertheless, the generators were not very good. An 1876 study for the Franklin Institute, by then high school teacher Elihu Thomson—later inventor of the wattmeter and of the jumping ring that will be discussed in the next chapter—showed that the Brush company's generator, at only 38%, was the most efficient then available. The low efficiency was largely due to losses within these devices: as discussed in the next section, unwanted eddy currents were induced because of the high conductivity of iron, and magnetic energy losses were caused while cycling the iron magnetization back and forth. Once recognized, the eddy current problem was quickly solved, and efficiencies rose to nearly 80%. By

1878, arc lamps had begun to appear on the streets of both Philadelphia and Boston. The hysteresis problems took somewhat longer to solve, but by 1890 motors and generators were operating near 90% efficiency. Modern motors and generators aren't much better than this, but they are much lighter, cheaper, and more reliable.

In the mid-19th century, industrial machines were driven by the turning power of steam engines or nearby rivers. However, by the early 1880s, it became possible for smaller companies to operate electrical motors with electricity purchased from a power company. Electricity-powered streetcars (with overhead lines for electric power produced by generators driven by coal-powered steam engines) became practical: indeed, they cost only one-tenth as much as horse-drawn streetcars. There was a rapid conversion to electrical power of streetcars and elevated trains, with a resultant skyrocketing in the rate of equine unemployment.

It was quickly realized that electric power could be transported over large distances, so that power from the Niagara Falls could be sent to New York (or power from the Columbia River could be sent to Seattle). Again, from the same source as the quotation at the chapter head:

> Electricity, in its important applications to machinery . . . is merely a convenient and easily manageable agency . . . by which mechanical power may be transferred from an ordinary prime motor, as a steam engine or a water wheel, to a secondary motor—it may be at a great distance—which is employed to do the work.

Once electrical power became more commonly available, new electrical devices, such as the carbon filament lamp, developed around 1880, made the world a different place. The electric power transformation, which included labor-saving devices such as the washing machine, was more noticeable than even the computer transformation occurring today. The following two chapters touch upon its effect in the area of communications: signal generation, transmission, detection, and manipulation. The present chapter concentrates on electric power.

13.2 Breakthroughs in Efficiency of Motors and Generators

It is of interest to indicate how the inefficient pre-1870's motors and generators evolved into the efficient—albeit bulky—ancestors of modern motors and generators. As indicated earlier, there were two major advances.

1. Elimination of eddy current losses. The original motor and generator designs had large amounts of conductor in their metal casing (for structural strength) and in their iron cores (to confine the magnetic field). When subjected to time-varying magnetic fields, these conducting materials had emfs induced in them, by Faraday's law. Hence, by Ohm's law, currents were induced. Such currents were not needed in the casing and the core, so the resulting Joule heating constituted a huge waste of energy. By the late 1870s, engineers were aware of this problem, and improved designs minimized it. The most obvious solution was to use less solid metal in the casing, leading to a "squirrel-cage" frame. See Figure 13.3(a). Such eddy currents, following

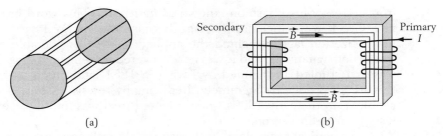

Figure 13.3 (a) Squirrel-cage frame for motor casing. Being mostly hollow, the induced currents and Joule heating losses are small compared to a solid frame. (b) Laminated transformer core. The induced currents circulate in the same direction as the current in the primary; the laminating gives the induced currents a high-resistance path and thus a small amplitude.

Lenz's law, circulated about the direction of the magnetic field. In the iron core, by using laminations or a packed set of much thinner iron rings, the path of the eddy currents could be given a much smaller cross-sectional area, thus increasing the electrical resistance and decreasing the rate of Joule heating. See Figure 13.3(b).

Figure 13.4 illustrates a square, before and after it has been cut in four, and the associated eddy currents due to magnetic field change normal to the page. Before cutting, let the effective resistance be R_0, and let the emf due to a time-varying magnetic field normal to the page be $\mathcal{E}_0$. The total rate of eddy current heating is then $\mathcal{E}_0^2/R_0$. Figure 13.4(b) shows that the shape of the eddy current path is the same after cutting as it was before. Each quarter square is subject to an emf $\mathcal{E}_0/4$ and has resistance $R = R_0$. (Although $R = \rho l/A$ of Chapter 7 does not hold here, it captures the correct ideas. For each quarter square the eddy current length in the plane decreases by a factor of two relative to the original length, as does the length in the plane associated with the current-carrying area. However, the current-carrying direction normal to the page does not change. Hence, effectively, both l and A halve, so the eddy current resistance is the same for both the big square of Figure 13.4a and the quarter squares of Figure 13.4b.) The total heating rate of all four quarter squares is $4\mathcal{E}^2/R = \mathcal{E}_0^2/4R_0$, one-fourth of the original rate.

2. **Hysteresis losses.** In ac motors and generators, the iron in the electromagnet caused unnecessary heating because the applied magnetic field cycled the

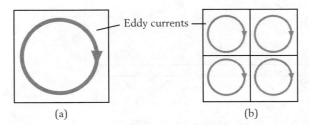

Figure 13.4 (a) Eddy currents for a square. (b) Eddy currents for a square that has been broken up into subsquares.

iron in a very lossy fashion, known as *hysteresis*. (This word has its origins in a Greek word meaning "to lag," because the energy loss is due to the magnetization lagging behind the applied field.) Considerable energy was lost in demagnetizing, and it was not recovered on remagnetizing. This effect was first studied and named by Ewing in 1881, and much useful work in this area was done by Steinmetz. Reducing hysteresis losses was a materials problem. In the early 1900s, grain-oriented steel was found to be much less lossy than ordinary iron.

A mechanical example of hysteresis. In old-fashioned windup clocks, energy is extracted from a spring as the clock mechanism goes through a period of its motion. In addition to energy that is extracted by choice (useful energy), there is energy wasted because the mechanism has friction (energy loss to heat) or is noisy (energy loss to sound). For example, it would take more force F to stretch a clock-spring from .09 mm to .10 mm than to relax it from .10 mm to .09 mm. The energy wasted during a cycle of the clock mechanism would be called a hysteresis loss. The magnetic analogy would be that it would take a greater applied magnetic field B to increase a magnet's magnetization M from 900 A/m to 1000 A/m than to decrease M from 1000 A/m to 900 A/m.

Let us now turn to a simple model that will serve to explain the behavior of both dc motors and generators. The next chapter discusses ac motors and generators.

13.3 Simple Model for DC Motor and Generator— The Linear Machine

Consider a circuit as in Figure 13.5, where a conducting bar of mass m can slide along a fixed conducting rail at the velocity v. (We present a top view of the circuit.) A uniform magnetic field $\vec{B}$ points into the page. The bar's position can be limited to a desirable range by the use of microswitches, which cause the direction of the current to reverse, and thus cause the direction of the force on the bar to reverse. We include an external emf $\mathcal{E}_0$, and let an external force

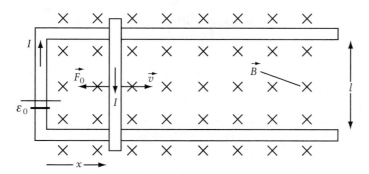

Figure 13.5 Simple model for a dc motor and generator. A conducting rod can slide along a fixed conducting rail, to which a constant emf is attached.

$\vec{F}_0$ act on the bar. For simplicity, we neglect the self-inductance of the circuit. However, we include the resistance R of the circuit, and a frictional resistance force of the bar against the rail, which for simplicity we assume to take the form $-m\vec{v}/\tau_f$ (just as in Chapter 7, for electrons in a wire). The time τ_f represents a frictional relaxation time. The shorter the τ_f, the greater the friction. A motor that slides to rest in less time has more friction; oiling that motor decreases the friction and lets it slide for a longer time. τ_f^{-1}, the inverse of τ_f, has the units of s^{-1} and is a relaxation rate. (A constant sliding frictional force could be written as $\vec{F}_{slide} = -\hat{v}F_{slide}$, with a direction $-\hat{v}$ that opposes the velocity $\vec{v}$, but we will not consider this case.)

13.3.1 *Operation as a Motor*

In running this linear machine as a motor, we will consider that the external power source, or input, is the constant $\mathcal{E}_0$. The output is a mechanical load force. There are a number of possible motor applications, and thus different types of mechanical load forces:

1. In lifting a weight W, the load is $\vec{F}_0$ and is due to W. The mass $M = W/g$ of the object also must be included, along with the mass m of the bar.

2. In pulling a boat through water, addition of the boat increases the mass m and increases the frictional resistance (so that τ_f decreases). Such a velocity-dependent drag force leads to a load force that is not constant. $F_0 = 0$ here.

3. In compressing a spring (for energy storage), the load force F_0 is proportional to the displacement x: $F_0 = -Kx$. Again, the load force is not constant.

13.3.2 *Operation as a Generator*

In running this linear machine as a generator, we take the external power source, or input, to be the force $\vec{F}_0$. As for the motor, for the generator there are a number of applications and load emfs.

1. If a battery of emf $\mathcal{E}_0$ and internal resistance r is to be recharged, the electrical load is both an increase in the electrical resistance by r and a constant emf $\mathcal{E}_0$.

2. If a lightbulb of resistance R_L is to be operated, the electrical load is a resistance R_L, with $\mathcal{E}_0 = 0$.

3. The electrical load also could be a capacitor that is being charged, or an inductor that is given a current.

For both the generator and the motor, there are even more complex possibilities, but their detailed study properly belongs to a specialized course, not to an introduction.

13.4 Equations Describing the Linear Machine

Our task is to solve for both the motion of (i.e., the position x), and the current I through, the bar. The equations we employ are (1) Newton's law for motion, subject to the load force and the magnetic force on the current-carrying bar; and (2) Ohm's law, with emfs produced by both the dc source $\mathcal{E}_0$ and by the motional emf due to the bar. This yields two equations for the two unknowns, the current I through the circuit, and the position x of the bar. If it is true, as the great inventor Edison is said to have stated in court, that Ohm's law is the basis of electrical engineering, then surely it is just as true that Newton's second law of motion is the basis of mechanical engineering.

13.4.1 *Newton's Law: The Basis of Mechanical Engineering*

Mechanical engineering is associated with the motion of machinery, described by Newton's second law of motion. What changes from problem to problem is the nature of the forces that cause the motion.

Let us first consider the magnetic force. In Figure 13.5, the magnetic field points into the paper, and the battery discharges ($I > 0$) when its current flows clockwise. Thus, for the bar a positive current flows downward. The magnetic force on the bar is given by

$$\vec{F}_B = I\vec{l} \times \vec{B}. \tag{13.1}$$

With $\vec{l}$ pointing along ↓ (i.e., $-\hat{y}$), and $\vec{B}$ pointing along ⊗, $\vec{l} \times \vec{B}$ points to the right, and is of magnitude $|\vec{F}_B| = I|\vec{l} \times \vec{B}| = IlB \sin 90° = IlB$. Hence

$$\vec{F}_B = IlB\hat{x}. \tag{13.2}$$

In addition, let there be an applied (or load) force $\vec{F}_L$ opposing this motion, with

$$\vec{F}_L = -F_0\hat{x}. \tag{13.3}$$

Finally, we will include a frictional force $\vec{F}_f$ proportional to the velocity, that opposes the motion, in the form

$$\vec{F}_f = -\frac{mv}{\tau_f}\hat{x}, \qquad v = \frac{dx}{dt}. \tag{13.4}$$

This drag force is like that of (7.39).

The net effect is that, including each of the forces (13.2), (13.3), and (13.4), Newton's second law of motion along $\hat{x}$ becomes

$$m\frac{dv}{dt} = (F_x)_{net} = IlB - F_0 - \frac{mv}{\tau_f}. \tag{13.5}$$

If we think of F_0 as the (leftward) applied force, then IlB may be thought of as a *back force* because it opposes the applied force.

13.4.2 *Ohm's Law: The Basis of Electrical Engineering*

Electrical engineering is associated with the motion of electric current, described by Ohm's law. What changes from problem to problem is the nature of the emfs that cause the motion. Let us now consider the emfs that act in the present case.

In Figure 13.5, the battery emf tends to drive current clockwise. Furthermore, there is an induced emf. By Lenz's law, if the bar moves to the right, it gains flux into the paper. Hence the induced emf must produce a flux that is out of the paper, so the induced emf must be counterclockwise. Let us compute it explicitly. Consider the motional emf

$$\mathcal{E}_{mot} = \oint \vec{v} \times \vec{B} \cdot d\vec{s}, \tag{13.6}$$

where $d\vec{s}$ circulates in a clockwise sense to produce a positive emf. This is nonzero only for the part of the circuit that includes the bar. Since it moves to the right, and the field is into the paper, $\vec{v} \times \vec{B}$ points along $\hat{y}$. This corresponds to a counterclockwise circulation, as obtained by Lenz's law. Moreover, since $\vec{v}$ and $\vec{B}$ are normal to each other, $|\vec{v} \times \vec{B}| = vB$. With $d\vec{s} = dy\hat{y}$, (13.6) becomes

$$\mathcal{E}_{mot} = \int_{l}^{0} vB\hat{y} \cdot dy\hat{y} = \int_{l}^{0} vB dy = -vBl. \tag{13.7}$$

In agreement with our qualitative discussion, the motional emf tends to drive current counterclockwise to oppose the increase of flux from the rightward motion.

With both the battery $\mathcal{E}_0$ and the motional emf, Ohm's law becomes

$$I = \frac{\mathcal{E}}{R} = \frac{\mathcal{E}_0 - vBl}{R}. \tag{13.8}$$

If self-inductance were included, then $-LdI/dt$ would be added to the numerator of (13.8).

If we think of $\mathcal{E}$ as the (clockwise) applied emf, then vBl may be thought of as a *back emf* because it opposes the applied emf.

Example 13.1 **Back emf theory**

Find the back emf for $v = 1$ m/s, $B = 0.1$ T, and $l = 0.1$ m.

Solution: Equation (13.7) gives $vBl = 0.01$ V.

Example 13.2 **Back emf measurement**

Often the back emf can be measured. Let a rotational motor have a 6 A current pass through it when a dc emf of 12 V is switched on, but only a 2 A current passes through it when it is in steady operation. Consider that L/R is so short that the current immediately reaches the dc value before the motor can start to turn. Find the resistance and back emf.

Solution: At $t = 0^+$ there is no motion, so there is no back emf. Then $I = \mathcal{E}_0/R$ yields $R = 12$ V/6 A $= 2\ \Omega$. When the motor is in steady operation, use of $I = (\mathcal{E}_0 - \mathcal{E}_{back})/R$ gives $\mathcal{E}_{back} = \mathcal{E}_0 - IR = 12$ V $- (2$ A$)(2\ \Omega) = 8$ V.

13.5 Solving the Equations

Our goal now is to solve (13.5) and (13.8) simultaneously for v and I. Substituting (13.8) into (13.5) yields

$$m\frac{dv}{dt} = \left(\frac{Bl\mathcal{E}_0}{R} - F_0\right) - mv\left(\frac{1}{\tau_f} + \frac{B^2l^2}{mR}\right). \tag{13.9}$$

We define the net relaxation time τ and the magnetic relaxation time τ_B via the inverse of the net relaxation rate

$$\frac{1}{\tau} \equiv \frac{1}{\tau_f} + \frac{1}{\tau_B}, \qquad \tau_B \equiv \frac{mR}{B^2l^2}, \tag{13.10}$$

and the effective force F_{eff} via

$$F_{eff} = \frac{Bl\mathcal{E}_0}{R} - F_0. \tag{13.11}$$

Clearly the net relaxation rate of (13.10) is the sum of a mechanical relaxation rate and an electromechanical relaxation rate. Likewise, the total force consists of the applied force F_0 and the magnetic *back force*, in this case given by $Bl\mathcal{E}_0/R$.

Example 13.3 **Magnetic relaxation time and magnetic back force**

For $m = 0.01$ kg, $R = 10^{-2}\ \Omega$, $B = 0.1$ T, and $l = 0.1$ m, find τ_B and the magnetic back force.

Solution: Equation (13.10) gives $\tau_B = 1$ s. Thus magnetic drag can be dominant over mechanical drag because, for a low-friction surface, τ_f can be hundreds of seconds. For $\mathcal{E}_0 = 10$ V, by (13.11) the magnetic back force is $Bl\mathcal{E}_0/R = 10$ N.

By (13.11), the constant part of the current, induced by the constant emf, produces a constant rightward magnetic force of $Bl\mathcal{E}_0/R$. Using (13.10) and (13.11), Equation (13.9) can be written in the more compact form

$$m\frac{dv}{dt} = F_{eff} - \frac{mv}{\tau}. \tag{13.12}$$

Once (13.12) is solved for v, (13.8) will yield I.

We first consider the initial response of the motor, and then its steady state. Section 13.6 considers how it attains that steady-state behavior.

13.5.1 *Initial Response of Motor*

Initially, the current I and velocity v are taken to be zero. However, in the absence of self-inductance, which is a measure of the magnetic field "inertia" (and therefore of the current, to which the magnetic field is proportional), the current can build up very quickly. On the other hand, the bar's mass inertia will prevent the velocity from building up immediately. Hence (13.8) and (13.2) yield an initial current and magnetic force

$$I_0 = \frac{\mathcal{E}_0}{R}, \qquad F_B(t = 0) = I_0 l B = \frac{\mathcal{E}_0 l B}{R}. \tag{13.13}$$

This can be a large current and a large force. For startup, a large force is often desirable. However, a large current cannot be sustained for a long time because the Joule heating ($I^2 R$) can cause the insulation on the windings to melt. On the other hand, the larger the current, the larger the startup force, and the shorter the time it takes for the motor to start operating.

13.5.2 *Steady State of Motor*

After a long time, the velocity reaches its maximum value, the terminal velocity v_∞, where $dv/dt = 0$. Equation (13.12), with its left-hand side set to zero, yields $v_\infty = F_{eff} \tau / m$. Equation (13.11) then yields

$$v_\infty = \frac{F_{eff} \tau}{m} = \left(\frac{Bl\mathcal{E}_0}{R} - F_0 \right) \frac{\tau}{m}. \tag{13.14}$$

In (13.14), increasing all types of load tends to decrease v_∞: for a more massive load, the mass m in the denominator increases; a larger load force F_0 is subtracted; and for a larger drag force, the relaxation time τ in the numerator decreases. Equation (13.14) also has some interesting consequences as a function of F_0:

1. If the load F_0 increases, the velocity v and the *back emf*, in this case given by vBl, decrease. Then, by Ohm's law, as in (13.8), the current increases. When we fix $v = 0$, meaning that the motor is held in place, the full startup current will pass through the motor, causing permanently large $I^2 R$ heating. This explains why, when motors "freeze up," they stop working: too much heat is produced, so the insulation on the coils melts or burns up, and the motor shorts out.

2. By (13.14), if the load force $F_0 > Bl\mathcal{E}_0/R$, the motor will run backwards! (In other words, it will run as a generator.)

Equations (13.14) and (13.8) yield the current I_∞ at long times. Using (13.10) for τ_B, we obtain

$$\begin{aligned}
I_\infty &= \frac{\mathcal{E}_0 - Blv_\infty}{R} = \frac{\mathcal{E}_0}{R} - \left(\frac{Bl\mathcal{E}_0}{R} - F_0 \right) \frac{\tau}{m} \frac{Bl}{R} \\
&= \frac{\mathcal{E}_0}{R} \left(1 - \frac{\tau}{\tau_B} \right) + \frac{Bl\tau}{mR} F_0 = \frac{\mathcal{E}_0}{R} \frac{\tau_B}{\tau_f + \tau_B} + \frac{Bl\tau}{mR} F_0.
\end{aligned} \tag{13.15}$$

Clearly, the larger the load F_0, the greater the current that must flow. In addition, the smaller the frictional time τ_f, the greater the current.

13.6 Efficiency and Load

A given device, either electrical or mechanical, provides power to a *load*. There are numerous types of load.

13.6.1 *Motors: Types of Load*

For a motor, the input energy is provided by the emf $\mathcal{E}_0$, and the electrical power input is the rate of discharge of that emf:

$$P_{input} = \mathcal{E}_0 I_\infty. \qquad \text{(motor input)} \qquad (13.16)$$

In pulling a boat through water, the load is the drag force mv_∞/τ_f, so we may set $F_0 = 0$. Then the useful mechanical power is

$$P_{useful} = \frac{mv_\infty}{\tau_f} v_\infty = \frac{mv_\infty^2}{\tau_f}. \qquad \text{(drag load)} \qquad (13.17)$$

The efficiency is given by (13.17) divided by (13.16).

For a lifting device, the load may be taken to be $F_0 \neq 0$. In that case, the useful mechanical power is

$$P_{useful} = F_0 v_\infty. \qquad \text{(lifting load)} \qquad (13.18)$$

The efficiency is given by (13.18) divided by (13.16).

13.6.2 *Generators: Types of Load*

For a generator, the input energy is provided by the force F_0, and the mechanical power input is the rate of work by that force:

$$P_{input} = F_0 v_\infty. \qquad \text{(generator input)} \qquad (13.19)$$

We often consider generators for the case when the power source causes steady motion. In that case, (13.15) provides the current. We then interpret F_0 as coming from the power source, and $\mathcal{E}_0$ as being due to a battery that is being recharged.

For a generator used to recharge an emf, the useful electrical power is

$$P_{useful} = \mathcal{E}_0 I_\infty, \qquad \text{(recharging emf)} \qquad (13.20)$$

and the efficiency is given by (13.20) divided by (13.19). Here the recharging emf is the load.

For a generator used to light a bulb, the useful electrical power is

$$P_{useful} = I_\infty^2 R, \qquad \text{(powering lightbulb)} \qquad (13.21)$$

and the efficiency is given by (13.21) divided by (13.19). Here the resistor is the load.

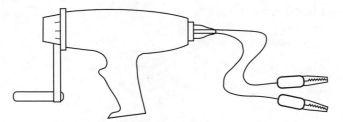

Figure 13.6 Genecon generator. Turning the handle produces a voltage difference across the leads. Likewise, connecting the leads to an applied voltage difference causes the handle to turn.

One implication of the relaxation time τ_B can be seen if we "short out" a generator ($\mathcal{E}_0 = 0$) by connecting its external leads to each other. Then the resistance is due only to its own wires and is relatively small. By (13.10), this leads to a very short time τ_B, so $\tau \approx \tau_B$, and the back force is approximately given by mv/τ_B, which can be surprisingly large. A rotational analog can be seen in the inexpensive and commonly used Genecon generator. See Figure 13.6. Comparison between the effort needed to turn the handle when it is disconnected ($R = \infty$), when it is connected to a flashlight bulb (moderate R), and when it is shorted (small R) reveals that the last case requires by far the largest effort.

By connecting one Genecon to another, we can turn the handle on one (so it is a generator), and that causes the handle on the other to turn (so it is a motor). This is just what the workman did in the quotation at the chapter head.

> **Example 13.4** **Force to drive a linear generator**
>
> A linear generator is used to provide the 2 A current required to run a lightbulb with a resistance of 20 Ω. Assuming that the generator is perfectly efficient, what force must be provided to the generator if $v_\infty = 15$ m/s?
>
> ***Solution:*** By (13.21), $\mathcal{P}_{useful} = I_\infty^2 R = (2)^2(20) = 80$ W. Since all the power input to the generator is converted to useful power, $\mathcal{P}_{useful} = \mathcal{P}_{input}$. By (13.19), $\mathcal{P}_{input} = F_0 v_\infty$, so $F_0 = \mathcal{P}_{input}/v_\infty = 80/15 = 5.3$ N. Note that for a rotational generator to generate this power at angular velocity 2 turns/s, or $\omega = 12.56$ radians/s, the torque τ would be given by $\mathcal{P}_{input} = \tau_0 \omega_\infty$, which leads to a torque of $80/12.56 = 6.37$ N-m.

13.7 Transients

When an electrical or mechanical device starts up or shuts down, or when there is a power surge, the system is subject to *transients*.

13.7.1 *Motor Startup*

On startup, or on a change in the load, the system must adjust. We consider the case where there is a load F_0, and we start up the motor. Thus we must solve

(13.12), which can be rewritten as

$$\frac{dv}{dt} = a - \frac{v}{\tau},\qquad(13.22)$$

where

$$a = \frac{F_{eff}}{m},\qquad(13.23)$$

and F_{eff} is given in (13.11) and τ is given in (13.10).

Equation (13.22) is very like our old friend, the charge-up of the capacitor in an RC circuit with an emf $\mathcal{E}$. Chapter 8 showed that the charge Q on a capacitor satisfies the equation

$$\frac{dQ}{dt} = \frac{\mathcal{E}}{R} - \frac{Q}{RC}.\qquad(13.24)$$

When started from $Q = 0$ at $t = 0$, the solution was found to be given by

$$Q = C\mathcal{E}[1 - \exp(-t/\tau_{RC})],\qquad \tau_{RC} = RC.\qquad(13.25)$$

With $(Q, \mathcal{E}/R, RC)$ of (13.24) $\to (v, a, \tau)$ of (13.22), and noting that $C\mathcal{E} = (\mathcal{E}/R)(RC)$, so $C\mathcal{E} \to a\tau$, the solution to (13.24) with $v = 0$ at $t = 0$ is

$$v = a\tau[1 - \exp(-t/\tau)].\qquad(13.26)$$

Note that $a\tau$ is the same as v_∞ of (13.14). See Figure 13.7 for a sketch of (13.26). From (13.26) and (13.8), we can find I if desired. Thus the characteristic relaxation time is τ of (13.10), a combination of the frictional time τ_f associated with

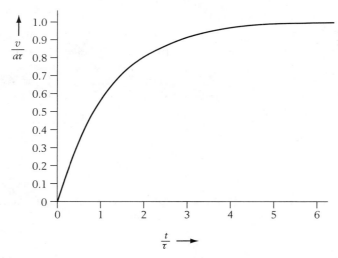

Figure 13.7 Velocity versus time for the linear motor of Figure 13.5, with the constant force F_{eff} of (13.11).

the bar or load, and the magnetic relaxation time τ_B associated with dissipation of induced currents in the resistance. The shorter of these times dominates τ. An efficient motor will have $\tau_B \ll \tau_f$ so that there is relatively little frictional drag.

13.7.2 *Generator Disconnection*

In this case, there is no emf, so $\mathcal{E}_0 = 0$, and there is no constant force, so $F_0 = 0$. By (13.11), we then have $F_{eff} = 0$. The motion can be found by solving (13.12) with $F_{eff} = 0$.

This is similar to the equation for the discharge of an RC circuit:

$$I = \frac{dQ}{dt} = \frac{\Delta V}{R} = \frac{-Q/C}{R}, \qquad \text{or} \qquad \frac{dQ}{dt} = -\frac{1}{RC}Q. \tag{13.27}$$

The solution of this equation, with the initial condition $Q = Q_0$ at $t = 0$, is

$$Q = Q_0 \exp(-t/\tau_{RC}), \qquad \tau_{RC} = RC. \tag{13.28}$$

With (Q, Q_0, RC) of (13.28) $\to (v, v_0, \tau)$ of (13.27), the solution to (13.27) for $F_{eff} = 0$, with the initial condition $v = v_0$ at $t = 0$, is

$$v = v_0 \exp(-t/\tau). \tag{13.29}$$

Rotational machines. Most commercial motors involve rotation. Therefore, instead of $F = ma$, $\tau = \mathcal{I}\alpha$ applies, where $\mathcal{I}$ is the moment of inertia and α is the angular acceleration. For such machines, there is a moving structure and a fixed structure. The *stator* (from *static*) is fixed in place, usually to the outer frame; the *rotor* (from *rotate*) is free to rotate. Independently of these structures, there are also electrical windings. The winding in which voltage is induced is called the *armature winding*, and the winding in which the field is produced is called the *field winding*. (Sometimes permanent magnets are used instead of field windings.) The detailed study of electrical machinery is a vast field.

13.8 Eddy Currents and MAGLEV

Optional

As discussed earlier, eddy currents often are a nuisance. But not always. In some cases, they are used for braking purposes (this is analogous to the drag effect discussed before, when a generator was disconnected). In other cases, they are used for *magnetic levitation*, or MAGLEV. In general, it is not easy to determine the eddy current distribution, although it can be determined when a monopole— one end of a long magnet—falls down the center of a long conducting tube (see Problem 12-G.7).

13.8.1 *General Results*

When a magnet moves parallel to the surface of a fixed conductor, it generates induced currents in the conductor, and it feels a force due to them. There is a drag component and a lift component. (For the loop pulled through the field (13.9) with $\mathcal{E}_0 = 0$, $F_0 = 0$ and no mechanical drag yields a magnetic drag force that is proportional to the velocity.) The drag component tries to bring the magnet to rest; the lift component tries to push the magnet away. Both of these serve to decrease the rate of change of the magnetic flux seen by the conductor, in agreement with Lenz's law.

There is a very simple way to think of the effect of moving a magnet. Since any magnet can be thought of as a linear combination of monopoles, for simplicity let us consider the magnet to be a monopole. After understanding a monopole, we can consider more general types of magnet. Let us also restrict ourselves to the case of a conductor with a flat surface.

First consider the response on the sudden creation of a monopole above the conductor, as appears to happen when the monopole is moving very quickly. (First the monopole is not above a given point on the surface, and then all of a sudden, it is there.) In that case, eddy currents are set up in the conductor, and these succeed in opposing the change in magnetic flux due to the monopole. To an observer above the conductor, it is as if there were an image monopole beneath the conductor. See Figure 13.8(a) for the image monopole and Figure 13.8(b) for the eddy currents. This is precisely the situation discussed in Section 11.11, where Figure 11.25(a) depicts a monopole above a perfect diamagnet, and Figure 11.27(a) depicts the corresponding surface currents. (In Section 11.11, however, the surface currents do not die off with time because perfect diamagnets are superconductors.) The force on the monopole due to its image is

$$F_I = \frac{k_m (q_m)^2}{(2s)^2}. \qquad \text{(conductor with flat surface)} \qquad (13.30)$$

Thus, in the limit of high velocity v (where "high" is relative to a velocity v_0 to be defined shortly), the force on the monopole is a *lift* force F_L, and

$$\lim_{v/v_0 \to \infty} F_L = F_I. \qquad \text{(conductor with flat surface)} \qquad (13.31)$$

Let us now give more careful consideration to what happens when the monopole moves. As noted by Maxwell, motion of a monopole from A to B can be thought of as the simultaneous superposition of an anti-monopole at A (thus canceling the existing monopole at A) and a new monopole at B. (This is

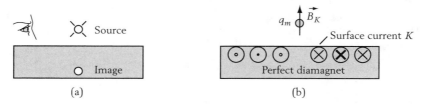

Figure 13.8 (a) Image and source. (b) Monopole appearing above a perfect diamagnet, where above the perfect diamagnet the field can be described as being due to the source and image of part (a).

very much like the "teleportation" that occurs in the television show *Star Trek*—prompting Captain Kirk's immortal words "Beam me up, Scotty!") The eddy current response to this motion is obtained by superposing the eddy current response to the field change from adding both the anti-monopole at *A* and the monopole at *B*. See Figure 13.9. This gives a qualitative sense of the nature of the eddy currents that will be set up. Eddy currents set up at a previous time, of course, die down, so the most recent eddy currents dominate.

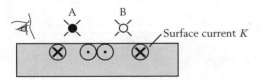

Figure 13.9 Surface currents caused by moving a monopole to the right. The change is equivalent to creating a monopole–antimonopole pair, the monopole at the new position and the anti-monopole (in dark) at the old position.

13.8.2 *Maxwell's Receding Image Construction*

We are now going to write down a remarkable result. As Maxwell did in his first paper on the subject, we will present it without proof. This is not the preferred approach in physics courses, which tend to derive everything from first principles. However, because the result is so simple, and the proof is so complex, Maxwell did not follow normal procedure when he first published the result, and neither shall we. [For details, see the *American Journal of Physics*, Vol. 60, p. 693 (1992).]

The result is true only for thin, flat sheets of nonmagnetic conductor (like aluminum or copper, but not iron). Here *thin* means that the distance of the magnet from the sheet is much greater than the thickness d of the sheet, which is taken to have conductivity σ. Here is the result.

The eddy currents set up by the image poles create magnetic fields whose future behavior is the same as the magnetic fields set up by image poles that move away from the sheet with what we call the *Maxwell recession velocity*

$$v_0 = \frac{1}{2\pi k_m \sigma d}. \tag{13.32}$$

Thus, if a monopole suddenly appears, so does an image monopole, with the image monopole moving away at velocity v_0. See Figure 13.10(a) for the situation as viewed from above, and Figure 13.10(b) for the situation as viewed from below. (Note that the conductor is now a thin sheet, as opposed to the more general case depicted in the previous figures.) We call the theory based upon (13.32) *Maxwell's receding image construction*.

On Science Fiction

It is a pity that such an interesting literary device as teleportation violates some of the most fundamental laws of physics. When an object dematerializes in the teleportation device, the energy associated with that object cannot suddenly disappear, but must smoothly get transported from the interior of the spaceship to the desired place. How such a huge amount of energy can pass through the hull of the spaceship is not addressed.

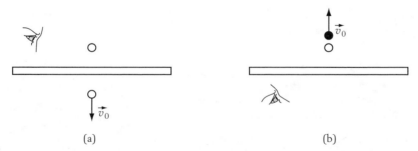

Figure 13.10 Maxwell's receding image construction on the sudden appearance of a monopole above the conducting sheet: (a) as seen from above, (b) as seen from below.

When the source monopole moves at velocity v parallel to the conductor's surface, as in Figure 13.9 the change at each step may be thought of as creation of a monopole at the new position of the source monopole, *and* as creation of an antimonopole at the old position of the source monopole. In this way, the new situation is a superposition of the original situation and the change. The response of the conducting sheet to the change is thus to produce a monopole and an antimonopole beneath the conducting sheet, with a horizontal separation that is proportional to v.

As the velocity v grows, the most recent image becomes more dominant because the horizontal separation between images increases. Compare Figures 13.11(a) and 13.11(b). From Figure 13.11(b), it should be clear that for high velocities (i.e., $v \gg v_0$), the force on the monopole is repulsive, with magnitude given by the image force of (13.30); it is thus a lift force.

In fact, (13.31) applies to any magnet moving at velocity v parallel to the sheet, with the appropriate image force. Of course, the image force depends on the magnet, and for a monopole only is given by (13.30). In this limit, the self-inductance of the sheet, which wants to oppose all changes in flux, dominates.

At low velocities, self-inductance is less important than resistance. Consider a related example. If in (13.9) we set $dv/dt = 0$, $\mathcal{E}_0 = 0$, and $1/\tau_f = 0$, then the

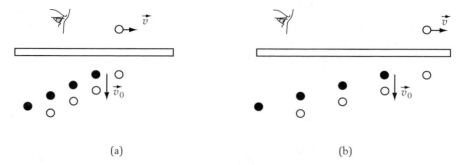

Figure 13.11 Maxwell's receding image construction for a monopole above the conducting sheet, moving rightward. The monopoles and antimonopoles (in dark) move downward at velocity v_0. (a) A slowly moving monopole. (b) A quickly moving monopole.

magnetic force F_B cancels the applied force F_0. This situation corresponds to the hand providing power, so it is a generator. Explicitly, the magnetic force is then given by $(B^2 l^2 / R)v$, which is the same as obtained in the previous chapter for a circuit pulled across a magnetic field. This is a drag force, and for thin sheets the proportionality to v holds in general at low velocities. Since F_I and v_0 are the natural units of force and velocity, at low velocities we thus expect that

$$\lim_{v/v_0 \to 0} F_D = \alpha F_I \frac{v}{v_0}, \tag{13.33}$$

where α is some constant that depends upon the details of the magnet.

13.8.3 *Lift–Drag Relationship*

We now derive a result, based on the receding image construction, that will enable us to relate the lift force F_L and the drag force F_D when a magnet moves at any given velocity parallel to the sheet. Then, from our knowledge of the high velocity limit of F_L—given by (13.31)—we can determine the high velocity limit of F_D. Moreover, from our knowledge of the low velocity limit of F_D—given by (13.33)—we can determine the low velocity limit of F_L. This will permit us to extrapolate the behavior of both F_L and F_D for all velocities, and thus give us a feeling for the behavior of eddy current MAGLEV systems.

First, note that for a magnet to move at a constant velocity parallel to the sheet, it must be subject to zero net force. Hence, in addition to the lift and drag forces, some equal and opposite forces must be acting. Let them be provided by your hand. The rate at which your hand does work against the drag force is given by

$$\mathcal{P}_{hand} = F_D v. \tag{13.34}$$

By energy conservation, this must equal the power that goes into eddy currents in the conducting sheet, which in turn must be the same as the power going into pushing the image charges away from the surface. Since the image charges all move with velocity v_0, and since, by action and reaction, the total force on them must be equal and opposite to the lift force, we obtain

$$\mathcal{P}_{images} = \Sigma_i (F_i v_0) = (\Sigma_i F_i) v_0 = F_L v_0. \tag{13.35}$$

Equating (13.35) and (13.34) yields the important result that

$$F_D v = F_L v_0. \tag{13.36}$$

13.8.4 *Theory of Eddy Current MAGLEV*

Combining (13.36) and (13.31) (for high velocity) yields

$$\lim_{v/v_0 \to \infty} F_D = F_I \frac{v_0}{v}. \tag{13.37}$$

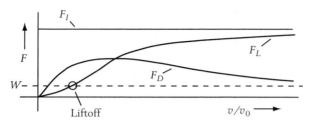

Figure 13.12 Force versus velocity for a magnet moving along a conducting sheet. Force is in units of the image force F_I, and velocity is in units of the Maxwell recession velocity v_0 of (13.32).

Hence the drag force decreases as v^{-1} at high velocities. Combining (13.36) and (13.33) (for low velocity) yields

$$\lim_{v/v_0 \to 0} F_D = \alpha F_I \left(\frac{v}{v_0} \right)^2 , \qquad (13.38)$$

so that the lift force varies as v^2 at low velocities.

Figure 13.12 sketches the behavior of the lift force F_L as a function of velocity v. A constant force W (for the weight of a train) is given. At low velocities, the lift force F_L is inadequate to support W, but at higher velocities, F_L can support W, and lift-off can occur. At even higher velocities, the drag starts to decrease, and the lift saturates at the image force. (When lift-off occurs, the lift force will tend to decrease, due to the increasing separation, so this figure, based on a constant separation, is a bit misleading. Once lift-off occurs, the height adjusts until $W = F_L$.) You are now an expert in eddy current MAGLEV.

And so the chapter ends. The next chapter will work out more of the implications of the laws of electromagnetism. Besides studying ac generators and transformers, and ac power, it also considers how eddy currents are produced when a uniform magnetic field oscillates with its direction in the plane of an infinite sheet of conductor. In this way, we will learn how eddy currents lead to electromagnetic shielding. (We will use the terminology *electrostatic screening* for the static screening of electric fields within a conductor, and *electromagnetic shielding* for the dynamic screening of both electric and magnetic fields within a conductor.)

Problems

13-1.1 Verify from the magnetic fields and the magnetic forces that Figure 13.1 gives the correct direction for the motion of the magnet of the motor on the left and for the wire of the motor on the right.

13-1.2 For both of the motors in Figure 13.1, the velocity is proportional to the force acting on them, and their acceleration is negligible. This is similar to what happens for electrons carrying current in a wire. (a) Discuss why the velocity is proportional to the force. (b) What role, besides being a good electrical conductor, does the liquid mercury play?

13-2.1 Which would make the best transformer core and why: solid plastic, plastic rods, solid iron, iron rods?

13-2.2 (a) Explain why the eddy current heating for Figure 13.4(b) should be one-fourth that for Figure 13.4(a). (b) If the loop of Figure 13.4(a) initially has an eddy current heating rate of 0.45 W and it is now broken into nine equivalent subloops, what is the new eddy current heating rate?

13-3.1 If a motor charges up a generator, what is considered to be the external power source, and what is considered to utilize the power?

13-3.2 If a generator drives a motor, what is considered to be the external power source, and what is considered to utilize the power?

13-4.1 During startup of a motor, at 120 V emf, the current is 12 A. Once the motor gets moving, the current is a steady 2 A. Find (a) the resistance and (b) the back emf.

13-4.2 For a linear motor with $B = 0.035$ T and $l = 24$ cm, the back force is 0.84 N. Find the current.

13-4.3 A linear motor with mass 0.58 kg, length 14 cm, and resistance 0.65 Ω, in a field of magnitude 0.46 T, is driven by a 60 V emf. In steady-state operation, it has velocity 2.3 m/s. (a) Determine its initial current. (b) Determine the initial magnetic force and the corresponding acceleration. (c) Determine the back emf when it is operating. (d) Determine the current when it is operating. (e) Determine τ_f. (Neglect the force F_0.)

13-4.4 A linear generator with mass 0.58 kg, length 14 cm, and resistance 0.65 Ω, in a field of magnitude 0.46 T, is driven by a 24 N force. There is no constant emf $\mathcal{E}_0$. In steady-state operation, it has a current of magnitude 36 A. (a) Determine its steady-state velocity. (b) Determine τ_f.

13-5.1 Consider a circuit with a moving arm of mass 46 g and length 12 cm, in a field of magnitude 0.08 T. (a) If the resistance is 0.86 Ω, determine the magnetic relaxation time. (b) If a 6.3 V battery is added to the circuit, find the initial current and the initial magnetic force.

13-5.2 Consider a circuit with a moving arm of mass 125 g and length 16 cm, in a field of magnitude 0.08 T. (a) If the resistance is 2.4 Ω, determine the magnetic relaxation time. (b) If it is driven by a 4.6 V emf, and there is no driving force, find

its steady-state velocity for $\tau_f = 120$ s. (c) Find its steady-state current.

13-5.3 Consider a circuit with a moving arm of mass 68 g and length 14 cm, in a field of magnitude 0.08 T. (a) If the resistance is 0.42 Ω, determine the magnetic relaxation time. (b) If it is driven by a 25 N force, and there is no driving emf, find its steady-state velocity for $\tau_f = 120$ s. (c) Find its steady-state current.

13-5.4 Consider a conducting rod of mass m and length l, initially at rest, but free to slide down two guide wires making an angle θ to the horizontal. The guide wires are connected at the bottom, so the rod and the guide wires form a complete circuit. There is a vertical field $\vec{B}$. See Figure 13.13. In the limit where self-inductance dominates over resistance (i.e., retain LdI/dt but neglect IR in the emf equation), derive the equation of motion for the position z along the guide wires and for the current. Solve for the position and current as a function of time, assuming that the position and current initially are zero. Take L to be constant.

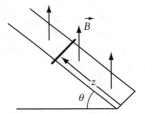

Figure 13.13 Problem 13-6.4.

13-6.1 A blender motor driven by a 120 V emf uses a 12 A current. If it operates at 60% efficiency, with a moment of inertia $\mathcal{I} = 0.0042$ Kg-m^2, at 14.5 Hz, determine the rotational drag time τ_f.

13-6.2 A lifting motor driven by a 250 V emf uses a 16 A current. If it operates at 80% efficiency, and lifts a weight of 417 N, at what velocity does it lift the weight?

13-6.3 A linear motion generator is driven by a 12 N force at velocity 2.3 m/s. It provides a current of 3.6 A at 6.8 V. Find the efficiency.

13-6.4 A rotational motion generator is driven by a 2 N-m torque at angular velocity 4.5 s^{-1}. If it is 70% efficient, find the current it provides at output voltage 5.6 V.

13-7.1 A circuit with a moving arm of mass 48 g and length 15 cm is in a field of magnitude 0.12 T. The resistance is 0.54 Ω. (a) Determine the magnetic relaxation time. (b) If there is a 2.4 V emf in the circuit, determine the initial force and acceleration. (c) If $\tau_f = 86$ s, determine the steady-state velocity and the velocity after 1.2 s.

13-7.2 A circuit with a moving arm of mass 212 g and length 24 cm, is in a field of magnitude 0.086 T. The resistance is 0.94 Ω. Let $\tau_f = 78$ s, and let there be a 4.5 V emf in the circuit. Determine (a) the magnetic relaxation time, (b) the steady-state velocity, (c) the velocity 3.4 s after the emf has been removed.

13-8.1 Compute v_0 of (13.32) for Cu with $d = 10$ nm, 100 nm, 100 μm, and 1 mm.

13-8.2 The lift and drag forces at a velocity of 0.5 m/s are 2.4 N and 15.8 N. The lift force at a velocity of 45 m/s is 38 N. Find v_0 and the drag force at velocity 45 m/s.

13-8.3 Using Figure 13.11(a), appropriate for $v \ll v_0$, show that $F_D \sim v$ for $v \ll v_0$. (*Hint:* There is no drag force for the nearest image, but for all pairs of images at the same horizontal position, partial cancellation of the force along the surface occurs. Show that for each pair this leads to a force proportional to v.)

13-8.4 Determine the current density K generated if a monopole suddenly materializes at a height h above *any* flat conductor. (b) For a thin conducting sheet, use the receding image construction to determine how K varies with time. *Hint:* Refer to Section 11.11.

13-8.5 Treat a conducting sheet as if it were a perfect diamagnet. Use Oersted's right-hand rule in what follows. (a) If a current-carrying wire is parallel to the sheet, how should the image current flow? See Figure 13.14. (b) If wire carries current into the

sheet, perpendicularly, how should the image current flow?

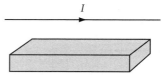

Figure 13.14 Problem 13-9.5.

13-G.1 Consider a box containing a fluid. It has two holes cut in it through which the fluid can empty. (a) Is the rate of relaxation bigger when both are open than when only one is open? (b) Should the relaxation rates be added, or should the relaxation times be added?

13-G.2 (N. Gauthier.) Consider a cylinder of radius a and length $l \gg a$. It is coated with a fixed charge density σ on its round outside and has moment of inertia $\mathcal{I}$. Wrapped around its outside is a massless string, one end attached to the outer surface of the cylinder, and the other end attached to a mass M. See Figure 13.15. (a) Find the acceleration of the mass for $\sigma = 0$. (b) Find the acceleration of the mass for finite σ. *Hint:* A time-varying rotation of the charged cylinder causes a time-varying magnetic field along the axis.

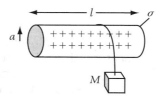

Figure 13.15 Problem 13-G.2.

13-G.3 Using electronic switches, capacitors are often added in parallel to motor circuits during startup, but are shortly removed. Explain in qualitative terms why they are added, why they are removed, and how they operate.

"I know not, but I wager that one day your government will tax it."

—Michael Faraday,
on the British prime minister's inquiring of the use of Faraday's dynamo (generator) (1831)

"No great law in Natural Philosophy has ever been discovered for its practical applications, but the instances are innumerable of investigations apparently quite useless in this narrow sense of the word which have led to the most valuable results."

—William Thomson, Lord Kelvin (1858)

Chapter 14

Alternating Current Phenomena: Signals and Power

Chapter Overview

Section 14.1 provides an introduction to, and motivation for, this chapter. Section 14.2 discusses the *LC* circuit, which has a natural resonance frequency. Section 14.3 studies the effect that a resistor *R* has on the *transient*, or temporary, response of such a circuit. Section 14.4 discusses an ac generator, and Section 14.5 discusses the effect of ac voltage on the individual circuit elements *R*, *L*, and *C*. With this as background, we consider the ac response of various circuits: *RC* and *LR* in Section 14.6 (with applications to signal filtering in Section 14.7), and *RLC* in Section 14.8. Section 14.9 discusses the principles of amplification, and Section 14.10 discusses the power factor, which is a measure of how efficiently power is used. Section 14.11 discusses an idealized version of the ac transformer, and the dramatic example of a Tesla coil, which employs two sets of transformers. Section 14.12 shows how ac power can be used to cause dc motion, and how ac power relates to 60-cycle and 120-cycle hum. Section 14.13 closes the chapter with a discussion of electromagnetic shielding by electrical conductors. ■

14.1 Introduction

Your local electric company—which, unless it is municipally owned, pays taxes—sells its customers an electrical signal at an oscillating, or alternating voltage. In the United States, the frequency of oscillation f is, to a few parts in 10^4, 60 cycles per second (cps), or 60 Hz (or Hertz). Related to the frequency f in Hz is the frequency in radians per second ω, where

$$\omega = 2\pi f. \quad \text{(frequency)} \quad (14.1)$$

The period of the oscillation T is given by

$$T = \frac{1}{f} = \frac{2\pi}{\omega}. \quad \text{(period)} \tag{14.2}$$

For power at $f = 60$ Hz, (14.1) gives $\omega = 376.99 \approx 377$ s^{-1}, and (14.2) gives $T = 1/60 = 0.01667$ s.

The voltage difference across the terminals of an electrical outlet in a house or in a factory oscillates with time. It may be represented in the form

$$\Delta V = V_m \sin \omega t, \quad \text{(oscillating voltage)} \tag{14.3}$$

where V_m is the *maximum* value of the voltage. Voltage is usually specified in terms of the *root mean square*, or *rms*, voltage V_{rms}. To obtain V_{rms} from V_m, first square the voltage. Second, time average the square over a period. (For a time-varying function g, we will write its time average over a period of the oscillation as $\overline{g}$.) Third, take the square root.

We time average the square of (14.3) by noting that, over a period, $\overline{\sin^2 \omega t} = \overline{\cos^2 \omega t}$, because $\sin^2 \omega t$ and $\cos^2 \omega t$ are simply shifted versions of each other. Since $\sin^2 \omega t + \cos^2 \omega t = 1$, we also have $\overline{\sin^2 \omega t} + \overline{\cos^2 \omega t} = 1$. Hence $\overline{\sin^2 \omega t} = \overline{\cos^2 \omega t} = \frac{1}{2}$. Thus

$$V_{rms} = \sqrt{\overline{\Delta V^2}} = \sqrt{V_m^2 \overline{\sin^2 \omega t}} = \sqrt{V_m^2 \frac{1}{2}} = \frac{V_m}{\sqrt{2}}. \quad \text{(rms ac voltage)} \tag{14.4}$$

In the United States, power is usually provided at about $V_{rms} = 120$ V, which corresponds to $V_m = \sqrt{2} V_{rms} = 169.7 \approx 170$ V. Although the electric company provides power that is specified in terms of an oscillating, or alternating *voltage*, it is conventional to use the term *alternating current*, or *ac*, to describe this power. With these details established, let's get to the big picture.

There are at least two practical reasons we study ac phenomena. First, power is transmitted as ac, and it is therefore important to know how this power is produced, transmitted, transformed, and utilized. Second, our communications are based upon modulations of signals at much higher frequencies, and it is therefore of importance to know how such signals are manipulated and amplified. Indeed, there is a fundamental mathematical theorem, discovered in 1807 by Fourier during his classic studies of heat flow, that any time-dependent quantity can be written as a linear combination of sinusoidal oscillations if enough oscillation frequencies are included. (Actually, Fourier discussed how to represent a spatially varying quantity—the temperature—as a linear combination of waves if enough wavelengths are included.) In the next chapter, we will finally discuss the means by which such communications take place: electromagnetic radiation.

This chapter primarily considers oscillating voltages, either when turning on or off a circuit that has a natural period of oscillation or when subjecting another circuit to an oscillating voltage. We consider only frequencies sufficiently low that the response of the circuit is essentially instantaneous. The next chapter will show that electromagnetic signals propagate at the speed of light $c \approx 3.0 \times 10^8$ m/s.

Because of this finite velocity, the circuit—of characteristic dimension d—mustn't be so large that the time delay across the circuit is comparable to the period $T = f^{-1}$. (Otherwise, it would be like having a chorus in a large room, where even a choir director can't prevent the time delays that make sounds from distant singers arrive at different times.) Specifically,

$$\frac{d}{c} \ll T = \frac{1}{f}. \tag{14.5}$$

A course in electricity and magnetism is a means to learn how radios and televisions work. The present chapter deals with how to tune a signal once it has been received by an antenna, how to manipulate that signal (treble and bass), and how to amplify it. It also discusses how to decipher the acoustic signal from the much higher frequency that is sent by the radio transmitter; this deciphering is different for AM and FM. Chapter 15 discusses how radio waves are generated by a radio station, how they are propagated through space, and how they are received by a receiving antenna. Recall that Chapter 10 indicated how to take an electrical signal from a tuner and use its interaction with a magnet to drive a speaker cone. In this way, it produces sound, the ultimate end product of the radio station's activities.

Although the circuit elements we study are simple—much simpler than the complicated electronics devices used in many high-tech applications—they illustrate certain fundamental principles. Today, vacuum tube diodes are *passé*, but there was a time when they were the most advanced nonlinear devices. Similarly, today's MOSFETS (metal oxide semiconductor field effect transistors) will give way to faster nonlinear devices. However, resistors, inductors, and capacitors, which are linear devices (double the input and you double the output), have not been replaced, and are unlikely to be replaced. Moreover, a hundred years from now, the technology will have changed, but the principles will remain the same.

14.2 *LC* Resonance

A circuit consisting only of an inductance L and a capacitance C has no resistance, and thus cannot dissipate energy. Hence the total energy E, which is shared by both capacitor and inductor, must be a constant. Thus

$$E = \frac{1}{2}\frac{Q^2}{C} + \frac{1}{2}LI^2 = \text{const.} \tag{14.6}$$

LC Circuit

The behavior of a circuit with a capacitance C and inductance L was first treated theoretically by W. Thomson (Lord Kelvin) in 1853. If time travel were possible, I would ask Maxwell why he didn't include the LC circuit in his great *Treatise* of 1873. The answer would probably have been that the technology of the time was not yet capable of measuring the relatively high resonance frequency of a typical LC circuit. Nevertheless, it was with an LC circuit that electromagnetic radiation—a major new prediction by Maxwell himself—was first detected by Hertz, in 1888.

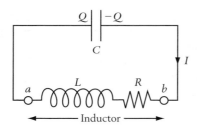

Figure 14.1 Circuit with capacitor C and inductor, modeled by inductance L in series with resistor R.

To obtain the circuit equation in this case, first assume that the inductor has a finite resistance R. That gives us an RLC circuit, to be considered in Section 14.3. We will then take the limit where $R \to 0$. See Figure 14.1.

Since there is zero net voltage change around the circuit, from Figure 14.1 we have

$$\Delta V_L + \Delta V_C = 0, \tag{14.7}$$

where $\Delta V_L = V_b - V_a$ and $\Delta V_C = V_a - V_b$. We have taken the high-voltage side of the inductor to drive current to the low-voltage side. For the inductor, Ohm's law takes the form

$$I = \frac{\mathcal{E} + \Delta V_L}{R} = \frac{\mathcal{E} - \Delta V_C}{R} = \frac{-L dI/dt - Q/C}{R}. \tag{14.8}$$

Note that if $dI/dt > 0$, the induced emf tends to drive the current opposite to the direction of positive current. For finite I in Equation (14.8), as the numerator $R \to 0$, the denominator must also approach zero. In the limit,

$$L\frac{dI}{dt} + \frac{Q}{C} = 0. \tag{14.9}$$

With $I = dQ/dt$, (14.9) becomes

$$L\frac{d^2Q}{dt^2} = -\frac{Q}{C}. \tag{14.10}$$

14.2.1 *Review of the Harmonic Oscillator*

Those who have become adept at recognizing old friends will remember that (14.10) is like the equation describing a harmonic oscillator. There a mass M (units of kg) is constrained to move only along one axis (i.e., the x-axis). It is attached to a spring, of spring constant K (units of N/m), with equilibrium position $x = 0$. When $x \neq 0$, $F_x = -Kx$, so that if the spring is to the left, then the force is to the right, thus tending to return the mass to the origin. Using F_x in Newton's law of motion yields

$$M\frac{d^2x}{dt^2} = -Kx. \tag{14.11}$$

This is a *second-order differential equation* (second derivative is the highest that appears). It has two initial conditions, and two undetermined constants are needed to match these initial conditions. You may have solved (14.11) in your mechanics course by a rotating circle construction. Let's now solve it by another method. Since the motion is expected to be oscillatory (but of unknown frequency ω_0), let's try the form

$$x = A\cos(\omega_0 t + \phi_0), \quad \text{so } \cos(\omega_0 t + \phi_0) = \frac{x}{A}. \tag{14.12}$$

Here A and ϕ_0 are constants determined by the initial conditions, and ω_0 has yet to be determined. (By convention, we take $A > 0$.) Placing (14.12) for x into (14.11), and using

$$\frac{d^2 \cos(\omega_0 t + \phi_0)}{dt^2} = -\omega_0^2 \cos(\omega_0 t + \phi_0),$$

we find that $K/M = \omega_0^2$, so

$$\omega_0 = \sqrt{\frac{K}{M}}. \tag{14.13}$$

Note that the stiffer the spring (larger K), the higher the ω_0; and the lighter the mass (smaller M), the higher the ω_0, as expected.

From x of (14.12), the velocity v is given by

$$v = \frac{dx}{dt} = -A\omega_0 \sin(\omega_0 t + \phi_0), \quad \text{so } \sin(\omega_0 t + \phi_0) = \frac{v}{\omega_0 A}. \tag{14.14}$$

Here's how to obtain A and ϕ_0. Let the initial ($t = 0$) values of position and velocity be x_0 and v_0. Using (14.12) and (14.14), the trig identity $1 = \cos^2(\omega_0 t + \phi_0) + \sin^2(\omega_0 t + \phi_0)$ becomes

$$1 = \frac{x^2}{A^2} + \frac{v^2}{\omega_0^2 A^2}. \tag{14.15}$$

Evaluating (14.15) at $t = 0$ then gives

$$A^2 = x_0^2 + \left(\frac{v_0}{\omega_0}\right)^2. \tag{14.16}$$

Again using (14.12) and (14.14), the trig identity $\tan(\omega_0 t + \phi_0) = \sin(\omega_0 t + \phi_0)/\cos(\omega_0 t + \phi_0)$ becomes

$$\tan(\omega_0 t + \phi_0) = -\frac{v}{\omega_0 x}. \tag{14.17}$$

Evaluating (14.17) at $t = 0$ then gives

$$\tan \phi_0 = -\frac{v_0}{\omega_0 x_0}. \tag{14.18}$$

Note that the inverse tangent function always yields a ϕ_0 in either the first or fourth quadrant. With $A > 0$, such a ϕ_0 always yields $x_0 > 0$. To accommodate $x_0 < 0$, a phase of π rad $= 180°$ must be added by hand to the inverse tangent of (14.18). Hence, if $x_0 = -3$ cm, $v_0 = 2$ cm/s, and $\omega_0 = 0.5$ s^{-1}, then $A = 5$ cm/s and $\phi_0 = (0.927 + \pi)$ rad $= 4.07$ rad $= (53 + 180)° = 233°$, which is in the third quadrant. Then (14.12) gives $x_0 = -3$ cm, and (14.14) gives $v_0 = 2$ cm/s, as desired.

14.2.2 *Return to the LC Circuit*

By analogy to the harmonic oscillator, replace (x, M, K) by $(Q, L, 1/C)$, to obtain the solution for the LC circuit equation of (14.10) as

$$Q = A\cos(\omega_0 t + \phi_0), \qquad \omega_0 = \sqrt{\frac{1}{LC}}. \qquad (LC\,\text{resonance}) \qquad (14.19)$$

The current is given by

$$I = \frac{dQ}{dt} = -\omega_0 A \sin(\omega_0 t + \phi_0). \qquad (14.20)$$

Here ω_0 is the frequency of oscillation, and A (in units of coulombs) and ϕ_0 (in units of radians or degrees) are constants determined by the initial values of the charge (Q_0) and the current (I_0). By analogy to (14.16) and (14.18),

$$A^2 = Q_0^2 + \left(\frac{I_0}{\omega_0}\right)^2, \qquad \tan\phi_0 = -\frac{I_0}{\omega_0 Q_0}. \qquad (14.21)$$

Equations (14.19) and (14.20) show that when the capacitor is fully charged (and correspondingly, when a spring is at maximum expansion or compression), the inductor carries no current (correspondingly, the mass has zero velocity). Moreover, when the inductor carries its maximum current (and correspondingly, when the mass has its maximum velocity), the capacitor has zero charge (correspondingly, the spring has zero displacement). Pursuing the analogy further, the potential energy of the spring corresponds to the energy stored by the capacitor's electric field, and the kinetic energy of the mass corresponds to the energy stored in the magnetic field of the inductor.

Example 14.1 Finding the amplitude and phase

(a) If the initial conditions are that $Q = Q_0$ and $I = 0$, then find ϕ_0 and A.
(b) Repeat for initial conditions $Q = 0$ and $I = I_0$.

Solution: (a) If the initial conditions are $Q = Q_0$ and $I = 0$, then (14.19) and (14.20) are satisfied by $\phi = 0$ and $A = Q_0$. (b) If the initial conditions are that $Q = 0$ and $I = I_0$, then (14.19) and (14.20) are satisfied by $\phi_0 = -\pi/2$ and $A = I_0/\omega_0$. This can be seen by performing the substitution.

Example 14.2 **Properties of an *LC* circuit**

Let $L = 5$ mH and $C = 2$ μF. Find (a) ω_0; (b) $f_0 = \omega_0/2\pi$; (c) $T_0 = f_0^{-1}$; (d) how many oscillations occur within 0.15 s.

Solution: (a) By (14.19), $\omega_0 = 10^4$ s^{-1}; (b) $f_0 = 1592$ Hz; (c) $T_0 = 6.283 \times 10^{-4}$ s; (d) $0.15/T_0 = 239$ oscillations.

On substituting (14.19) and (14.20) into (14.6), we find that the total energy is indeed constant. Explicitly, with $\phi = \omega_0 t + \phi_0$ and $\omega_0 = (LC)^{-1/2}$,

$$E = \frac{1}{2C} A^2 \cos^2 \phi + \frac{L}{2} A^2 \omega_0^2 \sin^2 \phi$$

$$= \frac{A^2}{2C} (\cos^2 \phi + \sin^2 \phi) = \frac{A^2}{2C}. \tag{14.22}$$

In the case of the harmonic oscillator, the energy is alternately stored as potential or kinetic energy; in the case of the *LC* circuit, the energy is alternately stored as electrical or magnetic energy.

The *LC* circuit is the heart of the tuning circuit for radio or TV. Turning a dial (or, in these digital days, pushing a button) adjusts either L or C until the desired station, with its own ω_0, is picked up.

14.3 *RLC* Circuit Transients

14.3.1 *Overview*

When a resistance R is added to a resonating *LC* circuit, the oscillations will eventually die out. See Figure 14.1, which illustrates the voltage $\Delta V = IR$ across the resistor. Assume that a constant emf is switched on at time $t = 0$.

Small R. If the resistance R is small, we expect that the current and voltage should look something like that in Figure 14.2(a), with many oscillations that slowly decay in amplitude.

Large R. If the resistance R is large, there will be an initial transient, with short characteristic time

$$\tau_{LR} = \frac{L}{R}, \tag{14.23}$$

Self Resonance

Just as electric charge on the surface of a resistor produces the voltage drop across the resistor (see Section 8.10), so too does electric charge on the surface of an inductor produce the voltage drop across the inductor (see Section 12.11.2). This means that, just as a resistor has a parasitic capacitance in parallel, so too does an inductor have a parasitic capacitance in parallel. Hence an inductor can be expected to self-resonate; indeed, this is a well-known laboratory phenomenon.

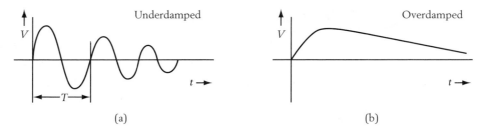

Figure 14.2 RLC circuit transients: (a) voltage across resistor R for low resistance (underdamping); (b) voltage across resistor R for high resistance (overdamping).

due to LR buildup of the current. It will be followed by a much longer decay, with characteristic time

$$\tau_{RC} = RC, \tag{14.24}$$

due to RC decay of the current flow. See Figure 14.2(b). In this case, there is no oscillation at all.

By considering the case of small resistance, and increasing it, it is possible to obtain a qualitative understanding of what happens for intermediate values of the resistance. In addition to the decrease in amplitude depicted in Figure 14.2(a), increasing the resistance increases the period T of each oscillation. This is because there is a resistive drag that tends to oppose the current, thereby slowing its motion. If a circuit oscillates with a slow decay, it is called *underdamped*. Thus the frequency of the motion should decrease with increasing R, until a critical resistance R_c is reached at which the frequency goes to zero. For larger resistances, there are no oscillations, only a rise and then a fall in the current, as in Figure 14.2(b). If the circuit doesn't oscillate at all, the circuit is called *overdamped*.

There is a simple way to estimate the dependence of R_c on C and L: we equate the two relaxation times $\tau_{LR} = L/R$ of (14.23) and $\tau_{RC} = RC$ of (14.24). Then

$$R_c \sim \sqrt{\frac{L}{C}}. \tag{14.25}$$

For $R \ll R_c$ the circuit is underdamped; for $R \gg R_c$ the circuit is overdamped.

Having described the physics of the situation, we must now extract it from the mathematics. With $I = dQ/dt$, differentiating (14.8) with respect to t yields

$$\frac{dI}{dt} = -\frac{L}{R}\frac{d^2 I}{dt^2} - \frac{1}{RC}I, \tag{14.26}$$

where we have used $I = dQ/dt$. This may be rewritten as

$$\frac{d^2 I}{dt^2} + \frac{R}{L}\frac{dI}{dt} + \frac{1}{LC}I = 0. \tag{14.27}$$

Mathematical Details

Rather than bore you with algebraic manipulations, we simply present the final results. The details can be done as problems. The general results will be given, which will allow any initial conditions. First, the critical resistance is given by

$$R_c = 2\sqrt{\frac{L}{C}}. \qquad (14.28)$$

This has precisely the form indicated in (14.25), the proportionality constant taking the specific value 2.

Example 14.3 Type of *RLC* circuit

Let $L = 5$ mH and $C = 2$ μF, as specified in the previous example. Also take $R = 10$ ohms. (a) Find R_c; (b) determine the qualitative behavior of the circuit.

Solution: (a) Equation (14.28) gives $R_c = 100$ ohms. (b) Since $R \ll R_c$, this is an underdamped circuit.

Small R. For $R < R_c$, the charge is given by

$$Q = Q_\infty + A \cos\left(\Omega t + \phi_0\right) \exp\left[-\frac{t}{2L/R}\right], \qquad (14.29)$$

where Q_∞ is the charge after a long time (given by $Q_\infty = C\mathcal{E}_0$ if there is a constant emf $\mathcal{E}_0$), and

$$\Omega = \sqrt{\frac{1}{LC} - \left(\frac{R}{2L}\right)^2} = \sqrt{\frac{1}{LC}}\sqrt{1 - \left(\frac{R}{R_c}\right)^2}. \qquad (14.30)$$

The current is given by $I = dQ/dt$. The form of (14.29) corresponds to Figure 14.2(a). The quantities A (in units of coulombs) and ϕ_0 (in radians) are determined by the two initial conditions, which are the values of Q and $I = dQ/dt$ at $t = 0$. We say that this is the *underdamped* case. For $R = 0$ and $\mathcal{E}_0 = 0$, (14.29) and (14.30) reduce to (14.19). Problem 14-3.3 asks you to show that (14.29) and (14.30) satisfy (14.27).

The Quality Factor Q Measures How Long a Circuit Will "Ring" A dimensionless measure of how long the circuit will oscillate is the number of radians of oscillation that occur during the exponential decay time L/R. This is called the *quality factor Q*. The higher the Q, the more the circuit will "ring"; for $R \ll R_c$, the circuit oscillates many times before the capacitor completely discharges. If $R < \frac{1}{2}R_c$, then Ω of (14.30) approximately equals ω_0 of (14.19). Then, with (14.28), Q is given by

$$Q = \text{number of radians of oscillation} = \frac{\Omega L}{R} \approx \frac{\omega_0 L}{R} = \frac{1}{\sqrt{LC}}\frac{L}{R} = \frac{R_c}{2R}. \qquad (14.31)$$

Example 14.4 **The quantity factor**

Consider the previous example, where $L = 5$ mH, $C = 2$ μF, and $R_c = 100$ ohms. (a) If $R = 10$ ohms, find Q; (b) estimate the number of cycles of oscillation.

Solution: (a) Equation (14.31) gives $Q \approx 5$. (b) This corresponds to $5/2\pi = 0.796$ of a full cycle of oscillation.

Another way to express the "quality" of an RLC circuit is to determine the ratio of the total energy E stored in the circuit, given by (14.22), to the energy loss per cycle, given by $\int_0^T I^2 R \, dt$. In the limit where $R \ll R_c$, we may use (14.20) for I. The period is given by (14.2) with $\omega = \omega_0$. Then, since the time average of $\sin^2 \omega_0 t$ over a period is $1/2$,

$$\int_0^T I^2 R \, dt = \frac{1}{2} \frac{2\pi}{\omega_0} A^2 \omega_0^2 R = \pi A^2 \omega_0 R. \tag{14.32}$$

Then, using (14.22), (14.32), $\omega_0 = (LC)^{-1/2}$, (14.28), and (14.31),

$$\frac{\text{energy stored}}{\text{energy loss per cycle}} = \frac{E}{\int_0^T I^2 R \, dt} = \frac{A^2/2C}{\pi A^2 \omega_0 R} = \sqrt{\frac{L}{C}} \frac{1}{2\pi R} = \frac{1}{\pi} \frac{R_c}{4R} = \frac{Q}{2\pi}. \tag{14.33}$$

Large R. For $R > R_c$, the current is given by

$$I = A_+ \exp\left(-\frac{t}{\tau_+}\right) + A_- \exp\left(-\frac{t}{\tau_-}\right), \tag{14.34}$$

where

$$\frac{1}{\tau_+} = \frac{R}{2L} + \sqrt{\left(\frac{R}{2L}\right)^2 - \frac{1}{LC}} = \frac{R}{2L}\left[1 + \sqrt{1 - \left(\frac{R_c}{R}\right)^2}\right],$$

$$\frac{1}{\tau_-} = \frac{R}{2L} - \sqrt{\left(\frac{R}{2L}\right)^2 - \frac{1}{LC}} = \frac{R}{2L}\left[1 - \sqrt{1 - \left(\frac{R_c}{R}\right)^2}\right] \tag{14.35}$$

and the quantities A_+ and A_- are constants with units of amperes that are determined by the initial conditions. The form of (14.34) corresponds to Figure 14.2(b). Note that for $R \gg R_c$ the shorter time τ_+ nearly equals τ_{LR}, and that for $R \ll R_c$ the longer time τ_- nearly equals τ_{RC}. We say that this is the *overdamped* case. When $R = R_c$, we have the case of *critical damping*. There are many mechanical analogs of the RLC circuit, such as a swinging door or a mass on a spring. Problem 14-3.4 asks you to show that (14.33) and (14.34) satisfy (14.27).

The purpose of displaying the solutions is to show that our initial qualitative discussion was correct. You are encouraged to remember only the qualitative introductory discussion associated with Figure 14.2.

14.4   AC Generator—Rotate Loop in Uniform $\vec{B}$ Field

We begin by discussing how it is that the electric company can generate ac power. The principle is as follows.

14.4.1   *The EMF of an AC Generator*

Consider a planar coil of area A and N turns, with normal $\hat{n}$, in a uniform $\vec{B}$ field. That could be due to a large magnet or solenoid. See Figure 14.3. The magnetic flux for all N turns is given by

$$\Phi_B = N \int \vec{B} \cdot \hat{n} \, dA = NBA \cos\theta, \qquad (14.36)$$

where θ is the angle between $\hat{n}$ and $\vec{B}$. If $\vec{B}$ and $\hat{n}$ point in the same direction at $t = 0$, and if the coil (or the field) is rotating at the uniform rate ω, then $\theta = \omega t$, so

$$\Phi_B = NBA \cos\omega t. \qquad (14.37)$$

Now apply Faraday's law, with positive emf along the direction of circulation of $d\vec{s}$ (which is counterclockwise to the observer in Figure 14.3, on applying the circuit-normal right-hand rule with the thumb along $\hat{n}$). Then the generator emf is given by

$$\mathcal{E}_g = \oint \vec{E} \cdot d\vec{s} = -\frac{d\Phi_B}{dt} = \omega NBA \sin\omega t. \qquad \text{(induced emf from rotation)}$$

$$(14.38)$$

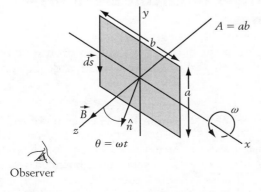

Figure 14.3 A simple ac generator: a rotating rectangular loop in a uniform magnetic field.

In practice, the electric company employs hydroelectric and steam turbine generators. There, what is called the *rotor* (containing either the coils or the magnet) turns relative to the *stator* (containing either the magnet or the coils), thus producing an emf, by (14.38). The frequency is determined by the turning rate of the rotor, which is kept nearly constant at 60 Hz. It cannot be kept exactly constant because of variations in the rate at which water or steam flows past the turbine. Furthermore, the larger the current I, the larger the magnetic moment $\vec{\mu}$ ($|\vec{\mu}| = NIA$) and the larger the back torque $\vec{\mu} \times \vec{B}$ that tends to slow down the turbine.

Equation (14.38) shows that a generator provides a time-dependent source of ac emf

$$\mathcal{E}_g = \mathcal{E}_m \sin \omega t, \qquad (14.39)$$

where $\mathcal{E}_m = \omega NBA$ for our particular generator. $\mathcal{E}_g$ peaks at a time t given by a quarter period, or $T/4$, since $\omega T = 2\pi$ and $\omega T/4 = \pi/2$. Another interpretation of how this emf comes about is that it arises from a time-varying mutual inductance M that is proportional to $\cos\theta = \cos\omega t$. We can also interpret $\mathcal{E}_g$ as a motional emf, via $\mathcal{E} = \oint \vec{E} \cdot d\vec{s}$ with $\vec{E} = \vec{v} \times \vec{B}$.

14.4.2 *The Voltage Drop across an AC Generator*

Now apply a generator to an arbitrary unknown circuit; the latter may be thought of as a "black box." See Figure 14.4. We choose positive current to flow in the direction of positive generator emf $\mathcal{E}_g$, and we choose positive terminal voltage $\Delta V_g = V_b - V_a$ to oppose the generator emf. This follows the convention applied earlier to a battery, where the terminal voltage opposes the battery emf. Then Ohm's law applied to the generator resistance R_g yields

$$I = \frac{\mathcal{E}_g - \Delta V_g}{R_g}. \qquad (14.40)$$

In the limit $R_g \to 0$ of an ideal generator, for finite I, (14.40) yields

$$\Delta V_g = \mathcal{E}_g. \qquad (14.41)$$

By (14.39) and (14.41),

$$\Delta V_g = \mathcal{E}_g = \mathcal{E}_m \sin \omega t. \qquad (14.42)$$

14.5 Response to AC Power of Circuit Elements— Impedance and Phase

The next few sections work out the consequences, for various circuits, of using an ac generator, with emf as in (14.39), rather than a dc battery, to drive

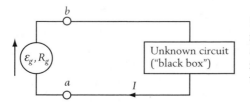

Figure 14.4 A generator of emf $\mathcal{E}_g$ and resistance R_g, driving current I through an unknown circuit ("black box"). An ideal generator is obtained in the limit where $R_g \to 0$.

current around a circuit. We cannot solve for all circuits at once: our approach will be to solve for individual circuit elements, and then build up more complex circuits from these individual circuit elements (the black box of Figure 14.4).

As early as 1879, the voltage and current of ac circuits (i.e., circuits driven by ac power) were analyzed mathematically using complex numbers, with measureable quantities obtained by taking the real part. An equivalent but more graphical viewpoint was developed in 1893 by Steinmetz, an electrical engineer well trained in mathematics and physics. It is called the method of *phasors* and is related to the rotating circle construction to describe the harmonic oscillator. Favorably received then by a large audience of practicing electrical engineers, it is still used today.

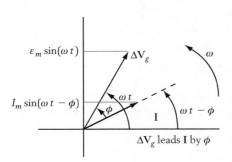

Figure 14.5 Phasor diagram. Instantaneous values of the voltage and current for a device are given by the y-components of the counterclockwise rotating vectors $\Delta\mathbf{V}_g$ and $\mathbf{I}$. Respectively, they have amplitudes $\mathcal{E}_m$ and I_m, and phase angles relative to the x-axis of ωt and $\omega t - \phi$. If ω, t, and I_m are considered to be known, then $\mathcal{E}_m$ and ϕ must be determined.

The idea of a phasor is that in an ac circuit the voltage or current for any device can be represented as the y-component of a vector (called the *phasor*) that rotates at frequency ω. For example, the generator voltage of (14.42) can be represented by a voltage phasor $\Delta\mathbf{V}_g$ of length $\mathcal{E}_m$, which rotates counterclockwise at the frequency ω, and coincides with the x-axis at $t = 0$. We denote phasors by boldface. See Figure 14.5, where $\mathcal{E}_m \sin \omega t$ of (14.42) is the y-component of $\Delta\mathbf{V}_g$. (Some authors choose to use the x-component of phasors; either choice will work, as long as that choice is made consistently.) $\Delta\mathbf{V}_g$ peaks at time $t_{\mathcal{E}}^{max} = T/4$.

For a given input emf, the time-dependent current I constitutes the response of the system. It can be written in the general form

$$I = I_m \sin\left(\omega t - \phi\right), \qquad I_m = \frac{\mathcal{E}_m}{Z},$$

(definition of impedance and phase shift)   (14.43)

where the presently unknown angle ϕ is called the *phase shift*, and the presently unknown quantity Z, with dimensions of ohms, is called the *impedance*. I peaks at $t_I^{max} = (T/4) + (\phi/\omega)$. If the current peaks before (after) the emf, then $\phi < 0$ ($\phi > 0$). For $\phi = 0°$, current and voltage are completely in phase, and for $\phi = 180°$, current and voltage are completely out of phase.

For a given circuit, to obtain the actual current at any time, we must thus determine both ϕ and Z. Figure 14.5 also represents the current phasor $\mathbf{I}$, where $I_m \sin(\omega t - \phi)$ is the y-component of $\mathbf{I}$. Note that for Figure 14.5 the voltage across the generator peaks before the current through the circuit so that in this case $\phi > 0$.

The remainder of this chapter considers currents of the form (14.43) and emfs of the form (14.39). Hence, in all phasor diagrams, the phase angle for I is $\omega t - \phi$. Section 14.10 shows that power-absorbing circuits have ϕ in either the first or fourth quadrant, and that power-providing circuits have ϕ in either the second or third quadrants.

Example 14.5 **Determining the impedance and phase shift**

Consider a circuit for which $\mathcal{E}_m = 20$ V and $\omega = 50$ s^{-1}. A measurement yields $I_m = 0.1$ A, and that the time interval $\Delta t = t_{\mathcal{E}}^{max} - t_I^{max}$ between the maxima in $\mathcal{E}$ and I is -0.004 s, corresponding to the current peaking later. Find Z and ϕ.

Solution: Equation (14.43) immediately yields $Z = 20/0.1 = 200$ Ω. Since the current peaks later, $\phi > 0$. Now note that $|\Delta t|/T = |\phi|/2\pi$. Since Δt and ϕ have opposite signs, $\Delta t/T = -\phi/2\pi$. Then $\phi = -(2\pi/T)\Delta t = -\omega\Delta t$. This gives $\phi = 0.2$ rad, or $\phi = 11.46°$. Alternatively, $\Delta t = t_{\mathcal{E}}^{max} - t_I^{max} = T/4 - (T/4 + \phi/\omega) = -\phi/\omega$, so $\phi = -\omega\Delta t$ again.

In what follows, we successively represent the voltage across circuits consisting of a resistor R, an inductor L, and a capacitor C, in terms of the current through the circuit. For more complex series circuits, we add the voltages of each circuit element to obtain the net voltage. From this the current (and thus the impedance Z and the phase ϕ) can be determined. The more complex circuits will be RC, LR, and RLC. The first two can be used for filtering out unwanted signals, and the last can be used for tuning to desired frequencies.

Since current and voltage have different units, in the figures (e.g., Figure 14.5), the relative lengths of current and voltage phasors are not relevant. The phase angle ϕ between $\Delta\mathbf{V}_g$ and $\mathbf{I}$ is defined in a local (rotating) coordinate system $x'y'$. With the x'-axis fixed on the rotating current phasor $\mathbf{I}$, and $\Delta V_{x'}$ and $\Delta V_{y'}$ the components of $\Delta\mathbf{V}_g$ along x' and y', we define

$$\tan\phi = \frac{\Delta V_{y'}}{\Delta V_{x'}}. \qquad (14.43')$$

14.5.1 *Resistor Voltage*

Consider a resistor, as in Figure 14.6(a). The current I across a resistor R with voltage drop $\Delta V_R = V_b - V_a$ is, by Ohm's law, given by

$$I = \frac{\Delta V_R}{R}. \qquad (14.44)$$

On representing the current as in (14.43), Equation (14.44) yields

$$\Delta V_R = IR = I_m R \sin(\omega t - \phi). \qquad \text{(ac voltage across resistor)} \quad (14.45)$$

ΔV_R is represented as a phasor $\Delta\mathbf{V}_R$ by a rotating vector that points along the current, with length $I_m R$. See Figure 14.6(b). We say that the voltage across the resistor is in phase with the current.

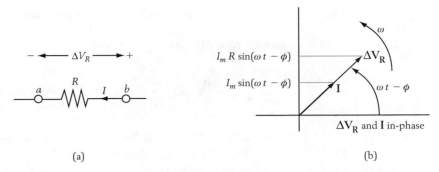

Figure 14.6 Current I and voltage drop ΔV_R across a resistor R. (a) Circuit diagram. (b) Phasor diagram. For a resistor, I and ΔV_R are in phase.

Now consider a circuit containing only a resistor and a generator. That is equivalent to the black box of Figure 14.4 containing only a resistor. How do we find ϕ and Z of (14.43)? The resistor is subject to the full voltage drop across the generator, so $\Delta V_g = \Delta V_R$. Then, by (14.41), $\Delta V_R = \Delta V_g = \mathcal{E}_g = \mathcal{E}_m \sin \omega t$. Hence, using $I_m = \mathcal{E}_m/Z$, (14.45) is satisfied by phase $\phi = 0$ and impedance $Z = R$. *Thus the resistance R is the impedance of a resistor alone.*

14.5.2 *Inductor Voltage*

Now apply Ohm's law to find the voltage ΔV_L across an inductor L of resistance $R_L \to 0$. As usual, I enters the high-voltage side. See Figure 14.7(a). With $\Delta V_L = V_b - V_a$ representing the voltage across the inductor, we write

$$I = \frac{\mathcal{E}_L + \Delta V_L}{R_L} = \frac{1}{R_L}\left(-L\frac{dI}{dt} + \Delta V_L\right). \tag{14.46}$$

In the limit as $R_L \to 0$, (14.46) gives $\Delta V_L = LdI/dt$. Then (14.43) yields

$$\Delta V_L = L\frac{dI}{dt} = I_m\omega L \cos(\omega t - \phi) = I_m\omega L \sin(\omega t - \phi + \pi/2), \tag{14.47}$$

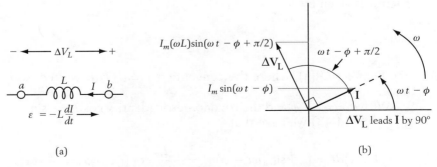

Figure 14.7 Current I and voltage drop ΔV_L across an inductor L. (a) Circuit diagram. (b) Phasor diagram. For an inductor, $\Delta V_L = LdI/dt$ "sees the future" of I, and thus "leads" I by 90°.

where with $\theta = \omega t - \phi$ we used the trigonometric identity $\cos\theta = \sin(\theta + \pi/2)$.

Equation (14.47) may be rewritten as

$$\Delta V_L = I_m X_L \sin(\omega t - \phi + \pi/2), \qquad X_L \equiv \omega L,$$
$$\text{(ac voltage across inductor)} \quad (14.48)$$

where X_L is called the *inductive reactance*. It has the same units as resistance, or ohms. Observe that X_L is small at low frequencies, but large at high frequencies.

ΔV_L is represented as a phasor $\Delta \mathbf{V_L}$ by a rotating vector of length $I_m \omega L = I_m X_L$, with a phase angle 90° greater than that of the current. See Figure 14.7(b). We say that the voltage across the inductor leads the current (by 90°). Since $\Delta V_L = L dI/dt$, the inductor voltage "sees" the future (the slope) of the current.

Now consider a circuit containing only an inductor and a generator. That is equivalent to the black box of Figure 14.4 containing only an inductor. How do we find ϕ and Z of (14.43)? The inductor is subject to the full voltage drop across the generator, so $\Delta V_g = \Delta V_L$. Then, by (14.41), $\Delta V_L = \Delta V_g = \mathcal{E}_g = \mathcal{E}_m \sin\omega t$. Hence, with $I_m = \mathcal{E}_m/Z$, we reproduce (14.48) for phase $\phi = \pi/2$ and impedance $Z = \omega L$. *Thus the inductive reactance $X_L = \omega L$ is the impedance of an inductor alone.*

Example 14.6 Inductive reactance

For $L = 5$ mH, find X_L at $\omega = 10^2$ s^{-1} and $\omega = 10^8$ s^{-1}.

Solution: Equation (14.48) gives $X_L = 0.5$ ohm at $\omega = 10^2$ s^{-1}, and $X_L = 5 \times 10^7$ ohms at $\omega = 10^8$ s^{-1}. Note how X_L increases with increasing frequency ω.

14.5.3 *Capacitor Voltage*

For a capacitor C with charge Q, the voltage drop $\Delta V_C = V_b - V_a$ across a capacitor is given by

$$\Delta V_C = \frac{Q}{C}. \qquad (14.49)$$

See Figure 14.8(a), where the convention $I = dQ/dt$ is used, so that $Q = \int I(t)dt$.

With $\theta = \omega t - \phi$ and the trigonometric identity $\cos\theta = -\sin(\theta - \pi/2)$, for I given by (14.43) we have that

$$Q = \int I(t)dt = I_m \int \sin(\omega t - \phi)dt = -\frac{I_m}{\omega}\cos(\omega t - \phi) = \frac{I_m}{\omega}\sin\left(\omega t - \phi - \frac{\pi}{2}\right).$$
$$(14.50)$$

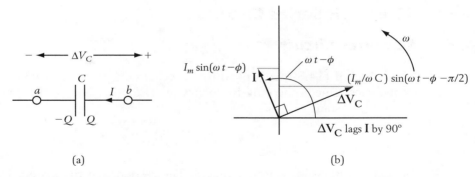

Figure 14.8 Current I and voltage drop ΔV_C across a capacitor C. (a) Circuit diagram. (b) Phasor diagram. For a capacitor, $\Delta V_C = Q/C$ "sees the past" of I ($Q = \int I\,dt$), and thus "lags" I by $90°$.

Hence the voltage drop across the capacitor is given by

$$\Delta V_C = \frac{Q}{C} = \frac{I_m}{\omega C} \sin\left(\omega t - \phi - \frac{\pi}{2}\right). \tag{14.51}$$

This may be rewritten as

$$\Delta V_C = I_m X_C \sin\left(\omega t - \phi - \frac{\pi}{2}\right), \qquad X_C = \frac{1}{\omega C},$$
$$\text{(ac voltage across capacitor)} \quad (14.52)$$

where X_C is called the *capacitive reactance*. It has the same units as resistance, or ohms. Observe that X_C is large at low frequencies, and small at high frequencies.

ΔV_C is represented as a phasor $\Delta \mathbf{V_C}$ by a rotating vector of length $I_m/\omega C$, with a phase angle $90°$ less than that of the current. See Figure 14.8(b). We say that the capacitor voltage lags the current (by $90°$). Since $\Delta V_C = Q/C = \int I\,dt/C$, ΔV_C "sees" the past of the current (the sum over the past).

Now consider a circuit containing only a capacitor and a generator. That is equivalent to the black box of Figure 14.4 containing only a capacitor. How do we find ϕ and Z of (14.43)? The capacitor is subject to the full voltage drop across the generator, so $\Delta V_g = \Delta V_C$. Then, by (14.41), $\Delta V_C = \Delta V_g = \mathcal{E}_g = \mathcal{E}_m \sin \omega t$. Hence, with $I_m = \mathcal{E}_m/Z$, we reproduce (14.52) for phase $\phi = -\pi/2$ and impedance $Z = (1/\omega C)$. *Thus the capacitative reactance $X_C = (\omega C)^{-1}$ is the impedance of a capacitor alone.*

Example 14.7 Capacitive reactance

For $C = 2\ \mu F$, find X_C for $\omega = 10^2\ s^{-1}$ and $\omega = 10^8\ s^{-1}$.

Solution: Equation (14.52) gives $X_C = 5000$ ohms at $\omega = 10^2\ s^{-1}$, and $X_C = 0.005$ ohm at $\omega = 10^8\ s^{-1}$. Note how X_C decreases with increasing frequency ω.

With this as background, we now analyze some more complex circuits.

14.6 *RC* and *LR* Series Circuits

14.6.1 *RC Series Circuit*

When an emf (14.39) is applied to an RC circuit, there are two characteristic frequencies. One is the emf frequency ω. The other is τ_{RC}^{-1}, the inverse of the RC time constant

$$\tau_{RC} = RC \tag{14.53}$$

for this circuit, studied in Chapter 8. The response of the system will depend on the ratio of these two frequencies, ω/τ_{RC}^{-1}, or $\omega\tau_{RC}$. For low frequencies, the capacitor, with the larger low-frequency impedance $(\omega C)^{-1}$, will dominate the system. For high frequencies, the resistor, with the larger high-frequency impedance R, will dominate the system. We now derive the response of the RC circuit given in Figure 14.9(a).

How do we find ϕ and Z of (14.43)? First, we relate ΔV_R and ΔV_C to $\mathcal{E}_m$ of (14.42). The voltage drop across the generator equals the sum of the voltage drops across the resistor and the capacitor, so

$$\Delta V_g = \Delta V_R + \Delta V_C. \tag{14.54}$$

Here, $\Delta V_g = V_b - V_a$, $\Delta V_R = V_c - V_a$, and $\Delta V_C = V_b - V_c$. Use of (14.41) and (14.39) then gives

$$\mathcal{E}_g = \mathcal{E}_m \sin \omega t = \Delta V_g = \Delta V_R + \Delta V_C. \tag{14.55}$$

We will solve for ϕ and Z in two ways.

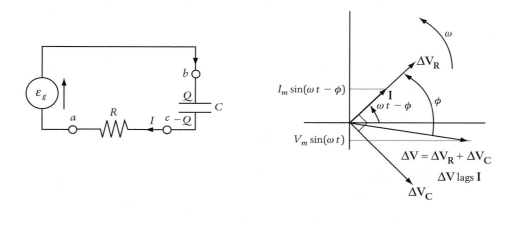

(a)   (b)

Figure 14.9 *RC* series circuit driven by a generator $\mathcal{E}_g$. (a) Circuit diagram. (b) Phasor diagram. The phase angle is intermediate between those for a resistor $(0°)$ and a capacitor $(-90°)$.

14.6.2 *Trigonometric Method*

Using $\sin(A + B) = \sin A \cos B + \sin B \cos A$, with $A = \omega t - \phi$ and $B = \phi$, so $A + B = \omega t$, the left-hand side of (14.55) can be rewritten as

$$\mathcal{E}_m \sin \omega t = \mathcal{E}_m[\sin(\omega t - \phi)\cos \phi + \cos(\omega t - \phi)\sin \phi]. \qquad (14.56)$$

With (14.45), (14.51), and $\sin(\omega t - \phi - \pi/2) = -\cos(\omega t - \phi)$, the right-hand side of (14.55) becomes

$$\Delta V_R + \Delta V_C = I_m\left[R\sin(\omega t - \phi) - \frac{1}{\omega C}\cos(\omega t - \phi)\right]. \qquad (14.57)$$

By (14.55), the coefficients of $\sin(\omega t - \phi)$ and $\cos(\omega t - \phi)$ in (14.56) and (14.57) must be the same. This leads to

$$\mathcal{E}_m \cos \phi = I_m R, \qquad \mathcal{E}_m \sin \phi = -\frac{I_m}{\omega C} = -I_m X_c. \qquad (14.58)$$

Using $\sin \phi / \cos \phi = \tan \phi$, the ratio of the left-hand sides in (14.58) yields

$$\tan \phi = -\frac{1/\omega C}{R} = -\frac{X_c}{R}. \qquad (14.59)$$

Thus ϕ lies in the fourth quadrant, intermediate between 0 (R alone) and $-\pi/2$ (C alone).

The sum of the squares of the two terms in (14.58) gives, with $\sin^2 \phi + \cos^2 \phi = 1$,

$$\mathcal{E}_m^2 = I_m^2\left[R^2 + \left(\frac{1}{\omega C}\right)^2\right] = I_m^2\left(R^2 + X_c^2\right), \qquad (14.60)$$

so

$$Z = \frac{\mathcal{E}_m}{I_m} = \sqrt{R^2 + \left(\frac{1}{\omega C}\right)^2} = R\sqrt{1 + \left(\frac{1}{\omega \tau_{RC}}\right)^2}. \qquad (14.61)$$

14.6.3 *Phasor Method*

In terms of the phasor representation in Figure 14.9(b), by (14.55) we must add the voltage phasors $\Delta \mathbf{V}_R$ and $\Delta \mathbf{V}_C$ across the resistor and the capacitor to obtain the total voltage phasor $\Delta \mathbf{V}_g$ across the generator. Since $\Delta \mathbf{V}_R$ and $\Delta \mathbf{V}_C$ are 90° out of phase, they add vectorially. By the Pythagorean theorem, the resultant phasor has a length squared $I_m^2 R^2 + I_m^2 X_C^2$, which must equal $\mathcal{E}_m^2$. As expected, this is the same as (14.60). We obtain the phase angle ϕ from the fact that its tangent is the ratio of the phasor voltage component along the local (rotating) y'-axis to the phasor voltage component along the local (rotating) x'-axis. By definition, the voltage $I_m R$ is along the local (rotating) x'-axis, so the voltage $-I_m/\omega C = -I_m X_c$ is along the local (rotating) y'-axis. The ratio, by (14.43'), gives (14.59). From (14.59) for ϕ and (14.61) for Z, the total current at any time can be obtained using (14.43).

Example 14.8 An *RC* circuit

Consider the values $R = 10$ ohms and $C = 2$ μF, for which Example 14.7 applies. Which circuit element dominates at low frequencies, and which dominates at high frequencies?

Solution: Since (14.52) gave $X_C = 5000$ ohms at $\omega = 10^2$ s^{-1}, and $X_C = 0.005$ ohm at $\omega = 10^8$ s^{-1}, the capacitor dominates at the lower frequency, and the resistor dominates at the higher frequency. The reasons are simple: at low frequencies the capacitor has time to charge up, thus blocking current from flowing; at high frequencies the capacitor does not have enough time to charge up, so the resistor dominates.

14.6.4 *LR Series Circuit*

When an emf (14.39) is applied to an LR circuit, there are two characteristic frequencies. One is the frequency ω. The other is τ_{LR}^{-1}, the inverse of the LR time constant

$$\tau_{LR} = \frac{L}{R} \tag{14.62}$$

for this circuit, studied in Chapter 12. The response of the system will depend on the ratio of these two frequencies, ω/τ_{LR}^{-1}, or $\omega\tau_{LR}$. For low frequencies the resistor, with the larger low-frequency impedance R, will dominate the system. For high frequencies the inductor, with the larger high-frequency impedance ωL, will dominate the system. We now derive the response of the LR circuit given in Figure 14.10(a).

How do we find ϕ and Z of (14.43)? First we relate ΔV_R and ΔV_L to $\mathcal{E}_m$ of (14.42). The voltage drop across the generator equals the sum of the voltage

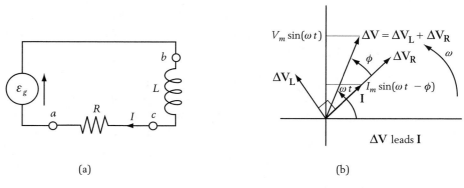

(a) (b)

Figure 14.10 LR series circuit driven by a generator $\mathcal{E}_g$. (a) Circuit diagram. (b) Phasor diagram. The phase angle is intermediate between those for a resistor (0°) and an inductor (90°).

drops across the resistor and the inductor, so

$$\Delta V_g = \Delta V_R + \Delta V_L. \tag{14.63}$$

Here $\Delta V_g = V_b - V_a$, $\Delta V_R = V_c - V_a$, and $\Delta V_L = \Delta V_b - \Delta V_a$. Use of (14.41) and (14.39) then gives

$$\mathcal{E}_g = \mathcal{E}_m \sin \omega t = \Delta V_g = \Delta V_R + \Delta V_L. \tag{14.64}$$

We present only the phasor solution. Here we must add the voltage drops across the resistor and the inductor. Since these are at $90°$ to each other, by (14.64) and (14.43), the phasors add vectorially to give a quantity of length

$$\mathcal{E} = I_m Z = I_m \sqrt{R^2 + (\omega L)^2} = I_m \sqrt{R^2 + X_L^2}. \tag{14.65}$$

Thus

$$Z = \sqrt{R^2 + (\omega L)^2} = R\sqrt{1 + (\omega \tau_{LR})^2}. \tag{14.66}$$

See Figure 14.10(b). We obtain the phase angle ϕ from the fact that its tangent is the ratio of the phasor voltage component along the local (rotating) y'-axis to the phasor voltage component along the local (rotating) x'-axis. By definition, the voltage $I_m R$ is along the local (rotating) x'-axis, so the voltage $I_m \omega L = I_m X_L$ is along the local (rotating) y'-axis. Thus, by (14.43′),

$$\tan \phi = \frac{\omega L}{R} = \frac{X_L}{R}. \tag{14.67}$$

From (14.66) for ϕ and (14.67) for Z, the total current at any time can be obtained using (14.43).

Example 14.9   An *LR* circuit

Consider the values $R = 10$ ohms and $L = 5$ mH, for which Example 14.6 applies. Which circuit element dominates at low frequencies, and which dominates at high frequencies?

Solution: Since (14.48) gave $X_L = 0.5$ ohm at $\omega = 10^2$ s^{-1}, and $X_L = 5 \times 10^7$ ohms at $\omega = 10^8$ s^{-1}, the resistor dominates at the lower frequency, and the inductor dominates at the higher frequency. The reasons are simple: at low frequencies, the current is changing only very slowly, so the inductor is not very effective at producing an opposing emf $-L dI/dt$; at high frequencies, the current is changing very rapidly, so the inductor is effective.

Food for Thought: How would you treat a resistor and inductor in parallel? A resistor and capacitor in parallel? A capacitor and an inductor in parallel? The last two cases are relevant to parasitic capacitance.

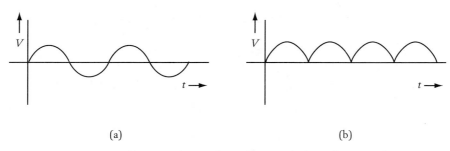

(a)   (b)

Figure 14.11 (a) Oscillating voltage. (b) Fully rectified oscillating voltage.

14.7 Rectifying and Filtering AC Voltages

Often, only ac voltage is available in situations where dc voltage is desired. Thus, somehow the ac voltage must be *rectified*, or converted to a dc voltage. Amplifiers use dc voltage as part of the input to a non-ohmic (i.e., nonlinear) device. An additional weak signal voltage can then produce a relatively large change in the output current. Certain motors, such as those used in battery-operated toys, also use dc power.

As discussed in the previous chapter, one way to produce a dc voltage from an ac voltage, as in Figure 14.11(a), is to use a *commutator*, which switches the connections every half-cycle. Another way is to use a set of diodes. Either way leads to an output that, in the ideal case (where the contacts do not cause any complications), consists of the absolute value of sine curves, as in Figure 14.11(b).

Such output can more closely approximate dc power if we employ a more complicated set of coils and commutators. If there are twice as many coils, then the signal will be picked up in its positive phase every 90° (rather than every 180°). That is an improvement, but it is still far from dc. The Gramme dynamo, or generator, of the 1870s, used at least 16 coils and commutators (i.e., every $180°/16 = 11.25°$), thus producing a much smoother voltage. (At relatively low power, rectification can be performed with electronic tubes or transistor devices called diodes.) To obtain an even clean or dc voltage, with a minimum of so-called *ripple* voltage, we *filter out* the time-varying parts of the voltage. We speak of *low-pass* and *high-pass* filters, by which we mean that the filter passes either low-frequency signals or high-frequency signals.

Filters also eliminate unwanted signals. For example, when we tune in a radio or TV, we don't want to hear two stations (or more) at the same time. A third common use of filtering is to eliminate unwanted static or hiss. Bass and treble controls, or the more sophisticated graphic equalizers, found on many stereo systems, are filtering devices. (Five frequency ranges may be associated with a graphic equalizer, each frequency range defined by a combination of a high-pass and a low-pass filter.) Note that *signal* really indicates a *signal voltage*.

14.7.1 *Inverse Relationship between Response in Frequency and in Time*

Because of (14.2), which shows that the period of an oscillation equals the inverse of its frequency, it should not be surprising to learn that the response of a circuit to a given frequency is related to its response within the inverse of a given time. That is, if a circuit responds well at high frequencies, it will respond well to signals that change significantly only within a short amount of time; if a circuit responds well at low frequencies, it will respond well to signals that change significantly only over a long amount of time. We will use this result in what follows.

14.7.2 *RC Circuit as Filter*

Consider Figure 14.9(a). By (14.45), the voltage $\Delta V_R = IR$ across a resistor is proportional to the current at that instant of time. By (14.49) the voltage $\Delta V_C = Q/C = \int I \, dt/C$ across a capacitor in series with R is proportional to the charge at that instant of time. *Because the charge is the time integral of the current, the capacitor largely sees the past.* The capacitor voltage has a certain inertia, because it cannot change instantly. (This is a general statement that includes the fact that, for ac voltages, the capacitor voltage ΔV_C lags the current I entering and leaving the plates of C.) By putting a voltage across the resistor and capacitor in series, and then picking up only the voltage across one, it is possible to emphasize either the present or the past.

For signals at a low frequency, the capacitor has time to charge up, and its voltage will dominate the circuit. Thus, picking up the signal from the capacitor in this circuit would smooth out a rippled dc signal from a generator.

For signals at a high frequency, the capacitor does not have enough time to charge up, so it will not develop much voltage. Thus, picking up the signal from the resistor in this circuit would decrease the amount of low-frequency noise, such as "leakage" from a 60 Hz power line or unwanted dc voltage.

> ### Picking Up a Signal
>
> When we say "pick up a signal," we mean that we use a measuring device of very high resistance (or very high impedance), so that it draws negligible current. Otherwise, we would have to include the measuring device as part of the circuit.

Example 14.10 **An *RC* circuit**

In an RC circuit subject to a maximum voltage $\mathcal{E}_m$, find the maximum voltage ΔV_{Cm} across the capacitor.

Solution: Use of (14.52) for ΔV_C, (14.43) for I_m, and (14.61) for Z gives, with $X_C/Z = X_C/\sqrt{R^2 + X_C^2} = 1/\sqrt{1 + (R/X_C)^2}$,

$$\Delta V_C = \mathcal{E}_m \frac{1}{\sqrt{1 + (\omega RC)^2}} \sin(\omega t - \phi - \pi/2).$$

Thus $\Delta V_{Cm} = \mathcal{E}_m/\sqrt{1 + (\omega RC)^2}$. Since ΔV_{Cm} is finite at low ω, but goes to zero at high ω, the capacitor picks up the voltage at low frequencies, but not at high frequencies. The capacitor is thus a low-pass filter for the RC circuit. Correspondingly, you may show that the resistor is a high-pass filter for the RC circuit, with a maximum voltage given by $\Delta V_{Rm} = \mathcal{E}_m(\omega RC)/\sqrt{1 + (\omega RC)^2}$. The phase ϕ is given by (14.59).

14.7.3 *LR Circuit as Filter*

By (14.45), the voltage $\Delta V_R = IR$ across a resistor is proportional to the current at that instant of time. By (14.47), the voltage $\Delta V_L = L(dI/dt)$ across an ideal inductor is proportional to the rate of change of the current at that instant of time. In a certain sense the inductor sees the (immediate) future. (This is a general statement that includes the fact that, for ac signals, the voltage ΔV_L leads the current I going into L.) By putting a voltage across the resistor and inductor in series, and then picking up only the voltage across one, it is possible to emphasize either the present or the future.

For signals at a low frequency, the resistor voltage IR will dominate the circuit. Thus, taking the signal off the resistor in this circuit would smooth out a rippled dc signal from a generator.

For signals at a high frequency, the inductor voltage LdI/dt will dominate the circuit. Thus, taking the signal off the inductor in this circuit would decrease the amount of low-frequency noise.

Example 14.11 An *LR* circuit

In an LR circuit subject to a maximum voltage $\mathcal{E}_m$, find the maximum voltage ΔV_{Lm} across the inductor.

Solution: Use of (14.48) for ΔV_L, (14.43) for I_m, and (14.66) for Z gives, with $X_L/Z = X_L/\sqrt{R^2 + X_L^2} = (X_L/R)/\sqrt{1 + (X_L/R)^2}$,

$$\Delta V_L = \mathcal{E}_m \frac{(\omega L/R)}{\sqrt{1 + (\omega L/R)^2}} \sin(\omega t - \phi + \pi/2).$$

Thus $\Delta V_{Lm} = \mathcal{E}_m(\omega L/R)/\sqrt{1 + (\omega L/R)^2}$. Since ΔV_L is finite at high ω, but goes to zero at low ω, this illustrates that the inductor picks up the voltage at high frequencies, but not at low frequencies. The inductor is thus a high-pass filter for the LR circuit. Correspondingly, you may show that the resistor is a low-pass filter for the LR circuit, with a maximum voltage given by $\Delta V_{Rm} = \mathcal{E}_m/\sqrt{1 + (\omega L/R)^2}$. The phase ϕ is given by (14.67).

14.8 *RLC* Resonance: Tuning AC

We shall show later in this section that an RLC circuit, as in Figure 14.1, has its largest response at the resonant frequency

$$\omega_0 = \sqrt{\frac{1}{LC}} \tag{14.68}$$

of a circuit with only a capacitor and an inductor. Section 14.10 will show that by varying either L or C (or both), we thus can tune the frequency of the circuit to pick up a range of frequencies around the resonant frequency. The range of frequencies $\Delta\omega$ for which the circuit responds well is proportional to

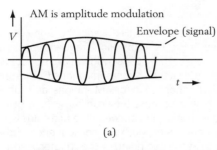

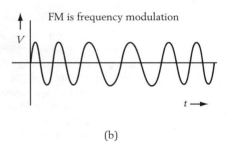

(a)

(b)

Figure 14.12 (a) Amplitude modulated signal (AM). The frequency is constant at the transmitter frequency (e.g., 1210 KHz), but the amplitude is modulated at much lower acoustic frequencies (up to about 10 KHz). (b) Frequency modulated signal (FM). The amplitude is constant, but the frequency is modulated around the transmitter frequency (e.g., 89.1 MHz), by perhaps 20 KHz.

the resistance R, as we will derive in (14.76). For small R, the response is very narrow about ω_0, and the circuit is very sensitive to the signals that are picked up. For large R, the response is very broad about ω_0, and the circuit is not as sensitive. One therefore might think that it is always advantageous to have a small resistance. That is not the case. The actual signal is contained in modulations of the carrier signal, and this can be described as a range of frequencies around the carrier frequency. Let the RLC circuit be used as a radio tuner. If the tuner response is too narrow in frequency, it will be able to pick up very weak signals, but it will not pick up enough of the high-frequency part of the radio signal. If the tuner response is too broad, then it might simultaneously pick up two stations. Note that there are two ranges, or bands, of commercial radio signals, *AM* and *FM*.

Let us now give a more detailed discussion of the RLC series circuit, depicted in Figure 14.13(a).

We present only the phasor solution. Rather than, for example, (14.55) for the RC circuit, or (14.64) for the LR circuit, for the RLC circuit

$$\mathcal{E}_g = \mathcal{E}_m \sin \omega t = \Delta V_R + \Delta V_L + \Delta V_C. \tag{14.69}$$

In Figure 14.13(b), the phasors for the capacitor and inductor point opposite to

AM Stands for Amplitude Modulation

The AM band corresponds to carrier frequencies running from about 55 KHz to 1600 KHz. Here, the frequency of the carrier is fixed, and its amplitude is modulated by the audio signal (at much lower frequencies): the stronger the audio signal, the greater the amplitude of the carrier signal; the higher the frequency of the audio signal, the faster the modulation. See Figure 14.12(a). A radio receiver uses a tuned RLC circuit to filter out frequencies not within a narrow frequency range centered about a particular carrier frequency. It then feeds the remaining signal through a circuit that responds too slowly to detect the carrier but fast enough to detect the *envelope* of the carrier (i.e., the audio signal). A rectifier, as in Figure 14.11(b), coupled with a low-pass filter, as in Example 14.10, will perform this function.

FM Stands for Frequency Modulation

The FM band runs from about 88 MHz to 108 MHz. Here, the signal is carried by modulating the carrier frequency. See Figure 14.12(b). High-frequency peaks are closer together in time than low-frequency peaks. The same considerations about narrowness and breadth hold for FM as for AM. The Federal Communications Commission (FCC) has reserved a larger bandwidth for FM stations than for AM stations; as a consequence, the sound quality of FM stations is higher because they can carry more of the higher frequencies than the AM stations. This relationship between the faithfulness, or *fidelity,* of sound reproduction, and the frequency bandwidth is a consequence of the mathematical theorem of Fourier, referred to at the beginning of the chapter. Another reason FM sound quality is higher is that there is much less "static" (due to local atmospheric disturbances) in the FM band than in the AM band.

one another. By the Pythagorean theorem, adding all three phasors vectorially gives a net phasor of length

$$\mathcal{E}_m = I_m Z = I_m \sqrt{R^2 + \left(\omega L - \frac{1}{\omega C}\right)^2}, \tag{14.70}$$

so that

$$Z = \sqrt{R^2 + \left(\omega L - \frac{1}{\omega C}\right)^2} = \sqrt{R^2 + (X_L - X_C)^2}.$$

$$\text{(impedance of } RLC \text{ circuit)} \quad (14.71)$$

Here X_L and X_C are defined in (14.48) and (14.52). We obtain the phase angle ϕ as follows. By definition, the voltage $I_m R$ is along the local (rotating) x'-axis, the voltage $I_m/\omega C$ is along the local (rotating) $-y'$-axis, and the voltage $I_m(\omega L)$

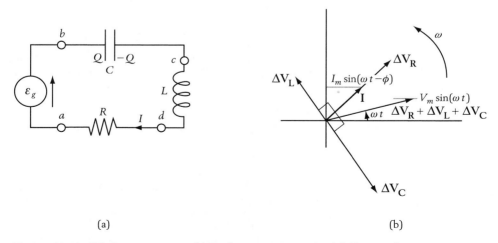

(a) (b)

Figure 14.13 RLC series circuit driven by a generator $\mathcal{E}_g$. (a) Circuit diagram. (b) Phasor diagram. The phase angle is intermediate between those for an inductor (90°) and a capacitor (−90°).

is along the local (rotating) y'-axis. Thus, by (14.43'),

$$\tan\phi = \frac{\omega L - 1/\omega C}{R} = \frac{X_L - X_C}{R}. \quad \text{(phase shift of } RLC \text{ circuit)} \quad (14.72)$$

See Figure 14.13(b). With (14.71) for Z and (14.72) for ϕ, the total current may be obtained from (14.43). For $\omega L \to 0$, we obtain the results for an RC circuit; for $(\omega C)^{-1} \to 0$, we obtain the results for an LR circuit. At the resonance frequency $\omega = \omega_0$, $X_L - X_C = 0$, so $Z = R$, the minimum possible value for Z. By (14.72), $\phi = 0$ on resonance.

For a given input voltage (which might represent the signal from a radio antenna), the maximum current occurs when the impedance is at a minimum. This is at resonance, where $Z = R$, so the full input voltage is across the resistor. Nevertheless, for a circuit of low resistance, at resonance the voltage maximum $I_m/\omega_0 C = I_m X_C$ across the capacitor (and its negative across the inductor) is usually much larger than the voltage maximum $I_m R$ across the resistor.

Example 14.12 **An *RLC* circuit**

Consider a circuit with the familiar values $L = 5$ mH and $C = 2$ μF, and with $R = 10$ Ω, driven at the frequency $\omega = 0.5 \times 10^4$ s^{-1}. Consider the power source to provide a maximum emf of $\mathcal{E}_m$. Find ω_0, R_c, Q, X_L, X_C, Z, ϕ, the time between voltage and current peaks, and the maximum voltage across each circuit element.

Solution: Equation (14.68) yields the resonant frequency $\omega_0 = 10^4$ s^{-1}, and (14.28) yields the critical resistance $R_c = 100$ ohms. Note that ω is half the resonance frequency ω_0, so the capacitor should dominate the inductor. By (14.31), the circuit has $Q = 5$. By (14.48), $X_L = \omega L = 25$ ohms, and by (14.52), $X_C = 1/\omega C = 100$ ohms. Then, by (14.70), $Z = \sqrt{10^2 + (25 - 100)^2} = 75.7$ ohms. Hence the capacitor dominates the circuit, but the other circuit elements certainly contribute. By (14.72), $\tan\phi = (X_L - X_C)/R = 25 - 100/10 = -7.5$, so $\phi \approx -82°$, or -1.438 radians. Let us set $\omega t^* = \phi$. By (14.39) and (14.43), the applied voltage peaks at a time $t^* = \phi/\omega = -1.438 \times 10^{-4}$ s before the current peaks; here, the negative sign means that the current peaks before the voltage, as expected if the capacitor dominates. Note that at this frequency $\Delta V_{Rm} = (R/Z)\mathcal{E}_m = 0.132\mathcal{E}_m$, $\Delta V_{Lm} = (\omega L/Z)\mathcal{E}_m = 0.33\mathcal{E}_m$, and $\Delta V_{Cm} = (1/\omega CZ)\mathcal{E}_m = 1.32\mathcal{E}_m$. Thus the maximum voltage across the capacitor is larger than the maximum applied voltage; however, the capacitor and inductor voltages oppose one another.

14.9 Principles of Amplification

Optional

Tuners would not be useful if, after a signal were received and tuned in, it could not also be *amplified*. This makes a good place to discuss amplifiers.

The first electronic amplifiers were called *valves* because their principle of operation was very similar to those for valves associated with water flow. There, for one valve setting—off—the valve prevents any water from passing. For another setting, all the water gets through. For an intermediate setting, a small change in the valve setting causes a very large change in the output.

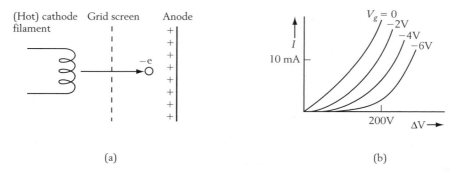

Figure 14.14 Amplification using a triode. (a) Electrons are emitted from the hot cathode. The circuit is completed by the positively charged anode (in modern semiconductor language, the "collector"). The current is sensitive to the voltage V_g of the grid screen. (b) Current versus anode voltage, for a family of grid screen voltages, illustrating the sensitivity of the current to the grid voltage.

A standard and easily visualized electrical amplification device is the triode. See Figure 14.14(a). There, within a vacuum tube, a negatively charged cathode emits electrons, which travel to the positively charged anode. Between them is a screen called a *grid*, which serves as the valve. Let the cathode-to-anode voltage ΔV be fixed, as in Figure 14.14.(b). (ΔV can be hundreds or even thousands of volts.) To prevent a current I so large that the triode burns out, the grid voltage V_g is given a negative value, which somewhat suppresses the current because the electrons are then repelled by the grid. Together, ΔV and V_g define the *operating point* of the triode. For a triode, the input consists of a small change ΔV_g in the grid voltage. The output consists of the change in IR voltage across an external resistor R through which the anode current I flows. Amplification occurs because small changes ΔV_g in the grid voltage cause large changes ΔI in the anode current. The amplification is proportional to $\Delta I / \Delta V_g$ and can correspond to a very large conductance (i.e., inverse resistance). (Small changes in ΔV do not cause large changes in I.) In the 1880s, Edison noticed that heating the cathode increases the current, but heating the anode has no such effect. (Edison didn't know this, but heating the cathode helps the charge carriers, electrons, escape; heating the anode has no such effect. This was an early clue that the charge carriers had negative charge.) Modern semiconductor electronics use a variety of devices whose ancestor was the triode.

14.10 Power and Power Factor

Let us now return to our discussion of tuners, in the broader context of the efficiency of power utilization.

When the electric company sends power to a factory, it wants to ensure that a significant portion of that power is actually used by the factory. Otherwise, all that happens is that the power company uses energy in the wires ($I^2 R$ heating) that go to and from the factory. For example, if the factory were simply a large stupid capacitor, then the capacitor would simply charge and discharge, with no actual use of power. Clearly, the electric company must characterize the effectiveness, or efficiency, of power utilization. It measures this with a quantity

Black-and-White Television and Computer Monitors

Year 2000 television screens and computer monitors typically use 25 keV electron beams (e.g., $\Delta V = 2500$ V) produced by the cathode in a vacuum tube to "paint" a succession of images on the screen, which is covered with phosphor spots that emit light when hit by the electron beam. Because electrons and electronics respond much more quickly than the eye, which can only resolve times on the order of 1/30 of a second, images are painted on the screen every 1/30 of a second. The image consists of two "interleaved" partial images of horizontal lines (odd numbered and even numbered) that are successively painted, each in 1/60 of a second. The horizontal sweep, with about 440 dots per line, is controlled by the magnetic field produced by electric current in a set of field coils above and below the tube. The vertical sweep (which enables the electron beam to go from one horizontal line to the next) is controlled by the magnetic field produced by electric current in a set of field coils on each side of the tube. Both of these sweeps are automatic, no matter what the television signal. Brightness is controlled by the grid voltage. The first 20 lines of each partial image are not visible because they correspond to the time it takes to go from the last line in one partial image to the first line in the next partial image. The final result is an image with a total of 485 visible lines from both partial images.

called the *power factor*, which is the ratio of the power used by the factory to the maximum power that it could use.

In this case, the average power absorbed by the system is given by the time average of the product of the input voltage $V_{in} = V_m \sin \omega t$ and the current $I = I_m \sin(\omega t - \phi)$. Thus, with an overbar denoting the time average,

$$\overline{\mathcal{P}} = I_m V_m \overline{\sin \omega t \sin(\omega t - \phi)}$$

$$= I_m V_m \overline{\sin^2 \omega t \cos \phi - \sin \omega t \cos \omega t \sin \phi} = \frac{I_m V_m}{2} \cos \phi. \qquad (14.73)$$

Here we have used $\overline{\sin^2 \omega t} = 1/2$ and $\overline{\sin \omega t \cos \omega t} = (1/2)\overline{\sin 2\omega t} = 0$ for the averages over a period. We call

$$\frac{\overline{\mathcal{P}}}{\overline{\mathcal{P}}_{max}} = \cos \phi, \qquad \left(\overline{\mathcal{P}}_{max} \equiv \frac{I_m V_m}{2} \right) \qquad \text{(power factor)} \qquad (14.74)$$

Color Television and Computer Monitors

These use three separate electron beams, one for the red phosphors, one for the greens, and one for the blues (hence RGB). To help ensure that the red electron beam doesn't hit the blue phosphors, a shadow mask separates the phosphors on the screen. This mask must conduct electricity (from the electron beam's charge), tolerate high temperatures (from the electron beam's energy), and remain aligned with the phosphors as it heats up and cools down. The most suitable material for this mask is *Invar,* so named because its length is nearly *invariant* to temperature changes—its coefficient of thermal expansion is very small. Invar has the difficulty that it can be magnetized, which can affect the focusing of the electron beams. For this reason, many color television sets and monitors have *degaussing coils* to generate a magnetic field that automatically demagnetizes the shadow mask.

> ### Power Factors
>
> For power to be absorbed, the power factor must be positive, so ϕ must lie in the first or fourth quadrant. Circuits such as those we have studied all have such a ϕ. If the power factor is negative, then the circuit must be providing power (in which case, we say that it is *active*), and ϕ lies in the second or third quadrant. Factories that provide, rather than use, electrical power have negative power factors.

the *power factor* for the system. (Here $\overline{\mathcal{P}}_{max}$ is the maximum power utilization possible.) The power factor is the efficiency of power utilization. Note that $\cos\phi = 0$ for either a capacitor ($\phi = -\pi/2$) or an inductor ($\phi = \pi/2$). If I_m is out of phase ($\phi = \pm\pi/2$), the power factor is zero; if I_m is in phase ($\phi = 0$), the power factor is unity.

14.10.1 *Power Absorbed by an RLC Circuit*

Let us now return to the problem of a radio tuner, and of the power absorbed by an RLC circuit as a function of frequency for fixed V_m. By (14.74), with $I_m = V_m/Z$ and $\cos\theta = R/Z$, and (14.71) for Z,

$$\overline{\mathcal{P}} = \frac{1}{2}V_m^2\frac{R}{R^2 + (X_L - X_C)^2}. \tag{14.75}$$

This has a maximum when $X_L = X_C$, which occurs for $\omega = \omega_0 = (LC)^{-\frac{1}{2}}$. At resonance, $Z = R$ and (14.75) shows that $\overline{\mathcal{P}}$ reaches its maximum value $\overline{\mathcal{P}}_{max} = V_m^2/2R$. If the frequency is off-resonance, either above or below, the power absorption is reduced. See Figure 14.15, which plots $\overline{\mathcal{P}}/\overline{\mathcal{P}}_{max}$ versus frequency for a circuit with the familiar values $L = 5$ mH and $C = 2\ \mu$F. Recall from Examples 14.2 and 14.3 that these lead to a resonant frequency $\omega_0 = \sqrt{1/LC} = 10^4\ \text{s}^{-1}$, and a critical resistance $R_c = 2\sqrt{L/C} = 100$ ohms. Three different values for the resistance R are considered. As expected, the larger the resistance, the broader the resonance.

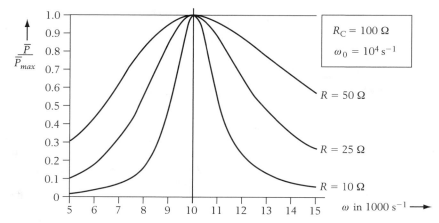

Figure 14.15 Resonance curve for power absorption by an RLC circuit. The larger the resistance R, the broader the resonance curve.

When the resonance is sufficiently narrow, the frequency shift needed to reduce the power absorption by a factor of two is the same on both sides of the resonance. It is given by requiring that $X_L - X_C = R$. Let us set $\omega = \omega + \Delta\omega$, and use $\omega^{-1} \approx \omega_0^{-1} - \omega_0^{-2}\Delta\omega$, which holds for $\Delta\omega \ll \omega_0$. Since $X_L - X_C = 0$ when $\omega = \omega_0$, for ω *near* ω_0 we have

$$X_L - X_C = \omega L - \frac{1}{\omega C} \approx (\Delta\omega)L + \frac{1}{\omega_0^2 C}(\Delta\omega) = 2(\Delta\omega)L.$$

Equating this to R yields $2(\Delta\omega)L = R$, or

$$\Delta\omega = \frac{R}{2L}. \tag{14.76}$$

For this frequency shift (either above or below resonance), the power absorbed is half the power absorbed at resonance. This assumes that V_m does not vary with frequency. $BW \equiv 2\Delta\omega$ is called the *half-power bandwidth*.

With (14.76), we may rewrite the quality factor of (14.31) as

$$Q = \frac{\omega_0 L}{R} = \frac{\omega_0}{2\Delta\omega}. \tag{14.77}$$

This may also be written as $Q = \omega_0/BW$.

Example 14.13   Bandwidth

Using values from Example 14.12, find the ω'_s at which the power is at half-maximum, and the half-power bandwidth $2\Delta\omega$.

Solution: Equation (14.77) gives $\Delta\omega = 10/(2 \cdot 5 \times 10^{-3}) = 0.1 \times 10^4 \, \text{s}^{-1}$. Hence, for $\omega = \omega_0 \pm \Delta\omega = 10^4 \pm 0.1 \times 10^4 \, \text{s}^{-1}$, giving $\omega_- = 0.9 \times 10^4 \, \text{s}^{-1}$ and $\omega_+ = 1.1 \times 10^4 \, \text{s}^{-1}$, the power is at half-maximum. The half-power bandwidth is $0.2 \times 10^4 \, \text{s}^{-1}$.

14.10.2   *Net Power Factor Unity and Resistance Matching for Maximum Power Transfer*

The previous discussion considered the power source to have negligible impedance. In some applications, this is not a valid approximation. In particular, to obtain maximum power transfer, we must include the impedance of the power source. For a power supply in the laboratory, with impedance Z_g and phase shift ϕ_g, and a system with impedance Z_S and phase shift ϕ_S, we can pose the question of when the maximum power transfer will occur to the system.

Rather than derive the answer in general, we will discuss one specific system, and show what the two conditions must be. Consider that the power source is an LR circuit with an emf and resistance R_g, and that the system is an RC circuit with resistance R_S. This combined circuit is equivalent to an RLC circuit with a resistance $R = R_g + R_S$. For an RLC circuit, the maximum current occurs on resonance, when the inductive and capacitive reactances are equal. This corresponds to a power factor of unity. Once this first condition is met, we obtain maximum power transfer when a second condition, $R_g = R_S$, is met, as discussed in Chapter 8. These two conditions—net power factor unity and equal resistances—must hold for maximum power transfer from a power source to an ac circuit.

14.11 Transforming AC

Electrical transformers use the mutual inductance between two circuits to *transform ac voltages*, either to higher or to lower values. However, the ac frequency is unaffected; 60 Hz remains 60 Hz. After electricity is generated at an electrical power plant, it's then *stepped up* to a high voltage for transmission along power lines, of resistance R. The extra cost of insulation for high voltages is outweighed by the decreased $I^2 R$ heating in the power lines. The electrical signal is then *stepped down* in voltage for usage in the home or factory.

Fluorescent tubes contain what is called a *ballast*, which contains a step-up transformer that produces the high ac voltage required to initiate startup of the tube. Doorbells and external house lighting usually are powered by step-down transformers, which provide a lower and safer voltage in the event of a short in the circuit.

14.11.1 *The AC Transformer*

Our discussion of transformers will neglect both eddy current loss and hysteresis loss. As discussed in Chapter 13, the latter can be made relatively small. We also neglect the resistive losses in the circuit of the primary, thus assuming that $R_P = 0$. Hence the only source of loss is the resistance R_S in the secondary. The primary is the side associated with the power source $\mathcal{E}_g$, and the secondary is the side associated with the electrical load R_S. See Figure 14.16. Let N_P turns of the primary and N_S turns of the secondary be wrapped around the iron of the transformer core. Also, we assume that the iron core perfectly retains the magnetic flux so that the flux (i.e., the number of field lines) crossing a turn of the primary is the same as the flux crossing a turn of the secondary. (Note: The mode of operation of an ac transformer is a consequence of Faraday's law. Thus it applies only to ac voltages and current, with their associated time-varying magnetic fields, not to dc voltages and currents, with their associated constant magnetic fields.)

The voltages V_P and V_S across the primary and secondary are caused by the induced emfs $\mathcal{E}_P$ and $\mathcal{E}_S$. Neglecting the resistance in the wire of the primary

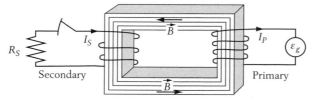

Figure 14.16 An ac transformer, with generator $\mathcal{E}_g$ powering the primary. The primary and secondary typically have different numbers of turns; their relative voltage is proportional to their turns ratio.

and secondary, then just as for the ac generator of (14.41), the voltages and emfs are instantaneously opposite. In terms of their amplitudes, they are the same: $V_P = \mathcal{E}_P$ and $V_S = \mathcal{E}_S$.

Now note that the same magnetic flux passes through each turn of both primary and secondary. (This observation makes it possible to avoid explicitly introducing the self-inductance and mutual inductance of the primary and secondary.) By Faraday's law, the emf is proportional to the rate of change of the magnetic flux. Thus the voltage per turn is the same in both primary and secondary, or

$$\frac{V_P}{N_P} = \frac{V_S}{N_S}.$$ (14.78)

When the secondary has fewer (more) turns, it is called a step-down (step-up) transformer. Equation (14.78) holds for any value of the load resistance R_S in the secondary.

Example 14.14 **Step-down transformer**

It is desired to step down 480 V power at 50 Hz to 12 V. (a) Find the turns ratio between primary and secondary. (b) Find the frequency of the voltage in the secondary.

Solution: (a) Equation (14.78) gives $N_P/N_S = V_P/V_S = 480/12 = 40$. (b) Both primary and secondary voltage are at 50 Hz.

Application 14.1 **Why High Voltage Power Lines?**

Via an ideal transformer, let a power station provide ac power with maximum current I_{Pm} through wires of resistance R_P to a load resistor R_S carrying maximum current I_{Sm}. The efficiency is $I_{Sm}^2 R_S/(I_{Sm}^2 R_S + I_{Pm}^2 R_P)$, or $[1 + (I_{Pm}/I_{Sm})^2 (R_P/R_S)]^{-1}$. [See (8.11).] For fixed secondary properties R_S and I_{Sm}, the efficiency is maximized both by minimizing R_P/R_S—thick wires—and by minimizing I_{Pm}/I_{Sm}—the power lines carry a relatively low current. The next subsection shows that, for power transfer by a transformer, the high voltage side has the lower current. Hence, power lines should be on the high voltage side.

14.11.2 ***The Effective Resistance of an AC Transformer***

Optional

An important relationship holds between the voltage changes V_P and V_g across the power source and the primary. Since we treat the primary circuit as having no resistance, $V_P = V_g$. This result also holds for any value of the load resistance R_S in the secondary.

Now consider the case where the switch in the secondary is not connected. Then $R_S \to \infty$, and $I_S = 0$. In that case, there is no way to dissipate energy, so the power factor of (14.73) for the primary must be zero. This means that the current I_P and the voltage V_P associated with the primary must be out of phase by 90°. Our earlier analysis of an inductor connected to an ac power source yields, by (14.48), that I_P has an amplitude $I_{Pm} = V_{Pm}/\omega L_P = V_{gm}/\omega L_P$.

Next consider what happens when the switch in the secondary is connected. Now a current i_S is induced in the secondary, so there will be resistive losses in the secondary. Because of this, and because of energy conservation, the primary must provide an amount of energy equal to that dissipated in R_S. From energy conservation and (14.78), we can determine i_S, without having to obtain a detailed solution for the two coupled circuits.

When the secondary is connected, so R_S is finite, by Ohm's law the current through it is driven by V_S:

$$i_S = \frac{V_S}{R_S}. \tag{14.79}$$

Since i_S and V_S are in phase, the power factor is unity, so by (14.73) the rms power loss is given by

$$\mathcal{P}_S = \frac{1}{2} i_{Sm} V_{Sm}. \tag{14.80}$$

The primary is a bit more complicated to describe. In addition to the out-of-phase current (relative to V_P), I_P, which does not provide power, there is an in-phase current i_P that provides an rms power

$$\mathcal{P}_P = \frac{1}{2} i_{Pm} V_{Pm}. \tag{14.81}$$

Enforcing energy conservation ($\mathcal{P}_S = \mathcal{P}_P$) by equating (14.80) and (14.81) yields

$$i_{Pm} V_{Pm} = i_{Sm} V_{Sm}. \tag{14.82}$$

Combining (14.82) and (14.78) gives

$$i_{Pm} = i_{Sm} \frac{V_{Sm}}{V_{Pm}} = i_{Sm} \frac{N_S}{N_P}. \tag{14.83}$$

With (14.79) and (14.78), (14.83) becomes

$$i_{Pm} = i_{Sm} \frac{N_S}{N_P} = \frac{V_{Sm}}{R_S} \frac{N_S}{N_P} = \frac{V_{Pm}}{R_S} \left(\frac{N_S}{N_P} \right)^2. \tag{14.84}$$

Hence, on connecting the secondary, the primary develops an effective resistance associated with the in-phase current i_{Pm} given by

$$R_P^{eff} = \frac{V_{Pm}}{i_{Pm}} = R_S \left(\frac{N_P}{N_S} \right)^2. \tag{14.85}$$

Thus the primary for step-up (step-down) transformers has less (more) resistance than the secondary. The above analysis holds as long as $L_P/R_P \gg \omega^{-1}$ or $X_{LP} \gg R_P$. That is, the inductive time constant for the primary must be much greater than the period of the ac power source, or the inductive reactance must be much greater than the resistance. Since we have taken $R_P = 0$, this inequality certainly holds. Although it is not obvious, the in-phase current I_P is not affected by the presence of the secondary, and thus it continues to satisfy $I_{Pm} = V_{Pm}/\omega L_P$.

Example 14.15	**Designing a transformer**

A transformer converts from the high output impedance (784 Ω) of an audio amplifier to the low impedance (4 Ω) of a speaker coil. If the primary has 112 turns, find the number of turns in the secondary.

Solution: By (14.85), $N_P/N_S = (R_P^{eff}/R_S)^{\frac{1}{2}} = 14$. Then $N_S = 112/14 = 8$ turns.

Besides the more obvious uses of transformers, they can also be used for impedance matching two devices with different resistances, such as an audio amplifier and a loudspeaker. Note that transformer efficiencies have not changed much in the past 100 years. However, their reliability and size (volumes smaller by a factor of 100!) have improved enormously.

14.11.3 The Tesla Coil
Optional

Tesla coils produce dramatic high-frequency, high-voltage electric sparks. They utilize the principles of transformers, ac driven circuits, and RLC transients.

A Tesla coil circuit uses two transformers. The first consists of a step-up iron core transformer at 120 V and 60 Hz, which drives the secondary to 25 KV and 60 Hz. See Figure 14.17. This charges up a low C capacitor and a spark gap (which, when it is not sparking, may be thought of as a capacitor), in series with an inductor L. Including the resistance in the wires, this is an RLC circuit. Because 60 Hz is a relatively low frequency, the voltage associated with the

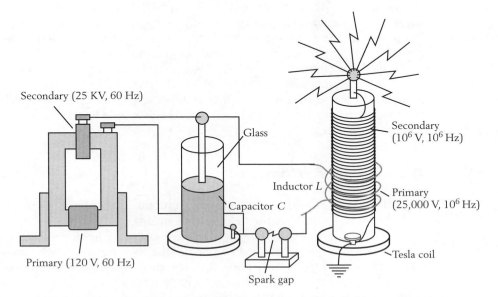

Figure 14.17 Tesla coil. The 60 Hz power source on the left drives an LC circuit that, when the voltage is high enough, causes sparking across the spark gap. This drives the primary on the right into oscillation at its high resonance frequency (on the order of a MHz). The Tesla coil, coupled weakly to this oscillation, draws off energy, at a high voltage and at the frequency of the resonance.

inductor L is not very high, so the voltages across the capacitor and the spark gap are both nearly 25 KV. A glass sheath keeps the capacitor from sparking when subjected to such high voltage, but nothing prevents the spark gap from sparking. When that occurs, charge rushes across the spark gap, thus taking this RLC circuit out of equilibrium. It then undergoes transients, as discussed in Section 14.3, which consist of a damped RLC resonance at a frequency on the order of a MHz. In contrast to what happens for the low-frequency 60Hz voltage from the power transformer, for the high-frequency voltage associated with the LC circuit, the inductor couples to the step-up air transformer. This raises the high-frequency voltage of the latter to on the order of a MV. See Figure 14.17. The many-turn secondary coil of this second transformer, which can be thought of as a parasite off the main RLC circuit, is what is meant by the Tesla coil.

The top of the Tesla coil gives a high voltage (MV), but sparks from it are not harmful. Some textbooks attribute this to a short penetration, or *skin depth*, of the magnetic field into the human body, as the body tries to set up eddy currents that keep out the rapidly changing magnetic field. This is not correct. As shown in Section 14.13, at MHz the skin depth is quite long. Sparks at this high voltage are not harmful primarily because at such high frequencies the current does not transfer much charge.

14.12 Getting DC Force from AC Power

Optional

Because ac voltage is so commonly available, motors using ac voltage are important. Commutators and electronic switches are employed to ensure that the torque or force that causes motion is always in the same direction. However, there are other ways to produce a time-averaged, or dc, force using ac power. We now show how this can occur. Before presenting any detailed arguments, we present a qualitative discussion based upon Lenz's law.

Consider a conducting loop that is in a time-varying, nonuniform magnetic field $\vec{B}$. By the motion statement of Lenz's law, the loop will "want" to go to regions where the time-varying $\vec{B}$ is as small as possible. This is possible only if the spatial profile of $\vec{B}$ is nonuniform; it is the principle upon which the following examples operate.

14.12.1 *A Linear AC Motor*

Consider a loop of resistance R and self-inductance L. It has dimensions a and b, with dimension x and b in the field, so that an area bx is in the oscillating field $\vec{B}_{ac}$. Moreover, the leftmost part of the circuit is not in *any* field. See Figure 14.18(a). This situation can be produced by placing the loop between two turns of a very large solenoid whose axis is normal to the page. From what we have just said, we expect there to be a time-averaged force tending to push the loop out of the field region. Let us show this explicitly.

For $\hat{n}$ and $\vec{B}_{ac}$ into the page, the magnetic flux associated with $\vec{B}_{ac}$ is

$$\Phi_B^{ac} = \int \vec{B}_{ac} \cdot d\vec{A} = \int \vec{B}_{ac} \cdot \hat{n} dA = B_{ac}bx. \tag{14.86}$$

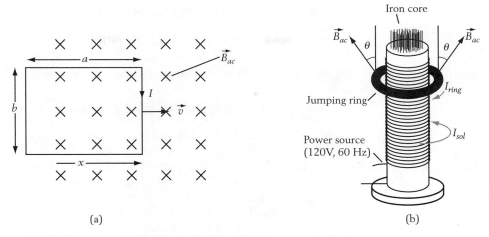

(a) (b)

Figure 14.18 (a) A conducting rectangular loop partially within an oscillating magnetic field. (b) A conducting ring in a flaring and oscillating magnetic field.

We let positive emf $\mathcal{E} = \oint \vec{E} \cdot d\vec{s}$ and current I circulate clockwise. Thus $d\vec{s}$ circulates clockwise, and by the circuit-normal right-hand rule, the normal $\hat{n}$ is indeed into the paper. Then, including the effect of self-inductance,

$$\mathcal{E} = -\frac{d\Phi_B^{ac}}{dt} - L\frac{dI}{dt} = -\frac{d}{dt}(B_{ac}bx) - L\frac{dI}{dt} = -\frac{dB_{ac}}{dt}bx - B_{ac}b\frac{dx}{dt} - L\frac{dI}{dt}.$$
$$(14.87)$$

If B_{ac} is changing rapidly relative to the motion, the motional emf $B_{ac}b(dx/dt)$ can be neglected. Then, taking

$$B_{ac} = B_m \cos \omega t, \qquad (14.88)$$

(14.87) becomes

$$\mathcal{E} = \omega B_m bx \sin \omega t - L\frac{dI}{dt}. \qquad (14.89)$$

Using (14.89) in Ohm's law then gives

$$I = \frac{\mathcal{E}}{R} = \frac{1}{R}\left(\omega B_m bx \sin \omega t - L\frac{dI}{dt}\right), \qquad (14.90)$$

or

$$I = \frac{\mathcal{E}_m}{R}\sin \omega t - \frac{L}{R}\frac{dI}{dt}, \qquad \mathcal{E}_m = \omega B_m bx. \qquad (14.91)$$

This is a rewritten version of equation (14.64), describing an ac driven LR circuit. The impedance Z and phase shift ϕ are given by (14.66) and (14.67). I then can be obtained from (14.43).

The motion along x is driven by the magnetic force on the right arm,

$$\vec{F}_{mag} = I\vec{l} \times \vec{B}_{ac} = IbB_{ac}\hat{x}. \tag{14.92}$$

With (14.88) and (14.43), Newton's equation of motion along x yields

$$m\frac{dv}{dt} = IbB_{ac} = I_m \sin(\omega t - \phi)bB_m \cos\omega t$$
$$= I_m b B_m[\sin\omega t \cos\omega t \cos\phi - \cos^2\omega t \sin\phi]. \tag{14.93}$$

Since (14.67) implies that $\sin\phi > 0$, the time-averaged force is to the left, thus pushing the loop out of the field; the loop "wants" to leave the region of the field. (This time-averaged force suggests how to devise a motor based on ac power.) In addition, there is an oscillating force, at frequency 2ω. If a capacitor is added to the circuit, and if the capacitor dominated the phase shift ϕ at the driving frequency, causing ϕ to be negative, so $\sin\phi < 0$, then the time-averaged force would be to push the loop to the right (i.e., farther into the field).

14.12.2 *The Jumping Ring*

We now analyze Elihu Thomson's jumping ring demonstration (1887), which first established that an ac power source can produce a time-averaged force. (Thomson's inventions, which include the wattmeter, are nearly as numerous as Edison's. His company, like Edison's, was swallowed up in the formation of General Electric. Thomson was later a president of MIT.) Consider Figure 14.18(b). An analysis of the force on a loop of radius a in a flaring B field at an angle θ was done in Chapter 10, yielding (10.29). The flaring field means that the higher the ring, the weaker the field, so from our earlier considerations, there should be a tendency for the ring to be thrown out of the field region. Let us show this explicitly.

Let $B = B_{ac}$ be due to the ac current I_{sol} in the solenoid, and let $I = I_{ring}$ be the current in the ring. When the currents are in the same direction, the force is downward (attractive), which is taken to be positive. Because the ring "tries" to produce a flux that *opposes* that of the solenoid, I_{ring} and I_{sol} will be in opposite directions, and the force will be upward (repulsive). We will not write down the circuit equation in this case, but merely note that, just as in the previous case, it is essential to include the self-inductance L as well the resistance R. Hence (14.64–14.67) apply.

Using (10.29), (14.50), and (14.88) yields vertical force component

$$F = IB(2\pi a)\sin\theta = I_{ring}B_{ac}(2\pi a)\sin\theta$$
$$= -I_m \sin(\omega t - \phi)B_m \cos\omega t(2\pi a)\sin\theta$$
$$= -I_m B_m(2\pi a)\sin\theta[\sin\omega t \cos\omega t \cos\phi - \cos^2\omega t \sin\phi]. \tag{14.94}$$

Since (14.67) implies that $\sin\phi > 0$, the time-averaged force is upward, or out of the field. In addition, there is an oscillating force, at frequency 2ω. On adding a capacitor to the circuit, if the capacitor were to dominate the phase shift ϕ at the driving frequency, making ϕ negative, then the time-averaged force would push the ring farther into the field: a "sucking" ring.

A more complete analysis would express I_m and ϕ in terms of L, R, and $\mathcal{E}_m$. It would also relate $\mathcal{E}_m$ to B_{ac}. However, the latter relationship can be obtained only approximately (qualitatively, $\mathcal{E}_m \sim \omega B_{ac} A_{ring} \cos\theta$). These details are not needed to demonstrate that we can produce a time-averaged force using ac power.

As the preceeding two examples make clear, phase is not an abstract concept; by suitable design a circuit can produce a repulsive force or an attractive force, according to the phase.

14.12.3 *Hum: 60 Hz and 120 Hz*

So-called 60-cycle hum occurs when there is an interaction between a permanent magnet and a power source at 60 Hz, such as for the speaker coil problem of Chapter 10, if the amplifier "leaks" signal at 60 Hz. For the solenoid associated with the jumping ring, there is also a hum, but it is not at 60 Hz. It is due to the interaction of the rods—so chosen to minimize eddy currents—that constitute the iron core. Half every cycle they are all pointing up, and a half-cycle later they are pointing down. This produces a magnetic repulsion twice every cycle, leading to 120-cycle hum. It is like the $\cos^2\omega t$ terms in (14.93) and (14.94), which have nonzero time averages.

Any circuit (even the "sucking" ring) is, for short times, dominated by its self-inductance, and this leads to a repulsive force on turn-on. However, this force is too weak to be observed for the sucking ring. Moreover, for the jumping ring, it is dominated by the repulsive force due to the ac voltage. This becomes clear if you power up the solenoid, pull the ring onto the solenoid, and then release the solenoid—the ring will still jump up.

14.12.4 *Induction Motor*

Once it was realized that ac power could lead to a net force, the road was clear for the development of motors based upon ac power. The induction motor, first developed by Tesla (1888), is commonly used today. It is based upon the idea that ac power can be used to simulate rotation. By Lenz's law, a rotating magnet just above a conducting disk produces eddy currents in that disk. The interaction between the magnet and the eddy currents in the disk then tends to bring the disk into rotation. (Recall the magnetic drag effect in Figure 12.10.) Similarly, a simulated rotating magnetic field, obtained by sequentially powering a set of coils placed at $120°$ to one another, brings a conducting disk into rotation. In the 19th century, the use of three-phase ac power facilitated this. Presently, electronics can arrange the timing of ordinary ac power to simulate a rotating field.

14.13 Electromagnetic Shielding—Skin Depth

Optional

14.13.1 *Overview*

Chapter 12 discussed eddy currents as an example of how a conductor "tries" to prevent changes in a magnetic field. Chapter 13 presented, without proof, Maxwell's receding image construction, which represents the eddy current

Table 14.1 Characteristic skin depths

Material	Conductivity	60 Hz (power)	100 MHz (FM)	10^{10} Hz (microwave)
Copper	6.0×10^7 ohm^{-1}-m^{-1}	0.0085 m	6.6×10^{-6} m	6.6×10^{-7} m
Salt water	4.4 ohm^{-1}-m^{-1}	31.4 m	0.0244 m	0.00244 m

response of a thin conducting sheet. The present section discusses the eddy current response of a semiinfinite conducting sheet, for magnetic fields that lie in the plane of the sheet. In this case, the derivation, although not simple, is easier than for Maxwell's receding image construction. This derivation is closely related to the derivation in the next chapter for electromagnetic radiation; you may want to compare them after reading both.

The central result of this section is that, because of eddy currents, an ac magnetic field will only be able to penetrate a conducting sheet to a limited extent, given by what is called the *skin depth*

$$\delta = \sqrt{\frac{2}{\mu_0 \sigma \omega}}. \tag{14.95}$$

This means that the electric and magnetic fields—and the eddy currents—within the sheet decrease exponentially away from the surface, with the characteristic decay length δ.

From Table 14.1, at power line frequencies, electric and magnetic fields are shielded out by a couple of cm of copper, but are not shielded out by the human body; at FM frequencies, they are shielded out by a thin layer of copper, but they can penetrate a couple of cm into the human body; at microwave frequencies, they are shielded out by a thin layer of copper, but they can penetrate a couple of mm into the human body. Thus, the electromagnetic radiation from a poorly protected microwave oven will not quite reach the retina (but we do not recommend that you work around such a microwave oven!). From Table 14.1 and (14.95), by proportionality of δ to $1/\sqrt{\omega}$, we deduce that if $f = 1$ MHz, then $\delta = 24.4$ cm. Hence the electric field from a 1 MHz Tesla coil will penetrate your body. The frequencies used in MRI range from 1.5 to 10 MHz; at the upper end of 10 MHz, $\delta = 7.71$ cm, which is enough to penetrate an arm or a leg, but not the human torso.

14.13.2 *Derivation*

To derive (14.95) for the skin depth, we apply Ampère's law and Faraday's law to rectangular circuits whose planes include the normal to the sheet.

Consider a semi-infinite system of uniform conductivity σ that occupies the half-space $x > 0$. Let an incident ac magnetic field $\vec{B}$, with frequency ω, point along the z-axis. We expect that an electric current and electric field pointing along the y-axis will be induced. See Figure 14.19. This current will produce its own magnetic field along the z-axis that will oppose the change in the incident magnetic field, thus preventing the net magnetic field from penetrating the conductor. Our goal is to determine the equations describing how the magnetic field is kept out. We will assume that $\vec{B}$ and $\vec{E}$ depend only on x and t.

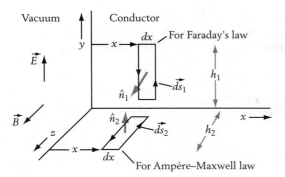

Figure 14.19 Geometry for electromagnetic shielding of an ac electric field coming in on the left, and being absorbed by the conductor on the right $(x > 0)$. Two imaginary circuits are drawn, one for use with Faraday's law (involving magnetic flux), and the other for use with Ampère's law (involving electric current).

14.13.3 Use of Faraday's law

Because $\vec{B}$ is along z, it produces a flux along z. Let us therefore apply Faraday's law to a small rectangular circuit whose normal $\hat{n}_1$ is along z, with dimension h_1 along y and dx along x. See Figure 14.19. In that case, by the circuit-normal right-hand rule, in computing the circulation of $\vec{E}$ we must take $d\vec{s}_1$ such that it circulates as in Figure 14.19.

For the Faraday's law circuit, the electric circulation is given by

$$\oint \vec{E} \cdot d\vec{s}_1 = [E_y(x + dx) - E_y(x)]h_1$$

$$\approx \left[E_y(x) + \frac{dE_y}{dx}dx - E_y(x)\right]h_1 \approx \frac{dE_y}{dx}(h_1 dx). \quad (14.96)$$

The associated magnetic flux is

$$\Phi_B = \int \vec{B} \cdot d\vec{A} = \int \vec{B} \cdot \hat{n}_1 dA = B_z(h_1 dx). \quad (14.97)$$

The negative rate of change of the associated magnetic flux is thus

$$-\frac{d\Phi_B}{dt} = -\frac{dB_z}{dt}(h_1 dx). \quad (14.98)$$

Using Faraday's law to equate (14.96) and (14.98) yields

$$\frac{dE_y}{dx} = -\frac{dB_z}{dt}. \quad (14.99)$$

14.13.4 Use of Ampère's law

Because $\vec{E}$ is along y, by (7.7) it generates in the sheet an electric current density $\vec{J} = \sigma\vec{E}$ along y. Let us therefore apply Ampère's law to a small rectangular circuit whose normal $\hat{n}_2$ is along y, with dimension h_2 along z and dx along x. See Figure 14.19. In that case, by the circuit-normal right-hand rule, in computing the circulation of $\vec{B}$ we must take $d\vec{s}_2$ such that it circulates as in Figure 14.19. For the Ampère's law circuit, the magnetic circulation is given by

$$\oint \vec{B} \cdot d\vec{s}_2 = [-B_z(x+dx) + B_z(x)]h_2$$

$$\approx \left[-B_z(x) - \frac{dB_z}{dx}dx + B_z(x) \right]h_2 \approx -\frac{dB_z}{dx}(h_2 dx). \quad (14.100)$$

The associated electric current is

$$\int \vec{J} \cdot d\vec{A} = \int \vec{J} \cdot \hat{n}_2 dA = J_y(h_2 dx) = \sigma E_y(h_2 dx). \quad (14.101)$$

Using Ampère's law to equate (14.100) to $4\pi k_m$ times (14.101) yields

$$-\frac{dB_z}{dx} = 4\pi k_m \sigma E_y = \mu_0 \sigma E_y. \quad (14.102)$$

14.13.5   *The Field Diffusion Equation and Its Solution*

Taking the space derivative of (14.102), and eliminating dE_y/dx via (14.99),

$$\frac{d^2 B_z}{dx^2} = 4\pi k_m \sigma \frac{dB_z}{dt} = \mu_0 \sigma \frac{dB_z}{dt}. \quad (14.103)$$

This equation, when solved, leads to (14.95) for the skin depth. It is called a *diffusion* equation. (Fourier was the first person to obtain a diffusion equation; for heat, he obtained a diffusion equation with the temperature T in place of B_z, and constants that depended upon the specific heat and the thermal conductivity.) Taking the space derivative of (14.99), and eliminating dB_z/dx via (14.102), yields an equation like (14.103), but now for E_y. When Maxwell obtained (14.103), he was already familiar with related equations, so he knew how to solve it. We expect that, in the conductor, B_z will oscillate in time at the frequency ω, and decay in space, so that the signal gets weaker on moving into the conductor. In fact, there is also an oscillation in space. Rather than try to solve the equation directly, we first write down the answer, and then indicate how to verify that it is the correct solution. For B_z satisfying $B_z(0,t) = B_m \sin \omega t$, the solution is

$$B_z(x,t) = B_m \sin(\omega t - kx) \exp(-qx), \qquad k = q = \frac{1}{\delta} = \sqrt{\frac{\mu_0 \sigma \omega}{2}}. \quad (14.104)$$

To establish that this is correct, note that on applying d^2/dx^2 to B_z in (14.103) there are three terms. One comes from d^2/dx^2 acting on the sine term, giving $-k^2 B_z(x,t)$, and one comes from d^2/dx^2 acting on the exponential term, giving $q^2 B_z(x,t)$. These two cancel if $q = k$. The third term involves twice the product of d/dx acting once on the sine term (converting it to a cosine times $-k$), and once on the exponential term (converting it to an exponential times $-q = -k$).

Thus

$$\frac{d^2 B_z}{dx^2} = 2k^2 B_m \cos(\omega t - kx) \exp(-kx). \tag{14.105}$$

This is now equated to the right-hand side of (14.103). With d/dt on B_z in (14.103) converting the sine to a cosine, we obtain

$$\mu_0 \sigma \frac{d B_z}{dt} = \mu_0 \sigma \omega B_m \cos(\omega t - kx) \exp(-kx). \tag{14.106}$$

Equating (14.105) and (14.106), and solving for $k \equiv 1/\delta$ in (14.105) leads to (14.95) for the skin depth.

Equations (14.102) and (14.104) determine $E_y(x,t)$. Both B_z and E_y decay on moving into the conductor, as desired. Since $J_y = \sigma E_y$, the eddy currents also decay on moving into the conductor. There is no magic about electromagnetic screening; the eddy currents are responsible for it. Note that if the input power is too high, the eddy currents can produce so much Joule heating that they cause the conductor to heat up and, perhaps, to melt!

Problems

14-1.1 Consider a signal that varies linearly from 1 V to −1 V in time $T/2$, then from −1 V to 1 V in time $T/2$, and so on. For this triangle-shaped signal, find the rms and average values of the voltage.

14-1.2 (a) Is there a unique value for the average voltage across an oscillating circuit if the time average is taken to be only a quarter-period? (b) A half-period? (c) A full-period? (d) A billion-and-a-quarter periods?

14-1.3 For an oscillating circuit it takes 0.02 s to go through a quarter-period. (a) Find the time it takes to go through a radian and through a full-period. (b) Find the inverse of these times. (c) Which (if any) of these corresponds to the frequency f and the angular frequency ω?

14-1.4 Consider a signal that is V_m for time $T/2$, then is zero for $T/2$, and so on. For this step function–shaped signal, find the rms and average values of the voltage.

14-1.5 Consider a signal that varies linearly from zero to V_m in $T/2$ and linearly back to zero in $T/2$, and so on. For this triangle-shaped signal, find the rms and average values of the voltage.

14-1.6 Consider a signal that goes linearly from zero to V_m in time T, then suddenly goes to zero for time T, and so on. For this sawtooth-shaped signal, find the rms and average values of the voltage.

14-1.7 Consider a signal that varies linearly from $-V_m$ to V_m in $T/2$ and linearly back to $-V_m$ in $T/2$, and so on. For this triangle-shaped signal, find the rms and average values of the voltage.

14-1.8 An ac signal has a period of 2.4 ms. Find (a) its frequency, (b) its angular frequency, and (c) the time it takes to go from zero to a maximum.

14-2.1 In an LC circuit, it is desired to double the frequency of the oscillation. You can change only one circuit element. What are your options?

14-2.2 An inductor is replaced in an LC circuit, and the oscillation frequency changes from 1250 Hz to 1950 Hz. What can you say about the replacement inductor relative to the original inductor?

14-2.3 The AM radio frequency broadcast band ranges from 550 kHz to 1550 kHz. (a) For a 40 μH inductor, find the capacitance range needed to tune across the AM band. (b) For an rms voltage 50 mV across the capacitor, find the energy stored in the circuit for signals at 550 kHz and 1550 kHz.

14-2.4 The FM radio frequency broadcast band ranges approximately from 88 MHz to 108 MHz. (a) For a 60 μF capacitor, find the inductance range needed to tune across the FM band. (b) For an rms voltage 50 mV across the inductor, find the energy stored in the circuit for signals at 88 MHz and 108 MHz.

14-2.5 A coil with $L = 145$ nH, to be used for nuclear magnetic resonance, resonates at 103.7 MHz. Find the value of the parasitic capacitance.

14-3.1 An underdamped RLC circuit initially has zero charge and zero current, and is connected to a battery at $t = 0$. (a) Sketch the current as a function of time. (b) Repeat for the charge.

14-3.2 An overdamped RLC circuit initially has zero charge and zero current, and is connected to a battery at $t = 0$. (a) Sketch the current as a function of time. (b) Repeat for the charge.

14-3.3 For the underdamped RLC circuit, show that (14.29) with (14.30) leads to a current that satisfies (14.27).

14-3.4 For the overdamped RLC circuit, show that (14.34) with (14.35) satisfies (14.27).

14-3.5 An RLC circuit has $L = 24$ mH, $C = 67$ nF, and $R = 5\Omega$. (a) Find the resonance frequency and the critical resistance. (b) Sketch the voltage across the capacitor if a constant emf is suddenly switched into the circuit.

14-3.6 An RLC circuit has $R = 5$ Ω, a resonance frequency of 3.76 MHz, and a critical resistance of 48 Ω. (a) Find its capacitance and inductance. (b) Sketch the voltage across the resistor if a constant emf is suddenly switched into the circuit.

14-4.1 It is desired to produce a 850 V maximum voltage from a generator turning 240 times per second in a 0.48 T magnetic field. It has 400 coils. (a) Find the area of each coil. (b) If the field falls to 0.24 T, find how many turns of coil are needed to raise the voltage back to 850 V.

14-4.2 An ac generator yields a 2 A current when the voltage across it is 4 V. (a) If the generator is ideal, find the emf. (b) If the generator has a 0.02 Ω resistance, find the emf. (c) In Figure 14.4, if the voltage $\Delta V_g = V_b - V_a$ is positive, does that tend to drive current up or down the left arm?

14-5.1 A black box has $\Delta V_{max} = 85$ V and $I_{max} = 14.5$ A, and the voltage leads the current in time by 0.65 ms when subjected to an ac emf at 240 Hz. (a) Find its impedance and phase. (b) Is energy absorbed or provided?

14-5.2 A high-voltage ac power line has two wires, carrying equal and opposite current. From measurements of voltage difference and current, how would you determine the direction of average power flow? *Hint:* Consider the phase for the case where power flows to a resistor.

14-5.3 Circuit A is connected to circuit B. It is found that the phase angle of (14.43) for circuit B, thought of as a single circuit element, is 125°. (a) Which circuit is providing power, and which is receiving power? (b) What is the phase angle for circuit A?

14-5.4 For an inductor connected to a 40 V rms ac voltage with a period of 0.02 s, the rms current is 0.08 A. (a) Find X_L, L, and ϕ. (b) Find the time average energy stored by the inductor.

14-5.5 For a capacitor connected to a 40 V rms ac voltage with a period of 0.02 s, the rms current is 0.08 A. (a) Find X_C, C, and ϕ. (b) Find the time average energy stored by the capacitor.

14-5.6 For a resistor connected to a 40 V rms ac voltage with a period of 0.02 s, the rms current is 0.08 A. (a) Find R and ϕ. (b) Find the time average energy dissipated by the resistor.

14-5.7 (a) Is voltage a scalar or a vector? (b) How is voltage "scalarized" in the phasor construction?

14-5.8 Explain why, for an ideal inductor, the voltage difference between its ends opposes the self-induced emf.

14-5.9 Capacitors and inductors behave like short circuits or open circuits in the appropriate limits of either low or high frequencies. (a) Which limits give which behavior for capacitors? Explain. (b) For inductors? Explain.

14-6.1 For an LR circuit the maximum generator emf is 160 V, and the maximum current is 4 A. The frequency is 60 Hz, and the current lags the voltage by 5 ms. Find Z, ϕ, R, X_L, and L.

14-6.2 For an RC circuit the maximum generator emf is 160 V, and the maximum current is 4 A. The frequency is 60 Hz, and the current leads the voltage by 5 ms. Find Z, ϕ, R, X_C, and C.

14-6.3 An LR circuit with $R = 20$ Ω and $L = 5$ mH is connected to a generator with rms emf of 60 V and frequency 50 Hz. Find X_L, Z, ϕ, the time by which the net voltage leads the current, and the maximum current.

14-6.4 An RC circuit with $R = 20$ Ω and $C = 5$ nF is connected to a generator with rms emf of 60 V and frequency 250 Hz. Find X_C, Z, ϕ, the time by which the net voltage lags the current, and the maximum current.

14-6.5 An unknown resistor is in series with a 12 nF capacitor. At 440 Hz, the net voltage lags the current by 0.075 ms. Find X_C, R, Z, and ϕ.

14-6.6 An unknown capacitor is in series with a 40 Ω resistor. At 440 Hz, the net voltage lags the current by 0.075 ms. Find X_C, C, Z, and ϕ.

14-6.7 An unknown resistor is in series with a 12 μF capacitor. At 440 Hz, the maximum current is 0.08 A for a maximum emf of 12.8 V. Find Z, X_C, R, ϕ, and the time by which the net voltage lags the current.

14-6.8 An unknown resistor is in series with a 25 mH inductor. At 880 Hz, the maximum current is 0.08 A for a maximum emf of 12.8 V. Find Z, X_L, R, ϕ, and the time by which the net voltage leads the current.

14-6.9 An unknown inductor is in series with a 12 Ω resistor. At 440 Hz, the net voltage leads the current by 0.055 ms. Find ϕ, X_L, L, and Z.

14-6.10 An unknown resistor is in series with a 12 μF inductor. At 440 Hz, the net voltage leads the current by 0.075 ms. Find ϕ, X_L, R, and Z.

14-6.11 A 120 V rms, 60 Hz generator is placed across a resistor and inductor in series, with $R = 24$ Ω and $L = 15$ mH. (a) Find the rms current through each of these circuit elements. (b) Find the rms voltage across each of these circuit elements. (c) Find the average rate at which power is lost to R and to L. (d) Find ϕ and the associated time delay.

14-6.12 A 120 V rms, 60 Hz generator is placed across a resistor and capacitor in series, with $R = 24$ Ω and $C = 940$ nF. (a) Find the rms current through each of these elements. (b) Find the rms voltage across each of these circuit elements.

(c) Find the average rate at which power is lost to R and to C. (d) Find ϕ and the associated time delay.

14-6.13 A 120 V rms, 60 Hz generator is placed across a circuit containing a resistor $R = 20$ Ω and another circuit element (either a capacitor or an inductor). The phase angle is 63°. (a) Identify the other element, give its reactance, and give its capacitance or inductance. (b) Find the average rate at which power is dissipated.

14-6.14 Derive the solution for the LR circuit using the trigonometric method.

14-7.1 Consider a resistor and an inductor that, at a frequency of 3500 Hz, have the same impedance of 200 Ω. When they are in series, a voltage pulse applied across them rises from 0 V to 6000 V and falls back to 0 V within a characteristic time of 8 μs. (a) Estimate the corresponding frequency. (b) Find the impedance of the resistor and of the inductor at this frequency. (c) Explain why inductors, rather than resistors, are often used to limit current associated with power surges.

14-7.2 An LR circuit is to be used as a low-pass filter for an input voltage V_{in}. (a) Across which element would you measure the output voltage V_{out}? (b) Repeat for an LR circuit to be used as a high-pass filter. (c) For $R = 17$ Ω, $f = 1.97$ kHz, $L = 4$ mH, and $V_{in}^{max} = 25$ V, find V_{out}^{max} for both R and L. (d) Repeat (c) for $f = 19.7$ kHz.

14-7.3 An RC circuit is to be used as a low-pass filter for an input voltage V_{in}. (a) Across which element would you measure the output voltage V_{out}? (b) Repeat for an RC circuit to be used as a high-pass filter. (c) For $R = 17$ Ω, $f = 1.97$ kHz, $C = 458$ nF, and $V_{in}^{max} = 25$ V, find V_{out}^{max} for both R and C. (d) Repeat (c) for $f = 19.7$ kHz.

14-7.4 Design two different circuits with $R = 24$ Ω that can be used as low-pass filters with a "roll-off" frequency of 4500 Hz, at which the ratio of the voltage across the resistor to the signal voltage is $1/\sqrt{2}$.

14-7.5 Show that, if an ac emf of amplitude $\mathcal{E}_m$ is applied to an RC circuit, then the maximum voltage across the resistor is given by $\Delta V_{Rm} = \mathcal{E}_m(\omega RC)/\sqrt{1 + (\omega RC)^2}$.

14-7.6 Show that, if an ac emf of amplitude $\mathcal{E}_m$ is applied to an LR circuit, then the maximum voltage across the resistor is given by $\Delta V_{Rm} = \mathcal{E}_m/\sqrt{1 + (\omega L/R)^2}$.

14-8.1 Consider an RLC circuit. Is it possible for the voltage across a circuit element to exceed the source voltage when the circuit element is (a) the inductor? (b) The resistor? (c) The capacitor?

14-8.2 You are told that an RLC circuit is not operating properly at high frequencies. Which circuit element would you suspect? Repeat if the problem is at low frequencies.

14-8.3 Consider an RLC circuit with $L = 2.5$ mH, $R = 4\ \Omega$, $C = 500\ \mu F$, driven by a 24 V maximum emf at a frequency of 400 Hz. (a) Find X_L, X_C, Z, and ϕ. (b) Find the maximum current and the maximum voltage across L, R, and C. (c) Find the time by which the driving voltage leads (or lags) the current. (d) Find the maximum voltage across the combination R and C, and the time by which this voltage leads (or lags) the current. (e) Find the maximum voltage across the combination R and L, and the time by which this voltage leads (or lags) the current. (f) Find the maximum voltage across the combination L and C, and the time by which this voltage leads (or lags) the current.

14-8.4 In an RLC circuit, $L = 56$ mH, $R = 2.4\ \Omega$, and $C = 8.4\ \mu F$. (a) For what angular frequency ω will the maximum current have its maximum value? (b) For what angular frequencies ω_+ and ω_- will the current have half the maximum value? (c) What is the fractional width $|\omega_+ - \omega_-|/\omega$ of the resonance?

14-8.5 A 1.20 V rms generator of variable frequency drives an RLC circuit with $R = 24\ \Omega$, $C = 940$ nF, and $L = 15$ mH. (a) Find the resonance frequency of this circuit, and the rms current through it at that frequency. (b) Find the rms voltage across each circuit element at resonance. (c) At twice the resonance frequency, find the impedance, the phase angle, and the rms current. (d) At twice the resonance frequency, find the rms voltage across each circuit element. (e) Compute the critical resistance and discuss whether this is a broad resonance or a narrow resonance.

14-8.6 A 1.06 V rms generator of variable frequency drives an RLC circuit with $R = 4\ \Omega$, $C = 34\ \mu F$, and $L = 2.6$ mH. (a) Find the resonance frequency of this circuit, and the current through it at that frequency. (b) Find the voltage across each circuit element at resonance. (c) At half

the resonance frequency, find the impedance, the phase angle, and the current. (d) At half the resonance frequency, find the voltage across each circuit element. (e) Compute the critical resistance and discuss whether this a broad resonance or a narrow resonance.

14-8.7 Derive the solution for the RLC circuit using the trigonometric method.

14-9.1 Explain why the first electronic amplifiers were called valves.

14-9.2 Explain why heating the cathode increases the current, but heating the anode does not.

14-9.3 Explain why a small change in the grid voltage of a triode can have a large effect on the current.

14-10.1 Explain why it is not possible, on average, to transfer ac power to a capacitor or an inductor.

14-10.2 An ac voltage divider is made of a resistor R and an inductor L in series, with the input voltage applied across the two in series, but the output voltage across only R. (a) Determine the ratio of the rms output voltage to the rms input voltage. (b) Determine its efficiency from its power factor. (c) For a voltage divider using a resistor R' instead of L, find the ratio of the rms output voltage to the rms input voltage. (d) Determine its efficiency of power utilization. (e) Which is a more energy-efficient method to control the output voltage?

14-10.3 An oil refinery generates its own power, but it has wires connecting it to the power company grid, both to buy and sell electrical power. (a) As seen by the power company, what is the sign of the power factor when the refinery is buying power? (b) When the refinery is selling power?

14-10.4 Power company A charges by power usage, whereas power company B charges by power usage/power factor. They are both making an 8% yearly return. (a) If you are an efficient power user, which power company would you expect to provide you with lower prices? (b) If you are an inefficent power user?

14-10.5 At 60 Hz, the phase and impedance for a factory are 65° and 2300 Ω. (a) Find the power factor. (b) What nondissipative circuit element, and what would be its value, if on adding it in series the power factor equals 0.8? (c) What is the new impedance?

14-11.1 A generator provides 120 V to a primary with 25 turns. Find the number of turns needed to produce, in the secondary: (a) 30 V; (b) 300 V.

14-11.2 120 V ac power is available, but you would like to use 12 V ac power for outdoor lighting. You are given a transformer core with seven turns of wire already wrapped around one side. (a) How many turns of wire would you wrap around the other side of the transformer? (b) Which set of turns would you solder to the outdoor lighting? (Turn the power off while you solder.)

14-11.3 A transformer with 250 turns in the primary and 50 turns in the secondary provides 20 V to the secondary. Find the rms voltage of the primary.

14-11.4 A transformer is to be used to convert from the high output impedance (850 Ω) of an audio amplifier to the low impedance (8 Ω) of a speaker coil. If the secondary (speaker coil) has 12 turns, find the number of turns in the primary. Interpret a nonintegral number of turns.

14-11.5 A doorbell transformer operates at 12 V output for 120 V input. (a) Find the turns ratio. (b) Explain why, in a transformer, the wire is usually thicker on the low-voltage side.

14-11.6 An 8 Ω speaker produces 40 W rms when connected via a transformer to an amplifier with output impedance 1200 Ω. Find (a) the turns ratio, (b) the current and voltage of the secondary, and (c) the current and voltage of the primary.

14-11.7 (a) An ac generator produces 12 A rms at 400 V rms with power factor unity. Find the rms power produced by this generator. (b) The generator voltage gets boosted by a step-up transformer to 12 kV. Find the power after the step-up transformer, assuming no losses in the transformer. (c) The power then is transmitted to (and from) an electrical load with wires having resistance 8 Ω each way, until it reaches a step-down transformer. Determine the rms power loss in the wires. (d) Determine the power available to the load.

14-11.8 A transformer has 600 turns in its primary. It is driven at 120 V rms, and its secondary has taps that yield rms voltages of 2 V, 5 V, and 8 V. Find the number of turns for each part of the secondary.

14-11.9 Explain the operation of a Tesla coil.

14-11.10 Explain why the top of a Tesla coil, despite its high voltage, does not give dangerous shocks.

14-11.11 Automobile ignitions are so named because they ignite, within the chamber associated with the piston, an explosive gas–air mixture that converts chemical energy from the gasoline to mechanical energy of the engine. Such ignitions generate high voltages using only a $\mathcal{E} = 12$ V battery in series with a large inductor L (with resistance R) and a capacitor C that can be shunted out when a switch S is connected. (In "automobilese," the inductor L is known as the solenoid, and the capacitor C is known as the condenser. The carburetor takes the fluid gasoline and vaporizes it with air in just the right proportion.) Here is how it develops the spark. (1) With S on (no capacitor), the battery sends current through the LR circuit of the inductor, until a current $I = \mathcal{E}/R$ is attained in characteristic time L/R. **(a)** If $I = 12$ A, find R. (2) The switch S suddenly takes $\mathcal{E}$ out of the circuit and puts C into the circuit, thus converting a powered LR circuit into an unpowered RLC circuit that starts with $Q = 0$ and $I = \mathcal{E}/R$. (3) A quarter-cycle of the RLC resonance period later, the voltage across the inductor reaches its maximum value V_{Lm}. **(b)** If $V_{Lm} = 300$ V, find R_c/R. **(c)** If a quarter-cycle corresponds to 0.84 ms, find L and C. (4) Coupled to the primary by a mutual inductance M is a secondary. The secondary functions as a step-up transformer, with a maximum voltage of 30,000 V to 100,000 V. **(d)** Find the turns ratio needed to produce 30,000 V. *Note:* Such a large voltage causes a spark in the gap of the spark plug, which is part of the secondary circuit. This drains energy from the primary, whose Q decreases during the spark. The primary is then ready to start the process again for the next spark plug. Modern automobiles use sophisticated electronics in their engine control units to ensure that the timing of each spark plug matches the stroke of its piston.

14-12.1 A toaster and an electric motor are designed for 120 V ac power. Which is more likely to work for 120 V dc power?

14-12.2 Explain the difference between 60-cycle hum and 120-cycle hum.

14-12.3 A linear motor with $R = 12$ Ω and $L = 45.2$ mH is driven by 120 V, 60 Hz ac power, producing an ac field of amplitude 0.37 T. Take

$b = 8.2$ cm. (a) When $x = 3.5$ cm, find the amplitude of the emf acting on the loop of the linear motor. (b) Find Z and ϕ for the circuit of the linear motor. (c) Find the amplitude of the current. (d) Find the time-averaged force.

14-12.4 A jumping ring with $R = 1.4$ Ω, $L = 420$ nH, and radius $a = 3.8$ cm is in an ac field of amplitude 0.26 T and flaring angle of 15°, due to a solenoid of radius $b = 2.8$ cm, driven by 120 V, 60 Hz ac power. (a) Estimate the amplitude of the emf acting on the jumping ring. (b) Find Z and ϕ for the circuit of the jumping ring. (c) Estimate the amplitude of the current in the ring. (d) Estimate the time-averaged force on the ring.

14-12.5 Consider a linear ac motor when there is a capacitor C in series with the inductance L and the resistance R. (a) Find the equation describing the current. (b) Find the current. (c) Find the time-averaged force. (d) Discuss the qualitative difference between the time-averaged force on the circuit when the driving frequency ω is above and is below the resonance frequency ω_0.

14-12.6 Consider the jumping ring when there is a capacitor C in series with the inductance L and the resistance R. (a) Find the equation describing the current. (b) Find the current. (c) Find the time-averaged force. (d) Discuss the qualitative difference between the time-averaged force on the circuit when the driving frequency ω is above and is below the resonance frequency ω_0.

14-13.1 (a) Find the skin depth for aluminum at 12.4 MHz. (b) Will a 0.001-inch-thick sheet of foil "shield out" electromagnetic radiation?

14-13.2 Show that E_y satisfies an equation like (14.103) for B_z.

14-13.3 Show that, for any functions $f(x)$ and $g(x)$,

$$\frac{d^2(fg)}{dx^2} = \frac{d^2 f}{dx^2} g + 2\frac{df}{dx}\frac{dg}{dx} + f\frac{d^2 g}{dx^2}.$$

14-13.4 Discuss microwave heating for $d \gg \delta$ and for $d \ll \delta$, where d is the dimension of the object in the microwave field.

14-G.1 The farther away the generator from the power consumption, the more line loss. To compensate for this, one can increase the emf by increasing the frequency at which the generator turns. This has

misleadingly been called compensation for "loss of frequency." Comment.

14-G.2 There is a story that the initial discovery of the J-psi particle, at a high-energy particle accelerator, was made by a group of physicists with a measuring device for particle energies that was sensitive only over a broad range of energies. The signal was repeatedly missed by another group of physicists who had an energy-measuring device that was very sensitive to a narrow range of energies. Only after the first group noticed a signal did the second group find that signal, and then they could study it in much more detail than the first group. Comment on how this relates to broad-band and narrow-band radio receivers.

14-G.3 Consider an underdamped RLC circuit. An emf $\mathcal{E} = \mathcal{E}_m \sin \omega t$ commences at $t = 0$, when $I = 0$ and $Q = 0$. Including transients, find I and Q for all $t > 0$.

14-G.4 Consider an RC circuit. An emf $\mathcal{E} = \mathcal{E}_m \sin \omega t$ commences at $t = 0$, when $I = 0$ and $Q = 0$. Including transients, find I and Q for all $t > 0$.

14-G.5 Consider an LR circuit. An emf $\mathcal{E} = \mathcal{E}_m \sin \omega t$ commences at $t = 0$, when $I = 0$ and $Q = 0$. Including transients, find I and Q for all $t > 0$.

14-G.6 Consider a resistor R and a capacitor C that are in parallel, and subject to the same ac voltage. By adding the current phasors, find the net current phasor, and then determine the impedance and phase shift for this combination.

14-G.7 Consider a resistor R and an inductor L that are in parallel, and subject to the same ac voltage. By adding the current phasors, find the net current phasor, and then determine the impedance and phase shift for this combination.

14-G.8 Let loop 1, carrying I_1, be part of a rotatable coil, to which a needle is attached. Let loop 2, carrying I_2, be fixed in place, producing a magnetic field that causes a torque on the first loop. This is the basis of a *dynamometer*, whose needle deflection is proportional to the torque between two current loops. (a) If $I_1 = I_2 = I$, the scale markings are not linear in current, unlike the torque for a permanent magnet and a current loop carrying current I. How should the scale markings vary with current? (b) Explain how such a device can be used to measure ac current. (c) Consider a circuit that is being monitored both for current and voltage. If the output of the ammeter goes through the first loop, and the output of the voltmeter through the second loop, show that the deflection can give a measure of the power. This is a true wattmeter. [The meter at the back or side of your house is a small motor that turns a disk at a rate proportional to the power consumption. The reading on the dial is a measure of the electrical energy consumed (the time integral of the electrical power).]

14-G.9 Here is some general information about electric current and *electric shock*, approximately in order of increasing current. Interspersed with the basic information are two questions. (1) *Perception* without harm can occur at a few microamperes of ac current, and somewhat larger for dc. (2) The *startle reaction* can be elicited by 0.5 mA rms at 60 Hz current; this is the maximum leakage current allowed by appliances. Startle can be caused by a 2 mA dc current, but sudden changes in smaller dc currents can lead to startle. At frequencies of about 1 kHz and higher, the threshold of startle reaction is ap-proximately equal to 1 mA per kHz of frequency. **(a)** Explain why higher frequencies require higher currents to cause the startle reaction. *Hint:* It relates to charging of the body. (3) Currents above 5 mA at 60 Hz can cause rapid contraction and relaxation cycle called *muscle tetanization*, which lasts as long as the current flows. Like perception, tetanization occurs at a higher current threshold for dc and for higher frequencies. (4) *Ventricular fibrillation*, a dis-organized and not spontaneously reversible arrhyth-mic motion of the heart, can be triggered by a short dc current burst, but it typically requires larger cur-rents. For times less than about 0.1 second, the lim-its for ac and dc current are the same. For longer durations, dc current has higher limits. Ventricular fibrillation might be caused by a 20 mA rms cur-rent at 60 Hz; the corresponding dc current would be about 55 mA. (5) The onset of *burning* occurs at a current of about 70 mA rms, independent of fre-quency. **(b)** Explain this frequency independence, and why burns are more of a danger at high fre-quencies than are shocks.

"When I placed the primary conductor in one corner of a large lecture room 14 meters long and 12 meters wide, the sparks [in the secondary] could be perceived in the farthest parts of the room; the whole room seemed filled with the oscillations of the electric force."

—Heinrich Hertz,
discoverer of electromagnetic radiation, predicted in 1865 by Maxwell (1888)

"It seemed to me that if the radiation could be increased, developed, and controlled, it would be possible to signal across space for considerable distances."

—Guglielmo Marconi,
inventor of radio communications,
in his later life reflecting on the idea he had in 1894, at the age of 20

Chapter 15

Maxwell's Equations and Electromagnetic Radiation

Chapter Overview

Section 15.1 introduces the chapter. Section 15.2 presents a brief history of modern communications. Section 15.3 launches into the technical aspects of our subject, with a discussion of Maxwell's *displacement current*, which serves as a new source for the magnetic field in Ampère's law. This completes the equations describing the electromagnetic field—Maxwell's equations—which have solutions that correspond to electromagnetic waves. To prepare you for electromagnetic waves, the next two sections discuss waves on a string, Section 15.4 obtaining the wave equation for the string, and Section 15.5 solving it for both standing waves (as for a guitar) and traveling waves (as on a long taut string). Section 15.6 solves Maxwell's equations for a particularly simple form of electromagnetic wave, known as a plane wave. Section 15.7 points out that electromagnetic radiation covers a spectrum ranging from the low-frequency waves associated with power lines, on up to successively higher frequencies in microwaves, visible light, and gamma rays. Section 15.8 considers power flow associated with a plane wave, and Section 15.9 considers the associated momentum flow. Section 15.10 considers polarization, a consequence of the fact that in free space the electromagnetic field vectors $\vec{E}$ and $\vec{B}$ are normal to the direction of propagation. Section 15.11 considers electromagnetic standing waves, as in a microwave cavity. Section 15.12 considers traveling waves for microwave waveguides and coaxial cables. Section 15.13 shows how the electric and magnetic properties of matter can influence, through the dielectric constant and the magnetic permeability, the velocity of an electromagnetic wave. We then show how this influences the phenomena of reflection and refraction. Section 15.14 reviews the critical experiments whereby Hertz established that, for a circuit that produced microwave oscillations, it also produced radiation in space whose properties were much like that for visible light. Subsection 15.15.1 deals with energy flow for traveling waves on a string, and Subsection 15.15.2 deals with compressional waves. ∎

15.1 Introduction

We are now prepared to take the final step in our study of electromagnetism, and treat electromagnetic radiation. Before getting to the physics, we will remind you of what the world was like before there were communications based upon electrical signaling, either along wires or through space itself. We will discuss matters of history and culture, matters that some take for granted, and of which others are completely unaware. For those who still think that electromagnetic radiation is magic, we are about to shatter that illusion. For those who no longer believe it is magic, we would like to encourage you to regain a sense of awe, a feeling that perhaps there is just a little bit of magic at work when radio and television signals are generated, traverse the earth, and then are detected and amplified for our business or pleasure. Electromagnetic radiation, of course, is not magic. Nevertheless, it is an amazing fact that in a mere 100 years since its discovery, humans have harnessed its power and now are completely dependent on it.

15.2 A Brief History of Communications

What follows is a brief history of communications, especially communications using electrical and electromagnetic signals.

Today, we take it for granted that we will be able to communicate with one another using phones that use phone lines of copper wire or optical fiber, and may also include microwave transmission and satellite linkups. The late 1990s saw the first widespread use of e-mail and the World Wide Web. All of our communications methods use some part of the spectrum of electromagnetic radiation, based upon the principles laid down in the previous chapters, and one more—due to Maxwell—that we shall develop shortly.

Only in 1837, after a number of years of inventors trying to transmit electric signals over a wire to cause mechanical movement, did Morse put the first commercial telegraph into operation. Previous to the electric telegraph, smoke signals (ineffective in winds), lantern-flashing (ineffective in cloudy weather), carrier pigeons (ineffective in winds, cloudy weather, and during mating season), and messengers (ditto) had been employed. Nevertheless, Paris and Lille, separated by 150 miles, could convey messages within two minutes, using a flag semaphore system with many intermediate flagmen. By 1861, the overland telegraph had improved to the point where New York and San Francisco were in contact with each other.

15.2.1 *Telephony and Telegraphy*

In the mid-1850s, telegraph cable was first laid across the Atlantic—and failed, because the tried-and-true empirical methods that telegraph engineers had developed from nearly 20 years of experience with overland routes were inadequate to the underwater environment. It was then that the young William Thomson (1824–1907) applied his prodigious scientific talents to the Atlantic cable problem, both developing a theory for it and developing more accurate and more sensitive methods to generate and detect telegraph signals. (The baud rate for the original Atlantic cable was perhaps 1 bit/s. In the year 2000, computer

Telegraphy and Water

Recall that when a signal is sent along a wire, the wire is not in equilibrium, and cannot be considered to be an equipotential. Different parts of the wire are at different voltages, and thus a current flows in the wire, according to Ohm's law. Moreover, that voltage is due to charge, according to Volta's law. From these ideas, in 1853 Kelvin developed a theory of the telegraph. The main reason signaling by single-wire telegraph cable is more difficult underwater than in air is that the capacitance is much greater in water. This is because charge on a wire in water polarizes the water, and due to the large dielectric constant of water ($\kappa \approx 80$), a nearly equal and opposite amount of charge in the water surrounds the charge on the (insulated) wire. Hence, at a given charge on a wire, for a wire in air the voltages along the wire are much greater than for a wire in water. Faraday discovered this difference in capacitance at the beginning of 1853, about the same time as Kelvin's theoretical work.

A secondary reason signaling by single-wire telegraph cable is more difficult underwater than in air is that the effective cable resistance is greater with water. The time-varying electric currents induced in the water by the signal along the cable lead to additional Joule heating and increased dissipation of energy. This corresponds to an increased effective resistance for the cable.

modems attained nearly 56,000 bits/s, and Integrated Service Digital Network—ISDN—lines commonly attained nearly 128,000 bits/s.) For his contributions to telegraphy, Thomson became rich, and was made a baron and then a lord: Lord Kelvin, the name by which he is more commonly known today. The telephone, which to reproduce speech required much more faithful reproduction of electrical signals than the telegraph, was invented in 1875.

Kelvin's practical work in the field of communications was to be superseded by that of Maxwell on electromagnetic radiation. Although probably the first person with whom Maxwell shared his electromagnetic theory, Kelvin initially did not accept Maxwell's prediction of electromagnetic radiation. Unfortunately, Maxwell, who died in 1879 at the age of 48, did not live to see his 1865 predictions verified by Hertz, in 1887. Kelvin wrote the preface to the English translation of Hertz's book, *Electric Waves*.

15.2.2 *Maxwell's Contribution*

For a number of years, Maxwell worked to express, in mathematical form, the physical ideas of Faraday. One of them is that electric charge is associated with electric field lines—Gauss's law. A second is the law that there is no true magnetic charge (so the flux of $\vec{B}$ through any closed surface is always zero). A third is that the emf produced in a circuit is proportional to the rate at which magnetic flux lines cross the circuit—Faraday's law, first given mathematical form by F. Neumann in 1845. In addition, Maxwell tried to develop an analogy due to Faraday, between electric currents and magnetic fields. This appears in a letter to Kelvin, dated 1861. In the end, Maxwell dropped this analogy, but he put another in its place. He found that, along with current (both free and Ampèrian), there should be a new source term on the right-hand side of Ampère's law.

The laws of electromagnetism—called Maxwell's equations—appear to be complex. However, it should be kept in mind that Maxwell used a very different and even more complex-looking—but equivalent—formulation of the equations of electromagnetism than we use today. What we today call Maxwell's equations actually were first published by Heaviside in 1885 and, independently, by Hertz, in 1890.

We now turn to Maxwell's greatest scientific discovery, the one on which wireless telecommunications is based. It is also the basis of all cable communications, including both metal wire and fiber optics. Indeed, Heaviside studied Maxwell's theory in order to apply it to telegraphy and telephony.

15.3 Maxwell's New Term—The Displacement Current

It is reasonable to inquire, "If, in Faraday's law, a time-varying magnetic field can produce a circulating electric field, why can't, in Ampère's law, a time-varying electric field produce a circulating magnetic field?" Once this question is asked, the issue becomes to determine how much circulating magnetic field is produced, and in which direction. However, direct observation of the effect by, for example, the rapid discharge of a capacitor, is difficult, even with modern equipment. Indeed, even after Maxwell determined, by theoretical means, how large this term is, it took over 20 years before the most important effect it implied—electromagnetic radiation—was detected, by Hertz.

The key element for Maxwell was the realization that a new term was needed in order to make Ampère's law consistent with conservation of charge. We present two justifications, one very general and one using the specific example of the charging of a capacitor.

15.3.1 *General Argument*

We already have three of the equations that describe the electromagnetic field. To repeat, they are Gauss's law, or

$$\oint \vec{E} \cdot \hat{n} dA = 4\pi k Q_{enc}, \qquad \text{(Gauss's law)} \qquad (4.8)$$

the no-magnetic-pole law, or

$$\oint \vec{B} \cdot \hat{n} dA = 0, \qquad \text{(no-magnetic-pole law)} \qquad (9.21)$$

and Faraday's law, or

$$\oint \vec{E} \cdot d\vec{s} = -\frac{d}{dt} \int \vec{B} \cdot \hat{n} dA. \qquad \text{(Faraday's law)} \qquad (12.4)$$

The fourth law would appear to be Ampère's law, but there is something missing from it (as Maxwell noticed), related to charge conservation. Charge conservation states that the flux of electric current $\vec{J}$ leaving a closed surface must result in a decrease of the charge enclosed. Mathematically, this takes the form

$$\oint \vec{J} \cdot d\vec{A} = -\frac{dQ_{enc}}{dt}.$$ (15.1)

Taking the time derivative of (4.8) and using (15.1) leads to the result that

$$\oint \vec{J} \cdot \hat{n} dA = -\frac{1}{4\pi k} \oint \frac{d\vec{E}}{dt} \cdot \hat{n} dA.$$ (15.2)

Note that these are both integrals over *closed* surfaces.

Now consider Ampère's law,

$$\oint \vec{B} \cdot d\vec{s} = 4\pi k_m \int \vec{J} \cdot d\vec{A},$$ (15.3)

where the integral over $d\vec{A} = \hat{n} dA$ is over an *open* surface, and $d\vec{s} = \hat{s} ds$ and $d\vec{A}$ are related by the circuit-normal right-hand rule.

The integral on the right-hand side of (15.3) is very like the integral on the left-hand side of (15.2), except that in (15.3) the surface is not closed. Maxwell modified the right-hand side of (15.3) by adding in a term related to the right-hand side of (15.2). His modified Ampère's law reads

$$\oint \vec{B} \cdot d\vec{s} = 4\pi k_m \int \vec{J} \cdot d\vec{A} + \frac{k_m}{k} \frac{d}{dt} \int \vec{E} \cdot d\vec{A},$$

(Maxwell's modified Ampère's law) (15.4)

or

$$\oint \vec{B} \cdot d\vec{s} = 4\pi k_m \int (\vec{J} + \vec{J}_D) \cdot d\vec{A}. \qquad \vec{J}_D \equiv \frac{1}{4\pi k} \frac{d\vec{E}}{dt} = \epsilon_0 \frac{d\vec{E}}{dt}$$

(effect of displacement current) (15.5)

$\vec{J}_D$ was called by Maxwell the *displacement current density*. The actual current density $\vec{J}$, like the electric field $\vec{E}$, has sources and sinks (such as capacitor plates). However, rearranging (15.2) and using (15.5) reveals that $\vec{J} + \vec{J}_D$ has no sources or sinks: $\oint \vec{J} \cdot d\vec{A} = 0 = -\oint \vec{J}_D \cdot d\vec{A} = 0$, so $\oint(\vec{J} + \vec{J}_D) \cdot d\vec{A} = 0$.

Equations (4.8), (9.21), (12.4), and (15.4) are collectively known as *Maxwell's equations*. For our purposes $(k_m/k)^{-1} = 9 \times 10^{18}$ m^2/s^2, which is essentially c^2, where c is the velocity of light in free space. It thus should come as no surprise that light is somehow related to electricity and magnetism.

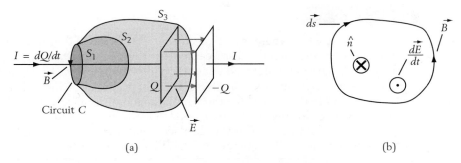

(a)   (b)

Figure 15.1 (a) Wire carrying current I to charge a pair of capacitor plates. The open surfaces S_1, S_2, and S_3 all have the same circuit C as their boundary. (b) On the same figure we represent, for C, a circuit element $d\vec{s}$ and a normal $\hat{n}$, and within the capacitor, the directions of $d\vec{E}/dt$ and $\vec{B}$.

15.3.2 *Displacement Current within a Capacitor*

The combination $\vec{J} + \vec{J}_D$ guarantees that, despite the ambiguity in determining which area to use for $\int \hat{n}\, dA$, the right-hand side of (15.4) is uniquely determined. Let us see how this works out for an electric circuit consisting of a long wire that is externally discharging the plates of a capacitor. At a given instant, let the current be I. Consider an Amperian circuit C that is a concentric circle surrounding the wire. It has a circulation given by the left-hand side of (15.4). To compute the right-hand side of (15.4), consider three surfaces associated with C. Surface S_1 is disk shaped, corresponding to the circle, through which a current I passes. See Figure 15.1(a).

The electric field is zero on S_1, so the right-hand side of (15.4) is given by

$$4\pi k_m \int \vec{J} \cdot d\vec{A} + \frac{k_m}{k}\frac{d}{dt}\int \vec{E} \cdot d\vec{A} = 4\pi k_m I + 0 = 4\pi k_m I. \qquad (15.6)$$

Surface S_2 is obtained by deforming the disk, as if it were completely extensible, while leaving the perimeter of the circle in place. The current I intersects S_2, as for S_1, and the electric field is zero on S_2, as for S_1, so the right-hand side of (15.4) is given by (15.6), as for S_1.

An even more extended version of S_2 is surface S_3, which has part of its surface cross between the plates of the capacitor. In that case, no current crosses the capacitor, but the electric field is nonzero within the capacitor. The electric field that crosses S_3 comes exclusively from the charge on the capacitor plate enclosed by S_3. Thus we can artificially close the surface S_3, and then use Gauss's law. Thus for S_3 the right-hand side of (15.4) is given by

$$4\pi k_m \int \vec{J} \cdot d\vec{A} + \frac{k_m}{k}\frac{d}{dt}\int \vec{E} \cdot d\vec{A} = 0 + \frac{k_m}{k}\frac{d}{dt}\int \vec{E} \cdot d\vec{A} \approx \frac{k_m}{k}\frac{d}{dt}\oint \vec{E} \cdot d\vec{A}$$

$$= \frac{k_m}{k}\frac{d}{dt}4\pi k Q_{enc} = 4\pi k_m I. \qquad (15.7)$$

Hence the right-hand side of (15.4) is the same for S_3 as it is for S_1 and S_2.

Application 15.1

In free space, where there is no true current, if the $\vec{E}$ field is into the page and starts to decrease, then the displacement current, by (15.5), will point out of the page. Let us apply Maxwell's extension of Ampère's law, (15.4), to include displacement current, and use Oersted's right-hand rule. Then a $\vec{B}$ field will circulate counterclockwise for the Amperian circuit in Figure 15.1(b). More formally, since clockwise $d\vec{s}$ corresponds to $\hat{n}$ into the page (and thus positive current into the page), both the displacement current (integrated over the cross-section) and the magnetic circulation are negative, as expected.

Hertz, in the introduction to his *Electric Waves*, noted that there are at least four ways to think about electricity: (1) action at a distance (Chapter 3); (2) the electric field produced by distant free charge (this is little more than action at a distance, as in Chapter 4); (3) the sum of the electric fields produced by distant free charge and by locally neutral polarization charge (e.g., in a dielectric, as in Chapter 7); (4) the electric field described solely by the local polarization of space $\vec{P}$, as embodied in the physical picture of flux tubes, and the mathematics of the displacement vector $\vec{D} \equiv \varepsilon_0 \vec{E} + \vec{P}$, whose only source is free charge. Hertz believed that Maxwell thought in the fourth fashion. Maxwell himself never really told us. He just left us the equations.

The presence of the displacement current is felt constantly; it is essential to electromagnetic radiation. In short, without Maxwell's new term, there would be no radio, no television, no sunlight—and no life.

15.4 Equation of Motion for a String under Tension

For Maxwell, immersed in the physics of his time, it was second nature to recognize a wave equation. A beginning student of physics may not be able to do this. It is important, therefore, that you learn how to recognize a wave equation and that you become aware of some of its properties. Because a string under tension supports waves, its equation of motion should support waves. For that reason, we study the motion of a string.

First, however, note that all waves satisfy the relationship

$$\lambda f = v, \tag{15.8}$$

where λ is the wavelength, f is the frequency in Hz (cycles per second), and v is the velocity of propagation. (Notational woe: We earlier used the symbol λ for charge per unit length, so we again have the problem of too many quantities to represent. In this chapter, λ will refer *exclusively* to wavelength.)

We now make a distinction between two classes of waves.

15.4.1 *Nondispersive and Dispersive Waves*

Nondispersive waves have the same velocity for all frequencies. This includes waves on a uniform string, sound waves in air, and electromagnetic waves in

> **Nondispersive Waves**
>
> Seismic waves on the solid surface of the earth are nondispersive. That is why distant measurements permit us to locate the position and intensity of distant earthquakes. For a given type of wave to be useful from the point of view of communication, it should be as nondispersive as possible. If water waves were nondispersive, we might be able to communicate from ship to distant ship by using them. They aren't, so we can't. We use our voices (sound waves) when near, and radio (electromagnetic radiation) when far.

free space. For each of these, there is a characteristic velocity of propagation that waves of all frequencies satisfy. As a consequence, if a complicated signal (like someone's voice) radiates outward, when it arrives at the listener, all the frequencies will arrive at the same time (i.e., nondispersively), and the sound will be intelligible.

Dispersive waves have a velocity that varies with frequency (or, equivalently, varies with wavelength). The most common example of this is waves on the surface of a body of water. These all travel at different velocities, the high frequencies traveling the fastest. A pebble dropped into a pool generates waves of many frequencies and, by (15.8), correspondingly many wavelengths. Hence, if a second pebble is dropped into the pool, a few seconds after the first, the high-frequency, short-wavelength waves from the second pebble will catch up to and outrun the low-frequency, long-wavelength waves from the first pebble. To a lesser extent, light in materials and sound are dispersive.

15.4.2 *Physical Picture*

We begin with some general considerations. For simplicity, neglect gravity. Attach the ends of a string of mass per unit length μ to two posts, such that the string is under tension F. (Notational woes again: Often the symbol T is used for *tension*, but consistent with our previous usage, T will be reserved for the *period*. The symbol T is also used for *temperature*.) The string then forms a straight line. See Figure 15.2(a). Rotating the string (a change in slope) does not cause it to vibrate. See Figure 15.2(b). Lifting the string up and down as a whole (a uniform displacement) does not cause it to vibrate. See Figure 15.2(c). Thus the acceleration of any part of the string does not depend on either the value of the vertical displacement y of the string or of its slope dy/dx, where x is the horizontal distance along the string.

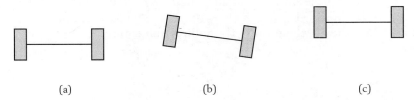

(a)   (b)   (c)

Figure 15.2 A string under tension, which, when straight, does not move: (a) original configuration, (b) rotated configuration, (c) vertically displaced configuration.

Moreover, if the string is wrapped on a curved frame, and the frame suddenly is removed, the string will not retain its shape: it will start to move. This tells us that curvature, which is proportional to d^2y/dx^2, can cause the string to accelerate. Since acceleration is given by d^2y/dt^2, we expect that the second time derivative is proportional to the second space derivative. In other words, *the acceleration is proportional to the curvature.*

Now consider the proportionality constant. The acceleration d^2y/dt^2 should be proportional to the force that acts on it (and this force should be proportional to the tension F), and inversely proportional to the string's mass (which should be proportional to the mass per unit length μ). Thus, if we are lucky about the dimensionality of F and μ, and doubly lucky about the arbitrary constant, we might guess that

$$\frac{d^2y}{dt^2} = \frac{F}{\mu}\frac{d^2y}{dx^2}. \tag{15.9}$$

In fact, this equation is correct. A check that it has the correct dimensionality can be done by observing that, because of the space and time dependences in the denominators of (15.9), F/μ should have the same dimension as the square of a velocity. To verify this, note that F has the same dimension as a force, or N, and that μ has the same dimension as a mass per unit length, or kg/m. Thus F/μ has the same dimension as N-m/kg = J/kg = (m/s)^2, which is indeed the square of a velocity.

This discussion neglects motion in and out of the page (i.e., along z). Because gravity has been neglected, motion along z will be similar to motion along y. Thus motion along z can be described by (15.9) with the y's replaced by z's.

A comment about notation: The variable y depends on both x and t. In (15.9), the d/dx's are taken at fixed t and the d/dt's are taken at fixed x. It is conventional to use another symbol for these *partial derivatives*, where only one independent variable is changed. (Think of independent variables like t as input, and dependent variables like x as output.) Thus, the mathematical cognescenti write $\partial/\partial x$ or ∂_x instead of d/dx, and $\partial/\partial t$ or ∂_t instead of d/dt. We will be a bit sloppy in what follows, but at least we have paid lip service to the mathematical conventions that you will see in more advanced courses.

15.4.3 *Derivation of the Wave Equation*

Let us now derive (15.9). It applies in the limit where the slope dy/dx is small. Consider a closeup diagram of the string. See Figure 15.3.

In a spatial region of small length dx, the length of string is $ds = \sqrt{dx^2 + dy^2} = dx\sqrt{1 + (dy/dx)^2} \approx dx$. Thus dx has mass $dm = \mu ds \approx \mu dx$. The mass times the acceleration in the y-direction is thus

$$dm\frac{d^2y}{dt^2} \approx \mu dx\frac{d^2y}{dt^2}. \tag{15.10}$$

Horizontal motion of the string would cause significant extension or compression, which would change the tension F. However, since the motion of the string is essentially vertical, such tension-changing effects are negligible. In that

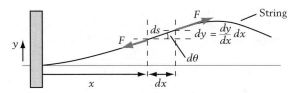

Figure 15.3 Closeup of string under tension F. In the absence of constraining forces, a curved string will move, even if it initially is at rest.

case, as can be seen in Figure 15.3, the net force dF_y on dm in the y-direction is given by the difference in the y components of the nearly constant tension F. Since at x the angle $\theta(x)$ that the string makes to the horizontal is small, we can set $\sin\theta \approx \tan\theta \equiv dy/dx$. Then

$$F_y = F\left[\sin\theta(x+dx) - \sin\theta(x)\right] \approx F\left[\tan\theta(x+dx) - \tan\theta(x)\right]$$

$$= F\left[\frac{dy}{dx}\bigg|_{x+dx} - \frac{dy}{dx}\bigg|_{x}\right] \approx F\frac{d^2y}{dx^2}dx. \tag{15.11}$$

The last approximate equality of (15.11) used the straight-line approximation (valid for small dx) that $f(x+dx) - f(x) \approx (df/dx)\,dx$, with $f(x) = dy/dx$. By Newton's second law of motion, we equate (15.10) to (15.11), thus obtaining (15.9).

Certainly, we believe that (15.9) has wave solutions because stringed instruments—like violins and violas and cellos and guitars and basses—support wave motion. Let us use that to guide us in finding wave solutions to (15.9). As a start, rewrite (15.9) as

$$\frac{d^2y}{dt^2} = v^2\frac{d^2y}{dx^2}, \qquad v^2 = \frac{F}{\mu} \qquad \text{(wave equation for a string)} \tag{15.12}$$

where v has the dimensions of a velocity.

We will discuss two types of solutions to (15.12), *standing waves* and *traveling waves*.

15.5 Waves on a String

Standing waves occur when there are fixed boundaries, as with stringed instruments. Traveling waves occur when there are open boundaries, or when we are far from the boundaries, as when we shout from the center of an immense auditorium.

15.5.1 *Standing Waves and the Theory of Stringed Instruments*

Now, what does a wave for a stringed instrument look like? If its ends are at $x = 0$ and $x = L$, then we may choose coordinates that make $y = 0$ at these ends. See Figure 15.4.

When a string on a stringed instrument is plucked or picked or pulled or bowed, it makes a characteristic sound. Analysis of this sound shows that it is dominated by a single frequency. Therefore, let us try a solution of the form

$$y(x, t) = g(x) \sin(\omega t + \phi). \tag{15.13}$$

Here there are many unknowns: the frequency ω in radians per second, the shape function $g(x)$, and the phase ϕ. (Recall that ω, the frequency in radians per second, equals 2π times f, the frequency in cycles per second.) The function $g(x)$ must satisfy the *boundary conditions* that $g(0) = 0$ and $g(L) = 0$, in order to make $y = 0$ at the ends. Substituting (15.13) into (15.12) yields

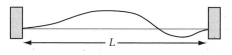

Figure 15.4 String that is fixed at its ends.

$$-\omega^2 g = v^2 \frac{d^2 g}{dx^2}, \qquad v^2 = \frac{F}{\mu}, \tag{15.14}$$

where we have factored out $\sin(\omega t + \phi)$.

This is a disguised version of the harmonic oscillator equation, (14.11). With K the spring constant, M the mass, and X the position coordinate, the harmonic oscillator equation is

$$-KX = M \frac{d^2 X}{dt^2}. \tag{15.15}$$

In (15.15), t is the independent variable and X is the dependent variable. Equation (15.15) has as its solution

$$X(t) = A \sin(\Omega t + \phi_0), \qquad \Omega = \sqrt{\frac{K}{M}}, \tag{15.16}$$

where ϕ_0 and A are determined by what are called the *initial conditions:* the initial values of X and dX/dt. [Don't be bothered by the use of Ω in (15.16), instead of ω_0 as in (14.12); we already have used the lookalike ω in (15.14).] Comparison shows that the set (K, X, M, t) of (15.15) is just like the set (ω^2, g, v^2, x) of (15.14). Since the solution to (15.15) is (15.16), we can obtain the solution to (15.14) by substituting the appropriate quantities. Replacing $\Omega = \sqrt{K/M}$ by $q = \sqrt{\omega^2/v^2} = \omega/v$, the solution for the shape function $g(x)$ of (15.14) is given by

$$g(x) = A \sin(qx + \phi_0), \qquad q = \frac{\omega}{v}. \tag{15.17}$$

We call q (yet another notational woe—q is not to be mistaken for a charge!) the *wavenumber.* In (15.17), q plays the same role as Ω does in (15.16), and as ω_0 did in (14.12). (We really have a problem with too many quantities for the number of symbols at our disposal. Other symbols often used for wavenumber are Q, K, and k. Each of them has its notational difficulties!)

Equation (15.17) still isn't a solution to our problem even though it satisfies the differential equation (15.14). The problem is that it doesn't yet satisfy the boundary conditions $g(0) = 0 = g(L)$. To satisfy $g(0) = 0$, set $\phi_0 = 0$ in (15.17). To satisfy $g(L) = 0$, set $\sin qL = 0$ in (15.17), which implies that

$$q = \frac{n\pi}{L}. \qquad (n \text{ a nonzero integer}) \qquad (15.18)$$

Hence, the boundary conditions restrict the allowed values of the wavenumber q. It is universally the case that boundary conditions impose restrictions. See Figure 15.5 for the modes corresponding to $n = 1$ and $n = 2$.

Let's now finish things up. Placing (15.17) into (15.13) yields the space and time variation of the displacement y of a string tied down at $x = 0$ and $x = L$. It is

$$y(x, t) = A \sin qx \sin(\omega t + \phi), \qquad q = \frac{n\pi}{L}. \qquad \text{(standing wave for string)} \qquad (15.19)$$

where, from (15.17) and (15.18), ω only takes on the values

$$\omega = vq = n\frac{\pi v}{L}, \qquad v = \sqrt{\frac{F}{\mu}}. \qquad \text{(vibrational frequencies for string)} \qquad (15.20)$$

This equation gives the natural vibrational frequencies, or *harmonics* of a string under tension. The $n = 1$ frequency is called the *fundamental*, or *first harmonic*. The $n = 2$ frequency is called the second harmonic, or first *overtone*. Equation (15.20) also contains the theory of the tuning of stringed instruments. For a given instrument the length L is fixed; thus the shorter instruments will have the higher frequencies. For a given length and tension, the lighter the string the higher the frequency. (Note: All strings on a given instrument have about the same tension; otherwise, some supports would be built stronger than others.) For a given length and mass per unit length, the higher the tension the higher the frequency. Any motion of the string is a superposition over a proper amount (with the proper phase) of each of the modes.

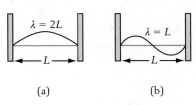

(a)   (b)

Figure 15.5 (a) Fundamental mode of a uniform string. (b) Second harmonic of a uniform string.

Waves satisfying (15.19) are called standing waves because they retain their shape (given by $\sin qx$) although their amplitude changes with time. We say that the string *resonates* at the *resonant frequencies* given by (15.20). The acoustic resonances of a room and the electromagnetic resonances of a microwave cavity are analogous to the resonance of a string.

> **Example 15.1** **Tuning a guitar string**
>
> A guitar string is 60 cm long and has a mass per unit length of 2.2 g/m. Find the tension it should be given so that its third harmonic has a frequency $f = 690$ Hz.
>
> **Solution:** Since $\omega = 2\pi f = 4335$ s^{-1}, $L = 60$ cm, and $n = 3$, the first part of (15.20) gives $v = \omega L/n\pi = 276$ m/s. The second part of (15.20) gives $F = \mu v^2 = 167.6$ N. This would lift a mass $m = F/g = 17.0$ kg under the earth's gravity.

15.5.2 *Traveling Waves*

Traveling waves, as seen by an observer at rest, do not repeat in time. They radiate energy away from the region in which they are generated. There is something special about the wave equation given by (15.12): its traveling waves do not change shape. (This is because the velocity is independent of the wavelength; it is *nondispersive*.) Thus, for an observer moving with the traveling wave, the wave will appear to be at rest. In this section, we derive $\lambda f = v$, and we show that (15.12) has solutions that correspond to waves with velocity v.

Consider a possible solution to (15.12) with the form

$$y(x, t) = g(x - vt). \tag{15.21}$$

Clearly, in (15.21), derivatives of g with respect to x are proportional to derivatives with respect to t. Indeed, $dy/dt = -v(dy/dx)$, and $d^2y/dt^2 = v^2(d^2y/dx^2)$, which is (15.12). Hence (15.21) is indeed a solution to (15.12). Moreover, $g(x - vt)$ represents a wave that travels to the right at velocity v because the $x = 0$, $t = 0$ value $g(0)$ is at $x = vt$ at time t. See Figure 15.6.

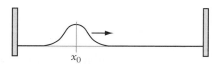

Figure 15.6 A rightward-moving traveling wave.

Note that, in a wave, the *material* doesn't travel rightward, only the position of the peak (and, more generally, the shape) of the wave. A chalk mark near the left end of the string doesn't travel to the right with a rightward wave, but rather it moves vertically. Similarly, we don't get carried to shore when we are floating on the water and a water wave goes by.

Now consider, far from the walls, the specific waveform

$$g(x - vt) = \sin[q(x - vt)], \tag{15.22}$$

where the wavenumber q is arbitrary. (We could also add an arbitrary phase ϕ, or use a cosine function.) We can relate q to both the frequency ω and the wavelength λ.

First, if at fixed x we wait a time period $t = T = 1/f = 2\pi/\omega$, then by definition, in (15.22) there will be a decrease in phase of 2π. Explicitly, in (15.22) this decrease in phase is $qvT = qv(2\pi/\omega) = 2\pi$, so

$$qv = \omega = 2\pi f. \tag{15.23}$$

Also, if at fixed t we displace ourselves by a wavelength λ, then, by definition, in (15.22) there will be an increase in phase of 2π. Explicitly, in (15.22) this increase in phase is $q\lambda = 2\pi$, so

$$q = \frac{2\pi}{\lambda}. \qquad \text{(wavenumber–wavelength relation)} \qquad (15.24)$$

Taking the ratio of (15.23) and (15.24) yields

$$\lambda f = v. \qquad \text{(frequency–wavelength relation)} \qquad (15.25)$$

As indicated earlier, this relationship is satisfied by all waves. If v is independent of frequency, as for nondispersive waves, the frequency and wavelength are inversely proportional to each other.

Note that the physical motion of the string is along y, whereas the wave propagates along x. Hence the motion of the string is *transverse* to the direction of propagation. In contrast, sound waves in air, if they travel along x, produce a displacement of the air along x, and thus the motion is *longitudinal* to the direction of propagation.

Example 15.2 **From frequency to wavelength**

Take the velocity of sound in air and in water to be $v_{air} = 340$ m/s, and $v_{water} = 1500$ m/s. For the guitar string of the previous example, $v = 276$ m/s. For a frequency $f = 300$ Hz, find the corresponding wavelengths.

Solution: Using (15.25), $f = 300$ Hz yields wavelengths $\lambda_{air} = 1.133$ m and $\lambda_{water} = 5.0$ m, and $\lambda_{guitar} = 0.92$ m.

You are now experts on the properties of wave equations, and you know how to recognize a wave equation—(15.12)—when you see one. We can now return to the problem of electromagnetic waves.

15.6 Electromagnetic Waves

Now consider, in three-dimensional space, an electromagnetic wave that has no y or z dependence. This is called a *plane wave*, since for any point on any plane $x = constant$, the wave has the same amplitude.

Our derivation for plane wave electromagnetic radiation closely follows the derivation of the last chapter for the skin depth. It employs the same rectangular circuits, and it employs the same laws, Ampère's law and Faraday's law, except that now we employ Maxwell's version of Ampère's law. The result for Faraday's law is *exactly* the same as in the previous chapter. If you are familiar with that section, don't bother to read the subsection on Faraday's law. The result for Ampère's law changes, however, because now, instead of the real current, there is the displacement current.

Consider empty space for the region $x > 0$. Let the $x = 0$ plane be a sheet of conductor that provides a time-varying current along the y-direction. Hence, at least within the sheet, by Ohm's law there is an electric field along the y-direction. We therefore assume that the electric field points along the y-direction. Moreover, from Ampère's right-hand rule, we expect a magnetic field near the sheet that points along the z-axis. We therefore assume that the magnetic field points along the z-direction. A plane wave with this form is said to be *linearly polarized* because as time goes by the electric field points along or against the same linear direction in space (and similarly for the magnetic field). In contrast, light from a lightbulb or from the sun has electric and magnetic field vectors whose direction in space changes nearly randomly (and rapidly) with time. Such light waves are said to be *unpolarized*.

Our goal is to determine the two equations describing how the electric and magnetic fields E_y and B_z vary in space and in time. We assume that the only spatial dependence comes from x.

15.6.1 *Use of Faraday's Law*

Because the magnetic field is along z, it produces a magnetic flux along z. Let us therefore apply Faraday's law to a small rectangular circuit whose normal $\hat{n}_1$ is along z, with dimension h_1 along y and dx along x. See Figure 15.7.

By the circuit-normal right-hand rule, $d\vec{s}_1$ must be taken to circulate as in the figure. Then the electric circulation is given by

$$\oint \vec{E} \cdot d\vec{s}_1 = [E_y(x + dx) - E_y(x)]h_1 \approx \frac{dE_y}{dx}(h_1 dx). \qquad (15.26)$$

The associated magnetic flux is

$$\Phi_B = \int \vec{B} \cdot d\vec{A} = \int \vec{B} \cdot \hat{n}_1 dA = B_z(h_1 dx). \qquad (15.27)$$

The negative rate of change of the associated magnetic flux is

$$-\frac{d\Phi_B}{dt} = -\frac{dB_z}{dt}(h_1 dx). \qquad (15.28)$$

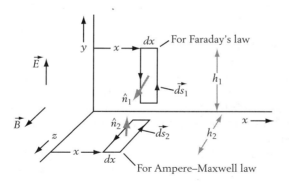

Figure 15.7 Geometry describing electromagnetic radiation flowing to the right, caused by an electric current on the $x = 0$ plane, oscillating along y. Two imaginary circuits are drawn, one for use with Faraday's law (involving magnetic flux), and the other for use with the Ampère–Maxwell law (involving electric flux).

Using Faraday's law to equate (15.26) and (15.28) yields

$$\frac{dE_y}{dx} = -\frac{dB_z}{dt}.$$

(15.29)

15.6.2 *Use of Ampère's Law, as Modified by Maxwell*

Because the electric field is along y, it produces an electric flux along y. Let us therefore apply Ampère's law to a small rectangular circuit whose normal $\hat{n}_2$ is along y, with dimension h_2 along z and dx along x. By the circuit-normal right-hand rule, we must take $d\vec{s}_2$ such that it circulates as in Figure 15.7. Then the magnetic circulation is given by

$$\oint \vec{B} \cdot d\vec{s}_2 = [-B_z(x+dx) + B_z(x)]h_2 \approx -\frac{dB_z}{dx}(h_2 dx).$$

(15.30)

For empty space, the true electric current is zero, but the displacement current is nonzero. From (15.5), it is given by

$$\int \vec{J}_D \cdot d\vec{A} = \int \vec{J}_D \cdot \hat{n}_2 dA = J_{Dy}(h_2 dx) = \frac{1}{4\pi k}\frac{dE_y}{dt}(h_2 dx).$$

(15.31)

Using Ampère's law to relate (15.30) and $4\pi k_m$ times (15.31) yields

$$-\frac{dB_z}{dx} = \frac{k_m}{k}\frac{dE_y}{dt} = \mu_0\epsilon_0\frac{dE_y}{dt}.$$

(15.32)

15.6.3 *The Electromagnetic Wave Equation*

Taking the time derivative of (15.29) and the x-derivative of (15.32), we can eliminate $d^2E_y/dtdx = d^2E_y/dxdt$, to obtain

$$\frac{d^2B_z}{dx^2} = \frac{k_m}{k}\frac{d^2B_z}{dt^2} = \mu_0\epsilon_0\frac{d^2B_z}{dt^2}.$$

(15.33)

A similar equation can be derived for E_y. Not only are $\vec{E}$ and $\vec{B}$ normal to each other, they are normal to the direction of propogation, which is along x.

Comparison to (15.12) shows that (15.33) is a wave equation, with velocity

$$c = \sqrt{\frac{k}{k_m}} = \frac{1}{\sqrt{\mu_0\epsilon_0}} \approx 3.0 \times 10^8 \frac{m}{s}.$$

(speed of light in vacuum)

(15.34)

Thus Maxwell's equations have, in free space, a solution corresponding to a coupled wave involving both the electric and magnetic fields, which propagate at a speed identical to the speed of light c in vacuum! Surely this is a hint that light is a form of electromagnetic radiation; indeed, the electromagnetic nature of light has been borne out by experiments and practical applications for over 100 years.

15.6.4   *Properties of Electromagnetic Waves*

Consider a traveling wave of the form

$$B_z(x,t) = A \sin(qx - \omega t). \tag{15.35}$$

Substitution into (15.32) yields, with (15.34),

$$\frac{dE_y}{dt} = -\frac{k}{k_m}\frac{dB_z}{dx} = -c^2 Aq \cos(qx - \omega t). \tag{15.36}$$

This integrates to

$$E_y = c^2 A\frac{q}{\omega}\sin(qx - \omega t) = cA\sin(qx - \omega t), \tag{15.37}$$

where we have used (15.23) with $v = c$. Comparison of (15.35) and (15.37) gives

$$E_y = cB_z, \quad \text{or} \quad |\vec{E}| = c|\vec{B}|, \quad \text{or} \quad E = cB, \tag{15.38}$$

for this rightward-traveling wave. A leftward-traveling wave would have the waveform $\cos(qx + \omega t)$, and $E_y = -cB_z$.

From this discussion we conclude that, in a vacuum, electromagnetic waves have the following properties:

1. They travel with the velocity of light c.

2. Their frequency ω is related to their wavenumber q by $\omega = cq$.

3. Like waves on a string, they are transverse. That is, the quantities that vary—$\vec{E}$ and $\vec{B}$—are normal to the direction of propagation $\hat{q}$. Then, just as $(\hat{i}, \hat{j}, \hat{k})$ is a right-handed triad, so is $(\vec{E}, \vec{B}, \vec{q})$ a right-handed triad. More specifically, $\vec{E} \times \vec{B}$ points in the direction of propagation $\hat{q}$, so

$$\vec{E} = -c\hat{q} \times \vec{B}, \qquad \vec{B} = \frac{1}{c}\hat{q} \times \vec{E}. \tag{15.39}$$

4. $|\vec{E}| = c|\vec{B}|$.

Example 15.3   **Properties of radiation**

Consider an electromagnetic wave that propagates along $\hat{y}$, with $\vec{E}$ instantaneously pointing along $\hat{z}$. $\vec{E}$ has a maximum amplitude E_m of 18 V/m. (a) Find the instantaneous direction of $\vec{B}$. (b) Find the maximum value of $|\vec{B}|$.

Solution: (a) Because $\vec{E}$, $\vec{B}$, and the direction of propagation $\hat{q}$ are mutually perpendicular, $\vec{B}$ must point along $\pm\hat{x}$. Specifically, $\vec{B}$ along $+\hat{x}$ satisfies (15.39). (b) By (15.38), the maximum value B_m of $|\vec{B}|$ is $B_m = E_m/c = 6.0 \times 10^{-8}$ T.

Finally, note that by Fourier's theorem, any shape can be reproduced as a sum over sines and cosines if enough wavelengths are included. Hence we can decompose, for example, a rightward-traveling localized pulse, into a sum (or integral) over waves with all wavelengths. Now, since in a vacuum each wavelength travels at the same velocity, that means that a localized pulse as a whole travels rightward at that same velocity, without any change in shape. This nondispersive behavior also holds for sound in air and for waves on a string. As discussed earlier, that is why sound and light are useful for communications.

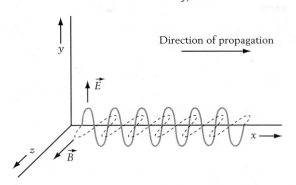

Figure 15.8 Representation of an electromagnetic wave that is traveling rightward. At any instant of time, its electric field and magnetic field are uniform along any plane defined by a constant value of x. In this case, $\vec{E}$ points along y and $\vec{B}$ points along z.

Figure 15.8 summarizes our results. This depicts two types of dependent vectors, the electric field vector $\vec{E}$, and the magnetic field vector $\vec{B}$, at a given instant of time. Properly, $\vec{E}$ and $\vec{B}$, having different units, do not really co-exist in same coordinate space; the x-axis represents the real space x-axis, the y-axis represents E_y space, and the z-axis represents B_z space. Nevertheless, we present Figure 15.8 to show how both $\vec{E}$ and $\vec{B}$ propagate together in time at the speed of light. As time goes by, the curves representing $\vec{E}$ and $\vec{B}$ would move rightward with velocity c.

15.7 The Full Electromagnetic Spectrum

Equation (15.34) is critical. It says that all electromagnetic waves in free space move with the same velocity. It also says that electromagnetic radiation can occur, in principle, at infinitely high and infinitesimally low frequencies, with corresponding short and long wavelengths. At the low-frequency end is ac power (60 Hz). Successively higher frequency and shorter wavelength give long radio waves, short radio waves, UHF, VHF, microwaves (produced by microwave tubes and by molecular rotations), infrared (produced by lasers and molecular vibrations), optical (produced by low-energy electron transitions), ultraviolet (produced by high-energy electron transitions), x-rays (produced by very high-energy electron transitions), and gamma rays (produced by high-energy nuclear transitions). Because of the Doppler effect, whereby the frequency of a wave increases (decreases) when the source approaches (recedes from) the observer, these forms of electromagnetic radiation are equivalent to one another, except for their different frequencies. See Figure 15.9.

Table 15.1 summarizes the wavelengths associated with visible electromagnetic radiation—light.

Table 15.1 Wavelengths of colors

Violet	Blue	Green	Yellow	Orange	Red
400–440 nm	440–480 nm	480–560 nm	560–590 nm	590–630 nm	630–700 nm

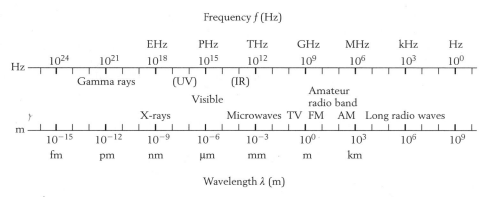

Figure 15.9 A representation of the electromagnetic spectrum. The upper line gives frequency f, and the lower line gives the corresponding wavelength λ, where $\lambda f = c$.

15.8 Electromagnetic Energy and Power Flow

An energy density is associated with an electromagnetic wave, and it travels at the velocity c of the wave. The total energy density u is the sum of the electric and magnetic energy densities, so from previous chapters,

$$u = u_E + u_B = \frac{E^2}{8\pi k} + \frac{B^2}{8\pi k_m}. \tag{15.40}$$

For the plane wave of Section 15.6, $E^2 = E_y^2$ and $B^2 = B_z^2$. Use of (15.34) then yields $c/k = (k_m c)^{-1}$, so with (15.38), Equation (15.40) becomes

$$u = \frac{E B c}{8\pi k} + \frac{E B}{8\pi c k_m} = \frac{E B}{8\pi c k_m} + \frac{E B}{8\pi c k_m} = \frac{E B}{4\pi c k_m}. \tag{15.41}$$

We now consider the flow of energy, using an argument similar to the one we used in Chapter 7 when we considered the flow of charge. Instead of charge density ne, we now consider energy density u; instead of drift velocity v_d, we now consider the speed of light c; and instead of electric current per unit area J, we now consider the power per unit area S.

Consider a small imaginary box of area A normal to x and thickness $dx \ll \lambda$. It has volume $A\,dx$ and contains an energy $u\,A\,dx$. The energy moves with velocity c so that this energy moves out of the box in a time $dt = dx/c$. Thus the energy per unit time, or power, crossing the box is $u\,A\,dx/dt = u\,Ac$. Hence the power per unit area S is given by $u\,Ac/A = uc$. Thus

$$S = uc = \frac{E B}{4\pi k_m} = \frac{E B}{\mu_0}. \qquad \text{(power per unit area)} \tag{15.42}$$

The energy flows in the direction of wave propagation. Another name for the power per unit area is the intensity, with symbol I.

A more general treatment shows that the energy flow and the magnitude of the intensity are both given by the *Poynting vector*

$$\vec{S} = \frac{\vec{E} \times \vec{B}}{4\pi k_m} = \frac{\vec{E} \times \vec{B}}{\mu_0}, \qquad \text{(EM power as a vector, the Poynting vector)}$$

$$(15.43)$$

Clearly, $\vec{S}$ is the vector form of S, so $S = |\vec{S}|$. The Poynting vector explains how power flows into a wire that is subject to Joule heating. The energy enters the sides of the wire, rather than along its axis, from the *electromagnetic field* to the wire. Of course, a voltaic cell or mechanical motion converted to an emf (by Faraday's law) is the ultimate source of energy for the electromagnetic field.

Example 15.4 Light from the sun at the earth

Visible radiation from the sun has an average intensity $\bar{S}$ of about 1500 W/m^2 at the earth's orbit, for which $R_E = 1.50 \times 10^{11}$ m. Find the characteristic value of the maximum electric and magnetic fields E_m and B_m incident on the earth.

Solution: Including a factor of one-half from averaging over an oscillation (as for ac circuits), (15.42) and (15.38) give

$$\bar{S} = \frac{E_m^2}{8\pi k_m c} = \frac{E_m^2}{2\mu_0 c}. \qquad (15.44)$$

Here E_m denotes a maximum electric field, averaged over all radiation frequencies. Solving for E_m gives $E_m = \sqrt{2\mu_0 c \bar{S}}$, which evaluates to $E_m = 1060$ V/m. Corresponding to this is $B_m = E_m/c = 3.54 \times 10^{-6}$ T.

Radiation by an isotropic spherical source. An *isotropic spherical source* is one for which the radiation intensity is the same in all directions, as for a lightbulb and for the sun. In this case, the average total power $\bar{\mathcal{P}}$ is obtained by multiplying the average intensity $\bar{S}$—a power per unit area—by the surface area $4\pi R^2$ of a sphere of radius R. Thus

$$\bar{\mathcal{P}} = \bar{S}(4\pi R^2). \qquad (15.45)$$

Since $\bar{\mathcal{P}}$ is independent of radius R, it is useful to rewrite (15.45) as

$$\bar{S} = \frac{\bar{\mathcal{P}}}{4\pi R^2}. \qquad (15.46)$$

By (15.44), the intensity $\bar{S}$ varies as E_m^2. Hence (15.46) implies that E_m falls off inversely with distance, and similarly for B_m.

Example 15.5 Light from the sun at Jupiter

Find $\bar{\mathcal{P}}_{sun}$, and $\bar{S}$, E_m and B_m at Jupiter. Take $R_J = 1.43 \times 10^{12}$ m.

Solution: Using values of $\bar{S}$ and R appropriate to the earth, from (15.45) we find that the sun radiates an average power $\bar{\mathcal{P}}_{sun} = 4.18 \times 10^{26}$ W! (Without this power, there would be no life on the earth.) Next, using $R_J = 1.43 \times 10^{12}$ m in (15.46), we deduce that, at Jupiter, $\bar{S} = 16.3$ W/m^2. Finally, using the values of E_m and B_m at the earth (Example 15.4), and the fact that E_m and B_m fall off inversely with radius, we deduce that $E_m = 111.2$ V/m and $B_m = 3.71 \times 10^{-7}$ T at Jupiter. Only terrestrial measurements of $\bar{S}$, R_E, and R_J were needed to obtain this information. (However, if the raw data on $\bar{S}$ is taken on the earth, compensation must be made for absorption and scattering by the earth's atmosphere.)

15.9 Momentum of Electromagnetic Radiation, Radiation Pressure

Optional

Just as the light wave carries energy, it also carries momentum—this despite the light having no mass! If it is any consolation, for light the ratio of momentum p to energy U is as low as possible: $p/U = 1/c$.

The discussion that follows shows that a charged particle can absorb energy and momentum from a linearly polarized light wave, and thus that light itself must possess both energy and momentum. Moreover, the ratio of energy absorption to momentum absorption is independent of the specific particle. However, that ratio is very much dependent on the fact that energy and momentum are absorbed from *light*.

Consider a charge q, of mass m, in a plane electromagnetic field like that in Figures 15.7 and 15.8, traveling to the right, with $\vec{E}$ along y, and $\vec{B}$ along z. The electric force on q is along y, and the magnetic force is normal to z. Explicitly, q feels the Lorentz force (11.35), or

$$\vec{F} = q(\vec{E} + \vec{v} \times \vec{B}) = q(E_y \hat{y} - v_x B_z \hat{y} + v_y B_z \hat{x}). \qquad (15.47)$$

Because $E_y = cB_z$, and $|\vec{v}| \ll c$, the first term dominates. That is, the force along $\hat{y}$ (the direction of $\vec{E}$) is the largest. Thus the particle is brought into motion along y by the electric field and absorbs energy at the rate

$$\mathcal{P} = F_y v_y = qE_y v_y. \qquad (15.48)$$

$\mathcal{P}$ thus is proportional to two oscillating terms, v_y and E_y. In the absence of any collisions, they are out of phase and average to zero (just as the product of a sine and cosine average to zero). However, when a small amount of damping due to collisions [e.g., a force $-(m/\tau)\vec{v}$] is included, the time average of $\mathcal{P}$ is nonzero. The specific form of the damping is immaterial.

Although y-momentum is transferred instantaneously, since the force along y oscillates, the net momentum transfer along y averages to zero. However, there is an average absorption of momentum along the direction of propagation x, coming from the magnetic force due to the motion caused by the electric field. The rate of absorption of x-momentum is given by the third term in (15.47), or

$$F_x = qv_y B_z. \qquad (15.49)$$

Because $B_z = E_y/c$, comparison of (15.48) and (15.49) shows that F_x equals $\mathcal{P}/c$. More explicitly, at any instant of time

$$\frac{\mathcal{P}}{F_x} = \frac{F_y v_y}{F_x} = \frac{q E_y v_y}{q v_y B_z} = \frac{E_y}{B_z} = c. \tag{15.50}$$

Thus the energy absorbed ($dU = \mathcal{P}dt$) and the momentum absorbed [$dp_x = (dp_x/dt)dt = F_x dt$] are proportional, with a coefficient that is independent of the absorber. Explicitly, (15.50) gives $dU = \mathcal{P}dt = cF_x dt = cdp_x$. Integration over time then gives

$$U = pc. \quad \text{(energy U and momentum p of EM wave)} \tag{15.51}$$

(Here we have written p for p_x.)

Consider a set of vanes, painted black on one side and silver on the other, that are free to rotate. In a high vacuum, they will, under uniform illumination, absorb more momentum from the silver side. This is because in the reflection process there is a momentum change that is twice as great as the actual incident momentum. However, inexpensive low-vacuum radiometers turn oppositely. This is because the momentum of the relatively massive air molecules (massive relative to light) dominates at atmospheric pressure, and thus the hot molecules that leave the black side of the vane give the vane more of a kick than do the light waves that reflect off the silvered side.

We might try to use the momentum of light to direct a spacecraft. Taking a burst of energy $U = 10$ W-hrs $= 3.6 \times 10^4$ J for a directed light source, the corresponding momentum is $p = 1.2 \times 10^{-4}$ kg-m/s. If this light completely reflects off a spacecraft of mass 10^3 kg, the spacecraft would gain twice this momentum, thereby increasing its velocity by 1.2×10^{-7} m/s. The momentum of light is low because its velocity c is high.

Since pressure P is force per unit area, or rate of change of momentum per unit area, by (15.51) we have

$$P = \frac{1}{A}\frac{dp}{dt} = \frac{1}{cA}\frac{dU}{dt} = \frac{S}{c}, \quad \text{(radiation pressure and power)} \tag{15.52}$$

since S is power per unit area. This is the *radiation pressure* of light. The radiation pressure of the sun is believed to be responsible for producing comet tails. At the surface of the sun, $\bar{S} = \bar{P}/4\pi R_{sun}^2 = 687 \times 10^7$ W/m^2, so (15.52) yields a radiation pressure there of $P_{rad} = 0.229$ Pa.

15.10 Index of Refraction and Snell's Law of Refraction

In a real material, the equations describing the electric and magnetic field must include the effect of electric and magnetic polarization of the material. As shown by (7.13), the electric field within a dielectric medium is decreased by the dielectric constant (or relative permittivity) κ, due to electric polarization. Thus

Table 15.2 Table of index of refraction (characteristic of optical frequencies)

Air (STP)	Water	Crown glass	Flint glass	Diamond	Ice	Benzene	Lucite	Salt
1.000293	1.333	1.52	1.66	2.42	1.31	1.501	1.491	1.544

$k = (4\pi\epsilon_0)^{-1} \to (4\pi\kappa\epsilon_0)^{-1}$. Alternatively, the permitivity ϵ_0 of free space is multiplied by κ.

Similarly, to obtain the magnetic field within a magnetic medium, the effect of the material's magnetic polarization, as shown by (10.23) and (10.24), the magnetic permeability of free space μ_0 is multiplied by the relative magnetic permeability μ_r. Ferrites, for example, have $\mu_r \approx 10^4$ at microwave frequencies. However, even for magnetic materials, $\mu_r \approx 1$ at optical frequencies, because the processes responsible for magnetic permeability, such as domain wall motion, *cannot* respond at optical frequencies, which are around 10^{15} Hz. On the other hand, since electrons are responsible for electric polarizability, and they *can* respond at optical frequencies, ϵ_r is not unity at optical frequencies.

The effect of nonzero polarizability and permeability on the propagation velocity of electromagnetic waves is that, when computing the velocity of light, in (15.34) we must make the replacements $k \to k/\epsilon_r$ ($\epsilon_0 \to \epsilon_0\epsilon_r$) and $k_m \to k_m\mu_r$ ($\mu_0 \to \mu_o\mu_r$). This leads to

$$v = \sqrt{\frac{k/\epsilon_r}{k_m\mu_r}} = \frac{c}{\sqrt{\epsilon_r\mu_r}} = \frac{c}{n}, \qquad n = \sqrt{\epsilon_r\mu_r}, \qquad \text{(speed of light in matter)}$$

$$(15.53)$$

where n is called the *index of refraction*. It depends upon the detailed properties of a given material. See Table 15.2. The origin of the word *refraction* only becomes clear when we discuss what happens when light passes from one material to another, as we do shortly.

For each material there is a characteristic frequency dependence. See Figure 15.10.

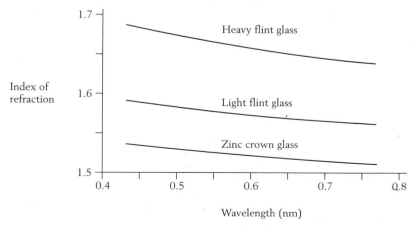

Figure 15.10 Index of refraction as a function of wavelength for three types of glass.

The dependence of n on frequency leads to dispersion; that is, the different colors of light in the medium do not travel with quite the same velocity. Dispersion must be minimized for optical fibers along which light signals are sent. In a quantitative sense, this dispersion of light in materials is minor compared to the dispersion of water waves.

15.10.1 *How Frequency and Wavelength Behave on Reflection and Refraction*

Consider a plane wave of frequency ω_1 impinging on the air–water surface at an angle θ_1 with respect to the normal. See Figure 15.11 (which features a brown pelican eyeing a rather large goldfish).

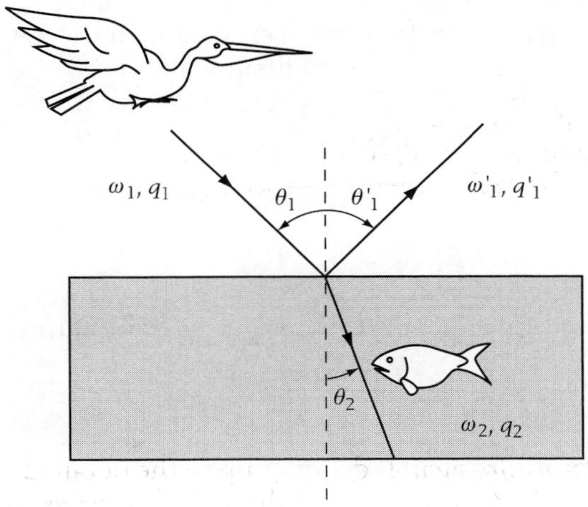

To learn how frequency changes on crossing the surface, consider a related problem, involving sound transmission and reflection. Let a drummer beat monotonously at the rate of once per second, with the sound in the air transmitted across the surface of a fishtank. The time interval between beats in both the air and the water is one beat per second. Moreover, the frequency of any sound reflected off the surface is one beat per second. In other words, the frequencies in each material are the same. This is also true for light. Hence

Figure 15.11 Reflection and refraction of light. A beam of light from behind the pelican is transmitted to the goldfish.

$$\omega_1 = \omega_1' = \omega_2. \quad \text{(frequency matching at interface)} \qquad (15.54)$$

By (15.23), this may be rewritten as

$$v_1 q_1 = v_1' q_1' = v_2 q_2. \qquad (15.55)$$

Since $v_1' = v_1$, by the first equality in (15.55), and by (15.24), we have

$$q_1' = q_1, \qquad \lambda_1' = \lambda_1. \qquad (15.56)$$

The transmitted (i.e., refracted) wave has the same frequency but a different velocity than the incident wave. Hence, by $\omega_1 = v_1 q_1 = \omega_2 = v_2 q_2$, and by $v_1 = c/n_1$, $v_2 = c/n_2$ [which follow from (15.53)], (15.55) shows that

$$\frac{q_2}{n_2} = \frac{q_1}{n_1}. \qquad (15.57)$$

From (15.24), (15.57) yields

$$\lambda_2 n_2 = \lambda_1 n_1. \quad \text{(wavelength matching at interface)} \quad (15.58)$$

Thus the wavelength changes under refraction, decreasing on going to a material of larger index of refraction.

15.10.2 *How Angle Behaves on Reflection and Refraction*

The incident wave propagates along the coordinate $y_1 = -y \cos\theta_1 + x \sin\theta_1$, so that it takes the form, analogous to (15.35), of $\cos(q_1 y_1 - \omega t)$. Along the surface $y = 0$, this becomes $\cos(q_1 \sin\theta_1 x - \omega t)$. Although the frequency is the same everywhere in space, the sound from the drummer (and the light reflecting off the pelican) does not arrive at the same time everywhere on the surface. In fact, along the interface $y = 0$, the phase of the incident wave varies as $q_1 \sin\theta_1 x$. To keep the same phase variations in the reflected and transmitted waves, $q_1' \sin\theta_1' x$ of the reflected wave and $q_2 \sin\theta_2 x$ of the transmitted wave must be the same. This leads to the conditions

$$q_1 \sin\theta_1 = q_1' \sin\theta_1' = q_2 \sin\theta_2. \quad (15.59)$$

Equation (15.56) and the first equality in (15.59) yield, for the reflected wave,

$$\theta_1' = \theta_1. \quad \text{(reflection angle)} \quad (15.60)$$

This is formalized in the familiar result that *the angle of reflection equals the angle of incidence.* Equation (15.55) and the second equality in (15.59) yield, for the transmitted wave,

$$\frac{\sin\theta_1}{v_1} = \frac{\sin\theta_2}{v_2}. \quad (15.61)$$

Then, using (15.53), we have *Snell's law.*

$$n_2 \sin\theta_2 = n_1 \sin\theta_1. \quad \text{(refraction angle)} \quad (15.62)$$

This implies, as is well known, that the transmitted light is bent, or *refracted.* The material with the larger index of refraction has the smaller angle to the normal. Equation (15.61) also describes how sound is refracted on going from one material to another. Note that on going from air to water, light bends toward the normal, whereas sound bends away from the normal. This is because light is slower in water (by a factor of 1.33), but sound is faster in water (by a factor of 4.3). Figure 15.11 accurately describes the path of light between the pelican and the goldfish, but the squawk of the pelican heard by the goldfish would take a more downward path, contacting the water to the left of the contact point for light.

| **Example 15.6** | **Refraction of light by water** |

Let $\theta_{pelican}$ be $45°$, with $n_{air} = 1$, and $n_2 = n_{water} = 1.33$. Find $\theta_{goldfish} = \theta_2$.

Solution: $\theta_{pelican} = \theta_1$, $n_1 = n_{air} = 1$, and $n_2 = n_{water} = 1.33$. Equation (15.62) yields $\theta_{goldfish} = \theta_2 = 32.1°$. This shows explicitly that light incident from the material with the smaller index of refraction is refracted to a smaller angle relative to the normal.

Correspondingly, light incident from the material with the larger index of refraction is refracted to a larger angle relative to the normal. This leads to the following interesting phenomenon that occurs when the refraction angle is $90°$, for then no light enters the material of smaller index of refraction.

15.10.3 *Critical Angle for Complete Internal Reflection*

What is called *complete internal reflection* can occur when the velocity in the second material is larger than, and thus the index of refraction is smaller than, in the first material. In that case, the transmitted angles θ_2 are larger in the second material. When θ_2 is $90°$, the incident light at angle θ_1 is at the *critical angle* θ_c. From (15.62), this occurs for

$$\sin \theta_c = \frac{n_2}{n_1}. \qquad \text{(critical angle in medium 2, for } n_2 < n_1) \qquad (15.63)$$

For angles exceeding θ_c, no light is transmitted into the second medium.

When the goldfish in Figure 15.11 looks up at the sky, complete internal reflection can occur. Since $n_{air} < n_{water}$, to apply (15.63) to Figure 15.11 we must interchange the meaning of 1 and 2. Then (15.63) gives $\sin \theta_c = n_1/n_2 = 1/1.33 = 0.75$, for which $\theta_c = 48.6°$. For $\theta > \theta_c$ there is complete internal reflection. **Note:** When the brown pelican in Figure 15.11 looks down at the water, complete internal reflection *cannot* occur.

| **Example 15.7** | **Complete internal reflection of sound** |

Mamie, listening for Buddy under water, can hear him only if she is sufficiently close to the normal, although she can always see him. Find the angle of complete internal reflection of his sounds.

Solution: The index of refraction for a medium is the ratio of the velocity of light in the standard medium air (properly, in vacuum) to the velocity of light in the medium. Equation (15.61) applies to sound, so with $\theta_2 = 90°$ and $\theta_1 = \theta_c$, (15.61) yields

$$\sin \theta_c = \frac{v_2}{v_1}.$$

Since $v_{water} = 4.3 v_{air}$, with water as medium 2, this gives $\theta_c = 13.4°$. Only within a cone of this angle, centered around Buddy, will Mamie hear Buddy.

15.10.4 *The Prism*

As indicated, the index of refraction depends upon frequency, so that light of different frequencies and wavelengths travels at different velocities, and therefore is dispersive. Moreover, the frequency dependence of the index of refraction implies that light of different colors bends slightly differently. This is the basis of operation of the prism, which separates the different pure colors of light. Isaac Newton (1666) was the first to study this effect systematically. The human eye cannot distinguish between a pure color and certain combinations of other pure colors. The study of color perception, which involves both physics and physiology, is still full of unanswered questions. See Figure 15.12, which shows incident white light that has been broken up into its component colors, due to their different indices of refraction (indicated in Figure 15.10).

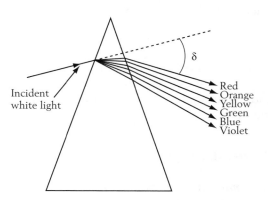

Figure 15.12 Separation of the pure colors of white light by a prism, due to the wavelength dependence of the index of refraction.

The least deflected color is red, which has the longest wavelength of the visible colors. Comparison with Figure 15.10, for common glasses, shows that red has an index of refraction that is closer to $n = 1$ than any of the other colors. The deflection angle δ would be zero for $n = 1$. Also see Color Plate 1 for a photo.

15.11 Transverse Nature of Electromagnetic Radiation; Polarization

An illustration of how the transverse nature of electromagnetic radiation is used can be seen in the different types of antennas that accompany television sets.

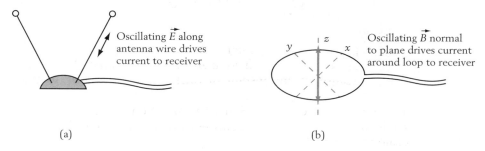

(a)   (b)

Figure 15.13 (a) Linear antenna, sensitive to the electric field of the incoming radiation. (b) Circular antenna, sensitive to the magnetic field of the incoming radiation.

Linear polarization   Random polarization

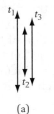

(a)

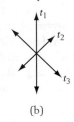

(b)

Figure 15.14 (a) Linear polarization of $\vec{E}$, where at every instant of time the electric field is along the same (fixed) axis normal to the direction of propagation. (b) Random polarization of $\vec{E}$, where at any instant of time the electric field is in a plane normal to the direction of propagation, but the direction can change over time within the plane.

Linear antennas (Figure 15.13a) pick up the signal of the electric field produced by a given TV station, causing current to flow along the direction of the antenna, into and out of the amplifier circuit of the TV. A linear antenna pointed along the direction of propagation will pick up no signal (unless there are reflections off buildings or cars, etc.). Circular antennas (Figure 15.13b) also pick up the electric field broadcast by the TV station. However, their operation is more easily understood by thinking of an incoming magnetic field that is normal to the plane of the circular antenna, and by its time variation inducing a circulating electric field within the antenna. Linear antennas are efficient for long-wavelength radiation, but they decrease in efficiency when the wavelength gets shorter than about half the length of the antenna. They work for VHF signals. Circular antennas pick up the higher-frequency, shorter-wavelength UHF signals.

For a given direction of propagation, at a given instant of time the electric field of electromagnetic radiation can point in any direction in the perpendicular plane. The way in which the electric field varies within that plane specifies its *state of polarization. Linear polarization* is the simplest type of polarization. The electric field of linearly polarized light propagating into the paper is depicted as a function of time in Figure 15.14(a). For very short time intervals, the electric field oscillates rapidly back and forth, represented by lines with arrows at both ends. For linearly polarized light, as time progresses the electric field amplitude can change in magnitude, but not in direction.

The light emitted from a lightbulb, as well as light from the sun, at any instant is likely to have its electric field in any direction within the perpendicular plane. This is because, at any instant of time, the light is likely to be emitted by a different atom in the light source, and the light emitted by a given atom lasts only on the order of 10^{-8}s. Hence the direction of the electric field changes very rapidly in time, the changes being essentially random to the observer. Such light is said to be *unpolarized*. It is depicted in Figure 15.14(b).

In principle, all forms of electromagnetic radiation can be linearly polarized. Microwave polarizers are parallel strips of metal, which preferentially along the field direction absorb the polarization. Similarly, long molecules absorb one polarization of light preferentially, as discovered by Edwin Land in 1928. For linearly polarized light, the electric field direction gives the direction of linear polarization, and the plane of the incident direction and the electric field gives the plane of polarization. (For historical reasons, the direction of linear polarization was first associated with the direction of the magnetic field, and the term *plane of polarization* was associated with the plane of the incident direction and the magnetic field. However, many authors now apply such usage to the electric field, and we will follow suit.)

If linearly polarized light has its electric field at an angle θ to the preferred axis of a linear polarizer, then the electric field can be decomposed into a fractional component $\cos\theta$ along the preferred axis, and a fractional component $\sin\theta$ normal to the preferred axis. The intensity associated with the absorption axis is completely absorbed. What remains is the electric field along the axis of the polarizer. The resultant intensity is lower by a factor of $\cos^2\theta$. If the incident intensity is I_0, and the reflected intensity is negligible, then the transmitted intensity is

$$I = I_0 \cos^2\theta. \qquad \text{(Malus's law for linear polarizer)} \qquad (15.64)$$

This is known as *Malus's law*. It was actually discovered in the context of polarization caused by transmission through biaxial crystals, a topic to be discussed in the next chapter.

When unpolarized light passes through an ideal linear polarizer, the intensity of the light that emerges is half that of the incident light, and the polarization is precisely that of the polarizer. This factor of two can be understood as follows. In (15.53), at any given instant of time the unpolarized light has a θ that can point anywhere in the transverse plane. Within a very short time, this direction changes randomly because another atom, uncorrelated with the first, now is radiating the light. Hence, averaging over θ gives $\overline{\cos^2} = \frac{1}{2}$. In Figure 15.15, we depict what happens when unpolarized light successively passes through two linear polarizers making an angle of θ with respect to each other. When a polarizer is used to study the polarization of light, it is called an *analyzer*.

Light can be *partially polarized* both by reflection from polished surfaces, and by scattering off molecules. Thus light that reflects off a mirror is partially

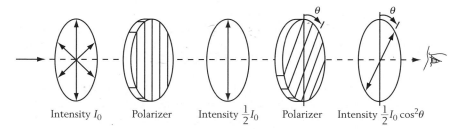

Intensity I_0 Polarizer   Intensity $\frac{1}{2}I_0$ Polarizer   Intensity $\frac{1}{2}I_0 \cos^2\theta$

Figure 15.15 Unpolarized light of intensity I_0 passing through a sequence of two linear polarizers, whose acceptance axes make an angle θ relative to each other. These polarizers work by completely absorbing electromagnetic radiation with its electric field along one direction, and completely transmitting the other component of electromagnetic radiation.

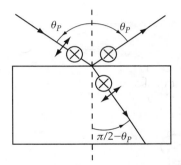

Figure 15.16 Geometry associated with complete polarization of reflected radiation (the Brewster angle). ⊗ represents polarization normal to the page.

polarized, as is light from a clear sky. Polarized sunglasses take advantage of polarization by reflection to reduce the glare from reflections and the background light from the sky.

There is a special angle of incidence for which reflected light is completely polarized normal to the plane defined by the direction of the incident wave and the normal to the interface (the *plane of incidence*.) This angle occurs when the angle of incidence from material 1 and the angle of refraction by material 2 are complementary, or $\theta_1 + \theta_2 = 90°$. See Figure 15.16.

For this angle of incidence, called the *Brewster angle*, and denoted by $\theta_1 = \theta_p$ (p for polarization), (15.62) gives $n_1 \sin \theta_p = n_2 \sin(\pi/2 - \theta_p) = n_2 \cos \theta_p$. Thus

$$\tan \theta_p = \frac{n_2}{n_1}. \qquad \text{(Brewster angle)} \qquad (15.65)$$

This effect can be observed using a set of polarizing sunglasses.

Figure 15.16 shows both the plane of incidence and light of both polarizations, for light incident at the Brewster angle. Light of both polarizations propagates within material 2. The reflected light must be caused by radiation from atoms of material 2 in the vicinity of the crystal surface (recall the planar antenna of Figure 15.7). Since the atoms in the crystal, in response to the electric field $\vec{E}$, develop dipole moments $\vec{p}$ along $\vec{E}$, the Brewster angle result indicates that only the component of $\vec{p}$ normal to the propagation direction produces radiation in the direction of propagation.

Radiation by induced dipole moments explains why sunlight, scattered by air molecules, is partially polarized. Figure 15.17 shows an observer viewing the sky at an angle of 90° to the sun. Only when the electric field has a component normal to the scattering plane (the plane of the page), or ⊗, will the air molecules develop dipole moments normal to the direction of propagation from the molecules to the observer (the vertical). Such light will be completely polarized, as can be observed using a set of polarizing sunglasses. In practice, due to multiple scattering, the polarization is not complete. At other observation angles, the light is only partially polarized.

15.12 Microwave Cavities—Standing Waves

Optional

Standing waves occur on a string only because forces at the ends keep the string from moving above or below the end, which is fixed in place. Standing waves occur in an acoustic cavity only because forces at the walls keep the air from moving into or out of the walls, which are fixed in place. Similarly, standing waves occur in a microwave cavity (such as a microwave oven) only because charges and currents at the walls of the cavity make the normal component of the electric field and the transverse component of the magnetic field take the appropriate values.

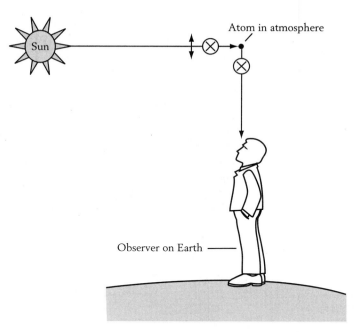

Figure 15.17 Unpolarized light from the sun being scattered off atoms in the atmosphere; for an observer at 90°, if there is no other scattering the light is completely polarized. ⊗ represents polarization normal to the page.

Within the walls of the microwave cavity, the electric and magnetic fields are zero, as a result of the so-called skin effect, discussed in Section 14.13. However, because of surface charges and surface currents, certain components of the electric and magnetic fields abruptly become nonzero just inside the cavities. Surface charge can make the component of the electric field perpendicular to the plane of the walls go abruptly from zero just within the walls to nonzero just within the cavity. However, the components of the electric field parallel to the plane of the wall cannot abruptly become nonzero just within the cavity; they are zero inside, and they remain zero just outside. Surface current can make the components of the magnetic field parallel to the plane of the wall go abruptly from zero within the walls to nonzero within the cavity. However, the component of the magnetic field perpendicular to the wall cannot abruptly become nonzero just within the cavity; it is zero inside, and it remains zero just outside. Without the charges and currents on the walls of the microwave cavity, there would be no microwaves within the cavity. See Figure 15.18.

Specifically, the boundary conditions on the electric and magnetic field are, with $\hat{n}$ the outward normal from the wall,

$$\vec{E} \times \hat{n} = \vec{0}, \qquad \vec{B} \cdot \hat{n} = 0. \tag{15.66}$$

At the surface, the relationships between the surface charge σ_S and the electric field, and between the surface current $\vec{K}$ and the magnetic field, are given by (4.22) and (11.43),

$$\vec{E} \cdot \hat{n} = 4\pi k \sigma_S, \qquad \vec{B} \times \hat{n} = -4\pi k_m \vec{K}. \tag{15.67}$$

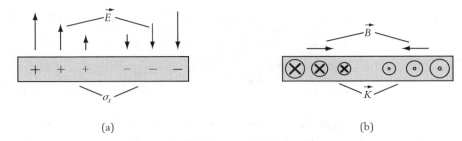

Figure 15.18 (a) A surface charge density σ_s at a conducting surface can change the normal component of the electric field $\vec{E}$ from zero inside the conductor to a nonzero value just outside the conductor. The transverse component is zero both inside and just outside the conductor. (b) A surface current density $\vec{K}$ at a conducting surface can change the transverse component of the magnetic field $\vec{B}$ from zero inside the conductor to a nonzero value just outside the conductor. The normal component is zero both inside and just outside the conductor.

Although (11.43), involving $\vec{B}$, was derived for the static behavior of a field-expelling perfect diamagnet, it is also valid for the dynamic behavior of a field-expelling conductor.

Equations (15.66) serve as the boundary conditions that determine the allowed wavenumbers of the microwave cavity, and Equations (15.67) tell us the surface charges and surface currents that cause the modes to occur. These surface charges and surface currents serve an analogous role to the forces exerted on a string by the contacts at the ends.

The character of the sound of a stringed instrument is determined by the way in which the strings are plucked; the closer to the end of the string, the greater the amplitude of the high-frequency modes relative to the low-frequency modes. Similarly, the radiation pattern within a microwave cavity depends very much upon the placement and shape of the waveguide that leads from the power tube to the microwave cavity. Within the structure of some microwave ovens, in order to more evenly distribute the radiation, a small rotating metal fan is used to scatter the radiation.

15.13 Wires, Co-axial Cables, and Waveguides—Traveling Waves

Optional

In 1857, Kirchhoff studied the problem of the transmission of electrical signals along a wire. He employed values for the capacitance per unit length and the inductance per unit length that were appropriate to the wire geometry.

Kirchhoff found that the velocity of propagation was given by

$$v = \sqrt{\frac{1}{\mathcal{LC}}},\tag{15.68}$$

where $\mathcal{C}$ is the capacitance per unit length—given for concentric cylinders by (7.6)—and $\mathcal{L}$ is the inductance per unit length—given for concentric cylinders by (13.38). Substituting these values, we obtain $v = \sqrt{k/k_m} = 3 \times 10^8$ m/s.

Kirchhoff's theoretical value, was too low by $\sqrt{2}$ because of an error in his calculation of the individual quantities C and $\mathcal{L}$.

In 1876, Heaviside studied this problem, but with unspecified values for the resistance per unit length, capacitance per unit length, and inductance per unit length. He obtained what has since been called the *telegrapher's equation*. This was a generalization of Kelvin's earlier theory, which neglected the effects of inductance.

On the experimental side of this question, in 1849, Fizeau measured the speed of light in air, finding a value of 3.15×10^8 m/s. In 1850, Fizeau and Gounelle determined the velocity of propagation along iron and copper wires to be about 10^8 m/s, and in 1875, Siemens found it to be, for iron wires, about 2.6×10^8 m/s.

In 1883, Heaviside began to apply Maxwell's theory to the problem of signal propagation along wires and co-axial cables, thus founding the modern theory of cable communications. When studying problems in radiation, Maxwell did not actually use the electric and magnetic fields, but rather another set of variables: the usual (scalar) electrical potential due to charges, and another (vector) potential due to currents. (In this introductory course, there is no need to study the vector potential.) To make the equations more tractable, Heaviside rewrote the equations that Maxwell used, giving us the theory in the form used today.

Figure 15.19 indicates the pattern of electric and magnetic fields, and the surface charges and currents, for a wave that is traveling leftward in a co-axial cable (leading, perhaps, to your television set). The cable has inner radius a and outer radius b. At a given position z along the cable, the electric field $\vec{E}$ points radially inward or outward. Let's say it points outward from the inner cylinder, as on the left part of Figure 15.19. Then the surface charge density σ_S must be positive where $\vec{E}$ leaves, and σ_S must be negative where $\vec{E}$ enters at the outer cylinder. For a leftward traveling wave, the magnetic field $\vec{B}$ must point such that the Poynting vector $\vec{E} \times \vec{B}$ points leftward. This determines in which direction $\vec{B}$ circulates around the inner cable. The surface currents $\vec{K}$ produce the magnetic

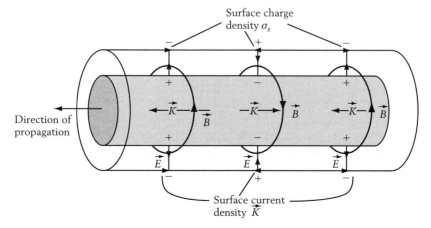

Figure 15.19 The surface charges and surface currents, and the electric and magnetic fields, for an electromagnetic wave propagating along a co-axial cable.

field, so from Oersted's right-hand rule we can determine the direction of the surface currents. Since $\vec{E}$ and $\vec{B}$ are related, by (15.67) the surface currents and the surface charge densities are related. For this type of mode, not only do $\vec{E}$ and $\vec{B}$ have components normal to the direction of propagation, but there is also a component of $\vec{E}$ along the direction of propagation, needed to produce a component of $\vec{K}$ along that direction.

Electromagnetic radiation along wires and co-axial cables is nondispersive. However, along waveguides, which have no inner wire, electromagnetic radiation is dispersive, not propagating at all below a geometry-related cut-off frequency ω_c. The narrower the waveguide, the higher the cut-off frequency. For frequencies just above ω_c, the waves are very slow and dispersive, whereas for frequencies very far above ω_c, the waves are nearly nondispersive. For communications purposes, usually it is best to work at higher frequencies, where the signals are nearly nondispersive and do not distort significantly.

15.14 Hertz's Studies of Electromagnetic Radiation

Optional

In 1886, Heinrich Hertz undertook to detect electromagnetic radiation, using a pair of matched spirals with spherical knobs at their ends (see Figure 15.20a). On exciting a spark between the knobs in one coil (the primary) by discharging a Leyden jar, the knobs of the other coil (the secondary) would spark—a dramatic demonstration of mutual electromagnetic induction. The spirals were so sensitive that they did not require a large bank of batteries to cause sparking. Indeed, "even the discharge of a small induction-coil would do, provided it had [enough energy] to spring across a spark gap." He then developed even more effective methods to generate and detect electrical sparks.

15.14.1 *The Spark-Excited Oscillating Antenna*

The spark from a primary excited electrical signals in any wire connected to the primary, much like hitting a hammer against a rod excites sound in a rod. Upon

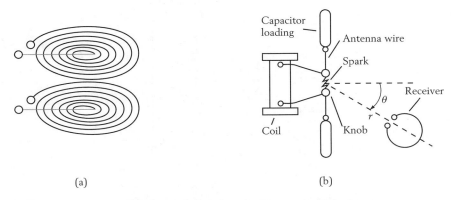

(a)   (b)

Figure 15.20 (a) A pair of coupled coils, which were used to demonstrate electomagnetic induction. (b) One of the geometries used by Hertz to detect electromagnetic radiation. The transmitting antenna is linear, with a spark gap, and is driven by a capacitor.

LC Oscillations

In the early 1840s Henry suspected, by their effect on magnetic needles, that capacitors discharge in an oscillatory fashion, and in 1847 Helmholtz noticed that H_2 and O_2 were produced at both electrodes when a Leyden jar was discharged through an electrolytic cell. Following Kelvin's theory of LC oscillations, it was realized that Leyden jars typically would oscillate at a frequency of about 10^4 Hz, and that open induction coils typically would oscillate at a frequency of about 10^6 Hz.

sparking, the gap resistance fell to a much lower value, where it remained while current flowed across the gap. Hertz attached a wire to each knob, so signals could travel from knob to wire, reflect off the end of the wire, return to the knob, and then cross the spark gap. Sometimes he "loaded" the wire by placing a capacitor at its end. See Figure 15.20(b). He used the resonance frequency formula $T = 1/f = 2\pi/\omega = 2\pi\sqrt{LC}$ to estimate that $T \approx 10^{-8}$ s, corresponding to a frequency $f = 1/T = 10^8$ Hz that was too high to measure at that time.

The wires attached to each knob were in a line. This made it easier, when studying mutual induction, to get good proximity with the secondary circuit, which was equipped with a micrometer spark gap (see Figure 15.21a). On increasing the gap until no sparking occurred, Hertz could obtain a measure of the intensity of the signals in the wire. By eliminating *electrostatic* induction as a source of sparking, he established that *electromagnetic* induction was its source. He suspected that the wire circuit (including the spark gap) was oscillating largely at a single frequency. Nevertheless, he had not *proved* that there were such oscillations.

15.14.2 *The Resonating Receiver*

Hertz reasoned that if the primary (the transmitter) was oscillating, and the secondary (the receiver) was brought into resonance with it, then the secondary

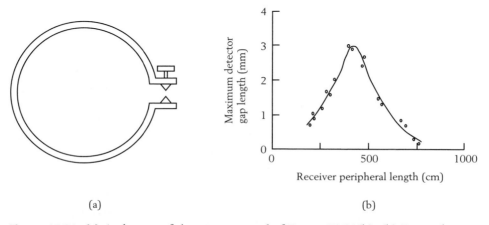

(a)   (b)

Figure 15.21 (a) A closeup of the receiving coil of Figure 15.20(b). (b) Data taken by Hertz, showing the increased sensitivity of the receiving coil in the vicinity of its resonance frequency.

would be a more effective detector. By varying the capacitance and inductance of each circuit, he could bring the two circuits in and out of resonance. [He changed the capacitance by connecting large spheres to the knobs or the ends of the antenna wire ("capacitive loading"), and he changed the inductance by changing the length of the circuit.] Figure 15.21(b) shows the largest gap across which a spark could be seen, as a function of the length of the secondary circuit (this is proportional to the self-inductance of the secondary). With his resonant receiving circuit, he mapped out the positions and orientation of the electric field produced in the air by his linear transmitting antenna. When the receiver was placed along the axis of the antenna ($\theta = 90°$), there was no signal for any receiver orientation. For other positions of the receiver, the maximum response occurred when the receiver was in the plane defined by the antenna and the line of sight, with the spark gap oriented perpendicular to the line of sight. See Figure 15.20(b).

15.14.3 *Waves in Wires, Waves in Space*

Hertz next studied the speed of propagation of electromagnetic waves along a long wire connected to his transmitter (the primary). Just outside the wire, he found regularly spaced null points which determined the wavelength λ. Estimating $f = 1/T$ from the theoretical value of T for the primary, he used the relation $v = f\lambda$ to obtain a speed of propagation v along wires; it was near the speed of light, and consistent with previous studies. Moving the receiver away from the wire, he saw complex patterns, which he attributed to interference between radiation traveling along the wire and the radiation in air. He concluded that the speed of electromagnetic radiation in air was finite, and on the order of the speed of light.

Hertz suspected that the complex patterns involved three sources: the wire, the direct radiation in air, and reflections off the walls of his laboratory. By eliminating the wires, he found a simple pattern of standing waves, confirming the presence of reflections from the walls.

15.14.4 *Further Experiments on Electromagnetic Radiation*

Hertz also showed that, when an electromagnetic signal propagated along a wire, the current only penetrated a small distance into the wire (the skin depth). Further, he found that even a cage of only four wires—provided that their separation was much less than the wavelength of the radiation—would provide electromagnetic shielding. (The metallic screens on the doors of microwave ovens use this effect. The screen permits short-wavelength light to escape, but retains microwave radiation, whose wavelength is much greater than the holes in the screen.)

For a shorter wavelength, Hertz made a focusing reflector out of a large concave parabolic mirror and used it to demonstrate straight-line propagation, polarization, and reflection. To demonstrate refraction he used a large wedge of tar.

Finally, by measuring quantities with amplitudes quadratic in the fields (as for the jumping ring, discussed in Section 14.12.2), first for the electric field and

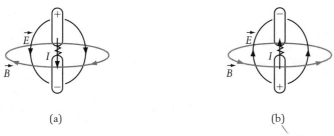

(a)   (b)

Figure 15.22 Closeup of the electric and magnetic fields in the immediate vicinity of the linear transmitting antenna of Figure 15.20(b): (a) when the ends of the antenna have maximum charge, (b) a half-period later.

then for the magnetic field, Hertz showed that *both* $\vec{E}$ and $\vec{B}$ are present in an electromagnetic wave.

15.14.5 *Theory of the Dipole Antenna*

Hertz also developed a theory of radiation by his dipole antenna. Near the antenna, the electric field is like that for a static distribution of charge, and the magnetic field is like that due to a long wire. See Figure 15.22, which depicts $\vec{E}$ and $\vec{B}$ at times a half-period apart.

Further from the antenna, the field pattern becomes more complicated. This is because, as time goes by, the directions of the electric and magnetic fields near the antenna must change. At a quarter-period the fields near the antenna are zero; this is where they must "pinch off" from the source and, like all other electromagnetic signals, propagate outward at the velocity c. See Figure 15.23, which depicts signals at $t = 0$, $t = T/4$, and $t = T/2$, where T is the period of oscillation of the antenna circuit.

Figure 15.24 depicts the field pattern for a few cycles of oscillation.

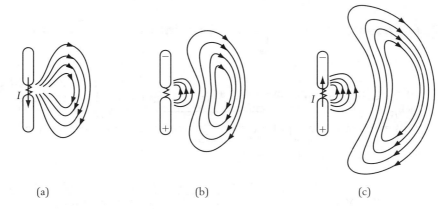

(a)   (b)   (c)

Figure 15.23 Electromagnetic radiation in the vicinity of the linear transmitting antenna of Figure 15.20(b): (a) the electric and magnetic fields before they "pinch" off, (b) a quarter-period later, (c) a half-period later.

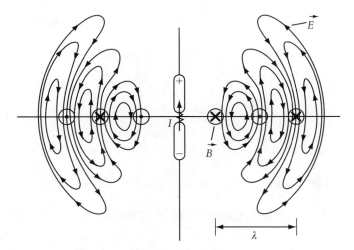

Figure 15.24 The electric field in the vicinity of the linear transmitting antenna of Figure 15.20(b) for nearly two periods of oscillation.

15.15 Supplementary Material

Optional

15.15.1 Energy Flow for Traveling Waves

Traveling waves transport energy. Since they travel with velocity v, the local rate of energy flow $\mathcal{P}$ (an energy per unit time) equals the energy per unit length ε times v:

$$\mathcal{P} = \varepsilon v. \tag{15.69}$$

If we can determine ε, we can determine $\mathcal{P}$.

Consider a string. The kinetic energy per unit length ε_{KE} is given by one-half the mass per unit length times the velocity squared, or

$$\varepsilon_{KE} = \frac{1}{2}\mu\left(\frac{dy}{dt}\right)^2. \tag{15.70}$$

There is also potential energy per unit length, ε_{PE}, associated with stretching of the string by the wave. Because the string has nearly constant tension F, the work done on the string in stretching it from length dx to length ds is $F(ds - dx)$. This goes into elastic potential energy, with the potential energy per unit length dx given by

$$F\frac{ds - dx}{dx} = F\frac{\sqrt{dx^2 + dy^2} - dx}{dx} = F\left(\sqrt{1 + \left(\frac{dy}{dx}\right)^2} - 1\right). \tag{15.71}$$

Since we assume that dy/dx is small, use of the straight-line approximation, as in (4.35) [$(1+x)^n \approx 1 + nx$ for small x, where we use dy/dx in place of x], gives

$$\sqrt{1 + \left(\frac{dy}{dx}\right)^2} \approx 1 + \frac{1}{2}\left(\frac{dy}{dx}\right)^2,$$

so

$$\varepsilon_{PE} = \frac{1}{2}F\left(\frac{dy}{dx}\right)^2. \tag{15.72}$$

We now show that, for a rightward traveling wave, described by (15.21), $\varepsilon_{KE} = \varepsilon_{PE}$.

By (15.21), $dy/dx = dg/dx$ and $dy/dt = -vdg/dx$. With $v^2 = F/\mu$ we then have

$$\varepsilon_{KE} = \frac{1}{2}\mu\left(\frac{dy}{dt}\right)^2 = \frac{1}{2}\mu v^2\left(\frac{dg}{dx}\right)^2 = \frac{1}{2}F\left(\frac{dy}{dx}\right)^2 = \varepsilon_{PE}. \quad \text{(traveling waves)} \tag{15.73}$$

This result holds both for leftward and rightward traveling waves.

Combining (15.69) and (15.73) we have, for traveling waves, that the rate of energy flow is

$$\mathcal{P} = \varepsilon v = v(\varepsilon_{KE} + \varepsilon_{PE}) = 2v\varepsilon_{KE} = v\mu\left(\frac{dy}{dt}\right)^2. \quad \text{(traveling waves)} \tag{15.74}$$

This same result can be obtained by another route. We can directly calculate the work per unit time dW/dt being done on the right part of a string at position x, due to the left part. Since the vertical force is $-Fdy/dx$ and the vertical displacement is dy, the work done is $dW = (-Fdy/dx)dy$, so

$$\frac{dW}{dt} = -F\frac{dy}{dx}\frac{dy}{dt}. \tag{15.75}$$

For a rightward wave, described by (15.21), $dy/dx = -v^{-1}dy/dt$, so with $v^2 = F/\mu$ (15.75) gives

$$\frac{dW}{dt} = \frac{F}{v}\left(\frac{dy}{dt}\right)^2 = v\mu\left(\frac{dy}{dt}\right)^2. \quad \text{(traveling waves)} \tag{15.76}$$

This equals $\mathcal{P}$ of (15.74). Equation (15.75) for dW/dt is more general than (15.76) and holds for any kind of motion of the string. Similarly, our first expressions for ε_{KE} and ε_{PE} hold for any kind of motion of the string.

15.15.2 *Compressional Motion (Longitudinal Waves)*

Sound waves are waves of compression. So are the waves you excite on hitting one end of a rod with a hammer. These involve not transverse motion, as for a string, but longitudinal motion. The x-coordinate of the object is changed, and this means we must distinguish it from the fixed x-coordinate of real space.

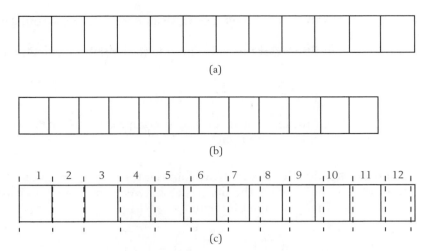

Figure 15.25 A rod subject to longitudinal distortions: (a) equilibrium, (b) uniform compression, (c) nonuniform longitudinal distortion.

For that reason, we will use u_x to denote the *deviation from equilibrium* of the x-coordinate of the object whose motion we are studying. However, the subscript x really is unnecessary here, so let's simplify our notation and use u.

Physical Picture for Compression of a Rod Consider a one-dimensional rod of cross-section A, and undistorted length l_0, that lies along the x-axis. See Figure 15.25(a), which represents the undistorted deviation $u(x) = 0$. Just as uniform vertical translations don't cause the string to come into motion, so uniform horizontal translations don't cause the rod to come into motion. In addition, uniform compression or expansion of the rod doesn't cause it to come into motion. See Figure 15.25(b), which represents the static uniform deviation of compression $u(x) = -0.08x$, for which $du/dx < 0$. (Even for the nonuniform case, $du/dx < 0$ corresponds to compression; correspondingly, $du/dx > 0$ corresponds to expansion.) Only *nonuniform* compressions or expansions can cause the rod to come into horizontal motion. This corresponds to $d^2u/dx^2 \neq 0$. It is very like what happens for the string, where $d^2y/dx^2 \neq 0$ causes vertical motion.

Figure 15.25(c) gives an example of a nonuniform compression. Here we have a snapshot of the rod at a specific time. We have arbitrarily broken up the rod into 12 slabs. The solid lines are the equilibrium positions and the dashed lines are the instantaneous positions. To determine the direction of the expected acceleration, consider two adjacent slabs. If the slabs have the same compression, their dashed separation line will not accelerate. If the slabs have different compressions, the separation line will move away from the slab that is more compressed. Thus the line separating 11 and 12 will accelerate to the left. (Keep in mind that this snapshot does not show the velocities at this instant of time.)

The force, or stress, associated with longitudinal motion is proportional to the strain du/dx. The proportionality constant is the product of the area A and Young's modulus Y (which has units of force per unit area). Typical values of Young's modulus for the elements are on the order of 10×10^{10} Pa, although for lead it is only 1.7×10^{10} Pa; atmospheric pressure is about 10^5 Pa, and a car tire is typically inflated to about three times this value; rubber has a Young's modulus of only 5×10^8 Pa.

Consider a position x of the rod. The side to the right of x feels a force

$$F_x = -YA\frac{du}{dx}. \tag{15.77}$$

The negative sign means that if the rod is compressed ($du/dx < 0$), then the force at its right end is to the right.

Equation of Motion for a Rod Now consider the force on a horizontal slice dx of the rod. The net force is the difference between the forces on its ends, at x and at $x + dx$. This gives a net force of

$$F_x = YA\left(\frac{du}{dx}\bigg|_{x+dx} - \frac{du}{dx}\bigg|_x\right) = YA\frac{d^2u}{dx^2}dx. \tag{15.78}$$

With mass per unit volume ρ, the mass contained by dx is $\rho A dx$. By Newton's second law of motion, (15.78) must equal the product of its mass and its horizontal acceleration d^2u/dt^2. Thus

$$\rho A dx\frac{d^2u}{dt^2} = YA\frac{d^2u}{dx^2}dx, \tag{15.79}$$

so

$$\frac{d^2u}{dt^2} = \frac{Y}{\rho}\frac{d^2u}{dx^2}. \tag{15.80}$$

Comparing with (15.12) of our earlier study of waves on a string, we conclude that (15.80) is a wave equation with velocity

$$v = \sqrt{\frac{Y}{\rho}}. \tag{15.81}$$

Note that when a rod is compressed it normally expands in the transverse direction, something we have not shown in Figure 15.25.

Longitudinal Waves in Bulk Matter One distinction between longitudinal waves in rods and longitudinal waves in a bulk material is that in bulk the proper variable is not displacement, but volume change. Not surprisingly, the compressibility determines the velocity of sound in this case. The compressibility is precisely the inverse of the bulk modulus B, a measure of the pressure increase to a fractional decrease in volume. Thus B has units of pressure. For water, $B = 2.05 \times 10^9$ Pa.

With this modification, a full analysis yields that the velocity of propagation is

$$v = \sqrt{\frac{B}{\rho}}. \tag{15.82}$$

For gases, $B = \gamma P$, where P is the pressure and γ is the dimensionless ratio of the specific heats at constant pressure and at constant volume: $\gamma = C_p/C_V$. For air, $\gamma = 1.4$.

Equation (15.82) applies both to traveling and to standing waves. One important example of standing waves is to sound waves in pipes. For closed pipes, the boundary conditions are that the velocities at the ends are both zero. This implies, by analogy to waves on a string, that sound waves resonate within a pipe when the length L of the pipe corresponds to a half-integral number of wavelengths λ, or

$$L = \frac{n}{2}\lambda. \qquad n = 1, 2, \ldots \qquad \text{(resonance of closed pipe)} \qquad (15.83)$$

For a pipe open at one end, the boundary condition at the closed end is still that the velocity is zero, for the sound can't move the rigid wall. On the other hand, at the open end, the velocity is completely unconstrained, which corresponds to it reaching its maximum value at the open end. Hence, for a maximum, in this case, the length L corresponds to a quarter-integral number of wavelengths, or

$$L = \frac{2n+1}{4}\lambda. \qquad n = 0, 1, \ldots \qquad \text{(resonance of pipe with one open end)}$$

$$(15.84)$$

In particular, for a given pipe length L, $n = 0$ of (15.84) gives $\lambda = 4L$, whereas $n = 1$ of (15.83) gives $\lambda = 2L$. Hence, for a given pipe length, a pipe with one open end gives a lower resonant frequency than for either a pipe closed at both ends or a pipe open at both ends, whose resonance wavelengths are also given by (15.83).

Problems

15-2.1 (a) For the Paris to Lille semaphore system, estimate the effective velocity of transmission. (b) Estimate the effective rate of transmission of a bit (flag up or flag down). (c) How long would it take to send a 100-word message, assuming 7 letters per word and 6 bits per letter, and a flagman who can send 1 bit per second? (d) If it takes 1 second for a flagman to receive a signal and 1 second to transmit it, estimate the number of flagmen between Paris and Lille.

15-2.2 Consider two concentric cylindrical shells of radii a and b, with $a < b$. Their interior is filled with de-ionized water, having relative dielectric permeability of 80. (a) Find the capacitance per unit length C. Evaluate it for $b = 0.25$ cm and $a = 0.36$ cm. (b) Let seawater fill the space between the shells, and let the inner shell be given charge Q. If $b - a \gg l$, where l is a "screening length" due to the ions (see the last section of Chapter 4), the field lines "end" on ions within a distance l. Estimate the capacitance per unit length for $l = 1.2$ nm. (c) If there is no outer conducting shell, estimate the capacitance per unit length for the inner shell.

15-3.1 Find the value of $|d\vec{E}/dt|$ that corresponds to a displacement current density (current per unit area) of 2.5×10^5 A/m^2.

15-3.2 For $|d\vec{E}/dt| = 4 \times 10^7$ V/m-s pointing to the left, find the displacement current density.

15-4.1 A string 1.2 m long has a mass of 0.252 kg and a tension of 46 N. If it has a vertical acceleration of 2.4 m/s^2 at point P, find d^2y/dx^2 at P.

15-4.2 A string 1.4 m long has a mass of 0.186 kg and a tension of 37 N. If $d^2y/dx^2 = 0.038$ m^{-1} at point P, find the vertical acceleration at P.

15-4.3 A circle of radius a, centered at $x = 0$, $y = a$, satisfies the equation $x^2 + (y - a)^2 = a^2$. (a) Compute dy/dx and d^2y/dx^2. (b) Evaluate d^2y/dx^2 at $x = 0$, which corresponds to the bottom of the circle. (c) Determine the relationship between d^2y/dx^2 at $x = 0$ and the radius of curvature a.

15-4.4 Explain why the acceleration of a string under tension does not depend on the string's overall height or slope, but does depend on its curvature.

15-5.1 A rope 3 m long has mass 0.05 kg and is at a tension of 240 N. Find the speed of propagation of transverse waves along it.

15-5.2 Sound propagates along a 200 m long wire at 60 m/s. It has mass 1.2 kg. Find its tension.

15-5.3 A steel wire 1.6 m long and of mass 0.012 kg is to be used to produce transverse waves with frequency 50 Hz and wavelength 0.4 m. What tension should it be given?

15-5.4 A string 1.2 m long and of mass 0.008 kg is fixed at one end. The other end passes over a pulley that supports a 0.25 kg mass. Find the velocity of transverse waves and how long it takes for a wave to pass from one end and back again.

15-5.5 A string telephone system consists of two paper cups with holes in their bottoms, connected by a taut string knotted around the outside of the holes. Speaking into a cup makes it vibrate, both along the direction of the string axis and along the two directions transverse to the string. In what follows, consider only the transverse motion. (a) Explain why the sound at the receiving end is louder than it would have been had the signal traveled through the air as a sound wave. (b) Can a transverse wave on a string travel faster than a sound wave in air? (c) For a 10 m long string of mass 0.284 kg, what tension will give a velocity equal to the velocity of sound (344 m/s)? (d) Will a paper cup support this tension?

15-5.6 A string is plucked on two occasions, for the same length of time. The second time it is plucked twice as hard as the first. (a) Compare the wave speeds in these two cases. (b) Compare the vertical velocities of a given point on the string.

15-5.7 For a leftward wave, show that the slope has magnitude given by the ratio of the particle speed dy/dt to the wave speed v.

15-5.8 Write down the equation describing a leftward vertical wave on a string, of amplitude 0.78 cm, frequency 248 Hz, and speed of 145 m/s. It satisfies $y = 0.52$ cm at $t = 0$ and $x = 1.4$ cm.

15-5.9 Sinusoidal water waves pass a dock at regular intervals of time. It takes 25 sec for eight waves to pass by; the wave peaks are 6.4 m apart, and the maximum water height is 0.36 m above what it is without the waves. (a) Find the wave speed. (b) Find the wave amplitude. (c) Find the instantaneous vertical velocity 0.56 s after a maximum. (d) Find the instantaneous vertical acceleration 0.56 s after a maximum.

15-5.10 A transverse wave on a string is described mathematically by the form $y(x, t) = (3.5$ cm$)\sin(0.085x - 1.46t + 0.75)$, where phase is measured in radians, x is measured in m, and t in s. Find (a) the amplitude, (b) the wavenumber, (c) the wavelength, (d) the angular frequency, (e) the frequency, (f) the wave velocity. (g) If the tension is 4.5 N, find the density. (h) At $t = 0$, find the position of the first maximum with $x > 0$. (i) At $x = 4$ cm, find the time of the third minimum after $t = 2.5$ s.

15-5.11 Leftward sinusoidal waves on a string have wave speed 8 m/s, amplitude 0.03 cm, and wavelength 2.5 cm. At $t = 4$ s and $x = 2$ cm, $y = -0.02$ cm. Find (a) the wavenumber, (b) the angular frequency, (c) the frequency, (d) the period, (e) the phase, (f) the mathematical expression for $y(x,t)$, (g) the value of y at $t = 8$ s and $x = 12$ cm.

15-5.12 A leftward transverse wave has amplitude 0.8 cm, velocity 250 m/s, and initial shape $\exp[-(x - 4)^2/a^2]$, with $a = 2$ cm and x in cm. (a) Write down an expression for the vertical displacement at a finite time t. (b) Evaluate it at $x = 0$ for $t = 0, 0.001$ and 0.002 sec. (c) Evaluate it at $x = -2$ cm for $t = 0, 0.001$ and 0.002 s.

15-5.13 A wave satisfies $y(x,t) = (2.2$ cm$)\cos(0.015x - 6.0t + 0.35)$, where phase is measured in radians, x is measured in cm, and t in s. (a) If the string has tension 48 N, find the mass density. (b) if the string has mass density 0.045 kg/m, find the tension.

15-5.14 Discuss why these wave shapes can or cannot be used to describe traveling waves: (a) $y = \sqrt{x + vt}$, (b) $y = A/[(x - vt)^2 + a^2]$.

15-5.15 A string of length 85 cm has a mass per unit length 6.45 g/m and tension 140 N. Determine (a) the wave speed, (b) the wavelengths of the first and the second harmonics, (c) the frequencies of the fundamental and the first overtone.

15-5.16 A string of length 74 cm and mass per unit length 12.5 g/m is to be used to produce a third harmonic of frequency 450 Hz. Find (a) the wavelength, (b) the sound velocity, (c) the tension in the string, (d) the wavelength of the corresponding frequency in air.

15-5.17 The fifth harmonic of a string satisfies the equation $y(x,t) = 0.36 \sin(0.84x) \cos(490t + 0.15)$, where phases are in radians, x and y are in meters, and t is in seconds. Find (a) the wavelength; (b) the length of the string; (c) the period; (d) the frequency; (e) the vertical displacement, velocity, and acceleration at $t = 0$ and $x = 0.12$.

15-5.18 At $t = 0$, a rightward wave satisfies $y(x, t = 0) = \exp[-(x/b)^2]$. It travels at speed v. Find $y(x,t)$ for $t > 0$.

15-5.19 A guitar string (along x) can produce sound in two distinct ways. Usually it is plucked so that the string moves parallel to the guitar face (along y). It also can be plucked by pulling it farther from the guitar face (along z). (a) Write down the equation of motion for the z-coordinate. (b) Find the wave speed for motion along z. (c) For a guitar length L, fing the lowest resonance frequency for motion along z.

15-6.1 For an EM wave that propagates along $+y$, if $|\vec{B}| = 0.04$ mT and $\vec{B}$ points along $+z$ at $y = 2$ cm and $t = 1.2$ s, find the direction and magnitude of $\vec{E}$ at $y = 2$ cm and $t = 1.2$ s.

15-6.2 For an EM wave that propagates along $+z$, if $|\vec{E}| = 52$ V/m and $\vec{E}$ points along $+x$ at $y = 3.4$ cm and $t = 2$ s, find the direction and magnitude of $\vec{B}$ at $y = 3.4$ cm and $t = 2$ s.

15-6.3 (a) For an electromagnetic wave propagating along x, if $E_y(x,t) = D\sin(qx - \omega t)$, find $B_z(x,t)$. (b) For an electromagnetic wave propagating along y, if $E_z(y,t) = D\sin(qy - \omega t)$, find $B_x(y,t)$.

15-6.4 In a microwave cavity filled with air, the nodes in the electric field are 3.2 cm apart. Find the wavelength, the wavenumber, the frequency f, and the angular frequency ω.

15-6.5 An electromagnetic wave from a cell phone travels along $+y$, its electric field given numerically by $\vec{E}(x,y,z,t) = 4.5 \times 10^4 (\text{V/m}) \hat{z} \sin[qx - \omega t]$, where its frequency of 3.2×10^{10} Hz lies in the microwave band. Find (a) the wavelength, (b) q, (c) ω, and (d) $\vec{B}(x,y,z,t)$.

15-6.6 An electromagnetic wave has $\vec{B}(x, y, z, t) = 1.5 \times 10^{-4} \text{T} \hat{z} \sin[qy - \omega t]$, where its wavelength of 0.24 mm lies in the infrared band. Find (a) the frequency f, (b) q, (c) ω, and (d) $\vec{E}(x,y,z,t)$.

15-7.1 Find the wavelength of (a) an AM station broadcasting at 990 kHz; (b) an FM station broadcasting at 89.1 MHz.

15-7.2 Find the frequencies f associated with the extremes of human vision, taken to be red (700 nm) and violet (400 nm).

15-8.1 (a) Assuming no energy loss, how must the amplitude of an outgoing spherical wave fall off with distance from the power source? (b) Repeat for an outgoing cylindrical wave.

15-8.2 A laser beam of 4 mW (rms), with wavelength 680 nm, propagates along the $+z$-direction. Its electric field is along the y-axis. Find (a) the maximum value of the electric and magnetic fields; (b) the frequency; and (c) explicit expressions for both $\vec{E}(x,y,z,t)$ and $\vec{B}(x,y,z,t)$, which are zero at the origin for $t = 0$.

15-8.3 Find the electric and magnetic field amplitudes needed to produce a time-averaged power of 4×10^5 W within an area of 120 cm^2.

15-8.4 Take the total intensity of starlight at the earth to be 10^{-9} the total intensity of light from the sun, at the earth. Find the distance from the sun where the sun's intensity at the earth equals that of starlight.

15-9.1 A flashlight of mass 0.18 kg and 4 Ω resistance is in outer space. It is equipped with two 1.5 V cells in series, each with an initial charge of 4 A-hr. The flashlight provides light with 100% efficiency at a constant rate until the batteries are totally discharged. Find (a) the rate at which it accelerates, (b) how long it accelerates, (c) the maximum speed it attains, and (d) the distance it travels while still lit.

15-9.2 A laser beam of 4 mW power (rms) and 1.4 mm^2 width, with wavelength 680 nm, propagates along the $+z$-direction. Its electric field is along the y-axis. Find (a) the electric and magnetic energy densities, (b) the Poynting vector, and (c) the radiation pressure.

15-9.3 An electromagnetic wave from a cell phone travels along $+y$, its electric field given numerically by $\vec{E}(x,y,z,t) = 4.5 \times 10^4 (\text{V/m})\hat{z} \sin[qy - \omega t]$, where its frequency of 3.2×10^{10} Hz lies in the microwave. Find (a) the electric and magnetic energy densities, (b) the Poynting vector, and (c) the radiation pressure.

15-9.4 An electromagnetic wave . has $\vec{B}(x,y,z,t) = 1.5 \times 10^{-4}\,\text{T}\hat{z}\sin[qy - \omega t]$, where its wavelength of 0.24 mm lies in the infrared. When $\vec{B}$ points along $+\hat{z}$, $\vec{E}$ points along $-\hat{x}$. Find (a) the electric and magnetic energy densities, (b) the Poynting vector, and (c) the radiation pressure.

15-9.5 Within a solenoid of radius R and n turns per unit length, carrying current I, there is an axial magnetic field $\vec{B}$. See Figure 15.26, which does not show the turns of wire. If the current is changing, there is also a tangential induced electric field. (a) Find the Poynting vector for $r < R$. (b) For $r < R$, find the power flow (per unit length along the axis), and give its direction when $dI/dt > 0$. (c) Find the net flow of power (per unit length along the axis) for an annulus from r to $r + dr$. (d) Find the rate of change of the magnetic field energy, and interpret your result.

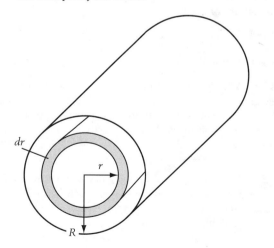

Figure 15.26 Problem 15-9.5.

15-9.6 Within a parallel-plate capacitor of radius R and plate separation $d \ll R$, with charge Q, there is an axial electric field $\vec{E}$. See Figure 15.27. If the charge varies, there is also a tangential induced magnetic field. (a) Find the Poynting vector for $r < R$. (b) For $r < R$, find the power flow (per unit length along the axis), and give its direction when $dQ/dt > 0$. (c) Find the net flow of power for an annulus from r to $r + dr$. (d) Compare with the rate of change of the electric field energy, and interpret your result.

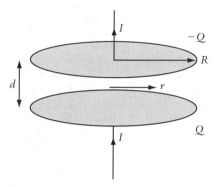

Figure 15.27 Problem 15-9.6.

15-10.1 Radiation at 67.5 MHz propagates with velocity 2.65×10^8 m/s along a plastic tube. (a) Find the wavelength. (b) Find the index of refraction. (c) If $\mu_r = 1$, find ε_r.

15-10.2 In a ferrite, radiation of 22.4 MHz propagates with velocity of 3.5×10^6 m/s. (a) Find the wavelength. (b) Find the index of refraction. (c) If $\varepsilon_r = 1.8$, find μ_r.

15-10.3 Light of wavelength 591.5 nm in air is incident on water at an angle of $23°$ from the normal. (a) Find the frequency of the light in air. (b) Find the frequency of the light in water. (c) Find the wavelength of the light in water. (d) Find the angles the reflected and refracted light make with the normal.

15-10.4 Sound of frequency 475 Hz in air is incident at $6°$ on a wall of rock for which the sound velocity is 2380 m/s. (a) Find the wavelength of sound in the air. (b) Find the wavelength of sound in the rock. (c) Find the frequency of sound in the rock. (d) Find the angles the reflected and refracted sound make with the normal.

15-10.5 (a) Find the critical angles for crown glass and for diamond. (b) Now consider two cut gems, one of crown glass and the other of diamond, with precisely the same geometry. From which is a randomly chosen light ray more likely to emerge?

(c) If the light intensities within the two stones are equal, which gem is likely to emit brighter light, when light does emerge?

15-10.6 Consider light that enters one side of a flint glass prism with corner angle of 90°. Let the light enter at an angle of 50° to the normal. See Figure 15.28. (a) Find the angle that the light emerges at. (b) Which will be deflected more, red light or blue?

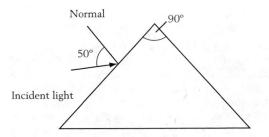

Figure 15.28 Problem 15-10.6.

15-11.1 Unpolarized light of intensity 4.6 W/m² is normally incident on a perfect linear polarizer whose preferred electric field axis is vertical. Find the intensity of the transmitted light, and describe its state of polarization.

15-11.2 Linearly polarized light of intensity 2.8 W/m² is normally incident on a perfect linear polarizer whose preferred electric field axis is vertical. The transmitted light intensity is 1.6 W/m². Describe the state of polarization of the incident light and the transmitted light.

15-11.3 The preferred electric field axes of linear polarizers 1 and 2 are along x and y, respectively. Their normals are along z. (a) If unpolarized light of intensity 0.4 W/m² is normally incident on 1, find the intensity and polarization of the light transmitted by 2. (b) If polarizer 2 is now rotated, so that its preferred electric field axis makes a counterclockwise angle of 35° to x, find the intensity and polarization of the transmitted light.

15-11.4 The preferred electric field axes of two linear polarizers are along x and y, respectively. Their normals are along z. Unpolarized light of intensity 1.5 W/m² is normally incident on the first. (a) If a third, middle polarizer is introduced between the two polarizers, its preferred electric field axis at a clockwise angle of 56° to y, find the intensity and polarization of the transmitted light. See Figure 15.29. (b) For the third polarizer, find the angle of the preferred electric field axis for which the intensity will be a maximum. (c) Find the intensity of the transmitted light, and describe its state of polarization.

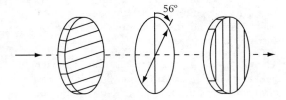

Figure 15.29 Problem 15-11.4.

15-11.5 A superposition of unpolarized light of intensity 2.6 W/m², and linearly polarized light of intensity 4.1 W/m², is normally incident on a perfect linear polarizer. If the polarizer's preferred electric field axis is vertical, and the polarization axis of the polarized light makes an angle of 56° to that, give the intensity and state of polarization of the emerging light.

15-11.6 Consider a superposition of unpolarized light and linearly polarized light. It has total intensity 4.6 W/m² and is normally incident on a perfect linear polarizer. When the polarizer's preferred electric field axis is vertical, the intensity of transmitted light is 3.7 W/m². When the polarizer's preferred electric field axis is horizontal, the intensity of transmitted light is 3.1 W/m². For the incident light, find (a) the intensity of unpolarized light, (b) the intensity of linearly polarized light, and (c) the angle that the electric field of the polarized light makes to the vertical.

15-11.7 (a) Find the Brewster angle for light passing from air into water. (b) Repeat for light passing from water into air.

15-11.8 As in Figure 15.17, an observer looks up at the sky, but now through a linear polarizer whose plane is normal to the incident light. The intensity of the transmitted light changes on rotating the preferred electric field axis of the polarizer, whose plane remains normal to the incident light. Let $\theta = 0$ correspond to maximum intensity. How should the intensity vary with θ? Assume that the light seen arises from only a single scattering of light by the atmosphere.

15-12.1 (a) Describe in your own words why the electric field can be normal to, but cannot be parallel to, the surface in a microwave cavity. (b) Describe in your own words why the magnetic field can be parallel to, but cannot be normal to, the surface in a microwave cavity.

15-12.2 At a given instant of time, the electric field at point P just outside the wall of a microwave cavity has magnitude 478 V/m, pointing into the wall. Determine the magnitude and sign of the surface charge density.

15-12.3 At a given instant of time, the magnetic field at point P just outside the wall of a microwave cavity has magnitude 3.62 mT, pointing along y. The outward normal is along z. Determine the magnitude and direction of the surface current density.

15-13.1 In Figure 15.19, verify that (a) the surface current density and the $\vec{B}$ field have the appropriate relationship, (b) the surface charge density and $\vec{E}$ have the appropriate relationship, (c) the propagation direction is as indicated.

15-13.2 Explain why it is preferable to use nondispersive waveguides when transmitting information using microwaves.

15-14.1 The end of a linear antenna normally cannot become very charged because it has a relatively low capacitance. Thus the antenna current, considered to be spatially varying (unlike the case in circuit theory) has two near-zeros. (a) For a linear antenna of length 1.2 m, what is the longest wavelength to which this corresponds in the linear antenna, considered as a resonator? (b) What is the wavelength of the radiation? [Note: For a guitar, the geometric resonances of the string determine the wavelength, and $v = \lambda f$ determines the frequency, which is then radiated to the air, with a sound velocity *different* from that of the string. For a linear antenna, the geometric resonances of the antenna determine the wavelength, and $v = \lambda f$ determines the frequency, which is then radiated to the air, with the *same* velocity (c) as that of the antenna. See the discussion following (15.68).]

15-14.2 Near a linear antenna, when the electric field is at a maximum, the magnetic field is zero, and when the magnetic field is at a maximum, the electric field is zero. On the other hand, far from the antenna the electric and magnetic field are in phase. What does this imply about phase relationships in the intermediate region?

15-14.3 Comparing Figure 15.21(b) to Figure 14.15, estimate R/R_c for Hertz's detector.

15-14.4 Two concentric circular disks of radius a, with seperation $d \gg a$, are centered along the x axis. Let Q and $-Q$ be uniformly distributed on the disks, centered at $x = \pm a/2$, respectively. A rod of resistance R, placed along the x-axis, connects the two disks. For a concentric circular loop, of radius $R \gg a$, find $\int (\vec{J} + \vec{J}_D) \cdot \hat{n} \, dA$, when the plane of the loop is inside and outside the disks.

15-15.1 A string has an unstretched length of L_0. It is now stretched to L. A standing wave $y(x, t) = A\sin(\pi x/L)\sin(\omega t)$ is set up. The maximum value for the additional length associated with the vertical displacement (the length of string is now $\int ds$, not $\int dx$) should be much less than the stretching length $L - L_0$; otherwise, the tension will not be nearly constant. (a) Find the additional length $\delta L = \int (ds - dx)$ due to this wave. (b) Setting $\delta L = L/64$, show that $A = L/4\pi \approx 0.08L$. This is a surprisingly large amplitude, yet it permits $L - L_0$ to be as small as $0.1L$ and still have $\delta L \ll L - L_0$.

15-15.2 A string 1.5 m long has a mass of 0.152 kg and a tension of 82 N. Find the amplitude of 250 Hz waves with time average transmitted power of 0.56 W.

15-15.3 A bar is made of aluminum (Young's modulus 7×10^{10} Pa, mass per unit volume 2.7×10^3 kg/m³). It is 5 m long and has a cross-sectional area of 4.8×10^{-4} m². Find: (a) the velocity of longitudinal waves; (b) the time it takes for a wave to pass from one end and back again.

15-15.4 For equal sound intensities at different frequencies, the kinetic energies of the air and thus the actual air velocities must be the same. (a) If two equal intensity sounds are at $f_1 = 125$ Hz and $f_2 = 1250$ Hz, which has the larger amplitude? (b) What does this say about the design of movable diaphragms for low-frequency speakers (doglike woofers) as opposed to high-frequency speakers (bird-like tweeters)?

15-15.5 A manufacturer produces radios with compact speakers that accurately reproduce low frequency sounds. Assume that a speaker contains a long wrapped up tube of length L that resonates at $\lambda = (1/2)L$ (a small hole permits sound to escape). Find the length of tube needed to reproduce the lowest piano key (27.5 Hz).

15-15.6 A tube resonates at 40 Hz when filled

with air. You are to add krypton gas to make it resonate at 27.5 Hz. (a) What fraction of krypton should be in the gas? (b) Does the pressure matter? (c) Could you use kryton and air at a lower pressure? (*Hint:* The velocity of sound in a gas is proportional to $\sqrt{k_B T/m}$, where k_B is Boltzmann's constant, T is the temperature, and m may be considered the average molecular mass.)

15-15.7 A tube that resonates at 27.5 Hz when filled with air is now filled with water. What is its new resonance frequency?

15-15.8 (a) Find the wavelengths (in air) associated with the extremes of human hearing, taken to be 20 Hz and 20,000 Hz. (b) Find the corresponding wavelengths in water.

15-G.1 Explain the principle of echo location. Apply it to find the distance to a building if your echo arrives with a 2.4 s delay. Take the sound velocity to be 344 m/s.

15-G.2 Develop a rule of thumb to estimate the distance between you and a source of lightning. When you see the lightning, start counting at about a rate of one per second, until the thunder arrives. (a) What integer would you multiply by to estimate the distance in miles? (b) In kilometers?

15-G.3 A rightward wave with velocity v is photographed using a strobe light with a time interval τ that is slightly more than half a period: $\tau = T/2 + \tau'$. (a) Show that the wave appears to be a leftward-moving wave with velocity $v' = v(T - \tau)/\tau$. (b) Show that the maximum apparent leftward velocity, is less than v. (c) If $v'/v = 0.28$, find τ/T. [*Hint:* Use the fact that $\sin qx \equiv \sin(qx + 2\pi)$.]

15-G.4 A uniform rope of length L and mass m hangs from the ceiling. (a) Show that at a distance y from the bottom of the rope, the local transverse wave velocity is given by $v = \sqrt{gy}$. (b) Show that the time it takes for the bottom of the rope to wiggle, if an earthquake shifts the ceiling, is $\tau = 2\sqrt{L/g}$.

15-G.5 Repeat Problem 15-G.4 when the rope supports a mass M attached to its bottom. (a) Show that $v = \sqrt{g}\sqrt{y + (M/m)L}$ and $\tau =$

$2\sqrt{L/g}(\sqrt{M/m + 1} - \sqrt{M/m})$. (b) Verify that this gives the expected results for $M/m \to 0$ and for $M/m \to \infty$.

15-G.6 Show that the time-derivative of the displacement vector $\vec{D} = \epsilon_0 \vec{E} + \vec{P}$ is the sum of the displacement current $\vec{J}_D$ and the polarization current $\partial \vec{P}/\partial t$.

15-G.7 Take a pen in your hands and bend it. (a) Is it under tension? (b) Is it curved? (c) Is it in equilibrium? (d) Does the argument about acceleration being proportional to curvature apply to a rigid object like a string? In the early 19th century, Mlle. Sophie Germain studied bending of a rigid beam, finding that the equation describing its motion required four spatial derivatives, rather than two, in order to produce an acceleration.

15-G.8 Kelvin developed a capacitance-and-resistance-based theory of telegraphy. Consider a co-axial cable where $\mathcal{R}$ and $\mathcal{C}$ are, respectively, the resistance per unit length and the capacitance per unit length. The current I, the charge per unit length λ, and the voltage V are all taken to vary with x along the axis, with the inner and outer parts assumed to always have equal and opposite charges. Consider a length dx. For slow variations of I and λ along x, show that: (a) the charge on dx changes at the rate $dQ/dt = I(x) - I(x + dx)$; (b) $d\lambda/dt = -dI/dx$; (c) $I = -\mathcal{R}^{-1}dV/dx$; (d) $dV/dx = \mathcal{C}^{-1}d\lambda/dx$; (e) $\mathcal{R}\mathcal{C}d\lambda/dt = d^2\lambda/dx^2$.

15-G.9 Before applying Maxwell's equations to telegraphy, Heaviside extended Kelvin's work by developing a theory of telegraphy where self-inductance dominates over resistance. This is important at high frequencies. Consider a co-axial cable made of a perfect conductor (no resistance; surface charge and surface currents). Denote by λ the charge per unit length and by I the current for the inner cylinder. The inner and outer wires are assumed to always have equal and opposite charges. (a) For slow variations along the axis show that charge conservation implies that $d\lambda/dt = -dI/dx$. (b) Next, again for slow variations along the axis, show that the emf equation for a resistance-less wire is given by $0 = \mathcal{C}^{-1}d\lambda/dx + \mathcal{L}dI/dt$. (c) From this deduce that the velocity of propagation is $v = \sqrt{1/\mathcal{L}\mathcal{C}}$.

"When two Undulations [waves], from Different Origins, coincide either perfectly or very nearly in direction, their joint effort is a Combination of the Motions belonging to each."

—Thomas Young,
stating the principle of interference in "Theory of Light and Colours," 1801

Chapter 16

Optics

Chapter Overview

Section 16.1 introduces the chapter. Section 16.2 introduces interference and diffraction, largely in the context of water waves. (However, these ideas apply to all types of waves, including sound waves and light waves.) Having introduced all these phenomena, the remainder of the chapter discusses light from a somewhat historical viewpoint.

Sections 16.3–16.5 discuss the early views of light, which tended to think of light as a particle emitted by a source. A more complete discussion—not the purpose of the present work—would also include the study of focusing by mirrors and lenses, and application of these focusing principles to optical instruments like the telescope and the microscope. Section 16.3 presents a brief history of optics to the end of the 17th century. It emphasizes two of the successes of the particle emission theory: explanations of the laws of reflection and refraction, and the formation of rainbows. Section 16.4 enumerates a number of 17th-century experiments on light, many of which are explainable only by the wave theory. Section 16.5 summarizes the sometimes contradictory theoretical developments of the 17th century.

Sections 16.6–16.8 consider the more modern view of light, which is based upon a wave viewpoint. (Both the particle and wave viewpoints were speculated on by the ancient Greeks, some 2000 years ago.) Section 16.6 introduces Young, applies his analysis of constructive and destructive interference in numerous contexts, and analyzes his famous two-slit interference experiment. Section 16.7 introduces Fresnel and gives his analysis of the observed intensity patterns both for Young's two-slit interference experiment and for single-slit diffraction. Section 16.8 discusses birefringent crystals, by which the phenomenon of polarization was discovered.

Sections 16.9 and 16.10 discuss the applications of diffraction, by which we analyze both the very large and the very small. Section 16.9 considers the diffraction grating, by which we determine the nature of the very distant stars. Section 16.10 considers diffraction by crystals of x-rays (like light, a form of electromagnetic radiation), by which we determine the positions of individual atoms within a crystal. ∎

16.1 Introduction

Man has become a celestial detective—his forensic tools the telescope and the diffraction grating—investigating the birth, evolution, and death of the stars.

1. The telescope, with a wide aperture, lets light of all pure colors pass through in nearly a straight line, intensified by focusing with a lens. This is an application

of *geometrical optics*, which includes the fundamental phenomena of reflection and refraction at surfaces.

2. The diffraction grating, with either many narrow lines for scattering or many narrow apertures for transmission, analyzes the light into its component colors. This is an application of *physical optics*, which includes *polarization* (discussed in Chapter 15) and the very general and closely related wave phenomena known as *interference* and *diffraction* (to be introduced in Section 16.2).

Chapter 15's discussion of light as a wave phenomenon depended heavily upon mathematical reasoning. The solution of Maxwell's equations yielded plane wave solutions of all frequencies and wavelengths, corresponding to a coupled oscillation of the electric and magnetic fields—electromagnetic radiation—traveling at the speed of light in vacuum. The identification of light as a small part of this spectrum was clear. However, even without knowing that light is an electromagnetic wave, by 1801 Thomas Young had established that light was a type of wave, with a specific range of wavelengths.

16.2 Interference and Diffraction

Although this chapter presents background material about optics in general, this section presents water wave experiments to illustrate the phenomena of interference and diffraction. These phenomena generalize to other types of waves, such as sound and light.

We can directly observe the medium—the water surface—associated with a water wave. As Young demonstrated in his lectures to the Royal Institution (1802–1803), when a plunger in a *ripple tank* containing water oscillates up and down at a fixed frequency f, there is an expanding wave of wavelength λ that has circular symmetry. At any instant of time there are maxima (peaks, or crests) and minima (troughs). Figure 16.1 presents a schematic of the top view of a

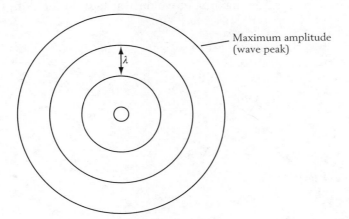

Maximum amplitude
(wave peak)

Figure 16.1 Schematic of the top view of a ripple tank containing a plunger that oscillates vertically at a fixed frequency f. The dark circles indicate, at a given instant of time, the maxima, or wave peaks (crests). The distance between the peaks is the wavelength λ. In water, $\lambda f = v$, where v is the wave velocity.

ripple tank. The circles denote the instantaneous locations of the wave peaks, which move radially outward with velocity $v = f\lambda$. Because water waves are dispersive, v varies with λ, rather than being a constant, as is the case for sound and light.

16.2.1 *Interference*

When two identical plungers within a ripple tank operate at the same frequency f, and therefore at the same wavelength λ, they may produce a pattern of maxima and minima on the water surface. If two maxima (or two minima) arrive at a point P at the same time (i.e., in phase), we say that there is *constructive interference*. See Figure 16.2(a). If a maximum and a minimum arrive at a point P at the same time (i.e., out of phase by 180°), we say that there is *destructive interference*. See Figure 16.2(b).

Let the two plungers be separated by the distance d and let the wavelength be λ. The overall pattern of maxima and minima depends on the relative phase of the plungers; for simplicity, consider only the case where at time t the plungers' vertical displacements $A(t)$ have the same amplitude a and phase ϕ: $A(t) = a\cos(\omega t + \phi)$. As will be shown, the pattern of maxima and minima depends on the ratio of λ to d and is independent of the common amplitude a and phase ϕ.

Consider that the frequency f can vary, and that initially it is very low, which means that λ is very large. In this case, the waves produced by the two plungers are very nearly in phase everywhere, giving a pattern much like that for a single plunger (Figure 16.1). Now let f increase, which decreases λ. This causes dephasing between the waves produced by the two plungers. Figure 16.3 depicts the case where $d \approx 0.7\lambda$, the open circles denoting the plungers. There, the two waves are in phase (denoted by closed dots, with an imaginary line C connecting the dots) only along the perpendicular bisector of the line connecting the plungers. As time passes, the maxima move outward, as do the dots. (The lines

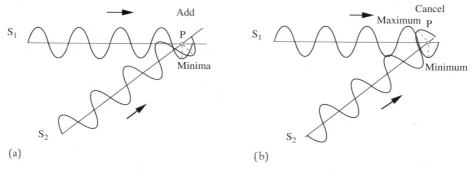

(a)   (b)

Figure 16.2 (a) Constructive interference between waves produced by two sources with the same amplitude, wavelength, and phase. Minima (and a half-period later, maxima) arrive at the same time, giving a doubling of the net wave amplitude. (b) Destructive interference between two waves of the same wavelength and from sources that are in phase with one another. Minima of one and maxima of the other arrive at the same time, giving zero net wave amplitude.

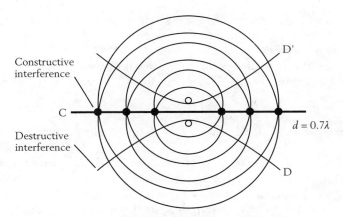

Figure 16.3 Interference between waves from two sources (the open circles) with the same amplitude, frequency, and phase, separated by $d = 0.7\lambda$. Constructive interference occurs only along the line C. Destructive interference occurs only along the curves D and D′.

D and D′ correspond to minima, or destructive interference, to be discussed shortly.)

If f is further increased, so that $d > \lambda$, there are additional curves along which constructive interference occurs. Figure 16.4 depicts the respective cases where $d = 1.5\lambda$ and $d = 4.5\lambda$. For $d = 1.5\lambda$, in addition to the perpendicular

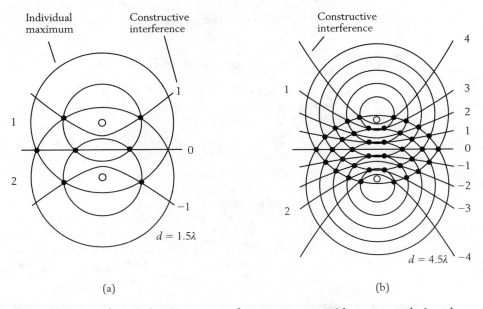

(a)   (b)

Figure 16.4 Interference between waves from two sources (the open circles) with the same amplitude, frequency, and phase. (a) Separation $d = 1.5\lambda$, where there are three curves along which constructive interference occurs. (b) Separation $d = 4.5\lambda$, where there are nine curves along which constructive interference occurs.

bisector there is an additional pair of curves, one to each side of the perpendicular bisector, where constructive interference occurs. For $d = 4.5\lambda$, there are four additional pairs of curves.

Let the distances from any point P to the plungers be r_1 and r_2. Constructive interference occurs when the separation $r_2 - r_1$ is an integral number of wavelengths, or

$$r_2 - r_1 = m\lambda, \qquad m = 0, \pm 1, \pm 2, \ldots \qquad \text{(constructive interference)} \quad (16.1)$$

By definition, when the distances between a point P and two foci differ by a constant, the curve traced out is a hyperbola. Thus, the locus of points where the maxima occur trace out a series of hyperbolae, defined by the integer m and the ratio d/λ. The larger the wavelength λ, or the smaller the source separation d, the larger the separation between maxima. This is seen in Figures 16.4(a) and 16.4(b), where the number to the right of each curve is the corresponding value of m.

Destructive interference occurs when the separation $r_2 - r_1$ is a half-integral number of wavelengths, or

$$r_2 - r_1 = \left(m + \frac{1}{2}\right)\lambda, \qquad m + \frac{1}{2} = \pm\frac{1}{2}, \pm\frac{3}{2}, \ldots \qquad \text{(destructive interference)}$$
$$(16.2)$$

Destructive interference corresponds to the local height of the water surface being undisturbed—still water. In Figure 16.3, where $d \approx 0.7\lambda$, the pair of curves D and D′ correspond to $m + \frac{1}{2} = \pm\frac{1}{2}$ in (16.2), where there are minima. The larger the wavelength, the larger the separation between minima. In Figures 16.4(a) and 16.4(b), between the maxima lie minima, which are not drawn.

Example 16.1   Interference maxima and minima

Two plungers are separated by $d = 25$ cm. (a) What is the largest λ for which there is a maximum corresponding to $n \neq 0$? (b) What is the largest λ for which there is destructive interference?

Solution: (a) If the wavelength λ exceeds the largest value of $r_2 - r_1$, which here is $d = 25$ cm, then (16.1) can be satisfied only for $m = 0$. Hence $\lambda \leq 25$ cm for a maxima with $n \neq 0$. When equality holds, only on the line defined by the two sources (and not in the region between the sources) is there constructive interference. (b) If one-half the wavelength λ exceeds the largest value of $r_2 - r_1$, which here is $d = 25$ cm, then (16.2) for destructive interference cannot be satisfied, even for $m = 0$. Hence, for destructive interference $\lambda \leq 50$ cm. When equality holds, so $r_2 - r_1 = (\frac{1}{2})\lambda$, only on the line defined by the two sources (and not in the region between the sources) is there destructive interference.

Example 16.2   Some details of an interference pattern

Two plungers are separated by $d = 25$ cm. Along their perpendicular bisector, a distance $D = 52$ cm away, there is a maximum at $y = 0$. Normal to the

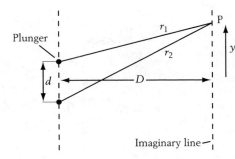

Figure 16.5 Schematic for analyzing the interference patterns of Figures 16.3 and 16.4. Point P lies along a line parallel to the line of centers of the two plungers, but a distance D away.

bisector at this distance, there is a second maximum at $y = 12$ cm. See Figure 16.5. (a) Find the wavelength λ. (b) Find the position y of the first minimum.

Solution: (a) Use $r_2 = \sqrt{D^2 + (y + d/2)^2}$ and $r_1 = \sqrt{D^2 + (y - d/2)^2}$. Placed in (16.1), with $m = 1$ and $y = 12$ cm, this yields $\lambda = 5.48$ m. (b) Using $\lambda = 5.48$ m in (16.2) with $m = 0$ gives

$$\sqrt{D^2 + (y + d/2)^2} - \sqrt{D^2 + (y - d/2)^2} = \frac{1}{2}\lambda = 2.74 \text{ m.}$$

The solution, found numerically, is $y = 5.90$ cm. This lies about halfway between the maxima at $y = 0$ and $y = 12$ cm.

16.2.2 *Diffraction*

A ripple tank also can be used to demonstrate *diffraction*. Consider a water wave with planar wavefront of wavelength λ incident on a barrier with an opening of width d. For $d \gg \lambda$ (i.e., for short wavelengths), as in Figure 16.6(a), the wave that passes through the barrier goes nearly in a straight line, as if the wave were a particle on a straight-line path. The related case of scattering by an obstacle of length d is given in Figure 16.6(b).

The case where $d \sim \lambda$ is depicted in Figure 16.7 both for a barrier with an opening and for an obstacle.

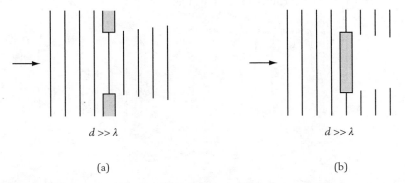

$d \gg \lambda$	$d \gg \lambda$
(a)	(b)

Figure 16.6 (a) Scattering by a short-wavelength wave incident on a barrier. (b) Scattering by a short-wavelength wave incident on the corresponding obstacle. The dark lines represent instantaneous maxima.

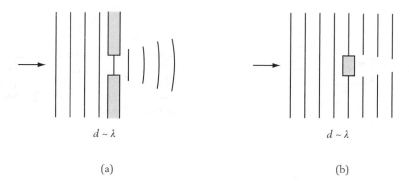

(a)   (b)

Figure 16.7 (a) Scattering by an intermediate-wavelength wave incident on a barrier. (b) Scattering by an intermediate-wavelength wave incident on the corresponding obstacle. The dark lines represent instantaneous maxima.

The case where $d \ll \lambda$ (i.e., for long wavelengths) is depicted in Figure 16.8(a) for a barrier with an opening. The wave that passes through the barrier spreads out uniformly, as if the opening were a point source. In the corresponding case for an obstacle, just behind it the wave amplitude builds up. See Figure 16.8(b). At large distances from the obstacle, and out of the incident beam, the diffracted pattern is the same as for the corresponding barrier. This identical diffraction pattern for both cases is an example of what is known in optics as *Babinet's principle*. This says that, for any pair of scatterers with complementary geometries (e.g., an opaque screen with a star cut out, and the corresponding opaque star), outside the geometrical shadow of either, the patterns of scattered light are the same.

To summarize, short wavelengths are associated with straight-line propagation, and long wavelengths are associated with a more spread out propagation. Related to this are two phenomena associated with sound: (1) a cricket, with its high-pitched (short-wavelength) chirp, can be localized more readily than a bullfrog, with its low-pitched (long-wavelength) croak. (2) When the corner of a building separates them from us, the low-pitched bullfrog can be heard more readily than the high-pitched cricket.

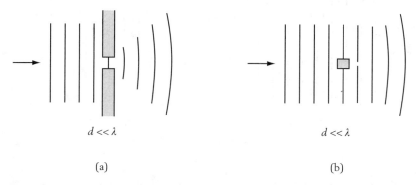

(a)   (b)

Figure 16.8 (a) Scattering by a long-wavelength wave incident on a barrier. (b) Scattering by a long-wavelength wave incident on the corresponding obstacle. The dark lines represent instantaneous maxima.

Example 16.3	Wave spreading (diffraction)

Radio waves of frequency 91.1 MHz and sound waves of frequency 440 Hz are incident on a metal wall with a circular hole of diameter 1.2 m. Take the speed of sound to be $v = 340$ m/s. Describe the extent of the spreading of the waves transmitted through the hole.

Solution: The relation $f\lambda = v$ implies that $\lambda = v/f$. For the radio wave, with $v = c = 3.0 \times 10^8$ m/s, this gives $\lambda = 3.29$ m. For the sound wave this gives $\lambda = 0.77$ m. The radio wave, with larger wavelength (3.29 m) than the 1.2 m diameter circular hole (as in Figure 16.8a), should exhibit mostly spreading, with some localization in the forward direction. The sound wave, with shorter wavelength (0.79 m) than the 1.2 m diameter circular hole, should exhibit mostly propagation in the forward direction, with some spreading (as in Figure 16.6a).
Sections 16.7.2 and 16.7.3 consider diffraction in more detail.

16.3 Optics to the End of the 17th Century

Surely primitive man knew that light travels (nearly) in straight lines, and that behind an illuminated object is a shadow. Euclid knew the law of reflection, as did the Romans, who used a metallic alloy called *speculum* for their mirrors. [This is the origin of the term *specular reflection* (e.g., off a mirror, or at a glancing angle for a less perfectly smooth surface, such as a piece of writing paper), as opposed to *diffuse reflection* (e.g., off a dully painted wall).] Ptolemy, some 1800 years ago, studied refraction, but he characterized it quantitatively only for small angles. Likewise, the ancients knew of the rainbow, and perhaps of the prism, but they had little understanding of these phenomena. Some viewed light as being generated by the eye, rather than taking the correct view, first stated by Al-Hasan (c. 1000), of the eye as a receiver of light whose source is elsewhere. Light here was considered to be particle-like, traveling in a straight line.

De Dominus, around 1590, used a large water-filled spherical glass vessel to reproduce the halo shape of a rainbow, with a bright bow at 42° to the incident sunlight. He showed that the primary (and secondary) rainbows involved one (and two) internal reflections in addition to the refractions on entering and exiting. Figure 16.9(a) presents a geometry where sunlight is incident from the left, directed along the x-axis. (The angle of incidence relative to the normal is θ, and the observation angle relative to the backward direction is ϕ.) De Dominus could not explain, however, how the halo shape occurred, or why the colors separate to have a continuous distribution of bows.

16.3.1 *The Law of Refraction (Snell's Law)*

Measured relative to the normal, let θ be the angle of incidence in air, and let θ' be the angle of refraction into another medium. The law of refraction, in a form equivalent to

$$\sin\theta = n\sin\theta', \tag{16.3}$$

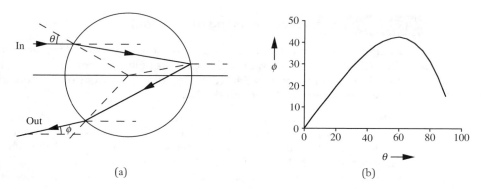

(a)   (b)

Figure 16.9 (a) Path of a ray incident on a water-filled spherical glass vessel. It undergoes refraction on entering, one reflection within, and refraction on leaving. (b) Exit angle ϕ as a function of incident angle θ. Not drawn are the ray reflected on entering, nor the transmitted ray associated with internal reflection, nor the reflected ray on leaving.

where the index of refraction n, was known to Snel (one "l") by 1621. Since $n > 1$, we have $\theta' < \theta$. The larger the n, the larger the magnitude of the angular deviation $\theta - \theta'$. For small angles, $\theta = n\theta'$, as known by Ptolemy.

16.3.2 *Theory of the Rainbow Angle*

From numerical ray tracing using Snell's law, and from experiments, Descartes (1637) studied the relationship between θ and ϕ in Figure 16.9(a). These results can be summarized by Figure 16.9(b). This is calculated on the basis of a ray subject to refraction on entering, an internal reflection, and a refraction on leaving, as in Figure 16.9(a). (Not drawn are the ray reflected on entering, nor the transmitted ray associated with the internal reflection, nor the reflected ray on leaving.) Figure 16.10 presents a detailed version of Figure 16.9(a), with light rays incident from the left, directed along the x-axis.

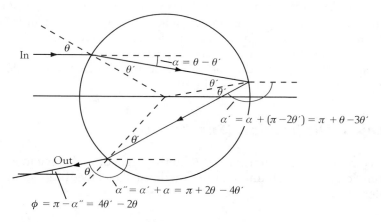

Figure 16.10 Details of reflection and refraction for Figure 16.9.

In Figure 16.10, a ray is incident on the raindrop at an angle θ relative to the normal, and refracts at an angle θ'. The refracted ray deflects from a straight path along the x-axis by a clockwise angle

$$\alpha = \theta - \theta'. \tag{16.4a}$$

The refracted ray travels along a straight path within the raindrop until it is internally reflected. The redirection on reflection can be described as an additional clockwise rotation by $\pi - 2\theta'$. Thus the internally reflected ray makes a clockwise angle relative to the x-axis of

$$\alpha' = \alpha + \pi - 2\theta' = \pi + \theta - 3\theta'. \tag{16.4b}$$

This internally reflected ray then travels in a straight line until it hits the internal surface of the raindrop at an angle θ' relative to the normal. There it refracts to an angle θ relative to the normal. This corresponds to a net deflection by an additional angle $\alpha = \theta - \theta'$ clockwise relative to the x-axis. The net effect is a clockwise angle of rotation by

$$\alpha'' = \alpha' + \alpha = \pi + 2\theta - 4\theta' \tag{16.4c}$$

relative to the x-axis. Viewed by an observer relative to the rainbow, this corresponds to a clockwise angle relative to the x-axis of

$$\phi = \pi - \alpha'' = 4\theta' - 2\theta. \tag{16.5}$$

The ϕ versus θ curve of Figure 16.9(b) is computed for $n = 4/3$, as appropriate for water. The maximum is near $42°$. No rays are deflected by more than $42°$, corresponding to the dark region observed outside a rainbow. Thus the primary rainbow is caused by a bunching up of rays at the maximum angle near $42°$. When raindrops are present, a bow can be seen in the part of the sky that corresponds to this angle. See Plate 2 for a primary and a secondary rainbow.

If we use (16.5) for ϕ and (16.3) relating θ' to θ, then setting $d\phi/d\theta = 0$ gives that ϕ has a maximum for

$$\sin\theta = \sqrt{\frac{4 - n^2}{3}}. \qquad \text{(rainbow angle)}$$

For $n = 4/3$ this yields $\sin\theta = 0.861$, so $\theta = 59.4°$, and then (16.1) gives $\sin\theta' = 0.645$, so $\theta' = 40.2°$. Finally, (16.5) gives $\phi = 42.0°$, in agreement with the measured result and with the graph of Figure 16.9(b).

16.3.3 *Reflection, Refraction, and the Principle of Least Time*

Descartes gave a "derivation" of Snell's law. Although elsewhere Descartes argued that light propagated instantaneously, his mechanically based derivation assumed that light has a finite velocity, whose component parallel to the surface is conserved on refraction. This reasoning yielded $v_1 \sin\theta_1 = v_2 \sin\theta_2$, rather than what we now know to be the correct result, given by (15.61):

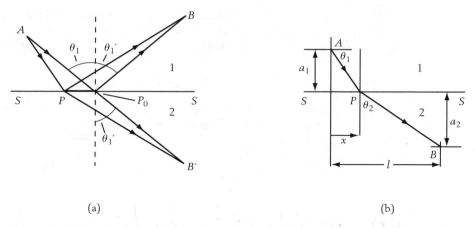

(a)   (b)

Figure 16.11 Reflection and refraction of light from a source at A, incident on the surface S. (a) Reflection to point B, where point B′ is the image of B. The path of least time is AP_0B'. (b) Refraction to point B. The general path APB is shown.

$\sin\theta_1/v_1 = \sin\theta_2/v_2$. To obtain agreement with experiment, Descartes had to assume that light travels more quickly in water and glass than in air.

Unsatisfied with Descartes's derivation, Fermat undertook to determine for a particle of light the straight-line path from A to B that takes the least time. In 1657, he showed that this ray satisfies a law of refraction equivalent to (15.61). (His derivation used a method he had developed earlier to find the tangents— i.e., the slopes—to curves. Newton generalized Fermat's method to yield what we now know as the differential calculus.) Since the velocity of light was not measured accurately until Fizeau (1850), it was not possible at that time to verify Fermat's form of the law of refraction. Following Fermat, we now apply the *principle of least time* to study reflection and refraction.

The law of reflection. Consider a point A in medium 1 where a ray of light is emitted, and a point B where light reflected from surface S is observed. See Figure 16.11(a). Draw the image point B′ associated with B, and a straight line from A to B′, intersecting S at P_0. Consider light that travels straight from A to a general point P on S, and then straight to B. The total distance that it travels is given by AP + PB, which equals AP + PB′. Figure 16.11(a) shows that the point P giving the shortest total distance is P_0. This corresponds to a straight line from A to B′, and an angle of reflection θ_1' that equals the angle of incidence θ_1:

$$\theta_1' = \theta_1. \quad \text{(law of reflection)} \quad (16.6)$$

Since the velocity is uniform in medium 1, this path is both the shortest path and the path of least time.

The law of refraction. Let point A be in medium 1, with light velocity v_1, and let point B be in medium 2, with light velocity v_2. We first consider a ray of light that goes straight from A in medium 1 to a point P on the surface S, and then straight in medium 2 from S to the point B. We then find the point for which the travel time T is least. See Figure 16.11(b).

Let the nearest distance from A to the surface S be a_1, the nearest distance from B to S be a_2, and let the distance along the surface from A to P be x. If the total distance along x from A to B is l, then the distance along x from P to B is $l - x$. Thus the total time T to go from A to B is the sum of the time $AP/v_1 = \sqrt{a_1^2 + x^2}/v_1$ to go from A to P and the time $BP/v_2 = \sqrt{a_2^2 + (l - x)^2}/v_2$ to go from P to B. Thus

$$T = \frac{\sqrt{a_1^2 + x^2}}{v_1} + \frac{\sqrt{a_2^2 + (l - x)^2}}{v_2}. \tag{16.7a}$$

The position x where T is minimized is the solution of $dT/dx = 0$. Noting that

$$\frac{d}{dx}\sqrt{a_1^2 + x^2} = \frac{x}{\sqrt{a_1^2 + x^2}} \quad \text{and} \quad \frac{d}{dx}\sqrt{a_2^2 + (l - x)^2} = -\frac{(l - x)}{\sqrt{a_2^2 + (l - x)^2}},$$

(16.7a) yields

$$\frac{dT}{dx} = \frac{1}{v_1}\frac{x}{\sqrt{a_1^2 + x^2}} - \frac{1}{v_2}\frac{l - x}{\sqrt{a_2^2 + (l - x)^2}} = \frac{1}{v_1}\sin\theta_1 - \frac{1}{v_2}\sin\theta_2, \tag{16.7b}$$

where Figure 16.11(b) has been used to obtain the sines. Setting (16.7b) to zero, the minimum time occurs for

$$\frac{\sin\theta_1}{v_1} = \frac{\sin\theta_2}{v_2}. \tag{16.8}$$

This is precisely the same as (15.61) and leads to the law of refraction in the form

$$n_1 \sin\theta_1 = n_2 \sin\theta_2, \qquad n_{1,2} = \frac{c}{v_{1,2}}. \qquad \text{(law of refraction)} \qquad \tag{16.9}$$

Like Descartes's explanation of the rainbow, Fermat's derivation of the laws of reflection and refraction was a great triumph of the particle viewpoint of light.

16.4 Late 17th-Century Discoveries about Light

Considerable experimental progress was made in the second half of the 17th century, much of it incomprehensible from the particle viewpoint.

1. Grimaldi (1616–1673) discovered and named diffraction (1665). Using a small hole through which sunlight passed, he noticed that (1) the shadow of objects on a nearby screen is slightly larger than would be expected geometrically; (2) on moving farther out of the geometrical shadow, there is a sequence of bright and dark fringes, getting narrower the farther they are from the shadow (within each bright fringe the center is white, the side nearest the shadow is bluish, and the side farthest from the shadow is reddish); (3) the center of the first bright fringe is noticeably brighter than the uniform illumination far outside the shadow. See Figure 16.12.

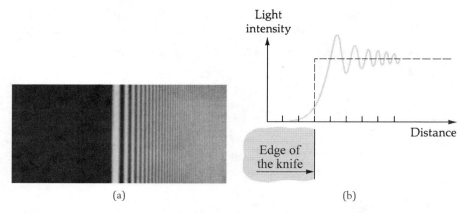

(a)   (b)

Figure 16.12 (a) Diffraction of a distant light around an edge, for an observer not so close that the geometrical shadow is completely obscured. Although first studied by Grimaldi, this is called Fresnel diffraction, after the man who first explained the phenomenon. (b) Intensity is plotted along the vertical and position along the horizontal.

2. Hooke (1635–1703), using glass plates pressed together, in his *Micrographia* (1665) gave the first published studies of the colors seen in light reflected off thin films and plates. He noticed that the colors were related to the thickness d of the gap between the plates; the harder together he pressed the plates, the farther apart the colored fringes. He suggested that perhaps reflection off the front and back surfaces could explain the effect, but he didn't develop the idea quantitatively. See Figure 16.13. His geometries included plates of variable thickness, and lenses placed one upon another. However, he could not measure

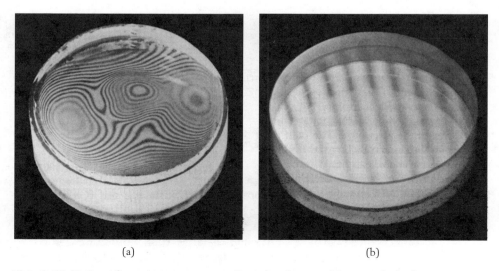

(a)   (b)

Figure 16.13 Interference patterns on reflection of monochromatic light from two lenses in near contact, first observed by Hooke. (a) Unpolished surfaces. (b) Surfaces polished flat to within a fraction of a wavelength, and slightly tilted relative to one another.

Figure 16.14 White light incident upon a prism, which separates the light into its pure colors (defined by their wavelengths), first studied by Newton.

the thickness of the films or plates that corresponded to each colored fringe, leaving that as a challenge for future workers.

3. Newton (1643–1727) performed experiments on the prism (1666), establishing that light was composed of a continuum of many pure colors, and that once a beam of pure color was isolated from other colors, it retained its integrity. See Figure 16.14 and Plate 1. He realized that, in many cases, light that appeared to be of a certain color (e.g., orange) would, when analyzed with the prism, separate into various pure colors (e.g., red and yellow). With this idea,

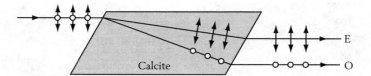

Figure 16.15 Unpolarized light refracted by a calcite crystal, which splits it into two beams, one satisfying Snell's law (the ordinary, or O, beam), and one not satisfying Snell's law (the extraordinary, or E, beam). Also given are the electric field polarizations of these beams (the open circles denote field normal to the page).

Newton explained the color separation of the rainbow as the consequence of the different colors having different indices of refraction in water. Newton also extended Hooke's work on the colors of thin films and plates, using a lens geometry that permitted him to measure the film thickness d. (We shall discuss this in more detail.)

4. Bartholinus (1625–1698) discovered *double refraction* (1669), which occurs only for crystals with a preferred axis (such as what was known to him as *Iceland spar*, now known as *calcite*, or $CaCO_3$). See Figure 16.15. When there is only one beam incident on the crystal—even if it is normally incident—typically there are two refracted beams. One of these (the ordinary, or O, beam) satisfies Snell's law; the other (the extraordinary, or E, beam) does not. (Figure 16.15 also indicates the directions of polarization, as now understood; at the time, scientists were not aware that the two exiting beams had different properties.) For one particular direction of propagation in the crystal (the *optic axis*), the O and E beams coincide.

5. Roemer (1675) showed that the times of eclipses of Jupiter's first moon, Io, depend on the motion of Jupiter relative to the earth (this is a form of the Doppler effect, where the frequency is determined by the period of Io's motion around Jupiter). See Figure 16.16, where in moving from A to B, Io's period decreases, and in moving from C to D, Io's period increases. This indicated that the speed of light is finite. From Roemer's data and the then best known value for the diameter of the earth's orbit about the sun (about two-thirds of its correct value), Huygens later estimated the speed of light c to be about 2×10^8 m/s.

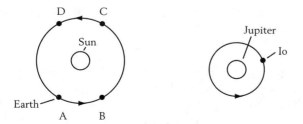

Figure 16.16 Schematic of the motion of the earth around the sun, and the motion of Jupiter's moon Io around Jupiter, with period T_{Io}. When the earth moves toward (from) Jupiter, the period decreases (increases), in proportion to the ratio of the earth's velocity to the velocity of light.

6. One more figure of this time period must be mentioned: Kepler, more well-known for establishing, by 1618, the three laws of planetary motion. His 1604 book *Astronomia Pars Optica* investigated image formation by the pinhole camera, explained vision as due to imaging by the lens of the eye on the retina, correctly described the causes of long-sightedness and short-sightedness, and explained how both eyes are used for depth perception. In 1608 Lippershey made a telescope with a converging objective lens and a diverging eye lens, which Galileo almost immediately improved upon and applied to terrestrial and astronomical studies, in 1610 discovering four satellites orbiting about Jupiter, and setting the stage for the Copernican revolution in our views of the universe. Kepler's 1611 book *Dioptrice* described total internal reflection for large angles, and founded modern geometrical optics using Ptolemy's small angle result $\theta = n\theta'$. It described real, virtual, upright and inverted images and magnification. It explained the principles of Lippershey's telescope, and also proposed what has become the modern telescope, with a converging objective and a converging eye lens. To Kepler, who had spent years trying to explain planetary orbits as circles upon circles, and then as ellipses, the straight lines of geometrical optics must have been a piece of cake.

Hence, by 1675, nearly all the basic phenomena of physical optics had been discovered. What lacked was a unifying explanation.

16.5 Late 17th-Century Views of Light

16.5.1 *Robert Hooke and Wave Pulses*

Hooke argued that light is associated with rapid, small-amplitude vibrations of an unspecified medium called the "ether." He employed an analogy to sound in air, where the medium for the vibrations is air. He also developed a theory of light propagation where the source produces pulses at regular time intervals (rather than the more modern view where the source produces a continuous oscillation whose period equals that regular time interval). Such pulses propagate outward, by analogy to water waves, "like Rays from the center of a Sphere ... all the parts of these spheres cut the Rays at right angles." He studied reflection and refraction using a ray-and-wavefront construction. (A *wavefront* or—to use Hooke's word, a *pulse*—gives the simultaneous position of many outgoing particles associated with different rays emitted at the same time. See Figure 16.17.) According to Hooke, the origin of colors lay in the pulse shape, and white light was a color unto itself. This view was not consistent with Newton's work with the prism, which showed that there is more than one way to obtain a given color.

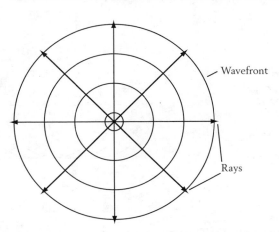

Figure 16.17 Rays and wavefront for a spherical source.

With two assumptions—(1) when a wavefront impinges on an interface, it initiates a new wavefront in the new medium; and (2) the new wavefront and the incident wavefront have the same velocities parallel to the interface—Hooke derived the same erroneous law of refraction as Descartes: $v_1 \sin\theta_1 = v_2 \sin\theta_2$.

16.5.2 *Christian Huygens and Huygens's Principle for the Addition of Wavelets*

Huygens, in his *Treatise on Light* (1678), developed a wave theory of light. "If, indeed, one looks for some other mode of accounting for the [uniform speed] of light, he will have difficulty in finding one better adapted than elasticity [of the ether]." Thinking of light waves in the sense of pulses in time (just as did Hooke), he developed what we now know as *Huygens's principle: If a wavefront is known at any instant of time, then at any future time the new wavefront is determined by the envelope of secondary wavelets that are produced at each point of the initial wavefront.* (The secondary wavelets are taken to travel at the same speed as the primary wave.)

Figure 16.18(a) depicts the propagation of light emitted as spherical waves at time $t = 0$ from position A. The dashed lines represent rays propagating radially outward from A, yielding the outgoing spherical primary wavefronts HH′ at time t, and JJ′ at time $2t$. (The value of t is immaterial; Huygens did not think of light as periodic either in space or in time.) At points C, D, and E on HH′ at time t, secondary wavelets are generated that travel at the same speed as the primary wave, yielding secondary wavelets cc′c″, dd′d″, and ee′e″ on JJ′ at time $2t$. The envelope of these secondary wavelets is a sphere that coincides with the primary wavefront. Similarly, Figure 16.18(b) depicts a primary wavefront that is a downward-directed plane at $t = 0$. It generates secondary wavelets whose envelope at time t coincides with the primary wavefront at time t. Hence, Huygens's construction permits spherical, planar, and even cylindrical wavefronts to maintain their shape. Considering only an envelope from the forward-moving

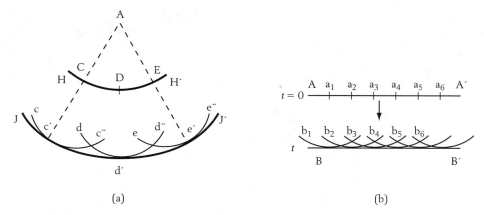

(a) (b)

Figure 16.18 Huygens's principle. (a) The point source at A produces a wavefront HH′. C, D, and E on HH′ are sources of secondary wavelets cc′c″, dd′d″, ee′e″, whose envelope forms the new spherical wavefront JJ′. (b) A planar wavefront AA′ has sources $a_1, \ldots$, producing the secondary wavelets $b_1, \ldots$, whose envelope forms the new planar wavefront BB′.

wavelets, Huygens's construction implicitly includes information about both the wave position and the wave velocity.

Because Huygens's principle was a wave theory, by analogy to sound he could immediately justify how light waves can pass through one another independently (inexplicable in a particle theory, except on assuming extraordinarily low particle densities). He could explain shadows, qualitatively, by arguing that within the shadow the secondary wavelets "do not combine at the same instant [i.e., in phase] to produce one single wave."

16.5.3 *Huygens's Theory of Reflection and Refraction*

Most important, for a planar wavefront incident on a planar interface between two transparent media, Huygens developed Hooke's arguments to obtain the laws of reflection and refraction. Like Fermat, Huygens took the velocity of light in glass to be less than in air, rather than greater than in air, as erroneously assumed by Hooke and Newton.

Consider Figure 16.19. This depicts two media with different indices of refraction (in fact, with $n_1 > n_2$, so $\theta_1 < \theta_2$, as occurs when light is incident from water to air). Incident rays in medium 1 yield both reflected rays in medium 1 and refracted rays in medium 2. The incident rays make an angle θ_1 to the normal N, so their wavefronts (e.g., the dashed line AA′) make an angle θ_1 to the interface SS. At $t = 0$, the wavefront AA′ is planar, with its edge at A just making contact with the interface. Spherical wavelets, depicted at time t, are generated at $t = 0$ in both media. Also at time t a second incident ray hits the interface at B, which begins to initiate another secondary wavelet. To find the time t envelope of wavelets in each medium, draw the common tangents to all the time t wavelets. Figure 16.19 shows only the infinitesimal radius wavelet (i.e., a point)

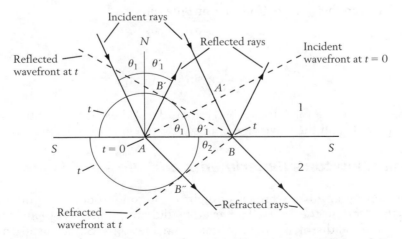

Figure 16.19 Application of Huygens's principle to reflection and refraction. Incident rays from 1 intersect surface S at A and B, producing secondary wavelets in both 1 and 2. The secondary wavelets originating at A are shown; at time t the wavefront has just reached B, so the secondary wavelets originating at B have zero radius.

at B, and the wavelet produced at A. In medium 1, the tangent to the wavelet produced at A is at B′. In medium 2, the tangent to the wavelet produced at A is at B″. Let us see how this leads to the laws of reflection and refraction.

Reflection. In Figure 16.19, consider the right triangles AA′B and BB′A, with common side AB normal to their common angle of 90°. Because sides AB′ and BA′ correspond to the same time interval, AB′ = BA′. Hence AA′B and BB′A are similar triangles. Therefore the angle θ_1' made by the outgoing wavefront to the interface equals the angle θ_1 made by the incoming wavefront to the interface, or

$$\theta_1' = \theta_1. \tag{16.10}$$

Because the angle between the normal and a ray equals the angle between the interface and a wavefront, we have thus established the *law of reflection: the angle of reflection equals the angle of incidence.* This has the same form found by Fermat, derived earlier as (16.6).

Refraction. In Figure 16.19, consider the right triangles BB′A and BB″A. These satisfy the relationships

$$AB'' = AB \sin \theta_2 \quad \text{and} \quad AB' = AB \sin \theta_1. \tag{16.11a}$$

Taking the ratios of the left- and right-hand sides of these equations then yields

$$\frac{AB'}{AB''} = \frac{\sin \theta_1}{\sin \theta_2}. \tag{16.11b}$$

The distances AB′ and AB″ are proportional to the velocities v_1 and v_2, so

$$\frac{AB'}{AB''} = \frac{v_1}{v_2}. \tag{16.11c}$$

Equating the right-hand sides of (16.11b) and (16.11c) yields

$$\frac{\sin \theta_1}{\sin \theta_2} = \frac{v_1}{v_2}, \tag{16.12}$$

or

$$n_1 \sin \theta_1 = n_2 \sin \theta_2, \qquad n_{1,2} \equiv \frac{c}{v_{1,2}}. \tag{16.13}$$

This is the same form found by Fermat, derived earlier as (16.9). Hence, both the particle and the wave viewpoints yield the laws of reflection and refraction.

16.5.4 *Isaac Newton's Experiments and Views of Light*
Optional

Newton began his researches on light in the mid-1660s and published papers on this work in the 1670s. However, he did not publish his *Opticks* until 1704, after the deaths of his rivals Hooke and Huygens. Some historians attribute this to his desire to avoid disagreements with these scientists, neither of whom initially accepted that Newton's prism experiments established that sunlight is a superposition of many pure colors.

Newton rejected the view that light is a wave because he could not see how it could lead to nearly straight-line propagation with pronounced shadows.

To avoid having his comprehensive work subject to misinterpretation, Newton carefully tried to distinguish experimental facts from theoretical hypotheses. Book I presents his experiments on the color separation of sunlight by a prism, and his *experimentum crucis* (crucial experiment) showing that these colors, once separated by a prism because of their (slightly) different indices of refraction, retain their integrity and their index of refraction afterward. He also showed that a beam of white light whose colors have been separated by a prism can be recombined by passing the separated beam through a second prism to produce another beam of white light, thus establishing that white light is composite. Book II extends Hooke's work on the colors of thin plates and thin films (e.g., oil slicks and soap bubbles). By studying the positions of the colors associated with a convex lens placed on flat glass (Newton's rings), Newton found that maxima occur in reflection when a certain condition held. Had he written the equation for the case of normal reflection, it would have read

$$d = \frac{mI}{2n}. \qquad \text{(fits of reflection)}$$

Here d is the thickness of the film, n is the index of refraction, m is an *odd* integer; the length I, which is specific to a given color, is what in 1704 Newton called the "interval of the fits" but in earlier work of 1675 he called "bigness." Book III presents his experiments and ideas on diffraction, which Newton called "inflection" because he thought of it as a disturbance in the paths of particles of light that pass by an edge, caused by forces associated with the edge.

Nevertheless, at least initially, his physical picture of light did not involve light particles alone. In the 1670s he wrote: "the rays of light . . . excite vibrations in the ether; . . . with various colours, according to their bigness and mixture; the biggest with the strongest colors, reds and yellows; the least with the weakest, blues and violets . . . much as nature makes use of several bignesses to generate sounds of divers tones." *Thus, to Newton, the vibrations of the "ether" had certain wave properties, even if the light rays themselves did not.* Newton indicated the "bigness" of light: "it is to be supposed that the ether is a vibrating medium like air, only the vibrations are far more swift and minute; those of air, made by a man's ordinary voice, succeeding one another at more than half a foot, or a foot distance; but of ether at a less distance than the hundred thousandth part of an inch." Because Newton could not draw on the word *wavelength*, his verbal definition of "interval of fits" to the modern reader could correspond to either a wavelength or half a wavelength. However, Newton states that yellow-orange has an interval of fits of 1/89,000 of an inch, or 285 nm, which is half of yellow-orange's 570 nm wavelength.

Newton studied sound and obtained a value for its velocity in air that was about 25% below the measured value. He also studied water waves with well-defined wavelengths, presenting an argument equivalent to $\lambda f = v$ [but more like $(\frac{1}{2}\lambda)(2 f) = v$]. Had he considered light to be a wave, then from his estimate of the bigness $\frac{1}{2}\lambda$ of light, and his own estimate of its speed v (700,000 times the speed of sound, or about 2.4×10^8 m/s), Newton could have estimated the frequency f of light to be on the order of 10^{15} Hz. That would have been consistent with the view of light as a rapid, small-amplitude oscillation.

16.6   Thomas Young—Interference

Newton's reputation continued to grow following his death, in large part because of the success of his *Principia*, relating force to motion, in stimulating quantitative scientific development. But to nontechnical readers like Benjamin Franklin, Newton's *Opticks* was the more accessible. Whoever was to carry optics forward was going to need the confidence to break with Newton's opposition to light as a wave. Around 1800, Thomas Young (1773–1829), educated broadly but by profession a physician specializing in the eye, entered the picture.

The Multi-faceted Thomas Young

Among other things, the polymath Young (known to his contemporaries as "The Phenomenon") explained visual *accommodation* (muscles attached to the lens change its shape, thus enabling the eye to focus at different distances); proposed the three-color theory of color vision (the first theory to propose that our perception of colors occurs *within* the eye, rather than being an intrinsic property of light); and made the first study of astigmatism (eyes that when viewed head-on have an elliptical shape also have two focal points, one for each axis of the ellipse). He also introduced the term *energy* (in the context of kinetic energy), was responsible for what is now known in elasticity as Young's modulus, and analyzed the effect of surface tension on the wetting angle of fluids. He knew over ten languages and made major contributions to deciphering the two unknown Egyptian languages accompanying the Greek on the then recently discovered Rosetta stone. Nevertheless, in his own judgment, Young's most important work was in establishing the wave nature of light. Young's works on light as a wave included two enduring demonstrations: (1) the previously discussed ripple tank, for demonstrating the interference of water waves; and (2) the two-slit experiment (to be discussed shortly), for demonstrating the interference of light waves. His work was not immediately recognized, ultimately due to his "unduly concise and obscure" presentation of mathematical detail.

16.6.1   *Light as a Wave*

In 1800, to argue in support of light as a wave, Young quoted Newton's analogy between the tones of sound and their wavelengths, and between the colors of light and their wavelengths. To argue that waves can be localized in direction (so as to produce shadows) Young countered Newton's view that sound waves "diverge equally in all directions" by noting that (1) the sound of cannons is much larger in the forward than the backward direction; and (2) "speaking trumpets" (megaphones) *by design* produce sound that is localized in direction. Further, to argue against "the Newtonian system" of light as an emitted particle, Young noted that (1) as observed by Huygens, it does not explain why the velocity of light is independent of the intensity of its source (the harder a ball—a type of particle—is thrown, the faster it goes); and (2) both partial reflection and refraction occur. Both of these phenomena are easily explained by the analogous behavior of sound waves.

In *Theory of Light and Colours* (1801), Young stated the *principle of interference*, quoted at the chapter heading: *When two Undulations [waves], from*

Different Origins, coincide either perfectly or very nearly in direction, their joint effort is a Combination of the Motions belonging to each.

The implications of this principle were worked out previously, in Section 16.2. Figure 16.2 illustrates the addition of two waves of the same amplitude and wavelength, when completely in phase and when completely out of phase. These cases correspond to (16.1) and (16.2).

16.6.2 *Coherence*

Young applied the principle of interference to two light waves only when they had the same wavelength and only when they had the same source, so they were in phase relative to one another over a time long compared to the resolution time of the eye. (This time is about 1/30 of a second, a fact used when moving pictures fool the eye into thinking that a rapid succession of individual photographs is a continuum of images.) Such "in-phase-ness" is known as *coherence*. A typical light source has a *coherence time* τ_c of about 10^{-9} s, corresponding to about 10^6 oscillations of the light, with characteristic frequency $f \sim 0.5 \times 10^{15}$ Hz. Since light travels at 3×10^8 m/s, this τ_c corresponds to what is called a *longitudinal coherence length* $l_c = c\tau_c$ of about 0.15 m. A "photograph" of such a wave would show that, starting from a given wave peak, within the longitudinal coherence length the other wave peaks would be approximately an integral number of wavelengths from the given wave peak, but 10^6 peaks away there would be significant dephasing. A light wave also has a transverse extent given by its *transverse coherence length*. Both types of coherence provide limitations on interference phenomena. Typical commercial lasers have τ_c's on the order of 10^{-8} s. Lasers with very well-defined frequencies can in principle have τ_c on the order of 10^2 s, but mechanical stability of the laser cavity reduces this to on the order of 10^{-5} s. By comparison, a bell with $f = 500$ Hz might have τ_c on the order of 2 s, corresponding to about 1000 oscillations, or a Q—see Section 14.3—of about 1000. Figure 16.20 gives schematics of wavefronts with good transverse and longitudinal coherence, with good longitudinal coherence only, and with good transverse coherence only.

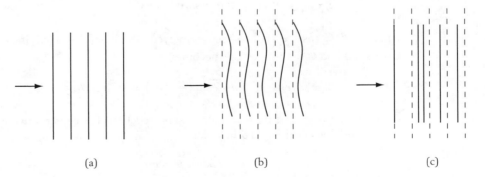

(a)   (b)   (c)

Figure 16.20 Examples of wavefronts with varying degrees of coherence: (a) good longitudinal and good transverse coherence, (b) good longitudinal coherence but weak transverse coherence, (c) good transverse coherence but weak longitudinal coherence. Dashed lines in (b) and (c) indicate, for comparison, good longitudinal and good transverse coherence. An even better representation would vary the thickness of the dark lines at the wave peak to indicate variations in amplitude.

16.6.3 *Colors of Striated Surfaces*

Young's first detailed application of the principle of interference was to the colors sometimes seen when light from a small distant light source is reflected off a surface inscribed with two nearby narrow, parallel grooves. He experimented with a micrometer for which each inscribed line was actually a pair of lines separated by 0.0001 inch, so $d = 2.54 \times 10^{-4}$ cm in Figure 16.21. (The pair-to-pair separation of 0.002 inch, which Young noted, was not relevant to the observed effects.) For sunlight at glancing incidence ($\alpha = 0$, or P_1-P_2 along the x-axis in Figure 16.21), he saw only bright red. With $\alpha = \pi/4 - \theta$, this corresponded to $\theta = \pi/4$. On rotation of the micrometer by α,

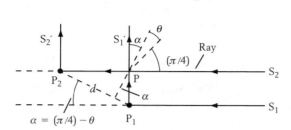

Figure 16.21 Normal intersection with the page at P_1 and P_2 of a pair of lines scratched on Young's micrometer. The incident wavefront S_1-S_2 scatters to the wavefront S_1'-S_2'. The point P is for purposes of comparison. In the experiment, S_1-S_2 and S_1'-S_2' were fixed, and P_2 was rotated about P_1.

thus decreasing θ, he also observed bright red for angles θ of 32°, 20.75°, and 10.25°.

Example 16.4 **Young's micrometer**

(a) For Young's micrometer experiment, relate the difference in path length $r_2 - r_1$ to the line separation d and the angle θ. (b) From the data, determine the wavelength λ. (To Young, the "length of an Undulation.")

Solution: (a) Consider a general rotation angle α, with points P_1 and P_2 separated by d. By Figure 16.21, the difference in path length $r_2 - r_1$ is given by

$$r_2 - r_1 = P_2 P - P_1 P = d(\cos\alpha - \sin\alpha) = d\sqrt{2}\left(\sin\frac{\pi}{4}\cos\alpha - \cos\frac{\pi}{4}\sin\alpha\right),$$

where $\sin\frac{\pi}{4} = \cos\frac{\pi}{4} = \frac{1}{\sqrt{2}}$ has been employed. With $\sin(A - B) = \sin A\cos B - \cos A\sin B$ and $\theta = \pi/4 - \alpha$, this becomes

$$r_2 - r_1 = d\sqrt{2}\sin\left(\frac{\pi}{4} - \alpha\right) = \sqrt{2}d\sin\theta.$$

(This gives $r_2 - r_1 = d$ for $\theta = \pi/4$, as expected.) (b) For constructive interference, using this $r_2 - r_1$ in (16.1) yields

$$\sqrt{2}d\sin\theta = m\lambda.$$

Taking $d = 2.54 \times 10^{-4}$ cm, and $m = 1, 2, 3, 4$ to correspond to 10.25°, 20.75°, 32°, and 45°, this gives $\lambda = 639$ nm, which indeed lies in the red. (Characteristically, Young merely stated without proof that the extra path difference was proportional to $d\sin\theta$.) Note that shorter wavelengths λ (e.g., toward the blue) give interference at smaller angles θ.

16.6.4 *Colors of Thin Films and Thin Plates*

Consider light of wavelength λ at near normal incidence on a plate of thickness d surrounded by air. In Figure 16.22, two reflected rays are represented, one from the front surface and the other from the back surface. Our goal is to understand the colors seen in reflection. As recognized by Newton, to the extent that light is not absorbed within the film, maxima in reflection correspond to minima in transmission, and vice versa.

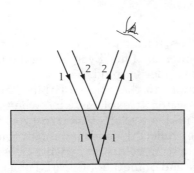

Figure 16.22 Paths of two rays reflected to the eye at near normal incidence.

In analyzing this situation, Young first noted that, for motion in one dimension, a light object colliding with a massive object will continue forward in the same direction, as in Figure 16.23(a), whereas a massive object colliding with a light object will bounce back (i.e., reverse direction), as in Figure 16.23(b). If the velocity is represented as a positive amplitude times a sine or cosine, this reversal in direction in the first case may be interpreted as a phase shift of 180°. Young then argued (correctly) that a similar phase shift occurs on reflection when light is incident on a more optically dense material (higher index of refraction). Thus, in Figure 16.22, of the two reflected rays, ray 1 involves a 180° phase shift and ray 2 does not. This phase shift corresponds to half a wavelength λ' in the medium. With n the index of refraction, (15.59) gives $\lambda'n = \lambda$, so $\lambda' = \lambda/n$.

Because of this extra phase shift, at near normal incidence the condition that the two reflected waves must yield constructive interference is that the extra path length $r_2 - r_1 = 2d$ must correspond to a half-integral number $m + \frac{1}{2}$ of wavelengths $\lambda' = \lambda/n$. Thus, for constructive interference on reflection,

$$2d = \left(m + \frac{1}{2} \right) \frac{\lambda}{n}, \qquad m = 1, 2, \ldots$$

(constructive interference in reflection, films) (16.14)

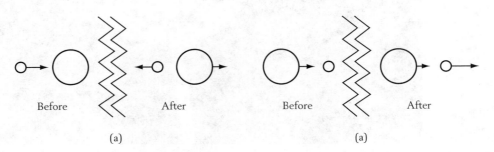

Before   After   Before   After

(a)   (a)

Figure 16.23 (a) Scattering of a light object off a heavy object. (b) Scattering of a heavy object off a light object.

Similarly, the condition for destructive interference is

$$2d = m\frac{\lambda}{n}. \qquad m = 0, 1, \ldots \qquad \text{(destructive interference in reflection, films)}$$

(16.15)

These equations are consistent with Newton's studies, but go beyond them in that wavelength appears, rather than interval of fits. Even Young did not use the word *wavelength*, but clearly he had that idea in mind.

Equations (16.14) and (16.15) apply to the colors of thin plates (Figure 16.13), soap films, and oil slicks on water. It explains why, when such films are very thin, so $d \ll \lambda$, they reflect no light at all [$m = 0$ in (16.15)], but at intermediate thicknesses the color they reflect depends upon film thickness. Interference of this sort explains the irridescent colors of the wing cases of beatles, of peacock and hummingbird feathers, and of mother-of-pearl. For a given order m, the larger the thickness d, the larger the wavelength λ needed for the characteristic interference maximum or minimum. Figure 16.24 illustrates a characteristic thin-film interference pattern when the thickness of the film varies, so the wavelength leading to enhanced reflection varies. Plate 3 shows the same figure in color. Plate 4 shows soap bubbles. Plate 5 shows peacock feathers. Enjoy.

From his values of the wavelengths of the different colors of light, Young was able to determine the corresponding frequencies (on the order of 10^{15} Hz), via $\lambda f = c$ (which did not appear in his paper), finding that "the absolute frequency expressed in numbers is too great to be distinctly conceived."

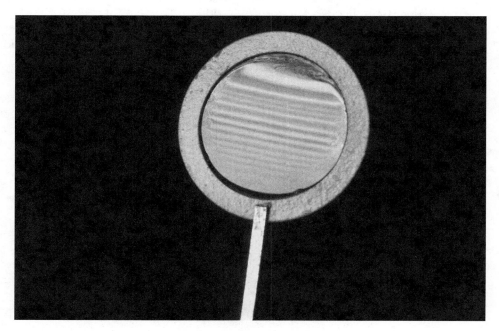

Figure 16.24 Oil slick, illustrating the different colors that can be seen on reflection as the film thickness changes.

Young also noted that if a material of intermediate index of refraction is placed between the air and the first medium, then both reflections involve phase shifts of 180°, and the conditions for constructive interference and destructive interference switch. (Placing sassafras oil, with index of refraction 1.535, between flint glass and crown glass, with indices of refraction 1.66 and 1.52, Young later established that such a phase shift does occur.)

Example 16.5 Soap film interference

Consider a soap film ($n = 1.33$) of thickness 540 nm, supported on a circular hoop. (a) For white light (from 400 nm to 700 nm) at normal incidence, find the reflected colors that are intensified and those that are weakened by interference. (b) Repeat for transmission, assuming that the light is neither scattered nor absorbed in the film. (c) Repeat for reflection, if the soap film is floating on a surface of sassafras oil ($n = 1.535$).

Solution: (a) For constructive interference (16.14) gives $m + \frac{1}{2} = 2dn/\lambda$, where in our case $2dn = 1436$ nm. For $\lambda = 400$ nm this gives $m + \frac{1}{2} = 3.59$, suggesting that $m = 3$ will give a solution for some value of λ in the visible. Indeed, $m = 3$ corresponds to 410 nm (violet). Further, $m = 2$ corresponds to 574 nm (yellow). (Larger m's correspond to λ's in the ultraviolet, and smaller m's to λ's in the infrared.) For destructive interference, (16.15) gives $m = 2dn/\lambda$. For $m = 3$ this gives $\lambda = 479$ nm (blue-green). (Again, larger m's correspond to λ's in the ultraviolet, and smaller m's lead to λ's in the infrared.) Thus, in reflection the film will appear to be a mixture of violet and yellow, with blue-green absent. (b) In the absence of absorption and scattering in the film, the colors that are reflected strongly will be missing from the transmitted light (thus violet and yellow are absent from transmission), and the colors that are least reflected will be enhanced in transmission (thus the film will appear to be blue-green in transmission). (c) With sassafras oil in place, there is an extra 180° phase shift, so $2d = m\lambda/n$ gives constructive interference, and $2d = (m + \frac{1}{2})\lambda/n$ gives destructive interference. In reflection, this will give an enhancement of blue-green and a weakening of violet and yellow. Thin-film antireflection coatings, having an index of refraction intermediate between that of air and glass, use this effect.

For thick films, the number of wavelengths giving constructive interference is large, and they are so close together in wavelength that, effectively, all wavelengths are observed. As a consequence, interference effects become less noticeable to the eye.

Example 16.6 Newton's rings: thickness versus radius

Young analyzed Newton's original data for thin films and thin plates. Figure 16.25(a) depicts a convex lens on a flat plate of glass. There are three reflections: off the flat top surface of the lens (not drawn), off the round bottom surface of the lens, and off the flat plate of glass. Only the latter two have reflected beams close enough to exhibit interference. The effective film thickness d of the air layer varies with distance from the center of the lens so that the colors change as this distance changes.

(a) For a given wavelength λ, find d as a function of the axial distance r from the center of the lens, of radius of curvature R. (b) Find the equations giving the positions of the maxima and minima.

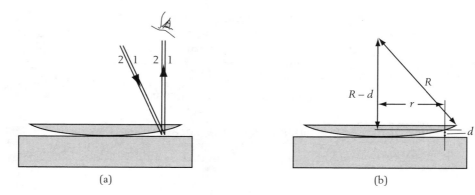

(a)   (b)

Figure 16.25 Newton's rings. (a) Paths of two rays reflected to the eye at near normal incidence. (b) Details of geometry. Newton measured r and deduced d. For small d, d varies quadratically with r.

Solution: (a) Figure 16.25(b) illustrates the lens geometry. At a distance r from the axis of a lens of radius of curvature R, the effective thickness d is given by

$$d = R - \sqrt{R^2 - r^2} = R - R\left(1 - \frac{r^2}{R^2}\right)^{\frac{1}{2}}.$$

For $r/R \ll 1$, $(1 - r^2/R^2)^{\frac{1}{2}} \approx 1 - r^2/2R^2$, from (4.35) with $x = -r^2/R^2$ and exponent $\frac{1}{2}$. Then

$$d \approx R - R\left(1 - \frac{r^2}{2R^2}\right) = \frac{r^2}{2R}.$$

Newton used this result to deduce the (for him) immeasureably small distance d that corresponds to the measureably large r. (b) According to Young, $2d$ must be set to $(m + \frac{1}{2})\lambda$ for maxima, and $m\lambda$ for minima. Note that $m = 0$ gives a minimum, corresponding to the dark spot at the center of Figure 16.26.

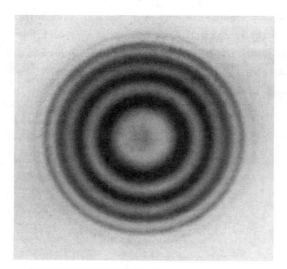

Figure 16.26 For a given wavelength λ, intensity observed on reflection in Newton's rings geometry of Figure 16.25, plotted as a function of position r.

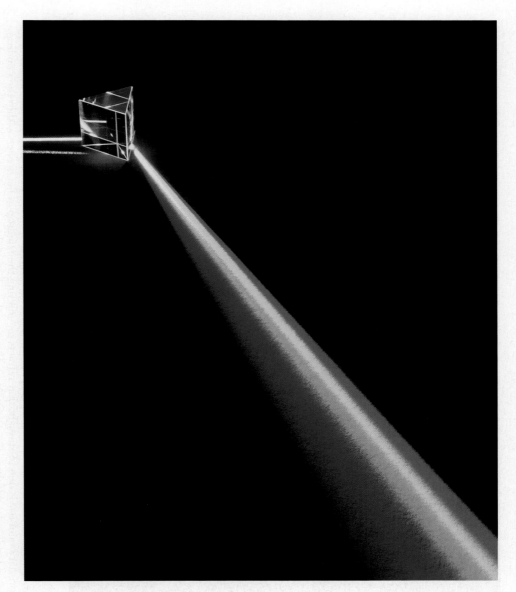

Plate 1 White light through a prism.

A beam of white light passed through an equilateral triangular prism. The white light is a composite of many pure colors with their own characteristic wavelengths and indices of refraction. On passing through the prism, each individual wavelength is refracted differently, thus splitting the beam into its component wavelengths, and revealing the full spectrum of visible light.

Plate 2 Double rainbow.

To see a rainbow, the observer must be between the Sun (behind the observer) and raindrops in the sky (in front of the observer). The primary rainbow is due to sunlight that refracts on entering the raindrop, reflects once within the raindrop, and then refracts on leaving the raindrop. See Figure 16.9. If all wavelengths had the same index of refraction, the rainbow would be seen as a sharp bright white ring, but due to the different indices of refraction, there is a spread of colors. The secondary rainbow is due to sunlight that refracts on entering the raindrop, reflects twice within the raindrop, and then refracts on leaving the raindrop. The colors are reversed in the secondary rainbow, relative to the primary rainbow.

Plate 3 Soap film.

The colors of soap films (and oil slicks) are basically due to interference between (1) light that reflects off the front surface, and (2) light that enters the soap film, reflects off the back surface, and then transmits back to the incident direction. The effect is sensitive to the thickness of the soap film and the path from the light source to the observer, so that the pattern can change if either the light source moves or if the observer moves. For very thin films, all colors are subject to destructive interference, so that the bubble appears to be black. For slightly thicker films, constructive interference can occur, but only for selective wavelengths. For thick films, the number of visible wavelengths that can interfere constructively is so large that all colors seem to be equally reflected. As water from the soap film evaporates, and the film thins, the soap film displays all three of these phenomena.

Plate 4 Soap bubble.

As for a soap film, the colors of a soap bubble are due to interference between (1) light that reflects off the front surface, and (2) light that enters the soap film, reflects off the back surface, and then transmits back to the incident direction. The effect is sensitive to the thickness of the soap film and the path from the light source to the observer, so that the pattern can change if either the light source moves or if the observer moves. Part of the different coloration seen here is due to the varying film thickness, and part to the varying observation angle.

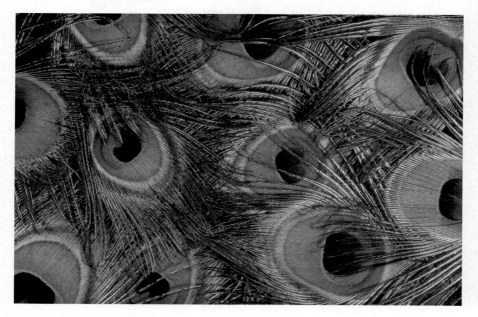

Plate 5 Peacock feather.

The colors of a peacock feather are due both to the ordinary coloration effect of preferential absorption by pigments and to the effects of interference between the front and back of the feather, thought of as a thin film. The latter effect leads to the shimmery irridescence of the peacock feather, because interference effects depend upon the path length of the light and thus the position of the observer.

Example 16.7   Newton's rings

Consider a Newton's ring apparatus with $R = 2.5$ cm that is illuminated with red light of $\lambda = 639$ nm. (a) Locate the first two minima and the first maximum. (b) Repeat assuming that the gap between the lens and the plate is filled with sassafras oil, whose index of refraction ($n = 1.535$) is intermediate between that of the lens and the plate.

Solution: (a) $r = 0$ corresponds to the central minimum on reflection. For the other minima, $2d = r^2/R = m\lambda$ gives $r = \sqrt{mR\lambda}$, so the next minimum occurs for $m = 1$, or $r_1^{max} = \sqrt{(1)(0.025)(639 \times 10^{-9})}$ m $= 1.264 \times 10^{-4}$ m. Similarly, for the maxima, $r = \sqrt{(m + \frac{1}{2})R\lambda}$. For $m = 0$ this gives $r_0^{min} = 0.894 \times 10^{-4}$ m. (b) Addition of sassafras oil, of index refraction n intermediate between that of the lens and the glass plate, interchanges the equations for minima and maxima. It also shifts the positions because the wavelength becomes $\lambda' = \lambda/n$. Thus $r = 0$ corresponds to the central maximum. Using $r = \sqrt{mR\lambda/n}$ with $m = 1$ gives $r = 1.020 \times 10^{-4}$ m for the second maximum. For the minima, $r = \sqrt{(m + \frac{1}{2})R\lambda/n}$ with $m = 0$ gives the first minimum at 0.721×10^{-4} m, and with $m = 1$ gives the second minimum at 1.249×10^{-4} m.

16.6.5   *Young's Two-Slit Experiment*

Like his ripple tank demonstration, the two-slit experiment appears only in Young's published lectures to the Royal Institution. It uses a single light source (to ensure temporal coherence) and three screens. The first screen contains a single slit narrow enough to ensure that it behaves like a line source; hence the wavefronts leaving the first screen are cylindrical in shape. The second contains two identical slits narrow enough to ensure that they behave like line sources driven by the light passing through the first screen. The third is the observation screen. See Figure 16.27. As Young writes, "In order that the effects of two portions of light may be thus combined, it is necessary that they be derived from the same origin, and that they arrive at the same point by different paths." When light of only a single pure color is used, the result on the observation screen is a

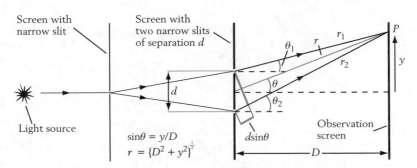

Figure 16.27 Geometry of Young's interference experiment. The first narrow slit, on the left, assures that the light from the light source is coherent (loosely, in-phase) as seen by the second set of narrow slits.

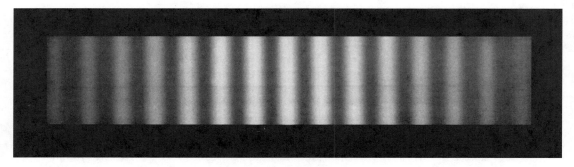

Figure 16.28 Interference pattern for Young's interference experiment of Figure 16.27.

series of light and dark bands, or fringes. See Figure 16.28. (If the light consists of many colors, the pattern on the screen is a superposition of the patterns of each color, weighted in proportion to their individual intensities.

These results can be explained in a manner similar to our earlier discussion of the constructive and destructive interference of water waves, summarized by (16.1) and (16.2). For the geometry of Figure 16.27, the slit separation is d and the distance between the slits and the observation screen is D. A distance y along the observation screen corresponds to the angle θ, at a distance $r = \sqrt{D^2 + y^2}$ from the midpoint between the slits. As in Example 16.2,

$$r_2 = \sqrt{D^2 + (y + d/2)^2} = \sqrt{D^2 + y^2 + yd + d^2/4},$$

but for $D \gg d$,

$$r_2 \approx \sqrt{D^2 + y^2 + yd} = \sqrt{r^2 + yd} = r\sqrt{1 + yd/r^2} \approx r + yd/2r.$$

In that same limit, $r_1 \approx r - yd/2r$. Hence $r_2 - r_1 \approx yd/r = d \sin\theta$. Placing this into (16.1) and (16.2) gives

$$d \sin\theta = m\lambda, \qquad m = 0, \pm 1, \pm 2, \ldots$$

(two-slit constructive interference)   (16.16)

$$d \sin\theta = \left(m + \frac{1}{2}\right)\lambda. \qquad m = 0, \pm 1, \pm 2, \ldots$$

(two-slit destructive interference)   (16.17)

For a given index m and fixed d, larger λ implies larger θ; for a given index m and fixed λ, larger d implies smaller θ. The results of a two-slit interference experiment, as in Figure 16.28, analyzed with (16.16) and (16.17), can be used to determine the wavelength of the incident light. Using these equations, Young found the wavelengths of the colors of the visible spectrum, in essential agreement with modern results. Two-slit interference cannot be explained from a particle viewpoint.

Example 16.8   Two-slit interference

Consider a double-slit interference experiment with a light source having $\lambda = 480$ nm, a slit separation of 0.08 mm, and an observation screen 42 cm away. (a) Find the number of maxima with $\theta \leq 30°$. (b) Find the angular position and the position on the screen of the third minimum. (c) Find the angular position and the position on the screen of the fourth noncentral maximum.

Solution: (a) In (16.16), setting $\theta = 30°$, so $\sin\theta = \frac{1}{2}$, yields $m = d/2\lambda = 166.7$. Hence there are 167 maxima out to $\theta = 30°$, from $m = 0$ to $m = 166$. (b) For $m = 2$, (16.17) yields $\sin\theta = 5\lambda/2d = 0.015$, so $\theta = 0.015$ rad $= 0.859°$. Since $y = D\tan\theta$, this gives $y = 42\tan 0.015 \approx 42(0.015) = 0.63$ cm. (c) For $m = 4$, (16.17) yields $\sin\theta = 4\lambda/d = 0.024$, so $\theta = 0.024$ rad $= 1.375°$. Since $y = D\tan\theta$, this gives $y = 42\tan 0.024 = 1.008$ cm.

16.6.6   *Young and Diffraction*

Young also performed two important diffraction experiments: (1) He showed that the width of the diffraction pattern produced by fibers (e.g., from cloth) is inversely proportional to the fiber diameter. (2) He studied the diffraction pattern produced by an obstacle, a strip of paper 1/30 inch wide, analogous to Figure 16.8(b). At large enough distances from the strip, in addition to Grimaldi's dark fringes outside the geometrical shadow, Young observed bright fringes within the shadow, and a bright fringe at the center of the shadow. See Figure 16.29.

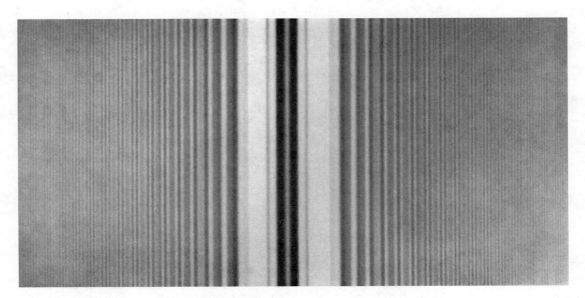

Figure 16.29 Diffraction fringes observed behind an opaque strip of paper.

Moreover, when with a card he cut off the light to one side of the strip, the bright fringes within the shadow disappeared, but the dark fringes outside the shadow remained, just as seen by Grimaldi. This effect of one edge on the pattern near the other edge was inexplicable from a particle emission viewpoint.

16.7 Augustin Fresnel—Theory of Interference and Diffraction Intensity

Fresnel's talents were evident at the age of nine; by systematic study of the best woods for popguns and archery bows, he made these toys so dangerous that they were banned by his friends' parents. As a civil engineer, beginning in 1814, Fresnel took time from his duties to undertake experimental and theoretical research on diffraction. Unaware of Young's work, he reproduced Young's experiments on diffraction by an obstacle. His first attempt at theory, like a similar attempt of Young, considered only interference between light bending around the far edge of the fiber and light reflected off the near edge. To obtain quantitative agreement with the data, Fresnel had to add, in an ad hoc manner, a half-wavelength phase shift at the edge. Moreover, this approach failed completely within the geometrical shadow. Fresnel's paper was read by Arago, who informed Fresnel of Young's prior work, but encouraged Fresnel to continue work on the problem.

By blackening one side of a knife (used in place of a fiber), Fresnel established that reflections off the diffracting edge had no effect on the diffraction pattern; hence the approach he and Young had taken was incorrect. In 1816, Fresnel went far beyond Young. Examination of his data showed that the position in space of each bright fringe, as the position of the observation screen was varied, traced out a hyperbola (as found by Young and Newton). Fresnel then showed that, in contradiction to the opinion expressed in Book III of Newton's *Opticks*, such a hyperbola could not be caused by particles of light being scattered by a short-range force associated with the edge of the diffracter. More important, using Huygens's principle and the principle of interference applied to the entire wavefront surrounding the edge (i.e., an infinite number of secondary wavelets), Fresnel showed that the wave theory *could* explain the observed hyperbola.

Poisson's (Unwanted) Spot

In 1818, the French Academy held a competition on the nature of light. Poisson, a reviewer who also was a zealous supporter of the particle nature of light, noted that according to the wave theory all the Huygens's wavelets originating at the edge of an opaque disk should arrive at the center of the shadow with the same phase. He thus came to the "absurd" conclusion that a bright spot should be at the center. (A century-old observation of this spot had been forgotten, and the analogy to Young's bright fringe at the center of the shadow of an opaque strip was not made.) Arago promptly observed the effect. This convinced most of the remaining doubters of the wave nature of light, and helped win for Fresnel the prize.

Fresnel's approach permits quantitative calculations of both interference patterns and of diffraction patterns. Diffraction viewed by an observer near the diffracter, as studied by Grimaldi, Newton, Young, and Fresnel (and called *near field*, or *Fresnel diffraction*), requires a relatively complex theory. We will consider only the more easily analyzed case of *far field*, or *Fraunhofer diffraction*, where the observer is relatively far from the diffracter. Fraunhofer diffraction is included as a limit of the more general and significantly more complex theory of Fresnel diffraction.

16.7.1 *Theory of Two-Slit Interference Pattern Intensity*

Consider light of frequency $\omega = 2\pi f$, where $\lambda f = c$, that passes through the two narrow, identical slits in Figure 16.27, separated by d, and proceeds to an observation screen at a distance D. Assume that on the observation screen the wave amplitudes (of the ether, according to Fresnel; however, we know them to be of the electric field) from each slit are approximately equal, so $d \ll D$. If the phase of the wave from the first slit is $\omega(t - r_1/c) = \omega t - r_1(\omega/c) = \omega t - 2\pi r_1/\lambda$, then the phase of the other is $\omega(t - r_2/c) = \omega t - 2\pi r_2/\lambda$, where r_1 and r_2 are the distances between the observation point and the slits. By Huygens's principle, the amplitude at the observation screen is proportional to the sum of the amplitudes from each slit. Thus the amplitude is proportional to

$$\cos \omega \left(t - \frac{r_1}{c} \right) + \cos \omega \left(t - \frac{r_2}{c} \right) = 2 \cos \omega \left[t - \frac{r_1 + r_2}{2c} \right] \cos \omega \left(\frac{r_2 - r_1}{2c} \right),$$

(16.18)

which follows from the trigonometric identity

$$\cos A + \cos B = 2 \cos \left[\frac{A + B}{2} \right] \cos \left[\frac{A - B}{2} \right].$$

(16.19)

This gives the expected amplitude doubling for $B = A$, and the expected zero amplitude for $B = A + \pi$.

Let each of these waves alone have a time-averaged intensity I_0 proportional to the square of the amplitude. Then the combination, with amplitude proportional to (16.18), has instantaneous intensity

$$I(t) = 8I_0 \cos^2 \omega \left[t - \frac{r_1 + r_2}{2c} \right] \cos^2 \left[\omega \left(\frac{r_2 - r_1}{2c} \right) \right].$$

(16.20)

This includes an extra factor of 2 that will be compensated by a factor of $\frac{1}{2}$, on time-averaging the first $\cos^2$ term in (16.20). Hence, denoting time averaging by an overbar,

$$\bar{I} = 4I_0 \cos^2 \left[\omega \left(\frac{r_2 - r_1}{2c} \right) \right].$$

(16.21)

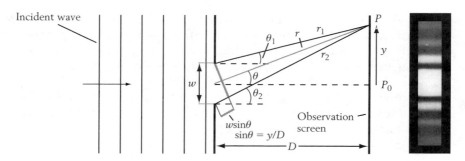

Figure 16.30 Geometry for single-slit diffraction by a distant light source. On the right is a characteristic diffraction pattern observed on the screen.

With $r_2 - r_1 \approx d \sin \theta$ and $\omega = 2\pi c / \lambda$, this becomes

$$\bar{I} = 4 I_0 \cos^2 \left[\frac{\pi d \sin \theta}{\lambda} \right]. \tag{16.22}$$

This has a pattern that oscillates from bright to dark as one moves across the observation screen, as observed in Figure 16.28. Moreover, averaged over the screen, the intensity is $2I_0$, double that for a single slit. Equation (16.22) reproduces the pattern of maximum and minima described by (16.16) and (16.17).

16.7.2 *Basic Theory of Diffraction by a Slit (Fraunhofer Diffraction)*

Diffraction may be thought of as due to interference between an infinite number of waves or wavelets.

For simplicity, consider diffraction by a narrow aperture (a slit), rather than diffraction by a narrow opaque body. Let light of wavelength λ be normally incident on a single slit of width w, as in Figure 16.30. Let r_1 and r_2 denote the distance from the edges to the observation point on the screen. At infinitesimally short wavelength ($\lambda \to 0$), the light travels in a straight line from the source, to the slit, to the observation screen. This yields a bright region corresponding to the slit, and a sharply defined geometrical shadow. At finite wavelength, the pattern expands and changes shape, due to diffraction, but is still centered about the position of the pattern for $\lambda \to 0$.

By Huygens's principle, the total amplitude at the screen equals the *integral* of the amplitudes from all the wavelets produced in each part ("slitlet") of the slit. For simplicity, we take the source at infinity. Thus, the incident light has the same phase at all points of the slit. (Alternatively, there can be a lens at the slit, of focal length equal to the distance to the source, so that the source is effectively at infinity.) We also assume that the distance D from the slit to the observation screen is much larger than the slit width w: $D \gg w$. Without doing any fancy mathematics, let's extract as much physics as possible.

1. If the slit width w is small enough (actually, less than a wavelength λ), then nowhere on the observation screen S will there be any perfectly dark fringes; wavelets from different parts of the slit are too much in phase to cancel the wavelets from other parts. Indeed, a narrow enough slit appears to be a line

source, which illuminates uniformly in all directions, as in Figure 16.8(a). (Similarly, a narrow hole appears to be a point source.)

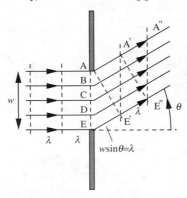

Figure 16.31 Illustration of how secondary wavelets produce zero net amplitude at certain observation angles θ. In this case, the sources from A to C produce secondary wavelets that completely cancel those from C to E.

2. As the slit width grows, becoming comparable in extent to a wavelength, it becomes possible to find points on S where phase cancellation occurs. Following Fresnel, for a wide enough slit ($w > \lambda$) let us locate where on the screen such cancellation occurs. These positions correspond to dark fringes, or minima, in the diffraction pattern.

For the diffracted beam making an angle θ to the forward direction, let the ends of the slit produce wavelets that are out of phase by a full wavelength, or $r_2 - r_1 = \lambda$; that is, in Figure 16.31 the distance from E to E' is λ. Then the slit can be decomposed into two equal half-slits, where any part of the near half-slit is half a wavelength out of phase with the corresponding part of the far half-slit. That is, on the observation screen the wavelets from C to E cancel those from A to C, so the sum is zero. (Equivalently, the area under a full period of a cosine is zero.)

More generally, there is a minimum of the total wave amplitude when

$$r_2 - r_1 = m\lambda, \qquad m = \pm 1, \pm 2, \ldots \qquad \text{(diffraction minima)} \quad (16.23)$$

For the geometry of Figure 16.31, $r_2 - r_1 \approx w \sin \theta$, so that

$$w \sin \theta = m\lambda. \qquad m = \pm 1, \pm 2, \ldots$$
$$\text{(far field, or Fraunhofer, diffraction minima)} \quad (16.24)$$

3. Approximately midway between the positions of the minima are the positions of the secondary maxima. *Caution:* Since the central maximum, rather than a minimum, corresponds to $m = 0$, there is no secondary maximum near $r_2 - r_1 = \lambda/2$. Near the secondary maxima, the largest phase difference between secondary wavelets is an integral number of half-wavelengths. For $m = 1$ this corresponds to $r_2 - r_1 = (3/2)\lambda$. Since 2/3 of the slit corresponds to a net dephasing of a wavelength, whose amplitudes sum to zero, only 1/3 of the slit (corresponding to a net dephasing of a half-wavelength) has an amplitude that sums to a nonzero value. The next maximum corresponds to 5/2 wavelengths. Here only 1/5 of the slit has an amplitude that sums to a nonzero value. Hence the secondary maxima become weaker on moving away from the central maximum.

Example 16.9 **Diffraction minima**

Yellow light ($\lambda = 580$ nm) shines on a narrow slit of width 1562 nm. (a) Find the angle for the second minimum. (b) Find the number of minima.

Solution: (a) By (16.24), for the second minimum $\sin\theta = 2\lambda/w = 0.743$, so $\theta = 0.837\ \text{rad} = 48.0°$. (b) Setting $\sin\theta = 1$ in (16.24) gives the largest value of m as $m = w/\lambda = 2.69$. Hence there are only two minima.

By Babinet's principle (see Section 16.2), (16.24) also applies to diffraction by an obstacle (e.g., a narrow opaque body, such as a fiber or a wire) outside the incident beam, as in Figure 16.8(b).

Example 16.10 **The eriometer**

For given m and λ, and small θ, the angle θ is inversely proportional to w, as observed by Young. He used this fact to construct a device he called an *eriometer*, initially used to measure the diameter of wool fibers, but later to measure the diameter of small particles within pus and blood. (Young was, after all, a physician.) (a) For $\lambda = 660\ \text{nm}$ (red), a fiber has its first minimum at $25°$. Estimate the fiber width. (b) For blue light, indicate qualitatively where the first minimum will be located. (c) For a wider fiber, indicate qualitatively where the first minimum will be located.

Solution: (a) By (16.24), for the first minimum we have a width $w = (1)(660)/\sin(25°) = 1562\ \text{nm}$. (b) For a shorter wavelength (e.g., in the blue), the first minimum would be at a smaller angle. (c) For a wider fiber, the first minimum would be at a smaller angle.

16.7.3 *Theory of Slit Diffraction Pattern Intensity*

By Huygens's principle, the total amplitude equals the *integral* of the amplitudes of all the wavelets produced in each part (slitlet) of the slit. Let $A_0(dy'/w)$ be the maximum wave amplitude due to a slitlet of width dy'. Then summing over all the slitlets, including their phase $\omega(t - R/c)$, yields a total amplitude

$$A = A_0 \frac{1}{w} \int_{-w/2}^{+w/2} dy' \cos\left(\omega t - \frac{2\pi R}{\lambda}\right), \tag{16.25}$$

where $R = \sqrt{D^2 + (y - y')^2}$ is the distance between the slit at y' to the observation screen at y.

Set $r = \sqrt{D^2 + y^2}$. Then, with the observer at an angle θ to the center of the screen, $R = \sqrt{D^2 + (y - y')^2} \approx \sqrt{r^2 - 2y'y} \approx r - yy'/r = r - y'\sin\theta$. Hence (16.25) becomes

$$A = A_0 \frac{1}{w} \int_{-w/2}^{+w/2} dy' \cos\left(\omega t - 2\pi \frac{r}{\lambda} + 2\pi \frac{y'\sin\theta}{\lambda}\right)$$

$$= A_0 \cos\left(\omega t - 2\pi \frac{r}{\lambda}\right) \frac{\sin(\pi w \sin\theta/\lambda)}{(\pi w \sin\theta/\lambda)}. \tag{16.26}$$

The time-averaged intensity, proportional to the square of (16.26), is

$$I = I_0 \left(\frac{\sin(\pi w \sin\theta/\lambda)}{(\pi w \sin\theta/\lambda)} \right)^2, \qquad (16.27)$$

where I_0 is the time-averaged intensity at the center of the slit.

Let us examine (16.27). First, if $w/\lambda < 1$, the trigonometric function in the numerator never goes to zero, so there is no minimum, as expected for a narrow slit. Indeed, for $w \ll \lambda$, there is no appreciable dependence on angle. Second, (16.27) yields minima at the expected positions, given by (16.24), in good agreement with the pattern shown on the right-hand side of Figure 16.30. Finally, the fact that $\sin x \to x$ as $x \to 0$ implies $I \to I_0$ as $\theta \to 0$, as expected.

Let $I_{m,m+1}$ be the amplitude of the secondary maximum between the m-th and $(m+1)$-th minima. By the previous subsection, the maxima are located approximately at $w\sin\theta/\lambda = m + \frac{1}{2}$, and thus (16.27) yields

$$I_{m,m+1} \approx \frac{I_0}{\left[\left(m + \frac{1}{2}\right)^2 \pi^2 \right]}. \qquad \text{(intensity of secondary maxima)}$$

This gives a quantitative measure of the falloff of intensity of the secondary maxima.

16.7.4 *The Rayleigh Criterion for Resolution*

Consider two distant point sources, with angular separation $\delta\theta_{sep}$ relative to an aperture leading to an observation screen. In the limit of small wavelength (i.e., the particle emission viewpoint), waves travel in straight lines. Hence the angular separation $\delta\theta_{sep}$ between the sources gives the same angular separation in the image on the observation screen. This is the principle of the pinhole camera.

For finite wavelength, light from a distant point source passing through an aperture spreads out, due to diffraction. Let $\delta\theta_R$ be the angular spread between the central diffraction peak and the first diffraction minimum. The *Rayleigh criterion* states that to resolve two sources their angular separation $\delta\theta_{sep}$ must be greater than or equal to the angular spread $\delta\theta_R$ of a single source due to diffraction, or $\delta\theta_{sep} \geq \delta\theta_R$. As is conventional in discussing the Rayleigh criterion, we call the slit width d rather than employ w of (16.24). With this modification, and the small angle approximation $\sin\theta \approx \theta$, (16.24) yields the angle between the central maximum and the first minimum ($m = 1$) to be

$$\delta\theta_{sep} \geq \frac{\lambda}{d}. \qquad \text{(Rayleigh observability criterion, slit width } d) \quad (16.28)$$

For a circular aperture of diameter d (e.g., a camera or eye lens), the effective width for diffraction effects is less than for a slit of width d. In this case (16.28)

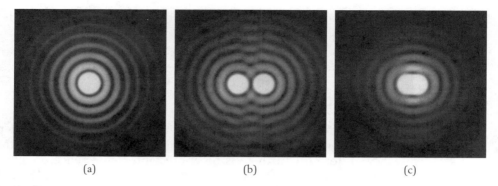

Figure 16.32 Diffraction pattern from two circular apertures: (a) pattern from a single aperture, (b) pattern from two well-separated apertures, (c) pattern from two apertures satisfying the Rayleigh criterion.

is replaced by

$$\delta\theta_{sep} \geq \delta\theta_R = 1.22\frac{\lambda}{d}.$$

(Rayleigh observability criterion, circular aperture of diameter d)   (16.29)

The factor of 1.22 comes from an analysis that generalizes (16.26), for a slit, to the case of a circular aperture. Larger telescopes give both smaller angular resolution $\delta\theta_{sep}$ and larger intensity $\sim d^2$. Figure 16.32 gives the diffraction pattern associated with a circular aperture, two well-separated circular apertures, and two apertures satisfying the Rayleigh criterion.

The spacial resolution of an image is determined by at least three factors: (1) time variations in the medium through which the light propagates, (2) the resolution of the receptors on the observation screen, and (3) the diffraction-limited resolution on passing through the aperture to the observation screen. The first factor is significant for astronomical observations through the earth's atmosphere, and is the major reason for having earth-orbiting telescopes. The second factor is significant both for film, where the photosensitive grains have a finite size, and for the eye, where the photosensitive rods and cones have a finite size. Whichever of the three factors produces the greatest angular spread dominates in determining the spatial resolution. In what follows, we shall assume that diffraction dominates. In the human eye, photoreceptor widths provide about the same limit on spacial resolution as does diffraction.

Example 16.11   **The telescope and the mouse**

An amateur telescope with diameter 0.16 m is focused on a mountain that is 3 km away. (a) For green light ($\lambda = 500$ nm), find the angular resolution of the central maximum. (b) Consider two beams of light reflected from a mouse, one from the tip of its nose and one from the tip of its tail 12 cm away. Find the angular separation of the two beams at the location of the telescope. (c) Determine if the mouse can be detected.

Solution: (a) By (16.29), $\delta\theta_R = 1.22\lambda/d = 3.81 \times 10^{-6}$ radians. (b) Think of each end of the mouse as a point source that sends a plane wave toward the telescope. Relative to the telescope, the angular separation of the two ends of the mouse is $\delta\theta_{sep} = 12$ cm/3 km $= 4 \times 10^{-5}$ radians, and this is also the angular separation on the screen between the diffraction patterns of the two ends of the mouse. (c) The angular separation $\delta\theta_{sep}$ is over ten times the diffraction-limited angular spread $\delta\theta_R$ of (16.29), so a 12 cm long mouse can be detected easily.

16.8 Polarization by Crystals

Crystal optics has an honored place in the history of physical optics in part because for many years it was only via crystals that polarization effects could be studied. The phenomena we are about to describe are complex, and we will not explain them in detail. Such explanations, however, would involve only the physics of polarization and phase shift, and the mathematics of algebra, geometry, and trigonometry. (One way to think about polarization is that it is a manifestation of the internal angular momentum of light, and thus it is a quantum phenomenon. It would be disappointing to find that the mathematics of a quantum phenomenon is completely trivial. Not to worry.) Crystal optics, straightforward in principle but complex in practice, is indispensible in modern optics for the manipulation of beams of light. Here the beams are very often nearly monochromatic, produced by a laser (light amplification by stimulated emission of radiation). The operation of a laser requires the introduction of so many new physical ideas that it is beyond the scope of the present work.

16.8.1 *Early Experiments*

In Bartholinus's original studies on birefringent (i.e., doubly refracting) crystals, for a single beam of sunlight normally incident on the crystal, two beams emerged. The ordinary (O) beam (which satisfies Snell's law) passed straight through, but the extraordinary (E) beam was deflected. See Figure 16.15. Huygens was the first to analyze the two emerging beams, by passing them through a second birefringent crystal (called an *analyzer,* the first crystal being called the *polarizer.*) In our discussion of such an experiment, the incident light and the polarizer will be fixed, but the analyzer will be rotated. We take the polarizer to be thinner than the analyzer; thus it gives a smaller deflection than the analyzer.

In Figure 16.33(a), the crystal axes of the polarizer and the analyser are aligned. Then the incident O and E beams, respectively, yield only O and E beams on passing through the second crystal. It is as if the combination is the same as a crystal of thickness equal to the sum of their thicknesses, with two beams emerging from the analyzer. (The X in the "intensity box" indicates the position of the O beam exiting the polarizer. The larger the dot, the larger the intensity.)

Now let the second crystal (the analyzer) be rotated about the propagation direction (here $\hat{x}$), the rotation being reckoned positive when counterclockwise as seen looking at the incoming beam. The beams emerging from the analyzer now become more complex; typically, each beam entering the analyzer splits in two, one of type E and the other of type O, so that a total of four beams emerge from the analyzer. For a rotation angle of 45°, see Figure 16.33(b). Note the

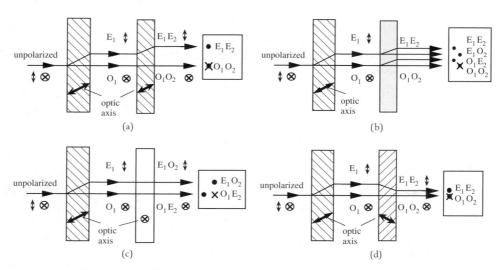

Figure 16.33 Polarization and analysis of unpolarized white light. The polarizer and analyzer have different thicknesses. The optic axis, in the page for (a), is rotated counterclockwise about the x-axis by 45° in (b), by 90° in (c), and by 180° in (d). The "intensity boxes" give the intensity patterns, where the X marks the position of an undeflected beam. The intensity patterns rotate successively by 45°, 90°, and 180°, while the intensities change and the states of polarization change.

deflections in the intensity box. For an arbitrary angle of rotation, the polarization of each outgoing beam is more complex than can be described with linear polarization. Note, however, that for all rotation angles there is an O_1O_2 beam. Figures 16.33(c) and 16.33(d) give the results for rotations of 90° and 180°.

For simplicity, first consider the case of a 180° rotation of the analyzer, as in Figure 16.33(d). Here the E beam bends back toward the O beam. It is as if the combination is the same as a crystal of thickness equal to the difference of their thicknesses, with two beams emerging from the analyzer.

For a 90° rotation of the analyzer, Figure 16.33(c) shows that the E and O beams interchange. This can be understood by arguing that the polarizer polarizes the O_1 and E_1 beams, but that to the rotated analyzer they appear to be E and O beams, respectively. The E_1O_2 beam is deflected only by the amount the E_1 beam is deflected, and the O_1E_2 beam is deflected only by the amount the E_2 beam is deflected.

Figure 16.33(b) presents the most complex case, where the polarizer has been rotated by 45°. In this case, both the incident O_2 and E_2 beams are, relative to the analyzer, mixtures of both types of beams, and this leads to two exit beams from each beam incident on the analyzer. Note that the E_1O_2 beam is deflected only by the polarizer, and the O_1E_2 beam is deflected only by the analyzer.

Huygens developed a wave theory that successfully described—but did not explain—the directions of both the O and E beams. In 1717, Newton interpreted these phenomena to be an indication that light has "sides"—his word for what we now call polarization.

16.8.2 *Eighteenth-Century Experiments*

Young's and Fresnel's wave theory was based on an analogy to sound, where there is only a single type of wave, associated with motion of the air along the direction of propagation—a *longitudinal* wave. However, the two beams seen in double refraction indicate more than just a single (longitudinal) degree of freedom. In the 1820's, die-hard advocates of the particle theory of light would not cease their criticism until the wave theory could explain the various crystal optics phenomena that were starting to accumulate.

In 1809, Malus discovered that ordinary light could be polarized by reflection, when he accidentally passed light, first reflected off window glass, through a birefringent crystal. The E and O waves had very different intensities, according to the angle of incidence on the window. Moreover, for one particular angle only one ray of light emerged. He named the associated property of light *polarization*. In 1815, Brewster discovered that, at the angle of complete polarization, the reflected and refracted rays are normal to one another. Combined with Snell's law, (16.9), this gives the *Brewster angle* θ_p of (15.65). In that same year, Brewster also discovered biaxial crystals, which are birefringent, but for which neither of the two emergent rays is "ordinary."

In 1816, Arago began to investigate the effect of polarization on diffraction. He first showed that a narrow light beam from a polarized source gave the same diffraction patterns as for ordinary light. Next he considered light from a single source, split by a birefringent crystal into two narrow light beams of different polarizations and directed to the opposite edges of a narrow opaque card. It yielded diffraction fringes at the edges of the shadow but no fringes near the center of the shadow. Further, he and Fresnel showed that two light beams with different sources would not produce interference patterns. These experiments suggested to Fresnel that light is a tranverse wave, and that the two beams are polarized in the two transverse directions. However, he could not convince Arago to include this idea in their joint publication.

Example 16.12   **Interference and polarization**

In a two-slit interference experiment using unpolarized light, let the time averaged intensities be $4I_0$ at the maxima, and zero at the minima. A linear polarizer is placed over slit 1, so that it accepts light polarized along $\hat{z}$, but it absorbs light polarized along $\hat{y}$. (a) Find the new intensities of the maxima and minima. (b) If the linear polarizer has thickness 150 nm and has index of refraction $n = 1.42$ for the transmitted light, discuss how the positions of the maxima and minima will shift. See Figure 16.34, where only up–down polarization leaves slit 1, but both types of polarization leave slit 2.

Solution: (a) Without the linear polarizer, the intensity averaged over the screen is $2I_0$, with I_0 coming from each type of polarization. With the polarizer, the average intensity of polarization along $\hat{y}$ will be halved because it comes only from slit 2. Hence the pattern on the screen will be the sum of uniform illumination of polarization along $\hat{y}$ from slit 2, with intensity $I_0/2$, and of the interference pattern from polarization along $\hat{z}$ from both slits, with intensity $2I_0$ at the maxima. Hence the new maxima and minima have intensities $5I_0/2$ and $I_0/2$. The average over

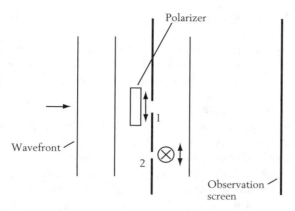

Figure 16.34 A two-slit interference experiment where a perfectly absorbing linear polarizer has been placed in front of one of the slits.

the screen is $3I_0/2$, as expected if one-fourth of the incoming light is absorbed by the linear polarizer. (b) Because the wavelength within the polarizer is λ/n, rather than λ in air, the beam passing through the polarizer is subject to a larger phase shift than the beam passing through air. We say that the light transmitted through slit 1 has a greater "optical thickness." On the observation screen, the central maximum shifts toward slit 1 so that the light from slit 2 can make up the extra phase shift gained by the light passing through slit 1.

Young learned of Arago's work and began to think that light involved transverse vibrations. He developed an analogy to the two transverse directions of vibration of a string (a string stretched along x can vibrate along both y and z). In 1821, Fresnel developed a theory for light in crystals, using the idea of transverse vibrations, which worked so well that it overcame nearly all objections to the idea that light is a wave. Nevertheless, a fundamental understanding of what was "waving" was still lacking.

That same year, Fresnel also proposed that sound in solids could have transverse waves. Navier then developed a dynamical set of equations describing solids. Refined by Cauchy in 1828, they were solved later that year by Poisson, who found three types of modes, one longitudinal and two transverse, in agreement with experiments on sound in solids. However, there was no evidence of a longitudinal mode for light in any material. Not until Maxwell's equations for the electromagnetic field was it understood how it is that for light there is no

Diffraction Gratings

Diffraction gratings are used to study the spectral components (i.e., the intensities of the different wavelengths) of light. This permits study of unknown light sources on the earth, sun, or stars. Diffraction is also used as a tool to study the structure of matter; diffraction of x-rays off crystals is used to determine the structural arrangements and identities of the atomic constituents. As indicated earlier, diffraction is a subset of interference where many waves or wavelets interfere.

longitudinal degree of freedom, and thus no longitudinal mode. Maxwell's theory of light, when crystal polarization and magnetization effects are included, also encompasses crystal optics.

16.9 Multiple Slits and Diffraction Gratings

Diffraction gratings consist of a series of equally spaced parallel lines that either reflect or transmit light. (Fraunhofer, in the 1820s, developed techniques to make fine and accurate gratings. This enabled more precise analysis of the

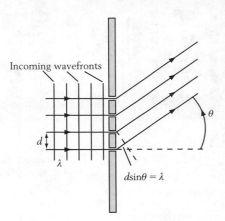

wavelengths of light than is possible by two-slit interference or by single-slit diffraction. Young's micrometer, of Example 16.4, can be thought of as a two-slit reflection grating.) Consider N narrow slits of width w, with center-to-center separation d. Let $w \ll \lambda$, so that each slit diffracts uniformly in all directions, and let $w \ll d$ to reduce the analysis to that of multiple slits with spacing d. Thus replace the single-slit of Figure 16.30 by a transmission diffraction grating, to obtain Figure 16.35. As for two-slit interference, take the light source to be either very far away or at the focal length of a lens just before or behind the slit. A typical pattern is shown on the observation screen in Figure 16.36.

Figure 16.35 Geometry of a diffraction grating (here, $N = 4$).

For a diffraction grating, the condition for constructive interference of light of wavelength λ is the same as for two-slit interference, or

$$d \sin \theta_m = m\lambda. \qquad m = 0, \pm 1, \pm 2, \ldots \qquad \text{(diffraction grating maxima)}$$
$$(16.30)$$

For each value of m, the angle that corresponds to the peak intensity is called the m-th order maximum. Peaks occur because the N scattered electric field components are in phase, yielding a net amplitude that is N times as large as for an individual vector. The peak intensity, which Fresnel correctly concluded is proportional to the square of the amplitude, is $N^2/4$ times larger than given by (16.22) for two slits. However, as seen in Figure 16.37, and as we derive shortly, the angular width of the maxima are much narrower than for two slits. This narrowing of the line is required by energy conservation; N lines give $N/2$ times as much overall intensity at the same angles as for two slits, but each line is $N^2/4$ times as intense and only $(N/2)^{-1}$ times as broad. The narrowness of their lines makes diffraction gratings a powerful diagnostic tool for the precise analysis of light. See Figure 16.38, which shows spectra of light emitted by various types of atoms and molecules, and has then passed through a diffraction grating.

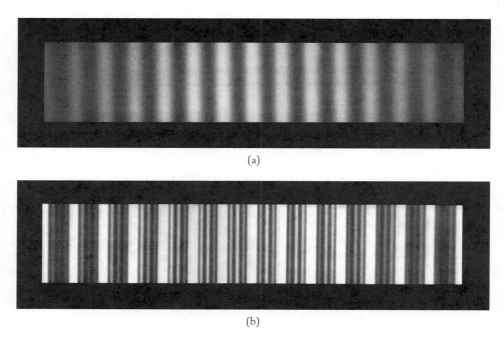

Figure 16.36 Experimental intensity pattern for a diffraction grating, showing the steepening in intensity and narrowing in angular spread of the maxima. (a) $N=2$. (b) $N=5$.

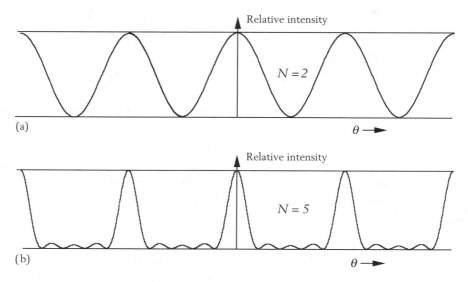

Figure 16.37 Theoretical intensity pattern for a diffraction grating, showing the steepening in intensity and narrowing in angular spread of the maxima. There are also weak secondary maxima of negligible intensity for large N. (a) $N=2$. (b) $N=5$.

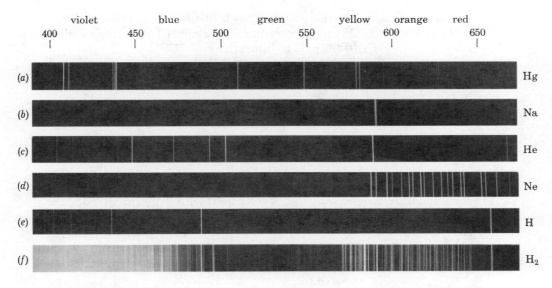

Figure 16.38 Spectra of light, for Hg, Na, He, Ne, H, and H_2, as analyzed with a diffraction grating. The lines are wider than the angular width of the grating. Thus we are observing the intrinsic linewidth rather than instrumental broadening.

Example 16.13 Diffraction maxima

Let a diffraction grating have slit separation 4000 nm. For a wavelength of 589 nm, find the angles that correspond to all the diffraction maxima.

Solution: By (16.30), the largest-order maximum occurs for the largest m satisfying $m \le d/\lambda$. Here $d/\lambda = 6.79$, so the largest-order maximum occurs for $m = 6$. Specifically, (16.30) gives $\theta_1 = 8.47°$, $\theta_2 = 17.1°$, $\theta_3 = 26.2°$, $\theta_4 = 36.1°$, $\theta_5 = 47.4°$, and $\theta_6 = 62.1°$.

There are three measures of the quality of a diffraction grating, only two of them independent: (1) the *angular width* $\Delta\theta_m$ of the m-th order maximum; (2) the *dispersion* $D = \Delta\theta_m/\Delta\lambda$, a measure of the angular spread $\Delta\theta_m$ of the m-th order maximum for a given separation $\Delta\lambda$ in wavelength of the incident light; (3) the *resolving power* $R = \lambda/\Delta\lambda$, a measure of the number of lines that can be observed distinctly. Each of these depends upon the order m of the maximum.

In what follows, when we become specific, we will consider a diffraction grating with $N = 500$, $d = 4000$ nm, and the so-called *sodium doublet*, which consists of two closely spaced spectral lines at 589.00 nm and 589.59 nm, so $\lambda = 589$ nm and $\Delta\lambda = 0.59$ nm. To the eye the sodium doublet appears yellow.

16.9.1 *Angular Width* $\Delta\theta_m$

Consider an angle $\Delta\theta_0$ slightly off the central maximum ($m = 0$), such that the first and N slits are out of phase by λ (thus for large N the first and second slits are out of phase by $\lambda/(N-1) \approx \lambda/N$). Summing over all the slits will give a net amplitude that is essentially zero. [This is like what happens for diffraction, where the integral (16.25) appears, rather than a sum.] The corresponding angle $\Delta\theta_0$ satisfies, not (16.30) for $m = 0$, but rather

$$d \sin \Delta\theta_0 = \frac{\lambda}{N}. \tag{16.31}$$

For small $\Delta\theta_0$, where $\sin \Delta\theta_0 \approx \Delta\theta_0$, we may rewrite (16.31) as $d\Delta\theta_0 \approx \lambda/N$, so

$$\Delta\theta_0 \approx \frac{\lambda}{Nd}. \tag{16.32}$$

Thus the central maximum is spread over an angular range that is inversely proportional to N, a narrowing that is consistent with experiment. For $N = 500$, $d = 4000$ nm, and for $\lambda = 589.00$ nm, (16.32) yields $\Delta\theta_0 = 2.945 \times 10^{-4}$ rad $= 0.01687°$.

More generally, for the m-th order maximum the angular width $\Delta\theta_m$ satisfies

$$d \sin(\theta_m + \Delta\theta_m) = m\lambda + \frac{\lambda}{N}. \tag{16.33}$$

Since, for small $\Delta\theta_m$, $\cos \Delta\theta_m \approx 1$ and $\sin \Delta\theta_m \approx \Delta\theta_m$, for small $\Delta\theta_m$ we have

$$\sin(\theta_m + \Delta\theta_m) = \sin \theta_m \cos \Delta\theta_m + \cos \theta_m \sin \Delta\theta_m \approx \sin \theta_m + \Delta\theta_m \cos \theta_m.$$

Thus (16.33) yields

$$d(\sin \theta_m + \Delta\theta_m \cos \theta_m) = m\lambda + \frac{\lambda}{N}. \tag{16.34}$$

Subtracting (16.30) from (16.34) yields

$$\Delta\theta_m d \cos \theta_m = \frac{\lambda}{N},$$

so

$$\Delta\theta_m = \frac{\lambda}{Nd \cos \theta_m}. \tag{16.35}$$

For $m = 0$, where $\cos \theta_0 = \cos 0 = 1$, (16.35) agrees with (16.32). For $N = 500$, $d = 4000$ nm, and $\lambda = 589.00$ nm, (16.35) yields $\Delta\theta_1 = 2.977 \times 10^{-4}$ rad $= 0.01706°$.

Equation (16.35) shows that to obtain a sharp line (narrow angular width), the diffraction grating should have a large total width Nd. Further, because of the cosine factor, lines at larger angles have a larger angular width $\Delta\theta_m$. Note that, for small θ_m, where $\cos \theta_m \approx 1$, $\Delta\theta_m$ is nearly independent of m.

16.9.2 *Dispersion D: Angular Width per Wavelength Interval*

Differentiation of (16.30) with respect to λ, at fixed m and d, gives

$$d \cos \theta_m \frac{d\theta_m}{d\lambda} = m,$$

which yields the *dispersion*

$$D \equiv \frac{d\theta_m}{d\lambda} = \frac{m}{d\cos\theta_m}. \tag{16.36}$$

The dispersion is a measure of the sensitivity of changes in angle θ_m of the m-th maximum to changes in wavelength λ. For an angular width $\Delta\theta_m$, the detectable wavelength interval $\Delta\lambda$ is given by

$$\Delta\lambda = \frac{d\lambda}{d\theta_m}\Delta\theta_m = \frac{\Delta\theta_m}{D}. \tag{16.37}$$

The larger the dispersion, the smaller the separation $\Delta\lambda$ between observable lines.

Example 16.14 **Dispersion of a diffraction grating**

Consider a diffraction grating with $N = 500$ and $d = 4000$ nm, and the sodium doublet, with $\lambda = 589.00$ nm, and $\Delta\lambda = 0.59$ nm. (a) Find the dispersion D for $m = 1$. (b) For $m = 1$ and $m = 2$, determine if the sodium doublet can be resolved.

Solution: (a) For $m = 1$, Example 6.13 gives $\theta_1 = 8.47°$. Then (16.36) yields $D = 2.5 \times 10^{-4}$ rad/nm. (b) Since $\Delta\theta_1 = 2.977 \times 10^{-4}$ rad $= 0.01706°$, by (16.37) a wavelength separation as large as $\Delta\lambda = 1.19$ nm can be measured. This is about twice the separation of the two lines of the sodium doublet, so study of the $m = 1$ maximum cannot resolve them. However, for $m = 2$, the doublet is just barely resolvable.

16.9.3 *Resolving Power R*

Consider two lines of wavelength λ and $\lambda + \Delta\lambda$ that are just barely resolvable. We define resolving power R via

$$\Delta\lambda = \frac{\lambda}{R}. \tag{16.38}$$

Resolving power R is a measure of the number of lines that can be distinctly observed within the vicinity $\Delta\lambda$ of a given wavelength λ. From (16.38), the larger the resolving power R, the smaller the separation $\Delta\lambda$ between observable lines. From (16.38),

$$R = \frac{\lambda}{\Delta\lambda}. \tag{16.39}$$

Applying (16.37) to introduce $\Delta\theta_m$, and then employing both (16.35) and (16.36) to eliminate $\Delta\theta_m$ and D yields

$$R = \frac{\lambda}{\Delta\theta_m}D = \frac{\lambda}{\lambda/(Nd\cos\theta)}\frac{m}{d\cos\theta} = Nm. \tag{16.40}$$

Example 16.15 **Resolving power of a diffraction grating**

Consider a diffraction grating with $N = 500$ and $d = 4000$ nm, and the sodium doublet, with $\lambda = 589.00$ nm, and $\Delta\lambda = 0.59$ nm. (a) Find the resolving power R of the grating, for $m = 1$. (b) Find the resolving power needed to resolve the sodium doublet.

Solution: (a) By (16.40), $R = Nm = 500$. (b) By (16.39), the necessary resolving power is $R = \lambda/\Delta\lambda = 589/0.59 = 1000$. This is about twice what is available from the $m = 1$ line. For $m = 2$, the R of (16.40) can be increased to 1000, making the line just barely resolvable, consistent with Example 16.14.

Equation (16.40) shows that, to be able to distinguish many lines, the grating should have a large number of parallel lines and large order m should be studied. Note that R is independent of d and λ.

16.10 X-Ray Scattering off Crystals

X-rays were discovered by Roentgen in 1895. He found that fresh photographic plates became exposed after elsewhere in his lab he accelerated high-energy cathode rays (now known to be electrons) onto a positively charged anode. X-rays were found to have no charge. It was suspected that x-rays were electromagnetic radiation of very short wavelength, and there were indications that x-rays had wavelengths that included the 0.1 nm range. At the same time, there was evidence that the distance between atoms was about 0.1 nm, and there were suspicions that crystals were regular arrays of atoms. Since x-rays primarily pass through most materials, the probability that an x-ray interacts with matter is relatively low (although the energy deposited can be significant when there is such an interaction).

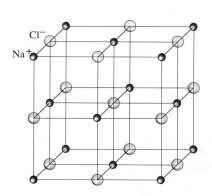

Figure 16.39 Schematic of NaCl crystal.

As shown in Section 16.2, if the distance between wave sources is large enough compared to the wavelength, interference can occur. In 1912, Laue and his colleagues Friedrich and Knipping scattered an x-ray beam off the atoms in a crystal, finding a pattern of spots that indicated a regular array of atoms in the crystal. W. L. Bragg quickly came up with an interpretation. A crystal can be thought of as containing planes of atoms with separation d that varies with the orientation of the plane. See Figure 16.39. Note that d typically is less than the separation between atoms. X-rays can scatter off each plane with an angle of incidence equal to the angle of reflection. There is momentum transfer but no energy transfer in such scattering.

We use the x-ray scattering convention that the angle is measured with respect to the plane (rather than with respect to the normal to the plane). Then, both

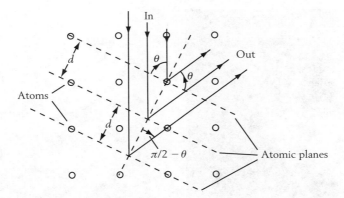

Figure 16.40 Geometry for Bragg scattering off a crystal. The crystal has an infinite number of planes and corresponding distances d.

coming in and going out, the extra distance on traversing an extra plane of atoms is $d\cos(\frac{1}{2}\pi - \theta) = d\sin\theta$. See Figure 16.40.

For constructive interference, this total extra distance must correspond to an integral number of wavelengths. This leads to the Bragg condition

$$2d\sin\theta = m\lambda. \qquad m = 1, 2, 3\ldots \qquad (16.41)$$

The angle θ is called the Bragg angle, and the scattering angle by which the scattered radiation deviates from the incident radiation is 2θ. Three methods are used to study crystals with x-rays, using (16.41).

1. Bragg's method uses monochromatic radiation incident on a single crystal, and rotates that crystal about a fixed axis. In Figure 16.40, that fixed axis can be considered to be normal to the page, so that rotation changes θ. For a given m, λ, and d, only for certain Bragg angles θ will scattering be observed.

2. Laue's method uses nonmonochromatic radiation incident on a single crystal, for which some wavelengths will satisfy the Bragg condition. See Figure 16.41(a). In terms of Figure 16.40 and Equation (16.41), the plane separation d and the Bragg angle θ are fixed, and for fixed m only at certain wavelengths λ will scattering be observed.

3. The Debye–Scherrer method uses monochromatic radiation incident on a powder of many identical crystals. In terms of Figure 16.40 and Equation (16.41), m and λ are fixed, but because each crystallite has its own orientation, only a fraction of them have their planes with separation d in the correct orientation for Bragg scattering. On the other hand, the ensemble of crystallites will scatter with symmetry about the axis of the incoming radiation so that there is a scattering cone. See Figure 16.41(b).

Example 16.16 **X-ray scattering**

Consider that x-rays with $\lambda = 0.4$ nm are scattered off a hypothetical crystal at a scattering angle of $130°$, and that this is the smallest scattering angle for this crystal. Find the planar separation within the crystal corresponding to this scattering angle.

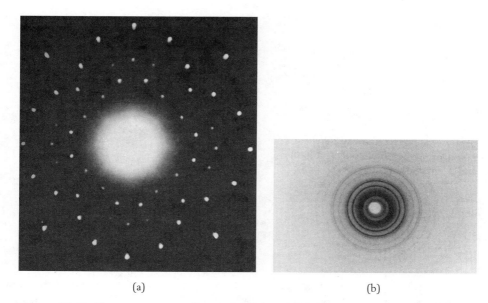

(a) (b)

Figure 16.41 X-ray scattering of a crystal: (a) Laue patterns (non-monochromatic radiation incident on a single crystal), (b) Debye–Scherrer patterns (monochromatic radiation incident on a powder of small crystals).

Solution: The Bragg angle is half the scattering angle, so $\theta = 65°$. Then (16.41), with $m = 1$, yields $d = 0.22$ nm. The atomic separation must be at least this large.

Application of (16.41) gives the location of atoms in the unit cell of a crystal. Detailed analysis of the intensity of each peak permits each atom to be identified. X-ray diffraction has been used to study the arrangement of atoms in crystals, even crystals consisting of complex biological molecules.

Acknowledgments

Figure 16.12a. Figure from *Atlas of Optical Phenomena* by Michel Cagnet, Maurice Francon & Jean Claude Thrierr. Copyright © 1962 by Springer-Verlag GmbH & Co. Used by permission of the publisher.

Figure 16.12b. Figure from *University Physics*, Second Edition by Alvin Hudson and Rex Nelson. Copyright © 1990 by Saunders College Publishing. Reproduced by permission of the publisher.

Figure 16.13. Figure from *College Physics*, fourth edition, by Franklin Miller, Jr., Harcourt Brace Jovanovich, Inc. Used by permission of the author.

Figure 16.24. Photograph by Larry Mulvehill. Used by permission of Photo Researchers, Inc.

Figure 16.26. Figure from *Atlas of Optical Phenomena* by Michel Cagnet, Maurice Francon & Jean Claude Thrierr. Copyright © 1962 by Springer-Verlag GmbH & Co. Used by permission of the publisher.

Problems

16-2.1 In a ripple tank, two plungers 80 cm apart pulse at the same frequency f and are in phase. On a plane 1.4 m away, the distance between the first and second maxima is 140 mm. (a) Find the wavelength. (b) For a wave speed of 22 m/s, find f.

16-2.2 (a) Consider a ripple tank where two plungers have the same frequency, but are out of phase by 180°. Find the condition on $r_2 - r_1$ for maxima and minima to occur. (b) Repeat for 90°. (c) Repeat for arbitrary relative phase ϕ.

16-2.3 Consider two plungers in a ripple tank, as in Figure 16.5, where $r_2^2 = D^2 + (y + d/2)^2$, and $r_1^2 = D^2 + (y - d/2)^2$. Show that Equation (16.1) leads to a set of hyperbolae if $|m|\lambda < d$. (*Hint:* Use coordinates $x = D$ and y.)

16-2.4 Two in-phase radar transmitters at wavelength 0.4 m are 3 m apart. The perpendicular bisector to the transmitters is normal to a plane that is 2.5 m from the midpoint between the transmitters. (a) Locate on this plane the positions of the first two maxima and the first two minima to one side of the central maximum. (b) On a circle of radius 2.5 m, centered at the midpoint, locate the maxima and minima.

16-2.5 Two speakers driven by the same amplifier are directed toward and are each 8 m from a wall, as in Figure 16.5. They are separated by 0.4 m. At $y = 0$, the intensity is a maximum. As y increases, the intensity decreases. At $y = 2.2$ m, the intensity is zero. Find the wavelength and frequency of the sound, for $v = 340$ m/s.

16-2.6 Two plungers, driven in phase at the same frequency, are separated by 2.5 cm. See Figure 16.5. Along an infinite line parallel to their connecting line, but 14 cm away, there are five maxima, one along the perpendicular bisector. Give the range of values that the wavelength can take.

16-2.7 (a) For two in-phase sources separated by $d = 4$ cm, if $\lambda = 18$ cm, how many curves giving constructive interference (maxima) are there? Curves giving destructive interference (minima)? (b) For $d = 4$ cm and $\lambda = 6$ cm, how many curves giving maxima are there? Curves giving minima? (c) For $d = 4$ cm and $\lambda = 1.4$ cm, how many curves giving maxima are there? Curves giving minima?

16-2.8 (a) Sketch the scattering pattern for sound waves in air, of frequency 238 Hz, incident on an obstacle of dimension 12 m. Take the sound beam to be much wider than 12 m. (b) Repeat for an obstacle of dimension 1.2 m. (c) Repeat for an obstacle of dimension 0.12 m.

16-2.9 (a) Sketch the scattering pattern for sound waves in air, of frequency 238 Hz, incident on a slit of dimension 12 m. Take the sound beam to be much wider than 12 m. (b) Repeat for a slit of dimension 1.2 m. (c) Repeat for a slit of dimension 0.12 m.

16-2.10 A water wave of wavelength 3.2 m is incident from the left on a barrier with two narrow gaps, separated by 0.46 m. (a) Describe the interference pattern of the transmitted waves, as seen on an imaginary line, as in Figure 16.5. (b) If the incident wave is incident at a 12° clockwise angle relative to the normal to the barrier, how does this change the interference pattern? *Hint*: What would be the path of a very short wavelength wave? (c) Find the new angle associated with the central maximum.

16-2.11 (a) Sound of frequency 240 Hz is normally incident on a wall with two slit-like holes separated by d. If on the other side of the wall at a distance D there is a plane for which the angular separation between the central maximum and the first minimum is 10°, determine the separation between the slits. Take $D \gg d$. *Hint*: Use the results of Example 16.2 in the limit where $D \gg y$. (b) If the incident wave now has frequency 960 Hz, find the the angular separation between the central maximum and the third minimum.

16-2.12 Two speakers, displaced by 2 m along the y-axis, emit sound at the same 180 Hz frequency, with the same intensity. The upper speaker peaks 35° before the lower speaker. (a) To what time does this correspond? (b) If the speakers were simply holes in the wall, driven by a single source, what angle would the source make to the normal to the holes? See Figure 16.5. (c) Find the angle that the first minimum makes with the normal.

16-3.1 Show that $\pi - 2\theta'$, used in deriving the angle change on reflection for the rainbow, gives the correct values for θ' given by 0° (normal incidence), $\pi/4$, and $\pi/2$ (glancing incidence).

16-3.2 Derive the law of reflection using the analytical method of Fermat (rather than the geometrical method employed in the text).

16-3.3 (a) Derive the total deflection angle and exit angle associated with the secondary rainbow, where there are *two* internal reflections. (b) Derive the stationary condition that gives the rainbow

angle for the secondary rainbow. (c) Determine the relation between the incident angle and the exit angle.

16-3.4 Snel expressed his result in terms of the secant of the complementary angles. Rewrite (16.9) in such terms. The extent to which, at the time of its discovery, Snell's law had been compared to actual data, is unknown. For finite angles, this problem had defied solution for some 1500 years.

16-4.1 Summarize Grimaldi's experiment and his discoveries about diffraction.

16-4.2 Summarize Hooke's experiment and his discoveries about the colors of thin films and plates.

16-4.3 Summarize Newton's experiment and his discoveries about the prism.

16-4.4 Summarize Bartholinus's experiment and his discoveries about double refraction.

16-4.5 Roemer could not accurately measure the period of a single revolution of Io. He could and did measure the average period for 40 successive revolutions. In Figure 16.16, let the position of Jupiter be at 0° in the circle representing the earth's orbit about the sun, and consider that Jupiter's position is nearly fixed. For an observer on the earth, at what angles in the earth's orbit about the sun will the period of Io's orbit show the maximum increase and decrease? (Roemer remarked that orbital irregularities and the motion of Jupiter relative to the earth were smaller effects than the relative motion of Io and the earth.)

16-4.6 In Figure 16.16, let the position of Jupiter be at 0° in the circle representing the earth's orbit. (a) How many periods of Io correspond to half an earth year? (b) Show that, if the motion of Jupiter is neglected, the extra distance that light travels from Jupiter as the earth goes from 0° to 180° equals the diameter d of the earth's orbit. (c) Show that the extra time associated with this motion equals d/c, where c is the speed of light. (d) By how much does this increase the apparent average period of Io about Jupiter?

16-4.7 What experiment, and specifically what aspects of it, would you cite to establish that shadows are not completely sharp?

16-4.8 White light reflecting off material A appears to be orange (O_A). White light (W) reflecting off material B also appears to be orange (O_B). One is a composite of red and yellow, and one is pure orange. How would you tell which is which?

16-4.9 Give a counterexample to the claim that all transparent crystals can be used to make cover sheets with a sharp image of what is underneath the sheet.

16-4.10 Give an example that argues against the claim that the period of the earth's moon about the earth is the same as seen from earth and from Mars.

16-5.1 Huygens actually wrote that one adds the effect of the primary wave and of the secondary wavelets. Thus when a primary wave passes through a medium, it excites secondary waves in that medium. (a) Argue that, as long as the primary wave and the secondary wavelets have the same velocity ($v_p = v_s$), the net wave has the same velocity as the primary wave. (b) What is the signal velocity if $v_p < v_s$? (c) If $v_p > v_s$?

16-5.2 What kind of Huygens construction would you use to represent an inward-moving spherical wavefront?

16-5.3 Explain why sound velocities, and the relative amounts of partial reflection and refraction, are intensity independent, using the fact that sound is a small disturbance in the ambient density and pressure.

16-5.4 Newton found the velocity of waves from an argument where he determined the period for a wave of a given wavelength (although he did not use that terminology). In Book 2 of his *Principia*, Newton gives Proposition 50, Problem 12: "*To find the distance between pulses.* In a given time, find the number of vibrations of the body by whose vibration the pulses are excited. Divide by that number the distance that a pulse could traverse in the same time, and the result is the length of one pulse." Let N be the number of vibrations, t the time for those vibrations, and D the distance a pulse can travel in that time. (a) Derive expressions for the frequency f and the wavelength λ. (b) Show that $\lambda f = v$ gives the velocity v. (Newton had neither the terminology nor the symbol for λ or for f.) [Answer: $\lambda = D/N$, $T = t/N$, $f = 1/T$, and $D = vt$ gives $\lambda = vt/N = vT = v/f$. In practice, Newton used $\frac{1}{2}\lambda$ and $2f$.]

16-6.1 A given musical note has many overtones, which are distinguished by subscripts. The guitar has six strings, whose thickest two have fundamentals at the frequencies E2 (82.4 Hz) and A2 (110 Hz). When the fifth fret of a string of length L is pressed, the effective string length is $L' = (3/4)L$. (a) For the tuned E2 string, to what frequency f'_E does $L' = (3/4)L$ correspond, and by what percentage is it out of tune with the frequency f_A of the tuned A2 string? (b) If f'_E and f_A don't quite match, and if they are struck at the same time, show that these two strings make sounds that cancel every $t = 1/(f'_E - f_A)$ seconds. This is called a *beat*. Hence, the more in tune are the two strings, the longer the time between beats. (c) Find the beat time between the E2 string at the fifth fret and the unfretted A2 string. In practice, if the lower-frequency E string has been tuned already, so f'_E is fixed, the tuner adjusts the tension on the A string until the time between beats is so long as to be unnoticeable.

16-6.2 Explain why Young employed a narrow slit in front of the two slits in his interference experiment.

16-6.3 Young argues that the shorter apparent wavelengths in interference from thin films of water and oil (as opposed to air films between glass plates) argues for a lower velocity of light in those media, relative to air. Justify this.

16-6.4 What role does coherence play in ordinary reflection? In reflection from thin films? From thick films?

16-6.5 (a) Explain why a two-slit interference pattern cannot be observed when light of optical frequency f_o from lightbulb 1 enters slit 1, and light of the same frequency f_0 from lightbulb 2 enters slit 2. (b) Explain why a two-slit interference pattern can be observed using sound from two speakers driven at the same acoustic frequency f_A by different acoustic oscillators.

16-6.6 (a) What is the minimum thickness d of a thin glass plate of index of refraction $n = 1.5$ for which constructive interference will be observed in air, for light of a given wavelength λ? (b) Repeat for the plate in water. (c) For incident intensity I_0 a film in air, of index of refraction n and thickness $d = \lambda/4n$, gives reflection intensity $I_1 = 0.9\,I_0$. For incident intensity I_0, estimate the reflected intensities for films of thickness $\lambda/8n$, $3\lambda/8n$, and $\lambda/2n$.

16-6.7 In a Young's two-slit experiment, one slit is covered with a blue filter and the other with a yellow filter. (a) Will there be an interference pattern on the observation screen? (b) Can light of two different colors be coherent?

16-6.8 At its thinnest part, a soap film on a wire hoop appears black on reflection, whereas the thinnest part of an oil slick on water appears bright on reflection. (a) Explain. (b) Relate to a Newton's ring experiment where the lenses are first in air and then in a fluid of large index of refraction.

16-6.9 Laura orients a hoop containing a 460 mm thick soap film ($n = 1.33$) normal to the sun's rays, which are incident from her back. (a) For such normal incidence, find the colors (from 400 nm to 700 nm) that are intensified on reflection. (b) Repeat for transmission.

16-6.10 White light (400 mm to 700 mm) at normal incidence on a soap film ($n = 1.33$) has an interference maximum only at 650 nm. It also has a single interference minimum. (a) Find the film thickness. (b) Find the wavelength of the interference minimum.

16-6.11 An oil slick ($n = 1.22$) of thickness 850 nm lies above water ($n = 1.33$). (a) Find the colors (from 400 nm to 700 nm) that are intensified on reflection at normal incidence. (b) Repeat for transmission.

16-6.12 For interference in a Newton's rings experiment, Young found experimentally that the first bright fringe moves out by a factor of six when air is replaced by water. Establish this theoretically.

16-6.13 White light at normal incidence on a soap film ($n = 1.33$) has interference maxima at 441 nm, 539 nm, and 693 nm. (a) Find the film thickness. (b) Find the wavelengths of the interference minima.

16-6.14 Two optically flat glass plates ($n = 1.5$) of width 6.4 cm are in contact at one end, and at the other end are separated by a thin wire of diameter $d = 0.02$ mm, forming a wedge of variable thickness. See Figure 16.42. Light of wavelength 620 nm is normally incident on the top plate. Consider interference between light reflected off the bottom of the top plate and off the top of the bottom plate. (a) Over the length of the wedge, find how many bright fringes are observed. Figure 16.13(b) show how the pattern would appear. (b) Repeat if oil with $n = 1.2$ is between the plates. (c) Repeat if oil with $n = 1.6$ is between the plates.

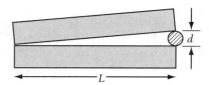

Figure 16.42 Problem 16-6.14.

16-6.15 Find the minimum thickness for a non-reflective coating ($n = 1.26$) upon glass ($n = 1.50$), if light of wavelength 570 nm (in air) at normal incidence has a minimum in reflected intensity.

16-6.16 Consider light incident from air at angle θ_1 on a thin film of index of refraction n and thickness d. Interference fringes are observed. See Figure 16.43. Note that the light that emerges at C travels a longer path than the light that is directly reflected, but that to compare the light reflected from A and the light transmitted at C, we must include an extra path length traveled until the light from A "catches up" to the light from C. Show that: (a) the extra path length in the film is $2d/\tan\theta_2$; (b) the extra path length in the air of the reflection off the top surface is $2d\tan\theta_2$; (c) the extra path length is such that the condition for constructive interference is $2dn\cos\theta_2 = \lambda(m + \frac{1}{2})$.

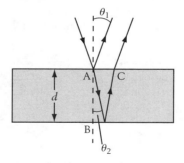

Figure 16.43 Problem 16-6.16.

16-6.17 For a wedge formed by two optically flat glass plates, at atmospheric pressure 1575 dark lines are observed across a thickness d for a 690.00 nm wavelength in air. See Figure 16.42. When the air pressure is increased by a factor of ten, 1519 dark lines are observed. (a) If the index of refraction n deviates from unity only by a term proportional to the pressure, find the index of refraction of air at atmospheric pressure. (b) Find the wavelength in vacuum. (c) Find d.

16-6.18 An oil slick ($n = 1.22$) of thickness 850 nm lies above water ($n = 1.33$). (a) Find the colors (from 400 nm to 700 nm) that will be intensified on reflection when the light is incident at an angle of 50° to the normal. (b) Repeat for transmission. *Hint*: See problem 16-6.16.

16-6.19 Michelson interferometer. See Figure 16.44. A partially silvered mirror P splits the beam

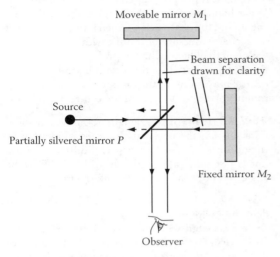

Figure 16.44 Problem 16-6.19.

in half, sends it along two paths to mutually perpendicular mirrors M_1 and M_2, which send it back to the partially silvered mirror where two of the now four beams recombine and go to the observer. All mirrors are very flat. As the distance between P and M_1 is adjusted, interference maxima and minima are successively observed. (a) Find the relationship between the number of maxima N, the wavelength λ, and the distance by which M_1 is moved. (b) If $N = 200$ maxima are observed when M_1 moves by 0.06 mm, determine the wavelength of the light.

16-6.20 A Young's two-slit apparatus with slit separation 0.67 mm is 1.8 m from a screen. (a) If red light, of wavelength 670 nm, is incident on the double slit, find for the zeroth- and first-order maxima their angular separation and their spatial separation on the screen. (b) Repeat for blue light, of wavelength 440 nm. (c) Repeat part (a) for the first- and second-order minima. (d) Find the highest order maximum that can be observed for red light; for blue light.

16-6.21 Monochromatic light is incident on a Young's two-slit apparatus with slit separation 0.250 mm, with a screen that is 1.1 m from the slits. The third order maximum occurs at $y = 5.4$ mm.

(a) Determine the wavelength of the light. (b) Determine the angle θ and position y on the screen of the fifth-order minimum. (c) Determine the highest-order maximum that can be observed.

16-6.22 Light from a helium-neon laser ($\lambda = 632.8$ nm) is incident upon a two narrow slits. (a) If the fourth-order minimum is observed at 18°, determine the slit separation. (b) Determine the largest order maximum that can be observed for this slit separation. (c) Repeat parts (a) and (b) for violet light of wavelength 440 nm.

16-6.23 Light of wavelength 490 nm is incident on a Young's two-slit apparatus. (a) Find the maximum slit separation for which there are no more than three maxima. (b) Find the maximum slit separation for which there are no more than a single pair of minima.

16-6.24 In a Young's two-slit experiment with light of wavelength 570 nm, the twelfth-order minima make an angle of 8.5° to the zero-th order maximum. (a) Determine the slit separation. (b) Determine the angle associated with the fourth-order maximum. (c) Determine the slit separation that would give a fifth-order maximum at 75°.

16-7.1 In 1816, Fresnel performed a two-source experiment with light from a source S and its reflection S′ (from a mirror), which recombined on the observing screen. See Figure 16.45. Show that for this geometry there is no possibility of what Newton called "inflection" effects, associated with scattering by edges. (Fresnel saw the same pattern as for two-slit diffraction. Thus the corpuscular theory failed to explain this experiment.)

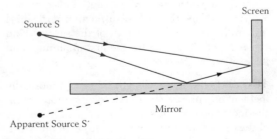

Figure 16.45 Problem 16-7.1.

16-7.2 On the observation screen for a Young's interference experiment, just off the central maximum the intensity falls off to 12% of the maximum. (a) Find the phase difference to which this corresponds. (b) For a slit separation of 0.35 mm and wavelength 620 nm, find the angle on the screen to which this corresponds.

16-7.3 In a Young's two-slit experiment, as in Figure 16.30, the distance to the screen is 62 cm, the slit separation is 0.23 mm, and the wavelength is 570 nm. (a) Find the distance y associated with the first-order maxima. (b) Find the distance y' above y where the intensity falls to 20% of the maximum.

16-7.4 (a) Compare Young's observation of a bright spot at the center of the geometrical shadow of a rectangular slip of paper to Arago's observation of a bright spot at the center of the geometrical shadow for a disk. (b) What would happen to the bright spot if the disk became irregular?

16-7.5 Give a qualitative argument to explain why, at the center of the shadow of an opaque strip, there should be a bright fringe. Repeat for an opaque disk.

16-7.6 White light is normally incident on a single slit of width 0.8 μm. (a) Determine the number of diffraction minima for blue light (460 nm) and for red light (680 nm). (b) Indicate the pattern on the observation screen within 30° of the central maximum.

16-7.7 For monochromatic light normally incident on a single slit of width 0.014 mm, the first diffraction minimum occurs at an angle of 1.5° from the central maximum. Find the wavelength of the light.

16-7.8 Light of wavelength 487 nm is normally incident on a narrow slit. On the observation screen, 1.24 m from the slit, the distance between the third minimum on either side of the central maximum is 1.58 cm. (a) Find the angle of diffraction of the third minimum. (b) Find the width of the slit. (c) Find the angle of diffraction for the first minimum, and its location on the screen.

16-7.9 Light of wavelength 624 nm is normally incident on a narrow slit. On the observation screen, the angle between the second minima on either side of the central maximum is 0.88°. (a) Find the width of the slit. (b) Find the angle of diffraction for first and third minima.

16-7.10 Light of wavelength 560 nm (yellow) and 640 nm (red) is incident on a slit. Estimate the angle on the observation screen of the first common maximum of these colors if the first minimum for red light is at 1.24°.

16-7.11 Light of wavelength 468 nm is normally incident on a slit of width 0.35 mm. (a) Find the angle of the first minimum. (b) If the screen is 24 cm from the slit, find the position on the screen of the first minimum.

16-7.12 (a) What happens to a diffraction pattern if the light is not normally incident on the slit? (b) What happens if the light is normally incident on the slit but the observation screen is tilted? (c) To get an accurate measure of the slit width, for a known wavelength of light, which is preferable to measure: the distance between the first minimum and the central maximum, or the distance between the first minima on opposite sides of the central maximum?

16-7.13 Light of wavelength 690 nm is incident on a slit. On the observation screen, 65 cm away, the distance between the first two diffraction minima is 4.86 mm. (a) Find the slit width. (b) Find the angle and positions of the second minima.

16-7.14 A laser of wavelength 620 nm is incident on a wire of diameter 0.86 mm. (a) Find the angle associated with the first diffraction minima. (b) To maintain wire diameter within a tolerance of 0.001 mm, determine to what accuracy the angle must be measured.

16-7.15 A laser of wavelength 633 nm is incident on a fiber that is 0.56 m from an observing screen. If the first minima on opposite sides are separated by 3.6 mm, determine the diameter of the fiber.

16-7.16 In a single-slit diffraction experiment with monochromatic light, the maximum between the second and third minima is at 12.4° and has intensity 0.68 W/m². If the minimum observable intensity is 0.21 W/m², identify the weakest observable maximum, and the approximate angle associated with it.

16-7.17 (a) Show that, in (16.27), the secondary maxima occur for $\tan\alpha = \alpha$, where $\alpha = \pi w \sin\theta/\lambda$. (b) Show that, when this condition holds, $I = I_0 \cos^2\alpha$.

16-7.18 A searchlight has wavelength 590 nm. (a) At what angle from the central beam is the first minimum? (b) For a person 240 m away, to what distance does this correspond?

16-7.19 Light of wavelength 590 nm passing through a slit can resolve two line sources with angular separation 1.5×10^{-4} radians. (a) Find the slit width. (b) For this slit width, find the limit of observable angular separation for x-rays with wavelength 0.059 nm. (c) Repeat for microwaves with wavelength 0.59 m.

16-7.20 A laser of wavelength 610 nm and beam diameter 0.8 mm is normally incident on an observing screen 1.3 m away. An opaque sphere is centered in the path of the laser, 0.9 m from the screen. If the distance on the screen from the undeflected beam to the first minimum is 1.4 mm, estimate the diameter of the sphere.

16-7.21 Light of wavelength 535 nm is emitted by two stars 84 light years away, and observed with a telescope of diameter 24 cm. (a) What minimum separation must the stars have to be distinguishable? (b) Compare this to the radius of the sun. (c) Repeat part (a) for a radio telescope with diameter 300 m at wavelength 21 cm.

16-7.22 At night, driver A sees a distant oncoming car B, which has two headlights 1.46 m apart. The nighttime diameter of the pupil of driver A's eye is 6.8 mm. For a wavelength of 550 nm, find (a) the minimum angular resolution at which driver A can distinguish the headlights of car B; (b) the distance between driver A and car B; (c) the time for the two cars to pass each other once driver A has resolved the headlights of car B (if each car moves at 60 mi/hr relative to the road).

16-7.23 The 19th-century French painter Georges Seurat (the inspiration for Stephen Sondheim's Broadway show *Sunday in the Park with George*) produced different colors not by mixing pigments, but with distinct, adjacent, dots of pigment that overlap in the eye due to diffraction. (a) Taking the diameter of the pupil of an eye in daylight to be 3.2 mm, for blue light (440 nm) find the diameter d at which two adjacent dots can just be resolved at 4 m. For distances closer (farther) than this, the dots can (cannot) be resolved. (b) Repeat for the comfortable focus near point distance of 25 cm. (c) Not all eyes are alike. For another person, with daytime pupil diameter 2.8 mm, find the distance s at which the dots can just be resolved, using d from part (a). (d) In dim light, the pupil diameter expands. For a pupil diameter of 4.6 mm in twilight, find s.

16-7.24 An acoustic speaker of diameter 12 inches, centered at the origin, is directed along the x-axis. If it emits sound at 120 Hz, find the smallest value of y for which the sound is a minimum at the point $(20$ m, y, $0)$.

16-8.1 When the optic axis is in the plane of a slab of birefringent crystal of thickness d, for normal incidence there is a single ray of light but there are two sets of interference fringes. (a) If the indices of refraction in this situation are n_O and n_E, derive the relationships for the two sets of interference maxima in reflection. (b) For $n_O = 1.4$ and $n_E = 1.5$, and $d = 650$ nm, find the colors from 400 nm to 700 nm that are most intense in reflection, for each polarization.

16-8.2 Give a qualitative explanation for the intensity patterns of Figure 16.33, including both position and relative intensity.

16-8.3 (a) In Example 16.12, find the additional "optical thickness" for light passing through the polarizer. (b) For a slit separation of 0.8 mm, find the shift in deflection angles of the interference pattern on the observation screen. (*Hint*: An extra path length is in some sense equivalent to the beam passing through the polarizer being farther away than the other beam. This corresponds to an angle that determines the shift in the interference pattern.)

16-8.4 A thin transparent wedge of maximum thickness 235 nm and index of refraction 1.53 is placed in front of one slit in a two-slit interference experiment with slit separation 0.5 mm, as in Figure 16.34. For light of wavelength 480 nm, determine the net shift in angle of the interference pattern on the screen, as the wedge moves all the way across the slit.

16-9.1 A grating has ten slits, with a slit separation of 2800 nm. (a) If it is illuminated with blue light (440 nm), determine the largest-order maximum. (b) Determine the angular positions of all the maxima. (c) For yellow light (580 nm), determine the largest-order maximum. (d) For $m = 1$ and yellow light, determine the angular width. (e) Determine the spread in wavelength about 440 nm that can be observed for $m = 1$.

16-9.2 A diffraction grating 15.2 mm wide has 8200 rulings. (a) Find the distance between adjacent rulings. (b) For a ruby laser (wavelength 690 nm), find the number of diffraction maxima. (c) Find the angles for the maxima.

16-9.3 Monochromatic light is incident on a diffraction grating with rulings separated by 5200 nm. The third maximum occurs at $17°14'$. (a) Find the wavelength of the light. (b) Find the number of maxima. (c) Find the angle for the sixth maximum.

16-9.4 Show that the intensity pattern for three narrow slits with uniform spacing d is

$$I = \frac{I_0}{9}(1 + 4\cos\alpha + 4\cos^2\alpha), \qquad \alpha = \frac{2\pi d \sin\theta}{\lambda},$$

and I_0 is the intensity for $\theta = 0$.

16-9.5 A grating has 240 rulings/mm. Find the observable range of visible wavelengths in the seventh order.

16-9.6 A grating has 270 rulings/mm. It has a principal maximum at $35°$. Find the possible visible wavelengths corresponding to such a maximum, and the order for each of these wavelengths.

16-9.7 For a grating that is 1.2 cm wide, a principal maximum at 526 nm is observed in third order at $32°$. Find the number of lines in the grating.

16-9.8 (a) To resolve two spectral lines at 487.45 nm and 487.62 nm, find the necessary resolving power. (b) For a grating with 1250 rulings, find the minimum order to give at least this resolution. (c) At this order, find the width of the grating that will give this resolution for an observation angle of $48°$.

16-9.9 Hydrogen (H_1) and its isotope deuterium (H_2) have lines at respective wavelengths of 656.45 nm and 656.27 nm. To determine their presence on the sun, diffraction measurements are made. Find the minimum number of slits needed to resolve these lines in third order.

16-9.10 (a) For a grating with 2600 rulings and ruling separation $d = 1800$ nm, and light of a wavelength 520 nm, find the angles that correspond to all diffraction maxima. For the second-order maximum, find (b) the angular width, (c) the dispersion, (d) the resolving power.

16-9.11 For a diffraction grating with grating separation d, it is implicit that the width w of each ruling (transmission or diffraction grating) is less than d. The text considered only the case $w \ll d$. (a) Show that there is no diffraction for $w = d$, and show that $w > d$ has no meaning. (b) Show that, when $d = 2w$, all the even orders are missing from a diffraction pattern. (*Hint*: Consider the diffraction pattern due to an individual slit of width w.) (c) When $2d = 3w$, identify the orders that are missing from a diffraction pattern. These results are independent of d and λ.

16-9.12 The rate of expansion of the universe is determined by analyzing diffraction grating experiments. When a star moves directly away from the earth with velocity v, its frequency f_0 is shifted to the lower value $f = f_0\sqrt{(1 - v/c)/(1 + v/c)}$. This is known as the relativistic Doppler shift. (a) How are the wavelengths shifted? (b) If the hydrogen line corresponding to 656.45 nm is identified, but it is shifted to 765.34 nm, what is the recession velocity of the star? It is found experimentally that the farther away a star is, the greater its recession velocity.

16-10.1 (a) For a Bragg angle of $10°$, determine the scattering angle. (b) Repeat for a Bragg angle of $80°$.

16-10.2 For scattering planes of separation 0.275 nm, find the scattering angles and the Bragg angles θ associated with second- and third-order Bragg scattering of radiation of wavelength 0.065 nm.

16-10.3 For x-rays of wavelength 0.085 nm, a Bragg angle of $15.4°$ is observed. Find the planar separation d to which this corresponds.

16-10.4 For a set of NaCl planes the interplanar spacing is 0.0796 nm. For an x-ray beam making a $26°$ angle to the normal to these planes, first-order Bragg reflection is observed. Find the wavelength of the x-rays.

16-10.5 A beam of x-rays contains wavelengths in the range 0.008 nm to 0.014 nm. It is incident at $36.7°$ relative to the normal to a family of planes with $d = 0.0284$ nm. Find which wavelengths are Bragg diffracted.

16-10.6 A crystal consists of identical atoms located on a simple cubic lattice, with nearest neighbor distance $a = 0.45$ nm. (a) What wavelength is needed to make a Bragg angle of $35°$ relative to a plane of atoms with plane separation a? (b) Repeat for a plane separation $a/\sqrt{2}$.

16-10.7 Consider a crystal with scattering planes of separation 0.342 nm. (a) Determine the wavelength and frequency of the radiation needed to produce a first-order Bragg angle of $17.4°$. (b) Repeat for a scattering angle of $46.5°$.

16-10.8 A simple cubic crystal with nearest neighbor distance $a = 0.45$ nm is ground into a powder and 0.34 nm x-rays are used to obtain a powder pattern. Find the Bragg angle and the scattering angle for the largest diameter ring that is observed.

16-10.9 For planes that go through only a few atoms in a crystal, the density of atoms in that plane must be relatively low. Explain why the scattering angles must be relatively large.

16-G.1 Discuss why the mechanism for color separation by a thin film—where each color has a different number of wavelengths that span the film—is different than the mechanism for color separation by a prism—where each color has a different index of refraction.

16-G.2 Derive the equation for the rainbow angle, following (16.5).

16-G.3 Derive (16.26).

16-G.4 Derive the angular dependence of the intensity profile for light of wavelength λ normally incident on two slits separated by d and each with slit width w.

16-G.5 Find the intensity profile for two slits with light normally incident from the same source, with wavelength λ. One slit has width w (net intensity I_0), and the other has width $w/2$ (net intensity $I_0/2$). Take their separation $d \gg w$.

16-G.6 Monochromatic light of wavelength λ is incident on a series of slits separated by d, as in Figure 16.46. (a) Show that the condition for a maximum is that $d(\sin\phi + \sin\theta) = m\lambda$. (b) Show that, for $\phi + \theta = 90°$, this reproduces the result for Young's micrometer. (c) Show that, for $\phi = 0$, this reproduces (16.30), the result for the usual geometry of a diffraction grating. (d) Show

that, for $\phi = \theta$, this reproduces (16.41), the result for the usual geometry of Bragg scattering.

16-G.7 (a) Using phasors, show that for N narrow slits, each of which transmits amplitude A_0, the net amplitude on transmission is given by

$$A = A_0 \left| \frac{\sin(N\alpha/2)}{\sin(\alpha/2)} \right|,$$

where $\alpha = 2\pi d \sin\theta/\lambda$. (b) Show that the time-averaged intensity is then given by

$$\bar{I} = I_0 \frac{\sin^2(N\alpha/2)}{\sin^2(\alpha/2)},$$

where I_0 is the time-averaged intensity of a single slit. (c) Show that for $N = 2$ this reduces to (16.22).

16-G.8 Consider N slits. When the finite width w of each slit is included, the total transmitted amplitude A_ω includes a factor that depends on the observer's orientation with respect to the slit, and depends on w. Use (16.26) for the amplitude of a single slit, but with r replaced by $r - kd\sin\theta$, where k is an index from 1 to N for the slit number. Show that A_ω equals A of Problem 16-G.6 multiplied by the factor $|(2\sin\beta/2)/\beta|$, where $\beta = 2\pi w \sin\theta/\lambda$, and that the intensity I_ω equals I of Problem 16-G.6 multiplied by this factor squared. Notice the modulation in Figure 16.36.

16-G.9 Water occupies the half-space $x \geq 0$. Air is to the left. See Figure 16.47. A light source is at $(d,0)$, and an observer is at $(-a,b)$. (a) Find expressions for the angles θ and ϕ in terms of the source and observer cordinates. Now consider that θ and ϕ are known. (b) For a ray from the source at an extra angle $d\theta$, find the extra angle $d\phi$. (c) Trace the rays seen by the observer at these two angles. Show that their intersection, which gives the position of the (virtual) image, corresponds to $d' = (d/n)(\cos\phi/\cos\theta)^3$, $s = d\tan\theta[1 - (\cos\phi/\cos\theta)^2]$. (d) For $\theta = 30°$, show that $d'/d = 0.611$ and $s/d = 0.1497$.

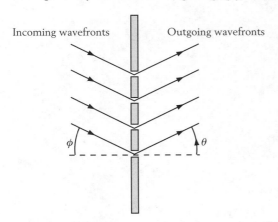

Figure 16.46 Problem 16-G.6.

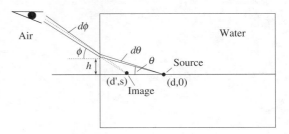

Figure 16.47 Problem 16-G.9.

Appendix A

General Mathematics Review

A.1 Simple Equations

(a) A straight line is represented by

$$y = mx + b,$$

where b is the intercept (the value of y for $x = 0$) and m is the slope. Another representation of a straight line is

$$y = m(x - x_0) + y_0,$$

where the straight line passes through the point (x_0, y_0).

(b) The quadratic equation is defined by

$$ax^2 + bx + c = 0.$$

Its solution is obtained as follows. Let $b' = b/a$ and $c' = c/a$. Then define d by

$$0 = a(x^2 + b'x + c') = a[(x + b'/2 + d)(x + b'/2 - d)].$$

Thus d satisfies $c' = (b'/2 + d)(b'/2 - d) = b'^2/4 - d^2$, so $d = \pm\sqrt{b'^2/4 - c'}$. This leads to the two solutions

$$x_\pm = -\frac{b'}{2} \pm d' = \frac{-b \pm \sqrt{b^2 - 4ac}}{2a}.$$

A.2 Scientific Notation and Powers

(a) Decimal notation is not to be taken for granted: it was not commonly used in Galileo's time, and Galileo did not use it for analyzing his experimental results. He might have approximated a number like 6.12 as the rational fraction $6\frac{1}{8}$, and performed calculations with that quantity.

For $n > 0$, 10^n means 10 multiplied by itself n times. Thus $10^3 = (10)(10)(10) = 1000$. For $n = 0$, $10^n = 10^0$ means 1. For $n < 0$, 10^n means $1/10^{-n}$. Thus $10^{-3} = 1/[(10)(10)(10)] = 1/1000 = 0.001$. Putting it all together, the number 78.24 means 7 10^1's plus 8 10^0's plus 2 10^{-1}'s plus 4 10^{-2}'s.

(b) The rules for multiplying powers of x are

$$x^m x^n = x^{m+n}.$$
$$(x^m)^n = x^{mn}.$$

As long as m and n are integers, it is clear how to calculate these quantities. The next section shows how to calculate these quantities for any real m and n.

We also have, for fractional powers or roots,

$$x^{\frac{1}{n}} \equiv \sqrt[n]{x}.$$

(c) We now define the logarithm and the antilogarithm. Let

$$x = a^y.$$

Here $a > 0$ is called the *base*. [In Section A.(22), above, the base was 10, and for computers typically it is 2.] Note that if $y = 0$, then $x = 1$, and if $y = 1$ then $x = a$.

Thinking of this as $x = f(y)$, and noting that this is a 1-to-1 function, a unique inverse function $y = f^{-1}(x)$ exists. (Inverse functions are obtained by plotting $y = g(x)$ and then reflecting the curve through the line $y = x$, so that x and y interchange.) This particular inverse function is called the logarithm, and it is written as

$$y = \log_a x.$$

Since $y = 0$ if $x = 1$, we deduce that $\log_a 1 = 0$ for all a. Since $y = 1$ if $x = a$, we deduce that $\log_a a = 1$ for all a.

Naturally, the inverse function of the logarithm ($y = \log_a x$), called the antilogarithm, is x itself:

$$x = \text{antilog}_a y = a^{\log_a x}.$$

Let $x_1 = a^{y_1}$ and $x_2 = a^{y_2}$, so $y_1 = \log_a x_1$ and $y_2 = \log_a x_2$. Then $x_1 x_2 = a^{y_1} a^{y_2} = a^{y_1 + y_2}$. Hence $y_1 + y_2 = \log_a(x_1 x_2)$. Thus

$$\log_a(x_1 x_2) = \log_a x_1 + \log_a x_2.$$

Similarly we can show that

$$\log_a \frac{1}{x_1} = -\log_a x_1.$$

$$\log_a x^n = n \log_a x.$$

Applying $\log_a$ to both sides of $b^{\log_b a} = a$ gives $(\log_b a)(\log_a b) = 1$, so

$$\log_b a = \frac{1}{\log_a b}.$$

To convert from logarithms in base b to logarithms in base a we apply $\log_b$ to both sides of $a^y = x$, which yields $y \log_b a = \log_b x$. Then with $y = \log_a x$

we have

$$\log_a x = \frac{\log_b x}{\log_b a}.$$

Natural logarithms involve the base $e = 2.718\ldots$, for which $\log_e 10 = 2.302585$. The proper definition of e is given in Section A.4. It makes the calculation of logarithms and antilogarithms (also known as exponentials) relatively simple, and thus it enables the calculation of quantities like x^m as $x^m = e^{\log_e x^m} = e^{m \log_e x}$.

Logarithms to the base 10 are the basis of the pre-electronic calculator device known as the *slide rule*. These involve rods calibrated with lengths along x given by the logarithm to the base 10 of x. Hence, on adding lengths from rods 1 and 2 (with the origin relative to rod 1), the logarithms are added (e.g., $\log_{10} x_1$ and $\log_{10} x_2$), giving a net length on rod 1 corresponding to $\log_{10}(x_1 x_2)$. The number on rod 1 corresponding to that length is $x_1 x_2$.

A.3 Arc Length and Trigonometry

(a) The arc length s of a circle is proportional to the radius r of the circle and the angle θ subtended by that arc (where we don't yet specify the angular units, which could be radians or degrees or whatever). Since s and r have the same units, for arbitrary angular units we write

$$s = \alpha \theta r,$$

where α is a constant that depends on our units. To determine α, we note that $s = 2\pi r$ for a complete circle. In radian measure we take $\alpha_{rad} = 1$, so $\theta = 2\pi$ radians determines a circle. In angular measure we have $2\pi r = \alpha_{deg} 360 r$, so $\alpha_{deg} = \pi/180$.

$s = r\theta$

Figure A.1 Arc length.

(b) The basis of trigonometry is the right triangle, where we will label the sides x and y and the angle θ. Note that x is the side adjacent to θ, and y is the side opposite to θ. The hypotenuse h is defined by the Pythagorean theorem

$$h^2 = x^2 + y^2.$$

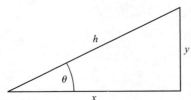

Figure A.2 Right triangle.

There are six dimensionless ratios of these sides, which define the six trigonometric functions sine, cosine, tangent, cosecant, secant, and cotangent. Specifically,

$$\sin \theta = \frac{y}{h}.$$

$$\cos \theta = \frac{x}{h}.$$

$$\tan \theta = \frac{y}{x}.$$

$$\csc \theta = \frac{h}{y} = \frac{1}{\sin \theta}.$$

$$\sec \theta = \frac{h}{x} = \frac{1}{\cos \theta}.$$

$$\cot \theta = \frac{x}{y} = \frac{1}{\tan \theta}.$$

Sometimes $\cot \theta$ is written as $\operatorname{ctn} \theta$.

For small θ, an arc is nearly a right triangle with $h \approx r$, $x \approx r$, and $y \approx s$. Hence, for small θ in radian measure, $\sin \theta \approx \tan \theta \approx \theta$ and $\cos \theta \approx 1$.

(c) The trigonometric functions satisfy the following relations (in degree measure):

$$\sin \theta = \cos(90° - \theta).$$

$$\cos \theta = \sin(90° - \theta).$$

$$\tan \theta = \cot(90° - \theta).$$

They have the following properties:

$$\sin(-\theta) = -\sin \theta.$$

$$\cos(-\theta) = \cos \theta.$$

$$\tan(-\theta) = -\tan \theta.$$

$$\sin(\theta + 180°) = -\sin \theta.$$

$$\cos(\theta + 180°) = -\cos \theta.$$

$$\tan(\theta + 180°) = \tan \theta.$$

(d) Using the Pythagorean theorem, an infinite number of trigonometric identities can be derived, the simplest of which are

$$\sin^2 \theta + \cos^2 \theta = 1.$$

$$\sec^2 \theta = 1 + \tan^2 \theta.$$

$$\csc^2 \theta = 1 + \cot^2 \theta.$$

(e) In addition, there are laws for the trigonometric functions of the sum of two angles:

$$\sin(A + B) = \sin A \cos B + \cos A \sin B.$$

From the properties of the sine function under $B \to -B$ we then have

$$\sin(A - B) = \sin A \cos B - \cos A \sin B.$$

From the relations between the sine and cosine functions we have that

$$\cos(A + B) = \cos A \cos B - \sin A \sin B.$$

From the properties of the cosine function under $B \to -B$ we then have

$$\cos(A - B) = \cos A \cos B + \sin A \sin B.$$

The sum of the relations for the sines yields

$$\sin A + \sin B = 2\sin\left[\frac{1}{2}(A + B)\right]\cos\left[\frac{1}{2}(A - B)\right].$$

The difference of the relations for the sines yields

$$\sin A - \sin B = 2\sin\left[\frac{1}{2}(A - B)\right]\cos\left[\frac{1}{2}(A + B)\right].$$

The sum of the relations for the cosines yields

$$\cos A + \cos B = 2\cos\left[\frac{1}{2}(A + B)\right]\cos\left[\frac{1}{2}(A - B)\right].$$

The difference of the relations for the cosines yields

$$\cos A - \cos B = 2\sin\left[\frac{1}{2}(A + B)\right]\sin\left[\frac{1}{2}(B - A)\right].$$

Note that, by use of the properties of the trigonometric functions, a knowledge of any of the above relations can be used to obtain any of the others. For example, in the last of these relations, letting $A \to 90° - A$ and $B \to 90° - B$ leads to the second relation.

Setting $A = B = \theta$ in two of the above equations yields

$$\sin 2\theta = 2\sin\theta\cos\theta.$$

$$\cos 2\theta = \cos^2\theta - \sin^2\theta.$$

From these relations a number of others can be derived, such as

$$\tan 2\theta = \frac{2\tan\theta}{1 - \tan^2\theta}.$$

$$\sin^2\left(\frac{\theta}{2}\right) = \frac{1}{2}(1 - \cos\theta).$$

$$\cos^2\left(\frac{\theta}{2}\right) = \frac{1}{2}(1 + \cos\theta).$$

$$\tan\left(\frac{\theta}{2}\right) = \pm\sqrt{\frac{1 - \cos\theta}{1 + \cos\theta}},$$

where $+$ holds for $\theta/2$ in the first or third quadrants, and $-$ holds otherwise.

(f) For the general triangle, with angles α, β, γ respectively opposite sides a, b, c, two classes of relations hold. (Note that, in degree measure, $\alpha + \beta + \gamma = 180$.) One is the Law of Sines:

$$\frac{a}{\sin \alpha} = \frac{b}{\sin \beta} = \frac{c}{\sin \gamma}.$$

They follow on finding, for each base, the altitudes in two different ways.

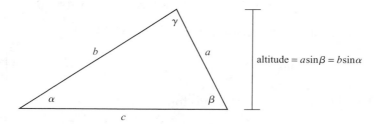

altitude $= a\sin\beta = b\sin\alpha$

Figure A.3 General triangle.

The other is the Law of Cosines:

$$a^2 = b^2 + c^2 - 2bc \,\cos\alpha.$$
$$b^2 = c^2 + a^2 - 2ca \,\cos\beta.$$
$$c^2 = a^2 + b^2 - 2ab \,\cos\gamma.$$

They are related by what is called *cyclic permutation*. (Can you see the relationships?) These equations follow on determining each side in two different ways. For example, the last of these equations follows on squaring both sides of $c = a \sin \beta + b \sin \alpha$, rearranging the right-hand side using the Law of Sines, and using $\alpha + \beta + \gamma = 180$.

A.4 Differential Calculus

(a) Differential calculus applied to $y = f(x)$ is merely finding the value of the slope at any point x. For example, to find the straight line that is tangent to $f(x)$ at (x_0, y_0), we compute $m = df/dx$ at x_0 and then use this in the second form of the straight line, given in Section A.1(a).

Here are some basic results about taking derivatives:

$$\frac{d}{dx}(fg) = \frac{df}{dx}g + f\frac{dg}{dx}. \qquad \text{(product rule)}$$

$$\frac{d}{dx}(f + g) = \frac{df}{dx} + \frac{dg}{dx}. \qquad \text{(sum rule)}$$

$$\frac{df}{dx} = \frac{df}{du}\frac{du}{dx}. \qquad \text{(chain rule)}$$

In this last case, $f = f(u)$ and $u = u(x)$.

(b) Here are some specific results:

$$\frac{d}{dx}x^n = nx^{n-1}.$$

The above result is proved straightforwardly in calculus texts for integer n. For rational $n = p/q$ with p and q integers it can be proved by writing $y = x^n = x^{p/q}$ in the form $y^q = x^p$, and then taking d/dx to both sides (i.e., implicit differentiation) to find dy/dx. Once the result is established for rational n it can be established, by a limiting process, for all real n.

In what follows, a is a constant, and ln means the natural logarithm (the logarithm in the base e, or $\log_e$).

$$\frac{d}{dx}\sin(ax) = a\cos(ax).$$

$$\frac{d}{dx}\cos(ax) = -a\sin(ax).$$

$$\frac{d}{dx}\tan(ax) = a\sec^2(ax).$$

$$\frac{d}{dx}\cot(ax) = -a\csc^2(ax).$$

$$\frac{d}{dx}\sec(ax) = a\tan(ax)\,\sec(ax).$$

$$\frac{d}{dx}\csc(ax) = -a\cot(ax)\,\csc(ax).$$

$$\frac{d}{dx}\ln(ax) = \frac{1}{x}.$$

$$\frac{d}{dx}e^{ax} = ae^{ax}.$$

The last result serves to define e.

(c) Here are some power series expansions that can be established by showing that the quantities on both sides of the equals sign have the same value at $x = 0$ and that the derivatives of both sides are the same.

The first expansion enables us to compute e:

$$e^x = 1 + x + \frac{x^2}{2!} + \frac{x^3}{3!} + \cdots$$

Applying this to the case $x = 1$, the left-hand side gives e, and including terms to x^6 the right-hand side gives ≈ 2.71806. This is more or less what a calculator does when it computes an exponential.

Next is the binomial theorem, applicable for $|x| < 1$:

$$(1 + x)^n = 1 + nx + \frac{n(n-1)}{2!}x^2 + \frac{n(n-1)(n-3)}{3!}x^3 + \cdots$$

(Here $n! = n(n-1)\ldots 1$, so $3! \equiv 3 \cdot 2 \cdot 1 = 6$.) When n is a positive integer, this is a series with n terms, the last of which is x^n. An important special case is $n = -1$:

$$\frac{1}{1+x} = (1+x)^{-1} = 1 - x + x^2 - x^3 + \cdots$$

The logarithm is given by

$$\ln(1+x) = x - \frac{x^2}{2} + \frac{x^3}{3} - \cdots$$

This is more or less what a calculator does when it computes a logarithm. You can verify that both sides agree for $x = 0$ and that the derivative of both sides gives both sides of the previous equation, for $(1+x)^{-1}$.

In radian measure for x, we have

$$\sin x = x - \frac{x^3}{3!} + \frac{x^5}{5!} - \cdots$$

$$\cos x = 1 - \frac{x^2}{2!} + \frac{x^4}{4!} - \cdots$$

$$\tan x = x + \frac{x^3}{3!} + \frac{2x^5}{15} + \cdots$$

Operating with d/dx on $\sin y = x$ gives $\cos y (dy/dx) = 1$. On employing $\cos y = \sqrt{1 - \sin^2 y} = \sqrt{1 - x^2}$ and the binomial theorem with $(x, n) \to (-x^2, -\frac{1}{2})$ we obtain

$$\frac{dy}{dx} = \frac{d}{dx}\sin^{-1} x = \frac{1}{\cos y} = \frac{1}{\sqrt{1 - x^2}} = 1 + \frac{1}{2}x^2 + \frac{3}{8}x^4 + \frac{5}{16}x^6 + \cdots$$

The power series for $y = \sin^{-1} x$ itself is

$$\sin^{-1} x = x + \frac{1}{6}x^3 + \frac{3}{40}x^5 + \frac{5}{112}x^7 + \cdots$$

You can verify that both sides agree for $x = 0$ and that the derivative of both sides gives both sides of the previous equation. For $x = \pi/6$, or $30°$, $\pi/6 = \sin^{-1}(\frac{1}{2})$. Including terms to x^7 gives $\pi \approx 3.141155$. This is more or less what a calculator does when it computes a sine function.

The inverse function for differentiation does not quite exist, because both $F(x)$ and $F(x) + a$ for any constant a have the same derivative $f(x)$, which we will call dF/dx. Nevertheless, we talk of the antiderivative, or inverse function $F(x)$, which is defined up to an arbitrary constant.

A.5 Integral Calculus

$$G(x) - G(a) = \int_a^x f(x)\,dx$$

is called the indefinite integral of the function $f(x)$ over the domain from a to x. It corresponds to the area under the curve $f(x)$ for that domain. Thus integration is just a form of addition, where an infinite number of very small terms are summed. It has the property that $dG/dx = f(x)$, so that $f(x)$ is the derivative of $G(x)$. Hence the integral is equal to the antiderivative, if the latter can be obtained (and is determined only up to an arbitrary constant.) In this way, many of the following integrals may be obtained from the corresponding derivatives:

$$\int x^n dx = \frac{x^{n+1}}{n+1}. \qquad n \neq -1$$

$$\int \frac{dx}{x} = \ln x.$$

$$\int e^{ax} dx = \frac{e^{ax}}{a}.$$

$$\int \sin(ax)\,dx = -\frac{\cos(ax)}{a}.$$

$$\int \cos(ax)\,dx = \frac{\sin(ax)}{a}.$$

$$\int \sec^2(ax)\,dx = \frac{\tan(ax)}{a}.$$

$$\int \csc^2(ax)\,dx = -\frac{\cot(ax)}{a}.$$

The following integrals can be obtained by a change of variable to $u = \cos x$ (and $u = \sin x$) and using the integral for the logarithm:

$$\int \tan(ax)\,dx = -\frac{\ln[\cos(ax)]}{a}.$$

$$\int \cot(ax)\,dx = \frac{\ln[\sin(ax)]}{a}.$$

This integral may be obtained by a change of variable to $u = a + bx$ and using the integral for the logarithm:

$$\int \frac{dx}{a+bx} = \frac{\ln(a+bx)}{b}.$$

Differentiation of the above with respect to b yields

$$\int \frac{dx}{(a+bx)^2} = -\frac{1}{b(a+bx)}.$$

Some inverse trigonometric functions are

$$\int \frac{dx}{a^2 + x^2} = \frac{1}{a} \tan^{-1} \frac{x}{a}.$$

$$\int \frac{dx}{\sqrt{a^2 - x^2}} = \sin^{-1} \frac{x}{a}.$$

Integration by parts sometimes permits integrals to be evaluated:

$$\int u\,dv = uv - \int v\,du,$$

which follows from integration on the chain rule in the form $d(uv) = u\,dv + du\,v$.

Here are some other useful integrals, the first three evaluated using integration by parts:

$$\int \ln(ax)\,dx = x\ln(ax) - x.$$

$$\int xe^{ax}\,dx = (ax - 1)\frac{e^{ax}}{a^2}.$$

$$\int x^2 e^{ax}\,dx = (a^2 x^2 - 2ax + 2)\frac{e^{ax}}{a^3}.$$

$$\int \frac{dx}{x^2 - a^2} = \frac{1}{2a}\ln\left(\frac{x - a}{x + a}\right). \quad (x^2 > a^2)$$

$$\int \frac{dx}{a^2 - x^2} = \frac{1}{2a}\ln\left(\frac{a + x}{a - x}\right). \quad (x^2 < a^2)$$

$$\int \frac{x\,dx}{a^2 + x^2} = \frac{1}{2}\ln(a^2 + x^2).$$

$$\int \frac{dx}{\sqrt{x^2 \pm a^2}} = \ln\left[x + \sqrt{x^2 \pm a^2}\right].$$

$$\int \frac{x\,dx}{\sqrt{a^2 - x^2}} = -\sqrt{a^2 - x^2}.$$

$$\int \frac{x\,dx}{\sqrt{x^2 \pm a^2}} = \sqrt{x^2 \pm a^2}.$$

Appendix B

Introduction to Spreadsheets

This introduction is given for those who are not already acquainted with spreadsheets. Each particular spreadsheet has its own symbols for certain operations, but all spreadsheets have the same basic operations. To be specific, we will use our spreadsheet to solve for the force on the charge q given in Figure 2.7.

A spreadsheet is a computer program (or *application*) that sets up rows and columns of *cells*, and permits mathematical manipulations on the entries in those cells. Because of this possibility of manipulation, we can think of spreadsheet use as *programming without programming*.

A cell can appear in either of two ways: either as one of many cells in the full *matrix* of cells, or as the entry cell. Figure B.1 shows the rows and columns of our spreadsheet. At the top is the entry bar, which happens to be that for cell E3. (We choose the entry bar by clicking on the corresponding cell in the spreadsheet, using the "mouse.") Below the entry bar are the rows and columns of the spreadsheet. In the entry bar, the boxes with the big X and big check are to reject or accept a typed-in entry.

We now describe the entry of the data and the arrangement of this particular application of the spreadsheet. (You may find a better way to set up your spreadsheet—what follows is merely an example.) Some of the entries simply define other entries. The numerical values are given just before Figure 2.7: $q = 2.0 \times 10^{-9}$ C, $q_1 = -4.0 \times 10^{-9}$ C and $q_2 = 6.0 \times 10^{-9}$ C, $R_1 = 0.2$ m, $R_2 = 0.3$ m, and $\theta_2 = 55$ degrees.

Cell A1—symbol k; cell B1—numerical value of k.

Cells A2, A3, and A4—symbols q, q_1, and q_2; cells B2, B3, and B4—numerical values for q, q_1, and q_2.

Cells C1 and D1—symbols x and y; cells C2 and D2—numerical values for x and y; cells C3 and D3—numerical values for x_1 and y_1; cells C4 and D4—numerical values for x_2 and y_2.

Cells E2, F2, and G2—symbols $x - x_n$, $y - y_n$, and R_n, to represent coordinate differences and relative distances $R_n = |\vec{r} - \vec{r}_n|$ of q_n ($n = 1, 2$) with respect to q.

Cells H2, I2, and J2—symbols F^*, F_x, and F_y, for the (signed) magnitudes of the forces, and their x and y components. [From (3.12′), for q_1 we have $F^* = kqq_1/R_1^2$.]

Here are some simple rules about spreadsheets.

1. If in cell X5 you want the number corresponding to the product of the numbers in cells X3 and X4, then in the entry bar for cell X5 you type $= $X3*X4. The $=$ sign tells the spreadsheet to perform the numerical

E3	✗	✓	=C2-C3						
	A	B	C	D	E	F	G	H	I	J
1	k	9E+9	x	y			-41.483	1.483E-6	1.112E-6	-9.83E-7
2	q	2E-9	0	0	$x-x_n$	$y-y_n$	R_n	F^*	F_x	F_y
3	q_1	-4E-9	0.2	0	-0.2	0	0.2	-1.8E-6	1.8E-6	0
4	q_2	6E-9	1.720E-1	2.457E-1	-1.72E-1	-2.45E-1	0.3	1.2E-6	-6.88E-6	-9.83E-7

Figure B.1 Example of a spreadsheet.

computation indicated in the entry bar for that cell; without it, only the *formula* X3*X4 will appear in the spreadsheet cell. Presently, cell E3 contains the entry = C2 − C3. We'll explain the meaning of the $ symbol shortly.

2. Spreadsheets have a powerful command called *fill*, which enables us to do repetitive calculations very easily. If cell E3 contains the product of E1 and E2 in the form = E1*E2, you can use the fill command to convert F3 to = F1*F2. If cell E3 contains this same product in the form = E2*E3, then the fill command will place = E2*F3 in F3; in other words, the $ symbol means that the next symbol does not get updated when one changes row or column. [At the top of the spreadsheet is a menu bar containing an entry called "Edit" or "Calculate." Clicking on this will expose a column of possible operations, including "Fill Down" and "Fill Right." To perform the above "fill," highlight both cells E3 and F3, and then click on "Fill Right."]

Let us now employ these rules. Although cell E3 contains the number −0.2, when we click on cell E3 we find in the entry bar the expression = C2 − C3. This represents the number in cell C2 minus the number in cell C3; that is, the difference in the x-coordinates of the charges q and q_1: $x − x_1$.

Cell G3 contains the formula = SQRT(E3^2 + F3^2), for the separation r_1 between q and q_1. Cell H3 contains the formula = B1*B2*B3/G3^2, for F^* due to q_1. Cell I3 contains the formula = B1*B2*B3*E3/G3^3, for the x-component of the force due to q_1. Cell J3 contains the formula = B1*B2*B3*F3/G3^3, for the y-component of the force due to q_1.

Cells G4, H4, I4, and J4 are similarly defined, but for q_2.

Cell I1 contains the formula = SUM(I3..I4), meaning the sum of the contents of cells I3 to I4, giving the total x-component of the force on q.

Cell J1 gives the total y-component of the force on q, and Cell H1 contains = SQRT(I1^2 + J1^2) for the magnitude of the total force on q. Cell G1 gives the angle θ (in radians) that the force makes with respect to the x-axis, computed from the entry = ATAN2(J1/I1).

To move a charge, we simply change its coordinates and the spreadsheet nearly instantly gives us the new answer. The fill command of the spreadsheet enables us to add in charges q_3, q_4, etc. with minimum difficulty.

Appendix C

The Periodic Table

Legend:

$$H \quad \text{— Symbol}$$
1 — Atomic number
1.008 — Atomic weight
Hydrogen — Name

( ) = Estimates

1 IA	2 IIA	3 IIIB	4 IVB	5 VB	6 VIB	7 VIIB	8	9 VIIIB	10	11 IB	12 IIB	13 IIIA	14 IVA	15 VA	16 VIA	17 VIIA	18 VIIIA
H 1 1.008 Hydrogen																	**He** 2 4.00 Helium
Li 3 6.94 Lithium	**Be** 4 9.01 Beryllium											**B** 5 10.81 Boron	**C** 6 12.01 Carbon	**N** 7 14.01 Nitrogen	**O** 8 16.00 Oxygen	**F** 9 19.00 Fluoride	**Ne** 10 20.18 Neon
Na 11 22.99 Sodium	**Mg** 12 24.31 Magnesium											**Al** 13 26.98 Aluminum	**Si** 14 28.09 Silicon	**P** 15 30.97 Phosphorus	**S** 16 32.07 Sulphur	**Cl** 17 35.45 Chlorine	**Ar** 18 39.95 Argon
K 19 39.10 Potassium	**Ca** 20 40.08 Calcium	**Sc** 21 44.96 Scandium	**Ti** 22 47.88 Titanium	**V** 23 50.94 Vanadium	**Cr** 24 52.00 Chromium	**Mn** 25 54.94 Manganese	**Fe** 26 55.85 Iron	**Co** 27 58.93 Cobalt	**Ni** 28 58.69 Nickel	**Cu** 29 63.55 Copper	**Zn** 30 65.39 Zinc	**Ga** 31 69.72 Gallium	**Ge** 32 72.61 Germanium	**As** 33 74.92 Arsenic	**Se** 34 78.96 Selenium	**Br** 35 79.90 Bromine	**Kr** 36 83.80 Krypton
Rb 37 85.47 Rubidium	**Sr** 38 87.62 Strontium	**Y** 39 88.91 Yttrium	**Zr** 40 91.22 Zirconium	**Nb** 41 92.91 Niobium	**Mo** 42 95.94 Molybdenum	**Tc** 43 (97.9) Technetium	**Ru** 44 101.01 Ruthenium	**Rh** 45 102.91 Rhodium	**Pd** 46 106.42 Palladium	**Ag** 47 107.87 Silver	**Cd** 48 112.41 Cadmium	**In** 49 114.82 Indium	**Sn** 50 118.71 Tin	**Sb** 51 121.76 Antimony	**Te** 52 127.60 Tellurium	**I** 53 126.90 Iodine	**Xe** 54 131.29 Xenon
Cs 55 132.91 Cesium	**Ba** 56 137.33 Barium	**La** 57 138.91 Lanthanum	**Hf** 72 178.49 Hafnium	**Ta** 73 180.95 Tantalum	**W** 74 183.85 Tungsten	**Re** 75 186.21 Rhenium	**Os** 76 190.2 Osmium	**Ir** 77 192.22 Iridium	**Pt** 78 195.08 Platinum	**Au** 79 197.97 Gold	**Hg** 80 200.59 Mercury	**Tl** 81 204.38 Thallium	**Pb** 82 207.2 Lead	**Bi** 83 208.98 Bismuth	**Po** 84 (209) Polonium	**At** 85 (210) Astatine	**Rn** 86 (222) Radon
Fr 87 223.02 Francium	**Ra** 88 226.03 Radium	**Ac** 89 227.03 Actinium	**Rf** 104 (261) Rutherfordium	**Db** 105 (262) Dubnium	**Sg** 106 (263) Seaborgium	**Bh** 107 (262) Bohrium	**Hs** 108 (265) Hassium	**Mt** 109 (266) Meitnerium	Unnamed discovery 110 Nov. 1994	Unnamed discovery 111 Nov. 1994	Unnamed discovery 112 1996		Unnamed discovery 114 1999		Unnamed discovery 116 1999		

Alkali metals — Group IA
Alkali earth metals — Group IIA
Halogens — Group VIIA
Noble gases — Group VIIIA

Lanthanides

Ce 58 140.12 Cerium	**Pr** 59 140.91 Praseodymium	**Nd** 60 144.24 Neodymium	**Pm** 61 (145) Promethium	**Sm** 62 150.36 Samarium	**Eu** 63 152.97 Europium	**Gd** 64 157.25 Gadolinium	**Tb** 65 158.93 Terbium	**Dy** 66 162.50 Dysprosium	**Ho** 67 164.93 Holmium	**Er** 68 167.26 Erbium	**Tm** 69 168.93 Thulium	**Yb** 70 173.04 Ytterbium	**Lu** 71 174.97 Lutetium

Actinides

Th 90 232.04 Thorium	**Pa** 91 231.04 Protactinium	**U** 92 238.03 Uranium	**Np** 93 237.05 Neptunium	**Pu** 94 (240) Plutonium	**Am** 95 243.06 Americium	**Cm** 96 (247) Curium	**Bk** 97 (248) Berkelium	**Cf** 98 (251) Californium	**Es** 99 252.08 Einsteinium	**Fm** 100 257.10 Fermium	**Md** 101 (257) Mendelevium	**No** 102 259.10 Nobelium	**Lr** 103 262.11 Lawrencium

Appendix D

Solutions to Odd-Numbered Problems

R-2.1 The ground wire provides an alternate current path that carries more current.

R-2.3 (a) 0 A, 0 W. (b) 2 A, rightward, 20 W. (c) 1 A, rightward, 5 W. (d) 1 A, leftward, 5 W. (e) 2 A, leftward, 20 W.

R-2.5 These precautions prevent current from flowing through the torso of your body.

R-3.1 Computers, VCRs, microwaves.

R-4.1 (a) 20 Ω. (b) 0.2 W. (c) 960 J. (d) 480 C.

R-5.1 (a) No. (b) The monitor can't work without power.

R-6.1 Her hair becomes charged and the strands repel each other.

R-6.3 The source provides a voltage, not a current.

R-6.5 You develop a charge by rubbing against the seat as you leave the car. Holding onto the outside surface as you exit allows the charge to leave gradually. Touching the outside surface only after you are outside forces the charge to leave all at once, shocking you.

R-7.1 (a) 3. (b) 3. (c) 3. (d) In the first case there is a collective effect; in the other cases there is an individual effect.

R-7.3 For an insulator a minus indicates an excess electron and a plus a deficit of an electron exactly at that spot. For a conductor they represent an excess or deficit in the average density of the conduction electrons in one mole.

R-7.5 (a) Assume that the extra charge carrier is a positive ion. The bulk of the liquid is neutral and all excess charge is due to ions distributed over its surface. (b) Assume that the extra charge carrier is an electron. The bulk is neutral and all excess charge, due to electrons, is distributed over its surface.

R-7.7 (a) Yes. (b) In pure water electric current is entirely due to H^+ and OH^- ion movement. In salt water the current is dominated by the movement

of Na^+ and Cl^- ions. In metal wire the current is due to the movement of electrons. In all cases a small average velocity is superimposed on the random motions of charge carriers.

R-8.1 Only (d) would receive full credit.

R-8.3 "Show that" problem.

R-8.5 $(-2, 1, 0)$ and a $180°$ rotation about the y-axis.

R-9.1 (a) "Show that" problem. (b) Use $\vec{a} = \hat{i}$, $\vec{b} = \hat{i}$, and $\vec{c} = \hat{j}$.

R-9.3 (a) "Show that" problem. (b) The left side equals -2; the right side equals 0; the two sides are unequal.

R-9.5 (a) "Show that" problem. (b) "Show that" problem. (c) "Show that" problem.

R-9.7 "Show that" problem.

R-9.9 (a) Counterclockwise rotation of $53.1°$. (b) $(-3.60, 5.20, -1)$. (c) $(-10.40, -2.20, 26)$. (d) $(-10.40, -2.20, 26)$. (e) They are the same. (f) 5.385, 6.403, 28.1. (g) $234.55°$. (h) Yes.

R-9.11 $(10, 5, -3)$ N-m.

R-9.13 $(4.8 \times 10^{-3}, 9.6 \times 10^{-3}, 0)$ N.

R-9.15 "Show that" problem.

R-9.17 "Show that" problem.

R-9.19 $(0, 0, -d/c)$, $(a/e, b/e, c/e)$, where $c = \sqrt{a^2 + b^2 + c^2}$.

R-9.21 (a) $\hat{n} \equiv (0.254, -0.381, 0.889)$. (b) $\frac{d\Phi_E}{dA} \equiv \vec{E} \bullet \hat{n} = -23.0$ volt/m. (c) $d\Phi_E = \frac{d\Phi_E}{dA} dA = -1.195 \times 10^{-5}$ volt-m.

1-2.1 See Figure 2.1 and the accompanying discussion of the amber effect.

1-2.3 They will repel if the positive ends are brought near each other but attract if the positive end of one is brought near the neutral end of the other.

1-3.1 The mechanical motion would be identical, but the induced charge would be opposite the previous induced charge.

1-3.3 Either end would be attracted to an electrically charged object due to electrostatic induction.

1-4.1 (a) In the 17th century, the most common way to charge an object was to rub it or to touch it to a previously charged object. Water and iron cannot be charged in this manner. (b) Put object on insulator, charge by induction and either grounding or sparking or contact.

1-4.3 Conductors: your body, a penny. Insulators: your clothing, plastic, your comb, a styrofoam cup.

1-4.5 When the person stands on the ground he is subject to the full voltage difference between the charge source and ground, which produces a large enough current to cause the shock.

1-4.7 Watson's view would predict identical shocks in the two cases. Franklin's view (the correct view) would predict a greater shock in the first case.

1-4.9 A is a conductor and B is an insulator.

1-5.1 (a) "A Penny Saved Is a Penny Earned" implies money conservation in a situation where there is a flow of resource, both in and out. (b) Decreasing their energy bill with thermal insulation has the same effect as increasing their revenue by the same amount.

1-5.3 After the first connection, A has 4.8 units and B has 3.2 units. After the second connection, A has 0.96 units and B has 0.64 units.

1-5.5 (a) The jar cannot lose charge to the ground, and losing it through the air could take hours or days. (b) As long as the top wire remains charged, it will attract charge to the bottom, and thus the Leyden Jar remains charged. (c) As long as the bottom wire remains charged it will attract charge to the top, and thus very little charge can be drawn off the top wire. (d) The bottom retains its charge, since it is insulated. Once the top wire is connected to ground, the bottom wire will attract charge to the top and the Leyden Jar would regain its strength.

1-6.1 If the tube were left in place, electrostatic induction would occur. Then the experiment would be less reproducible because the electrostatic induction would depend on the placement of the tube.

1-6.3 Rapid discharge of a nearby source causes a rapid decrease of the electrostatically induced charge in the human body.

1-6.5 The induced charge on each leaf will be of the same sign and hence the leaves will repel.

1-6.7 (a) Rubbing the balloons on clothing charges them up by friction. They are attracted to the wall by electrostatic induction and fall as they gradually lose their charge. (b) No. (c) The moisture in the air will draw off charge. The more moisture, the faster the charge will be drawn off.

1-6.9 When the negatively charged rod is brought near the grounded sphere, negative charge driven by electrostatic induction flows away from the sphere, leaving the sphere positively charged. When the ground connection is removed, the sphere remains positively charged, but also subject to electrostatic induction from the rod. When the rod is removed, the positive charge on the sphere redistributes.

1-7.1 $+2e$.

1-7.3 (a) Allowed. (b) Prohibited. (c) Allowed.

1-7.5 Electrons are transferred from the cloth to the rod. Protons generally cannot be transferred.

1-7.7 (a) The styrofoam bag discharges faster. (b) No. (c) The styrofoam still discharges faster, but it will start later than in (a).

1-8.1 People come in integer values; you are either alive or you are dead. There is no such thing as people conservation, as established by the phenomena of birth and death.

1-8.3 6.25×10^{19} electrons.

1-8.5 3.125×10^9.

1-9.1 "Show that" problem. A has units of C/m^3.

1-9.3 (a) The charge per unit length for rods 1 and 2 are q_1/l_1 and q_2/l_2, respectively. (b) $(q_1 + q_2)/(l_1 + l_2)$. (c) $(q_1 + q_2)/l_2$. (d) $(q_1 + q_2)/2l_2$.

1-9.5 (a) C has units of C/m^3 and B has units of C/m^4. (b) $(C + Br)(4\pi r^2 dr)$. (c) $(4\pi C)(a^3/3) + (4\pi B)(a^4/4)$. (d) $C + (3/4)Ba$.

1-9.7 (a) C has units of C/m^3 and B has units of C/m^5. (b) $(C + Bz^2)\pi a^2 dz$. (c) $\pi a^2(Cl + Bl^3/3)$. (d) $C + Bl^2/3$.

1-9.9 "Show that" problem.

1-10.1 When the tapes are pulled apart they develop a charge. Your finger is neutral. By electrostatic induction each will be attracted to your finger.

1-10.3 Use a versorium. It will respond at a greater separation to the tape with more charge.

1-10.5 (a) See Figure 4.24. (b) Smaller. (c) The same.

2-2.1 (a) "Show that" problem. (b) It decreases by a factor of $1/\sqrt{2}$. Yes. (c) 4.07×10^{-7} C. (d) $m_1 \rightarrow m_1 + M/2$.

2-3.1 4.77×10^{-6} N, attractive.

2-3.3 0.626 mC.

2-3.5 "Show that" problem. The maximum force will have magnitude $kQ^2/4r^2$.

2-3.7 68.6 μC.

2-3.9 (a) 3.33×10^{-10} C. (b) 4.80×10^{-10} sC.

2-4.1 $\theta = 60°$, $q_{max} = 2l \sin\theta \sqrt{\frac{mg \tan\theta}{k}}$, $q_{max} = 4.76 \times 10^{-7}$ C.

2-4.3 θ.

2-4.5 (a) $\frac{kqQ}{4R^2 \sin^2(\theta/2)}$. (b) $\frac{kqQ|\cos(\theta/2)|}{4R^2 \sin^2(\theta/2)}$.

2-5.1 (a) 0 N. (b) 28.4 m. (c) 3.59×10^{-6} C.

2-5.3 $F = 9.53N$, $\theta = -19°$.

2-5.5 "Show that" problem. $F_x = -1.102 \times 10^{-6}$ N, $F_y = -0.995 \times 10^{-6}$ N, $F = 1.484 \times 10^{-6}$ N, $\theta' = -137.9°$.

2-6.1 Rotating the rod about its perpendicular bisector does not change the configuration so it should not change F_y, but this rotation should reverse F_y. We conclude $F_y = 0$. Yes.

2-7.1 (a) $F_x = \frac{kqQ}{a(a+l)}$, $F_y = F_z = 0$. (b) $F_x \rightarrow \frac{kqQ}{a^2}$.

2-7.3 The y component of the force due to an arbitrary infinitesimal charge dQ between y and $y + dy$ is exactly cancelled by an equal charge dQ between $-y$ and $-y - dy$. Thus $F_y = 0$.

2-7.5 $F_x = \frac{kqQ}{a\sqrt{L^2+a^2}}$, $F_y = -\frac{kqQ}{L}[\frac{1}{a} - \frac{1}{\sqrt{L^2+a^2}}]$.

2-7.7 $F_x = F_z = 0$, $F_y = -\frac{2kqQ}{\pi a^2}$.

2-7.9 The force will point directly away from the center of the arc, at an angle of $\alpha/2$. The force will have magnitude $F = \frac{2kqQ}{\alpha a^2} \sin(\frac{\alpha}{2})$.

3-2.1 (a) $F_e = 2.003 \times 10^{-17}$ N. (b) $F_g = 2.935 \times 10^{-25}$ N. The force of gravity is much smaller than the electrostatic force.

3-2.3 (a) 6.12×10^{11} N/C. (b) 1.76×10^{13} m/s^2.

3-2.5 (a) $\theta = \tan^{-1}(\frac{|Q|E}{mg})$. (b) $T = mg \sec \times [\tan^{-1}(\frac{|Q|E}{mg})]$. (c) $\theta = 4.75 \times 10^{-4}$ degrees, $T = 0.412$ N.

3-2.7 No.

3-2.9 $E = 2500$ N/C, $|Q| = 0.11$ nC.

3-2.11 Scalar fields: pressure, density, temperature. Vector fields: flow velocity.

3-2.13 (a) Gravity is always attractive, so the field due to m_2 will point toward m_2, pulling m_1 toward m_2. (b) $\vec{g} = -\sum_i \frac{Gm_i}{R_i^2} \hat{R}_i$.

3-3.1 (a) 85,000 N/C $\hat{x}$. (b) $-85,000$ N/C $\hat{x}$.

3-4.1 Drag-dominated.

3-4.3 (a) The advantage is that a ball of charge creates field lines inside. (b) The disadvantage is that there are too many arrowheads if you want to sketch the field quickly. (c) The grass seed method has two problems: the grass seeds can align with the field in either direction, and the density of the grass seeds gives only a qualitative representation of magnitude.

3-5.1 (a) 175 N/C $\downarrow$. (b) 1100 N/C $\uparrow$. (c) 425 N/C $\downarrow$.

3-5.3 (a) Up $\uparrow$. (b) $\frac{kQ}{(a)^2}(1 + \frac{a}{Q})(1 + 2\cos\theta)\uparrow$.

3-5.5 (a) 105° clockwise. $E'_x = 22.6$ N/C, $E'_y = 54.7$ N/C. (b) This calculation was relatively simple. Doing the calculation from scratch would have involved 23 separate calculations and then the addition of 23 vectors, a very involved calculation.

3-5.7 (a) $\vec{E} = kQ[\frac{1}{(x+a)^2} + \frac{1}{(x-a)^2} - \frac{2}{x^2}]\hat{x}$. (b) $\vec{E} \rightarrow \frac{6a^2}{x^4}\hat{x}$.

3-5.9 "Show that" problem.

3-6.1 (a) $Q = \lambda L$. (b) $\vec{E} = \frac{k\lambda L}{x(x-L)}\rightarrow$. (c) $\vec{E} = \frac{k\lambda(2x-L)}{x(L-x)}\rightarrow$.

3-6.3 $\vec{E} = -\frac{kqQ(3-\sqrt{5})}{2a^2\sqrt{10}}\hat{x}$.

3-6.5 (a) Down. (b) $\vec{E} = -\frac{4k\lambda}{a}\hat{y}$.

3-6.7 (a) 0. (b) $\vec{p} = \frac{\pi}{2}\alpha R^2\hat{x}$. (c) $\vec{E} = -\frac{k\pi\alpha}{2R}\hat{x}$.

3-6.9 $\vec{E} = -\frac{1}{4}k\sigma(1 - \cos(2\alpha))\hat{z}$.

3-6.11 (a) $\vec{E} = \frac{4k\lambda y}{y^2+a^2}\hat{y}$. (b) $\vec{E} = -\frac{4k\lambda a}{y^2+a^2}\hat{x}$.

3-6.13 (a) Place q at $(14.7, 0, 0)$ m. (b) Place λ in the xy plane, parallel to the y-axis, and intersecting the x-axis at $x = -288$ m. (c) Impossible.

3-6.15 (a) 20 N/C downward. (b) The upper sheet has charge density 0.442 nC/m^2, and the lower sheet has charge density -0.0884 nC/m^2.

3-7.1 Along.

3-7.3 $\vec{E}$ is zero 2/3 of the way from 2λ to λ.

3-7.5 Within an insulator we can arrange the charge however we please, but within a conductor the like charges would repel each other toward the surface disrupting the uniform volume charge distribution.

3-8.1 (a) 3.6×10^{-25} J. (b) $90°$, 1.8×10^{-25} N-m.

3-8.3 (a) 2.5×10^{-25} N-m. (b) 1.17×10^{-25} N-m.
(c) -2.207×10^{-25} J.

3-8.5 (a) $\vec{F} = q(\frac{2kp}{x^3})\hat{\rightarrow}$. (b) $\vec{F} = -q(\frac{2kp}{x^3})\hat{\rightarrow}$. (c)
$p = 7.11 \times 10^{-25}$ C-m.

3-9.1 (a) $\vec{F} = pA\hat{x}$. (b) 2.37×10^{-9} N $\hat{x}$. (c)
2.37×10^{-9} N $\hat{x}$.

3-10.1 (a) -2.59×10^6 m/s. (b) 4.64×10^{-8} s.

3-10.3 (a) -728 N/C. (b) 6.44×10^{-9} C/m².
(c) 4.3×10^6 m/s, $21.8°$.

3-10.5 2.01×10^{-5} m.

3-10.7 (a) The velocity v does not depend on the
radius r. (b) $T = 2\pi r \sqrt{\frac{m_e}{2ek\lambda}}$.

4-2.1 (a) The flux will have the opposite sign.
(b) -34 N-m²/C. (c) $\Phi_E = -\frac{1}{\varepsilon_o} Q_{enc} = -4\pi k Q_{enc}$.

4-2.3 -7.68×10^{-6} N-m²/C.

4-2.5 $E\pi R^2$.

4-2.7 $2\pi kQ[1 - \frac{d}{\sqrt{R^2+d^2}}]$.

4-3.1 (a) 0. (b) 679 N-m²/C. (c) 0.

4-3.3 (a) $4\pi kQ$. (b) $(2/3)\pi kQ$.

4-3.5 (a) $(0.8, 0.48, -0.36)$. (b) 8.12 N/C.
(c) 5.64 N/C. (d) $\pm46°$. (e) 1.128×10^{-3} N-m²/C.
(f) 9.97×10^{-15} C.

4-4.1 1.77 nC/m².

4-4.3 -3.76×10^{-9} C.

4-5.1 (a) -1.94×10^{-12} C. (b) 2.59×10^5 N/C.

4-5.3 (a) 1.30×10^{-10} C/m². (b) 2.66×10^9 N/C.

4-5.5 (a) $\rho = \frac{\lambda}{\pi(b^2-a^2)}$. (b) 0, $\frac{2k\lambda(r^2-a^2)}{(b^2-a^2)r}$, $2k\lambda/r$.

4-5.7 (a) $\lambda_1 = -0.72$ nC/m, $\lambda_2 = 0.85$ nC/m.
(b) 19.47 N/C.

4-5.9 (a) $\vec{g} = \frac{\vec{F}_{grav}}{M}$, $\oint \vec{g} \bullet \hat{n} dA = -4\pi GM_{enc}$.
(b) $\vec{g} = -\frac{4\pi rG\rho}{3}\hat{r}$. (c) $T = 2\pi\sqrt{\frac{3}{4\pi\rho G}}$. (d) 5070 s.

4-6.1 No.

4-6.3 (a) 25 V. (b) The dipole moment on the
conductor will point away from the point charge.
The point charge will be attracted to the conductor.

4-6.5 (a) The inner surface has $-Q$ and the outer
surface has $+Q$. (b) The inner surface has $-Q$ and
the outer surface has zero charge.

4-6.7 Connecting the slats makes them behave

almost like a solid conductor, which can screen out
an external field.

4-7.1 0.212 nC/m².

4-7.3 "Explain in your own words" problem.

4-7.5 (a) Starting to the left and proceeding
to the right, the electric fields are $10\pi k\sigma\rightarrow$,
$18\pi k\sigma\rightarrow$, $2\pi k\sigma\rightarrow$, and $10\pi k\sigma\leftarrow$. (b) Starting with
the left side of #1, going rightward, the charge
densities are $-(5/2)\sigma$ and $(9/2)\sigma$, $-(9/2)\sigma$ and
$(1/2)\sigma$, $-(1/2)\sigma$ and $-(5/2)\sigma$.

4-7.7 (a) -7.96 nC/m², (b) -4.0×10^{-11} C,
(c) $\vec{E} = 0$. (d) $E_r = -400$ N/C (radially inward).

4-7.9 (a) 0, kQ/r^2, $-kQ/r^2$ (radial component
of field). (b) $Q_a^{inner} = 0$, $Q_a^{outer} = Q$. (c) $Q_b^{inner} = -Q$, $Q_b^{outer} = -Q$.

4-7.11 (a) 0, $4k\lambda/r$, $-2k\lambda/r$ (radial component of
field). (b) $\lambda_a^{inner} = 0$, $\lambda_a^{outer} = 2\lambda$. (c) $\lambda_b^{inner} = -2\lambda$,
$\lambda_b^{outer} = -\lambda$.

4-8.1 (a) All of the charge resides on the cup's
outer surface. (b) -3μC. (c) 2μC. (d) -0.4μC.

4-9.1 $2\pi(1 - \frac{s}{\sqrt{s^2+d^2/4}})$.

4-10.1 3×10^3 C/m².

4-11.1 $E_x = 0$, $E_y = -\frac{4k\lambda}{x^2+b^2}$, $\sigma_z = -\frac{\lambda b}{4(x^2+b^2)}$.

4-11.3 (a) $\frac{kQ^2}{8\pi r^4}$. (b) $(\frac{1}{b} - \frac{1}{a})\frac{kQ^2}{2}$.

5-2.1 81.2 N.

5-3.1 (a) 2×10^{-8} J. (b) -2×10^{-8} J. In the first
case we raise the electrical potential energy; in
the second case we lower the electrical potential
energy.

5-3.3 -2 V/cm, -2.2 V/cm, -2.4 V/cm.

5-3.5 2.7484 V.

5-4.1 5.7×10^{-14} m.

5-4.3 (a) 35.33 V, the electron is heading toward
the higher potential. (b) 883.25 N/C $\hat{y}$.

5-4.5 (a) 4.8×10^{-5} J. (b) 20,000 V, the starting point is at a lower potential than the endpoint.
(c) 2500 V/cm.

5-4.7 (a) The 6 V plate, 1.78×10^6 m/s. (b)
1.44×10^{-18} J. (c) -1.44×10^{-18} J. (d) 9 V.

5-4.9 0.058%.

5-4.11 (a) 8.88×10^5 V. (b) 0.63×10^{-7} C.
(c) 3.94×10^{-13} C/m².

5-5.1 The field is largest near the two bottom

corners and smallest on the line directly between the two side plates.

5-5.3 The field is largest at the upper right and at the lower left corners and smallest at the center.

5-5.5 (a) "Show that" problem. (b) $-Q$.

5-5.7 (a) Spheres centered at the charge. (b) Yes. (c) Yes. (d) No.

5-5.9 The field must be zero at the crossing point.

5-6.1 -2.0 V.

5-6.3 $(2kq^2/a)(2 + 1/\sqrt{2})$.

5-6.5 (a) 48 kV. (b) 192 kV. (c) 0 J.

5-6.7 (a) Halfway between them. (b) Yes. Take $V_\infty = -\frac{4kq}{a}$.

5-6.9 5.56×10^{-10} C.

5-6.11 -442 nC/m^2.

5-6.13 -1.753 V.

5-7.1 (a) 0. (b) Yes. (c) $-\frac{a}{2}x^2 - bx$.

5-7.3 (a) $-b^2$. (b) $+b^2$. (c) $2b^2$. (d) No.

5-8.1 (a) The potential at the center is the same as on the surface. (b) The center.

5-8.3 2.02 nC/m.

5-8.5 $-(1/3)Ar^3$, the equipotential surfaces are concentric cylinders centered on the z-axis.

5-8.7 $2\pi k\sigma(\sqrt{z^2 + a^2} - z)$.

5-9.1 (a) $V(z) = \frac{kQ}{\sqrt{R^2 + z^2}}$. (b) $E_z = -\frac{kQz}{(R^2 + z^2)^{3/2}}$. (c) No. (d) E_x and E_y will in general change under rearrangement.

5-9.3 Sketch not provided.

5-9.5 (a) 0. (b) $V(x) = ka[x\ln(\frac{x+l/2}{x-l/2}) - l]$. (c) "Show that" problem. (d) "Show that" problem.

5-9.7 (a) $V = -kQ/b$. (b) $V(a) = -0.9kQ/b < 0$. (c) For $q > 0.1 Q$.

5-10.1 (a) 24.85 V, 29.25 V. (b) -22 V/m $\hat{x}$. (c) -22 V/m $\hat{x}$.

5-10.3 (a) $V(x) = \frac{kQ}{l}[\ln|x+l| - \ln|x|]$. (b) $E_x = \frac{kQ}{l}[\frac{1}{x} - \frac{1}{x+l}]$. (c) $E_x = \frac{kQ}{l}[\frac{1}{x} - \frac{1}{x+l}]$.

5-10.5 "Show that" problem.

5-10.7 (a) 2.95 V, 8.05 V. (b) -25.5 V/m $\hat{r}$. (c) $-25r^3 \hat{r}$, -25 V/m $\hat{r}$.

5-10.9 $(-2y + 8x)\hat{x} + (-2x + 10y)\hat{y}$.

5-10.11 (a) $E_x = -\frac{4}{3}\frac{V_a}{d}(\frac{x}{d})^{1/3}$. (b) 0, -0.84×10^4 V/m, -1.06×10^4 V/m, -1.2×10^4 V/m, -1.33×10^4 V/m. (c) $-\frac{1}{9\pi k}(\frac{1}{xd^2})^{2/3}$.

Note: The total charge per unit area between the plates is -1.18×10^{-7} C/m^2.

5-11.1 (a) 6300 V, -1125 V. (b) 3.15×10^5 N/C outward, 2.81×10^4 N/C inward. (c) 6×10^{-9} C, 3×10^{-9} C, 1350 V, 1350 V. (d) 6.75×10^4 N/C outward, 3.375×10^4 N/C outward.

5-11.3 No.

5-11.5 $Q_{10R} = 10Q_R$, $\sigma_{10R} = \frac{1}{10}\sigma_R$, $V_R = V_{10R}$, $E_{10R} = \frac{1}{10}E_R$.

5-11.7 (a) 9.09 V. (b) 100 V. (c) $100 \times [1 - (\frac{10}{11})^n]$ V.

5-12.1 1.114×10^6 N/C $\hat{x}$.

5-12.3 1.323×10^{10} N/C.

6-2.1 (a) 23.6 nF. (b) 2.12×10^2 m.

6-2.3 Excess charge affects the structure of the solid sphere less than the structure of the shell.

6-2.5 4.5×10^{-8} m.

6-3.1 (a) 0.144 nF. (b) 1.25 nC.

6-3.3 (a) 5.79×10^{-11} F. (b) 1.736×10^{-12} C.

6-3.5 (a) By how close the plates can be kept without touching each other. (b) By how large the plates can be without making the capacitor unusable.

6-3.7 (a) 1250 V. (b) 0.0905 m^2. (c) 6.25×10^5 N/C. (d) 5.52×10^{-6} C/m^2.

6-3.9 This problem involves dielectrics, which are not discussed until Section 6.5. (a) 4.734 cm. (b) 14.25 nF. (c) 1.083×10^{-4} C.

6-3.11 (a) 19.23 pF. (b) 4.62 nC. (c) 6.92×10^3 N/C.

6-4.1 (a) Make three parallel arms, each with three 2μF capacitors in series. (b) Put two units from part (a) in parallel.

6-4.3 $A/4\pi k(d_1 + d_2)$. $C = Q/\Delta V$ has ΔV increase by the factor $(d_1 + d_2)/d$.

6-4.5 (a) 3.43 μF, 14 μF. (b) 41.16 μC, 5.145 V, 6.86 V. (c) 12 V, 96 μC, 72 μC.

6-4.7 154 nF.

6-4.9 4 nF.

6-4.11 5 μC.

6-4.13 9 μC on each capacitor, $\Delta V_1' = 3$V, $\Delta V_2' = 1.5$ V. No. No.

6-5.1 $d = 2.67 \times 10^{-4}$ m, A = 242 m^2.

6-5.3 3.91.

6-5.5 Above about 1 V, electrolysis occurs at the

plates and an ion current would flow. The capacitor then would not be able to hold charge.

6-5.7 (a) 33.25 cm^2. (b) 120 pF, 3.5 kV.

6-5.9 "Show that" problem.

6-5.11 "Show that" problem.

6-5.13 (a) 0.28×10^{-17} F $= 2.8$ aF. (b) 0.058 V.

6-5.15 $C_{total} = \frac{1+\kappa}{2} \frac{ab}{k(b-a)}$.

6-6.1 (a) 6.67×10^{-8} F, 3.32×10^{-6} m. (b) 4.8×10^{-4} J. (c) 3.61×10^7 V/m. (d) 5761 J/m^3.

6-6.3 4×10^{-6} F.

6-6.5 (a) $Q_A = Q_B = 160$ nC, $\Delta V_A = 4$ V, $\Delta V_B = 8$ V, $U_A = 0.32$ μJ, $U_B = 0.64$ μJ. (b) $\Delta V_A = \Delta V_B = 5.333$ V, $Q_A = 213.3$ nC, $Q_B = 106.7$ nC, $U_A = 0.568$ μJ, $U_B = 0.284$ μJ. (c) 0.108 μJ. (d) $\Delta V_A = \Delta V_B = 1.777$ V, $Q_A = 71.1$ nC, $Q_B = 35.6$ nC, $U_A = 0.063$ μJ, $U_B = 0.032$ μJ. (e) 0.757 μJ.

6-6.7 (a) $U = \frac{kQ^2}{2R}$. (b) $dU = -(kQ^2/2R^2) dR$, $P_{el} = \frac{E^2}{8\pi k}$.

6-6.9 (a) $Q_1 = Q_2 = Q_3 = Q_4 = 360$ μC, $\Delta V_1 = \Delta V_4 = 60$ V, $\Delta V_2 = \Delta V_3 = 30$ V. (b) 32.4 mJ. (c) $Q_1 = Q_4 = 240$ μC, $Q_2 = Q_3 = 480$ μC, $\Delta V_1 = \Delta V_2 = \Delta V_3 = \Delta V_4 = 40$ V. (d) 28.8 mJ. (e) -240 μC.

6-6.11 1.128×10^{12} J.

6-7.1 (a) $|\vec{E}_{diel}|/|\vec{E}_0| = 1/5$. (b) $V_{diel}/V_0 = 1/5$, $V_{diel} = 1.2$ V. (c) $Q_{diel}/Q_0 = 1$. (d) $C_{diel}/C_0 = 5$. (e) $U_{diel}/U_0 = 1/5$.

6-7.3 In the first case, U gives all of the energy. In the second case, the capacitor energy must be included.

6-7.5 "Show that" problem.

6-7.7 Take $C = 12$ μF initially. (a) $U_{cap} = (1/2)CV^2$, $\Delta U_{batt} = -CV^2$, $U_{heat} = (1/2)CV^2$; $U'_{cap} = CV^2$, $W'_{hand} = (1/2)CV^2$; $U''_{cap} = (1/2)CV^2$, $\Delta U''_{batt} = CV^2$, $U''_{heat} = (1/2)CV^2$. (b) $U'_{cap} = (1/4)CV^2$, $\Delta U'_{batt} = (1/2)CV^2$, $W'_{hand} = (1/4)CV^2$.

6-8.1 (a) 0.2233 μC. (b) 191.4 V. (c) $Q_a = 0.0106$ μC, $Q_b = 0.2127$ μC.

6-8.3 "Show that" problem. $V_1 - V_2 = Q_1(p_{11} - p_{21}) + Q_2(p_{12} - p_{22}) + Q_3(p_{13} - p_{23})$. The term $Q_3(p_{13} - p_{23})$ is the effect of Q_3.

6-8.5 (a) Charge will distribute over the surface of a conductor, so that the material is irrelevant and only the shape of the surface is important. Polarization of an insulator depends on the dielectric constant and thus the material. (b) The sphere's

induced dipole moment, due to polarization, is relatively small for $r \gg a$.

6-9.1 (a) Along the length of the molecule. (b) α would have to depend on the direction of $\vec{E}$. In most cases $\vec{E}$ and $\vec{p}$ would not even point in the same direction.

6-9.3 "Show that" problem.

6-10.1 "Show that" problem.

6-10.3 $\frac{k\lambda}{a}^2$.

7-1.1 (a) No current. (b) No current. (c) 5 mA.

7-2.1 10,800 C, 3.75×10^{19} electrons/s.

7-2.3 (a) 35,840 C. (b) 2.39×10^3 s, or about 66.4 hr.

7-2.5 5.09×10^{-11} A.

7-2.7 (a) 133.3 A. (b) 4.76×10^6 A/m^2. (c) $J_x = 4.47 \times 10^6$ A/m^2, $J_y = -1.629 \times 10^6$ A/m^2, $J_z = 0$. (d) 6.12 A.

7-2.9 No. Opposite directions.

7-2.11 995 A/m^2.

7-3.1 (a) 4 Ω, 5 Ω. (b) Non-ohmic.

7-3.3 (a) 2 A. (b) -2 A.

7-3.5 (a) 3.1296×10^{-5} Ω/m. (b) 0.125 Ω.

7-3.7 8.45 Ω-m.

7-3.9 31.2 mm.

7-3.11 (a) 7 mA. (b) 8.4 nV. (c) 20 nΩ.

7-3.13 (a) 240 Ω, 2.4 Ω. (b) Resistance of the bulb for home use.

7-3.15 (a) 5 kV. (b) 250 Ω.

7-3.17 (a) 3.6×10^4 J. (b) 0.833 A. (c) 6000 C.

7-3.19 (a) 28.8 Ω. (b) 28.8×10^6 /m.

7-3.21 (a) 23.9 horsepower. (b) 1000%. (c) No. It cannot exceed 100%.

7-4.1 (a) 0.28 V/m rightward. (b) 0.532 V. The current flows from the higher voltage left end.

7-4.3 (a) 2.79×10^{-8} Ω-m. (b) Aluminum.

7-4.5 (a) For $d = 0.04$ cm, $R_{Cu} = 0.01547$ Ω, $R_{steel} = 8.27 \times 10^{-6}$ Ω. (b) $\Delta V_{Cu} = 0.0310$ V, $\Delta V_{steel} = 1.654 \times 10^{-6}$ V.

7-4.7 Charge on the surface of the circuit, including the wire, makes an electric field that drives the current through the wire.

7-4.9 In previous chapters we were considering conductors in equilibrium. In this chapter, there is current flowing through the conductors, so they are not in equilibrium. Thus the electric field is not necessarily zero within the conductors.

7-5.1 (a) 13 Ω. (b) $\Delta V_2 = 27$ V, $\Delta V_1 = 12$ V. (c) 39 V. (d) 3 A. (e) 3 A.

7-5.3 (a) 6 Ω. (b) 3.5 A. (c) 35 V. (d) 3.5 A. (e) 10 Ω.

7-5.5 (a) 1.5 Ω. (b) Both are 12 V. (c) 12 V. (d) 11 A. (e) 1.09 Ω.

7-5.7 (a) $R_1 = 4$ Ω, $R_2 = 2.67$ Ω, $R_3 = 1.2$ Ω. (b) $I_3 = 10$ A, $I_2 = 6$ A. (c) 16 V. (d) 28 V. (e) 2.8 Ω.

7-5.9 One has a parallel combination of two resistors in series with another parallel combination of two resistors. The other has a series combination of two resistors in parallel with another series combination of two resistors.

7-5.11 96 Ω, 144 Ω, 100 W.

7-5.13 (a) $I_A = 3$ A, $P_A = 0.9$ W, $I_B = 0$ A, $P_B = 0$ W. (b) $I_A = I_B = 3$ A, $P_A = P_B = 0.9$ W. (c) $I_A = I_B = 1.5$ A, $P_A = P_B = 0.225$ W. (d) $I_B = 2$ A, $P_B = 0.4$ W, $I_A = I_C = 1$ A, $P_A = P_C = 0.1$ W.

7-5.15 (a) Resistors in parallel add as inverses, and capacitors in parallel add directly. (b) Resistors in series add directly, and capacitors in series add as inverses. (c) The formulas for total resistance and capacitance take the same form if $R \leftrightarrow 1/C$. $R = \Delta V/I$ and $1/C = \Delta V/Q$.

7-5.17 If a person stands on only one foot, then most current would pass through the foot touching the ground. If the person were to stand on both feet, there would be a path for current into the torso of the body where current is most dangerous.

7-6.1 (a) 1 mV. (b) 20,000 Ω/V.

7-6.3 (a) 1.98 kΩ. (b) 0.5526 Ω. (c) 84.0 mV.

7-6.5 (a) 7.02 V. (b) 7.4%.

7-7.1 The ion density is high near the plates, but low farther from the plates.

7-7.3 Starting a car four times a day consumes 0.667 % of the chemical charge every day; about the same amount as the non-current-producing sulfation reaction does.

7-7.5 1662 s = 27.7 min.

7-8.1 1.36 V, 0.3 Ω.

7-8.3 (a) The filling, aluminum foil and saliva in the mouth. (b) The chemical energy of the voltaic cell (the foil and filling are the electrodes and the saliva is the electrolyte).

7-9.1 (a) 43,2000 C. (b) 60 hr. (c) 24 hr.

7-9.3 (a) 2880 C = 0.8 A-hr. (b) 0.1 A. (c) 8 hr.

7-10.1 (a) Voltage gains of 0.6 V and 1.2 V across the electrodes, voltage losses $ir = 0.3$ V and $iR = 1.5$ V across the resistances. (b) 0.6 hr. (c) 7776 J.

7-10.3 (a) 2 A. (b) Voltage loss of 0.3 V and voltage gain of 1.3 V across the electrodes, voltage losses $ir = 0.6$ V and $iR = 0.4$ V across the resistances.

7-11.1 (1) $\vec{F}_- = 1.59 \times 10^{-17} N\,\hat{y}$, $\vec{F}_+ = -1.61 \times 10^{-17}$ N $\hat{y}$. (2) $\vec{v}_- = 3.18 \times 10^{-6}$ m/s $\hat{y}$, $\vec{v}_+ = -3.22 \times 10^{-6}$ m/s $\hat{y}$.

7-11.3 (a) "Show that" problem. (b) 11.2×10^6 s. (c) 9.54×10^{-7} m.

7-12.1 1.846×10^6 s.

7-12.3 $n = 3.84 \times 10^{13}$ /m³, $v_d = 9000$ m/s.

7-12.5 $dI = \omega \Sigma r dr$, $I = \omega \Sigma a^2/2$.

7-12.7 2.415 μC/m².

7-12.9 "Show that" problem.

7-13.1 Insulators, semiconductors, conductors.

7-13.3 $J = nev$. If the critical velocities v_c are not too different for semiconductors and metals, then the densities n mostly determine the J_c's.

8-3.1 (a) $\mathcal{E} = 24$ V. (b) $R = 0.024$ Ω. (c) 105.0 A-hr.

8-3.3 (a) 30 days. (b) 23.1 days.

8-3.5 (a) $\mathcal{E} = 10.72$ V. (b) $r = 0.0245$ Ω.

8-3.7 (a) $I = 5.2$ A charging the battery. (b) 1.352 W heating. (c) 10.72 W charging. (d) 88.98% efficiency.

8-3.9 (a) $R = 0.75$ Ω. (b) $I = 3$ A.

8-4.1 (a) 40 hr. (b) 12 cents.

8-5.1 (a) $\mathcal{E} = 0.6$ V, $r = 600$ Ω. (b) 4.02%. (c) 5.14%.

8-5.3 $I = 2.25$ A.

8-5.5 (a) $R = 3.62$ Ω. (b) $\Delta V = 261$ V. (c) $M = 1.190 \times 10^4$ kg. (d) 78.6%.

8-5.7 (a) $R = 2.4$ Ω. (b) 99.59%. (c) 96%.

8-6.1 (a) $I_1 = I + I_2$. (b) $I_1 = -\frac{\Delta V}{R_1}$, $I_2 = \frac{\Delta V}{R_2}$, $I = \frac{\Delta V - \mathcal{E}}{r}$.

8-6.3 $R = 788$ Ω.

8-6.5 (a) $0.05 \le r/R \le 0.55$. (b) $0.033 \le r/R \le 0.367$.

8-6.7 (a) $R = 15$ Ω, $\mathcal{E} = 120$ V.

8-7.1 (a) $I = I_1 = -I_2 = 22.2$ A. (b) $\Delta V_1 = -\Delta V_2 = 11.78$ V. (c) $P_1 = 266.4$ W discharge, $P_2 = -222.0$ W charge.

8-7.3 (a) Reverse the direction of positive I_2 relative to Figure 8.13(b). Then $I_1 = I + I_2$, $I_1 = \frac{\mathcal{E}_1 - \Delta V}{r_1} = 600 - 100\Delta V$, $I_2 = \frac{\mathcal{E}_2 + \Delta V}{r_2} = 500 - 50\Delta V$, $I = \frac{\Delta V}{R}$. (b) $\Delta V = 2.946$ V. (c) $I = 294.6$ A, $I_1 = 305.4$ A, $I_2 = 10.8$ A. (d) $P_R = 867.9$ W, $P_{r_1} = 932.7$ W, $P_{r_2} = 140.0$ W, $P_1 = 1832.4$ W discharge, $P_2 = 108.0$ W discharge.

8-7.5 (a) B_1 off, B_2 off. (b) B_1 dim, B_2 dim. (c) B_1 bright, B_2 off. (d) B_1 bright, B_2 off.

8-7.7 $R = (-1 + \sqrt{3})30$ Ω $= 21.96$ Ω.

8-7.9 $0.5R$.

8-7.11 $R = 7.2$ Ω.

8-7.13 "Show that" problem.

8-7.15 "Show that" problem.

8-7.17 (a) All positive currents and ΔV are as in Figure 8.39, so that $I_1 = \frac{\mathcal{E} - \Delta V}{r_1}$, etc. Also, $I = \Delta V / R$ and $I = I_1 + I_2 + I_3$. (b) $\Delta V = 4.963$ Ω. (c) $I = 496.3$ A, $I_1 = 103.7$ A, $I_2 = 251.8$ A, $I_3 = 140.7$ A. (d) Current is conserved.

8-8.1 (a) $Q_2 = 0, I_1 = 0, I = 6$ A. (b) $I_2 = 0, I = I_1 = 1.5$ A.

8-8.3 (a) $Q = 0$, $I_1 = 4/3$ A, $I_2 = 2/3$ A, $I_3 = 2/3$ A, $I_4 = 4/3$ A. $I = I_1 - I_2 = 2/3$ A goes to the capacitor. (b) $I_1 = I_3 = 0.8$ A, $I_2 = I_4 = 1.0$ A, $I = 0$, $Q = 14.4$ μC.

8-8.5 (a) $I_R = 0$. (b) $I_R = I_0$.

8-9.1 (a) $\tau_{RC} = 12$ s. (b) $I_0 = 0.84$ μA, $I_\infty = 0$. (c) $Q_0 = 0$, $Q_\infty = 10.01$ μC. (d) $U_{heat} = 4.536$ μJ. (e) $U_{heat} = 6.048$ μJ.

8-9.3 (a) "Show that" problem. (b) 96 μJ.

8-9.5 (a) $R = 5.1 \times 10^{15}$ Ω. (b) $\tau_{RC} = 2.04 \times 10^7$ s. (c) $C = 1.96 \times 10^{-13}$ F.

8-10.1 More turns through small angles require less surface charge but are more expensive to make. Two 45° might be adequate.

8-10.3 (a) $r/R \to \infty$. (b) Let $R_{eq}^{-1} = r^{-1} + R^{-1}$. Then $Q_{C_p} = (\mathcal{E}/R)(r + R)C_p[1 - e^{-t/R_{eq}C}]$. (c) $I_R = Q_{C_p}/RC_p$.

8-10.5 (a) $2\Sigma_s$. (b) Charge will actually flow from one part of the surface through the bulk to another part of the surface. (c) A combination of bulk and surface current would provide the least resistance, but in practice the resistance to surface current is very large because of the associated small cross-sectional area.

8-10.7 (a) If $\sigma_1 < \sigma_2$, then the electric field is larger in material 1 than in material 2, so a positive surface charge will on the material 1 side increase the field and will on the material 2 side decrease the field. (b) $\Sigma = J(\sigma_1 - \sigma_2)/4\pi k \sigma_1 \sigma_2$.

8-11.1 Many choices are possible. For $R_1 = R_3 = 0$, (9.65′) gives $I_1 = I_3 = 0$, whereas (9.65) gives nonzero I_1 and I_3.

8-12.1 $n = 1.241 \times 10^{24}$ /m^3.

9-2.1 $q_{m1} = 26.8$ A-m, $q_{m2} = 13.4$ A-m.

9-2.3 $\mu = 1.64$ A-m^2.

9-2.5 $B = 0.0259$ T.

9-2.7 $\vec{F} = q_m \vec{B}$, q_m due to one pole of a long magnet and $\vec{B} = k_m[-\vec{\mu} + 3(\vec{\mu} \cdot \hat{R})\hat{R}]/R^3$ due to a short magnet. The force on the distant pole of the long magnet is neglected.

9-2.9 (a) The magnet is strongly attracted to the soft iron rod when the soft iron is brought up to the magnet's poles, but only weakly attracted when the rod is brought up to the magnet's center. (b) The soft iron is strongly attracted to the magnet when the magnet is brought to any part of the soft magnet.

9-2.11 "Show that" problem.

9-3.1 $q_m = 5.12$ A-m, $\sigma_m = 3.2 \times 10^5$ A/m, $\mu = 0.256$ A-m^2.

9-3.3 $M = 3.99 \times 10^5$ A/m.

9-3.5 (a) $r \ll a$, no r-dependence. (b) $a \ll r \ll l$, $|\vec{B}| \sim r^{-2}$. (c) $l \ll r$, $|\vec{B}| \sim r^{-4}$.

9-4.1 Discontinuity in $\vec{B}$ is $\mu_0(\alpha - 1)\vec{M}$.

9-5.1 Put a rod of soft magnet in a line between a pole of a permanent magnet and the region where the field is to be intensified.

9-5.3 When field lines are expelled (concentrated), the object that is their source is repelled (attracted).

9-5.5 (a) $H = 1481$ A/m, $\chi = 1.688$. (b) $M_{emu} = 2.50$ mmu/cm^3, $B_{emu} = 50$ G, $H_{emu} = 18.6$ Oe.

9-6.1 M_r and M_s are too far in value; the magnet would not retain its magnetization.

9-7.1 (a) $|\vec{H}| = 485$ A/m, $|\vec{B}| = 0.98$ T. (b) Small, except perhaps near the poles.

9-7.3 By keeping in the field lines, the keeper

magnet lets the magnetization of one pole of the magnet magnetize the other end, and vice versa.

9-8.1 In Figure 9.19(a) there are no poles because $\vec{M}$ is normal to $\hat{n}$; hence $\vec{H} \approx \vec{0}$. In Figure 9.19(b) the poles produce a demagnetization field $\vec{H} \approx -\vec{M}$ that makes $\vec{B} = \mu_0(\vec{H} + \vec{M}) \approx \vec{0}$.

9-9.1 $|\vec{B}| = 6.63 \times 10^{-6}$ T.

9-9.3 72.9° dip angle.

9-9.5 "Show that" problem.

9-10.1 If the Fe and Nd interaction were ferromagnetic, the net magnetic moment would be larger than when they are antiferromagnetic: $7\mu_{Fe} + \mu_{Nd}$ rather than $7\mu_{Fe} - \mu_{Nd}$.

10-1.1 (a) $\vec{B} = -0.056\hat{x}$ T. (b) $\vec{B} = 0.056\hat{x}$ T.

10-1.3 (a) $B_{1x} = -0.037$ T. $B_{1y} = 0.015$ T. (b) $B_{2x} = -0.074$ T. $B_{2y} = -0.030$ T. (c) $B_x = 0.111$ T. $B_y = -0.015$ T.

10-2.1 (a) South pole up. (b) Rightward force. (c) Leftward force on magnet. (d) No torque on magnet.

10-2.3 (a) Left magnet has moment into page, right has moment out of page. (b) Attracted to loop on left.

10-2.5 The wire moves downward.

10-2.7 (a) In units of A-m², $\vec{\mu} = -0.1731\hat{i} - 0.0261\hat{j} + 0.3385\hat{k}$.

10-2.9 (a) Attractive. (b) Repulsive. (c) Wires carrying parallel currents will attract.

10-3.1 (a) $\vec{\mu} = -4.52 \times 10^{-4}\hat{z}$ A-m². (b) $\vec{\tau} = -1.808 \times 10^{-6}\hat{y}$ N-m. (c) $\vec{\tau} = -1.664 \times 10^{-6}\hat{y}$ N-m.

10-3.3 (a) $\mu = 1.024 \times 10^{-4}$ A-m². $q_m = 8.53 \times 10^{-4}$ A-m. (b) $|\vec{B}| = 94.8$ nT. (c) 0.521 nT. (d) $|\vec{B}| = 10.24$ nT.

10-3.5 (a) $M = 9.27 \times 10^6$ A/m. (b) For a good magnet, $M \approx 10^5 - 10^6$ A/m.

10-4.1 (a) $|\vec{F}| = 81.6$ μN. (b) Reversing the current or field reverses the force, pumping the blood the opposite way. (c) Reversing both current and field does not change the direction of the force.

10-4.3 Compress.

10-4.5 $\vec{F} = -2a|\vec{B}|\hat{y}$.

10-4.7 (a) $\vec{F} = 2\pi Na^2 AI\hat{z}$. (b) $\vec{F} = 0.253\hat{z}$ N.

10-4.9 (a) "Show that" problem. (b) $\vec{F} = 0.049\hat{y}$.

10-4.11 (a) $\vec{F}_b = 0.009\hat{y}$ N. $\vec{F}_t = -0.018\hat{y}$ N. (b) $\vec{F}_l = 0.0225\hat{x}$ N. $\vec{F}_r = -0.0225\hat{x}$ N. (c) $\vec{F}_{net} = -0.009\hat{y}$ N, compress.

10-4.13 (a) $a = 10^7$ m/s². (b) $|\vec{F}| = 1.4 \times 10^5$ N. (c) $I = 2.8 \times 10^6$ A.

10-4.15 $\vec{F} = [2\pi k_m q_m I a^2/(r^2 + a^2)^{3/2}]\hat{y}$.

10-5.1 (a) Force is out of the page. (b) $|\vec{F}| = 2.206 \times 10^{-17}$ N. (c) $a = 247g$, where $g = 9.8$ m/s².

10-5.3 $|\vec{B}| = 9.42 \times 10^{-17}$ T.

10-5.5 $\vec{F} = (2.611 \times 10^{-20}\hat{x} - 7.658 \times 10^{-20}\hat{y} - 10.27 \times 10^{-20}\hat{z})$ N, $|\vec{F}| = 1.307 \times 10^{-19}$ N.

10-5.7 $\vec{B} = (0.0286\hat{j} - 0.0755\hat{k})$ T.

10-5.9 "Show that" problem.

10-6.1 (a) Electron moves on a semicircle that bends left, and comes back out of the field region. (b) Proton moves on a semicircle that bends right, and comes back out of the field region.

10-6.3 (a) It deflects along $-\hat{y}$. (b) $v = 2.11 \times 10^8$ m/s. Using $m = m_0(1 - v^2/c^2)^{-\frac{1}{2}}$ gives $v = 1.726 \times 10^8$ m/s.

10-6.5 The period and radius are independent. The student is wrong.

10-6.7 (a) $T = 5.45 \times 10^{-7}$ s. (b) $|\vec{B}| = 0.241$ T. (c) 41.7 V.

10-6.9 Deuteron, 2.55 cm; triton, 3.12 cm; ^{3}He nucleus, 1.56 cm; ^{4}He nucleus, 1.80 cm.

10-6.11 (a) 26.1 T. (b) 30.1 T.

10-6.13 (a) $\vec{F}$ is perpendicular to $\vec{B}$. (b) $v_\perp^2 = v^2 - v_\parallel^2 = \sqrt{2mE} - v_\parallel^2$, and both E and $v_\parallel$ do not change. (c) "Show that" problem.

10-6.15 (a) $\theta = 124.6°$. (b) $R \doteq 8.35 \times 10^{-4}$ cm. (c) $p = 3.57 \times 10^{-3}$ m.

10-6.17 (a) Circle centered at $x = z = mv_0/qB$, $y = 0$, with radius $R = \sqrt{2}mv_0/qB$. (b) Penetration is $(\sqrt{2} - 1)mv_0/qB$. (c) Time in field is $(\pi/2)(m/qB)$. (d) Exits at $x_0 = 2mv_0/qB$.

10-6.19 (a) "Show that" problem. (b) "Show that" problem. (c) The two formulas coincide.

10-6.21 (a) "Show that" problem. (b) "Show that" problem.

10-7.1 Because the charge carriers move to the side of the wire, the force does act on the charge carriers themselves (i.e., the current). Thus Maxwell was in error here—"to err is human."

10-7.3 (a) Negative. (b) $v_d = 0.882$ mm/s.

10-7.5 (a) $R_H = 7.35 \times 10^{-11}$ m³/C. (b) $R_t = -10.5 \times 10^{-9}$ Ω. (c) $l = 1.22\ \mu$m.

10-7.7 (a) Top is positive. (b) $\vec{E}_{mot} = -vB\hat{y}$, $\vec{E}_{es} = vB\hat{y}$. (c) $\Delta V = vBl$. (d) $\Delta V = vBl\cos\theta$.

10-8.1 $dW_{emf} = -dW_{pmf} = -0.084$ J.

11-2.1 $d\vec{B} = 2.72 \times 10^{-10}(2\hat{i} + \hat{k})$ T.

11-3.1 "Show that" problem.

11-4.1 (a) 5.29 m. (b) 281 turns.

11-4.3 2.81 mm.

11-4.5 (a) $d\vec{B} = (k_m I ds/a^2)\hat{\otimes}$. (b) $\vec{B} = (\pi k_m I/a)\hat{\otimes}$.

11-4.7 3.67 mA.

11-4.9 $B_x = 6 \times 10^{-5}$ T.

11-5.1 (a) $\vec{B} = \vec{0}$, $\vec{B} = 4\pi k_m K\hat{i}$, $\vec{B} = 12\pi k_m K\hat{i}$. (b) $|\vec{B}| = 0.00314$ T.

11-5.3 (a) Two close coils behave like a single coil, with a single maximum at their midpoint. (b) Two distant coils behave like two independent coils, with a local minimum at their midpoint. (c) $B_y = 2\pi k_m I R^2[((y + s/2)^2 + R^2)^{-3/2} + ((y - s/2)^2 + R^2)^{-3/2}]$. (d) $s = R$.

11-5.5 "Show that" problem.

11-5.7 "Show that" problem.

11-5.9 (a) $\vec{B} = -168.4\hat{z}\ \mu$T. (b) $\vec{B} = -31.0\hat{z}\ \mu$T. (c) $\vec{B} = 49.1\hat{z}\ \mu$T.

11-5.11 $\vec{B} = -2k_m K \ln(\frac{x+w/2}{x-w/2})\hat{y}$.

11-5.13 (a) $K = M$, counterclockwise. (b) $\vec{B} = k_m I[\frac{a^2}{(z^2+a^2/4)\sqrt{z^2+a^2/2}} + \frac{b^2}{(z^2+b^2/4)\sqrt{z^2+b^2/2}}I]\hat{z}$.

11-5.15 (a) $B_x = \frac{N}{L}2\pi k_m I[\frac{x+L/2}{\sqrt{(x+L/2)^2+a^2/4}} - \frac{x-L/2}{\sqrt{(x-L/2)^2+a^2/4}}]$. (b) For $L/a \to 0$, $B_x \to \frac{2\pi k_m NI a^2}{(x^2+a^2)^{3/2}}$. For $L/a \to \infty$, $B_x \to 4\pi k_m NI/L$.

11-5.17 (a) $dI/dr = \sigma\omega r$. (b) $I = \omega(a^2/2)$. (c) $B_x = 2\pi k_m \sigma\omega[\sqrt{x^2 + a^2} - 2x + \frac{x^2}{\sqrt{x^2+a^2}}]$.

11-6.1 (a) $\vec{F} = 3.04 \times 10^{-21}\hat{y}$ N. (b) $\vec{F} = -3.04 \times 10^{-21}\hat{x}$ N. (c) $\vec{F} = \vec{0}$.

11-6.3 $\vec{F} = \frac{2evk_m I}{r}\hat{i}$.

11-6.5 (a) $|\vec{F}| = 2k_m I^2\ln(a/r)$, to the left. (b) $|\vec{F}| = 9.15 \times 10^{-5}$ N.

11-6.7 (a) $\vec{B} = \frac{2\pi k_m I_2}{b}(-\hat{x})$. (b) $\vec{F} = \frac{2\pi^2 k_m I_1 I_2 na^2}{b}\hat{x}$. (c) $\vec{B} = -2.51 \times 10^{-4}\hat{x}$ T, $\vec{F} = 1.18 \times 10^{-5}\hat{x}$ N.

11-6.9 $F/l = k_m I^2/a$.

11-7.1 (a) $\Gamma_B = 1.72 \times 10^{-5}$ T-m. (b) $\Gamma_B = 0$.

11-7.3 (a) $I_{enc} = 19{,}900$ A. (b) $\Gamma_B = 0.05$ T-m. (c) $\Gamma_B = 0$ T-m.

11-7.5 (a) Field circulates counterclockwise. (b) $\Gamma_B/s = 0.04$ T. (c) $I_{enc} = 636.6$ A. (d) Field points out of the page.

11-8.1 "Show that" problem. Deformation doesn't affect circulation if no current passes through the deforming circuit.

11-8.3 (a) And. (b) For both a physical circuit and an Ampèrian circuit, $d\vec{s}$ is defined, but only for a physical circuit does it point along the local current direction.

11-9.1 Take $d\vec{s}$ to be clockwise, so $I_{enc} > 0$ is into the page. (a) $\Gamma_B = -120$ T-m. (b) $I_{enc} = -9.55 \times 10^7$ A (out of page). (c) $\vec{J} \cdot \hat{n} = 7.96 \times 10^6$ A/m².

11-9.3 (a) $\Gamma_B = 2ydydx$. (b) $I_{enc} = ydydx/2\pi k_m$. (c) $I/A = y/2\pi k_m$.

11-9.5 (a) $\Gamma_B = 5.6 \times 10^{-6}$ T-m. (b) $B = 3.56 \times 10^{-5}$ T. (c) $I_{enc} = 4.45$ A into the page. (d) $I_{enc}/A = 2266$ A/m².

11-10.1 (a) $|\vec{B}| = 9.60$ mT. (b) $|\vec{B}| = 15$ mT. (c) Counterclockwise.

11-10.3 (a) $r < a$ and $a < r < b$ counterclockwise, $c < r$ clockwise. (b) Concentric circle of radius $r > c$. (c) $\Gamma_B = 2\pi r|\vec{B}|$. (d) $\Gamma_B = 12\pi k_m I$. (e) $|\vec{B}| = 6k_m I/r$.

11-10.5 Take I_{inner} to be out of the page, and $\hat{\phi}$ to indicate the counterclockwise tangent. (a) $J_{core} = \frac{I}{\pi a^2}$, $J_{sheath} = \frac{I}{\pi(c^2-b^2)}$. (b) For $r < a$, $\vec{B} = \frac{2k_m I}{a^2}r\hat{\phi}$; for $a < r < b$, $\vec{B} = \frac{2k_m I}{r}\hat{\phi}$; for $b < r < c$, $\vec{B} = \frac{2k_m I}{r}(1 - \frac{r^2-b^2}{c^2-b^2})\hat{\phi}$; for $c < r$, $\vec{B} = \vec{0}$.

11-10.7 (a) "Show that" problem. (b) $d\vec{B}' = -k_m(nIz)(d\vec{s} \times \vec{\rho})/R^3$. (c) "Show that" problem.

11-11.1 (a) $\vec{F} = \frac{k_m q_m^2}{4h^2}\hat{y}$. (b) $h = (q_m/2)\sqrt{k_m/Mg}$.

11-11.3 (a) "Show that" problem. (b) $B_y = \frac{k_m q_m}{(-y+h)^2}$ for $y < 0$. (c) $F = q_m B = \frac{k_m q_m^2}{(2h)^2}$.

11-12.1 (a) Induced surface current K circulates clockwise as seen from above. (b) $K = \frac{q_m}{2\pi}\frac{\rho}{(\rho^2+h^2)^{3/2}}$. (c) $\vec{F} = \frac{k_m q_m^2}{(2h)^2}(-\hat{z})$.

11-13.1 See Section 11.13.

11-14.1 (a) $B_{gap} = 17.2$ mT. (b) $B_{in} = B_{gap} = 17.2$ mT. (c) $A_{gap}/A = 45.5$.

11-14.3 (a) $B_{in} = 1.29$ T. (b) $B_{gap} = 1.29$ T.

12-3.1 (a) Calico is an insulator. (b) The greater the current, the greater the deflection. (c) Counterclockwise.

12-3.3 (a) Counterclockwise. (b) To increase the response.

12-3.5 (a) Smaller. (b) Clockwise.

12-3.7 (a) Source of magnetic field. (b) Counterclockwise.

12-4.1 (a) Increases out of the page. (b) Into the page. (c) Clockwise. (d) Clockwise. (e) Up. (f) Compress. (g) No tendency to rotate.

12-4.3 No current.

12-4.5 If R increases, (a) decrease. (b) increase. (c) same direction as primary. (d) same direction as primary. (e) loops attract and expand. (f) If R decreases, all answers reverse.

12-4.7 (a) Increase out of the page. (b) Into the page. (c) Clockwise. (d) Clockwise. (e) Leftward, compressive. (f) If the field is tilted, the effects will decrease.

12-4.9 (a) counterclockwise. (b) move foil away and compress the foil.

12-4.11

observer	$d\vec{B}_{ext}/dt$	B_{ind}	$\mathcal{E}_{ind}, I_{ind}$	$\vec{\tau}$	compress or expand
reader	$\otimes$	$\odot$	counterclockwise	opposite $\vec{\omega}$	expand

12-4.13

observer	$d\vec{B}_{ext}/dt$	B_{ind}	$\mathcal{E}_{ind}, I_{ind}$	$\vec{F}_{net}$	compress or expand
from above	$\otimes$	$\odot$	counterclockwise	$\odot$	compress

12-5.1 (a) $\vec{E}$ and $\vec{B}$ are vectors; $\mathcal{E}$, Φ_B, and $d\Phi_B/dt$ are scalars. (b) $\vec{E}$, $\mathcal{E}$, and $d\Phi_B/dt$ do not change; $\vec{B}$ and Φ_B reverse.

12-5.3 (a) $\Phi_B = -(16t + 32t^2) \times 10^{-7}$ Wb. (b) $d\Phi_B/dt = -(16 + 64t) \times 10^{-7}$ Wb/s. (c) $\mathcal{E} = (16 + 64t) \times 10^{-7}$ V. (d) $|\vec{F}| = 2.56(1 + 4t + 2t^2 + 8t^3) \times 10^{-11}$ N. (e) 3.03 s.

12-5.5 (a) −2.39 mV. (b) −2.47 mV. (c) −0.65 mA.

12-5.7 (a) 0.008 V counterclockwise. (b) 0.32 mA counterclockwise. (c) 5.12×10^{-5} N pushing the loop into the field region.

12-5.9 (a) 0.096 T into the page. (b) 0.00144 Wb/s. (c) 0.036 mA counterclockwise. (d) 6.912×10^{-8} N drawing the circuit into the solenoid.

12-5.11 (a) $2NBA$. (b) $2NBA/R$.

12-5.13 (a) $(2\pi r)(dr/dt)B^2/R$. (b) 144 N/m.

12-6.1 (a) 0.2 H. (b) 27 A/s.

12-6.3 $-(32.4 + 32.4t)$ mV.

12-6.5 (a) $2k_m N_1 N_2 ha/\rho$. (b) $2k_m N_1 N_2 ha/\rho$.

12-6.7 (a) $\frac{2\pi^2 k_m a^2 b^2}{R^3}$. (b) 2.56×10^{-11} H.

12-6.9 (a) $4\pi^2 k_m N_c N_s I_s a^2/l$. (b) Counterclockwise. (c) $d\Phi_B/dt = 4\pi^2 k_m N_c N_s (dI_s/dt)a^2/l$, $\mathcal{E} = -4\pi^2 k_m N_c N_s (dI_s/dt)a^2/l$. (d) 0.126 mH.

12-7.1 (a) $-0.08\hat{y}$ V/m. (b) Bottom is higher, by 18.35 mV.

12-7.3 (a) $\Delta V_{right} = 0$. (b) $\Delta V_{left} = 0$. (c) $\Delta V_{top} = 0$.

12-7.5 East wing is higher by 0.135 V.

12-7.7 1.42 μV, clockwise viewed from above.

12-7.9 (a) $\omega r Bb$. (b) $b/\sigma cd$. (c) Force $r\omega B^2 bcd\sigma$ opposing motion, torque $r^2 \omega B^2 bcd\sigma$ opposing motion. (d) "Show that" problem.

12-8.1 (a) Middle arm. (b) Before, $I_J = 0.048$ A, $I_w = 1.2$ A, both right to left. (c) After, I_w is unchanged and $I_J = 1.2$ A, left to right. (d) $\Delta V_w = 24$ V, $\Delta V_J = 600$ V, both clockwise, and $\Delta V_L = 624$ V counterclockwise. (e) See Figure 12.21.

12-9.1 2.703 mH.

12-9.3 −61,300 A/s, the sign meaning that I_2 and I_1 are changing in opposite senses.

12-9.5 For a pure inductor, $\Phi_B = \int(\vec{B}_1 + \vec{B}_2) \cdot d\vec{A}_1$ conserved. Thus an increase in $\vec{B}_1$ is accompanied by a decrease in $\vec{B}_2$. This corresponds to out-of-phase magnets, which will repel.

12-10.1 (a) 72 turns. (b) −1.55 mV.

12-10.3 (a) $N^2 L$. (b) 0.138 mH.

12-10.5 $L = 2k_m N^2 a \ln[(b+a)/a]$.

12-11.1 (a) $dI/dt = 2.21 \times 10^5$ A/s, $V_R = 0.966$ V, $V_L = 1.434$ V. (b) $I = 6.52$ A, $dI/dt = 1.38 \times 10^5$ A/s, $V_R = 1.5$ V.

12-11.3 (a) $dI/dt = 5.78 \times 10^5$ A/s. (b) 4.8 V. (c) 37.25 ns.

12-11.5 (a) 6.73 mH. (b) 0.410 ms. (c) 19.8 mA.

12-11.7 (a) At $t = 0^+$, $V_L = 12$ V. At $t = 25$ μs,

$V_L \approx 0$. (b) At $t = 25$ μs, $V_{R_L} = 0.293$ V, and $V_R = 11.707$ V. (c) At $t = 0^+$, the electric field is electromagnetically induced. At $t = 25$ μs, the field is electrostatic.

12-12.1 (a) 0.1568 J. (b) 2.31×10^8 J/m^3. (c) 24.1 T.

12-12.3 0.416 μH/m.

12-12.5 (a) 0.008 s. (b) See Figure 12.31. (c) At $t = 0$, $I = 0$ and $dI/dt = 3 \times 10^3$ A/s. At $t = 0.002$ s, $I = 5.31$ A and $dI/dt = 2.34 \times 10^3$ A/s. At $t = \infty$, $I = 24$ A and $dI/dt = 0$. (d) Battery provides 0, 63.7 W, and 288 W. (e) Resistor uses 0, 14.1 W, and 288 W. (f) Inductor builds up energy at rate 0, 49.7 W, and 0. (g) Except for negligible parasitic capacitance, there is no electrical energy. (h) $\mathcal{P}_{batt} = \mathcal{P}_L + \mathcal{P}_R$.

12-13.1 (a) 1.1 N/C. (b) 1.76 N/C.

12-13.3 (a) "Show that" problem. (b) 0.00398. (c) $|E_\theta| = 0.0329$ V/m, $E_z = 0.131$ mV/m, $\Delta V_L = 0.0262$ mV. (d) E_θ is electromagnetically induced. (e) E_z is electrostatic.

12-13.5 Let $\tilde{R}^2 = R_r R_l + R_r R_m + R_l R_m$. If $\mathcal{E}_r > 0$, then $I_l = \mathcal{E}_r R_m / \tilde{R}^2$ goes up the left arm and $I_r = \mathcal{E}_r (R_m + R_l)/\tilde{R}^2$ goes down the right arm. (a) $I_r R_r$. (b) $I_l R_l$. (c) $\mathcal{E}_r = -0.397$ V, so $I_r R_r = -0.283$ V (voltmeter bottom is positive) and $I_l R_l = -0.0567$ V (voltmeter top is positive).

12-13.7 (a) 4 A clockwise. (b) 8 V across left, 0 V across top, 4 V across right, 0 V across bottom. (c) $V_A = -5$ V, $V_B = -2$ V, $V_C = -3$ V, $V_D = 0$ V. (d) 5 V, -3 V, 1 V, -3 V.

12-14.1 (a) $I_1 = I_2 = 6$ A, right to left. (b) $I = 1.7$ A in both arms, circulating clockwise. (c) $I = 0.96$ A in both arms, circulating clockwise.

12-14.3 (a) $I_1 = I_2 = 0$. (b) $I_1 = I_2 = 0$. (c) $I_1 = 0.545$ A, $I_2 = 0.15$ A. (d) $I = 0.346$ A, circulating counterclockwise. (e) 68.6 μs, 0.193 A.

13-2.1 "Verify that" problem.

13-3.1 From best to worst: iron rods, iron, plastic (either rods or solid). Iron rods have a large magnetization, but a relatively large resistance.

13-4.1 The motor is an external source of mechanical power, and the generator uses the electrical power.

13-5.1 (a) 10 Ω. (b) 100 V.

13-5.3 (a) 92.31 A. (b) 0.59 N, 1.02 m/s^2. (c) 0.015 V. (d) 92.3 A. (e) 2.25 s.

13-6.1 (a) 4293 s. (b) 7.32 A. (c) 0.0703 N.

13-6.3 (a) 227 s. (b) 2.89×10^4 m/s. (c) 1.461×10^3 A.

13-7.1 0.0403 s.

13-7.3 88.7%.

13-8.1 (a) 80 s. (b) 0.08 N, 1.667 m/s^2. (c) 69.1 m/s^2, 1.97 m/s.

13-9.1 2.65×10^6 m/s, 2.65×10^5 m/s, 2.65×10^2 m/s, 2.65×10^1 m/s.

13-9.3 "Show that" problem.

13-9.5 (a) Opposite source current. (b) Same as source current.

14-1.1 $V_{rms} = 1/\sqrt{3}$, $\bar{V} = 0$.

14-1.3 (a) 0.0127 s per radian and 0.08 s per period. (b) 12.5 s^{-1} and 78.5 rad/s. (c) 12.5 s^{-1} is f and 78.5 rad/s is ω.

14-1.5 (a) $V_{rms} = V_m/\sqrt{3}$. (b) $\overline{\Delta V} = V_m/2$.

14-1.7 (a) $V_{rms} = V_m/\sqrt{3}$. (b) $\overline{\Delta V} = 0$.

14-2.1 Decrease either L or C by a factor of 4.

14-2.3 (a) 2.093×10^{-9} F to 0.264×10^{-9} F. (b) 2.62×10^{-12} J and 0.329×10^{-12} J.

14-3.1 (a) Current starts at zero but with finite slope, going through some damped oscillations until it is zero at long times. (b) Charge starts at zero and with zero slope, going through some damped oscillations until it is finite at long times.

14-3.3 "Show that" problem.

14-3.5 (a) $\omega_0 = 2.49 \times 10^4$ rad/s, $f_0 = 3.97 \times 10^3$ Hz, $R_c = 1197$ Ω.

14-4.1 (a) 29.4 cm^2. (b) 800 turns.

14-5.1 (a) 5.86 Ω, 56.1°. (b) Energy is absorbed.

14-5.3 (a) B provides power, A receives power. (b) 55°.

14-5.5 (a) 500 Ω, 0.628 F, $-90°$. (b) 502.6 J.

14-5.7 (a) Scalar. (c) The y-component of the rotating vector, or phasor, gives a scalar that equals the voltage.

14-5.9 (a) At low f capacitors serve as open circuits (finite voltage), at high f capacitors serve as closed (or short) circuits (zero voltage). (b) At low f inductors serve as closed (or short) circuits (zero voltage), at high f inductors serve as open circuits (finite voltage).

14-6.1 $Z = 500$ Ω, $\phi = 0.188$ rad $= 10.8°$, $R = 39.4$ Ω, $X_L = 7.50$ Ω, $L = 0.199$ H.

14-6.3 $X_L = 1.57 \ \Omega$, $Z = 20.6 \ \Omega$, $\phi = 0.0784 \ \text{rad} = 4.49°$, $t_{lead} = 0.0398$ ms, $I_m = 3.0$ A.

14-6.5 $X_C = 3.02 \times 10^4 \ \Omega$, $R = 1.430 \times 10^5 \ \Omega$, $Z = 1.461 \times 10^5 \ \Omega$, $\phi = -11.9°$.

14-6.7 $Z = 160 \ \Omega$, $X_L = 30.14 \ \Omega$, $R = 157.1 \ \Omega$, $\phi = -0.1894 \ \text{rad} = -10.85°$, $t_{lag} = 6.85 \times 10^{-5}$ ms.

14-6.9 $\phi = 0.188$ rad, $X_L = 2.29 \ \Omega$, $L = 0.911$ mH, $Z = 12.22 \ \Omega$.

14-6.11 (a) $I_{rms} = 3.52$ A for all circuit elements. (b) $\Delta V_{R,rms} = 85.4$ V, $\Delta V_{L,rms} = 20.1$ V. (c) $\mathcal{P}_{R,rms} = 300.6$ W, $\mathcal{P}_{L,rms} = 0$. (d) $\phi = 0.23$ rad, $t_{lag} = 0.61$ ms.

14-6.13 (a) Inductor with $L = 0.1$ H. (b) $\bar{\mathcal{P}} = 35.45$ W.

14-7.1 (a) $F = 0.0625$ Hz. (b) $Z_R = R = 200 \ \Omega$, $Z_L = X_L = 3.56 \times 10^{-3} \ \Omega$. (c) Power surges have high frequencies, for which inductors have a high impedance.

14-7.3 (a) Capacitor. (b) Resistor. (c) $V_{R,m} = 2.39$ V, $V_{C,m} = 24.9$ V. (d) $V_{R,m} = 17.35$ V, $V_{C,m} = 18.00$ V.

14-7.5 "Show that" problem.

14-8.1 (a) Yes. (b) No. (c) Yes.

14-8.3 (a) $X_L = 157.1 \ \Omega$, $X_C = 0.796 \ \Omega$, $Z = 156.3 \ \Omega$, $\phi = 88.53°$. (b) $I_m = 0.1535$ A, $V_{L,m} = 24.1 \ \Omega$, $V_{R,m} = 0.614 \ \Omega$, $V_{C,m} = 0.122 \ \Omega$. (c) Voltage leads current by 6.15×10^{-4} s. (d) 0.626 V, voltage lags current by 7.82×10^{-5} s. (e) $157.1 \ \Omega$, voltage leads current by 6.15×10^{-4} s. (f) $156.3 \ \Omega$, voltage lags current by 6.25×10^{-4} s.

14-8.5 (a) 1340 Hz, $I_{rms} = 0.05$ A. (b) $V_{L,rms} = 6.315$ V, $V_{R,rms} = 1.20$ V, $V_{C,rms} = 6.32$ V. (c) $Z = 191 \ \Omega$, $\phi = 82.8°$, $I_{rms} = 0.00688$ A. (d) $V_{L,rms} = 1.587$ V, $V_{R,rms} = 0.1508$ V, $V_{C,rms} = 0.3968$ V. (e) $R_c = 252.6 \ \Omega$. Definitely a resonance, but rather broad, since $\mathcal{Q} = 5.625$ is fewer than the 6.26 radians that correspond to a full period.

14-8.7 "Show that" problem.

14-9.1 For water, a valve permits a small force to control a large flow of water; for electricity, a valve permits a small voltage to control a large flow of electricity.

14-9.3 See Figure 14.14(b). The grid voltage determines whether or not electrons are drawn off the cathode; once off the cathode, they go to the anode.

14-10.1 Neither of them can dissipate energy, and on average neither of them can store more than one cycle's worth of energy.

14-10.3 (a) $\mathcal{P} > 0$ when refinery is buying power. (b) $\mathcal{P} < 0$ when refinery is selling power.

14-10.5 (a) $\cos\phi_0 = 0.423$. (b) Add capacitor with $C = 1.958 \ \mu$F. (c) $Z = 1215 \ \Omega$.

14-11.1 (a) 6.25 turns. (b) 62.5 turns.

14-11.3 100 V.

14-11.5 (a) $N_s/N_p = 1/10$. (b) Lower voltage side has higher current; thicker wire decreases the Joule heating rate.

14-11.7 (a) $\mathcal{P}_{gen} = 4800$ W. (b) $\mathcal{P} = \mathcal{P}_{gen} = 4800$ W. (c) $\mathcal{P}_{wires} = 2.56$ W. (d) $\mathcal{P}_{load} = 4797.44$ W.

14-11.9 See Section 14.11.3.

14-11.11 (a) $R = 1 \ \Omega$. (b) $R_c/R = \pi$. (c) $L = 0.84$ mH, $C = 0.34$ mF. (d) $N_s/N_p = 100$.

14-12.1 Toaster.

14-12.3 (a) $\mathcal{E}_m = 4.88$ V. (b) $Z = 20.8 \ \Omega$, $\phi = 52.8°$. (c) $I_m = 5.76$ A. (d) $F = 0.105$ N.

14-12.5 (a) $IR + LdI/dt + Q/C = \mathcal{E}_m \sin\omega t$, $\mathcal{E}_m = \omega B_m l x$. (b) $I = \mathcal{E}_m/Z \sin\omega t$, Z and ϕ as in (14.71) and (14.72). (c) $\bar{F} = \mathcal{E}_m b B_m/2Z$.

14-13.1 (a) 23.9 μm. (b) $d = 0.001$ inch is 25.4 μm, so that the field will penetrate, but down by a factor of $e^{-d/\delta} = 0.346$.

14-13.3 "Show that" problem.

15-2.1 (a) 2011 m/s. (b) 2 min. (c) 70 min. (d) 60 flagmen.

15-3.1 2.8×10^{16} N/C-s.

15-4.1 0.011/m.

15-4.3 (a) $dy/dx = x/(a-y)$, $d^2y/dx^2 = a^2/(a-y)^3$. (b) $1/a$. (c) $d^2y/dx^2 = 1/R$, where $R = a$ is the radius of curvature.

15-5.1 120 m/s.

15-5.3 3 N.

15-5.5 (a) Energy flowing along the string is more concentrated. (b) Yes. (c) 3361 N. (d) It corresponds to hanging a mass of 342.6 kg; unlikely.

15-5.7 "Show that" problem.

15-5.9 (a) 2.048 m/s. (b) 0.36 m. (c) −0.65 m/s. (d) −0.626 m/s².

15-5.11 (a) 251 m^{-1}. (b) 2010 s^{-1}. (c) 320 Hz. (d) 0.003125 s. (e) −8045.75 rad, which is equivalent to 3.01 rad. (f) $y(x, t) = (0.03 \text{ cm})$ $\sin[251x + 201t + 3.01]$. (g) 0.00296 cm.

15-5.13 (a) 3.0 kg/m. (b) 0.72 N.

15-5.15 (a) 147 m/s. (b) 170 cm and 85 cm. (c) 86.5 Hz and 172.9 Hz.

15-5.17 (a) 7.48 m. (b) 22.44 m. (c) 0.0128 s. (d) 78 Hz. (e) −2.65 m/s. (f) −8600 m/s².

15-5.19 (a) No. (b) No. (c) No. (d) $d^2y/dx^2 = v^2(d^2y/dt^2)$ and $d^2z/dx^2 = v^2(d^2z/dt^2)$.

15-6.1 $-12,000\hat{x}$ V/m.

15-6.3 (a) $(D/c)\sin(qx - \omega t)$. (b) $(D/c)\sin \times (qy - \omega t)$.

15-6.5 (a) 0.009375 m. (b) 670.2 m⁻¹. (c) 2.01 s⁻¹. (d) $0.15\sin(qy - \omega t)$ mT.

15-7.1 (a) 303 m. (b) 3.367 m.

15-8.1 (a) r^{-1}. (b) $r^{-1/2}$.

15-8.3 (a) 1.59×10^5 V/m. (b) 0.53 mT.

15-9.1 (a) 4.17×10^{-8} m/s². (b) 19,200 s. (c) 8×10^{-4} m/s. (d) 7.68 m.

15-9.3 (a) $u_E = u_B = 8.95 \sin^2(qx - \omega t)$ mJ/m³. (b) $\vec{S} = 5.37\hat{y}\sin^2(qx - \omega t)$. (c) $\mathcal{P} = 0.0179 \sin^2 (qx - \omega t)$ N/m².

15-9.5 (a) $\vec{S} = -2\pi k_m n^2 r I \frac{dI}{dt}\hat{r}$. (b) $\mathcal{P}/l = 4\pi^2 n^2 r^2 I|dI/dt|$. (c) $d\mathcal{P}/l = 8\pi^2 n^2 r dr I|dI/dt|$. (d) $d(dU_B/dt)/l = 8\pi^2 n^2 r dr I|dI/dt|$. This equals $d\rho/l$, so $d\rho = d(dU_B/dt)$.

15-10.1 (a) 3.93 m. (b) 1.132. (c) 1.28.

15-10.3 (a) 5.07×10^{14} Hz. (b) 5.07×10^{14} Hz. (c) 443.7 nm. (d) $\theta_{refl} = 23°$. (e) $\theta_{refr} = 17.04°$.

15-10.5 (a) $\theta_c^{crown \, glass} = 41.14°$, $\theta_c^{diamond} = 24.41°$. (b) Crown glass. (c) Diamond.

15-11.1 2.3 W/m², polarized vertically.

15-11.3 (a) 0. (b) 0.134 W/m², at 35° to the x axis.

15-11.5 2.58 W/m², polarized vertically.

15-11.7 (a) 53.1°. (b) 36.9°.

15-12.1 (a) See Section 15.12. (b) See Section 15.12.

15-12.3 $-2881\hat{x}$ A/m.

15-13.1 "Show that" problem.

15-14.1 (a) 2.4 m. (b) 2.4 m.

15-14.3 $R/R_c \approx 0.25$. This corresponds to $Q = R_c/2R \approx 2.0$ radians, or 0.32 of a full oscillation. It damps out very quickly.

16-2.1 (a) 76.6 mm. (b) 287 Hz.

16-2.3 $(4d^2 - 4m^2\lambda^2)y^2 - 4m^2\lambda^2 D^2 = m^2\lambda^2$ $(d^2 - m^2\lambda^2)$.

16-2.5 (a) 0.212 m. (b) 1604 Hz.

16-2.7 (a) One line of maxima, no lines of minima. (b) One line of maxima, two lines of minima. (c) Five lines of maxima, six lines of minima.

16-2.9 (a) See Figure 16.6(a). (b) See Figure 16.7(a). (c) See Figure 16.8(a).

16-2.11 (a) 4.09 m. (b) 12.50°.

16-3.1 "Show that" problem.

16-3.3 (a) $\phi = 2\pi + 2\theta - 6\theta'$. (b) $\sin^2 \theta = \frac{1-(n/3)^2}{1-(1/3)^2}$. (c) $\phi - 180° = 50.97°$.

16-4.1 See Section 16.4.

16-4.3 See Section 16.4.

16-4.5 At 90° (top of circle) get maximum increase in period; at 270° (bottom of circle) get maximum decrease in period.

16-4.7 Grimaldi's experiment showing bright and dark fringes inside and outside the geometrical shadow.

16-4.9 Birefringent crystals, unless oriented properly, will give two images.

16-5.1 (a) "Show that" problem. (b) v_s. (c) v_p.

16-5.3 (a) For small-amplitude disturbances the equations are linear, which leads to an amplitude-independent sound velocity. (b) The relative amounts of reflection and refraction are amplitude-independent.

16-6.1 (a) 109.87 Hz. (b) Tone at $(f_A - f_E')^{-1}$. (c) 7.5 s.

16-6.3 $\lambda'/\lambda = v/c$. Since the experiments indicate that λ' in the film is less than λ in air, we deduce that $v < c$.

16-6.5 (a) The sources have a coherence time much shorter than the measuring time of the eye. (b) The sources have a coherence time longer than the measuring time of the ear.

16-6.7 (a) No. (b) No.

16-6.9 (a) 489.44 mm. (b) 611.8 nm and 407.9 nm.

16-6.11 (a) 691.3 nm, 518.5 nm, and 414.8 nm. (b) 592.4 nm and 460.0 nm.

16-6.13 (a) 912 nm. (b) 606.5 nm, 485.2 nm, and 404.3 nm.

16-6.15 113.1 nm.

16-6.17 (a) $n_1 = 1.000293$. (b) $\lambda_{vac} = 690.2$ nm. (c) $d = 0.4998$ m.

16-6.19 (a) $N = 2\Delta D/\lambda$. (b) 600 nm.

16-6.21 (a) 409 nm. (b) 8.1 mm. (c) 611.

16-6.23 (a) 490 nm. (b) 245 nm.

16-7.1 "Show that" problem.

16-7.3 Break up the illuminated part of the plane into tiny strip-shaped sources parallel to the finite opaque strip. Break up the illuminated part of the plane into tiny annulus-shaped sources concentric with the finite opaque disk. Because of the symmetry of the observation point at the center of the geometrical shadow, wavelets from all strips (or annuli) add in phase, giving an interference maximum.

16-7.5 366.5 nm.

16-7.7 (a) 16.27 μm. (b) 2.20°, 6.60°.

16-7.9 (a) 7.68°. (b) 3.24 cm.

16-7.11 (a) 18.47 μm. (b) 4.28°, 4.87 cm.

16-7.13 19.7 μm.

16-7.15 (a) "Show that" problem. (b) "Show that" problem.

16-7.17 (a) 0.393 mm. (b) 1.5×10^{-7} radians. (c) 15 radians—unresolvable.

16-7.19 (a) 2.36×10^{12} m. (b) 3400 times the radius of the sun. (c) 6.79×10^{14} m.

16-7.21 (a) 0.671 mm. (b) 0.0419 mm. (c) 3.5 m. (d) 5.75 m.

16-8.1 (a) $2d = (m + \frac{1}{2})\lambda/n_{O,E}$. (b) 520 nm ($m = 3$) and 404.4 nm ($m = 4$). (c) 557.1 nm ($m = 3$) and 433.3 nm ($m = 4$).

16-8.3 (a) 0.063 nm. (b) 4.503°.

16-9.1 (a) Maxima out to $m = 6$. (b) $m = 0$, 0°; $m = 1$, ±9.04°; $m = 2$, ±18.32°; $m = 3$, ±28.13°; $m = 4$, ±38.94°; $m = 5$, ±51.79°; $m = 6$, ±70.54°. (c) Maxima out to $m = 4$. (d) 0.0212 rad = 1.2°. (e) 43.9 nm.

16-9.3 (a) 513.4 nm. (b) $m_{max} = 10$. (c) $\theta_6 = 36.3°$.

16-9.5 400 nm $\leq \lambda \leq$ 595 nm.

16-9.7 4030 lines.

16-9.9 1220 slits.

16-9.11 (a) "Show that" problem. (b) "Show that" problem. (c) Multiples of three.

16-10.1 (a) 20°, 80°. (b) 160°, 10°.

16-10.3 0.16 nm.

16-10.5 $m = 4$ gives $\lambda = 0.01139$ nm, $m = 5$ gives $\lambda = 0.00911$ nm.

16-10.7 (a) 0.205 nm, 1.47×10^{18} Hz. (b) 0.270 nm, 1.111×10^{18} Hz.

16-10.9 $2d \sin \theta = m\lambda$ and $n_V = n_A/d = $ constant relates the volume and area densities and the plane separation. Higher n_A means smaller d and thus larger θ and larger scattering angle 2θ.

Index

Some Standard Abbreviations for Units

ampere A
angstrom $\text{Å} = 10^{-8}$ m
atmosphere atm
British thermal unit Btu
coulomb $C = \text{A-s}$
degree Celsius °C
degree Fahrenheit °F
farad $F = C/V$
foot ft
gauss $G = 10^{-4}$ T
gram g
henry $H = \text{V-s/A}$
hertz $Hz = \text{cycle/s}$
horsepower hp
hour h
inch in.

kelvin K
joule $J = \text{kg-m}^2/\text{s}^2$
kilocalorie kcal (Cal)
kilogram kg
meter m
minute min
mole mol
newton N
ohm Ω
pascal Pa
pound lb
second s
tesla $T = \text{N/A-m}$
volt V
watt $W = \text{J-s}$
weber $Wb = \text{T-m}^2$

Some Mathematical Formulas (See Also Appendix A)

Algebra

If $ax^2 + bx + c = 0$, then $x = -\frac{1}{2a}[b \mp \sqrt{b^2 - 4ac}]$.
If $x = a^y$, then $y = \log_a x$. $\log_a(xy) = \log x + \log y$.

Geometry

Triangle with base b and height h has Area $bh/2$.
Circle of radius r has Circumference $2\pi r$ and Area πr^2.
Sphere of radius r has surface Area $4\pi r^2$ and Volume $\frac{4\pi}{3}r^3$
Cylinder of radius r and height h has Area of curved surface $2\pi rh$ and Volume $= \pi r^2 h$.

Trigonometry

2π radians $= 360$ degrees
$\sin(90° - \theta) = \cos\theta$, $\cos(90° - \theta) = \sin\theta$
$\sin(-\theta) = -\sin\theta$, $\cos(-\theta) = \cos\theta$
$\sin^2\theta + \cos^2\theta = 1$
$\sin 2\theta = 2\sin\theta\cos\theta$, $\cos 2\theta = \cos^2\theta - \sin^2\theta$
$\sin(A \pm B) = \sin A \cos B \pm \cos A \sin B$, $\cos(A \pm B) = \cos A \cos B \mp \sin A \sin B$
Law of cosines: $a^2 = b^2 + c^2 - 2bc\cos\alpha$
Law of sines: $\frac{1}{a}\sin\alpha = \frac{1}{b}\sin\beta = \frac{1}{c}\sin\gamma$
Expansions for $x \ll 1$ (and, where appropriate, radian measure for x):

$(1 + x)^n = 1 + nx + \frac{n(n-1)}{2!}x^2 + \frac{n(n-1)(n-3)}{3!}x^3 + \cdots$
$\frac{1}{1+x} = (1 + x)^{-1} = 1 - x + x^2 - x^3 + \cdots$
$\ln(1 + x) = x - \frac{x^2}{2} + \frac{x^3}{3} - \cdots$
$\sin x = x - \frac{x^3}{3!} + \frac{x^5}{5!} - \cdots$
$\cos x = 1 - \frac{x^2}{2!} + \frac{x^4}{4!} - \cdots$
$\tan x = x + \frac{x^3}{3!} + \frac{2x^5}{15} + \cdots$